Selected physical properties of metals (Continued)

Metal		Atomic Number	Crystal Structure	Lattice Parameter (Å)	Atomic Mass (g/g · mole)	Density (g/cm³)	Melting Temperature (°C)
Mercury	Hg	80	rhomb		200.59	13.546	− 38.9
Molybdenum	Mo	42	BCC	3.1468	95.94	10.22	2610
Nickel	Ni	28	FCC	3.5167	58.71	8.902	1453
Niobium	Nb	41	BCC	3.294	92.91	8.57	2468
Osmium	Os	76	HCP	$a = 2.7341$ $c = 4.3197$	190.2	22.57	2700
Palladium	Pd	46	FCC	3.8902	106.4	12.02	1552
Platinum	Pt	78	FCC	3.9231	195.09	21.45	1769
Potassium	K	19	BCC	5.344	39.09	0.855	63.2
Rhenium	Re	75	HCP	$a = 2.760$ $c = 4.458$	186.21	21.04	3180
Rhodium	Rh	45	FCC	3.796	102.99	12.41	1963
Rubidium	Rb	37	BCC	5.7	85.467	1.532	38.9
Ruthenium	Ru	44	HCP	$a = 2.6987$ $c = 4.2728$	101.07	12.37	2310
Selenium	Se	34	hex	$a = 4.3640$ $c = 4.9594$	78.96	4.809	217
Silicon	Si	14	FCC	5.4307	28.08	2.33	1410
Silver	Ag	47	FCC	4.0862	107.868	10.49	961.9
Sodium	Na	11	BCC	4.2906	22.99	0.967	97.8
Strontium	Sr	38	FCC BCC	6.0849 4.84	87.62 (> 557°C)	2.6	768
Tantalum	Ta	73	BCC	3.3026	180.95	16.6	2996
Technetium	Tc	43	HCP	$a = 2.735$ $c = 4.388$	98.9062	11.5	2200
Tellurium	Te	52	hex	$a = 4.4565$ $c = 5.9268$	127.6	6.24	449.5
Thorium	Th	90	FCC	5.086	232	11.72	1755
Tin	Sn	50	FCC	6.4912	118.69	5.765	231.9
Titanium	Ti	22	HCP BCC	$a = 2.9503$ $c = 4.6831$ 3.32	47.9 (> 882°C)	4.507	1668
Tungsten	W	74	BCC	3.1652	183.85	19.254	3410
Uranium	U	92	ortho	$a = 2.854$ $b = 5.869$ $c = 4.955$	238.03	19.05	1133
Vanadium	V	23	BCC	3.0278	50.941	6.1	1900
Yttrium	Y	39	HCP	$a = 3.648$ $c = 5.732$	88.91	4.469	1522
Zinc	Zn	30	HCP	$a = 2.6648$ $c = 4.9470$	65.38	7.133	420
Zirconium	Zr	40	HCP BCC	$a = 3.2312$ $c = 5.1477$ 3.6090	91.22 (> 862°C)	6.505	1852

Note to the Student

Dear Student,

   If you winced when you learned the price of this textbook, you are experiencing what is known as "sticker shock" in today's economy. Yes, textbooks are expensive, and we don't like it any more than you do. Many of us here at PWS-KENT have sons and daughters of our own attending college, or we are attending school part-time ourselves. However, the prices of our books are dictated by cost factors involved in producing them. The costs of typesetting, paper, printing, and binding have risen significantly each year along with everything else in our economy.

   The prices of college textbooks have increased less than most other items over the past fifteen years. Compare your texts sometime to a general trade book, i.e., a novel or nonfiction book, and you will easily see substantial differences in the quality of design, paper, and binding. These quality features of college textbooks cost money.

   Textbooks should not be considered only as an expense. Other than your professors, your textbooks are your most important source for what you learn in college. What's more, the textbooks you keep can be valuable resources in your future career and life. They are the foundation of your professional library. Like your education, your textbooks are one of your most important investments.

   We are concerned, and we care. We pledge to do everything in our power to keep our textbook prices under control, while maintaining the same high standards of quality you and your professors require.

Wayne A. Barcomb
President
PWS-KENT Publishing Company

The PWS Civil Engineering Series List

THE SCIENCE AND ENGINEERING OF MATERIALS

Second Edition

Donald R. Askeland

University of Missouri—Rolla

PWS-KENT Publishing Company

Boston

PWS-KENT
Publishing Company

Editor: J. Donald Childress, Jr.
Production Editor: Carolyn Ingalls
Production: Greg Hubit Bookworks
Copyeditor: Patricia Harris
Interior and Cover Design: Carolyn Ingalls
Cover Photograph: The Image Bank/Pete Turner. Used by permission.
Art Studio: Art by Ayxa
Manufacturing Coordinator: Margaret Sullivan Higgins
Typesetting: The Alden Press
Cover Printing: New England Book Components
Printing and Binding: R. R. Donnelley & Sons Company

PWS-KENT Publishing Company is a division of Wadsworth, Inc.

Printed in the United States of America

3 4 5 6 7 8 9 — 93 92 91 90

LIBRARY OF CONGRESS CATALOGING-IN-PUBLICATION DATA

Askeland, Donald R.
 The science and engineering of materials / Donald R. Askeland.—
2nd ed.
     p.   cm.
  Includes index.
  ISBN 0-534-91657-0
  1. Materials.   I. Title.
TA403.A74   1989
620.1′1—dc19

88-7944
CIP

Dedicated to Clarence E. Askeland

PREFACE

Our understanding of the relationship between structure and properties provides the basis for both our selection of existing materials and our development of new materials. For instance, we are able to engineer the electronic and atomic structure of materials to produce miniaturized electronic and optical devices. By manipulating molecular structure, we have produced a vast spectrum of polymers capable of operating under extreme conditions, including elevated temperatures or wear. Controlling microstructure has led to many new metal alloys and ceramics, including lightweight aerospace alloys and ceramic superconductors. And we've played wizard, juggling the structure of composite materials to produce unique properties.

We rely on such scientific understanding to shape materials into useful products, with our materials processing both depending on and influencing the structure and properties of the materials we use. New materials processing techniques enable us better to manufacture both conventional and newly developed materials to take economic advantage of these properties. Production of tough ceramic materials for use in engines, metals that have a memory, composites tailored to withstand specific forces, semiconductor devices whose structures are built almost on an atom-by-atom sequence, and ultimately processing in the space environment are all examples of the sophisticated materials manufacturing techniques.

Finally, the environment present during processing and use affects the behavior of a material. By melting and pouring aluminum alloys in air, for example, we may produce gas pores in the finished casting. Most high-strength metal alloys rapidly lose their properties when exposed to high temperatures. Materials may oxidize, corrode, or suffer radiation damage in other adverse environments.

As in the first edition, this book presents the three-way relationship between structure, properties, and processing of materials. The text can serve, first, those engineering students who are formally introduced to materials in only one course and who do not continue in this field. Such students need a basic understanding of materials behavior, including references to specific materials and the processing of materials, so that they can appropriately select materials. Second, this text can introduce the science of materials, the types of materials available, and the application and processing of materials to the materials-oriented engineering student. Armed with this preview, these students will later go on to study in more advanced courses the details of materials structure and properties.

Part I of the text introduces atomic and crystal structure, the foundation for understanding the mechanical and physical behavior of materials. Part II explains how we control structure and mechanical properties by examining the most common strengthening mechanisms; in general, this section uses metals and alloys as examples of these mechanisms, although reference to other materials is made whenever possible. This part also explores the structure-property relationship by considering processing techniques such as casting, forming, powder processing, joining, and heat treatment—as well as control of composition. Part III describes the common metal alloys, ceramics, polymers, and composite materials, showing how the mechanical properties of each class can be controlled and again pointing out both similarities and differences among groups. Part IV examines the physical properties of materials, such as electronic, magnetic, optical, and thermal behavior, and how these properties can be controlled and utilized in a practical manner. Finally, Part V describes the way materials perform during service, with an emphasis on preventing and analyzing corrosion and mechanical failure.

A number of changes and additions were made in preparing this second edition. The chapter on deformation of materials was moved earlier in the text, permitting easier reference to mechanical behavior and allowing the chapters that include solidification to follow one another without interruption. In addition, the chapter on ferrous metal alloys was moved ahead to follow immediately the chapter on phase transformations. This better connects the introduction of the phase transformations occurring in steels to the more detailed coverage of steel heat treatments. Some additional material was moved; for example, fracture mechanics is now considered in mechanical testing, while much of the discussion of ceramic crystal structures has been incorporated in the chapter on atomic arrangement.

In addition, new topics have been introduced into most of the chapters; the Arrhenius relationship, mechanical testing of brittle materials, the unary phase diagram, rapid solidification processing, aluminum-lithium alloys, dual phase steels, advanced ceramic materials (including superconductors), and new thermoplastic polymers are included. A more thorough coverage of fiber-reinforced composites, electronic devices, and photonic phenomena has been introduced in what is surely a futile attempt to keep pace with these rapidly developing fields. Finally, a number of new examples, photographs, and drawings are included, and virtually all of the practice problems at the end of each chapter have been replaced.

The text is intended for students at the sophomore or junior level who almost surely have had sufficient exposure to chemistry, physics, and mathematics. The author recognizes that the entire text cannot be covered in a one-semester course; instead, by appropriate selection of topics, an instructor can emphasize metals, provide a general overview of materials, concentrate on mechanical behavior, or emphasize physical properties. The text also provides the student with a useful reference book for subsequent manufacturing, materials, or materials selection courses.

ACKNOWLEDGMENTS

I am indebted to the many people who have provided the assistance, encouragement, and constructive criticism leading to the preparation of both the first and this second edition.

My colleagues at UMR—Robert Wolf, Harry Weart, Ron Kohser, and especially Fred Kisslinger—have used the text in our introductory courses and have provided invaluable suggestions.

Mike Norberg, Scott Miller, Chris Ramsay, Greg Lynch, and Darren Washausen have provided assistance with gathering figures.

Advice from numerous colleagues at other institutions, including Maynard Bauleke and his materials class at the University of Kansas, Alan Wolfenden and the ME EN 222 class at Texas A&M, and Jeff Giacomin at Texas A&M, has been especially valuable.

Special thanks are extended to the hundreds of Met 121 and AE 241 students at UMR; they have provided me with the inspiration to complete this text.

I wish to acknowledge the thoughtful reviews of the manuscript offered by Gerald Liedl, Purdue University; Elmer Schwartz, University of South Carolina; Marion L. Shepard, Duke University; Peter A. Thrower, Pennsylvania State University; James G. Vaughan, University of Mississippi; and more than twenty other instructors who have used the text.

I wish to express my appreciation to the editorial staff at PWS-KENT for wielding a large enough club to keep me working at this project.

Finally, and most importantly, I am deeply indebted to my wife Mary and son Per.

Donald R. Askeland

CONTENTS

1

INTRODUCTION TO MATERIALS

1-1 INTRODUCTION

All engineers are involved with materials on a daily basis. We manufacture and process materials, design and construct components or structures using materials, select materials, analyze failures of materials, or simply hope the materials we are using perform adequately.

As responsible engineers, we are interested in improving the performance of the product we are designing or manufacturing. Electrical engineers want integrated circuits to perform properly, switches in computers to react instantly, and insulators to withstand high voltages even under the most adverse conditions. Civil and architectural engineers wish to construct strong, reliable structures that are aesthetic and resistant to corrosion. Petroleum and chemical engineers require drill bits or piping that survive in abrasive or corrosive conditions. Automotive engineers desire lightweight yet strong and durable materials. Aerospace engineers demand lightweight materials that perform well both at high temperatures and in the cold vacuum of outer space. Metallurgical, ceramic, and polymer engineers wish to produce and shape materials that are more economical and possess improved properties.

The intent of this text is to permit the student to become aware of the types of materials available, to understand their general behavior and capabilities, and to recognize the effects of the environment and service conditions on the material's performance.

1-2 TYPES OF MATERIALS

We will classify materials into several groups—metals, ceramics, polymers, semiconductors, and composite materials (Table 1-1). Materials in each of these groups often possess different structures and properties. The differences

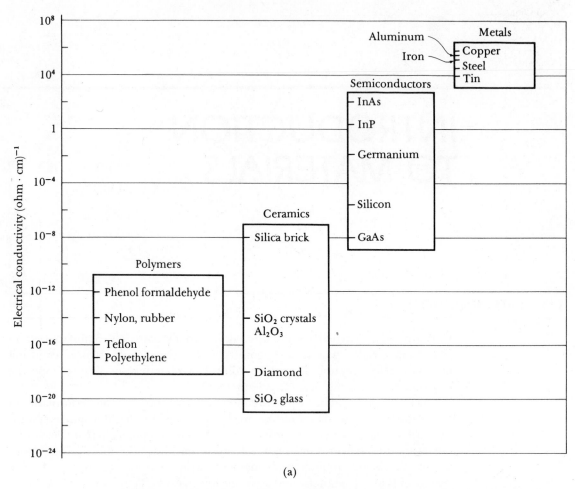

(a)

FIGURE 1-1(a)   Representative electrical conductivities for the various categories of materials. Particularly large differences are observed.

in electrical conductivity and strength for these materials are shown in Figures 1-1(a) and (b).

Metals.   Metals and alloys, which include steel, aluminum, magnesium, zinc, cast iron, titanium, copper, nickel, and many others, have the general characteristics of good electrical and thermal conductivity, relatively high strength, high stiffness, ductility or formability, and shock resistance. They are particularly useful for structural or load-bearing applications. Although pure metals are occasionally used, combinations of metals called *alloys* are normally designed to provide improvement in a particular desirable property or permit better combinations of properties. The section through a jet engine shown in Figure 1-2 illustrates the use of several metal alloys for a very critical application.

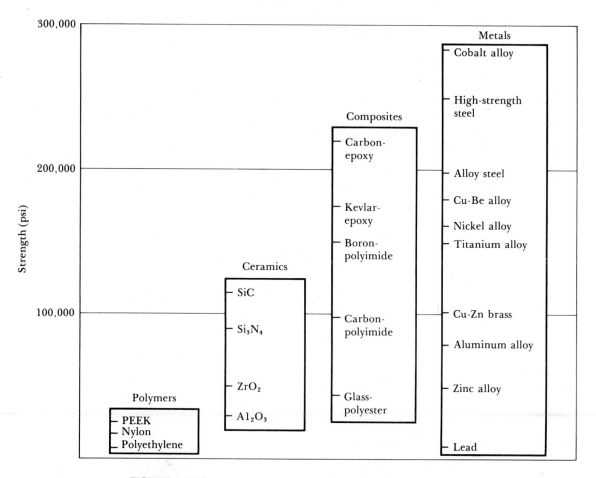

FIGURE 1-1(b) Representative strengths of the various categories of materials.

Ceramics. Ceramics, such as brick, glass, tableware, refractories, and abrasives, have low electrical and thermal conductivities and consequently are often used as insulators. Ceramics are strong and hard but also are very brittle. New processing techniques are being developed to allow ceramics to be used in load-bearing applications, such as impellers in turbine engines (Figure 1-3).

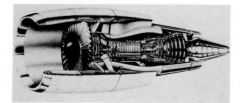

FIGURE 1-2 A section through a jet engine. The forward compression section operates at low to medium temperatures, and titanium parts are often used. The rear combustion section operates at high temperatures, and nickel-base superalloys are required. The outside shell sees low temperatures, and aluminum and composites are satisfactory. (Courtesy of GE Aircraft Engines.)

TABLE 1-1   Representative examples, applications, and properties for each category of materials

	Applications	Properties
Metals		
Copper	Electrical conductor wire	High electrical conductivity, good formability
Gray cast iron	Automobile engine blocks	Castability, machinability, vibration damping
Alloy steels	Wrenches	Good strengthening by heat treatment
Ceramics		
SiO_2–Na_2O–CaO	Window glass	Good optical properties and thermal insulation
Al_2O_3, MgO, SiO_2	Refractories for containing molten metal	Thermal insulation, high melting temperature, relatively inert to molten metal
Barium titanate	Transducers for stereo record players	Piezoelectric behavior converting sound to electricity
Polymers		
Polyethylene	Food packaging	Easily formed into thin, flexible, airtight film
Epoxy	Encapsulation of integrated circuits	Good electrical insulation and moisture resistance
Phenolics	Adhesives to join plies in plywood	Strength and moisture resistance
Semiconductors		
Silicon	Transistors and integrated circuits	Unique electrical behavior
GaAs	Fiber-optic systems	Converts electrical signals to light
Composites		
Graphite-epoxy	Aircraft components	High strength-to-weight ratio
Tungsten carbide-cobalt	Carbide cutting tools for machining	High hardness yet good shock resistance
Titanium-clad steel	Reactor vessels	Low cost and high strength of steel with good corrosion resistance of titanium

Ceramics have excellent resistance to high temperatures and certain corrosive media and have a number of unusual optical and electrical properties that are used in constructing integrated circuits, fiber-optic systems, and a variety of sensing devices.

Polymers.   Polymers include rubber, plastics, and many types of adhesives. They are produced by creating large molecular structures from organic molecules in a process known as *polymerization*. Polymers have low electrical and thermal conductivities, have low strengths, and are not suitable for use at high temperatures. *Thermoplastic* polymers, in which the long molecular chains are not rigidly connected, have good ductility and formability; *thermosetting* polymers are stronger but more brittle because the molecular chains are tightly linked (Figure 1-4). Polymers are used in many applications, including electronic devices (Figure 1-5).

FIGURE 1-3 A variety of complex ceramic components, including impellers and blades, which allow turbine engines to operate more efficiently at higher temperatures. (Courtesy of Certech, Inc.)

Semiconductors.   Although silicon, germanium, and a number of compounds such as GaAs are very brittle, they are essential for electronic, computer, and communication applications. The electrical conductivity of these materials can be controlled so that they can be used in electronic devices such as transistors, diodes, and integrated circuits (Figure 1-6). Information is now being transmitted by light through fiber-optic systems; semiconductors, which convert electrical signals to light and vice versa, are essential components in these systems.

Composite Materials.   Composites are formed from two or more materials, producing properties that cannot be obtained by any single material. Concrete, plywood, and fiberglass are typical, although crude, examples of composite materials (Figure 1-7). With composites we can produce lightweight, strong, ductile, high temperature-resistant materials that are otherwise unobtainable, or we can produce hard yet shock-resistant cutting tools that would otherwise

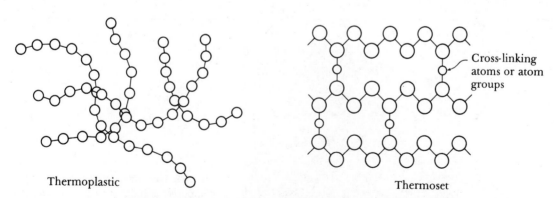

Thermoplastic

Cross-linking atoms or atom groups

Thermoset

FIGURE 1-4 Polymerization occurs when small molecules combine to produce larger molecules, or polymers. The polymer molecules can have a chainlike structure (thermoplastics) or can form three-dimensional networks (thermosets).

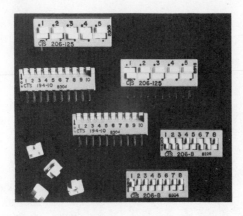

FIGURE 1-5 *Polymers are used in a variety of electronic devices, including these computer dip switches, where moisture resistance and low conductivity are required. (Courtesy of CTS Corporation.)*

shatter. Advanced aircraft and aerospace vehicles rely heavily on composites such as carbon-fiber-reinforced polymers (Figure 1-8).

◇ | **Example 1-1**

You wish to select the materials needed to carry a current between the components inside an electrical "black box." What materials would you select?

Answer:

The material that actually carries the current must have a high electrical conductivity. Thus, we need to select a *metal* wire. Copper, aluminum, gold, or silver might all serve. However, the metal wire must be insulated from the rest of the "black box" to prevent short circuits or arcing. Although a ceramic coating would be an excellent insulator, ceramics are brittle; the wire could not be bent without the ceramic coating breaking off. Instead we would select a *polymer* or plastic coating with good insulating characteristics yet good ductility.

FIGURE 1-6 *Integrated circuits for computers and other electronic devices rely on the unique electrical behavior of semiconducting materials. (Courtesy of Rogers Corporation.)*

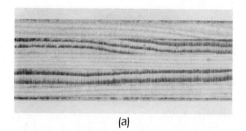

(a)

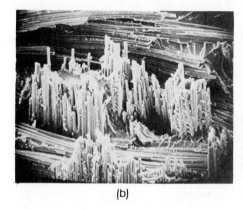

(b)

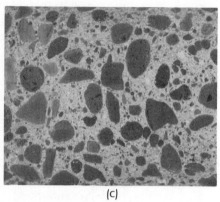

(c)

FIGURE 1-7 Some examples of composite materials. (a) Plywood is a laminar composite composed of layers of wood veneer. (b) Fiberglass is a fiber-reinforced composite containing stiff, strong glass fibers in a softer polymer matrix ($\times$ 175). (c) Concrete is a particulate composite containing coarse sand or gravel in a cement matrix (reduced 50%).

FIGURE 1-8 The X-wing for advanced helicopters relies on a composite composed of a carbon-fiber-reinforced polymer. Courtesy of Sikorsky Aircraft Division—United Technologies Corporation.

Example 1-2

What materials are used to make coffee cups? What particular property makes these materials suitable?

Answer:

Coffee cups are normally made of ceramic or plastic materials. Both ceramics and polymers have excellent thermal insulation due to their low thermal conductivity. Disposable expanded polystyrene cups are particularly effective, since they contain many gas bubbles which further improve insulation. Actually, we could consider the disposable cups to be made from a composite of a polymer and gas!

Metal cups, however, are seldom used because the high thermal conductivity permits the heat to be transferred, burning our hands.

1–3 STRUCTURE-PROPERTY-PROCESSING RELATIONSHIP

We are interested in producing a component that has the proper shape and properties, permitting the component to perform its task for its expected lifetime. The materials engineer meets this requirement by taking advantage of a complex three-part relationship between the internal structure of the material, the processing of the material, and the final properties of the material (Figure 1-9). When the materials engineer changes one of these three aspects of the relationship, either or both of the others also change. We must therefore determine how the three aspects interrelate in order to finally produce the required product.

Properties.   We can consider the properties of a material in two categories—mechanical and physical (Table 1-2).

Mechanical properties describe how a material responds to an applied force. The most common mechanical properties are strength, ductility, and stiffness (modulus of elasticity). However, we are often interested in how a material behaves when it is exposed to a sudden, intense blow (impact), continually cycled through an alternating force (fatigue), exposed to high temperatures (creep), or subjected to abrasive conditions (wear). Mechanical properties also determine the ease with which a material can be deformed into a useful shape. A metal part formed by forging must withstand the rapid application of a force without breaking and must have a high enough ductility to deform to the proper shape. Often, small structural changes have a profound effect on the mechanical properties of a material.

Physical properties, which include electrical, magnetic, optical, thermal, elastic, and chemical behavior, depend on both structure and processing of a material. Even tiny changes in composition cause profound changes in the electrical conductivity of many semiconducting materials. For example, high firing

TABLE 1-2   Typical examples of the properties of materials

Mechanical Properties	Physical Properties

Creep
    Creep rate
    Stress-rupture properties
	Chemical
	    Corrosion
	    Refining

Ductility
    % Elongation
    % Reduction in area
	Density

	Electrical
	    Conductivity
	    Dielectric (insulation)
	    Ferroelectricity and piezoelectricity

Fatigue
    Endurance limit
    Fatigue life	

Hardness
    Scratch resistance
    Wear rates
	Magnetic
	    Ferrimagnetic
	    Ferromagnetic

Impact
    Absorbed energy
    Toughness
    Transition temperature
	Optical
	    Absorption and color
	    Diffraction
	    Lasing action
	    Photoconduction
	    Reflection, refraction, and transmission

Strength
    Modulus of elasticity
    Tensile strength
    Yield strength
	Thermal
	    Heat capacity
	    Thermal conductivity
	    Thermal expansion

temperatures may greatly reduce the thermal insulation characteristics of ceramic brick; small amounts of impurities change the color of a glass or polymer.

◇ | **Example 1-3**

Describe some of the key mechanical and physical properties we would consider in selecting a material for an airplane wing.

Answer:

First, let's look at mechanical properties. We obviously want the material to have a high strength to support the forces acting on the wing. We must also recognize that the wing is exposed to a cyclical application of a force as well as vibration—this suggests that fatigue properties are important. In supersonic flight, the wing may become very hot, so resistance to creep is critical.

Important physical properties are density and corrosion resistance. The wing should be as light as possible, so the material should have a low density. If the wing is exposed to a marine atmosphere, corrosion resistance also may be important.

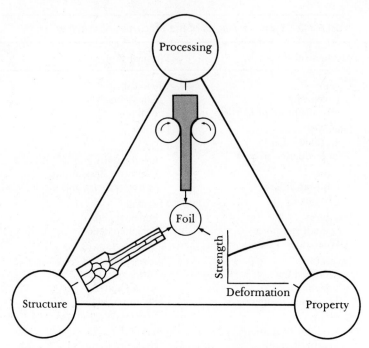

FIGURE 1-9   The three-part relationship between structure, properties, and processing method. When aluminum is rolled into foil, the rolling process changes the metal's structure and increases its strength.

◇ │   **Example 1-4**

What properties are required for a solar cell used to generate electricity for a satellite?

Answer:

The electrical and optical properties are most important in this case! The materials for solar cells, such as silicon, must interact with radiation, or light, to change the electron configuration of the atom. This interaction and change in structure in turn produce the electrical current that is desired.

Structure.   The structure of a material can be considered on several levels, all of which influence the final behavior of the product (Figure 1-10). At the finest level is the structure of the individual atoms that compose the material. The arrangement of the electrons surrounding the nucleus of the atom significantly affects electrical, magnetic, thermal, and optical behavior and may also influence corrosion resistance. Furthermore, the electronic arrangement influences

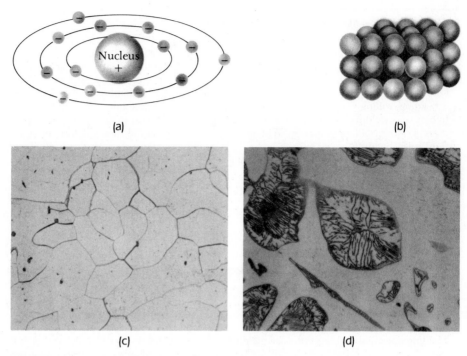

FIGURE 1-10 *Four levels of structure in a material. (a) Atomic structure, (b) crystal structure, (c) grain structure in iron (× 100) (d) multiple-phase structure in white cast iron (× 200).*

how the atoms are bonded to one another and helps determine the type of material—metal, ceramic, semiconductor, or polymer.

At the next level, the arrangement of the atoms in space is considered. Metals, semiconductors, many ceramics, and some polymers have a very regular atomic arrangement, or crystal structure. The crystal structure influences the mechanical properties of metals. Other ceramic materials and many polymers have no orderly atomic arrangement—these amorphous, or glassy, materials behave very differently from crystalline materials. For instance, glassy polyethylene is transparent, whereas crystalline polyethylene is translucent. Imperfections in either type of atomic arrangement may be controlled to produce profound changes in properties.

A grain structure is found in most metals, semiconductors, and ceramics and occasionally in polymers. The size and shape of the grains play a key role at this level. In some cases, as with silicon chips for integrated circuits or metals for jet engine parts, we wish to produce a material containing only one grain, or a single crystal.

Finally, in most materials, more than one phase is present, with each phase having its unique atomic arrangement and properties. Control of the type, size, distribution, and amount of these phases within the main body of the material provides an additional way to control properties.

TABLE 1-3 Typical materials processing techniques

Metals

Casting: sand, die cast, permanent mold, investment, continuous casting
Liquid metal is poured or injected into a solid mold to produce a desired shape.

Forming: forging, wire drawing, deep drawing, bending, rolling
Solid metal is deformed by high pressure, often while hot, into useful shapes.

Joining: gas welding, resistance welding, brazing, arc welding, soldering, friction welding, diffusion bonding
Several pieces of metal are joined together, using liquid metal, deformation, or high pressures and temperatures to provide bonding.

Machining: turning, drilling, milling, cutting
Metal is removed by a cutting operation, leaving a finished shape.

Powder metallurgy
Metal powders are compacted at high pressures into a useful shape, then heated at high temperatures to permit the particles to join together.

Ceramics

Casting, including slip casting
Liquid or slurries of liquid plus solid ceramics are poured into a desirable shape.

Compaction: extrusion, pressing, isostatic forming
Solid or viscous slurries of liquid and solid ceramics are compacted into a useful shape.

Sintering
Compacted solids are heated at high temperatures to cause the solids to bond together.

Polymers

Molding: injection molding, transfer molding
Hot or even liquid polymer is forced into a mold; this resembles the casting process.

Forming: spinning, extrusion, vacuum forming
Heated polymer is forced through a die opening or around a pattern to produce a shape.

Semiconductors

Crystal growing
A liquid is frozen to produce a single crystal.

Chemical vapor decomposition
A solid is condensed from a gas onto a second substrate material.

Composites

Casting, including infiltration
A liquid surrounds one of the constituents to produce the completed composite.

Forming
A soft constituent is forced by pressure to deform around a second constituent of the composite.

Joining: adhesive bonding, explosive bonding, diffusion bonding
The two constituents are joined together by gluing, deformation, or high-temperature processes.

Compaction and sintering
Powdered constituents are pressed into shapes, then heated to cause the powders to join.

Processing. Materials processing produces the desired shape of a component from the initial formless material (Table 1-3). Metals can be processed by pouring liquid metal into a mold (casting), joining individual pieces of metal (welding, brazing, soldering, adhesive bonding), forming the solid metal into useful shapes using high pressures (forging, drawing, extrusion, rolling, bending), compacting tiny metal powder particles into a solid mass (powder metallurgy), or removing excess material (machining). Similarly, ceramic materials can be formed into shapes by related processes such as casting, forming, extrusion, or compaction, often while wet, and heat treatment at high temperatures to drive off the fluids and to bond the individual constituents together. Polymers are produced by injection of softened plastic into molds (much like casting), drawing, and forming. Often a material is heat treated at some temperature below its melting temperature to effect a desired change in structure. The type of processing we use depends, at least partly, on the properties, and thus the structure, of the material.

Often a variety of materials are used during processing to produce a final component. One example of this is the evaporative pattern casting (lost foam) process for making aluminum castings illustrated in Figure 1-11. In this manufacturing process, a *polymer* pattern is made by expanding polystyrene beads into a die, forming a shape nearly identical to the intended part. The pattern is then coated with a thin layer of *ceramic* and backed up by loose sand grains (sand, or silica, is another *ceramic*). Finally, molten aluminum, a *metal*, is poured into the mold; the polymer pattern vaporizes as the metal takes its place. Solidification of the molten aluminum produces the final cast shape.

◇ | **Example 1-5**

The first step in the manufacture of tungsten filaments for light bulbs is accomplished by powder metallurgy rather than casting. Explain.

Answer:

One of the physical properties of tungsten is its high melting temperature, 3410° C. In order to make a casting, the tungsten must be heated to an exceptionally high temperature. Powder metallurgy processing, by which powdered tungsten is compacted into a solid mass, is done at much lower temperatures.

1-4 ENVIRONMENTAL EFFECTS ON MATERIAL BEHAVIOR

The structure-property-processing relationship is also influenced by the surroundings to which the material is subjected.

Temperature. Changes in temperature dramatically alter the properties of materials (Figure 1-12). The strength of most materials decreases as the temperature increases. Furthermore, sudden catastrophic changes may occur when heating above critical temperatures. Metals that have been strengthened by certain heat treatments or forming techniques may suddenly lose their strength when heated. Very low temperatures may cause a metal to fail in a brittle manner even though the applied loads are low. High temperatures can also change the structure of ceramics or cause polymers to melt or char.

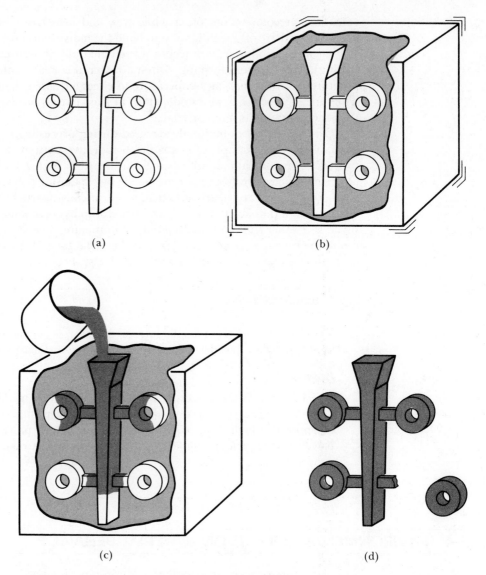

(a)

(b)

(c)

(d)

FIGURE 1-11   The evaporative pattern casting process involves the use of polymers and ceramics to produce a metal part. (a) An expanded polystyrene pattern is coated with ceramic, (b) the coated pattern is surrounded by vibrated sand, (c) molten metal vaporizes and displaces the Styrofoam, and (d) the solidified castings are removed from the tree and cleaned.

The design of materials with improved resistance to temperature is essential in many technologies, as illustrated by the increase in operating temperatures of aircraft and aerospace vehicles (Figure 1-13). As faster speeds are obtained, more heating of the vehicle skin occurs due to friction with the air; in addition, engines operate more efficiently at higher temperatures. In order to obtain

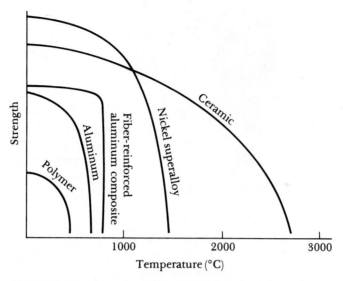

FIGURE 1-12 *Increasing temperature normally reduces the strength of a material. Polymers are suitable only at low temperatures. Some composites, special alloys, and ceramics have excellent properties at high temperatures.*

higher speeds and better fuel economy, new materials have gradually increased allowable skin and engine temperatures. But materials engineers are continually faced with new challenges. The "Orient Express," an advanced aircraft intended to carry passengers across the Pacific Ocean in less than three hours, will require

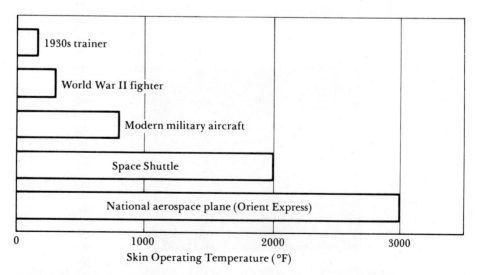

FIGURE 1-13 *Skin operating temperatures for aircraft have increased with the development of improved materials. (After M. Steinberg, **Scientific American**, October, 1986.)*

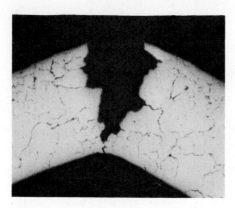

FIGURE 1-14 When hydrogen dissolves in tough pitch copper, steam is produced at the grain boundaries, thus creating thin voids. The metal is then weak and brittle and fails easily ($\times$ 50).

the development of even more exotic materials and processing techniques in order to meet the higher temperatures that will be encountered.

Corrosion.   Most metals and polymers react with oxygen or other gases, particularly at elevated temperatures. Metals and ceramics may catastrophically disintegrate (Figure 1-14); polymers may become brittle. Materials are also attacked by a variety of corrosive liquids. A metal may be uniformly or selectively consumed or may develop cracks or pits, leading to premature failure (Figure 1-15). Ceramics can be attacked by other liquid ceramics, and solvents can dissolve polymers. The materials engineer faces the challenge of developing new materials or coatings that will prevent these reactions and permit materials to operate in more extreme environments.

Radiation.   High-energy radiation, such as neutrons produced in nuclear reactors, can affect the internal structure of all materials, producing a loss of

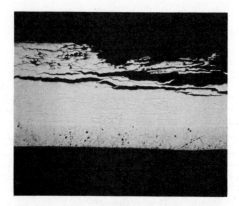

FIGURE 1-15 Attack of an aluminum fuel tank by bacteria in contaminated jet fuel causes severe corrosion, pitting, and eventual failure ($\times$ 10).

strength, embrittlement, or critical alteration of physical properties. External dimensions may also change, causing swelling or even cracking.

 Example 1-6

What precautions might have to be taken when joining titanium by a welding process?

Answer:

During welding, the titanium is heated to a high temperature. The high temperature may cause detrimental changes in the structure of the titanium, even eliminating some of the strengthening mechanisms by which the properties of the metal were obtained. Furthermore, titanium reacts rapidly with oxygen, hydrogen, and other gases at high temperatures. A welding process must supply a minimum of heat while protecting the metal from the surrounding atmosphere. Special gases, such as argon, or even a vacuum may be needed.

1–5 MATERIALS SELECTION

When selecting a material for a given application, care must be taken to select (1) a material that can develop the desired physical and mechanical properties, (2) a material that can be processed or manufactured into the desired shape, and (3) a material and process that are economical. In satisfying these three requirements, trade-offs may have to be made in order to produce a satisfactory yet marketable product.

As one example, we may decide to produce a complex metal fastener by investment casting rather than by forging. In forging, a heated metal blank is deformed into shape by the application of a force; the forging may then be machined into the final shape. In investment casting, several steps are required, including producing an aluminum die, injecting molten wax into the die to produce a pattern, coating the solidified wax with ceramic, heating the ceramic mold to melt out the wax and strengthen the mold, and, finally, pouring the molten metal into the mold to produce a part that is almost completely finished. Although the forging process itself may be much less expensive than the complicated casting process, the additional cost of machining may make the final forged part more expensive.

As another example, material cost is normally based on cost per pound. We must consider the *density* of the material, or its weight per unit volume, in our design and selection (Table 1-4). Aluminum may cost more per pound than steel, but it is only one-third the weight of steel. Although parts made from aluminum may have to be thicker, the aluminum part may be less expensive than the one made from steel, due to the weight difference.

TABLE 1-4   Strength-to-weight ratio of various materials

Material	Strength (lb/in^2)	Density (lb/in^3)	Strength-to-weight ratio (in.)
Polyethylene	1,000	0.03	0.03×10^6
Pure aluminum	6,500	0.098	0.07×10^6
Pure copper	30,000	0.32	0.09×10^6
Low-carbon steel	57,000	0.28	0.21×10^6
Pure titanium	35,000	0.16	0.22×10^6
Al_2O_3	30,000	0.114	0.26×10^6
Nylon	11,000	0.04	0.28×10^6
Epoxy	15,000	0.05	0.30×10^6
High-carbon steel	89,000	0.28	0.32×10^6
Si_3N_4	70,000	0.114	0.61×10^6
Heat-treated alloy steel	240,000	0.28	0.86×10^6
Heat-treated aluminum alloy	86,000	0.098	0.88×10^6
Carbon-carbon composite	60,000	0.065	0.92×10^6
Heat-treated titanium alloy	170,000	0.16	1.06×10^6
Kevlar-epoxy composite	65,000	0.05	1.30×10^6
Carbon-epoxy composite	80,000	0.05	1.60×10^6

In some instances, particularly in aerospace applications, weight is critical, since additional vehicle weight increases the fuel consumption and reduces the range. By using materials that are lightweight but very strong, aerospace vehicles can be designed to improve fuel utilization. Many advanced aerospace vehicles use composite materials instead of aluminum. These composites, such as carbon-epoxy, are more expensive than the traditional aluminum alloys; however, the fuel savings due to the higher *strength-to-weight ratio* of the composite (Table 1-4) may more than offset the higher initial cost of the aircraft.

◇   **Example 1-7**

Suppose we wish to make a metal cable that can support a force of 1000 kg. The strength of aluminum is $2000 \, kg/cm^2$, while that of steel is $3000 \, kg/cm^2$. The densities of the two metals are $2.7 \, g/cm^3$ for aluminum and $7.8 \, g/cm^3$ for steel. The costs of the metals are \$1.10/kg for aluminum and \$0.66/kg for steel. Compare the volume, weight, and cost of the cable per meter for each material.

Answer:

Let's find the cross-sectional area of the cable needed for each material, since the strength is in terms of kg/cm^2.

$$\text{Area of Al cable} \; = \; \frac{1000 \, \text{kg force}}{2000 \, \text{kg/cm}^2 \, \text{strength}} \; = \; 0.5 \, \text{cm}^2$$

$$\text{Area of steel cable} \; = \; \frac{1000 \, \text{kg force}}{3000 \, \text{kg/cm}^2 \, \text{strength}} \; = \; 0.33 \, \text{cm}^2$$

The volume of metal per meter of cable is

$$\text{Volume of Al cable/m} = (0.5\,\text{cm}^2)(100\,\text{cm/m}) = 50\,\text{cm}^3/\text{m}$$
$$\text{Volume of steel cable/m} = (0.33\,\text{cm}^2)(100\,\text{cm/m}) = 33\,\text{cm}^3/\text{m}$$

The weight of metal per meter of cable is

$$\text{Weight of Al cable/m} = (50\,\text{cm}^3/\text{m})(2.7\,\text{g/cm}^3) = 135\,\text{g/m}$$
$$\text{Weight of steel cable/m} = (33\,\text{cm}^3/\text{m})(7.8\,\text{g/cm}^3) = 258\,\text{g/m}$$

The cost of the cable per meter is

$$\text{Cost of Al cable/m} = (135\,\text{g/m})\left(\frac{1}{1000}\,\frac{\text{kg}}{\text{g}}\right)(\$1.10/\text{kg}) = \$0.15/\text{m}$$

$$\text{Cost of steel cable/m} = (258\,\text{g/m})\left(\frac{1}{1000}\,\frac{\text{kg}}{\text{g}}\right)(\$0.66/\text{kg}) = \$0.17/\text{m}$$

The aluminum cable, although larger in volume and diameter, is lighter in weight and less expensive than the steel cable.

SUMMARY

The procedure of selecting a material, processing the material into a useful shape, and obtaining the needed properties is a complicated process involving knowledge of the structure-property-processing relationship. The remainder of this text is intended to introduce the student to the wide variety of materials available. As we do so, we will come to understand the fundamentals of the structure of materials, how the structure affects the behavior of the material, and the role that processing and the environment play in shaping the relationship between structure and properties.

If you are ready, let's begin.

GLOSSARY

Alloys. Combinations of metals which enhance the general characteristics of metals.

Ceramics. A group of materials characterized by good strength and high melting temperatures, but poor ductility and electrical conductivity. Ceramic raw materials are typically compounds of metallic and nonmetallic elements.

Composites. A group of materials formed from combinations of metals, ceramics, or polymers in such a manner that unusual combinations of properties are obtained.

Metals. A group of materials having the general characteristics of good ductility, strength, and electrical conductivity.

Polymerization. The process by which organic molecules are joined into giant molecules, or polymers.

Polymers. A group of materials normally obtained by joining organic molecules into giant molecular chains or networks.

Polymers are characterized by low strengths, low melting temperatures, and poor electrical conductivity.

Semiconductors. A group of materials having intermediate electrical conductivity and other unusual physical properties.

Thermoplastics. A special group of polymers that are easily formed into useful shapes. Normally, these polymers have a chainlike structure.

Thermosets. A special group of polymers that are normally quite brittle. These polymers typically have a three-dimensional network structure.

PRACTICE PROBLEMS

1. The Stealth aircraft is designed so that it will not be detected by radar. What physical property should the materials used in the Stealth plane possess to meet this design requirement?

2. Certain materials, such as tungsten carbide, are compounds consisting of both metallic and nonmetallic elements. To which category of materials does tungsten carbide belong?

3. From which category of materials would you select a material best suited for building a vessel to contain liquid steel?

4. Consider a cement wall reinforced with steel bars. Into which category of materials would you place reinforced concrete?

5. Boron nitride (BN) and silicon carbide (SiC) are important materials in abrasive grinding wheels. In what category of materials do BN and SiC belong? BN and SiC are in the form of small particles and are often embedded in a polymer to produce the grinding wheel. In what category of materials does the entire wheel belong?

6. Silicon carbide (SiC) fibers are sometimes mixed with liquid aluminum. After the mixture freezes, a fiber-reinforced composite results. Would you reinforce aluminum with high-strength polyethylene fibers in the same manner? Explain your answer.

7. The nose of the space shuttle is composed of graphite (carbon). Based on this information, what type of material would graphite be?

8. Suppose we would like to make a porous metal filter to keep the oil in our automobile engine clean. Which one of the metal processing techniques listed in Table 1-3 might be used to produce these filters?

9. Sintering is listed in Table 1-3 as a ceramic processing technique. In which one of the metal processing techniques would you expect sintering also to be used?

10. Which of the three ceramic processing methods mentioned in Table 1-3 do you think is used to produce glass bottles?

11. By which one of the four methods of producing composite materials listed in Table 1-3 would you expect plywood to be made?

12. Injection molding to produce plastic parts most closely resembles which one of the metals processing methods?

13. The *Voyager* is an experimental aircraft that flew around the world nonstop on a single tank of fuel. What type of material do you think made up most of the aircraft? Explain why this type of material was selected.

14. United States coinage, such as the quarter, appear silvery on the face, but close inspection reveals a reddish color at the edges. Based on your observations, to which one of the five categories of materials should a quarter belong? Explain.

15. Relays in electrical circuits open and close frequently, causing the electrical contacts to wear. MgO is a very hard, wear-resistant material. Why would this material not be suitable for use as contacts in a relay?

16. What mechanical properties would

you consider most important when selecting a material to serve as a spring for an automobile suspension? Explain.

17. The devices used for memory in personal computers typically contain an integrated circuit, electrical leads, a strong, nonconducting base, and an insulating coating. From what material should each of these four basic components be made? Explain your selections.

18. Automobile bumpers might be made from a polymer material. Would you recommend a thermoplastic or a thermosetting polymer for this application? Explain.

19. Sometimes a nearly finished part is *coined*. During coining, a force is applied to deform the part into its final shape. For which of the following could this be done without danger of breaking the part—brass, Al_2O_3, thermoplastic polymers, thermosetting polymers, silicon?

20. A scrap metal processor would like to be able to identify different materials quickly, without resorting to chemical analysis or lengthy testing. Describe some possible testing techniques based on the physical properties of materials.

Part I

ATOMIC STRUCTURE, ARRANGEMENT, AND MOVEMENT

The electronic structure of the atom determines the nature of the atomic bonding, which in turn imparts certain general properties to metals, ceramics, and polymers. Furthermore, the electronic structure plays a predominant role in determining the physical properties, such as electrical conductivity, dielectric and magnetic behavior, and optical and thermal characteristics, of the material.

The arrangement of atoms into a crystalline or amorphous structure influences the physical properties and in particular the mechanical behavior of materials. Imperfections in the normal atomic arrangement play a critical role in understanding the deformation and mechanical properties of materials.

Finally, the movement of atoms, known as diffusion, is important for many heat treatments and manufacturing processes, and for both physical and mechanical properties of materials.

In the next four chapters, we will examine atomic structure, atomic arrangement, imperfections in the atomic arrangement, and atom movement. This examination will lay the groundwork needed to understand the structure and behavior of materials which we will discuss later.

2

ATOMIC STRUCTURE

2-1 INTRODUCTION

The structure of a material may be divided into four levels—atomic structure, atomic arrangement, microstructure, and macrostructure. Although the main thrust of this text is to understand and control the microstructure and macrostructure of various materials, we must first understand the atomic and crystal structures.

Atomic structure influences how the atoms are bonded together, which in turn helps us to categorize materials as metals, semiconductors, ceramics, and polymers and permits us to draw some general conclusions concerning the mechanical properties and physical behavior of these four classes of materials.

2-2 THE STRUCTURE OF THE ATOM

Most of us know that an atom is composed of a nucleus surrounded by electrons. The nucleus contains neutrons and positively charged protons and thus carries a net positive charge. The negatively charged electrons are held to the nucleus by an electrostatic attraction. The electrical charge q carried by each electron and proton is 1.60×10^{-19} C. Because the numbers of electrons and protons in the atom are equal, the atom as a whole is electrically neutral.

The *atomic number* of an element is equal to the number of electrons or protons in each atom. Thus, an iron atom, which contains 26 electrons and 26 protons, has an atomic number of 26.

Most of the mass of the atom is contained within the nucleus. The mass of each proton and neutron is about 1.67×10^{-24} g, but the mass of each electron is only 9.11×10^{-28} g. The *atomic mass M*, which is equal to the average number of protons and neutrons in the atom, is the mass of the *Avogadro number* N_A of atoms.[1] $N_A = 6.02 \times 10^{23}\,\text{mol}^{-1}$ is the number of atoms or molecules in a g·mole. Therefore, the atomic mass has units of g/g·mole. An alternate unit for atomic mass is the atomic mass unit, or amu, which is $\frac{1}{12}$ the mass of carbon 12.

Atoms of the same element that contain a different number of neutrons in the nucleus are called *isotopes* and thus have a different atomic mass. The atomic mass used for such an element is an average value of those of the different isotopes and thus the atomic mass may not be a whole number.

◇ Example 2-1

In a collection of nickel atoms, 70% of the atoms contain 30 neutrons and 30% of the atoms contain 32 neutrons. The atomic number for nickel is 28. Calculate the approximate average atomic mass of nickel.

Answer:

The atomic mass of the atoms containing 30 neutrons is M = number of protons + number of neutrons = 28 + 30 = 58 g/g·mole. These atoms are the Ni^{58} isotope.

The atomic mass of the atoms containing 32 neutrons is M = 28 + 32 = 60 g/g·mole, or the atomic mass of Ni^{60}.

The average atomic mass of nickel is

$$
\begin{aligned}
M \text{ of Ni} &= (0.7)(M \text{ of } Ni^{58}) + (0.3)(M \text{ of } Ni^{60}) \\
&= (0.7)(58) + (0.3)(60) = 58.6 \text{ g/g·mole}
\end{aligned}
$$

2–3   THE ELECTRONIC STRUCTURE OF THE ATOM

Electrons occupy discrete energy levels within the atom. Each electron possesses a particular energy, with no more than two electrons in each atom having the same energy. This also implies that there is a definite energy difference between each electron.

Quantum Numbers.   The energy level to which each electron belongs is determined by four quantum numbers. The number of possible energy levels is determined by the first three quantum numbers.

1. The *principal quantum number n* is assigned integral values 1, 2, 3, 4, 5, . . . , that refer to the quantum shell to which the electron belongs (Figure 2-1). Often the quantum shells are assigned a letter rather than a number; the shell for $n = 1$ is designated K, for $n = 2$ is L, for $n = 3$ is M, and so on.

---

[1] Often the atomic mass is called the *atomic weight*.

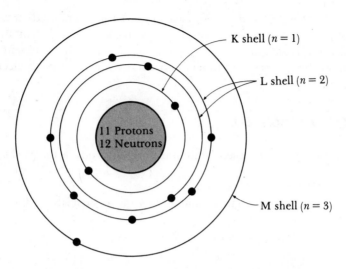

FIGURE 2-1   *The atomic structure of sodium, atomic number 11, showing the electrons in the K, L, and M quantum shells.*

2. The number of energy levels in each quantum shell is determined by the *azimuthal quantum number l* and the *magnetic quantum number* m_l. The azimuthal quantum numbers may also be assigned numbers: $l = 0, 1, 2, \ldots, n - 1$. If $n = 2$, then there are also two azimuthal quantum numbers, $l = 0$ and $l = 1$. The azimuthal quantum numbers are often designated by lowercase letters,

$$s \text{ for } l = 0 \qquad d \text{ for } l = 2$$
$$p \text{ for } l = 1 \qquad f \text{ for } l = 3$$

The magnetic quantum number m_l gives the number of energy levels, or orbitals, for each azimuthal quantum number. The total number of magnetic quantum numbers for each l is $2l + 1$. The values for m_l are given by whole numbers between $-l$ and $+l$. For $l = 2$, there are $2(2) + 1 = 5$ magnetic quantum numbers, with values -2, -1, 0, $+1$, and $+2$.

◇ | **Example 2-2**

Calculate the number of possible ortibals in the L shell, where $n = 2$.

Answer:

If $n = 2$, then $l = 0, 1$.

For $l = 0$, there are $2(0) + 1 = 1$ magnetic quantum numbers, so $m_l = 0$.

For $l = 1$, there are $2(1) + 1 = 3$ magnetic quantum numbers, so $m_l = -1, 0, +1$.

Consequently, there are a total of four orbitals possible in the L shell.

3. The *Pauli exclusion principle* specifies that no more than two electrons, each with opposing electronic spins, may be present in each orbital. The *spin quantum number m_s* is assigned values $+\frac{1}{2}$ and $-\frac{1}{2}$ to reflect the different spins. Figure 2-2 shows the quantum numbers and energy levels for each electron in a sodium atom.

◇ Example 2-3

Determine the maximum number of electrons in the M shell of an atom.

Answer:

The principal quantum number for the M shell is $n = 3$. If $n = 3$, then $l = 0$, 1, 2.

s level, $l = 0$, $m_l = \quad 0$, $m_s = +\frac{1}{2}, -\frac{1}{2}$ $\qquad$ 2 electrons

p level, $l = 1$, $m_l = -1$, $m_s = +\frac{1}{2}, -\frac{1}{2}$

$\qquad = \quad 0$, $m_s = +\frac{1}{2}, -\frac{1}{2}$ $\qquad$ 6 electrons

$\qquad = +1$, $m_s = +\frac{1}{2}, -\frac{1}{2}$

d level, $l = 2$, $m_l = -2$, $m_s = +\frac{1}{2}, -\frac{1}{2}$

$\qquad = -1$, $m_s = +\frac{1}{2}, -\frac{1}{2}$

$\qquad = \quad 0$, $m_s = +\frac{1}{2}, -\frac{1}{2}$ $\qquad$ 10 electrons

$\qquad = +1$, $m_s = +\frac{1}{2}, -\frac{1}{2}$

$\qquad = +2$, $m_s = +\frac{1}{2}, -\frac{1}{2}$

Thus, a total of 18 electrons may be present in the M shell.

If we continue the exercises in Examples 2-2 and 2-3, we could represent the maximum number of electrons in each energy shell by the pattern in Table 2-1.

The shorthand notation frequently used to denote the electronic structure of an atom combines the numerical value of the principal quantum number, the lowercase letter notation for the azimuthal quantum number, and a superscript showing the number of electrons in each orbital. Thus, the shorthand notation for the electronic structure of germanium, which has an atomic number of 32, is

$$1s^2 2s^2 2p^6 3s^2 3p^6 3d^{10} 4s^2 4p^2$$

The electronic configurations for the elements are summarized in Table 2-2.

Deviations from Expected Electronic Structures. The orderly building up of the electronic structure is not always followed, particularly when the atomic number is large and the d and f levels begin to fill. For example, we would expect

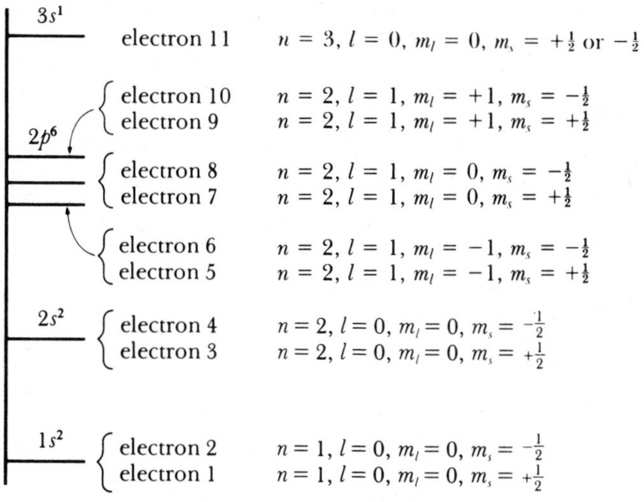

$3s^1$

electron 11 $n = 3, l = 0, m_l = 0, m_s = +\frac{1}{2} \text{ or } -\frac{1}{2}$

$2p^6$

electron 10 $n = 2, l = 1, m_l = +1, m_s = -\frac{1}{2}$
electron 9 $n = 2, l = 1, m_l = +1, m_s = +\frac{1}{2}$

electron 8 $n = 2, l = 1, m_l = 0, m_s = -\frac{1}{2}$
electron 7 $n = 2, l = 1, m_l = 0, m_s = +\frac{1}{2}$

electron 6 $n = 2, l = 1, m_l = -1, m_s = -\frac{1}{2}$
electron 5 $n = 2, l = 1, m_l = -1, m_s = +\frac{1}{2}$

$2s^2$

electron 4 $n = 2, l = 0, m_l = 0, m_s = -\frac{1}{2}$
electron 3 $n = 2, l = 0, m_l = 0, m_s = +\frac{1}{2}$

$1s^2$

electron 2 $n = 1, l = 0, m_l = 0, m_s = -\frac{1}{2}$
electron 1 $n = 1, l = 0, m_l = 0, m_s = +\frac{1}{2}$

FIGURE 2-2 *The complete set of quantum numbers for each of the 11 electrons in sodium.*

the electronic structure of iron, atomic number 26, to be

$$1s^2 2s^2 2p^6 3s^2 3p^6 \boxed{3d^8}$$

However, from Table 2-2, we find that the actual structure is

$$1s^2 2s^2 2p^6 3s^2 3p^6 \boxed{3d^6 4s^2}$$

The unfilled $3d$ level causes the magnetic behavior of iron, as we will see in Chapter 18. Many other examples of this behavior can be found in Table 2-2.

Valence. The *valence* of an atom is related to the ability of the atom to enter into chemical combination with other elements and is often determined by the number of electrons in the outermost combined sp level. Examples of the

TABLE 2-1 The pattern used to assign electrons to energy levels

		$l = 0$ (s)	$l = 1$ (p)	$l = 2$ (d)	$l = 3$ (f)	$l = 4$ (g)	$l = 5$ (h)
$n = 1$	(K)	2					
$n = 2$	(L)	2	6				
$n = 3$	(M)	2	6	10			
$n = 4$	(N)	2	6	10	14		
$n = 5$	(O)	2	6	10	14	18	
$n = 6$	(P)	2	6	10	14	18	22

Note: 2, 6, 10, 14, . . . , refer to the number of electrons in the energy level.

TABLE 2-2   The electronic configuration for each of the elements

Atomic Number	Element	K 1s	L 2s	2p	M 3s	3p	3d	4s	N 4p	4d	4f	O 5s	5p	5d	P 6s	6p
1	Hydrogen	1														
2	Helium	2														
3	Lithium	2	1													
4	Beryllium	2	2													
5	Boron	2	2	1												
6	Carbon	2	2	2												
7	Nitrogen	2	2	3												
8	Oxygen	2	2	4												
9	Fluorine	2	2	5												
10	Neon	2	2	6												
11	Sodium	2	2	6	1											
12	Magnesium	2	2	6	2											
13	Aluminum	2	2	6	2	1										
14	Silicon	2	2	6	2	2										
15	Phosphorus	2	2	6	2	3										
16	Sulfur	2	2	6	2	4										
17	Chlorine	2	2	6	2	5										
18	Argon	2	2	6	2	6										
19	Potassium	2	2	6	2	6		1								
20	Calcium	2	2	6	2	6		2								
21	Scandium	2	2	6	2	6	1	2								
22	Titanium	2	2	6	2	6	2	2								
23	Vanadium	2	2	6	2	6	3	2								
24	Chromium	2	2	6	2	6	5	1								
25	Manganese	2	2	6	2	6	5	2								
26	Iron	2	2	6	2	6	6	2								
27	Cobalt	2	2	6	2	6	7	2								
28	Nickel	2	2	6	2	6	8	2								
29	Copper	2	2	6	2	6	10	1								
30	Zinc	2	2	6	2	6	10	2								
31	Gallium	2	2	6	2	6	10	2	1							
32	Germanium	2	2	6	2	6	10	2	2							
33	Arsenic	2	2	6	2	6	10	2	3							
34	Selenium	2	2	6	2	6	10	2	4							
35	Bromine	2	2	6	2	6	10	2	5							
36	Krypton	2	2	6	2	6	10	2	6							
37	Rubidium	2	2	6	2	6	10	2	6			1				
38	Strontium	2	2	6	2	6	10	2	6			2				
39	Yttrium	2	2	6	2	6	10	2	6	1		2				
40	Zirconium	2	2	6	2	6	10	2	6	2		2				
41	Niobium	2	2	6	2	6	10	2	6	4		1				
42	Molybdenum	2	2	6	2	6	10	2	6	5		1				
43	Technetium	2	2	6	2	6	10	2	6	6		1				
44	Ruthenium	2	2	6	2	6	10	2	6	7		1				
45	Rhodium	2	2	6	2	6	10	2	6	8		1				
46	Palladium	2	2	6	2	6	10	2	6	10						
47	Silver	2	2	6	2	6	10	2	6	10		1				
48	Cadmium	2	2	6	2	6	10	2	6	10		2				
49	Indium	2	2	6	2	6	10	2	6	10		2	1			

TABLE 2-2 (continued)

Atomic Number	Element	K	L		M			N				O			P	
		1s	2s	2p	3s	3p	3d	4s	4p	4d	4f	5s	5p	5d	6s	6p
50	Tin	2	2	6	2	6	10	2	6	10		2	2			
51	Antimony	2	2	6	2	6	10	2	6	10		2	3			
52	Tellurium	2	2	6	2	6	10	2	6	10		2	4			
53	Iodine	2	2	6	2	6	10	2	6	10		2	5			
54	Xenon	2	2	6	2	6	10	2	6	10		2	6			
55	Cesium	2	2	6	2	6	10	2	6	10		2	6		1	
56	Barium	2	2	6	2	6	10	2	6	10		2	6		2	
57	Lanthanum	2	2	6	2	6	10	2	6	10	1	2	6		2	
⋮	⋮	⋮	⋮	⋮	⋮	⋮	⋮	⋮	⋮	⋮	⋮	⋮	⋮	⋮	⋮	⋮
71	Lutetium	2	2	6	2	6	10	2	6	10	14	2	6	1	2	
72	Hafnium	2	2	6	2	6	10	2	6	10	14	2	6	2	2	
73	Tantalum	2	2	6	2	6	10	2	6	10	14	2	6	3	2	
74	Tungsten	2	2	6	2	6	10	2	6	10	14	2	6	4	2	
75	Rhenium	2	2	6	2	6	10	2	6	10	14	2	6	5	2	
76	Osmium	2	2	6	2	6	10	2	6	10	14	2	6	6	2	
77	Iridium	2	2	6	2	6	10	2	6	10	14	2	6	9		
78	Platinum	2	2	6	2	6	10	2	6	10	14	2	6	9	1	
79	Gold	2	2	6	2	6	10	2	6	10	14	2	6	10	1	
80	Mercury	2	2	6	2	6	10	2	6	10	14	2	6	10	2	
81	Thallium	2	2	6	2	6	10	2	6	10	14	2	6	10	2	1
82	Lead	2	2	6	2	6	10	2	6	10	14	2	6	10	2	2
83	Bismuth	2	2	6	2	6	10	2	6	10	14	2	6	10	2	3
84	Polonium	2	2	6	2	6	10	2	6	10	14	2	6	10	2	4
85	Astatine	2	2	6	2	6	10	2	6	10	14	2	6	10	2	5
86	Radon	2	2	6	2	6	10	2	6	10	14	2	6	10	2	6

valence are

$$\text{Mg:} \quad 1s^2 2s^2 2p^6 \boxed{3s^2} \qquad\qquad \text{valence} = 2$$

$$\text{Al:} \quad 1s^2 2s^2 2p^6 \boxed{3s^2 3p^1} \qquad\qquad \text{valence} = 3$$

$$\text{Ge:} \quad 1s^2 2s^2 2p^6 3s^2 3p^6 3d^{10} \boxed{4s^2 4p^2} \qquad\qquad \text{valence} = 4$$

The valence also depends on the nature of the chemical reaction. The electronic structure of phosphorus is

$$1s^2 2s^2 2p^6 \boxed{3s^2 3p^3}$$

Phosphorus has the expected valence of five when it combines with oxygen. But the valence of phosphorus is only three—the electrons in the 3p level—when it reacts with hydrogen. Manganese may have a valence of 2, 3, 4, 6, or 7!

Atomic Stability. If an atom has a valence of zero, no electrons enter into chemical reactions and the element is inert. An example is argon, which has the electronic structure

$$1s^2 2s^2 2p^6 \boxed{3s^2 3p^6}$$

Other atoms also prefer to behave as if their outer sp levels are either completely full, with eight electrons, or completely empty. Aluminum with the electronic structure

$$1s^2 2s^2 2p^6 \boxed{3s^2 3p^1}$$

has three electrons in its outer sp level. An aluminum atom readily gives up its outer three electrons to empty the $3sp$ level. The nature of the atomic bonding and the chemical behavior of aluminum is determined by the mechanism through which these three electrons interact with surrounding atoms.

On the other hand, chlorine, with an electronic structure

$$1s^2 2s^2 2p^6 \boxed{3s^2 3p^5}$$

contains seven electrons in the outer $3sp$ level. The reactivity of chlorine is caused by its desire to fill its outer energy level by accepting an electron.

Electronegativity.   *Electronegativity* describes the tendency of an atom to gain an electron. Atoms with almost completely filled outer energy levels, like chlorine, are strongly electronegative and readily accept electrons. However, atoms with nearly empty outer levels, such as sodium

$$1s^2 2s^2 2p^6 \boxed{3s^1}$$

readily give up electrons and are strongly *electropositive*. High atomic number elements also have a low electronegativity; because the outer electrons are at a greater distance from the positive nucleus, electrons are not as strongly attracted to the atom. Electronegativities for some elements are shown in Fig. 2-3.

◇ Example 2-4

Using the electronic structures, compare the electronegativities of calcium and bromine.

Answer:

The electronic structures, obtained from Table 2-2, are

Ca: $1s^2 2s^2 2p^6 3s^2 3p^6 \boxed{4s^2}$

Br: $1s^2 2s^2 2p^6 3s^2 3p^6 3d^{10} \boxed{4s^2 4p^5}$

Calcium has two electrons in its outer $4s$ orbital and bromine has seven electrons in its outer $4s4p$ orbital. Calcium tends to give up electrons and is strongly electropositive, but bromine tends to accept electrons and is strongly electronegative.

◇

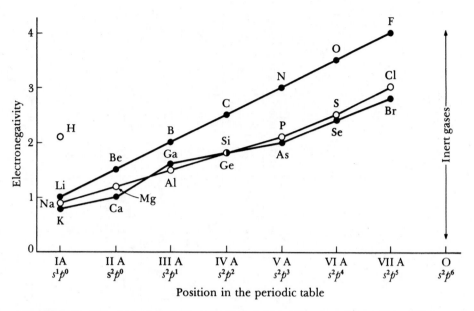

FIGURE 2-3 *The electronegativities of selected elements versus the position of the element in the periodic table.*

2-4   THE PERIODIC TABLE

The construction of the familiar periodic table (Figure 2-4) is based on the electronic configuration of the elements. Consequently, the periodic table can give us some clues to the behavior of the elements.

The IA to VIIA Elements. The rows in the periodic table correspond to quantum shells, or principal quantum numbers [Figure 2-5(a)]. For example, the elements lithium through neon contain electrons in the L shell ($n = 2$), whereas sodium through argon contain electrons in the M shell ($n = 3$).

The columns refer to the number of electrons present in the outermost sp energy level and correspond to the most common valence [Figure 2-5(b)]. Thus lithium, sodium, and potassium in column IA have a valence of one, whereas fluorine, chlorine, and bromine in column VIIA have a valence of seven. The column O on the far right represents the inert gases, which have the outer sp level full. Normally, the elements in each column have similar properties and behavior.

The IIIB to VIIIB Elements. In each of these rows, an inner energy level is progressively filled [Figure 2-5(c)]. The elements in the fourth row, scandium through zinc, are the *transition* elements and contain valence electrons in the N shell. However, the inner $3d$ level in the M shell is not full. The electronic structures of the transition elements are given in Table 2-3.

Similar situations arise in subsequent rows; the $4d$ level is filled in the

IA	IIA	IIIB	IVB	VB	VIB	VIIB	VIIIB	VIIIB	VIIIB	IB	IIB	IIIA	IVA	VA	VIA	VIIA	O
1 **H** 1.00797																	2 **He** 4.003
3 **Li** 6.939	4 **Be** 9.012											5 **B** 10.81	6 **C** 12.011	7 **N** 14.007	8 **O** 15.9994	9 **F** 19.00	10 **Ne** 20.183
11 **Na** 22.99	12 **Mg** 24.31											13 **Al** 26.98	14 **Si** 28.09	15 **P** 30.974	16 **S** 32.064	17 **Cl** 35.453	18 **Ar** 39.948
19 **K** 39.102	20 **Ca** 40.08	21 **Sc** 44.96	22 **Ti** 47.90	23 **V** 50.94	24 **Cr** 52.00	25 **Mn** 54.94	26 **Fe** 55.85	27 **Co** 58.93	28 **Ni** 58.71	29 **Cu** 63.54	30 **Zn** 65.37	31 **Ga** 69.72	32 **Ge** 72.59	33 **As** 74.92	34 **Se** 78.96	35 **Br** 79.909	36 **Kr** 83.80
37 **Rb** 85.47	38 **Sr** 87.62	39 **Y** 88.905	40 **Zr** 91.22	41 **Nb** 92.91	42 **Mo** 95.94	43 **Tc** 98	44 **Ru** 101.1	45 **Rh** 102.90	46 **Pd** 106.4	47 **Ag** 107.87	48 **Cd** 112.4	49 **In** 114.82	50 **Sn** 118.69	51 **Sb** 121.75	52 **Te** 127.60	53 **I** 126.90	54 **Xe** 131.30
55 **Cs** 132.905	56 **Ba** 137.34	57 **La** 138.91	72 **Hf** 178.49	73 **Ta** 180.95	74 **W** 183.85	75 **Re** 186.2	76 **Os** 190.2	77 **Ir** 192.2	78 **Pt** 195.09	79 **Au** 196.97	80 **Hg** 200.59	81 **Tl** 204.37	82 **Pb** 207.19	83 **Bi** 208.98	84 **Po** 210	85 **At** 210	86 **Rn** 222
87 **Fr** 223	88 **Ra** 226	89 **Ac** 227															

← Lanthanide series
← Actinide series

58 **Ce** 140.12	59 **Pr** 140.91	60 **Nd** 144.24	61 **Rm** 147	62 **Sm** 150.35	63 **Eu** 152	64 **Gd** 157.25	65 **Tb** 158.92	66 **Dy** 162.50	67 **Ho** 164.93	68 **Er** 167.26	69 **Tm** 168.93	70 **Yb** 173.04	71 **Lu** 174.97
90 **Th** 232.04	91 **Pa** 231	92 **U** 238.03	93 **Np** 237	94 **Pu** 242	95 **Am** 243	96 **Cm** 247	97 **Bk** 247	98 **Cf** 251	99 **Es** 254	100 **Fm** 253	101 **Md** 256	102 **No** 254	103 **Lw** 257

FIGURE 2-4   The periodic table of the elements.

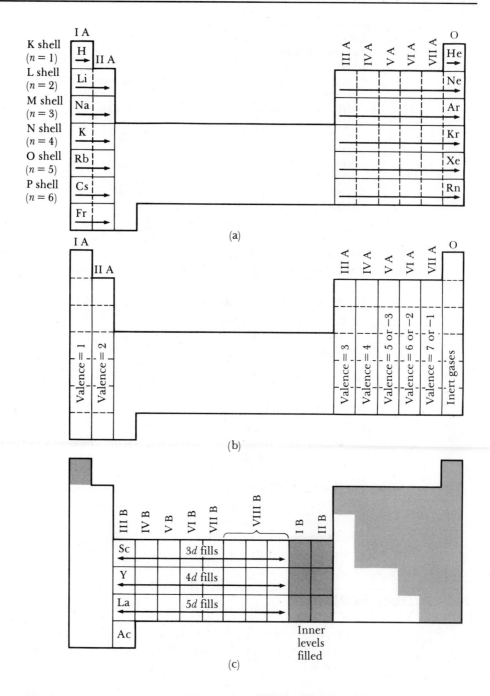

FIGURE 2-5   In the periodic table, (a) rows correspond to principal quantum numbers and show the change in electronegativity, (b) columns correspond to similar electronegativities, and (c) transition elements have partly filled inner energy levels. The lightly shaded regions in (c) denote nonmetallic elements.

TABLE 2-3 The electronic configuration of the transition elements

Group	Element	Electronic Configuration
IIIB	Sc	$\ldots 3s^2 3p^6 \boxed{3d^1} 4s^2$
IVB	Ti	$\ldots 3s^2 3p^6 \boxed{3d^2} 4s^2$
VB	V	$\ldots 3s^2 3p^6 \boxed{3d^3} 4s^2$
VIB	Cr	$\ldots 3s^2 3p^6 \boxed{3d^5} 4s^1$
VIIB	Mn	$\ldots 3s^2 3p^6 \boxed{3d^5} 4s^2$
VIIIB	Fe	$\ldots 3s^2 3p^6 \boxed{3d^6} 4s^2$
VIIIB	Co	$\ldots 3s^2 3p^6 \boxed{3d^7} 4s^2$
VIIIB	Ni	$\ldots 3s^2 3p^6 \boxed{3d^8} 4s^2$

yttrium series, the $5d$ level in the lanthanide series, and the $5f$ level in the actinide series. Elements within each of these series tend to have similar behavior.

The IB and IIB Elements.   These elements, which include copper, silver, and gold, have complete inner shells and one or two valence electrons. We might compare copper with potassium, in Group IA.

K: $1s^2 2s^2 2p^6 3s^2 3p^6 \boxed{4s^1}$

Cu: $1s^2 2s^2 2p^6 3s^2 3p^6 \boxed{3d^{10} 4s^1}$

The filled $3d$ level in copper helps keep the valence electrons tightly held to the inner core; copper, as well as silver and gold, are consequently very stable and unreactive. In potassium, however, the valence electron is not tightly held by the inner $3s3p$ shell and is very reactive.

Distribution of Metals and Nonmetals.   The lightly shaded regions in Figure 2-5(c) show the elements that typically behave as nonmetals; the remaining elements generally display a metallic nature. The nonmetallic elements, in particular boron, carbon, nitrogen, and oxygen, often combine with metallic elements to produce ceramic materials.

2–5 ATOMIC BONDING

There are four important mechanisms by which atoms are bonded in solids. In three of the four mechanisms, bonding is achieved when the atoms fill their outer s and p levels.

The Metallic Bond.   The metallic elements, which have a low valence, give up their valence electrons to form a "sea" of electrons surrounding the atoms (Figure 2-6). Aluminum, for example, gives up its three valence electrons, leaving behind a core consisting of the nucleus and inner electrons. Since three negatively charged electrons are missing from this core, the core becomes an ion

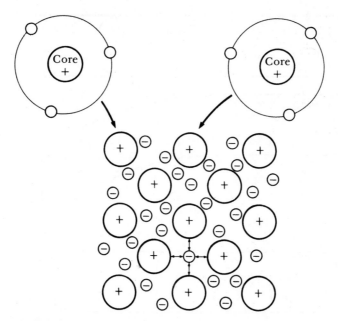

FIGURE 2-6   *The metallic bond forms when atoms give up their valence electrons, which then form an electron sea. The positively charged atom cores are bonded by mutual attraction to the negatively charged electrons.*

with a positive charge of three. The valence electrons, which are no longer associated with any particular atom, move freely within the electron sea and become associated with several atom cores. The positively charged atom cores are held together by mutual attraction to the electron, thus producing the strong metallic bond.

Metallic bonds are *nondirectional*; the electrons holding the atoms together are not fixed in one position. When a metal is bent and the atoms attempt to change their relationship to one another, the direction of the bond merely shifts, rather than the bond breaking [Figure 2-7(a)]. This permits metals to have good ductility and to be deformed into useful shapes.

The metallic bond also allows metals to be good electrical conductors. Under the influence of an applied voltage, the valence electrons move [Figure 2-7(b)], causing a current to flow if the circuit is complete. Other bonding mechanisms require much higher voltages to free the electrons from the bond.

The Covalent Bond.   Covalently bonded materials share electrons among two or more atoms. For example, a silicon atom, which has a valence of four, obtains eight electrons in its outer energy shell by sharing its electrons with four surrounding silicon atoms (Figure 2-8). Each instance of sharing represents one covalent bond; thus each silicon atom is bonded to four neighboring atoms by four covalent bonds.

In order for the covalent bonds to be formed, the silicon atoms must be

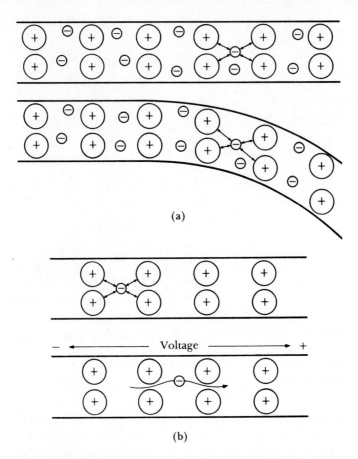

(a)

(b)

FIGURE 2-7 (a) Atoms joined by the metallic bond can shift their relative positions when the metal is deformed, permitting metals to have good ductility. (b) When a voltage is applied to a metal, the electrons in the electron sea can easily move and carry a current.

arranged so the bonds have a fixed *directional* relationship with one another. In the case of silicon, this arrangement produces a *tetrahedron*, with angles of about 109° between the covalent bonds (Figure 2-9). There is a much higher probability that electrons are located near these covalent bonds than elsewhere around the atom core.

Although covalent bonds are very strong, materials bonded in this manner have poor ductility and poor electrical conductivity. When a silicon rod is bent, the bonds must break if the silicon atoms are to permanently change their relationships to one another. Furthermore, for an electron to move and carry a current, the covalent bond must be broken, requiring high temperatures or voltages.

Thus, covalently bonded materials are brittle rather than ductile and behave as electrical insulators instead of conductors. Many ceramic, semi-

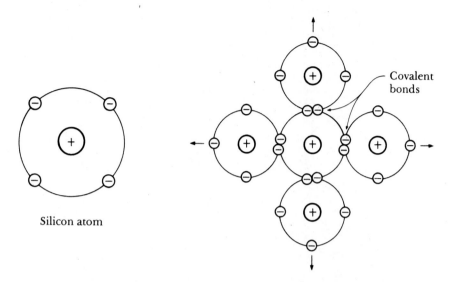

FIGURE 2-8 Covalent bonding requires that electrons be shared between atoms in such a way that each atom has its outer **sp** orbital filled. In silicon, with a valence of four, four covalent bonds must be formed.

conductor, and polymer materials are fully or partly bonded by covalent bonds, explaining why glass shatters when dropped and why bricks are good insulating materials.

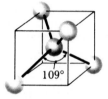

FIGURE 2-9 Covalent bonds are directional. In silicon, a tetrahedral structure is formed, with angles of about 109° required between each covalent bond.

◇ | **Example 2-5**

Describe how covalent bonding joins oxygen and silicon atoms in silica (SiO_2).

Answer:

Silicon has a valence of four and shares electrons with four oxygen atoms, thus giving a total of eight electrons for each silicon atom. However, oxygen has a valence of six and shares electrons with two silicon atoms, giving oxygen a total of eight electrons.

Figure 2-10 shows one of the possible structures. As in silicon, a tetrahedral structure is produced.

The Ionic Bond. When more than one type of atom is present in a material, one atom may donate its valence electrons to a different atom, filling the outer energy shell of the second atom. Both atoms now have filled (or empty) outer energy levels but both have acquired an electrical charge and behave as ions. The atom that contributes the electrons is left with a net positive charge and is a *cation*, while the atom that accepts the electrons acquires a net negative charge

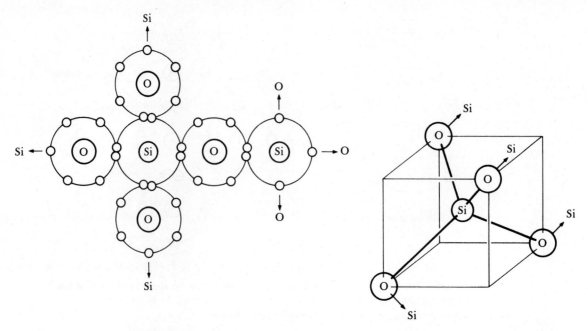

FIGURE 2-10   The tetrahedral structure of silica (SiO_2), which contains covalent bonds between silicon and oxygen atoms.

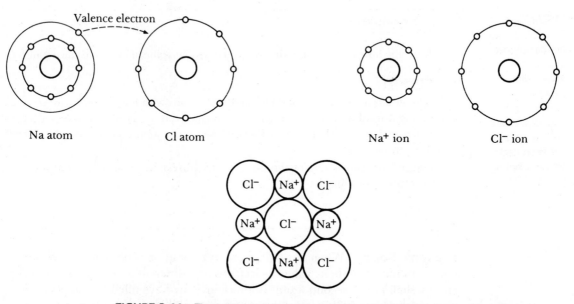

FIGURE 2-11   The ionic bond is created between two unlike atoms with different electronegativities. When sodium donates its valence electron to chlorine, each becomes an ion, attraction occurs, and the ionic bond is formed.

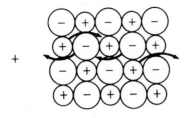

FIGURE 2-12 When a voltage is applied to an ionic material, entire ions must move to cause a current to flow. Ion movement is slow and the electrical conductivity is poor.

and is an *anion*. The oppositely charged ions are then attracted to one another and produce the ionic bond. For example, attraction between sodium and chloride ions (Figure 2-11) produces sodium chloride, or table salt, NaCl.

When a force is applied to a sodium chloride crystal, the electrical balance between the ions is upset. Partly for this reason, ionically bonded materials behave in a brittle manner. (A more detailed explanation for this brittle behavior will be offered in a later chapter.) Electrical conductivity is also poor; the electrical charge is transferred by the movement of entire ions (Figure 2-12), which do not move as easily as electrons.

◇ | **Example 2-6**

Describe the ionic bonding between magnesium and chlorine.

Answer:

The electronic structures and valences are

$$\text{Mg:} \quad 1s^2 2s^2 2p^6 \boxed{3s^2} \qquad \text{valence} = 2$$
$$\text{Cl:} \quad 1s^2 2s^2 2p^6 \boxed{3s^2 3p^5} \qquad \text{valence} = 7$$

Each magnesium atom gives up its two valence electrons, becoming a Mg^{2+} ion. Each chlorine atom accepts one electron, becoming a Cl^- ion. To satisfy the ionic bonding, there must be twice as many chlorine atoms as magnesium atoms present, and a compound, $MgCl_2$, is formed.

◇ | **Example 2-7**

As a general rule, would you expect anions or cations to have the larger ionic radius?

Answer:

Because the cation gives up its electron whereas the anion accepts an electron, we find that anions tend to be larger than cations.

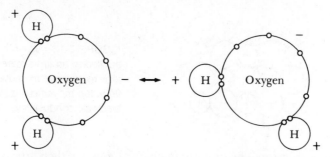

FIGURE 2-13 The Van der Waals bond is formed due to polarization of molecules or groups of atoms. In water, electrons in the oxygen tend to concentrate away from the hydrogen. The resulting charge difference permits the molecule to be weakly bonded to other water molecules.

Van der Waals Bonding.   Van der Waals bonds join molecules or groups of atoms by weak electrostatic attractions. Many plastics, ceramics, water, and other molecules are permanently *polarized*; that is, some portions of the molecule tend to be positively charged, while other portions are negatively charged. The electrostatic attraction between the positively charged regions of one molecule and the negatively charged regions of a second molecule weakly bond the two molecules together (Figure 2-13).

Van der Waals bonding is a *secondary bond*, but the atoms within the molecule or group of atoms are joined by strong covalent or ionic bonds. Heating water to the boiling point breaks the Van der Waals bonds and changes water to steam, but much higher temperatures are required to break the covalent bonds joining oxygen and hydrogen atoms together.

Van der Waals bonds can dramatically change the properties of materials. Since polymers normally have covalent bonds, we would expect polyvinyl chloride (PVC plastic) to be very brittle. However, polyvinyl chloride contains many long, chainlike molecules (Figure 2-14). Within each chain, bonding is covalent, but individual chains are bonded to one another by Van der Waals bonds. Polyvinyl chloride can be deformed significantly by breaking only the Van der Waals bonds as the chains slide past one another.

Mixed Bonding.   In most materials, bonding between atoms is a mixture of two or more types. Iron, for example, is bonded by a combination of metallic and covalent bonding; we will see in the next chapter that the directional nature of the covalent portion may prevent atoms in iron from packing as efficiently as we might expect.

Compounds formed from two or more metals (*intermetallic compounds*) may be bonded by a mixture of metallic and ionic bonds, particularly when there is a large difference in electronegativity between the elements. Because lithium has an electronegativity of 1.0 and aluminum has an electronegativity of 1.5, we would expect AlLi to have a combination of metallic and ionic bonding. On the other hand, because both aluminum and vanadium have electronegativities of 1.5, we would expect Al_3V to be bonded primarily by metallic bonds.

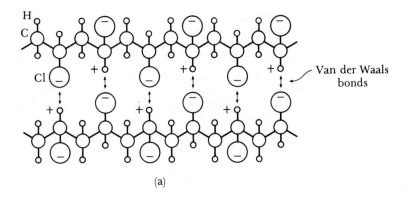

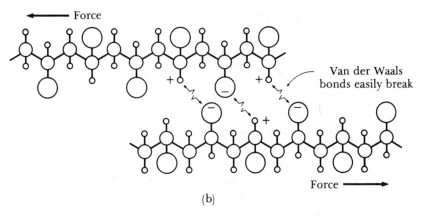

FIGURE 2-14   (a) In polyvinyl chloride, the chlorine atoms attached to the polymer chain have a negative charge and the hydrogen atoms are positively charged. The chains are weakly bonded by Van der Waals bonds. (b) When a force is applied to the polymer, the Van der Waals bonds are broken and the chains slide past one another.

Many ceramic and semiconducting compounds, which are combinations of metallic and nonmetallic elements, have a mixture of covalent and ionic bonding. As the electronegativity difference between the atoms increases, the bonding becomes more ionic. The fraction of bonding that is covalent can be estimated from the equation

$$\text{fraction covalent } = \text{ exp } (-0.25\Delta E^2) \tag{2-1}$$

where ΔE is the difference in electronegativities.

◇ | **Example 2-8**

We used SiO_2 as an example of a covalently bonded material. What fraction of the bonding is covalent?

Answer:

From Figure 2-3, we estimate the electronegativity of silicon to be 1.8 and that of oxygen to be 3.5. The fraction of the bonding that is covalent is

$$\text{fraction covalent} = \exp\left[-0.25(3.5 - 1.8)^2\right] = \exp(-0.72) = 0.486$$

Although the covalent bonding represents only about half of the bonding, the directional nature of these bonds plays an important role in the eventual structure of SiO_2.

2–6   BINDING ENERGY AND INTERATOMIC SPACING

Interatomic spacing is the equilibrium distance between atoms and is caused by a balance between repulsive and attractive forces. In the metallic bond, for example, the attraction between the electrons and the atom core is balanced by the repulsion between atom cores. Equilibrium separation occurs when the total energy of the pair of atoms is at a minimum, or when no net force is acting to either attract or repulse the atoms (Figure 2-15).

The interatomic spacing in a solid metal is equal to the atomic diameter, or twice the atomic radius r. We cannot use this approach for ionically bonded materials, however, since the spacing is the sum of the two different ionic radii. Atomic and ionic radii for the elements are listed in Appendix B and will be used in the next chapter.

The minimum energy in Figure 2-15 is the *binding energy*, or the energy required to create or break the bond. Consequently, materials having a high binding energy also have a high strength and a high melting temperature. Ionically bonded materials have a particularly large binding energy, due to the large difference in electronegativities between the ions (Table 2-4); metals have lower binding energies, since the electronegativities of the atoms are similar.

Other properties can be related to the force-distance and energy-distance expressions in Figure 2-15. For example, the *modulus of elasticity* of a material,

TABLE 2-4   *Binding energies for the four bonding mechanisms*

Bond	Binding Energy (kcal/mol)
Ionic	150–370
Covalent	125–300
Metallic	25–200
Van der Waals	< 10

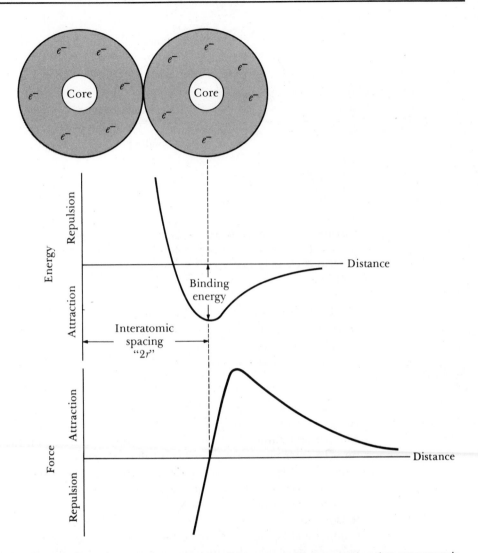

FIGURE 2-15 Atoms or ions are separated by an equilibrium spacing that corresponds to the minimum energy of the atoms or ions (or when zero force is acting to repel or attract the atoms or ions).

which is the amount that a material will stretch when a force is applied, is related to the slope of the force-distance curve (Figure 2-16). A steep slope, which correlates with a higher binding energy and a higher melting point, means that a greater force is required to stretch the bond; thus, the material has a high modulus of elasticity. In a similar manner, the *coefficient of thermal expansion*, which describes how much a material expands or contracts when its temperature is changed, can be related to the energy-distance curve. When the material is heated, additional energy is supplied to the material, causing the atoms to separate. When the binding energy is large, the additional energy causes a

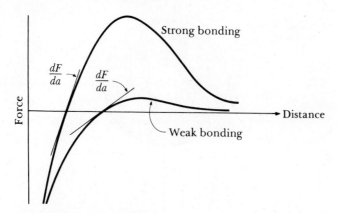

FIGURE 2-16 The force-distance curve for two materials, showing the relationship between atomic bonding and the modulus of elasticity. A steep **dF/da** slope gives a high modulus.

smaller change in the separation, or the material has a lower coefficient of thermal expansion.

SUMMARY

The electronic structure of an atom may be characterized by examining the energy levels to which each electron is assigned by the four quantum numbers. The periodic table of the elements is constructed based on the electronic structure. In later chapters, we will find that the energies of the electrons play an important role in determining many of the physical properties of a material.

We have found that the electronic structure plays an important role in determining the bonding between atoms, permitting us to assign general characteristics to each type of material. Metals have good ductility and electrical conductivity because of the metallic bond. Ceramics, semiconductors, and many polymers have poor ductility and electrical conductivity because of the covalent and ionic bonds. Van der Waals bonds are responsible for good ductility in certain polymers.

GLOSSARY

Anion. A negatively charged ion produced when an atom, usually of a nonmetal, accepts one or more electrons.

Atomic mass. The mass of the Avogadro number of atoms, g/g · mole. Normally, this is the average number of protons and neutrons in the atom. Also called the atomic weight.

Atomic number. The number of protons or electrons in an atom.

Avogadro number. The number of atoms or molecules in a g · mole. The Avogadro number is 6.02×10^{23} per mole.

Binding energy. The energy required to separate two atoms from their equilibrium spacing to an infinite distance apart. Alternately, the binding energy is the strength of the bond between two atoms.

Cation. A positively charged ion produced when an atom, usually of a metal, gives up its valence electrons.

Covalent bond. The bond formed between two atoms when the atoms share their valence electrons.

Electronegativity. The relative tendency of an atom to accept an electron and become an anion. Strongly electronegative atoms readily accept electrons.

Interatomic spacing. The equilibrium spacing between the centers of two atoms. In solid elements, the interatomic spacing equals the apparent diameter of the atom.

Ionic bond. The bond formed between two different atom species when one atom (the cation) donates its valence electrons to the second atom (the anion). An electrostatic attraction binds the ions together.

Isotopes. Isotopes of an element have the same atomic number but a different number of neutrons, thus giving a different atomic mass.

Metallic bond. The electrostatic attrac-tion between the valence electrons and the positively charged cores of the atoms.

Pauli exclusion principle. No more than two electrons in a material can have the same energy. The two electrons have opposite magnetic spins.

Polarized molecule. A molecule whose structure causes portions of the molecule to have a negative charge while other portions have a positive charge, leading to electrostatic attraction between the molecules.

Quantum numbers. The numbers that assign electrons in an atom to discrete energy levels. The four quantum numbers are the principal quantum number n, the azimuthal quantum number l, the magnetic quantum number m_l, and the spin quantum number m_s.

Valence. The number of electrons in an atom that participate in bonding or chemical reactions. Usually, the valence is the number of electrons in the outer sp energy level.

Van der Waals bond. A weak electrostatic attraction between polar molecules. The polar molecules have concentrations of positive and negative charges at different locations.

PRACTICE PROBLEMS

1. Silicon, which has an atomic number of 14, is composed of three isotopes: 92.21% of the Si atoms contain 14 neutrons, 4.7% contain 15 neutrons, and 3.09% contain 16 neutrons. Estimate the atomic mass of silicon.

2. Titanium, which has an atomic number of 22, is composed of five isotopes: 7.93% of the Ti atoms contain 24 neutrons, 7.28% contain 25 neutrons, 73.94% contain 26 neutrons, 5.51% contain 27 neutrons, and 5.34% contain 28 neutrons. Estimate the atomic mass of titanium.

3. Bromine, which has an atomic number of 35 and an atomic mass of $79.909 \, g/g \cdot mole$, contains two isotopes—Br^{79} and Br^{81}. Determine the percentage of each isotope of bromine.

4. Silver, which has an atomic number of 47 and an atomic mass of $107.87 \, g/g \cdot mole$, contains two isotopes—Ag^{107} and Ag^{109}. Determine the percentage of each isotope of silver.

5. Tin, with an atomic number of 50, has all of its inner energy levels filled except the $4f$ level, which is empty. From its electronic structure, determine the expected valence of tin.

6. Mercury, with an atomic number of 80, has all of its inner energy levels filled except the $5f$ and $5g$ levels, which are empty. From its electronic structure, determine the expected valence of mercury.

7. Calculate the number of atoms in 100 g of silver. Assuming that all of the valence electrons can carry an electrical current, calculate the number of these charge carriers per 100 g.

8. Suppose there are 8×10^{13} electrons in 100 g of germanium that are free to move and carry an electrical current. (a) What fraction of the total valence electrons are free to move? (b) What fraction of the covalent bonds must be broken? (On average, there are one covalent bond per germanium atom and two electrons in each covalent bond.)

9. Compare the number of atoms in 1 g of uranium with the number of atoms in 1 g of boron. Then, using the densities of each (see Appendix A), calculate the number of atoms per cubic centimeter in uranium and boron.

10. Suppose you collect 5×10^{26} atoms of nickel. Calculate the mass in grams and the volume in cubic centimeters represented by this number of atoms. See Appendix A for the density.

11. Calculate the volume in cubic centimeters occupied by one mole of gold. See Appendix A for the necessary data.

12. Suppose you have 15 moles of iron. Calculate the number of grams and the volume in cubic centimeters occupied by the iron. See Appendix A for the necessary data.

13. A decorative steel item having a surface area of 150 in^2 is plated with a layer of chromium 0.005 in. thick. Calculate the number of atoms required to produce the plating.

14. Examine the elements in the IVB to VIIIB columns of the periodic table. As you go to a higher atomic number in each column (as from Ni to Pd to Pt), how does the melting temperature change? Would you expect this, based on the atomic structure?

15. Examine the elements in the IA column of the periodic table. As you go to a higher atomic number, how does the melting temperature change? Would you expect this, based on the atomic structure? Is this behavior different from what was observed in the elements in Problem 14? Can you explain this difference?

15. Determine the formulas of the compounds formed when each of the following metals reacts with oxygen: (a) calcium, (b) aluminum, (c) germanium, (d) potassium.

17. Would you expect Al_2O_3 or aluminum to have the higher modulus of elasticity? Explain.

18. Would you expect silicon or nickel to have the higher coefficient of thermal expansion? Explain.

19. The compound GaAs is an important semiconductor material in which the atoms are joined by mixed ionic-covalent bonding. What fraction of the bonding is ionic?

20. The compound InP is an important semiconductor material in which the atoms are joined by mixed ionic-covalent bonding. If the fraction of covalent bonding is found to be 0.914, estimate the electronegativity of indium. Does your calculated value compare well with what you might expect, based on Figure 2-3?

21. Would you expect bonding in the intermetallic compound Ca_2Mg to be predominantly ionic or metallic? Explain.

22. The electronegativities of both copper and nickel are 1.8. Would you expect bonding in the intermetallic compound Ni_2Mg to be more or less metallic than in $CuAl_2$? Explain.

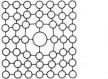

3

ATOMIC ARRANGEMENT

3–1 INTRODUCTION

Atomic arrangement plays an important role in determining the microstructure and behavior of a solid material. In metals, some arrangements permit exceptional ductility, whereas others permit exceptional strength. Certain physical properties of ceramics rely on the atomic arrangement; transducers used to produce the electrical signal in a stereo record player rely on an atomic arrangement that produces a permanent displacement of electrical charge within the material. The diverse behavior of polymers, such as rubber, polyethylene, and epoxy, is caused by differences in the atomic arrangement.

In this chapter, we will describe typical atomic arrangements in perfect solid materials and develop the nomenclature used to characterize this arrangement. We will then be prepared to see how imperfections in the atomic arrangement permit us to understand both deformation and strengthening of many solid materials.

3–2 SHORT-RANGE ORDER
VERSUS LONG-RANGE ORDER

If we neglect imperfections in materials, there are three levels of atomic arrangement (Figure 3-1).

No Order. In gases such as argon, the atoms have no order; argon atoms randomly fill up the space to which the gas is confined.

Short-Range Order. A material displays short-range order if the special arrangement of the atoms extends only to the atom's nearest neighbors. Each water molecule in steam has a short-range order due to the covalent bonds between the hydrogen and oxygen atoms; that is, each oxygen atom is joined to

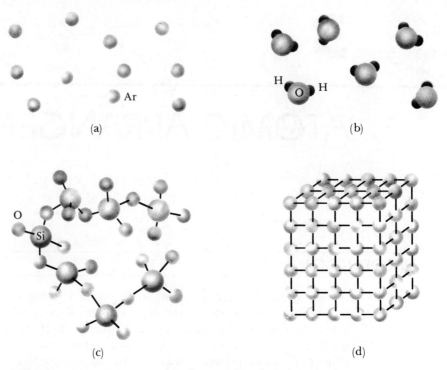

(a)

Ar

(b)

H   O   H

(c)

O

Si

(d)

FIGURE 3-1 *The levels of atomic arrangement in materials. (a) Inert gases have no regular ordering of atoms. (b) and (c) Some materials, including steam and glass, have ordering only over a short distance. (d) Metals and many other solids have a regular ordering of atoms that extends through the material.*

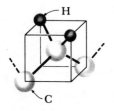

H

C

FIGURE 3-2
The tetrahedral
structure in
polyethylene.

two hydrogen atoms, forming an angle of about 104.5° between the bonds. However, the water molecules have no special arrangement but instead randomly fill up the space available to them.

A similar situation exists in ceramic glasses. In Example 2-5, we described the tetrahedral structure in silica, which satisfies the requirement that four oxygen atoms be covalently bonded to each silicon atom. Because the oxygen atoms must form angles of about 109° to satisfy the directionality requirements of the covalent bonds, a short-range order results. However, the tetrahedral units may be joined together in a random manner.

Polymers also display short-range atomic arrangements and, in fact, may closely resemble the silica glass structure. Polyethylene is composed of chains of carbon atoms, with two hydrogen atoms attached to each carbon. Because carbon has a valence of four and the carbon and hydrogen atoms are bonded covalently, a tetrahedral structure is again produced (Figure 3-2). The tetrahedral units can be joined together in a random manner to produce the polymer chains.

Ceramics and polymers having only this short-range order are often characterized as *amorphous* materials. *Glasses,* which form in both ceramic and polymer systems, are amorphous materials often having good strength and stiffness and unique physical properties but brittle behavior. A few specially prepared metals and semiconductors may also possess only short-range order.

Long-Range Order. Metals, semiconductors, many ceramics, and even some polymers have a crystalline structure in which the atoms also display long-range order; the special atomic arrangement extends throughout the entire material. The atoms form a regular repetitive, gridlike pattern, or lattice. The *lattice* is a collection of points, called *lattice points*, which are arranged in a periodic pattern so that the surroundings of each point in the lattice are identical. One or more atoms are associated with each lattice point. Consequently, each atom has both a short-range order, since the surroundings of each lattice point are identical, and a long-range order, since the lattice extends periodically throughout the entire material.

The lattice differs from material to material in both shape and size, depending on the size of the atoms and the type of bonding between the atoms. The *crystal structure* of a material refers to the size, shape, and atomic arrangement within the lattice.

3-3 UNIT CELLS

The *unit cell* is a subdivison of the crystalline lattice that still retains the overall characteristics of the entire lattice. A unit cell is shown in the lattice in Figure 3-3 by the heavy lines. By stacking identical unit cells, the entire lattice can be constructed.

We identify 14 types of unit cells, or *Bravais lattices*, grouped in seven crystal structures (Figure 3-4 and Table 3-1). Lattice points are located at the corners of the unit cells and, in some cases, at either faces or the center of the unit cell. Let's look at some of the characteristics of a lattice or unit cell.

Lattice Parameter. The lattice parameters, which describe the size and shape of the unit cell, include the dimensions of the sides of the unit cell and the angles between the sides (Figure 3-5). In a cubic crystal system, only the length of one of the sides of the cube is necessary to completely describe the cell (angles of 90° are assumed unless otherwise specified). This length, measured at room temperature, is the lattice parameter a_0. The length is often given in angstrom

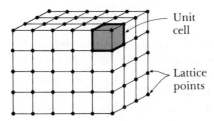

Unit cell

Lattice points

FIGURE 3-3 A lattice is a periodic array of points that define space. The unit cell (heavy outline) is a subdivision of the lattice that still retains the characteristics of the lattice.

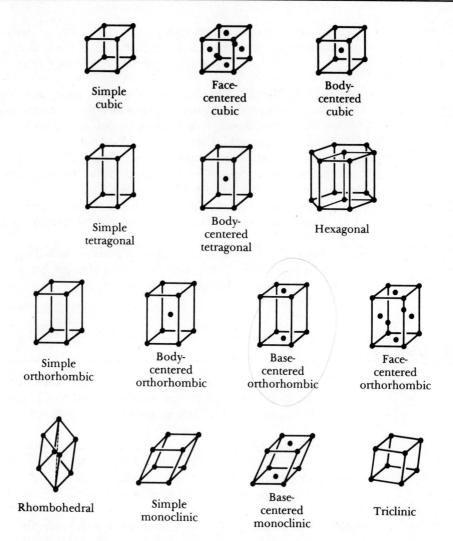

FIGURE 3-4 *The seven crystal systems and 14 Bravais lattices. Characteristics of the crystal systems are summarized in Table 3-1.*

units or nanometers, where

$$1 \text{ angstrom } (\text{Å}) = 10^{-8}\text{cm} = 10^{-10}\text{m}$$

$$1 \text{ nanometer } (\text{nm}) = 10^{-7}\text{cm} = 10^{-9}\text{m} = 10 \text{Å}$$

Several lattice parameters are required to define the size and shape of complex unit cells. For an orthorhombic unit cell, we must specify the dimensions of all three sides of the cell, a_0, b_0, and c_0. Hexagonal unit cells require two dimensions, a_0 and c_0, and the angle of 120° between the a_0 axes. The most complicated cell, the triclinic cell, is described by three lengths and three angles.

TABLE 3-1   Characteristics of the seven crystal systems

Structure	Axes	Angles between Axes
Cubic	$a = b = c$	All angles equal 90°
Tetragonal	$a = b \neq c$	All angles equal 90°
Orthorhombic	$a \neq b \neq c$	All angles equal 90°
Hexagonal	$a = b \neq c$	Two angles equal 90° One angle equals 120°
Rhombohedral	$a = b = c$	All angles are equal and none equals 90°
Monoclinic	$a \neq b \neq c$	Two angles equal 90° One angle (β) not equal to 90°
Triclinic	$a \neq b \neq c$	All angles are different and none equals 90°

Number of Atoms per Unit Cell.   A specific number of lattice points defines each of the unit cells. For example, the corners of the cells are easily identified, as are body- and face-centered positions (Figure 3-4). When counting the number of lattice points belonging to each unit cell, we must recognize that lattice points may be shared by more than one unit cell. A lattice point at a corner of one unit cell is shared by seven adjacent unit cells; only $\frac{1}{8}$ of each corner belongs to one particular cell. Thus, the number of lattice points from the corner positions in one unit cell is

$$\left(\frac{1}{8} \frac{\text{lattice point}}{\text{corner}} \right) \left(8 \frac{\text{corners}}{\text{cell}} \right) = 1 \frac{\text{lattice point}}{\text{unit cell}}$$

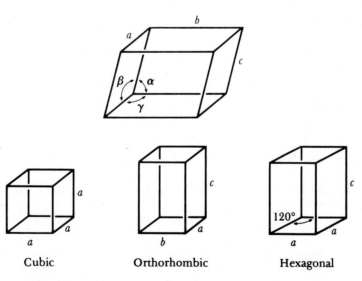

FIGURE 3-5 Definition of the lattice parameters and their use in three crystal systems.

Simple cubic   Body-centered cubic   Face-centered cubic

FIGURE 3-6 *The models for simple cubic (SC), body-centered cubic (BCC), and face-centered cubic (FCC) unit cells, assuming only one atom per lattice point.*

Corners contribute $\frac{1}{8}$ of a point, faces contribute $\frac{1}{2}$, and body-centered positions contribute a whole point.

The number of atoms per unit cell is the product of the number of atoms per lattice point and the number of lattice points per unit cell. In most metals, one atom is located at each lattice point, so the number of atoms is equal to the number of lattice points. The structures of simple cubic (SC), face-centered cubic (FCC), and body-centered cubic (BCC) unit cells, with one atom per lattice point, are shown in Figure 3-6. In more complicated structures, particularly compounds and ceramic materials, several or even hundreds of atoms may be associated with each lattice point, forming very complex unit cells.

◇ | **Example 3-1**

Determine the number of lattice points per cell in the cubic crystal systems.

Answer:

In the SC unit cell, lattice points are located only at the corners of the cube:

$$\frac{\text{lattice points}}{\text{unit cell}} = (8 \text{ corners})\left(\frac{1}{8}\right) = 1$$

In BCC unit cells, lattice points are located at the corners and the center of the cube:

$$\frac{\text{lattice points}}{\text{unit cell}} = (8 \text{ corners})\left(\frac{1}{8}\right) + (1 \text{ center})(1) = 2$$

In FCC unit cells, lattice points are located at the corners and faces of the cube:

$$\frac{\text{lattice points}}{\text{unit cell}} = (8 \text{ corners})\left(\frac{1}{8}\right) + (6 \text{ faces})\left(\frac{1}{2}\right) = 4$$

Figure 3-7 shows the contribution that each lattice point makes to the individual unit cell.

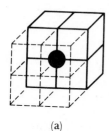

(a)

(b)

FIGURE 3-7 (a) Corner atoms are shared by eight unit cells. (b) Face-centered atoms are shared by two unit cells.

Atomic Radius versus Lattice Parameter.   In simple structures, particularly those with only one atom per lattice point, we can calculate the relationship between the apparent size of the atom and the size of the unit cell. We must locate the direction in the unit cell along which atoms are in continuous contact. These are the *close-packed directions*. By geometrically determining the length of the direction relative to the lattice parameters, and then adding the number of atomic radii along this direction, we can determine the desired relationship.

◇ | **Example 3-2**

Determine the relationship between the atomic radius and the lattice parameter in SC, BCC, and FCC structures.

Answer:

If we refer to Figure 3-8, we find that atoms touch along the edge of the cube in SC structures. The corner atoms are centered on the corners of the cube, so

$$a_0 = 2r \tag{3-1}$$

In FCC structures, atoms touch along the face diagonal of the cube, which is $\sqrt{2}a_0$ in length. There are four atomic radii along this length—two radii from the face-centered atom and one radius from each corner, so

$$a_0 = \frac{4r}{\sqrt{2}} \tag{3-2}$$

In BCC structures, atoms touch along the body diagonal, which is $\sqrt{3}a_0$ in length. There are two atomic radii from the center atom and one atomic radius from each of the corner atoms on the body diagonal, so

$$a_0 = \frac{4r}{\sqrt{3}} \tag{3-3}$$

◇

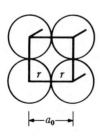
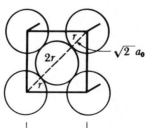

Simple cubic   Face-centered cubic   Body-centered cubic

FIGURE 3-8   The relationship between the atomic radius and the lattice parameter in cubic systems. (See Example 3-2.)

◇ | **Example 3-3**

The atomic radius of iron is 1.24 Å. Calculate the lattice parameters of BCC and FCC iron.

Answer:

For BCC iron,

$$a_0 = \frac{4r}{\sqrt{3}} = \frac{(4)(1.24)}{\sqrt{3}} = 2.86 \, \text{Å} = 2.86 \times 10^{-8} \, \text{cm}$$

For FCC iron,

$$a_0 = \frac{4r}{\sqrt{2}} = \frac{(4)(1.24)}{\sqrt{2}} = 3.51 \, \text{Å} = 3.51 \times 10^{-8} \, \text{cm}$$

Coordination Number.   The number of atoms touching a particular atom, or the number of nearest neighbors, is the *coordination number* and is one indication of how tightly and efficiently atoms are packed together. In simple crystal structures containing only one atom per lattice point, we find that the atoms have a coordination number related to the lattice structure. By inspecting the unit cells in Figure 3-9, we see that each atom in the SC lattice has a coordination number of six, while each atom in the BCC lattice has eight nearest neighbors. In Section 3-5, we will show that each atom in the FCC lattice has a coordination number of 12, which is the theoretical maximum.

Packing Factor.   The *packing factor* is the fraction of space occupied by atoms, assuming that atoms are hard spheres. The general expression for the packing

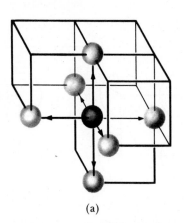
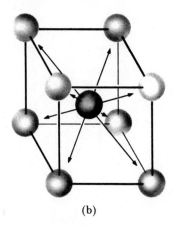

(a)   (b)

FIGURE 3-9 Illustration of coordination in SC and BCC unit cells. Six atoms touch each atom in SC, while the eight corner atoms touch the body-centered atom in BCC.

factor is

$$\text{Packing factor} = \frac{(\text{number of atoms/cell})(\text{volume of each atom})}{\text{volume of unit cell}} \qquad (3\text{-}4)$$

◇ | **Example 3-4**

Calculate the packing factor for the FCC cell.

Answer:

There are four lattice points per cell; if there is one atom per lattice point, there are also four atoms per cell. The volume of one atom is $4\pi r^3/3$ and the volume of the unit cell is a_0^3.

$$\text{Packing factor} = \frac{(4\ \text{atoms/cell})(\frac{4}{3}\pi r^3)}{a_0^3}$$

Since for FCC unit cells, $a_0 = 4r/\sqrt{2}$,

$$\text{Packing factor} = \frac{(4)(\frac{4}{3}\pi r^3)}{(4r/\sqrt{2})^3} = 0.74$$

In metals, the packing factor of 0.74 in the FCC unit cell is the most efficient packing possible. BCC cells have a packing factor of 0.68 and SC cells have a packing factor of 0.52. Materials may have low packing factors as a consequence of atomic bonding. Metals with only the metallic bond are packed as efficiently as possible. Metals with mixed bonding may have unit cells with less than the maximum packing factor. No common engineering metals have the SC structure, although this structure is found in ceramic materials.

Density.   The theoretical density of a metal can also be calculated using the properties of the crystal structure. The general formula is

$$\text{Density}\ \rho = \frac{(\text{atoms/cell})(\text{atomic mass of each atom})}{(\text{volume of unit cell})(\text{Avogadro number})} \qquad (3\text{-}5)$$

◇ | **Example 3-5**

Determine the density of BCC iron, which has a lattice parameter of 2.866 Å.

Answer:

$$\text{Atoms/cell} = 2$$

Atomic mass $=$ 55.85 g/g · mole

Volume of unit cell $= a_0^3 = (2.866 \times 10^{-8})^3 = 23.55 \times 10^{-24}$ cm³/cell

Avogadro number $\mathcal{N}_A = 6.02 \times 10^{23}$ atoms/g · mole

$$\rho = \frac{(2)(55.85)}{(23.55 \times 10^{-24})(6.02 \times 10^{23})} = 7.879 \text{ g/cm}^3$$

The measured density is 7.870 g/cm³. We will explain the slight discrepancy between the theoretical and measured density in the next chapter when we discuss vacancies.

The Hexagonal Close-Packed Structure.   A special form of the hexagonal lattice, the hexagonal close-packed structure (HCP), is shown in Figure 3-10. The unit cell is the skewed prism, shown separately. The HCP structure has one lattice point per cell—one from each of the eight corners of the prism—but two atoms are associated with each lattice point. One atom is located at a corner, while the second atom is located within the unit cell.

In ideal HCP metals, the a_0 and c_0 axes are related by the ratio $c/a = 1.633$. Most HCP metals, however, have c/a ratios that differ slightly from the ideal value, due to mixed bonding. Because the HCP structure, like the FCC structure, has a very efficient packing factor of 0.74 and a coordination number of 12, a number of metals possess this structure. Table 3-2 summarizes the characteristics of the most important crystal structures in metals.

3-4   ALLOTROPIC OR POLYMORPHIC TRANSFORMATIONS

Materials that can have more than one crystal structure are called *allotropic* or *polymorphic*. The term allotropy is normally reserved for this behavior in pure

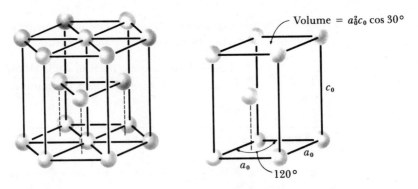

FIGURE 3-10   The hexagonal close-packed (HCP) lattice and its unit cell.

TABLE 3-2 Characteristics of common metallic crystals

Structure	a_0 versus r	Atoms per Cell	Coordination Number	Packing Factor	Typical Metals
Simple cubic (SC)	$a_0 = 2r$	1	6	0.52	None
Body-centered cubic (BCC)	$a_0 = 4r/\sqrt{3}$	2	8	0.68	Fe, Ti, W, Mo, Nb, Ta, K, Na, V, Cr, Zr
Face-centered cubic (FCC)	$a_0 = 4r/\sqrt{2}$	4	12	0.74	Fe, Cu, Al, Au, Ag, Pb, Ni, Pt
Hexagonal close-packed (HCP)	$a_0 = 2r$ $c_0 = 1.633a_0$	2	12	0.74	Ti, Mg, Zn, Be, Co, Zr, Cd

elements, while polymorphism is a more general term. You may have noticed in Table 3-2 that some metals, such as iron and titanium, have more than one crystal structure. At low temperatures iron has the BCC structure, but at higher temperatures iron transforms to an FCC structure. These transformations provide the basis for the heat treatment of steel and titanium.

Many ceramic materials, such as silica (SiO_2), also are polymorphic. A volume change may accompany the transformation during heating or cooling; if not properly controlled, this volume change causes the material to crack and fail.

◇ | **Example 3-6**

Calculate the change in volume that occurs when BCC iron is heated and changes to FCC iron. At the transformation temperature, the lattice parameter of BCC iron is 2.863 Å and the lattice parameter of FCC iron is 3.591 Å.

Answer:

$$\text{Volume of BCC cell} = a^3 = (2.863)^3 = 23.467 \, \text{Å}^3$$
$$\text{Volume of FCC cell} = a^3 = (3.591)^3 = 46.307 \, \text{Å}^3$$

But the FCC unit cell contains four atoms and the BCC unit cell contains only two atoms. Two BCC unit cells, with a total volume of $2(23.467) = 46.934 \, \text{Å}^3$, will contain four atoms. Thus, the volume change is

$$\text{Volume change} = \frac{46.307 - 46.934}{46.934} \times 100 = -1.34\%$$

This indicates that iron contracts on heating.

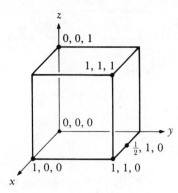

FIGURE 3-11 Coordinates of points in the unit cell. The numbers refer to numbers of lattice parameters.

3–5 POINTS, DIRECTIONS, AND PLANES IN THE UNIT CELL

Coordinates of Points. We can locate certain points, such as atom positions, in the lattice or unit cell by constructing the right-handed coordinate system in Figure 3-11. Distance is measured in terms of the number of lattice parameters we must move in each of the x, y, and z coordinates to get from the origin to the point in question. The coordinates are written as the three distances, with commas separating the numbers.

Directions in the Unit Cell. Certain directions in the unit cell are of particular importance. Metals deform, for example, in directions along which atoms are in closest contact. Properties of a material may depend on the direction in the crystal along which the property is measured. *Miller indices* for directions are the shorthand notation used to describe these directions. The procedure for finding the Miller indices for directions is as follows:

(a) Using a right-handed coordinate system, determine the coordinates of two points that lie on the direction.

(b) Subtract the coordinates of the "tail" point from the coordinates of the "head" point to obtain the number of lattice parameters traveled in the direction of each axis of the coordinate system.

(c) Clear fractions and/or reduce the results obtained from the subtraction to lowest integers.

(d) Enclose the numbers in square brackets [   ]. If a negative sign is produced, represent the negative sign with a bar over the number.

◇ | **Example 3-7**

Determine the Miller indices of directions A, B, and C in Figure 3-12.

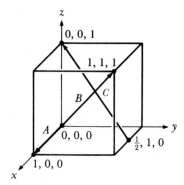

FIGURE 3-12 Crystallographic directions and coordinates required for Example 3-7.

Answer:

Direction A

(a) Two points are 1, 0, 0, and 0, 0, 0

(b) 1, 0, 0 − 0, 0, 0 = 1, 0, 0

(c) No fractions to clear or integers to reduce

(d) [100]

Direction B

(a) Two points are 1, 1, 1 and 0, 0, 0

(b) 1, 1, 1 − 0, 0, 0 = 1, 1, 1

(c) No fractions to clear or integers to reduce

(d) [111]

Direction C

(a) Two points are 0, 0, 1 and $\frac{1}{2}$, 1, 0

(b) 0, 0, 1 − $\frac{1}{2}$, 1, 0 = $-\frac{1}{2}$, −1, 1

(c) $2(-\frac{1}{2}, -1, 1) = -1, -2, 2$

(d) $[\bar{1}\,\bar{2}\,2]$

Several points should be noted about the use of Miller indices for directions.

1. A direction and its negative are not identical; [100] is not equal to [$\bar{1}$00]. They represent the same line but opposite directions. Going east isn't the same as going west!

2. A direction and its multiple are identical; [100] is the same direction as [200]. We just forgot to reduce to lowest integers.

3. Certain groups of directions are equivalent; they have their particular indices primarily because of the way we construct the coordinates. For example, in a cubic system a [100] direction is a [010] direction if we redefine the

TABLE 3-3 Directions of the form $\langle 110 \rangle$ in cubic systems

$$\langle 110 \rangle = \begin{cases} [110] & [\bar{1}\bar{1}0] \\ [101] & [\bar{1}0\bar{1}] \\ [011] & [0\bar{1}\bar{1}] \\ [1\bar{1}0] & [\bar{1}10] \\ [10\bar{1}] & [\bar{1}01] \\ [01\bar{1}] & [0\bar{1}1] \end{cases}$$

coordinate system as shown in Figure 3-13. We may refer to groups of equivalent directions as *directions of a form*. The special brackets $\langle \ \ \rangle$ are used to indicate this collection of directions. All of the directions of the form $\langle 110 \rangle$ are shown in Table 3-3.

Another way of characterizing equivalent directions is by the distance between lattice points along the direction. For example, we could examine the [110] direction in an FCC unit cell (Figure 3-14); if we start at the 0, 0, 0 location, the next lattice point is at the center of a face, or a 1/2, 1/2, 0 site. The distance between lattice points is therefore one-half of the face diagonal, or $\sqrt{2}a_0/2$. In copper, which has a lattice parameter of 3.6151 $\times$ 10^{-8} cm, the *repeat distance* is 2.556 $\times$ 10^{-8} cm.

The *linear density* is the number of lattice points per unit length along the direction. In copper, there are two repeat distances along the [110] direction in each unit cell; since this distance is $\sqrt{2}a_0 = 5.1125 \times 10^{-8}$ cm, then

$$\text{Linear density} = \frac{2 \text{ repeat distances}}{5.1125 \times 10^{-8}} = 3.9 \times 10^7 \text{ lattice points/cm}$$

Note that the linear density is also the reciprocal of the repeat distance.

Finally, we could compute the *packing fraction* of a particular direction, or the fraction actually covered by atoms. For copper, in which one atom is located at each lattice point, this fraction is equal to the product of the linear density

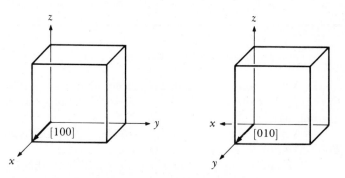

FIGURE 3-13 Equivalency of crystallographic directions of a form in cubic systems.

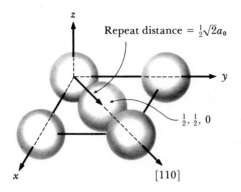

FIGURE 3-14 Determining the repeat distance, linear density, and packing fraction for a [110] direction in FCC copper.

and twice the atomic radius. For FCC copper, the atomic radius $r = \sqrt{2}a_0/4 = 1.2781 \times 10^{-8}$ cm. Therefore, the packing fraction is

$$
\begin{aligned}
\text{Packing fraction} &= (\text{linear density})(2r) \\
&= (3.9 \times 10^7)(2)(1.2781 \times 10^{-8}) \\
&= 1.0
\end{aligned}
$$

In this case, atoms lie continuously along the [110] direction, as expected, since the [110] direction is close packed in FCC metals. All of the directions of the form $\langle 110 \rangle$ in copper will have the same linear density and packing fraction.

◇ | **Example 3-8**

Calculate the repeat distance, linear density, and packing fraction for the [111] direction in FCC copper.

Answer:

If we start at the lattice point at 0, 0, 0, we do not encounter another lattice point until 1, 1, 1. This distance (Figure 3-15) is

$$\text{Repeat distance} = \sqrt{3}a_0 = \sqrt{3}(3.6151 \times 10^{-8}) = 6.262 \times 10^{-8} \text{ cm}$$

The linear density is the reciprocal of the repeat distance, or

$$\text{Linear density} = \frac{1}{6.262 \times 10^{-8}} = 1.597 \times 10^7 \text{ lattice points/cm}$$

The packing fraction is the linear density times $2r$, or

$$\text{Packing fraction} = (1.597 \times 10^7)(2)(1.278 \times 10^{-8}) = 0.408$$

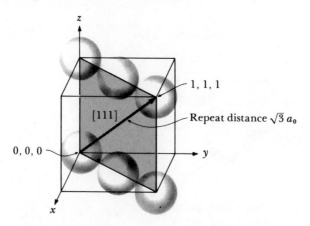

FIGURE 3-15   Determining the repeat distance, linear density, and packing fraction for a [111] direction in FCC copper (Example 3-8).

The linear density and packing fraction are lower than along the [110] direction because the [111] direction is not close packed.

Planes in the Unit Cell.   Certain planes of atoms in a crystal are also significant; for example, metals deform along planes of atoms that are most tightly packed together. Miller indices can be used as a shorthand notation to identify these important planes, as described in the following procedure.

(a) Identify the points at which the plane intercepts the x, y, and z coordinates in terms of the number of lattice parameters. If the plane passes through the origin, the origin of the coordinate system must be moved!

(b) Take reciprocals of these intercepts.

(c) Clear fractions but do not reduce to lowest integers.

(d) Enclose the resulting numbers in parentheses (   ). Again, negative numbers should be written with a bar over the number.

◇ | **Example 3-9**

Determine the Miller indices of planes A, B, and C in Figure 3-16.

Answer:

Plane A

(a) $x = 1, y = 1, z = 1$

(b) $\dfrac{1}{x} = 1, \dfrac{1}{y} = 1, \dfrac{1}{z} = 1$

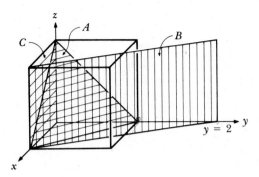

$\frac{1}{2}, 1, 0$

$2, 1, 0$

FIGURE 3-16 Crystallographic planes and intercepts for Example 3-9.

(c) No fractions to clear

(d) $(1\,1\,1)$

Plane B

(a) The plane never intercepts the z axis, so $x = 1, y = 2$, and $z = \infty$.

(b) $\dfrac{1}{x} = 1, \dfrac{1}{y} = \dfrac{1}{2}, \dfrac{1}{z} = 0$

(c) Clear fractions: $\dfrac{1}{x} = 2, \dfrac{1}{y} = 1, \dfrac{1}{z} = 0$

(d) $(2\,1\,0)$

Plane C

(a) We must move the origin since the plane passes through $0, 0, 0$. Let's move the origin one lattice parameter in the y-direction. Then $x = \infty, y = -1$, and $z = \infty$.

(b) $\dfrac{1}{x} = 0, \dfrac{1}{y} = -1, \dfrac{1}{z} = 0$

(c) No fractions to clear

(d) $(0\,\bar{1}\,0)$

Several important aspects of the Miller indices for planes should be noted.

1. Planes and their negatives are identical (this was not the case for directions). As an example, consider Figure 3-17. The shaded plane has the indices (020) if the x, y, and z coordinates are used but has the indices $(0\bar{2}0)$ if the x', y', and z' coordinates are used. But we are considering the same plane! Therefore, $(020) = (0\bar{2}0)$.

2. Planes and their multiples are not identical (again, this is the opposite of what we found for directions). We can show this by defining planar densities and planar packing fractions. The *planar density* is the number of atoms per unit area

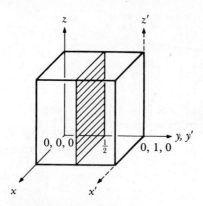

FIGURE 3-17 A plane and its negative are identical.

whose centers lie on the plane; the *packing fraction* is the fraction of that plane actually covered by these atoms. Example 3-10 shows how these can be calculated.

◇ | **Example 3-10**

Calculate the planar density and planar packing fraction for the (010) and (020) planes in simple cubic polonium, which has a lattice parameter of 3.34×10^{-8} cm.

Answer:

The two planes are drawn in Figure 3-18. On the (010) plane, atoms are centered at each corner of the cube face, with $\frac{1}{4}$ of each atom actually in the face of the unit cell. Thus, the total atoms in each face is one. The planar density is

$$\text{Planar density (010)} = \frac{\text{atoms per face}}{\text{area of face}} = \frac{1 \text{ atom per face}}{(3.34 \times 10^{-8})^2}$$

$$= 8.96 \times 10^{14} \text{ atoms/cm}^2$$

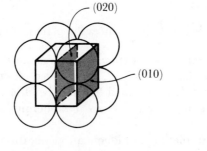
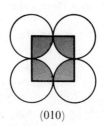
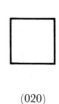

(010)

(020)

FIGURE 3-18 The planar densities of the (010) and (020) planes in SC unit cells are not identical. (See Example 3-10.)

The planar packing fraction is given by

$$\text{Packing fraction (010)} = \frac{\text{area of atoms per face}}{\text{area of face}} = \frac{(1 \text{ atom})(\pi r^2)}{(a_0)^2}$$

$$= \frac{\pi r^2}{(2r)^2} = 0.79$$

However, no atoms are centered on the (020) planes. Therefore, the planar density and the planar packing fraction are both zero. The (010) and (020) planes are not equivalent!

3. In each unit cell, *planes of a form* represent groups of equivalent planes that have their particular indices because of the orientation of the coordinates. We represent these groups of similar planes with the notation { }. The planes of the form {110} in cubic systems are shown in Table 3-4.

TABLE 3-4   Planes of the form {110} in cubic systems

$$\{110\} \begin{cases} (110) \\ (101) \\ (011) \\ (1\bar{1}0) \\ (10\bar{1}) \\ (01\bar{1}) \end{cases}$$

Note: The negatives of the planes are not unique planes.

4. In cubic systems, a direction that has the same indices as a plane is perpendicular to that plane. Figure 3-19 shows a unit cell containing a (100) plane and a [100] direction and clearly indicates this property. This is not always true for noncubic cells.

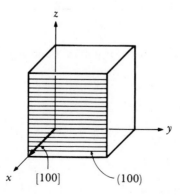

FIGURE 3-19   A direction in a cubic unit cell is perpendicular to a plane with the same indices.

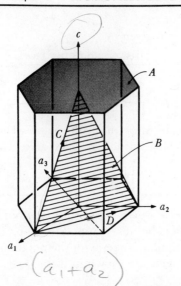

FIGURE 3-20 Miller-Bravais indices are obtained for crystallographic planes in HCP unit cells by using a four-axis coordinate system. (See Example 3-11.)

$$a_3 = -(a_1 + a_2)$$

Miller Indices for Hexagonal Unit Cells.   A special set of *Miller-Bravais indices* has been devised for hexagonal unit cells because of the unique symmetry of the system (Figure 3-20). The coordinate system uses four axes instead of three, with the a_3 axis being redundant. The procedure for finding the indices of planes is exactly the same as before, but four intercepts are required, giving indices of the form $(hkil)$. Because of the redundancy of the a_3 axis and the special geometry of the system, the first three integers in the designation, corresponding to the a_1, a_2, and a_3 intercepts, are related by the equation $h + k = -i$.

Directions in HCP cells are denoted with either the three-axis or four-axis system. With the three-axis system, the procedure is the same as for conventional Miller indices; examples of this procedure are shown in Example 3-11. A more complicated procedure, by which the direction is broken up into four vectors, is needed for the four-axis system. We then determine the number of lattice parameters we must move in each direction to get from the "tail" to the "head" of the direction, while for consistency still making sure that $h + k = -i$. This is illustrated in Figure 3-21, showing that the [010] direction is the same as the $[\bar{1}2\bar{1}0]$ direction.

◇ | **Example 3-11**

Determine the Miller-Bravais indices for planes A and B and directions C and D in Figure 3-20.

Answer:

Plane A

(a) $a_1 = a_2 = a_3 = \infty, c = 1$

(b) $\dfrac{1}{a_1} = \dfrac{1}{a_2} = \dfrac{1}{a_3} = 0, \dfrac{1}{c} = 1$

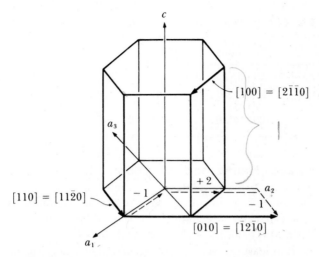

FIGURE 3-21 Typical directions in the HCP unit cell using both three- and four-axis systems. The dashed lines show that the $[\bar{1}2\bar{1}0]$ direction is equivalent to a $[010]$ direction.

(c) No fractions to clear

(d) (0001)

Plane B

(a) $a_1 = 1, a_2 = 1, a_3 = -\frac{1}{2}, c = 1$

(b) $\dfrac{1}{a_1} = 1, \dfrac{1}{a_2} = 1, \dfrac{1}{a_3} = -2, \dfrac{1}{c} = 1$

(c) No fractions to clear

(d) $(11\bar{2}1)$

Direction C

(a) Two points are 0, 0, 1 and 1, 0, 0

(b) 0, 0, 1 − 1, 0, 0 = − 1, 0, 1

(c) No fractions to clear or integers to reduce

(d) $[\bar{1}01]$

Direction D

(a) Two points are 0, 1, 0 and 1, 0, 0

(b) 0, 1, 0 − 1, 0, 0 = − 1, 1, 0

(c) No fractions to clear or integers to reduce

(d) $[\bar{1}10]$

TABLE 3-5   Close-packed planes and directions

Structure	Directions	Planes
SC	$\langle 100 \rangle$	None
BCC	$\langle 111 \rangle$	None
FCC	$\langle 110 \rangle$	$\{111\}$
HCP	$\langle 100 \rangle$, $\langle 110 \rangle$	(0001), (0002)

Close-Packed Planes and Directions.   In examining the relationship between atomic radius and lattice parameter, we looked for close-packed directions, where atoms are in continuous contact. We can now assign Miller indices to these close-packed directions, as shown in Table 3-5.

We can also examine FCC and HCP unit cells more closely and discover that there is at least one set of close-packed planes in each. A close-packed plane is shown in Figure 3-22. Notice that a hexagonal arrangement of atoms is produced in two dimensions. The close-packed planes are easy to find in the HCP unit cell; they are the (0001) and (0002) planes of the HCP structure and are given the special name *basal planes*. In fact, we can build up an HCP unit cell by stacking together close-packed planes in an . . . *ABABAB . . . stacking sequence* (Figure 3-22). Atoms on plane *B*, the (0002) plane, fit into the valleys between atoms on plane *A*, the bottom (0001) plane. If another plane identical in orientation to plane *A* is placed in the valleys of plane *B*, the HCP structure is created. Notice that all of the possible close-packed planes are parallel to one another. Only the basal planes—(0001) and (0002)—are close packed.

We can easily determine the coordination number of the atoms in the HCP structure from Figure 3-22. We find that the center atom in a basal plane is touched by six other atoms in the same plane. Three atoms in a lower plane and three atoms in an upper plane also touch the same atom. The coordination number is 12.

In the FCC structure, close-packed planes are of the form $\{111\}$ (Figure 3-23). When stacking parallel (111) planes, atoms in plane *B* fit over valleys in plane *A* and atoms in plane *C* fit over valleys in both planes *A* and *B*. The fourth

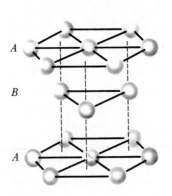

FIGURE 3-22   The *ABABAB* stacking sequence of close-packed planes produces the HCP structure.

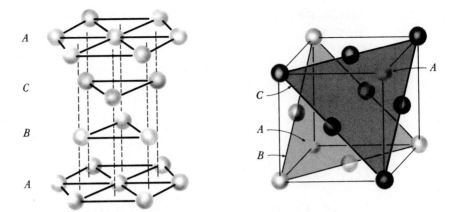

FIGURE 3-23   The *ABCABCABC* stacking sequence of close-packed planes produces the FCC structure.

plane fits directly over atoms in plane *A*. Consequently, a stacking sequence . . . *ABCABCABC* . . . is produced using the (111) plane. Again we find that each atom has a coordination number of 12.

Unlike the HCP unit cell, there are four sets of nonparallel close-packed planes—(111), (11$\bar{1}$), (1$\bar{1}$1), and ($\bar{1}$11)—in the FCC cell. This difference between the FCC and HCP unit cells—the presence or absence of intersecting close-packed planes—significantly affects the behavior of metals with these structures.

Anisotropic Behavior.   Because of differences in atomic arrangement in the planes and directions within a crystal, the properties also vary with direction. A material is *anisotropic* if its properties depend on the crystallographic direction along which the property is measured. If the properties are identical in all directions, the crystal is *isotropic*. The anisotropic behavior of the modulus of elasticity is illustrated in Table 3-6 for several materials. A few materials, including tungsten (W), are isotropic.

TABLE 3-6   Variation of modulus of elasticity ($\times 10^6$ psi) with crystal direction

Material	[100]	[111]	Random
Al	9.2	11.0	10.0
Cu	9.7	27.8	18.1
Fe	19.1	40.4	30.0
Nb	22.0	11.8	14.9
W	59.2	59.2	59.2
MgO	35.6	48.7	45.0
NaCl	6.3	4.7	5.3

Interplanar Spacing. The distance between two adjacent parallel planes of atoms with the same Miller indices is called the *interplanar spacing* d_{hkl}. The interplanar spacing in cubic materials is given by the general equation

$$d_{hkl} = \frac{a_0}{\sqrt{h^2 + k^2 + l^2}} \tag{3-6}$$

where a_0 is the lattice parameter and h, k, and l represent the Miller indices of the adjacent planes being considered.

◇   **Example 3-12**

Calculate the distance between adjacent (111) planes in gold, which has a lattice parameter of 4.0786 Å.

Answer:

$$d_{111} = \frac{4.0786}{\sqrt{1^2 + 1^2 + 1^2}} = \frac{4.0786}{\sqrt{3}} = 2.355 \text{ Å}$$

3–6   INTERSTITIAL SITES

In any of the crystal structures that have been described, there are small holes between the usual atoms into which smaller atoms could be placed. These locations are called *interstitial sites*. We will find that these interstitial sites are often partly filled; small impurity atoms, for instance, may be located in these positions. In some cases, we intentionally add small atoms to create metal alloys; this is the case when small carbon atoms are added to iron to produce steels. Finally, the structure of many ceramic materials can be understood by utilizing these interstitial sites. For instance, in ionically bonded compounds, cations may be located at normal lattice points, while anions may be inserted into the interstitial positions.

An atom, when placed into an interstitial site, will touch two or more of the usual atoms in the unit cell. Therefore, these interstitial atoms will have a coordination number equal to the number of atoms it is touching. Figure 3-24 shows the interstitial locations for the SC, FCC, BCC, and HCP structures. The *cubic* interstitial site, with a coordination number of eight, occurs in the SC structure. *Octahedral* sites give a coordination number of six, while *tetrahedral* sites give a coordination number of four. Examples of the octahedral and tetrahedral positions in BCC, FCC, and HCP unit cells are shown in the figure. As an example, the octahedral sites in BCC unit cells are located at faces of the cube; a small atom placed in the octahedral site will touch the four atoms at the

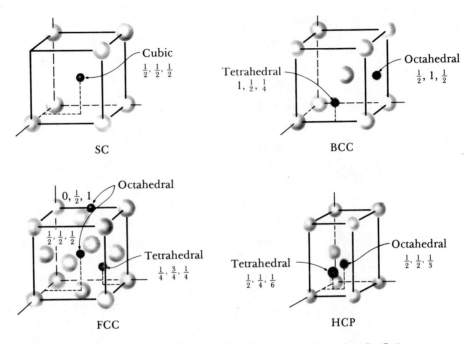

FIGURE 3-24 *The location of the interstitial sites in common unit cells. Only representative sites are shown.*

corners of the face, the atom in the center of the unit cell, plus another atom at the center of the adjacent unit cell, giving a coordination number of six. In FCC unit cells, octahedral sites occur at the center of each edge of the cube as well as in the center of the unit cell.

◇ | **Example 3-13**

Write the coordinates for all of the tetrahedral intersitial sites in the FCC crystal structure.

Answer:

Tetrahedral sites: $\frac{1}{4}, \frac{1}{4}, \frac{1}{4}$ $\frac{1}{4}, \frac{1}{4}, \frac{3}{4}$ $\frac{3}{4}, \frac{1}{4}, \frac{1}{4}$ $\frac{3}{4}, \frac{1}{4}, \frac{3}{4}$

$\frac{1}{4}, \frac{3}{4}, \frac{1}{4}$ $\frac{1}{4}, \frac{3}{4}, \frac{3}{4}$ $\frac{3}{4}, \frac{3}{4}, \frac{1}{4}$ $\frac{3}{4}, \frac{3}{4}, \frac{3}{4}$

◇ | **Example 3-14**

Calculate the number of octahedral sites that uniquely belong to one FCC unit cell.

Answer:

The octahedral sites include the 12 edges of the unit cell, with the coordinates

$\frac{1}{2}, 0, 0 \quad \frac{1}{2}, 1, 0 \quad \frac{1}{2}, 0, 1 \quad \frac{1}{2}, 1, 1$

$0, \frac{1}{2}, 0 \quad 1, \frac{1}{2}, 0 \quad 1, \frac{1}{2}, 1 \quad 0, \frac{1}{2}, 1$

$0, 0, \frac{1}{2} \quad 1, 0, \frac{1}{2} \quad 1, 1, \frac{1}{2} \quad 0, 1, \frac{1}{2}$

plus the center position, $\frac{1}{2}, \frac{1}{2}, \frac{1}{2}$. Each of the sites on the edge of the unit cell is shared between four unit cells, so only $\frac{1}{4}$ of each site belongs uniquely to each unit cell. Therefore, the number of sites belonging uniquely to each cell is

(12 edges) ($\frac{1}{4}$ per cell) + 1 center location = 4 octahedral sites

We can calculate the size of each interstitial site in terms of the size of atoms at the regular lattice positions. Example 3-15 shows how the radius of the interstitial site is related to the atom size for the cubic and octahedral sites.

◇ **Example 3-15**

Calculate the radius of an atom that will just fit into (a) the cubic site and (b) the octahedral site. The radius of the atoms in the normal lattice positions is R.

Answer:

(a) Figure 3-25(a) shows the arrangement of the atoms when the smaller atom just fits into the center of a cube.

$$2R + 2r = 2R\sqrt{3}$$
$$r = \sqrt{3}R - R = (\sqrt{3} - 1)R$$
$$\frac{r}{R} = 0.732$$

(b) Figure 3-25(b) shows the arrangement of the atoms when the smaller atom just fits into the center of an octahedron.

$$2R + 2r = 2R\sqrt{2}$$
$$r = \sqrt{2}R - R = (\sqrt{2} - 1)R$$
$$\frac{r}{R} = 0.414$$

Interstitial atoms whose radii are slightly larger than the radius of the interstitial site may enter that site, pushing the surrounding atoms slightly apart. However, atoms whose radii are smaller than the radius of the hole are not allowed to fit into the interstitial site, or the ion will "rattle" around in the site.

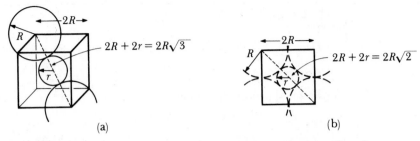

FIGURE 3-25   Illustration for Example 3-15 to determine the radius of an interstitial site for (a) a cubic hole and (b) an octahedral hole.

If the interstitial atom becomes too large, it prefers to enter a site having a larger coordination number (Table 3-7). Therefore, an atom whose radius ratio is between 0.225 and 0.414 will enter a tetrahedral site; if its radius is somewhat larger than 0.414, it will enter an octahedral site instead. When atoms have the same size, as in pure metals, the radius ratio is one and the coordination number is 12, which is the case for metals with the FCC and HCP structures.

TABLE 3-7   The coordination number and the radius ratio

Coordination Number	Location of Interstitial	Radius Ratio	Representation
2	Linear	0–0.155	
3	Corners of triangle	0.155–0.225	
4	Corners of tetrahedron	0.225–0.414	
6	Corners of octahedron	0.414–0.732	
8	Corners of cube	0.732–1.000	

◇   **Example 3-16**

Determine the expected coordination number for each ion in NiO.

Answer:

From Appendix B, the ionic radii are

$$r_{Ni} = 0.69 \, \text{Å} \qquad r_O = 1.32 \, \text{Å}$$

$$\frac{r_{Ni}}{r_O} = \frac{0.69}{1.32} = 0.523$$

From Table 3-7, we find that the coordination number should be six, since $0.414 < 0.523 < 0.732$.

3-7 IONIC CRYSTALS

Many ceramic materials contain ionic bonds between the negatively charged anions and the positively charged cations. These ionic materials must have crystal structures that assure electrical neutrality yet permit ions of different sizes to be efficiently packed.

Electrical Neutrality.   If the charges on the anion and the cation are identical, the ceramic compound has the formula AX, and the coordination number for each ion is identical to assure a proper balance of charge. As an example, each cation may be surrounded by six anions, while each anion is in turn surrounded by six cations. However, if the valence of the cation is $+2$ and that of the anion is -1, then twice as many anions must be present, and the formula is of the form AX_2. The structure of the AX_2 compound must assure that the coordination number of the cation is twice the coordination number of the anion. For example, each cation may have 8 anion nearest neighbors, while only 4 cations will touch each anion.

Ionic Radii.   The crystal structures of the ionically bonded compounds often can be described by placing the cations at the normal lattice points of a unit cell, with the anions then located at one or more of the interstitial sites described in Section 3-6. The ratio of the sizes of the ionic radii of the anion and cation influences both the manner of packing and the coordination number (Table 3-7). A number of common structures in ceramic compounds are described in the following paragraphs.

Cesium Chloride Structure.   Cesium chloride (CsCl) is simple cubic, with the "cubic" interstitial site filled by the Cl anion (Figure 3-26). The radius ratio, $r_{Cs}/r_{Cl} = 1.67/1.81 = 0.92$, dictates that cesium chloride have a coordination

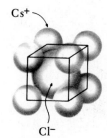

FIGURE 3-26
The cesium chloride structure, an SC unit cell with two ions per lattice point.

number of eight. We can characterize the structure as a simple cubic structure with two ions—one Cs and one Cl—associated with each lattice point. This structure is possible when the anion and the cation have the same valence.

◇   **Example 3-17**

For KCl, (a) verify that the compound may have the cesium chloride structure and (b) calculate the packing factor for the compound.

Answer:

(a) From Appendix B, $r_K = 1.33\,\text{Å}$ and $r_{Cl} = 1.81\,\text{Å}$, so

$$\frac{r_K}{r_{Cl}} = \frac{1.33}{1.81} = 0.735$$

Since $0.732 < 0.735 < 1.000$, the coordination number is eight and the CsCl structure is likely.

(b) The ions touch along the body diagonal of the unit cell, so

$$\sqrt{3}a_0 = 2r_K + 2r_{Cl} = 2(1.33) + 2(1.81) = 6.28\,\text{Å}$$

$$a_0 = 3.63\,\text{Å}$$

$$\text{Packing factor} = \frac{\dfrac{4\pi}{3}r_K^3(1\text{ K ion}) + \dfrac{4\pi}{3}r_{Cl}^3(1\text{ Cl ion})}{a_0^3}$$

$$= \frac{\dfrac{4\pi}{3}(1.33)^3 + \dfrac{4\pi}{3}(1.81)^3}{(3.63)^3} = 0.725$$

Sodium Chloride Structure.   The radius ratio for sodium and chloride ions is $r_{Na}/r_{Cl} = 0.97/1.81 = 0.536$; the sodium ion has a charge of $+1$, while the chloride ion has a charge of -1. Therefore, based on the charge balance and radius ratio, each anion and cation must have a coordination number of six. The FCC structure, with Na cations at FCC positions and Cl anions at the four octahedral sites, will satisfy these requirements (Figure 3-27). We can also consider this structure to be FCC with two ions—one Na and one Cl—associated with each lattice point. Many ceramics, including MgO, CaO, and FeO, have this structure.

◇   **Example 3-18**

Show that MgO can have the sodium chloride crystal structure and calculate the density of MgO.

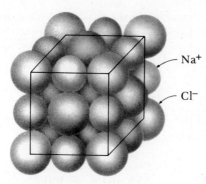

FIGURE 3-27 *The sodium chloride structure, an FCC unit cell with two ions per lattice point.*

Answer:

From Appendix B, $r_{Mg} = 0.66\,Å$ and $r_O = 1.32\,Å$, so

$$\frac{r_{Mg}}{r_O} = \frac{0.66}{1.32} = 0.50$$

Since $0.414 < 0.50 < 0.732$, the coordination number is six and the sodium chloride structure is possible.

The atomic weights are 24.3 and $16\,g/g \cdot mole$ for magnesium and oxygen, respectively. The ions touch along the edge of the cube, so

$$a_0 = 2r_{Mg} + 2r_O = 2(0.66) + 2(1.32) = 3.96\,Å$$

$$\rho = \frac{(4\,Mg\ ions)(24.3) + (4\,O\ ions)(16)}{(3.96 \times 10^{-8}\,cm)^3(6.02 \times 10^{23})} = 4.31\,g/cm^3$$

Zinc Blende Structure.   Although the Zn ions have a charge of $+2$ and S has a charge of -2, zinc blende (ZnS) cannot have the sodium chloride structure because $r_{Zn}/r_S = 0.74/1.84 = 0.402$. This radius ratio demands a coordination number of four, which in turn means that the sulfide ions will enter tetrahedral sites in a unit cell, as indicated by the small "cubelet" in the unit cell (Figure 3-28). The FCC structure, with Zn cations at the normal lattice points

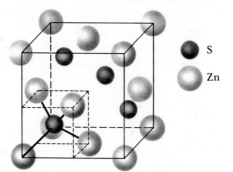

FIGURE 3-28 *The zinc blende unit cell.*

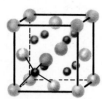

FIGURE 3-29
The fluorite unit cell.

and S anions at one-half of the tetrahedral sites, can accommodate the restrictions of both charge balance and coordination number.

Fluorite Structure.   The fluorite structure is FCC, with anions located at all eight of the tetrahedral positions (Figure 3-29). Thus, there are four cations and eight anions per cell and the ceramic compound must have the formula AX_2, as in calcium fluorite, or CaF_2. The coordination number of the calcium ions is eight but that of the fluoride ions is four, therefore assuring a balance of charge.

◇ | **Example 3-19**

Verify that CaF_2 can have the fluorite structure.

Answer:

From Appendix B, $r_{Ca} = 0.99\,Å$ and $r_F = 1.36\,Å$, so

$$\frac{r_{Ca}}{r_F} = \frac{0.99}{1.36} = 0.728$$

This predicts a coordination number of six, since $0.414 < 0.728 < 0.732$. However, this coordination number is not possible if the charge balance is to be maintained. There must be twice as many fluoride ions around the calcium ions as there are calcium ions around fluoride ions. The requirement for satisfying the charge balance offsets the ionic radius effect.

◇

120°

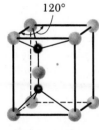

FIGURE 3-30
The wurtzite unit cell.

Wurtzite Structure.   A series of crystal structures are formed when ions are placed in the interstitial locations of the HCP structure. One example is the wurtzite structure, ZnS, which has half of the tetrahedral sites filled with anions (Figure 3-30).

◇ | **Example 3-20**

Based on charge balance and radius ratio, which of the cubic structures we have discussed would BeO most likely have?

Answer:

From Appendix B, the radius ratio is

$$\frac{r_{Be}}{r_O} = \frac{0.35}{1.32} = 0.265$$

This radius ratio requires a coordination number of four. Of the cubic structures we have examined, only the zinc blende structure, with half of the tetrahedral sites filled, satisfies both the coordination number and charge balance requirements.

◇

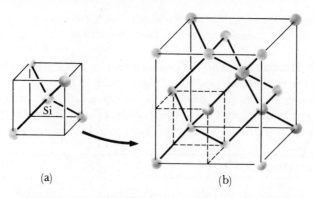

FIGURE 3-31   (a) Tetrahedron and (b) the diamond cubic (DC) unit cell. This open structure is produced because of the requirements of covalent bonding.

3-8   COVALENT STRUCTURES

Covalently bonded materials frequently must have complex structures in order to satisfy the directional restraints imposed by the bonding.

Diamond Cubic Structure.   Elements such as silicon, germanium, and carbon in its diamond form are bonded by four covalent bonds and produce a tetrahedron (Figure 3-31). The coordination number for each silicon atom is only four, due to the nature of the covalent bonding.

As these tetrahedral groups are combined, a large cube can be constructed [Figure 3-31(b)]. This large cube contains eight smaller cubes that are the size of the tetrahedral cube; however, only four of the cubes contain tetrahedra. The large cube is the *diamond cubic*, or DC, unit cell. The lattice is a special FCC structure. The atoms on the corners of the tetrahedral cubes provide atoms at each of the regular FCC lattice points. However, four additional atoms are present within the DC unit cell from the atoms in the center of the tetrahedral cubes. We can describe the DC lattice as an FCC lattice with two atoms associated with each lattice point. Therefore, there must be eight atoms per unit cell.

◇ | **Example 3-21**

Determine the packing factor for DC silicon.

Answer:

We find that atoms touch along the body diagonal of the cell (Figure 3-32). Although atoms are not present at all locations along the body diagonal, there

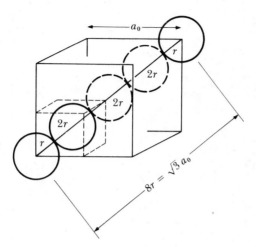

FIGURE 3-32 Determining the relationship between lattice parameter and atomic radius in diamond cubic.

are voids that have the same diameter as atoms. Consequently,

$$\sqrt{3}a_0 = 8r$$

$$\text{Packing factor} = \frac{(8 \text{ atoms/cell})(\frac{4}{3}\pi r^3)}{a_0^3}$$

$$= \frac{(8)(\frac{4}{3}\pi r^3)}{(8r/\sqrt{3})^3}$$

$$= 0.34$$

Crystalline Silica.   In a number of its forms, silica, or SiO_2, has a crystalline ceramic structure which is partly covalent and partly ionic. Figure 3-33 shows

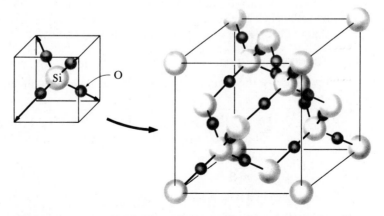

FIGURE 3-33 The silicon-oxygen tetrahedra and their combination to form the β-cristobalite form of silica.

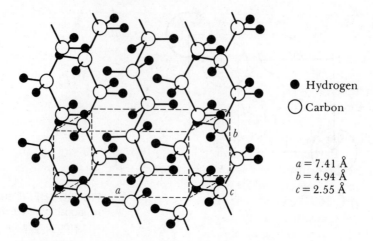

FIGURE 3-34 *The unit cell of crystalline polyethylene.*

the crystal structure of one of the forms of silica, β-cristobalite, which is a complicated FCC structure. The ionic radii of silicon and oxygen are 0.42 Å and 1.32 Å, respectively, so the radius ratio is $r_{Si}/r_O = 0.318$ and the coordination number is four.

Crystalline Polymers.   A number of polymers may form a crystalline structure. The dashed lines in Figure 3-34 outline the unit cell for the lattice of poly-ethylene. Polyethylene is obtained by joining C_2H_4 molecules together to produce long polymer chains that form an orthorhombic unit cell. Some polymers, including nylon, can have several polymorphic forms.

◇ | **Example 3-22**

How many carbon and hydrogen atoms are in each unit cell of crystalline polyethylene? There are twice as many hydrogen atoms as carbon atoms in the chain. The density of polyethylene is about 0.95 g/cm³.

Answer:

If we let x be the number of carbon atoms, then $2x$ is the number of hydrogen atoms.

$$\rho = \frac{(x)(12 \text{ g/g} \cdot \text{mole}) + (2x)(1 \text{ g/g} \cdot \text{mole})}{(7.41 \times 10^{-8})(4.94 \times 10^{-8})(2.55 \times 10^{-8})(6.02 \times 10^{23})}$$

$$0.95 = \frac{14x}{56.2}$$

$$x = 3.8 \simeq 4 \text{ carbon atoms per cell}$$

$$2x = 8 \text{ hydrogen atoms per cell}$$

SUMMARY

Atoms are arranged in solid materials with either a short-range or long-range order. Amorphous materials, such as glasses and many polymers, have only a short-range order, determined primarily by the restrictions of covalent bonding. Crystalline materials, including metals and many ceramics, have both a long- and short-range order. The long-range periodicity in these materials is described by the crystal structure.

Particularly in metals, the crystal structure is closely related to the general mechanical properties and behavior of the material. As we shall see in subsequent chapters, metals with FCC structures are normally soft and ductile, metals with BCC structures are much stronger, and metals with HCP structures tend to be relatively brittle. We will be able to explain the differences in behavior of metals by examining the role of crystal structure in deformation.

The crystal structures can become more complicated when atoms are situated at interstitial sites in the unit cell. Many ceramic structures can be interpreted in this way.

GLOSSARY

Allotropy. The characteristic of a material being able to exist in more than one crystal structure, depending on temperature and pressure.

Anisotropy. Having different properties in different directions.

Basal plane. The special name given to the close-packed plane in hexagonal close-packed unit cells.

Coordination number. The number of nearest neighbors to an atom in its atomic arrangement.

Crystal structure. The arrangement of the atoms in a material into a regular repeatable lattice.

Density. Mass per unit volume of a material, usually in units of g/cm^3.

Diamond cubic. A special type of face-centered cubic crystal structure found in carbon, silicon, and other covalently bonded materials.

Directions of a form. Crystallographic directions that all have the same characteristics, although their "sense" is different.

Glass. A solid, noncrystalline material that has only short-range order between the atoms.

Interplanar spacing. Distance between two adjacent parallel planes with the same Miller indices.

Isotropy. Having the same properties in all directions.

Lattice. A collection of points that divide space into smaller equally sized segments.

Lattice parameters. The lengths of the sides of the unit cell and the angles between those sides. The lattice parameters describe the size and shape of the unit cell.

Lattice points. Points that make up the lattice. The surroundings of each lattice point are identical anywhere in the material.

Linear density. The number of lattice points per unit length along a direction.

Long-range order. A regular repetitive arrangement of the atoms in a solid which extends over a very large distance.

Miller-Bravais indices. A special shorthand notation to describe the crystallographic planes in hexagonal close-packed unit cells.

Miller indices. A shorthand notation to describe certain crystallographic directions and planes in a material.

Packing factor. The fraction of space occupied by atoms.

Planar density. The number of atoms per unit area whose centers lie on the plane.

Planes of a form. Crystallographic planes that all have the same characteristics, although their orientations are different.

Polymorphism. Allotropy, or having more than one crystal structure.

Repeat distance. The distance from one lattice point to the adjacent lattice point along a direction.

Short-range order. The arrangement of the atoms is regular and predictable only over a short distance, usually one or two atom spacings.

Stacking sequence. The sequence in which close-packed planes are stacked. If the sequence is *ABABAB* a hexagonal close-packed unit cell is produced; if the sequence is *ABCABCABC* a face-centered cubic structure is produced.

Tetrahedron. The structure produced when atoms are packed together with a fourfold coordination.

Unit cell. A subdivision of the lattice that still retains the overall characteristics of the entire lattice.

PRACTICE PROBLEMS

1. How many lattice points are unique to the base-centered orthorhombic unit cell?

2. Why is there no base-centered tetragonal structure? Draw a lattice for this structure, then determine what the actual unit cell is.

3. Why is there no base-centered cubic structure? Draw a lattice for this structure, then determine what the actual unit cell is.

4. A material has a cubic unit cell with one atom per lattice point. If $a_0 = 4.0786\,\text{Å}$ and $r = 1.442\,\text{Å}$, determine the crystal structure.

5. A material has a cubic unit cell with one atom per lattice point. If $a_0 = 5.025\,\text{Å}$ and $r = 2.176\,\text{Å}$, determine the crystal structure.

6. Using the atomic radius data in Appendix B, calculate the packing factor for crystalline polyethylene.

7. The density of lead is $11.36\,\text{g/cm}^3$, its atomic mass is $207.19\,\text{g/g}\cdot\text{mole}$, and the crystal structure is FCC. Calculate (a) the lattice parameter and (b) the atomic radius for lead.

8. The density of tantalum is $16.6\,\text{g/cm}^3$, its atomic mass is $180.95\,\text{g/g}\cdot\text{mole}$, and the crystal structure is BCC. Calculate (a) the lattice parameter and (b) the atomic radius for tantalum.

9. How many unit cells are present in a cubic centimeter of face-centered cubic nickel? The atomic radius of nickel is $1.243\,\text{Å}$.

10. Calculate (a) the volume and (b) the mass of one million unit cells of body-centered cubic iron. The atomic radius of iron is $1.241\,\text{Å}$.

11. A material with a cubic structure has a density of $0.855\,\text{g/cm}^3$, an atomic mass of $39.09\,\text{g/g}\cdot\text{mole}$, and a lattice parameter of $5.344\,\text{Å}$. If one atom is located at each lattice point, determine the type of unit cell.

12. A material with a cubic structure has a density of $10.49\,\text{g/cm}^3$, an atomic mass of $107.868\,\text{g/g}\cdot\text{mole}$, and a lattice parameter of $4.0862\,\text{Å}$. If one atom is located at each lattice point, determine the type of unit cell.

13. Antimony has a hexagonal unit cell

#26,27,29,31,33,39,43,49,51,53,55,56
Due 9/27

with $a_0 = 4.307\,\text{Å}$ and $c_0 = 11.273\,\text{Å}$. If its density is $6.697\,\text{g/cm}^3$ and its atomic mass is $121.75\,\text{g/}$ $\text{g}\cdot\text{mole}$, calculate the number of atoms per cell.

14. One of the forms of plutonium has a face-centered orthorhombic structure, with $a = 3.159\,\text{Å}$, $b = 5.768\,\text{Å}$, and $c = 10.162\,\text{Å}$. The density of Pu is $17.14\,\text{g/cm}^3$ and the atomic mass is $239.052\,\text{g/g}\cdot\text{mole}$. Determine (a) the number of atoms per cell and (b) the number of atoms at each lattice point.

0.15 **15.** Prasiodymium has a special hexagonal structure with four atoms per unit cell; the lattice parameters are $a_0 = 3.6721\,\text{Å}$ and $c_0 = 11.8326\,\text{Å}$, while the atomic radius is $1.8360\,\text{Å}$. Calculate the packing factor of Pr.

16. Gadolinium has an HCP structure just below $1260°\,\text{C}$ with $a = 3.6745\,\text{Å}$ and $c = 5.8525\,\text{Å}$. Just above $1260°\,\text{C}$, Gd transforms to a BCC structure with $a = 4.06\,\text{Å}$. Calculate the percent volume change when Gd cools from the BCC to the HCP structure. Does the metal expand or contract during cooling?

-1.71 70 **17.** Lanthanum has an FCC structure just below $865°\,\text{C}$ with $a = 5.337\,\text{Å}$, but has a BCC structure with $a = 4.26\,\text{Å}$ just above $865°\,\text{C}$. Calculate the percent volume change when La heats from the FCC to the BCC structure. Does the metal expand or contract during heating?

18. Lanthanum has a special HCP structure just below $325°\,\text{C}$ and an FCC structure just above $325°\,\text{C}$. At $325°\,\text{C}$, the lattice parameters for the HCP structure are $a = 3.779\,\text{Å}$ and $c = 12.270\,\text{Å}$; the lattice parameter for the FCC structure at this temperature is $5.303\,\text{Å}$. Lanthanum has a density of $6.146\,\text{g/cm}^3$ and an atomic mass of $138.9055\,\text{g/g}\cdot\text{mole}$. (a) Calculate the number of atoms in the special HCP unit cell. (b) Calculate the percent volume change when the FCC form of La transforms to the HCP structure on cooling. Does the metal expand or contract during cooling?

19. At $1450°\,\text{C}$, thorium changes from one type of cubic unit cell to a different cubic cell, with a 0.5% decrease in volume during heating. Below $1450°\,\text{C}$, the lattice parameter is $5.187\,\text{Å}$,

Chpt. 3 # 7, 8
Chpt. 19 # 1, 2, 3, 4

while the lattice parameter of the higher temperature form is $4.11\,\text{Å}$. What is the ratio between the number of atoms in the unit cell of the high-temperature form to the number of atoms in the unit cell of the low-temperature form of Th?

20. α-Mn has a cubic structure with $a_0 = 8.931\,\text{Å}$ and a density of $7.47\,\text{g/cm}^3$. β-Mn has a different cubic structure with $a_0 = 6.326\,\text{Å}$ and a density of $7.26\,\text{g/cm}^3$. γ-Mn has a tetragonal structure with $a = 3.748\,\text{Å}$ and $c = 9.40\,\text{Å}$ and a density of $7.21\,\text{g/cm}^3$. The atomic mass of manganese is $54.9380\,\text{g/g}\cdot\text{mole}$. (a) Calculate the number of atoms in each of the three polymorphic forms of manganese. (b) Assuming that the radius of the Mn atom is $1.12\,\text{Å}$ in all three forms, calculate the packing factor for each of the three unit cells.

21. Determine the Miller indices for the directions in the cubic unit cell shown in Figure 3-35.

$A_0 = 1.12(2)$
$A_0 = \dfrac{1.12(4)}{\sqrt{2}}$
$A_0 = \dfrac{1.12(4)}{\sqrt{3}}$

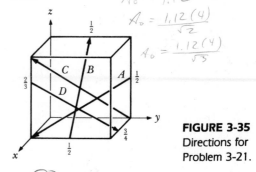

FIGURE 3-35
Directions for
Problem 3-21.

22. Determine the Miller indices for the directions in the cubic unit cell shown in Figure 3-36.

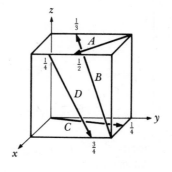

FIGURE 3-36
Directions for
Problem 3-22.

23. Determine the Miller indices for the planes in the cubic unit cell shown in Figure 3-37.

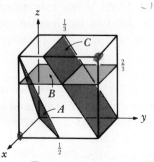

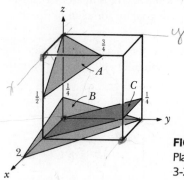

FIGURE 3-37
Planes for Problem 3-23.

24. Determine the Miller indices for the planes in the cubic unit cell shown in Figure 3-38.

FIGURE 3-38
Planes for Problem 3-24.

25. Determine the Miller indices for the directions in the hexagonal unit cell in Figure 3-39, using the three-digit system.

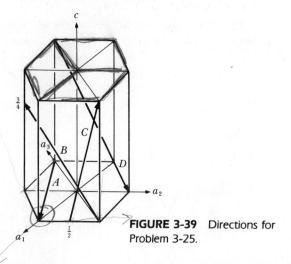

FIGURE 3-39 Directions for Problem 3-25.

26. Determine the Miller indices for the directions in the hexagonal unit cell in Figure 3-40, using the three-digit system.

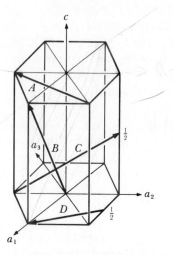

FIGURE 3-40 Directions for Problem 3-26.

27. Determine the Miller-Bravais indices for the planes in the hexagonal unit cell in Figure 3-41.

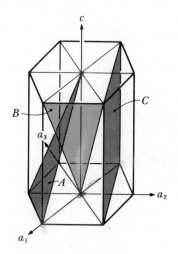

FIGURE 3-41 Planes for Problem 3-27.

28. Determine the Miller-Bravais indices for the planes in the hexagonal unit cell in Figure 3-42.

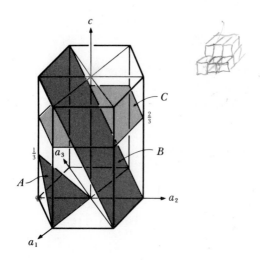

FIGURE 3-42   Planes for Problem 3-28.

29. Sketch the following directions and planes within a cubic unit cell.

(a) [112]   (b) [3$\bar{1}$0]   (c) [1$\bar{1}$1]   (d) [101]

(e) [041]   (f) [20$\bar{3}$]   (g) ($\bar{1}$01)   (h) (11$\bar{1}$)

(i) (0$\bar{1}$3)   (j) (1$\bar{2}$1)   (k) (201)   (l) (120)

30. Sketch the following directions and planes within a cubic unit cell.

(a) [123]   (b) [1$\bar{1}$0]   (c) [010]   (d) [131]

(e) [$\bar{1}$21]   (f) [134]   (g) (2$\bar{2}$0)   (h) (301)

(i) (1$\bar{1}$2)   (j) (011)   (k) (421)   (l) (1$\bar{4}$1)

31. Draw the (11$\bar{1}$) plane and identify the six ⟨110⟩ directions that lie in that plane in a cubic lattice.

32. Draw the ($\bar{1}$10) plane and identify the four directions of the form ⟨111⟩ that lie in that plane in a cubic lattice.

33. How many planes of the form {131} are found in a cubic system? Would you give the same answer if we used a tetragonal or ortho-rhombic system? Explain.

34. How many planes of the form {123} are found in a cubic system? What are the indices of the planes of the form {123} in a tetragonal system?

35. How many directions of the form ⟨123⟩ are found in a cubic system? Would you give the same answer if we used a tetragonal or orthorhombic system? Explain.

36. How many directions of the form ⟨221⟩ are found in a cubic system? What are the indices of the directions of the form ⟨221⟩ in a tetragonal system?

37. What are the indices of the planes of the form {412} in an orthorhombic system?

38. What are the indices of the directions of the form ⟨121⟩ in an orthorhombic system?

39. Determine whether the [101] direction in a tetragonal unit cell with a c/a ratio of 1.5 is perpendicular to the (101) plane. If it is not perpendicular, calculate the angle between the direction and the plane.

40. Determine whether the [110] direction in an orthorhombic unit cell with $a = 3\,\text{Å}$, $b = 4\,\text{Å}$, and $c = 5\,\text{Å}$ is perpendicular to the (110) plane. If it is not perpendicular, calculate the angle between the direction and the plane.

41. Draw the plane in a cubic system that passes through the coordinates 1, 1, 0; 0, 1, 1; and 0, 0, 1. What are the Miller indices of this plane?

42. Draw the plane in a cubic system that passes through the coordinates 1, 1, 0; 0, 0, 1; and 0, 1, 0. What are the Miller indices of this plane?

43. Draw the plane in a cubic system that passes through the coordinates 1, 0, 1; $\frac{1}{2}$, 0, 1; and 1, $\frac{1}{2}$, 0. What are the Miller indices of this plane?

44. Draw the plane in a cubic system that passes through the coordinates 1, 0, 0; 0, 0, 1; and $\frac{1}{2}$, 1, $\frac{1}{2}$. What are the Miller indices of this plane?

45. In the four-digit system for finding the indices for a direction in HCP unit cells, is the [110] equal to a [11$\bar{2}$0]? Show this by constructing the path from the tail to the head of the direction.

46. In the four-digit system for finding the indices for a direction in HCP unit cells, is the [100] equal to a [2$\bar{1}$$\bar{1}$0]? Show this by constructing the path from the tail to the head of the direction.

47. In the four-digit system for finding the

indices for a direction in HCP unit cells, is the $[011]$ equal to a $[\bar{1}2\bar{1}3]$? Show this by constructing the path from the tail to the head of the direction.

48. Is the $[1\bar{2}10]$ direction in an HCP unit cell perpendicular to the $(1\bar{2}10)$ plane? Draw each and verify your answer.

49. Calculate the linear density of a line in the $[111]$ direction in (a) simple cubic, (b) body-centered cubic, and (c) face-centered cubic unit cells, assuming a lattice parameter of $4.0\,\text{Å}$ in each case.

50. Calculate the linear density of a line in the $[110]$ direction in (a) simple cubic, (b) body-centered cubic, and (c) face-centered cubic unit cells, assuming a lattice parameter of $4.0\,\text{Å}$ in each case.

51. Calculate the packing fraction in the $[111]$ direction in (a) simple cubic, (b) body-centered cubic, and (c) face-centered cubic unit cells. In which, if any, of these structures is the $[111]$ direction a close-packed direction?

52. Calculate the packing fraction in the $[110]$ direction in (a) simple cubic, (b) body-centered cubic, and (c) face-centered cubic unit cells. In which, if any, of these structures is the $[110]$ direction a close-packed direction?

53. Calculate the packing fraction of a (111) plane in (a) a simple cubic, (b) a body-centered cubic, and (c) a face-centered cubic unit cell. In which, if any, of these structures is the (111) plane a close-packed plane?

54. Calculate the packing fraction of a (110) plane in (a) a simple cubic, (b) a body-centered cubic, and (c) a face-centered cubic unit cell. In which, if any, of these structures is the (110) plane a close-packed plane?

55. Calculate the planar density on a (111) plane in (a) simple cubic, (b) body-centered cubic, and (c) face-centered cubic unit cells, assuming a lattice parameter in each case of $4.0\,\text{Å}$.

56. Calculate the planar density of a (110) plane in (a) simple cubic, (b) body-centered cubic, and (c) face-centered cubic unit cells, assuming a lattice parameter in each case of $4.0\,\text{Å}$.

57. Calculate the linear densities in the $[110]$ and $[101]$ directions in a face-centered tetragonal unit cell with $a = 4.0\,\text{Å}$ and $c = 6.0\,\text{Å}$.

58. Calculate the planar densities in the (110) and (101) planes in a face-centered tetragonal unit cell with $a = 4.0\,\text{Å}$ and $c = 6.0\,\text{Å}$.

59. Calculate the linear densities in the $[100]$, $[010]$, $[110]$, and $[001]$ directions of a base-centered orthorhombic unit cell with $a = 3.0\,\text{Å}$, $b = 5.0\,\text{Å}$, and $c = 8.0\,\text{Å}$.

60. Calculate the planar densities in the (100), (010), (110), and (001) planes of a base-centered orthorhombic unit cell with $a = 3.0\,\text{Å}$, $b = 5.0\,\text{Å}$, and $c = 8.0\,\text{Å}$.

61. Calculate the interplanar spacing between the following planes in gold (see Appendix A for the lattice parameter).

(a) $(12\bar{4})$ (b) (201) (c) $(1\bar{1}2)$ (d) $(\bar{3}21)$

62. The interplanar spacing between (231) planes is found to be $0.89\,\text{Å}$. Calculate the lattice parameter if the material has a cubic crystal structure.

63. Show that the radius ratio for an atom or ion that just fits into a tetrahedral interstitial site without disturbing the surrounding atoms or ions is 0.225.

64. Show that the radius ratio for an atom or ion that just fits into a triangular interstitial site without disturbing the surrounding atoms or ions is 0.155.

65. List the coordinates for all six of the octahedral sites that lie in a BCC unit cell. How many of these sites belong uniquely to one BCC cell?

66. Using the ionic radii given in Appendix B, determine the coordination number expected for the following compounds.

(a) FeO (b) CaO (c) SiC (d) PbS (e) B_2O_3

67. Using the ionic radii given in Appendix B, determine the coordination number expected for the following compounds.

(a) Al_2O_3 (b) TiO_2 (c) MgO

(d) SiO_2 (e) $CuZn$

68. Based on the ionic radius ratio and the necessity for charge balance, which of the cubic structures discussed in the text would you expect CdS to possess?

69. Based on the ionic radius ratio and the necessity for charge balance, which of the cubic structures discussed in the text would you expect CoO to possess?

70. The compound NiO has the sodium chloride crystal structure. Based on the data in Appendix B, calculate (a) the lattice parameter, (b) the density, and (c) the packing factor for NiO.

71. The compound TeO_2 has the fluorite structure. Determine (a) the number of tellurium and oxygen ions in each unit cell, (b) the expected coordination number based on the radius ratio, (c) the lattice parameter, and (d) the packing factor for the compound.

72. One of the forms of BeO has the zinc blende structure at high temperatures. Determine (a) the lattice parameter, (b) the density, and (c) the packing factor for the compound.

73. Which of the cubic structures discussed in the text would you expect CsBr to possess, based on its expected coordination? Calculate (a) the lattice parameter, (b) the packing factor, and (c) the density of the compound.

74. Germanium has the diamond cubic structure with a lattice parameter of 5.6575 Å. Calculate the size of the germanium atom in the unit cell. Does this best match up with germanium's atomic radius or ionic radius?

75. Calculate the fraction of the [111] direction and the fraction of the (111) plane actually covered by atoms in the diamond cubic unit cell.

76. Calculate the fraction of the [111] direction and the fraction of the (111) plane actually covered by sodium ions in the NaCl unit cell.

77. How many oxygen and silicon ions are present in cristobalite? Using the ionic radii in Appendix B, calculate (a) the lattice parameter. (b) the packing factor, and (c) the density of cristobalite.

chpt. 3   59, 60, 62, 63, 64, 65

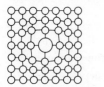

4

IMPERFECTIONS IN THE ATOMIC ARRANGEMENT

4–1 INTRODUCTION

All materials contain imperfections in the arrangement of the atoms which have a profound effect on the behavior of the material. By controlling lattice imperfections we create stronger metals and alloys, more powerful magnets, improved transistors and solar cells, glassware of striking colors, and many other materials of practical importance.

In this chapter we will introduce the three basic types of lattice imperfections—point defects, line defects (or dislocations), and surface defects. We must remember, however, that these imperfections represent only defects in the perfect atomic arrangement and should not suggest that the material itself is defective. Indeed, these "defects" are intentionally added to produce a desired set of mechanical and physical properties. In later chapters we will show how we control these defects through alloying, heat treatment, or processing techniques to produce improved engineering materials.

4–2 DISLOCATIONS

Dislocations are line imperfections in an otherwise perfect lattice. We can identify two types of dislocations—the screw dislocation and the edge dislocation. The *screw dislocation* (Figure 4-1) can be illustrated by cutting partway through a perfect crystal, then skewing the crystal one atom spacing. If we were to follow a crystallographic plane one revolution around the axis on which the crystal was skewed, traveling equal atom spacings in each direction, we would finish one atom spacing below our starting point. The vector required to complete the loop and return us to our starting point is the *Burgers vector* **b**. If we continued our rotation, we would trace out a spiral path. The axis, or line, around which we

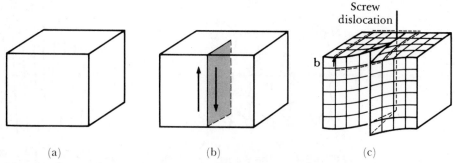

FIGURE 4-1   *The perfect crystal (a) is cut and sheared one atom spacing, (b) and (c). The line along which shearing occurs is a screw dislocation. A Burgers vector* **b** *is required to close a loop of equal atom spacings around the screw dislocation.*

trace out this path is the screw dislocation. We see that the Burgers vector is parallel to the screw dislocation.

An *edge dislocation* (Figure 4-2) can be illustrated by slicing partway through a perfect crystal, spreading the crystal apart, and partly filling the cut with an extra plane of atoms. The bottom edge of this inserted plane represents the edge dislocation. If we describe a clockwise loop around the edge dislocation by going an equal number of atom spacings in each direction, we would finish one atom spacing from our starting point. The vector that is required to complete the loop is again the Burgers vector. In this case, the Burgers vector is perpendicular to the edge dislocation. By introducing the dislocation, the atoms above the dislocation line are squeezed too closely together, while the atoms below the dislocation are stretched too far apart. The perfection of the lattice has been destroyed by the presence of the dislocation.

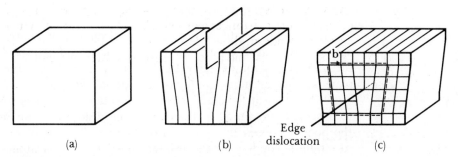

FIGURE 4-2   *The perfect crystal in (a) is cut and an extra plane of atoms is inserted (b). The bottom edge of the extra plane is an edge dislocation (c). A Burgers vector* **b** *is required to close a loop of equal atom spacings around the edge dislocation.*

◇ | **Example 4-1**

Suppose we have a BCC structure with a lattice parameter of 4.0 Å that contains the dislocation shown in Figure 4-3. Determine the direction and length of the Burgers vector.

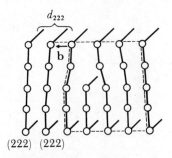

(222) (222)

FIGURE 4-3 The Burgers vector for Example 4-1 is perpendicular to the (222) planes and has a length equal to the interplanar spacing between (222) planes.

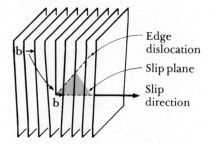

Edge dislocation

Slip plane

Slip direction

FIGURE 4-4 After the Burgers vector is translated from the loop to the dislocation line, a plane is defined.

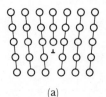

(a)

Shear stress

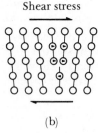

(b)

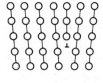

(c)

FIGURE 4-5 When a shear force is applied to the dislocation in (a), the atoms are displaced (b), until the dislocation moves one Burgers vector in the slip direction (c).

Answer:

The clockwise loop around the dislocation is closed by the vector **b**. Because **b** is perpendicular to the (222) planes, the Miller indices of direction **b** must be [222] or, reducing to lowest integers, [111]. The length of **b** is the distance between two adjacent (222) planes. From Equation (3-6),

$$d_{222} = \frac{a_0}{\sqrt{h^2 + k^2 + l^2}} = \frac{4}{\sqrt{2^2 + 2^2 + 2^2}} = 1.155 \, \text{Å}$$

The Burgers vector is a [111] direction that is 1.155 Å in length.

We could translate the Burgers vector from the loop to the edge dislocation, as shown in Figure 4-4. After this translation, we find that the Burgers vector and the edge dislocation define a plane in the lattice. The Burgers vector and the plane are helpful in explaining how materials deform.

When a shear force acting in the direction of the Burgers vector is applied to a crystal containing a dislocation, the dislocation can move by breaking the bonds between the atoms in one plane. The cut plane is shifted slightly to establish bonds with the original partial plane of atoms. This causes the dislocation to move one atom spacing to the side, as shown in Figure 4-5. If this process continues, the dislocation moves through the crystal until a step is produced on the exterior of the crystal (Figure 4-6). The crystal has now been deformed. If dislocations could be continually introduced into one side of the crystal and moved along the same path through the crystal, the crystal would eventually be cut in half.

The process by which a dislocation moves and causes a metal to deform is called *slip*. The direction in which the dislocation moves, the *slip direction*, is the direction of the Burgers vector for edge dislocations. During slip the edge

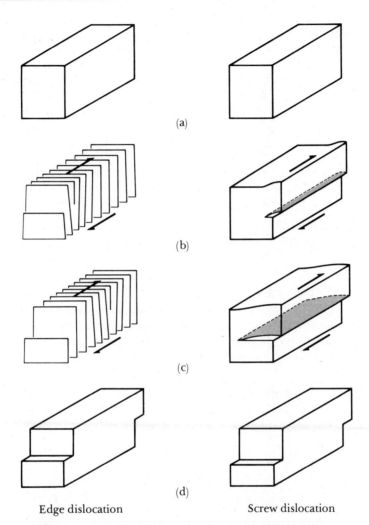

(a)

(b)

(c)

(d)

Edge dislocation   Screw dislocation

FIGURE 4-6   A shear force acting on a dislocation introduced into a perfect crystal (a) causes the dislocation to move through the crystal until a step is created (d). The crystal is now deformed.

dislocation sweeps out the plane formed by the Burgers vector and the dislocation; this plane is called the *slip plane*. The combination of slip direction and slip plane is the *slip system*. A screw dislocation produces the same result; the dislocation moves in a direction perpendicular to the Burgers vector, although the crystal deforms in a direction parallel to the Burgers vector.

During slip, the dislocation moves from one set of surroundings to an identical set of surroundings. To do this with the least expenditure of energy, the dislocation will move the shortest possible distance—in a direction in which the repeat distance is small or the linear density is large. The close-packed directions satisfy this criterion and are therefore the usual slip directions. In addition, slip

TABLE 4-1   *Slip planes and directions in metallic structures*

Crystal Structure	Slip Plane	Slip Direction
BCC	$\{110\}$ $\{112\}$ $\{123\}$	$\langle 111 \rangle$
FCC	$\{111\}$	$\langle 110 \rangle$
HCP	(0001) $\{11\bar{2}0\}$ $\{10\bar{1}0\}$ $\{10\bar{1}1\}$ } See note	$\langle 100 \rangle$ $\langle 110 \rangle$

Note: These planes are active in some metals and alloys or at elevated temperatures.

occurs between planes of atoms that are particularly smooth (so there are smaller "hills and valleys" on the surface) and far apart. Planes of atoms having a high planar density fulfill this requirement. Therefore, the slip planes are typically close-packed planes or those as closely packed as possible. The most common slip systems in metals are summarized in Table 4-1.

◇ Example 4-2

Calculate the length of the Burgers vector in copper.

Answer:

Copper is FCC with a lattice parameter of 3.6151×10^{-8} Å. The close-packed directions, or the directions of the Burgers vector, are of the form $\langle 110 \rangle$. The repeat distance along the $\langle 110 \rangle$ directions is one-half the face diagonal, since lattice points are located at corners and centers of faces.

$$\text{Face diagonal} = \sqrt{2}a_0 = (\sqrt{2})(3.6151 \times 10^{-8}) = 5.1125 \times 10^{-8} \text{ cm}$$

The length of the Burgers vector, or the repeat distance, is

$$\mathbf{b} = \tfrac{1}{2}(5.1125 \times 10^{-8}) = 2.5563 \times 10^{-8} \text{ cm}$$

◇ Example 4-3

The planar density of the (112) plane in BCC iron is 9.94×10^{14} atoms/cm^2. Calculate (a) the planar density of the (110) plane and (b) the interplanar spacings for both the (112) and (110) planes. On which type of plane would slip normally occur?

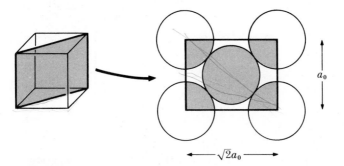

FIGURE 4-7 The atom locations on a (110) plane in a BCC unit cell (Example 4-3).

Answer:

The lattice parameter of BCC iron is 2.866×10^{-8} cm. The (110) plane is shown in Figure 4-7, with the portion of the atoms lying within the unit cell being shaded. Note that one-fourth of the four corner atoms plus the center atom lie within an area of a_0 times $\sqrt{2}a_0$. The planar density is

$$\text{Planar density } (110) \;=\; \frac{\text{atoms}}{\text{area}} \;=\; \frac{2}{\sqrt{2}\,(2.866 \times 10^{-8})^2}$$

$$=\; 1.72 \times 10^{15} \text{ atoms/cm}^2$$

The interplanar spacings are

$$d_{110} \;=\; \frac{2.866 \times 10^{-8}}{\sqrt{1^2 + 1^2 + 0}} \;=\; 2.0266 \times 10^{-8} \text{ cm}$$

$$d_{112} \;=\; \frac{2.866 \times 10^{-8}}{\sqrt{1^2 + 1^2 + 2^2}} \;=\; 1.17 \times 10^{-8} \text{ cm}$$

The planar density and interplanar spacing of the (110) plane are larger than those for the (112) plane; therefore, the (110) plane would be the preferred slip plane.

4–3 SIGNIFICANCE OF DISLOCATIONS

Although slip can occur in some ceramics and polymers, the slip process is particularly helpful to us in understanding the mechanical behavior of metals. First, slip explains why the strength of metals is much lower than the value predicted from the metallic bond. If we had to break an iron bar by breaking all of the metallic bonds across the cross section, as shown in Figure 4-8, we would have to exert a force of several million pounds per square inch. Instead, we could deform the bar by causing slip, during which only a tiny fraction of all

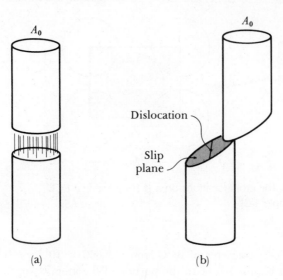

FIGURE 4-8 *Without dislocations (a), a material would fail by breaking all of the bonds across the surface A_0. However, when a dislocation slips (b), bonds are only broken along the line of the dislocation.*

the metallic bonds need be broken at any one time. Perhaps a force of only 10,000 lb/in^2 would be required to deform the iron bar by slip.

Second, slip provides ductility in metals. If no dislocations were present, the iron bar would be brittle; metals could not be shaped by the various metalworking processes, such as forging, into useful shapes.

Third, we control the mechanical properties of a metal or alloy by interfering with the movement of dislocations. An obstacle introduced into the crystal prevents a dislocation from slipping unless we apply higher forces. If we must apply a higher force, then the metal must be stronger!

Enormous numbers of dislocations are found in materials. The *dislocation density*, or total length of dislocations per unit volume, is usually used to represent the amount of dislocations present. Dislocation densities of 10^6 cm/cm^3 are typical of the softest metals, while densities up to 10^{12} cm/cm^3 can be achieved by deforming the material. Figure 4-9 shows dislocations at a very high magnification.

◇ | **Example 4-4**

A typical dislocation density in soft copper is 10^6 cm/cm^3. If the dislocations in 1000 g of copper were placed end to end, how many miles of dislocation would be available?

Answer:

The density of copper is 8.93 g/cm^3. Therefore

$$\text{Volume of copper} \ = \ \frac{1000 \, \text{g}}{8.93 \, \text{g/cm}^3} \ = \ 112 \, \text{cm}^3$$

$$\text{Length} = (112\,\text{cm}^3)(10^6\,\text{cm/cm}^3) = 1.12 \times 10^8\,\text{cm}$$

$$= \frac{(1.12 \times 10^8\,\text{cm})}{(2.54\,\text{cm/in.})(12\,\text{in./ft})(5280\,\text{ft/mile})}$$

$$= 696\,\text{miles}$$

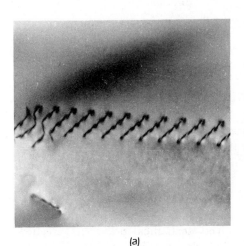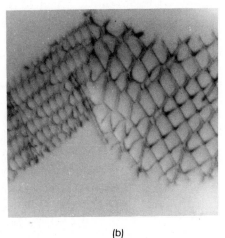

(a)	(b)

FIGURE 4-9 Electron photomicrographs of dislocations in Ti$_3$Al. (a) Dislocation pileups ($\times$ 36,500), (b) dislocation network ($\times$ 15,750). (Courtesy Gerald Feldewerth.)

4-4 SCHMID'S LAW

We can understand the differences in behavior of metals that have different crystal structures by examining the slip process. Suppose we apply a unidirectional force F to a cylinder of metal which is one single crystal (Figure 4-10). We can orient the slip plane and slip direction to the applied force by defining the angles λ and ϕ. λ is the angle between the slip direction and the applied force and ϕ is the angle between the normal to the slip plane and the applied force.

In order for the dislocation to move in its slip system, a shear force acting in the slip direction must be produced by the applied force. This resolved shear force F_r is given by

$$F_r = F \cos \lambda$$

If we divide the equation by the area of the slip plane, $A = A_0/\cos \phi$, we obtain *Schmid's law*

$$\tau_r = \sigma \cos \phi \cos \lambda \tag{4-1}$$

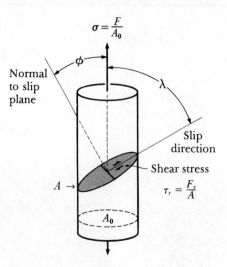

FIGURE 4-10 A resolved shear stress τ may be produced on a slip system, causing the dislocation to move on the slip plane in the slip direction.

where

$$\tau_r = \frac{F_r}{A} = \text{resolved shear } \textit{stress} \text{ in the slip direction}$$

$$\sigma = \frac{F}{A_0} = \text{unidirectional } \textit{stress} \text{ applied to the cylinder}$$

FIGURE 4-11
When the slip plane is perpendicular to the applied stress σ, the angle λ is 90° and no shear stress is resolved.

◇   **Example 4-5**

Suppose the slip plane is perpendicular to the applied stress σ, as in Figure 4-11. Then, $\phi = 0°$, $\lambda = 90°$, $\cos \lambda = 0$, and therefore $\tau_r = 0$. Even if the applied stress σ is enormous, no resolved shear stress develops along the slip direction and the dislocation cannot move. (You could perform a simple experiment to demonstrate this with a deck of cards. If you push on the deck at an angle, the cards slide over one another, as in the slip process. If you push perpendicular to the deck, however, the cards do not slide.) Slip cannot occur if the slip system is oriented so that either λ or ϕ is 90°.

◇

◇   **Example 4-6**

Calculate the resolved shear stress on the (111) $[\bar{1}01]$ slip system if a stress of 10,000 psi is applied in the [001] direction of an FCC unit cell. (See Figure 4-12.)

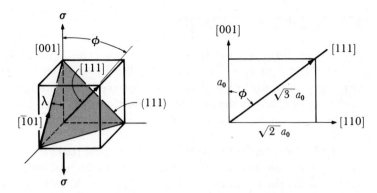

FIGURE 4-12 A normal stress σ is applied in the [001] direction of the unit cell. This produces an angle λ of 45° to the [$\bar{1}$01] slip direction and an angle ϕ of 54.76° to the normal to the (111) plane. (See Example 4-6.)

Answer:

By inspection, $\lambda = 45°$ and $\cos \lambda = 0.707$. The normal to the (111) plane must be the [111] direction. We can then calculate that

$$\cos \phi = \frac{1}{\sqrt{3}} = 0.577 \quad \text{or} \quad \phi = 54.76°$$

$$\tau_r = \sigma \cos \lambda \cos \phi = (10,000 \text{ psi})(0.707)(0.577) = 4079 \text{ psi}$$

The *critical resolved shear stress* τ_{crss} is the shear stress required to break enough metallic bonds in order for slip to occur. Thus slip occurs, causing the metal to deform, when the applied stress produces a resolved shear stress that equals the critical resolved shear stress.

$$\tau_r = \tau_{crss} \tag{4-2}$$

 Example 4-7

Consider a slip system in which $\lambda = 70°$ and $\phi = 30°$. Slip is found to begin when a stress of 5000 psi is applied. Calculate the critical resolved shear stress.

Answer:

$$\tau_{crss} = \tau_r \text{ when slip begins}$$
$$\tau_{crss} = \sigma \cos \lambda \cos \phi = (5000 \text{ psi})(\cos 70)(\cos 30) = 1481 \text{ psi}$$

4–5   INFLUENCE OF CRYSTAL STRUCTURE

We can use Schmid's law to compare the properties of metals having BCC, FCC, and HCP crystal structures. Table 4-2 lists three important factors that we can examine. We must be careful to note, however, that this discussion describes the behavior of nearly perfect single crystals. Real engineering materials are seldom single crystals and are never quite perfect.

Critical Resolved Shear Stress.   If the critical resolved shear stress in a metal is very high, the applied stress σ must also be high in order for τ_r to equal τ_{crss}. If σ is large, the metal must have a high strength! In FCC metals, which have close-packed planes, the critical resolved shear stress is low, about 50 to 100 psi in a perfect crystal; FCC metals are expected to have low strengths. On the other hand, BCC crystal structures contain no close-packed planes. Dislocations must move on nonclose-packed planes, such as $\{110\}$, $\{112\}$, and $\{123\}$ planes. Now we must exceed a higher critical resolved shear stress, on the order of 10,000 psi in perfect crystals, before slip occurs; therefore, BCC metals tend to have high strengths.

We would expect the HPC metals, because they contain close-packed planes, to have low critical resolved shear stresses. In fact, in HCP metals such as zinc that have a c/a ratio greater than or equal to the theoretical ratio of 1.633, the critical resolved shear stress is less than 100 psi, just as in FCC metals. However, in HCP titanium, the c/a ratio is less than 1.633; the close-packed planes are spaced too closely together. Slip now occurs on planes such as $(10\bar{1}0)$, or the faces of the hexagon, and the critical resolved shear stress is then as great as or greater than in BCC metals.

◇ | **Example 4-8**

Calculate the c/a ratios for HCP zinc and titanium and determine the likely slip processes in each. Will zirconium behave more like zinc or titanium?

Answer:

From Appendix A, we can determine the lattice parameters for each metal.

$$\text{Zinc:} \quad \frac{c}{a} = \frac{4.9470}{2.6648} = 1.856$$

$$\text{Titanium:} \quad \frac{c}{a} = \frac{4.6831}{2.9503} = 1.587$$

$$\text{Zirconium:} \quad \frac{c}{a} = \frac{5.1477}{3.2312} = 1.593$$

The c/a ratio for zinc is greater than 1.633; slip is expected to occur primarily on basal planes, with low critical resolved shear stresses. Both titanium and zirconium have c/a ratios less than ideal; high critical resolved shear stresses are expected for each.

◇

TABLE 4-2 Summary of factors affecting slip in metallic structures

Factor	FCC	BCC	HCP
Critical resolved shear stress (psi)	50–100	5,000–10,000	50–100[a]
Number of slip systems	12	48	3[b]
Cross-slip	Can occur	Can occur	Cannot occur[b]
Summary of properties	Ductile	Strong	Relatively brittle

[a] For HCP metals with $c/a = 1.633$.

[b] By alloying or heating to elevated temperatures, additional slip systems are active in HCP metals, permitting cross-slip to occur and thereby improving ductility.

Number of Slip Systems. If at least one slip system is oriented to give the angles λ and ϕ near 45°, then τ_r equals τ_{crss} at low applied stresses. Ideal HCP metals possess only one set of parallel close-packed planes, the (0001) planes, and three close-packed directions, giving three slip systems. Consequently, the probability of the close-packed planes and directions being oriented with λ and ϕ near 45° is very low. The HCP crystal may fail in a brittle manner without a significant amount of slip.

However, in HCP metals with a low c/a ratio, or when HCP metals are properly alloyed, or when the temperature is increased, other slip systems become active, making these metals less brittle than expected.

On the other hand, FCC metals contain four nonparallel close-packed planes of the form {111} and three close-packed directions of the form ⟨110⟩ within each plane, giving a total of 12 slip systems. At least one slip system is favorably oriented for slip to occur at low applied stresses, causing FCC metals to have low strengths but high ductilities.

Finally, BCC metals have as many as 48 slip systems that are nearly close packed. Several slip systems are always properly oriented for slip to occur—in fact, there are too many possible slip systems. Dislocations moving on one slip plane may interfere with movement of dislocations on other active slip planes. This interference leads to high strengths in BCC metals yet still permits at least some ductility as well.

Cross-Slip. Suppose a dislocation moving on one slip plane encounters an obstacle and is blocked from further movement. The dislocation can shift to a second intersecting slip system, also properly oriented, and continue to move. This is called *cross-slip* (Figure 4-13). In many HCP metals, no cross-slip can occur because the slip planes are parallel, not intersecting. Therefore, the HCP metals tend to remain brittle. Fortunately, additional slip systems become active when HCP metals are alloyed or heated, thus improving ductility. Cross-slip is possible in both FCC and BCC metals because a number of intersecting slip systems is present. Cross-slip consequently helps maintain ductility in these metals.

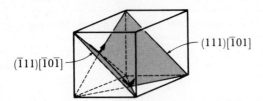

FIGURE 4-13 Cross-slip of a dislocation from a ($\bar{1}$11) [$\bar{1}$0$\bar{1}$] system to a (111) [$\bar{1}$01] system in an FCC crystal.

In summary, the most significant characteristics of the three important crystal structures in single crystals of metals are: FCC metals are ductile, BCC metals are strong, and HCP metals are relatively brittle. These general conclusions still tend to be followed even in real engineering metals and alloys which are not single crystals.

Other Factors.   Dislocations are more difficult to move in ionic and covalent materials. In covalent materials, such as silicon, the directionality of the bonds prevents easy movement of dislocations. Consequently, silicon behaves in a rather brittle manner.

In ionic materials, including most ceramics, dislocations are present and are observed to move. However, movement of a dislocation disrupts the charge balance around the anions and cations, requiring that bonds between anions and cations be broken. During slip, ions with a like charge must also pass close together, causing repulsion. Finally, the repeat distance along the slip direction, or the Burgers vector, is larger than in metals. Consequently, slip is much more difficult and ceramic materials normally display brittle behavior.

4–6 POINT DEFECTS

Point defects are localized disruptions of the lattice involving one or possibly several atoms. These imperfections, shown in Figure 4-14, may be introduced by movement of the atoms when they gain energy by heating, during processing of the material, by introduction of impurities, or intentionally through alloying.

Vacancies.   A *vacancy* is produced when an atom is missing from a normal site. Vacancies are introduced into the crystal during solidification, at high temperatures, or as a consequence of radiation damage. At room temperature, very few vacancies are present (Example 4-9), but this number increases exponentially as we increase the temperature.

$$n_v = n \exp\left(\frac{-Q}{RT}\right) \tag{4-3}$$

where n_v is the number of vacancies per cm^3; n is the number of lattice points per

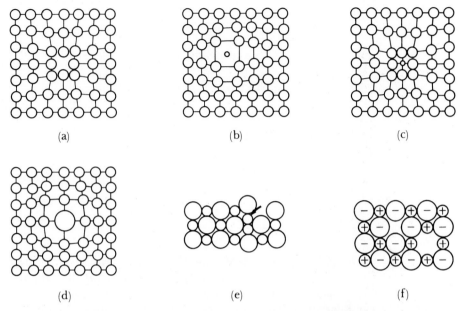

FIGURE 4-14   *Point defects: (a) vacancy, (b) interstitial atom, (c) small substitutional atom, (d) large substitutional atom, (e) Frenkel defect, and (f) Schottky defect. All of these defects disrupt the perfect arrangement of the surrounding atoms.*

cm^3; Q is the energy required to produce a vacancy, in cal/mole; R is the gas constant, 1.987 cal/mole $\cdot$ K; and T is the temperature in degrees Kelvin. Due to the large thermal energy near the melting temperature, there may be as many as one vacancy per 1000 lattice points.

◇ | **Example 4-9**

Calculate the number of vacancies per cubic centimeter and the number of vacancies per copper atom when copper is at (a) room temperature and (b) 1084°C (just below the melting temperature). About 20,000 cal/mole are required to produce a vacancy in copper.

Answer:

The lattice parameter of FCC copper is 3.6151×10^{-8} cm. The number of copper atoms or lattice points per cm^3 is

$$n = \frac{4 \text{ atoms/cell}}{(3.6151 \times 10^{-8})^3} = 8.47 \times 10^{22} \text{ copper atoms/cm}^3$$

(a) At room temperature, $T = 25 + 273 = 298$ K

$$n_v = (8.47 \times 10^{22}) \exp\left[\frac{-20,000}{(1.987)(298)}\right]$$

$$= 1.815 \times 10^8 \text{ vacancies/cm}^3$$

$$\frac{n_v}{n} = \frac{1.815 \times 10^8}{8.47 \times 10^{22}} = 2.14 \times 10^{-15}$$

(b) Just below the melting temperature, $T = 1084 + 273 = 1357\,\text{K}$

$$n_v = (8.47 \times 10^{22}) \exp\left[\frac{-20,000}{(1.987)(1357)}\right]$$

$$= 5.09 \times 10^{19} \text{ vacancies/cm}^3$$

$$\frac{n_v}{n} = \frac{5.09 \times 10^{19}}{8.47 \times 10^{22}} = 6.0 \times 10^{-4}$$

Note the tremendous increase in the number of vacancies as the temperature increases.

Interstitial and Substitutional Defects.   An *interstitial* defect is formed when an extra atom is inserted into the lattice structure at a site which is not a normal lattice point. A *substitutional* defect is introduced when an atom is replaced by a different type of atom. The substitutional atom remains at the original normal lattice point. Both interstitial and substitutional defects are present in materials as impurities and may also be intentionally introduced as alloying elements. The number of these defects is usually independent of temperature. Both again distort the surrounding lattice structure.

An *interstitialcy* is created when an atom identical to those at the normal lattice points is located in an interstitial position. Because this involves a great deal of distortion of the lattice, it is relatively rare.

Imperfection Pairs.   A *Frenkel* defect is a vacancy-interstitial pair formed when an ion jumps from a normal lattice point to an interstitial site, leaving behind a vacancy. A *Schottky* defect is a pair of vacancies in an ionically bonded material. In this case, both an anion and a cation must be missing from the lattice if electrical neutrality is to be preserved in the crystal.

A final important point defect occurs when an ion of one charge replaces an ion of a different charge. This might be the case when an ion with a valence of $+2$ replaces an ion with a valence of $+1$ (Figure 4-15). In this case, an extra positive charge is introduced into the structure. To maintain a charge balance, one solution would be to create a vacancy where a $+1$ cation normally would be located.

Point defects disturb the perfect arrangement of the surrounding atoms. When a vacancy or a small substitutional atom is present, the surrounding atoms collapse towards the point defect, stretching the bonds between the surrounding atoms and producing a tensile stress field. An interstitial, interstitialcy, or large substitutional atom pushes the surrounding atoms together, producing a compressive stress field. In either case, the effect is widespread, extending perhaps hundreds of atom spacings from the actual point defect. A dislocation

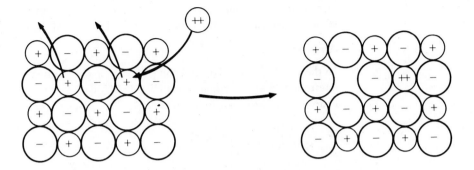

FIGURE 4-15 When a divalent atom replaces a monovalent cation, a second monovalent cation may also be removed, creating a vacancy.

moving through the general vicinity of a point defect encounters a lattice in which the atoms are not at their equilibrium positions. This disruption requires that a higher stress be applied to force the dislocation past the defect, therefore increasing the strength of the metal.

◇ **Example 4-10**

Iron has a measured density of 7.87 g/cm³. The lattice parameter of BCC iron is 2.866 Å. Calculate the percentage of vacancies in pure iron.

Answer:

From Equation (3-5),

$$\rho = \frac{(\text{atoms/cell})\,(55.85\text{ g/g} \cdot \text{mole})}{(2.866 \times 10^{-8})^3\,(6.02 \times 10^{23})} = 7.87\text{ g/cm}^3$$

$$\text{Atoms/cell} = \frac{(7.87)\,(2.866 \times 10^{-8})^3\,(6.02 \times 10^{23})}{55.85} = 1.998$$

There should be 2 atoms/cell in a perfect BCC iron crystal. Thus, the difference must be due to the presence of vacancies.

$$\text{Vacancies} = \frac{2 - 1.998}{2} \times 100 = 0.1\%$$

◇ **Example 4-11**

In FCC iron, carbon atoms are located at interstitial sites with coordinates $\frac{1}{4}, \frac{1}{4}, \frac{1}{4}$, whereas carbon atoms enter interstitial sites at $0, \frac{1}{2}, \frac{1}{4}$ in BCC iron. The lattice parameter is 3.571 Å for FCC iron and 2.866 Å for BCC iron. Carbon atoms have a radius of 0.71 Å. Would we expect a greater distortion of the lattice by an interstitial carbon atom for FCC or BCC iron?

Answer:

In the FCC crystal structure, the $\frac{1}{4}, \frac{1}{4}, \frac{1}{4}$ site is surrounded by four iron atoms in a tetrahedral arrangement. From Table 3-7, we know that the minimum radius ratio for fourfold coordination is

$$\frac{r \text{ (interstitial)}}{r \text{ (iron atom)}} = 0.225$$

$$r \text{ (interstitial)} = (0.225)\left(\frac{\sqrt{2}a_0}{4}\right) = \frac{(0.255)(\sqrt{2})(3.571)}{4} = 0.284 \text{ Å}$$

In the BCC iron, the $0, \frac{1}{2}, \frac{1}{4}$ site is shown in Figure 4-16.

$$R \text{ (iron atom)} = \frac{\sqrt{3}a_0}{4} = \frac{(\sqrt{3})(2.866)}{4} = 1.241 \text{ Å}$$

$$\left(\tfrac{1}{2}a_0\right)^2 + \left(\tfrac{1}{4}a_0\right)^2 = (r + R)^2$$

$$(r + R)^2 = 2.566$$

$$r \text{ (interstitial)} = 1.602 - 1.241 = 0.361 \text{ Å}$$

The carbon atom, 0.71 Å, is larger than either interstitial site. A compressive field is developed, which in this case will be larger in the FCC lattice than in the BCC lattice. Carbon atoms will likely disrupt slip more in the FCC iron.

4–7 SURFACE DEFECTS

Surface defects are the boundaries, or planes, that separate a material into regions, each region having the same crystal structure but different orientations.

Material Surface.   The exterior dimensions of the material represent surfaces at which the lattice abruptly ends. Each atom at the surface no longer has the proper coordination number; atomic bonding is disrupted and may even be incomplete. The exterior surface may also be very rough, may contain tiny notches, and may be much more reactive than the bulk of the material.

Grain Boundaries.   The microstructure of metals and many other solid materials consists of many grains. A *grain* is a portion of the material within which the arrangement of the atoms is identical. However, the orientation of the atom arrangement, or crystal structure, is different for each adjoining grain. Three grains are shown schematically in Figure 4-17; the lattice in each grain is identical but the lattices are oriented differently. A *grain boundary* is the surface that separates the individual grains and is a narrow zone in which the atoms are

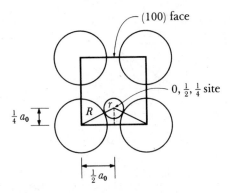

FIGURE 4-16 The location of the $0, \frac{1}{2}, \frac{1}{4}$ interstitial site in BCC metals, showing the arrangement of the normal atoms and the interstitial atom.

not properly spaced. The atoms are too close at some locations in the grain boundary, causing a region of compression, while in other areas the atoms are too far apart, causing a region of tension.

One method of controlling the properties of a material is by controlling the grain size. By reducing the grain size, we increase the number of grains and hence increase the amount of grain boundary. Any dislocation moves only a short distance before encountering a grain boundary, and the strength of the metal is increased. The *Hall-Petch* equation relates the grain size to the yield strength of the metal.

$$\sigma_y = \sigma_0 + Kd^{-1/2} \tag{4-4}$$

where σ_y is the yield strength or the stress at which the material permanently deforms, d is the average diameter of the grains, and σ_0 and K are constants for

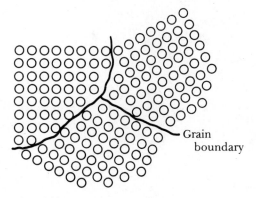

FIGURE 4-17 The atoms near the boundaries of the three grains do not have an equilibrium spacing or arrangement.

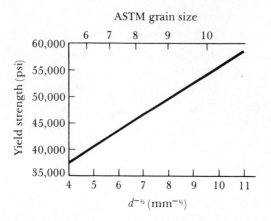

FIGURE 4-18 *The effects of grain size on the yield strength of a steel at room temperature.*

the metal. Figure 4-18 shows this relationship in steel. We will describe in later chapters how the grain size can be controlled through solidification, alloying, and heat treatment.

One technique by which grain size is specified is the ASTM (American Society for Testing & Materials) grain size number. The number of grains per square inch is determined from a photograph of the metal taken at magnification $\times 100$ (Figure 4-19). The number of grains per square inch N is entered into Equation (4-5) and the ASTM grain size number n is calculated.

$$N = 2^{n-1} \tag{4-5}$$

A large ASTM number indicates many grains, or a fine grain size, and correlates with high strengths.

FIGURE 4-19 *Photomicrograph showing the grain structure of nickel ($\times 100$).*

◇ **Example 4-12**

Suppose we count 16 grains per square inch at magnification $\times 100$ in a photomicrograph of a metal. Determine the ASTM grain size number.

Answer:

$$\mathcal{N} = 16 = 2^{n-1}$$
$$\log 16 = (n - 1) \log 2$$
$$1.204 = (n - 1)(0.301)$$
$$n = 5$$

 Example 4-13

Suppose we count 16 grains per square inch in a photomicrograph taken at magnification $\times 250$. What is the ASTM grain size number?

Answer:

If we count 16 grains per square inch at magnification $\times 250$, we must have at magnification $\times 100$

$$\mathcal{N} = \left(\frac{250}{100}\right)^2 (16) = 100 = 2^{n-1}$$
$$\log 100 = (n - 1) \log 2$$
$$2 = (n - 1)(0.301)$$
$$n = 7.64$$

Small Angle Grain Boundaries. A *small angle grain boundary* is an array of dislocations that produces a small misorientation between the adjoining lattices (Figure 4-20). Because the energy of the surface is less than that at a regular grain boundary, the small angle grain boundaries are not as effective in blocking slip. Small angle boundaries formed by edge dislocations are called *tilt boundaries* and those caused by screw dislocations are called *twist boundaries*.

 Example 4-14

Determine the angle θ across a small angle grain boundary in copper when the dislocations in the boundary are $1000\,\text{Å}$ apart.

Answer:

The grains are tilted one Burgers vector in each direction every $1000\,\text{Å}$. The Burgers vector in FCC copper is [110], so the length of the Burgers vector is the repeat distance in the [110] direction, or d_{110}.

$$d_{110} = \frac{a_0}{\sqrt{h^2 + k^2 + l^2}} = \frac{3.615}{\sqrt{2}} = 2.557\,\text{Å}$$

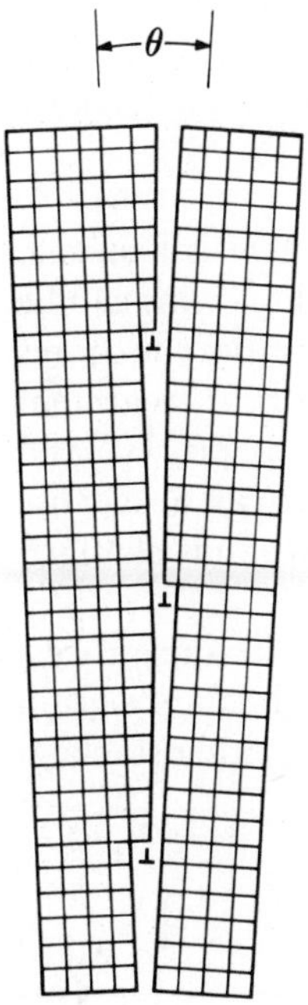

FIGURE 4-20
The small angle grain boundary is produced by an array of dislocations, causing an angular mismatch θ between the lattices on either side of the boundary.

$$\sin \frac{\theta}{2} = \frac{2.557}{1000} = 0.002557$$

$$\theta = 0.293°$$

Stacking Faults.   *Stacking faults* occur in FCC metals and represent an error in the stacking sequence of close-packed planes. Normally, a stacking sequence of *ABCABCABC* is produced in a perfect FCC lattice. But suppose the following sequence is produced:

   ABCABABCABC

In the portion of the sequence indicated, a type *A* plane is shown where a type *C* plane would normally be located. This small region, which has an HCP stacking sequence instead of the FCC stacking sequence, represents a stacking fault. Stacking faults interfere with the slip process.

Twin Boundaries.   A *twin boundary* is a plane across which there is a special mirror image misorientation of the lattice structure (Figure 4-21). Twins can be produced when a shear force, acting along the twin boundary, causes the atoms to shift out of position. Twinning occurs during deformation or heat treatment of certain metals. The twin boundaries interfere with the slip process and increase the strength of the metal. Movement of twin boundaries can also cause a metal to deform. Figure 4-21 shows that the formation of a twin has changed the shape of the metal.

   The effectiveness of the surface defects in interfering with the slip process can be judged from the surface energies (Table 4-3). The high-energy grain boundaries are much more effective in blocking dislocations than either stacking faults or twin boundaries.

TABLE 4-3   Energies of surface imperfections in selected metals[a]

Surface Imperfection (ergs/cm^2)	Al	Cu	Pt	Fe
Stacking fault energy	200	75	95	—[b]
Twin boundary energy	120	45	195	190
Grain boundary energy	625	645	1000	780

[a] After R. E. Reed-Hill, *Physical Metallurgy Principles*, 2nd Ed., D. Van Nostrand, 1973.
[b] Stacking faults do not occur in BCC metals such as iron.

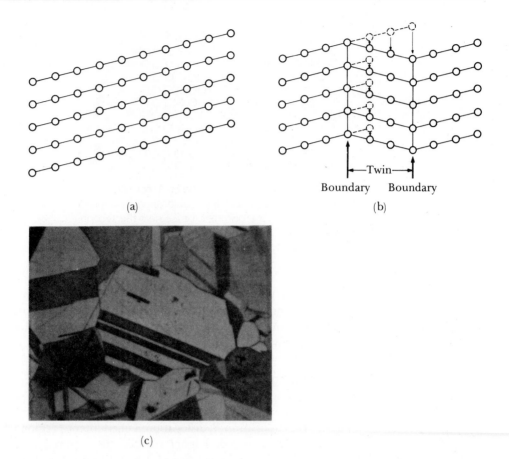

(a)

—Twin—
Boundary   Boundary

(b)

(c)

FIGURE 4-21   Application of a stress to the perfect crystal (a) may cause a displacement of the atoms (b), causing the formation of a twin. Note that the crystal has deformed as a result of twinning. (c) A photomicrograph of twins within a grain of brass (× 250).

4–8   CONTROL OF THE SLIP PROCESS

In a perfect crystal, the fixed, repeated arrangement of the atoms gives the lowest possible energy within the crystal. Any imperfection in the lattice raises the internal energy at the location of the imperfection. The local energy is increased because near the imperfection, the atoms either are squeezed too closely together (compression) or are forced too far apart (tension).

One dislocation in an otherwise perfect lattice can move easily through the crystal if the resolved shear stress equals the critical resolved shear stress. However, if the dislocation encounters a region where the atoms are displaced from their usual positions, a higher stress is required to force the dislocation past

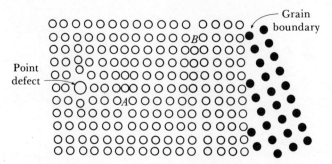

FIGURE 4-22   If the dislocation at point *A* moves to the left, it is blocked by the point defect. If the dislocation moves to the right, it interacts with the disturbed lattice near the second dislocation at point *B*. If the dislocation moves farther to the right, it is blocked by a grain boundary.

the region of high local energy. Thus, the material is stronger. We can, then, control the strength of a material by controlling the number and type of imperfections. Three common strengthening mechanisms are based on the three categories of lattice defects in crystals.

Strain Hardening.   Dislocations disrupt the perfection of the lattice. In Figure 4-22, the atoms below the dislocation line at point *B* are compressed, while the atoms above dislocation *B* are too far apart. If dislocation *A* moves to the right and passes near dislocation *B*, dislocation *A* encounters a region where the atoms are not properly arranged. Higher stresses are required to keep the second dislocation moving; consequently, the metal must be stronger. The more dislocations that are present in the lattice, the more likely it is that any one dislocation will be blocked. Therefore, increasing the number of dislocations increases the strength of the material. We will discuss this effect formally in Chapter 7 as *strain hardening*.

Solid Solution Strengthening.   Any of the point defects also disrupt the perfection of the lattice. If dislocation *A* moves to the left, it encounters a disturbed lattice caused by the point defect; higher stresses are needed to continue slip of the dislocation. By intentionally introducing substitutional or interstitial atoms, we can cause the material to become stronger by *solid solution strengthening*. This will be discussed further in Chapter 9.

Grain Size Strengthening.   Finally, surface imperfections such as grain boundaries disturb the lattice. If dislocation *B* moves to the right, it encounters a grain boundary and is blocked. By increasing the number of grains, or reducing the grain size, *grain size strengthening* is achieved. Control of grain size will be discussed in a number of later chapters.

SUMMARY

We have found in this chapter that the deformation of a metal can be explained in terms of the movement of certain lattice imperfections, namely dislocations, by a mechanism known as slip. The ease with which slip occurs depends first on the type of crystal structure and second on the number of other imperfections in the lattice. Based on this analysis, FCC metals should be ductile but weak, BCC metals should be strong with moderate ductility, and HCP metals should be relatively brittle with moderate strength. We can strengthen any of these structures by introducing larger numbers of lattice imperfections.

GLOSSARY

ASTM grain size number. A measure of the size of the grains in a crystalline material obtained by counting the number of grains per square inch at magnification × 100.

Burgers vector. The direction and distance that a dislocation moves in each step.

Critical resolved shear stress. The shear stress required to cause a dislocation to move and cause slip.

Cross-slip. A change in the slip system of a dislocation.

Dislocation. A line imperfection in the lattice of a crystalline material. Movement of dislocations helps explain how materials deform. Interference with the movement of dislocations helps explain how materials are strengthened.

Edge dislocation. A dislocation introduced into the lattice by adding an "extra half plane" of atoms.

Frenkel defect. A pair of point defects produced when an ion moves to create an interstitial site, leaving behind a vacancy.

Grain. A portion of a solid material within which the lattice is identical and oriented in only a single direction.

Grain boundary. A surface defect representing the boundary between two grains. The lattice has a different orientation on either side of the grain boundary.

Interstitial defect. A point defect produced when an atom is placed into the lattice at a site that is normally not a lattice point.

Schottky defect. A pair of point defects in ionically bonded materials. In order to maintain a neutral charge, both a cation and an anion vacancy must form.

Screw dislocation. A dislocation produced by skewing a crystal so that one atomic plane produces a spiral ramp about the dislocation.

Slip. Deformation of a material by the movement of dislocations through the lattice.

Slip direction. The direction in the lattice in which the dislocation moves. The slip direction is the same as the direction of the Burgers vector.

Slip plane. The plane swept out by the dislocation line during slip. Normally, the slip plane is a close-packed plane, if one exists in the crystal structure.

Slip system. The combination of the slip plane and the slip direction.

Small angle grain boundary. An array of dislocations causing a small misorientation of the lattice across the surface of the imperfection.

Stacking fault. A surface defect in FCC metals caused by the improper stacking sequence of close-packed planes.

Substitutional defect. A point defect produced when an atom is removed from a regular lattice point and replaced with a different atom, usually of a different size.

Tilt boundary. A small angle grain boundary composed of an array of edge dislocations.

Twin boundary. A surface defect across which there is a mirror image misorientation of the lattice. Twin boundaries can also move and cause deformation of the material.

Twist boundary. A small angle grain boundary composed of an array of screw dislocations.

Vacancy. A vacancy is created when an atom is missing from a lattice point.

PRACTICE PROBLEMS

1. Calculate, using the data in Appendix A, the c/a ratios for magnesium, cadmium, rhenium, and beryllium. In which of these metals is slip expected to occur easily on the basal planes? In which is slip expected to occur on the prismatic planes such as $(10\bar{1}0)$?

2. Determine the Miller indices of the slip directions on the (111) plane in an FCC unit cell.

3. Determine the Miller indices of the slip directions on the (101) plane in a BCC unit cell.

4. Determine the Miller indices of the slip planes in FCC unit cells that include the [110] slip direction.

5. Determine the Miller indices of the {110} slip planes in BCC unit cells that include the [111] slip direction.

6. Does the $[\bar{1}11]$ slip direction lie in the (112) slip plane in the BCC unit cell? Show by suitable drawings.

7. Does the $[\bar{1}\bar{1}1]$ slip direction lie in the (123) slip plane in the BCC unit cell? Show by suitable drawings.

8. Using the three-digit form for indices in the HCP unit cell, determine the slip directions in the $(01\bar{1}1)$ slip plane.

9. Using the three-digit form for indices in the HCP unit cell, determine the slip directions in the $(10\bar{1}0)$ slip plane.

10. Verify that the planar density of the {112} planes in BCC iron is 9.94×10^{14} atoms/cm^2, as used in Example 4-3.

11. Determine the length of the Burgers vector in BCC tungsten.

12. Compare the planar density and interplanar spacings for the {111} and {110} planes in FCC aluminum. Based on your calculations, on which planes would slip occur?

13. When BCC iron is in the softest possible condition, the dislocation density is about 10^6 cm/cm^3; large amounts of deformation of the iron increase the dislocation density to about 10^{12} cm/cm^3. How many grams of iron are necessary to produce 1000 miles of dislocation in (a) soft iron and (b) deformed iron?

14. The dislocation density in an aluminum sample is found to be 5×10^7 cm/cm^3. Calculate the total length of dislocations in 100 g of aluminum (see Appendix A for the necessary data). How many miles of dislocation are present in the sample?

15. The circumference of the earth is roughly 24,000 miles. If dislocations totalling this length were placed into one cubic centimeter, what would be the dislocation density?

16. Suppose that a single crystal of an FCC metal is oriented so that the [001] direction is parallel to an applied stress of 3000 psi. Calculate the resolved shear stress acting on the (111) slip plane in the $[\bar{1}10]$, $[01\bar{1}]$, and $[\bar{1}01]$ slip directions. Which slip system(s) will become active first?

17. Suppose that a single crystal of an FCC metal is oriented so that the [001] direction is parallel to the applied stress. If the critical resolved shear stress required for slip is 225 psi, calculate the magnitude of the applied stress required to cause slip to begin on the (111) plane in the $[1\bar{1}0]$, $[0\bar{1}1]$, and $[10\bar{1}]$ slip directions.

18. Suppose that a single crystal of a BCC metal is oriented so that the [001] direction is parallel to an applied stress of 12,000 psi. Calculate the resolved shear stress acting on the $(1\bar{1}0)$, $(0\bar{1}1)$, and $(\bar{1}01)$ planes in the [111] slip direction. Which slip system(s) will become active first?

19. Suppose that a single crystal of a BCC metal is oriented so that the [001] direction is parallel to the applied stress. If the critical resolved shear stress required for slip is 8500 psi, calculate the magnitude of the applied stress required to cause slip to begin in the $[\bar{1}11]$ direction on the (110), $(0\bar{1}1)$, and (101) slip planes.

20. Suppose that a single crystal of an FCC metal is oriented so that the [001] direction is parallel to an applied stress. When the applied stress is 6200 psi, a dislocation on the (111) plane just begins to move in the $[0\bar{1}1]$ direction. Calculate the critical resolved shear stress in this material. Based on this result, do you suspect that there are numerous lattice imperfections in the metal?

21. Suppose that a single crystal of a BCC metal is oriented so that the [001] direction is parallel to an applied stress. When the applied stress is 13,500 psi, a dislocation on the $(\bar{1}01)$ plane begins to move in the [111] direction. Calculate the critical resolved shear stress in this material. Based on this result, do you suspect that there are numerous lattice imperfections in the metal?

22. Suppose a single crystal of a hexagonal close-packed metal is oriented so that the [001] direction is parallel to an applied stress. Can slip occur in the basal plane? Can slip occur in the prismatic planes such as $(10\bar{1}0)$? Explain your answers to both questions.

23. FCC aluminum has a density of 2.695 g/cm³ and a lattice parameter of 4.04958 Å.

Calculate (a) the fraction of the lattice points that contain vacancies and (b) the total number of vacancies in a cubic centimeter of aluminum.

24. HCP magnesium has a density of 1.735 g/cm³ and lattice parameters of $a_0 = 3.2087$ Å and $c_0 = 5.209$ Å. Calculate (a) the average number of atoms per lattice point in the unit cell and (b) the total number of vacancies in a cubic centimeter of magnesium

25. The lattice parameter of BCC cesium is 6.13 Å. If one atom is missing from one out of 800 unit cells, calculate (a) the number of vacancies per cubic centimeter and (b) the density of cesium.

26. The lattice parameter of FCC strontium is 6.0849 Å. If one atom is missing for each 1500 strontium atoms, calculate (a) the density of strontium and (b) the number of vacancies per gram.

27. A BCC alloy of tungsten containing substitutional atoms of vanadium has a density of 16.912 g/cm³ with a lattice parameter of 3.1378 Å. Calculate the fraction of vanadium atoms in the alloy.

28. An FCC alloy of copper containing substitutional atoms of tin has a density of 7.717 g/cm³ and a lattice parameter of 3.903 Å. Calculate the fraction of tin atoms in the alloy.

29. Suppose that when one-third of the atoms in HCP magnesium are replaced by cadmium atoms, the lattice parameters of the alloy are $a_0 = 3.133$ Å and $c_0 = 5.344$ Å. Calculate the expected density of the alloy.

30. Body-centered cubic iron has a lattice parameter of 2.868 Å after a carbon atom enters one interstitial site in every twentieth unit cell. Estimate (a) the density of the iron-carbon alloy and (b) the packing factor for the structure, assuming that $r_{Fe} = 1.241$ Å and $r_C = 0.77$ Å.

31. Carbon atoms enter interstitial positions in FCC nickel, producing a lattice parameter of 3.5198 Å and a density of 8.955 g/cm³. Calculate (a) the atomic fraction of carbon atoms in the nickel and (b) the number of unit cells you would have to examine to find one carbon atom.

32. Would you expect a Frenkel defect to change the lattice parameter or density of MgO? Explain.

33. Suppose one Schottky defect occurred in every tenth unit cell of NaCl, producing a lattice parameter of 5.57 Å. Calculate the density of the sodium chloride.

34. Suppose the lattice parameter of CsCl is 4.0185 Å and the density is 4.285 g/cm³. Calculate the number of Schottky defects per unit cell.

35. Suppose the (111) plane is parallel to the surface of an FCC metal. What is the coordination number for each atom at the surface?

36. The ASTM grain size number for a metal is 6. How many grains would be observed per square inch in a photograph taken at a magnification of 100? How many actual grains are present per square inch?

37. Twelve grains per square inch are counted in a photograph taken at magnification × 500. Calculate the ASTM grain size number. Is this a coarse, medium, or fine grain size?

38. Eighteen grains per square inch are counted in a photograph taken at magnification × 75. Calculate the ASTM grain size number. Is this a coarse, medium, or fine grain size?

39. Figure 4-23 shows the microstructure of a material at magnification × 100. Estimate the ASTM grain size number.

40. Figure 4-24 shows the microstructure of a material at magnification × 500. Estimate the ASTM grain size number.

41. Calculate the angle θ of a small angle grain boundary in BCC iron when the dislocations are 7500 Å apart.

42. A small angle grain boundary is tilted 0.75° in FCC copper. Calculate the average distance between the dislocations.

43. Suppose that two Fe^{3+} ions are substituted for normal Fe^{2+} ions in FeO. What other changes in the atom arrangement would be required to maintain the proper charge balance?

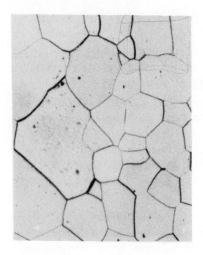

FIGURE 4-23 Microstructure of palladium, for Problem 4-39 (× 100). From **Metals Handbook**, Vol. 9, 9th Ed., American Society for Metals, 1985.

44. Suppose an Fe^{2+} ion is substituted for an Na^+ ion in NaCl. Explain why you would expect that the charge balance would be maintained by forming a vacancy rather than by adding another Cl^- ion to the lattice.

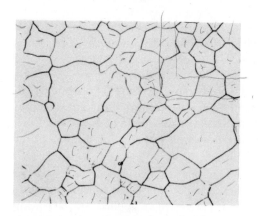

FIGURE 4-24 Microstructure of iron, for Problem 4-40 (× 500). From **Metals Handbook**, Vol. 9, 9th Ed., American Society for Metals, 1985.

5

ATOM MOVEMENT IN MATERIALS

5–1 INTRODUCTION

Diffusion is the movement of atoms within a material. Atoms move in an orderly fashion to eliminate concentration differences and produce a homogeneous, uniform composition. Atoms also can be forced to move by applying voltages or external forces to the material. In fact, atoms even move about randomly in pure metals when no external forces are applied or no concentration differences exist.

Movement of atoms is required for many of the treatments that we perform on materials. Diffusion is required for the heat treatment of metals, the manufacture of ceramics, the solidification of materials, the manufacture of transistors and solar cells, and even the electrical conductivity of many ceramic materials.

In this chapter, we will concentrate on understanding the diffusion of atoms in solid materials. We should recognize, however, that diffusion occurs in gases and liquids as well. But convection, mixing, and other phenomena assist the diffusion of atoms in the gaseous and liquid states; furthermore, diffusion is much more rapid in gases and liquids because of less efficient packing of the atoms.

5–2 STABILITY OF ATOMS

Chapter 4 showed that imperfections could be introduced into the lattice of a crystal. However, these imperfections, and indeed even the normal atoms in their lattice positions, are not stable or at rest. Instead, the atoms possess some thermal energy, and there is the possibility that they may move. For instance, an atom may move from a normal lattice point to occupy a nearby vacancy. An atom may move from one interstitial site to another. Vacancies or interstitialcy atoms may be created. Atoms may jump across a grain boundary, permitting the grain boundary to move.

The ability of atoms and imperfections to move is related to the temperature, or thermal energy, possessed by the atoms. Both theory and experiment have shown that the rate of movement is related to temperature or thermal energy by the *Arrhenius* equation

$$\text{Rate} = c_0 \exp\left(\frac{-Q}{RT}\right) \tag{5-1}$$

where c_0 is a constant, R is the gas constant ($1.987\ \text{cal/mol}\cdot\text{K}$), T is the absolute temperature (K), and Q is the *activation energy* (cal/mol) required to cause the imperfection to move. This equation is derived from a statistical analysis of the probability that the atoms will have the extra energy Q needed to cause movement. The rate is related to the number of atoms that move.

We can rewrite the equation by taking natural logarithms of both sides:

$$\ln\ (\text{rate}) = \ln\ (c_0) - \frac{Q}{RT} \tag{5-2}$$

By plotting ln (rate) of some reaction versus $1/T$ (Figure 5-1), the slope of the curve will be $-Q/R$, and consequently Q can be calculated. The constant c_0 is the intercept when $1/T$ is zero. One of the important consequences of the Arrhenius equation is the diffusion of atoms in a material.

◇ | **Example 5-1**

Suppose that interstitial atoms are found to move from one site to another at the rates of 5×10^8 jumps/s at 500°C and 8×10^{10} jumps/s at 800°C. Calculate the activation energy Q for the process.

Answer:

Figure 5-1 represents the data on a ln (rate) versus $1/T$ plot; the slope of this curve, as calculated in the figure, gives $Q/R = 14{,}000\ \text{K}^{-1}$, or $Q = 28{,}000\ \text{cal/mol}$. Alternately, we could write two simultaneous equations:

$$5 \times 10^8 = c_0 \exp\left[\frac{-Q}{(1.987)(500 + 273)}\right] = c_0 \exp\ (-0.000651Q)$$

$$8 \times 10^{10} = c_0 \exp\left[\frac{-Q}{(1.987)(800 + 273)}\right] = c_0 \exp\ (-0.000469Q)$$

Since

$$c_0 = \frac{5 \times 10^8}{\exp\ (-0.000651Q)},$$

then

$$8 \times 10^{10} = \frac{(5 \times 10^8)\exp\ (-0.000469Q)}{\exp\ (-0.000651Q)}$$

$$160 = \exp\ [(0.000651 - 0.000469)Q] = \exp\ (0.000182Q)$$

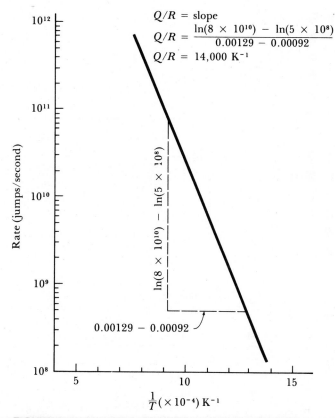

FIGURE 5-1   *The Arrhenius plot of ln (rate) versus 1/T can be used to determine the activation energy required for a reaction.*

$$\ln(160) = 5.075 = 0.000182Q$$

$$Q = \frac{5.075}{0.000182} = 28,000 \, \text{cal/mol}$$

5–3   DIFFUSION MECHANISMS

Even in absolutely pure solid materials, atoms move from one lattice position to another. This process, known as self-diffusion, can be detected by using radioactive tracers. Suppose we introduce a radioactive isotope of gold (Au^{198}) onto the surface of normal gold (Au^{197}). After a period of time the radioactive atoms move into the regular gold, and eventually the radioactive atoms are uniformly distributed throughout the entire gold sample. Although self-diffusion

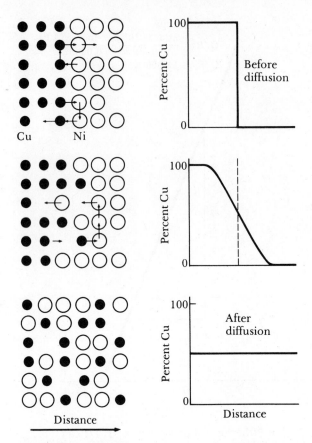

FIGURE 5-2 *Diffusion of copper atoms into nickel. Eventually, the copper atoms are randomly distributed throughout the nickel.*

occurs continually in all materials, the effect on the material's behavior is not significant.

Diffusion of unlike atoms in materials also occurs. If a nickel sheet is bonded to a copper sheet, nickel atoms gradually diffuse into the copper and copper atoms migrate into the nickel. Again, the nickel and copper atoms eventually are uniformly distributed (Figure 5-2).

There are several mechanisms by which atoms diffuse (Figure 5-3).

Vacancy Diffusion.   In self-diffusion and diffusion involving substitutional atoms, an atom leaves its lattice site to fill a nearby vacancy (thus creating a new vacancy at the original lattice site). As diffusion continues, we have a countercurrent flow of atoms and vacancies.

Interstitial Diffusion.   When a small interstitial atom is present in the crystal structure, the atom moves from one interstitial site to another. No vacancies are required for this mechanism to work.

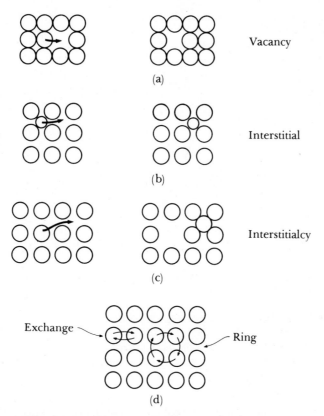

Vacancy

(a)

Interstitial

(b)

Interstitialcy

(c)

Exchange

Ring

(d)

FIGURE 5-3 *Diffusion mechanisms in materials. (a) Vacancy or substitutional atom diffusion, (b) interstitial diffusion, (c) interstitialcy diffusion, and (d) exchange and ring diffusion.*

Other Diffusion Mechanisms. Sometimes a substitutional atom leaves its normal lattice site and enters an interstitial position. This *interstitialcy* diffusion mechanism is uncommon because the atom does not easily fit into the small interstitial site. Atoms also move by a simple *exchange* mechanism or by a *ring* mechanism. However, the vacancy and interstitial mechanisms appear to be responsible for diffusion in most cases.

5–4 ACTIVATION ENERGY FOR DIFFUSION

A diffusing atom must squeeze past the surrounding atoms to reach its new site. In order to do this, energy must be supplied to force the atom to its new position. This is shown schematically for vacancy and interstitial diffusion in Figure 5-4. The atom is originally in a low-energy, relatively stable location. In order to move to a new location, the atom must pass over an energy barrier. The energy barrier is the *activation energy Q*. Heat supplies the atom with the energy needed to exceed this barrier.

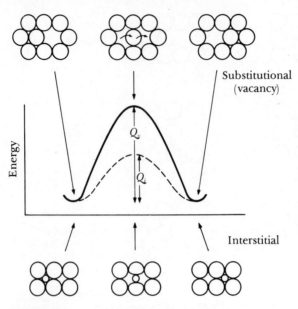

FIGURE 5-4 As atoms squeeze past one another during diffusion, a high energy is required. This energy is the activation energy **Q**. Generally more energy is required for a substitutional atom than for an interstitial atom.

Normally less energy is required to squeeze an interstitial atom past the surrounding atoms; consequently, activation energies are lower for interstitial diffusion than for vacancy diffusion. Typical values for activation energies are shown in Table 5-1; a low activation energy indicates easy diffusion.

5–5 RATE OF DIFFUSION (FICK'S FIRST LAW)

The rate at which atoms diffuse in a material can be measured by the *flux J*, which is defined as the number of atoms passing through a plane of unit area per unit time (Figure 5-5). *Fick's first law* explains the net flux of atoms,

$$J = -D \frac{\Delta c}{\Delta x} \tag{5-3}$$

where J is the flux (atoms/cm$^2 \cdot$ s), D is the diffusivity or *diffusion coefficient* (cm^2/s), and $\Delta c/\Delta x$ is the *concentration gradient* (atoms/cm$^3 \cdot$ cm). Several factors affect the flux of atoms during diffusion.

Concentration Gradient.   The concentration gradient shows how the composition of the material varies with distance; Δc is the difference in concentration over the distance Δx (Figure 5-6). We should note that the flux is initially high

TABLE 5-1   Diffusion data for selected metals

Diffusion Couple	Q (cal/mol)	D_0 (cm^2/s)
Interstitial diffusion		
C in FCC iron	32,900	0.23
C in BCC iron	20,900	0.011
N in FCC iron	34,600	0.0034
N in BCC iron	18,300	0.0047
H in FCC iron	10,300	0.0063
H in BCC iron	3,600	0.0012
Self-diffusion		
Au in Au	43,800	0.13
Al in Al	32,200	0.10
Ag in Ag	45,000	0.80
Cu in Cu	49,300	0.36
Fe in FCC iron	66,700	0.65
Pb in Pb	25,900	1.27
Pt in Pt	67,600	0.27
Mg in Mg	32,200	1.0
Zn in Zn	21,800	0.1
Ti in HCP Ti	22,900	0.4
Fe in BCC iron	58,900	4.1
Heterogeneous diffusion		
Ni in Cu	57,900	2.3
Cu in Ni	61,500	0.65
Zn in Cu	43,900	0.78
Ni in FCC iron	64,000	4.1
Au in Ag	45,500	0.26
Ag in Au	40,200	0.072
Al in Cu	39,500	0.045

Adapted from Y. Adda and J. Philibert, *La Diffusion dans les Solides*, Vol. 2, 1966.

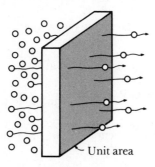

FIGURE 5-5   The flux during diffusion is defined as the number of atoms passing through a plane of unit area per unit time.

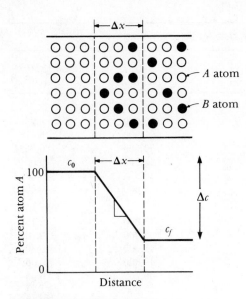

FIGURE 5-6   Illustration of the concentration gradient.

when the concentration gradient is high and gradually decreases as the gradient is reduced.

◇ | **Example 5-2**

One way to manufacture transistors, which amplify electrical signals, is to diffuse impurity atoms into a semiconductor material such as silicon. Suppose a silicon wafer 0.1 cm thick, which originally contains one phosphorus atom for every 10,000,000 Si atoms, is treated so that there are 400 P atoms for every 10,000,000 Si atoms at the surface. Calculate the concentration gradient (a) in atomic percent/cm and (b) in atoms/cm$^3 \cdot$ cm. The lattice parameter of silicon is 5.4307 Å.

Answer:

First, let's calculate the initial and surface compositions in atomic percent.

$$c_i = \frac{1 \text{ P atom}}{10,000,000 \text{ atoms}} \times 100 = 0.00001 \text{ at} \% \text{ P}$$

$$c_s = \frac{400 \text{ P atoms}}{10,000,000 \text{ atoms}} \times 100 = 0.004 \text{ at} \% \text{ P}$$

$$\frac{\Delta c}{\Delta x} = \frac{0.00001 - 0.004 \text{ at} \% \text{ P}}{0.1 \text{ cm}} = -0.0399 \text{ at} \% \text{ P/cm}$$

To find the gradient in terms of atoms/cm$^3 \cdot$ cm, we must find the volume of the unit cell.

$$V_{cell} = (5.4307 \times 10^{-8})^3 = 1.6 \times 10^{-22}\,cm^3/cell$$

The volume occupied by 10,000,000 Si atoms, which are arranged in a DC structure with 8 atoms/cell, is

$$V = \frac{10,000,000\ atoms}{8\ atoms/cell}(1.6 \times 10^{-22}\,cm^3/cell) = 2 \times 10^{-16}\,cm^3$$

The compositions in atoms/cm^3 are

$$c_i = \frac{1\ P\ atom}{2 \times 10^{-16}\,cm^3} = 0.005 \times 10^{18}\ P\ atoms/cm^3$$

$$c_s = \frac{400\ P\ atoms}{2 \times 10^{-16}\,cm^3} = 2 \times 10^{18}\ P\ atoms/cm^3$$

$$\frac{\Delta c}{\Delta x} = \frac{0.005 \times 10^{18} - 2 \times 10^{18}\ P\ atoms/cm^3}{0.1\ cm}$$

$$= -1.995 \times 10^{19}\ P\ atoms/cm^3 \cdot cm$$

 Example 5-3

A thick-walled pipe 3 cm in diameter contains a gas including 0.5×10^{20} N atoms per cm^3 on one side of a 0.001 cm-thick iron membrane. The gas is continuously introduced to the pipe. The gas on the other side of the membrane contains 1×10^{18} N atoms per cm^3. Calculate the total number of nitrogen atoms passing through the iron membrane at 700°C if the diffusion coefficient for nitrogen in iron is 4×10^{-7} cm^2/s.

Answer:

$$c_1 = 0.5 \times 10^{20}\ N\ atoms/cm^3 = 50 \times 10^{18}\ N\ atoms/cm^3$$

$$c_2 = 1 \times 10^{18}\ N\ atoms/cm^3$$

$$\Delta c = (1 - 50) \times 10^{18} = -49 \times 10^{18}\ N\ atoms/cm^3$$

$$\Delta x = 0.001\ cm$$

$$J = -D\frac{\Delta c}{\Delta x} = -\left(4 \times 10^{-7}\frac{cm^2}{s}\right)\left(\frac{-49 \times 10^{18}\ N\ atoms/cm^3}{0.001\ cm}\right)$$

$$J = 1.96 \times 10^{16}\ N\ atoms/cm^2 \cdot s$$

$$\text{Total atoms/s} = JA = J\left(\frac{\pi}{4}d^2\right) = (1.96 \times 10^{16})\left(\frac{\pi}{4}\right)(3)^2$$

$$= 1.39 \times 10^{17}\ atoms/s$$

Obviously, if the gas on the high nitrogen side of the membrane were not continuously replenished, that side would soon be depleted of nitrogen atoms.

Temperature and the Diffusion Coefficient. The diffusion coefficient D is related to temperature by an Arrhenius equation

$$D = D_0 \exp\left(\frac{-Q}{RT}\right) \tag{5-4}$$

where Q is the activation energy (cal/mol), R is the gas constant (1.987 cal/mol · °C), and T is the absolute temperature (K). D_0 is a constant for a given diffusion system. Typical values for D_0 are given in Table 5-1, while the temperature dependence of D is shown in Figure 5-7 for several materials.

When the temperature of a material increases, the diffusion coefficient and the flux of atoms increase as well. At higher temperatures, the thermal energy

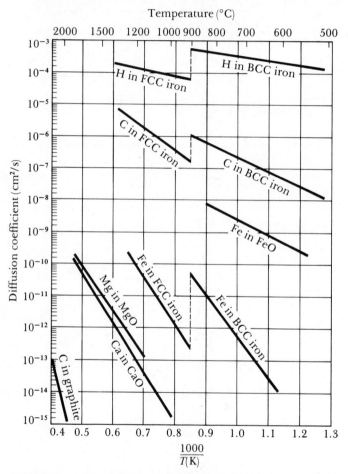

FIGURE 5-7 The diffusion coefficient **D** as a function of reciprocal temperature for several metals and ceramics. In this Arrhenius plot, **D** represents the rate of the diffusion process.

supplied to the diffusing atoms permits the atoms to overcome the activation energy barrier and more easily move to new lattice sites. At low temperatures, often below about 0.4 times the absolute melting temperature of the material, diffusion is very slow and may not be significant. For this reason, the heat treatment of metals and the processing of ceramics are done at high temperatures, where atoms move rapidly to complete reactions or to reach equilibrium conditions.

◇ | **Example 5-4**

The diffusion coefficient for aluminum in copper is found to be $2.5 \times 10^{-20}\,cm^2/s$ at 200°C and $3.1 \times 10^{-13}\,cm^2/s$ at 500°C. Calculate the activation energy for the diffusion of aluminum in copper.

Answer:

Let's make a ratio between the diffusion coefficients at the two temperatures, $200°C = 473\,K$ and $500°C = 773\,K$.

$$\frac{D_{773}}{D_{473}} = \frac{D_0 \exp\left(\dfrac{-Q}{(1.987)(773)}\right)}{D_0 \exp\left(\dfrac{-Q}{(1.987)(473)}\right)} = \frac{3.1 \times 10^{-13}}{2.5 \times 10^{-20}}$$

$$\exp\left[Q\left(\frac{-1}{773} - \frac{-1}{473}\right)\left(\frac{1}{1.987}\right)\right] = 1.24 \times 10^7$$

$$\exp\left(0.0004129Q\right) = 1.24 \times 10^7$$

$$0.0004129Q = \ln\left(1.24 \times 10^7\right) = 16.33$$

$$Q = 39{,}600\,cal/mol$$

Activation Energy and Diffusion Mechanism.   A small activation energy Q increases the diffusion coefficient and flux, since less thermal energy is required to overcome the smaller activation energy barrier. Interstitial diffusion, with a low activation energy, usually occurs an order of magnitude or more faster than vacancy, or substitutional, diffusion.

Activation energies are usually lower for atoms diffusing through open crystal structures compared with close-packed crystal structures. The activation energy for carbon diffusing in FCC iron is 32,900 cal/mol, but is only 20,900 cal/mol for carbon diffusing in BCC iron.

Activation energies are also lower for diffusion of atoms in materials with a low melting temperature (Figure 5-8) and are usually lower for small substitutional atoms compared with larger atoms.

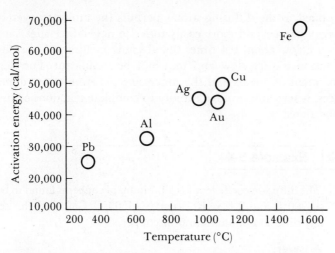

FIGURE 5-8 The activation energy for self-diffusion increases
as the melting point of the metal increases.

◇ │ **Example 5-5**

Compare the diffusion coefficients for hydrogen and nickel in FCC iron at
1000°C and explain the difference.

Answer:

From Table 5-1, at $T = 1000°C = 1273\,\text{K}$

$$D_H = 0.0063 \exp\left(\frac{-10{,}300}{RT}\right) = 0.0063 \exp\left[\frac{-10{,}300}{(1.987)(1273)}\right]$$

$$= 0.0063 \exp(-4.07) = (0.0063)(0.017) = 1.07 \times 10^{-4}\,\text{cm}^2/\text{s}$$

$$D_{Ni} = 4.1 \exp\left(\frac{-64{,}000}{RT}\right) = 4.1 \exp\left[\frac{-64{,}000}{(1.987)(1273)}\right]$$

$$= 4.1 \exp(-25.3) = (4.1)(1.03 \times 10^{-11}) = 4.2 \times 10^{-11}\,\text{cm}^2/\text{s}$$

Because hydrogen is a tiny interstitial atom and nickel is a larger substitutional
atom, the activation energy for diffusion of hydrogen in iron is small and the
rate of diffusion is seven orders of magnitude greater than that of nickel.

◇ │ **Example 5-6**

Compare the diffusion coefficients of carbon in BCC and FCC iron at the
allotropic transformation temperature of 910°C and explain the difference.

Answer:

From Table 5-1, at $T = 910°C = 1183\,\text{K}$

$$D_C(\text{BCC}) = 0.011\,\exp\left[\frac{-20{,}900}{(1.987)(1183)}\right] = 0.011\,\exp\,(-8.89)$$

$$= (0.011)(1.38 \times 10^{-4}) = 1.52 \times 10^{-6}\,\text{cm}^2/\text{s}$$

$$D_C(\text{FCC}) = 0.23\,\exp\left[\frac{-32{,}900}{(1.987)(1183)}\right] = 0.23\,\exp\,(-14.00)$$

$$= (0.23)(8.31 \times 10^{-7}) = 1.91 \times 10^{-7}\,\text{cm}^2/\text{s}$$

Diffusion is faster in the BCC iron than in the FCC iron, even at the same temperature, because the packing factor of BCC structures is less than that of FCC structures.

Types of Diffusion.   In *volume* diffusion, the atoms move through the crystal from one lattice or interstitial site to another. Because of the surrounding atoms, the activation energy is large and the rate of diffusion is relatively slow.

However, atoms can also diffuse along boundaries, interfaces, and surfaces in the material. Atoms diffuse more easily by *grain boundary diffusion* because the atom packing is poor in the grain boundaries. Because atoms can more easily squeeze their way through the disordered grain boundary, the activation energy is low. *Surface diffusion* is easier still. Consequently, the activation energy is lower and the diffusion coefficient is higher for grain boundary and surface diffusion (Table 5-2).

◇   **Example 5-7**

Consider a diffusion couple set up between pure tungsten and a tungsten-1 at % thorium alloy. After several minutes of exposure at 2000° C, a transition zone of 0.01 cm thickness is established. What is the flux of thorium atoms at this time if diffusion is due to (a) volume diffusion, (b) grain boundary diffusion, and (c) surface diffusion?

TABLE 5-2   *The effect of the type of diffusion for thorium in tungsten and for self-diffusion in silver*

Diffusion Type	Diffusion Coefficient	
	Thorium in Tungsten	Silver in Silver
Surface	$0.47\,\exp\,(-66{,}400/RT)$	$0.068\,\exp\,(-8{,}900/RT)$
Grain boundary	$0.74\,\exp\,(-90{,}000/RT)$	$0.24\,\exp\,(-22{,}750/RT)$
Volume	$1.00\,\exp\,(-120{,}000/RT)$	$0.99\,\exp\,(-45{,}700/RT)$

Answer:

The lattice parameter of BCC tungsten is about $3.165\,\text{Å}$. Thus, the number of tungsten atoms/cm^3 is

$$\frac{\text{W atoms}}{\text{cm}^3} = \frac{2\ \text{atoms/cell}}{(3.165 \times 10^{-8})^3\,\text{cm}^3/\text{cell}} = 6.3 \times 10^{22}$$

In the tungsten-1 at % thorium alloy, the number of thorium atoms is

$$c_{\text{Th}} = (0.01)(6.3 \times 10^{22}) = 6.3 \times 10^{20}\,\text{Th atoms/cm}^3$$

In the pure tungsten, the number of thorium atoms is zero. Thus, the concentration gradient is

$$\frac{\Delta c}{\Delta x} = \frac{0 - 6.3 \times 10^{20}}{0.01\ \text{cm}} = -6.3 \times 10^{22}\,\text{Th atoms/cm}^3 \cdot \text{cm}$$

(a) Volume diffusion

$$D = 1.0 \exp\left(\frac{-120{,}000}{(2273)(1.987)}\right) = 2.89 \times 10^{-12}\,\text{cm}^2/\text{s}$$

$$\mathcal{J} = -D\frac{\Delta c}{\Delta x} = -(2.89 \times 10^{-12})(-6.3 \times 10^{22})$$

$$= 18.2 \times 10^{10}\,\text{Th atoms/cm}^2 \cdot \text{s}$$

(b) Grain boundary diffusion

$$D = 0.74 \exp\left(\frac{-90{,}000}{(2273)(1.987)}\right) = 1.64 \times 10^{-9}\,\text{cm}^2/\text{s}$$

$$\mathcal{J} = -(1.64 \times 10^{-9})(-6.3 \times 10^{22}) = 10.3 \times 10^{13}\,\text{Th atoms/cm}^2 \cdot \text{s}$$

(c) Surface diffusion

$$D = 0.47 \exp\left(\frac{-66{,}400}{(2273)(1.987)}\right) = 1.9 \times 10^{-7}\,\text{cm}^2/\text{s}$$

$$\mathcal{J} = -(1.9 \times 10^{-7})(-6.3 \times 10^{22}) = 12 \times 10^{15}\,\text{Th atoms/cm}^2 \cdot \text{s}$$

Time.   Diffusion requires time; the units for flux are atoms/cm$^2 \cdot s$! If a large number of atoms must diffuse to produce a uniform structure, long times may be required, even at high temperatures. Times for heat treatments may be reduced by using higher temperatures or by making the *diffusion distances* (related to Δx) as small as possible.

We find that some rather remarkable structures and properties are obtained if we prevent diffusion. Steels quenched rapidly from high temperatures to prevent diffusion form nonequilibrium structures which provide the basis for sophisticated heat treatments.

5-6 COMPOSITION PROFILE (FICK'S SECOND LAW)

Fick's second law, which describes the dynamic, or nonsteady state, diffusion of atoms, is the differential equation $dc/dt = D(d^2c/dx^2)$, whose solution depends on the boundary conditions for a particular situation. One solution is

$$\frac{c_s - c_x}{c_s - c_0} = \text{erf}\left(\frac{x}{2\sqrt{Dt}}\right) \tag{5-5}$$

where c_s is a constant concentration of the diffusing atoms at the surface of the material, c_0 is the initial uniform concentration of the diffusing atoms in the material, and c_x is the concentration of the diffusing atom at location x below the surface after time t. These concentrations are illustrated in Figure 5-9. The function erf is the error function and can be evaluated from Figure 5-10.

The solution to Fick's second law permits us to calculate the concentration of one diffusing species near the surface of the material as a function of time and

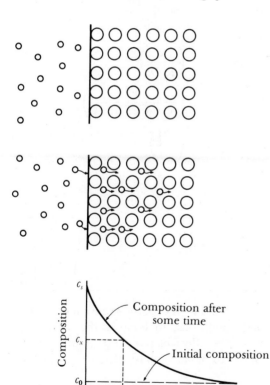

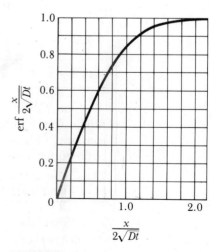

FIGURE 5-9 Diffusion of atoms into the surface of a material, illustrating the use of Fick's second law.

FIGURE 5-10 The error function.

distance, provided that the diffusion coefficient D remains constant and the concentrations of the diffusing atom at the surface c_s and within the material c_0 remain unchanged.

◇ Example 5-8

The surface of a 0.1% C steel is to be strengthened by carburizing. In carburizing, the steel is placed in an atmosphere that provides a maximum of 1.2% C at the surface of the steel at a high temperature. Carbon then diffuses into the surface of the steel. For optimum properties, the steel must contain 0.45% C at a depth of 0.2 cm below the surface. How long will carburizing take if the diffusion coefficient is $2 \times 10^{-7} \, \text{cm}^2/\text{s}$?

Answer:

From the problem,

$$c_s = 1.2\% \, \text{C} \qquad c_0 = 0.1\% \, \text{C} \qquad c_x = 0.45\% \, \text{C} \qquad x = 0.2 \, \text{cm}$$

From Fick's second law,

$$\frac{c_s - c_x}{c_s - c_0} = \frac{1.2 - 0.45}{1.2 - 0.1} = 0.68 = \text{erf}\left(\frac{x}{2\sqrt{Dt}}\right)$$

$$= \text{erf}\left(\frac{0.2}{2\sqrt{(2 \times 10^{-7})t}}\right)$$

$$0.68 = \text{erf}\left(\frac{224}{\sqrt{t}}\right)$$

From Figure 5-10, we find that

$$\frac{224}{\sqrt{t}} = 0.71$$

$$t = \left(\frac{224}{0.71}\right)^2 = 99{,}536 \, \text{s} = 27.6 \, \text{h}$$

One of the consequences of Fick's second law is that the same concentration profile can be obtained for different conditions, so long as the term Dt is constant. This permits us to determine the effect of temperature on the time required for a particular heat treatment to be accomplished.

◇ Example 5-9

We find that 10 h are required to cause carbon to diffuse 0.1 cm into the surface of a steel gear at 800°C. What time is required to achieve the same carbon depth at 900°C? The activation energy for the diffusion of carbon atoms in FCC iron is 32,900 cal/mol.

Answer:

The temperatures are $800 + 273 = 1073\,\text{K}$ and $900 + 273 = 1173\,\text{K}$. Since we want the carbon profile to remain the same at both temperatures,

$$D_{1073}t_{1073} = D_{1173}t_{1173}$$

$$t_{1173} = \frac{D_{1073}t_{1073}}{D_{1173}} = t_{1073}\frac{\exp\,(-Q/1073R)}{\exp\,(-Q/1173R)}$$

$$t_{1173} = (10\,\text{h})\frac{\exp\,(-32,900/(1.987)\,(1073))}{\exp\,(-32,900/(1.987)\,(1173))}$$

$$= 10\exp\,(-1.315) = (10)\,(0.268) = 2.68\,\text{h}$$

$\diamond$

Equation 5-5 requires that there be a constant composition c_0 at the interface; this is the case in a process such as carburizing of steel (Example 5-8), in which carbon is continuously supplied to the steel surface. In many cases, however, the surface concentration gradually changes during the process. In these cases, *interdiffusion* of atoms occurs, as shown in Figure 5-2, and Equation 5-5 is no longer valid.

Sometimes interdiffusion can cause difficulties. For example, when aluminum is bonded to gold at an elevated temperature, the aluminum atoms will diffuse faster into the gold than gold atoms diffuse into the aluminum. Consequently, more total atoms will eventually be on the original gold side of the interface than on the original aluminum side. This causes the physical location of the original interface to move towards the aluminum side of the diffusion couple. Any foreign particles originally trapped at the interface also move with the interface. This movement of the interface due to unequal diffusion rates is called the *Kirkendall effect*.

In certain cases, voids form at the interface as a result of the Kirkendall effect. In tiny integrated circuits, gold wire is welded to aluminum to provide an external lead for the circuit. During operation of the circuit, voids may form by coalescence of vacancies involved in the diffusion process; as the voids grow, the Au-Al connection is weakened and eventually may fail. Because the area around the connection discolors, this premature failure is called the *purple plague*. One technique to prevent this problem is to expose the welded joint to hydrogen. The hydrogen dissolves in the aluminum, fills the vacancies, and prevents self-diffusion of the aluminum atoms. This keeps the aluminum atoms from diffusing into the gold and embrittling the weld.

5-7 DIFFUSION AND MATERIALS PROCESSING

Diffusional processes become very important when materials are used or processed at elevated temperatures. Many important examples of this will be discussed in later chapters. In this section, three cases in which diffusion is important will be considered.

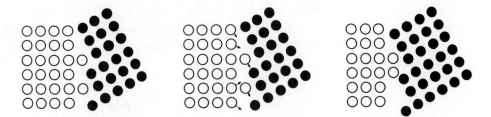

FIGURE 5-11   Grain growth occurs as atoms diffuse across the grain boundary from one grain to another.

Grain Growth.   A material composed of many grains contains a large number of grain boundaries, which represent a high-energy area because of the inefficient packing of the atoms. A lower overall energy is obtained in the material if the amount of grain boundary area is reduced by grain growth.

Grain growth involves the movement of grain boundaries, permitting some grains to grow at the expense of others. Diffusion of atoms across the grain boundary (Figure 5-11) is required for grain growth to occur. Consequently, the growth of the grain boundaries is related to the activation energy for an atom to jump across the boundary. High temperatures or low activation energies increase the size of the grains. Many heat treatments of metals, which include holding the metal at high temperatures, must be carefully controlled to avoid excessive grain growth.

We can impede grain growth by introducing obstacles at the grain boundaries that provide a "drag" force which prevents the grains from enlarging. We will find that we can introduce these particles in a variety of ways to control grain growth during high-temperature use and processing of many materials.

Diffusion Bonding.   *Diffusion bonding*, a method used to join materials, occurs in three steps (Figure 5-12). The first step forces the two surfaces together at a high pressure, flattening the surface, fragmenting impurities, and producing a high atom-to-atom contact area. Atomic bonding across the interface establishes a joint. Normally, the pressure is applied at high temperatures at which the material is softer and more ductile.

As the surfaces remain pressed together at high temperatures, atoms diffuse along grain boundaries to the remaining voids; the atoms condense and reduce the size of any voids in the interface. Because grain boundary diffusion is rapid, this second step may occur rapidly. Eventually, however, grain growth isolates remaining voids from the grain boundaries. For the third step—final elimination of the voids—volume diffusion, which is comparatively slow, must occur.

The diffusion bonding process is often used with some of the more exotic alloys, such as titanium, for joining dissimilar metals and materials, and even for joining ceramics.

Sintering.   A number of materials are manufactured into useful shapes by a process that requires consolidation of small particles into a solid mass. *Sintering*

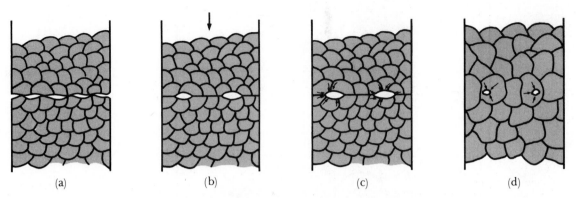

FIGURE 5-12 *The steps in diffusion bonding. (a) Initially the contact area is small, (b) application of pressure deforms the surface, increasing the bonded area, (c) grain boundary diffusion permits voids to shrink, and (d) final elimination of the voids requires volume diffusion.*

is the high-temperature treatment that causes the particles to join together and gradually reduces the volume of pore space between them. Sintering is a frequent step in the manufacture of ceramic components as well as in the production of metallic parts by powder metallurgy. A variety of composite materials are produced using the same techniques.

When a powdered material is compacted into a shape, the powder particles are in contact with one another at numerous sites, with a significant amount of pore space between the particles. In order to reduce the boundary energy, atoms diffuse to the boundaries, permitting the particles to be bonded together and eventually causing the pores to shrink. If sintering is carried out for a long time, the pores may be eliminated and the material becomes dense.

Surfaces that have a small radius of curvature grow rapidly. The points of contact have the smallest radius and thus grow first. Atoms diffuse to these points, while vacancies diffuse away from the interface. The net movement of vacancies permits the particles to move closer together (Figure 5-13), causing the pore size to decrease and the density to increase. The rate of sintering depends on the temperature, the activation energy and diffusion coefficient for diffusion, and the original size of the particles.

In powder metallurgy processes, summarized in Figure 5-14, metal powders with a diameter of 0.0001 to 0.03 cm are produced. Techniques for producing the powders include *atomization*, in which a jet of gas or liquid fragments a tiny metal stream into spherical droplets; *pulverization*, in which a brittle metal is crushed and ground to a fine size; and *chemical reduction*, in which metal compounds are chemically converted to metal particles. The powders are then compacted at high pressures in dies to produce intricate, precise green compacts. The compacts have sufficient strength to be handled and placed into a furnace, where they are sintered. Although the powder compact does shrink in size during sintering, small components requiring a minimum of machining often can be economically produced.

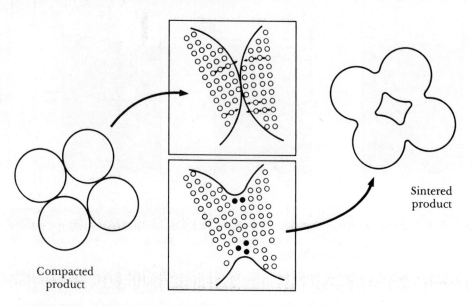

FIGURE 5-13 *Diffusion processes during sintering and powder metallurgy. Atoms diffuse to points of contact, creating bridges, and eventually filling the pores.*

5–8 DIFFUSION IN IONIC COMPOUNDS AND POLYMERS

In metals and alloys, atoms can move into any nearby vacancy or interstitial site. But in other materials, atom movement may be somewhat more restricted.

Ionic Materials.   In ionic materials, such as many ceramics, a diffusing ion only enters a site having the same charge. In order to reach that site, the ion must physically squeeze past adjoining ions, pass by a region of opposite charge, and move a relatively long distance (Figure 5-15). Consequently, the activation energies are high and the rates of diffusion are lower for ionic materials than for metals (see Figure 5-7).

We also find that cations (with a positive charge) often have higher diffusion coefficients than anions (with a negative charge). The cations, because they have given up their valence electrons, typically have a smaller size and thus can diffuse more easily than the larger anions. In sodium chloride, for instance, the activation energy for diffusion of chloride ions is about twice that for diffusion of sodium ions.

Diffusion of the ions also provides a transfer of electrical charge; in fact, the electrical conductivity of ionically bonded ceramic materials is related to temperature by an Arrhenius equation. As the temperature increases, the ions diffuse more rapidly, electrical charge is transferred more quickly, and the electrical conductivity is increased.

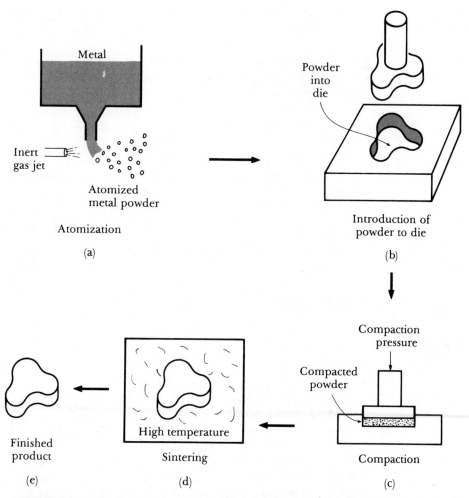

FIGURE 5-14 Powder metallurgy processing flow diagram: (a) molten metal is atomized by a jet of inert gas, (b) the powder is poured into a die, (c) the powder is pressed into a compact, and (d) the compact is sintered to produce the final product (e).

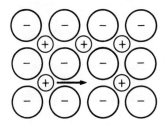

FIGURE 5-15 Diffusion in ionic compounds. Anions can only enter other anion sites.

Polymers.   Diffusion of atoms can occur from one location to another along a long polymer chain; however, strong covalent bonds must be broken for this to occur.

More often, we are concerned with the diffusion of atoms or small molecules between the long polymer chains. For example, polymer films are typically used as packaging to store food. If air diffuses through the film, the food may spoil. If air diffuses through the rubber inner tube of an automobile tire, the tire will deflate. Diffusion of some molecules into a polymer can cause swelling problems. On the other hand, diffusion is required to enable dyes to uniformly enter many of the synthetic polymer fabrics. Selective diffusion through polymer membranes is used to cause desalinization of water; while water molecules pass through the polymer membrane, the ions in the salt are trapped.

In each of these examples, the diffusing atoms, ions, or molecules penetrate between the polymer chains rather than moving from one location to another within the chain structure. Diffusion will be more rapid through this structure when the diffusing species is smaller or when larger voids are present between the chains. Diffusion through crystalline polymers, for instance, is slower than through amorphous polymers, which have no long-range order and consequently have a lower density.

SUMMARY

Atoms move through a solid material, particularly at high temperatures when a concentration gradient is present, by diffusion mechanisms. Atom diffusion is of paramount importance to us, because many of the materials processing techniques, such as sintering, powder metallurgy, and diffusion bonding, require diffusion. Furthermore, many of the heat treatments and strengthening mechanisms used to control structures and properties in materials are diffusion-controlled processes. The stability of the structure and the properties of materials during use at high temperatures depend on diffusion. Finally, many important materials are produced by deliberately preventing diffusion, thereby forming nonequilibrium structures. In the following chapters, we will encounter many instances in which diffusion plays a significant role.

GLOSSARY

Activation energy.   The energy required to cause a particular reaction to occur. In diffusion, the activation energy is related to the energy required to move an atom from one lattice site to another.

Concentration gradient.   The rate of change of composition with distance in a nonuniform material, expressed as atoms/$cm^3 \cdot$ cm or at %/cm.

Diffusion.   The movement of atoms within a material.

Diffusion bonding.   A joining technique in which two surfaces are pressed together at high pressures and temperatures. Diffusion of atoms to the interface fills in voids and produces a strong bond.

Diffusion coefficient. A temperature-dependent coefficient that is related to the rate at which atoms diffuse. The diffusion coefficient depends on the temperature and activation energy.

Diffusivity. Another term for diffusion coefficient.

Fick's first law. The equation relating the flux of atoms by diffusion to the diffusion coefficient and the concentration gradient.

Fick's second law. The partial differential equation that describes the rate at which atoms are redistributed in a material by diffusion.

Flux. The number of atoms passing through a plane of unit area per unit time. This is related to the rate at which mass is transported by diffusion in a solid.

Grain boundary diffusion. Diffusion of atoms along grain boundaries. This is faster than volume diffusion because the atoms are less closely packed in grain boundaries.

Grain growth. Movement of grain boundaries by diffusion in order to reduce the amount of grain boundary area. As a result, small grains shrink and disappear while other grains become larger.

Interstitialcy diffusion. A diffusion mechanism by which an atom leaves its regular lattice position to fill an interstitial position.

Interstitial diffusion. Diffusion of small atoms from one interstitial position to another in the crystal structure.

Sintering. A high-temperature treatment used to join small particles. Diffusion of atoms to points of contact causes bridges to form between the particles. Further diffusion eventually fills in any remaining voids.

Surface diffusion. Diffusion of atoms along surfaces, such as cracks or particle surfaces.

Vacancy diffusion. Diffusion of atoms when an atom leaves a regular lattice position to fill a vacancy in the crystal. This process creates a new vacancy and the process continues.

Volume diffusion. Diffusion of atoms through the interior of grains.

PRACTICE PROBLEMS

1. When a force is applied at 300°C, each inch of a metal stretches at the rate of 0.001 in./min; at 400°C, the rate of stretching is 0.00375 in./min; and at 500°C, the rate is 0.01 in./min. Calculate (a) the activation energy for the stretching process and (b) the constant c_0. What are the units for each?

2. The fraction of the lattice points containing vacancies in copper is 2.24×10^{-15} at 100°C but is 2.42×10^{-6} at 700°C. Calculate (a) the activation energy required to form a vacancy, (b) the fraction of the lattice points that contain vacancies 5°C below the melting point (1085°C), and (c) the number of vacancies per unit cell 5°C below the melting point.

3. The diffusion coefficient for Al in Al_2O_3 is 7.48×10^{-19} cm^2/s at 1000°C and is 2.48×10^{-10} cm^2/s at 1500°C. Calculate the activation energy and the diffusion constant D_0.

4. The diffusion coefficient for Ni in MgO is 1.23×10^{-12} cm^2/s at 1200°C and is 1.45×10^{-10} cm^2/s at 1800°C. Calculate the activation energy and the diffusion constant D_0.

5. Estimate the activation energy for self-diffusion in titanium.

6. Would you expect the activation energy for self-diffusion in silicon, which has the covalent bond, to follow the curve in Figure 5-8? Explain.

7. Suppose a 0.01 cm thick wafer of germanium contains one gallium atom per 10^8 Ge atoms on one surface and 1000 Ga atoms per 10^8 Ge atoms at the other surface. Calculate the concentration gradient (a) in atomic percent/cm and (b) in atoms/cm$^3 \cdot$ cm. The lattice parameter for Ge is 5.66 Å.

8. One surface of a 0.05 cm thick wafer of germanium contains 5 gallium atoms per 10^5 Ge atoms. How many gallium atoms per 10^5 Ge atoms must be introduced at the other surface to give a concentration gradient of -2×10^{19} atoms/cm$^3 \cdot$ cm? The lattice parameter for Ge is 5.66 Å.

9. A 0.001 cm thick foil of iron separates a gas containing 1×10^{22} H atoms/cm^3 from another chamber containing 6×10^{16} H atoms/cm^3. If the system is operating at 1000°C and the iron is FCC, calculate (a) the concentration gradient of hydrogen through the foil and (b) the flux of hydrogen atoms through the foil.

10. Nitrogen gas is held in a pressure chamber by a BCC iron foil only 0.02 cm thick. The concentration of nitrogen is 3×10^{20} atoms/cm^3 on one side of the foil but 5×10^{10} atoms/cm^3 on the other side. Calculate the flux of nitrogen atoms through the foil at 750°C.

11. Calculate the concentration gradient required to produce a flux of 10×10^5 hydrogen atoms/cm$^2 \cdot$ s through a BCC iron foil at 800°C.

12. An FCC iron structure is to be manufactured that will allow no more than 100 H atoms/cm^2 to pass through it in one minute at 950°C. The composition of the hydrogen is 1×10^{20} H atoms/cm^3 on one side of the foil and 5×10^{17} H atoms/cm^3 on the other side. Calculate (a) the concentration gradient and (b) the minimum thickness of the iron required.

13. Determine the maximum allowable temperature that will produce a flux of less than 950 N atoms/cm$^2 \cdot$ s through a BCC iron foil when the concentration gradient is -3×10^{16} atoms/cm$^3 \cdot$ cm.

14. Suppose a diffusion couple is produced between pure nickel and pure copper at 1000°C. Will the interface tend to move toward the pure nickel or the pure copper side? Explain, using suitable calculations.

15. A carburizing process is done to a 0.15% C steel by introducing 1.1% C at the surface at 1000°C, where the iron is FCC. Calculate the carbon content at 0.01 cm, 0.05 cm, and 0.10 cm beneath the surface after 1 h.

16. A carburizing process is done to a 0.0% C steel by introducing 1.0% C at the surface. Calculate the carbon content 0.1 cm beneath the surface after holding at 912°C for 1 h if (a) the iron is FCC and (b) the iron is BCC. Explain the difference that the structure makes.

17. We would like to produce 0.45% C at a distance of 0.25 cm beneath the surface of a steel part by carburizing. If the steel originally contains 0.15% C and 1.2% C is introduced to the surface, how long will carburizing take at 1100°C? Assume the iron is FCC.

18. We would like to produce 0.3% C at a distance of 0.15 cm beneath the surface of a steel part by carburizing. If the steel originally contains 0.10% C and 1.0% C is introduced to the surface, what carburizing temperature is required if the treatment is to be accomplished in 0.33 hours? Assume the iron is FCC.

19. We would like to produce 0.2% C at a distance of 0.10 cm beneath the surface of a 0.01% C steel part by carburizing. If we carburize at 975°C for 10 h, how much carbon must we expose to the surface of the part? Assume the steel is FCC during treatment.

20. We would like to produce 0.12% N at a distance of 0.40 cm beneath the surface of a steel containing 0.002% N by introducing 0.35% N to the surface of the steel. How long would this nitriding require if it were done at 700°C? Assume the iron is BCC.

21. During nitriding of a BCC steel containing 0.005% N, 0.45% N is introduced to the surface at 650°C for 4 h. Calculate the nitrogen content at a distance of 0.2 cm beneath the surface of the steel.

22. Calculate the nitriding temperature when a BCC steel containing 0.075% N is exposed to 0.25% N for 18 h, producing 0.12% N at a distance of 0.08 cm beneath the surface.

23. Decarburization of a steel occurs when carbon diffuses from the steel to the surface and enters the atmosphere. How long will it take for a 0.60% C steel surface to decarburize below 0.10% C for a depth of 0.2 cm if the FCC steel is held at 1250°C in an atmosphere containing zero carbon?

24. Calculate the distance below the surface of a 0.80% C steel that, after exposure to air for 100 h at 1200°C, is decarburized to 0.2% C or less. The steel is FCC during this process.

25. A carburizing heat treatment of an FCC steel normally can be successfully completed at 1250°C in 2 h. What time would be necessary if the temperature were lowered to 1100°C?

26. A nitriding heat treatment of a BCC steel normally requires 2 h at 600°C. What temperature would be required to reduce the heat treatment time to 1 h?

27. Suppose, during a diffusion bonding process used to join copper to nickel, that bonding was 90% complete after 1 h at 800°C. But at 900°C, only 5.9 min were required to obtain the same bonding. Assuming that the rate of bonding is related only to diffusion, calculate the activation energy for the process. Consulting Table 5-1, is it likely that bonding is controlled primarily by diffusion of Cu into the nickel or by diffusion of Ni into the copper? (Note that rate is the reciprocal of time.)

28. Suppose the grain size of copper doubles when the metal is held at 500°C for 5 h, but doubles at 800°C in only 8 min. Estimate the activation energy for grain growth. Does this correlate with the activation energy for self-diffusion of copper? Should it? (Note that rate is the reciprocal of time.)

29. Suppose we would like to join SiC to Si_3N_4 by a diffusion bonding process. What problems might we have during the first step in the bonding process? Can you think of any way that these problems might be minimized? Would you expect that diffusion bonding of these materials would take more or less time than if two metals were joined at the same temperature?

30. Suppose we were to produce a silver contact for an electrical relay using powder metallurgy. Sintering initially occurs very rapidly, then slows significantly at a later time. Explain why the rate decreases during the latter stages of the process.

31. At 680°C, the electrical conductivity of NaCl is 10^{-4} ohm$^{-1} \cdot$ cm^{-1}; at 420°C, the electrical conductivity is 10^{-8} ohm$^{-1} \cdot$ cm^{-1}. Which ion—Na or Cl—do you expect is transferring the greater portion of the charge? What is the activation energy for diffusion of this ion in NaCl?

32. A balloon filled with helium deflates more rapidly than a balloon filled with air. Explain.

Part II

CONTROLLING THE MICROSTRUCTURE AND MECHANICAL PROPERTIES OF MATERIALS

◇

In Part II, we will examine the methods used to control the microstructure and macrostructure of materials. We will find that by controlling the structure we in turn can control the mechanical properties.

There are six important mechanisms used to control structure and properties—grain size strengthening, solid solution strengthening, strain hardening, dispersion strengthening, age hardening, and phase transformations. All introduce barriers to slip. In the first three methods, we rely on the three types of lattice imperfections. By controlling surface defects such as grain boundaries, we obtain *grain size strengthening*. Controlling point defects such as substitutional atoms gives *solid solution strengthening*. Increasing the number of line defects, or dislocations, provides *strain hardening*.

We obtain strengthening in the other three mechanisms by introducing multiple phases, where each phase has a different composition or crystal structure. The boundaries between the phases can provide strengthening by interfering with the deformation mechanisms. *Dispersion strengthening* is a general term indicating strengthening by multiple phases. *Age hardening* is a special technique that provides an optimum, fine dispersion of phases. *Phase transformations* include more sophisticated treatments, often relying on allotropic transformations.

We will discuss the strengthening mechanisms, at least partly, from the point of view of processing of the material. In particular, we will examine solidification, alloying, deformation, and heat treatment. *Solidification* helps determine grain size, grain shape, and the fineness and distribution of phases in many multiple-phase alloys. *Alloying* produces solid solution strengthening and provides the basis for dispersion strengthening. *Deformation* processing produces strain hardening and helps control grain size and shape. *Heat treatment* permits us to perform the dispersion strengthening, age hardening, and phase transformation strengthening techniques.

Before looking at strengthening mechanisms and the processes used to control these mechanisms, we will first briefly examine the mechanical testing of materials and understand the results of these tests, which are the mechanical properties of a material.

6

MECHANICAL TESTING AND PROPERTIES

6–1 INTRODUCTION

We select materials for many components and applications by matching the mechanical properties of the material to the service conditions required of the component. The first step in the selection process requires that we analyze the application to determine the most important characteristics that the material must possess. Should the material be strong, or stiff, or ductile? Will it be subjected to repeated application of a high force, a sudden intense force, a high stress at elevated temperature, or abrasive conditions? Once we have determined the required properties, we can select the appropriate material using data listed in handbooks. However, we must know how the properties listed in the handbook are obtained, know what the properties mean, and realize that the properties listed are obtained from idealized tests that may not apply exactly to real-life engineering applications.

In this chapter we will study several tests that are used to measure how a material withstands an applied force. The results of these tests are the mechanical properties of the material.

Tensile Test

6–2 THE STRESS-STRAIN DIAGRAM

The *tensile test* measures the resistance of a material to a static or slowly applied force. A test setup is shown in Figure 6-1; a typical specimen has a diameter of 0.505 in. and a gage length of 2 in. The specimen is placed in the testing machine

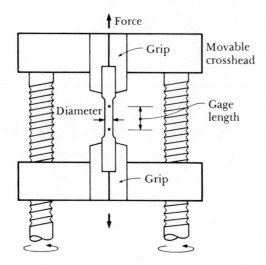

FIGURE 6-1 A unidirectional force is applied to a specimen in the tensile test by means of the moveable crosshead.

and a force F, called the *load*, is applied. A strain gage or extensometer is used to measure the amount that the specimen stretches between the gage marks when the force is applied. Table 6-1 includes the effect of the load on the gage length of an aluminum alloy test bar.

Figure 6-2 shows the load versus gage length for our test. Displaying the results of the test in this manner describes the behavior of this material when the diameter is 0.505 in. Unfortunately, this figure does not tell us the force required to produce a given amount of stretching if the diameter is larger or smaller.

Engineering Stress and Strain.   The results of a single test apply to all sizes and shapes of specimens for a given material if we convert the force to stress and the distance between gage marks to strain. *Engineering stress* and *engineering strain*

TABLE 6-1   The results of a tensile test of a 0.505-in. diameter aluminum alloy test bar

Load (lb)	Stress (psi)	Gage Length (in.)	Strain (in./in.)
0	0	2.000	0
1000	5,000	2.001	0.0005
3000	15,000	2.003	0.0015
5000	25,000	2.005	0.0025
7000	35,000	2.007	0.0035
7500	37,500	2.030	0.0150
7900	39,500	2.080	0.0400
8000	40,000	2.120	0.0600
8000 (max.)	40,000	2.160	0.0800
7600 (fracture)	38,000	2.205	0.1025

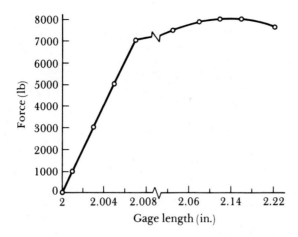

FIGURE 6-2 Graph of the load-gage length data from Table 6-1 for an aluminum alloy test specimen 0.505 in. in diameter.

are defined by the following equations.

$$\text{Engineering stress} = \sigma = \frac{F}{A_0} \tag{6-1}$$

$$\text{Engineering strain} = \varepsilon = \frac{l - l_0}{l_0} \tag{6-2}$$

where A_0 is the original cross-sectional area of the specimen before the test begins, l_0 is the original distance between the gage marks, and l is the distance between the gage marks after force F is applied. The conversions from load-gage length to stress-strain are also included in Table 6-1. The stress-strain curve (Figure 6-3) is usually used to record the results of a tensile test.

◇ **Example 6-1**

Convert the load-gage length data in Table 6-1 to engineering stress and strain and plot a stress-strain curve.

Answer:

For the 1000-lb load

$$\sigma = \frac{F}{A_0} = \frac{1000}{(\pi/4)\,(0.505)^2} = \frac{1000}{0.2} = 5000\,\text{psi}$$

$$\varepsilon = \frac{l - l_0}{l_0} = \frac{2.001 - 2.000}{2.000} = 0.0005\,\text{in./in.}$$

The results of similar calculations for each of the remaining loads are given in Table 6-1 and are plotted in Figure 6-3.

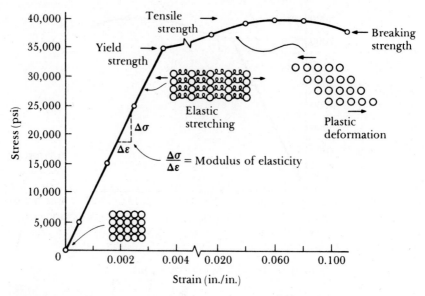

FIGURE 6-3   The stress-strain curve for an aluminum alloy from Table 6-1.

◇ | **Example 6-2**

Compare the force required to produce a stress of 25,000 psi in a 1-in. diameter bar and in a 2-in. diameter bar.

Answer:

$$F = \sigma A_0 = (25,000)\left(\frac{\pi}{4}\right)(1)^2 = 19,635\text{-lb force for a 1-in. bar}$$

$$F = \sigma A_0 = (25,000)\left(\frac{\pi}{4}\right)(2)^2 = 78,540\text{-lb force for a 2-in. bar}$$

The engineering strain tells us how much each inch of the metal will stretch for a given applied stress. If the metal part is 10 in. long, we can multiply the strain by 10 to determine the total amount that the part will stretch, assuming that the part stretches uniformly.

◇ | **Example 6-3**

Suppose a 5000-lb force is applied to a 0.505-in. diameter bar that is 50 in. long. The bar is made from the same aluminum alloy we have discussed previously. Determine the length of the bar when the force is applied.

Answer:

$$\sigma = \frac{F}{A_0} = \frac{5000}{(\pi/4)(0.505)^2} = 25{,}000 \text{ psi}$$

From Figure 6-3, $\varepsilon = 0.0025 \text{ in./in.}$

$$\frac{l - l_0}{l_0} = 0.0025$$

$$\frac{l - 50}{50} = 0.0025$$

$$l = 50 + (0.0025)(50) = 50.125 \text{ in.}$$

A 0.505-in. diameter is specified for the standard cylindrical test specimen because the original cross-sectional area is 0.20 in^2. We can convert force to stress simply by multiplying by five.

Units.   Many different units are used to report the results of the tensile test. The most common units for stress are pounds per square inch (psi) and megapascals (MPa). The units for strain include inch/inch, centimeter/centimeter, and meter/meter. The conversion factors for stress are summarized in Table 6-2. Because strain is really dimensionless, no conversion factors are required to change the system of units.

   Example 6-4

Determine the stress in megapascals when a 5000-lb force is applied to a 0.505-in. diameter bar.

Answer:

$$\sigma = \frac{F}{A_0} = \frac{5000}{(\pi/4)(0.505)^2} = 25{,}000 \text{ psi}$$

$$\sigma = (25{,}000 \text{ psi})(0.006895 \text{ MPa/psi}) = 172.4 \text{ MPa}$$

TABLE 6-2   Units and conversion factors for stress

1 psi = pounds per square inch
1 MPa = megapascal
$1 \text{ MN/m}^2 = 1 \text{ MPa}$ = meganewton per square meter
                          = newton per square millimeter
1 GPa = 1000 MPa = gigapascal
1 ksi = 1000 psi
1 ksi = 6.895 MPa
1 psi = 0.006895 MPa
1 MPa = 0.145 ksi = 145 psi

Elastic versus Plastic Deformation.   When a force is first applied to the specimen, the bonds between the atoms are stretched and the specimen elongates. When we remove the force, the bonds return to their original length and the specimen returns to its initial size. Stretching of the metal in this elastic portion of the stress-strain curve is recoverable.

6–3   PROPERTIES OBTAINED FROM THE TENSILE TEST

Information concerning the strength, stiffness, and ductility of a material can be obtained from a tensile test.

Yield Strength.   The *yield strength* is the stress at which slip becomes noticeable and significant. It therefore is the stress that divides the elastic and plastic behavior of the material. If we are designing a component that must support a force during use, we must be sure that the component does not plastically deform. We must therefore select a material that has a high yield strength, or we must make the component large so that the applied force produces a stress that is below the yield strength.

On the other hand, if we are manufacturing shapes or components by some deformation process, the applied stress must exceed the yield strength to produce a permanent change in the shape of the material.

◇ | **Example 6-5**

You are to design a cable that must support an elevator cab that weighs 10,000 lb. The cable is made from the aluminum alloy in Example 6-1. Calculate the minimum diameter of the cable required to support the cab without permanent deformation.

Answer:

We must not exceed the yield strength of 35,000 psi.

$$A_0 = \frac{F}{\sigma} = \frac{10,000}{35,000} = 0.286\,\text{in}^2$$

$$d = \sqrt{\frac{4}{\pi}A_0} = \sqrt{\frac{4}{\pi}(0.286)} = 0.603\,\text{in.}$$

◇

◇ | **Example 6-6**

You wish to bend an aluminum bar which is $\frac{1}{2}$ in. $\times$ 6 in. in cross section into a bracket by applying a tensile force. What is the minimum force that must be exerted by your forming equipment?

Answer:

We must exceed the yield strength of 35,000 psi.

$$F = \sigma A_0 = (35,000)\left(\tfrac{1}{2}\right)(6) = 105,000\text{-lb force}$$

The stress-strain curve often has a different shape from that shown in Figure 6-3. In some materials, the stress at which the material changes from elastic to plastic behavior is not easily detected. In this case, we may determine an *offset yield strength* [Figure 6-4(a)]. We decide that a small amount of permanent deformation, such as 0.2% or 0.002 in./in., might be allowable without damaging the performance of our component. We can construct a line parallel to the initial portion of the stress-strain curve but offset by 0.002 in./in. from the origin. The 0.2% offset yield strength is the stress at which our constructed line intersects the stress-strain curve.

◇ **Example 6-7**

Determine the 0.2% offset yield strength for gray cast iron [Figure 6-4(a)].

Answer:

By constructing a line starting at 0.002 in./in. strain, which is parallel to the elastic portion of the stress-strain curve, we find that the 0.2% offset yield strength is 40,000 psi.

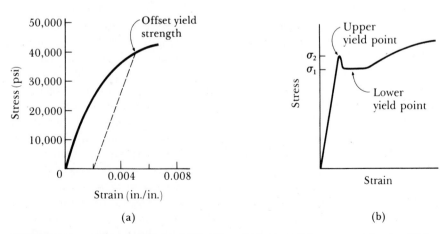

FIGURE 6-4   (a) Determing the 0.2% offset yield strength in gray cast iron and (b) upper and lower yield point behavior in a low-carbon steel.

On the other hand, the stress-strain curve for certain low-carbon steels displays a double yield point [Figure 6-4(b)]. The material is expected to plastically deform at stress σ_1. However, small interstitial atoms clustered around the dislocations interfere with slip and raise the yield point to σ_2. Only after we apply the higher stress σ_2 does the dislocation slip. After slip begins at σ_2, the dislocation moves away from the clusters of small atoms and continues to move very rapidly at the lower stress σ_1. We can easily determine the lower (σ_1) yield strength for materials having this type of stress-strain behavior.

Tensile Strength.   The *tensile strength* is the stress obtained at the highest applied force and thus is the maximum stress on the engineering stress-strain curve. In many ductile materials, deformation does not remain uniform. At some point, one region deforms more than other areas and a large local decrease in the cross-sectional area occurs (Figure 6-5). This locally deformed region is called a *neck*. Because the cross-sectional area becomes smaller at this point, a lower force is required to continue its deformation, and the engineering stress, calculated from the original area A_0, will decrease. The tensile strength is the stress at which necking begins in ductile materials.

Tensile strengths are often reported in handbooks because they are easy to measure; they are useful in comparing the behaviors of materials, and they permit us to estimate other properties which are more difficult to measure. However, the tensile strength is relatively unimportant for materials selection or materials fabrication—the yield strength determines whether the material will or will not deform.

Modulus of Elasticity.   The *modulus of elasticity*, or *Young's modulus*, is the slope of the stress-strain curve in the elastic region. This relationship is *Hooke's law*.

$$E = \frac{\sigma}{\varepsilon} = \text{modulus of elasticity} \tag{6-3}$$

The modulus is closely related to the forces bonding the atoms in the material, as was discussed in Chapter 2 (Figure 2-16). A steep slope in the force-distance graph at the equilibrium spacing indicates that high forces are required to separate the atoms and cause the metal to stretch elastically. Thus, the metal has a high modulus of elasticity. Binding forces, and consequently the modulus of elasticity, are higher for high melting point metals (Table 6-3).

The modulus is a measure of the *stiffness* of the material. A stiff material, with a high modulus of elasticity, maintains its size and shape even under an elastic load. If we are designing a shaft and bearing, we may need very close tolerances. But if the shaft deforms elastically, those close tolerances may cause excessive rubbing, wear, or seizing. Figure 6-6 shows the elastic behavior of iron and aluminum. If a stress of 30,000 psi is applied to the shaft, the steel deforms elastically 0.001 in./in. while, at the same stress, aluminum deforms 0.003 in./in. Iron has a modulus of elasticity three times greater than that of aluminum.

Force

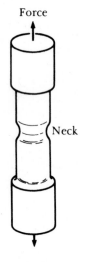

Neck

FIGURE 6-5
Localized deformation of a ductile material during a tensile test produces a necked region.

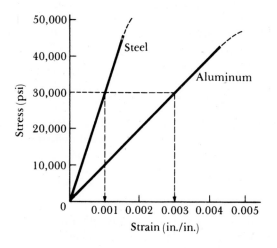

FIGURE 6-6 Comparison of the elastic behavior of steel and aluminum.

◇ **Example 6-8**

From the data in Example 6-1, calculate the modulus of elasticity of the aluminum alloy. Use the modulus to determine the length of a 50-in. bar to which a stress of 30,000 psi is applied.

Answer:

When a stress of 35,000 psi is applied, a strain of 0.0035 in./in. is produced. Thus

$$\text{Modulus of elasticity} = E = \frac{\sigma}{\varepsilon} = \frac{35,000}{0.0035} = 10 \times 10^6 \, \text{psi}$$

TABLE 6-3 Relationship between the modulus of elasticity and the melting temperature of metals

Metal	Melting Temperature (°C)	Modulus of Elasticity (psi)
Pb	327	2.0×10^6
Mg	650	6.5×10^6
Al	660	10.0×10^6
Ag	962	10.3×10^6
Au	1064	11.3×10^6
Cu	1085	18.1×10^6
Ni	1453	29.9×10^6
Fe	1538	30.0×10^6
Mo	2610	43.4×10^6
W	3410	59.2×10^6

From Hooke's law

$$\varepsilon = \frac{\sigma}{E} = \frac{30,000}{10 \times 10^6} = 0.003 \, \text{in./in.} = \frac{l - l_0}{l_0}$$

$$l = l_0 + \varepsilon l_0 = 50 + (0.003)(50) = 50.15 \, \text{in.}$$

◇

Poisson's ratio relates the longitudinal elastic deformation produced by a simple tensile or compressive stress to the lateral deformation that must simultaneously occur:

$$\mu = \frac{-\varepsilon_{\text{lateral}}}{\varepsilon_{\text{longitudinal}}} \tag{6-4}$$

For ideal materials, we can show that Poisson's ratio is $\mu = 0.5$. However, in real materials we find that less lateral strain develops than would be predicted based on conservation of volume; typically, Poisson's ratios of $\mu = 0.3$ are measured.

Ductility.   *Ductility* measures the amount of deformation that a material can withstand without breaking. There are two ways to express ductility. First, we can measure the distance between the gage marks on our specimen before and after the test. The % *elongation* describes the amount that the specimen stretches before fracture.

$$\% \, \text{Elongation} = \frac{l_f - l_0}{l_0} \times 100 \tag{6-5}$$

where l_f is the distance between gage marks after the specimen breaks.

A second approach is to measure the percent change in cross-sectional area at the point of fracture before and after the test. This % *reduction in area* describes the amount of thinning that the specimen undergoes during the test.

$$\% \, \text{Reduction in area} = \frac{A_0 - A_f}{A_0} \times 100 \tag{6-6}$$

where A_f is the final cross-sectional area at the fracture surface.

Ductility is important to both designers and manufacturers. The designer of a component would prefer a material that displays at least some ductility so that, if the applied stress is too high, the component deforms before it breaks. Fabricators want a ductile material so they can form complicated shapes without breaking the material in the process.

◇ **Example 6-9**

The aluminum alloy in Example 6-1 has a final gage length after failure of 2.195 in. and a final diameter of 0.398 in. at the fractured surface. Calculate the ductility of this alloy.

Answer:

$$\% \text{ Elongation} = \frac{l_f - l_0}{l_0} \times 100 = \frac{2.195 \times 2.000}{2.000} \times 100 = 9.75\%$$

$$\% \text{ Reduction in area} = \frac{A_0 - A_f}{A_0} \times 100$$

$$= \frac{(\pi/4)(0.505)^2 - (\pi/4)(0.398)^2}{(\pi/4)(0.505)^2} \times 100$$

$$= 37.8\%$$

The final gage length is less than 2.205 in. (see Table 6-1) since, after fracture, the elastic strain is recovered.

6–4   BRITTLE BEHAVIOR

Ductile materials display an engineering stress-strain curve that goes through a maximum at the tensile strength. In more brittle materials, the maximum load or tensile strength occurs at the point of failure. In extremely brittle materials, such as ceramics, the yield strength, tensile strength, and breaking strength are all the same (Figure 6-7).

In many brittle materials, particularly ceramics and certain composite materials, the normal tensile test cannot easily be performed due to the presence of flaws at the surface. Often, placing the brittle material in the grips of the tensile testing machine will cause these flaws to promote cracking, invalidating the test. Preparation of tensile specimens of brittle materials may also be expensive. One approach used to minimize these problems is the bend test (Figure 6-8). By applying the load at three points and causing bending, a tensile force

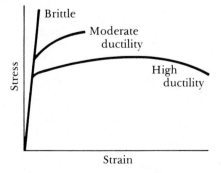

FIGURE 6-7   *The stress-strain behavior for brittle materials compared with that of more ductile materials.*

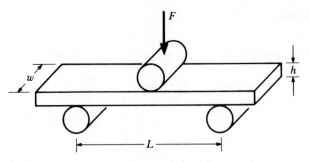

FIGURE 6-8 *The modulus of rupture test often used for measuring the strength of brittle materials.*

acts on the material opposite the middle point. Fracture begins at this location. The *flexural strength* or *modulus of rupture* (MOR), given by Equation 6-7, is used to describe the strength of the material.

$$\text{Flexural strength (psi)} = \frac{3FL}{2wh^2} \tag{6-7}$$

where F is the applied load in pounds, L is the distance between the two outer points, w is the width of the specimen, and h is the height of the specimen.

Since cracks and flaws tend to remain closed in compression, brittle materials are often designed so that only compressive stresses are acting on the part. Often, we find that brittle materials fail at much higher compressive stresses than tensile stresses (Table 6-4), although ductile materials such as metals may have tensile and compressive strengths that are nearly equal.

◇ | **Example 6-10**

The flexural strength of a composite material reinforced with glass fibers is 45,000 psi. The sample, which is 0.5 in. wide, 0.375 in. high, and 8 in. long, is supported between two rods 5 in. apart. What force is required to fracture the material?

Answer:

Based on the description of the sample, $w = 0.5$ in., $h = 0.375$ in., and $L = 5$ in. From Equation 6-7

$$45,000 = \frac{3FL}{2wh^2} = \frac{(3)(F)(5)}{(2)(0.5)(0.375)^2} = 106.7F$$

$$F = \frac{45,000}{106.7} = 422 \,\text{lb}$$

◇

TABLE 6-4 Comparison of the tensile, compressive, and flexural strengths of selected ceramic and composite materials

Material	Tensile Strength (psi)	Compressive Strength (psi)	Flexural Strength (psi)
Polyester—50% glass fibers	23,000	32,000	45,000
Polyester—50% glass fabric	37,000	27,000[a]	46,000
Al_2O_3	30,000	375,000	50,000
SiC	25,000	100,000	37,000

[a]A number of composite materials are quite poor in compression.

6–5   TEMPERATURE EFFECTS

The tensile properties are significantly affected by temperature (Figure 6-9). The yield strength, tensile strength, and modulus of elasticity decrease at higher temperatures, whereas the ductility, as measured by the amount of strain at failure, commonly increases. A materials fabricator may wish to deform a material at a high temperature (known as *hot working*) to take advantage of the higher ductility and lower required stress.

6–6   TRUE STRESS–TRUE STRAIN

The decrease in engineering stress beyond the tensile point occurs because of our definition of engineering stress. We used the original area A_0 in our calculations, which is not precise because the area continually changes. We define *true stress* and *true strain* by the following equations.

$$\text{True stress} = \sigma_t = \frac{F}{A} \tag{6-8}$$

$$\text{True strain} = \int \frac{dl}{l_0} = \ln\left(\frac{l}{l_0}\right) = \ln\left(\frac{A_0}{A}\right) \tag{6-9}$$

where A is the actual area at which the force F is applied. The expression $\ln(A_0/A)$ must be used after necking begins. The true stress-strain curve is compared with the engineering stress-strain curve in Figure 6-10. The true stress continues to increase after necking because, although the load required decreases, the area decreases even more.

We seldom require true stress and true strain. As soon as we exceed the yield strength, the metal begins to deform. Our component has failed because it no longer has the original intended shape. Furthermore, a significant difference

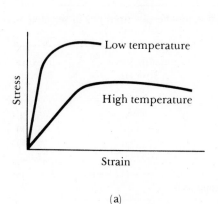

(a)

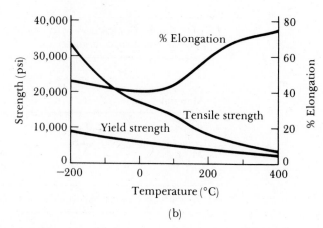

(b)

FIGURE 6-9 The effect of temperature (a) on the stress-strain curve and (b) on the tensile properties of an aluminum alloy.

develops between the two curves only when necking begins. But when necking begins, our component is grossly deformed and no longer satisfies its intended use.

◇ | **Example 6-11**

Compare the engineering stress and strain with the true stress and strain for the aluminum alloy in Example 6-1 at (a) the maximum load and (b) fracture. The diameter at maximum load is 0.497 in. and at fracture is 0.398 in.

Answer:

(a) At the tensile or maximum load

$$\text{Engineering stress} = \frac{F}{A_0} = \frac{8000}{(\pi/4)(0.505)^2} = 40,000\,\text{psi}$$

$$\text{True stress} = \frac{F}{A} = \frac{8000}{(\pi/4)(0.497)^2} = 41,237\,\text{psi}$$

$$\text{Engineering strain} = \frac{l - l_0}{l_0} = \frac{2.120 - 2.000}{2.000} = 0.060\,\text{in./in.}$$

$$\text{True strain} = \ln\left(\frac{l}{l_0}\right) = \ln\left(\frac{2.120}{2.000}\right) = 0.058\,\text{in./in.}$$

(b) At fracture

$$\text{Engineering stress} = \frac{F}{A_0} = \frac{7600}{(\pi/4)(0.505)^2} = 38,000\,\text{psi}$$

$$\text{True stress} = \frac{F}{A} = \frac{7600}{(\pi/4)(0.398)^2} = 61,090\,\text{psi}$$

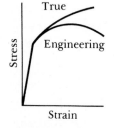

FIGURE 6-10
The relationship between the true stress–true strain diagram and the engineering stress-strain diagram.

$$\begin{aligned}
\text{Engineering strain} &= \frac{l - l_0}{l_0} = \frac{2.205 - 2.000}{2.000} = 0.1025 \,\text{in./in.} \\
\text{True strain} &= \ln\left(\frac{A_0}{A}\right) \ln\left[\frac{(\pi/4)(0.505)^2}{(\pi/4)(0.398)^2}\right] \\
&= \ln(1.610) = 0.476 \,\text{in./in.}
\end{aligned}$$

The true stress becomes much greater than the engineering stress only after necking begins.

Impact Test

6–7   NATURE OF THE IMPACT TEST

The tensile test is normally performed at a low strain rate, at which the specimen is very slowly elongated. When a material is subjected to a sudden, intense blow, in which the strain rate is extremely rapid, the material may behave in a much more brittle manner than is observed in the tensile test. An *impact test* is often used to evaluate the brittleness of a material under these conditions. Many test procedures have been devised, including the *Charpy* test and the *Izod* test (Figure 6-11). The Izod test is often used for nonmetallic materials. The test specimen may be either notched or unnotched; V-notched specimens better measure the resistance of the material to crack propagation.

In the test, a heavy pendulum which starts at an elevation h_0 swings through its arc, strikes and breaks the specimen, and reaches a lower final elevation h_f. By knowing the initial and final elevations of the pendulum, the difference in potential energy can be calculated. This difference is the *impact energy* absorbed by the specimen during failure. The energy is usually expressed in foot · pounds (ft · lb) or joules (J), where 1 ft · lb = 1.356 J. The ability of a material to withstand an impact blow is often referred to as the *toughness* of the material.

6–8   PROPERTIES OBTAINED FROM THE IMPACT TEST

The results of a series of impact tests performed at various temperatures are shown in Figure 6-12. At high temperatures, a large absorbed energy is required to cause the specimen to fail, whereas at low temperatures even a relatively ductile material may fail with little absorbed energy.

Transition Temperature.   At high temperatures, the material behaves in a ductile manner, with extensive deformation and stretching of the specimen

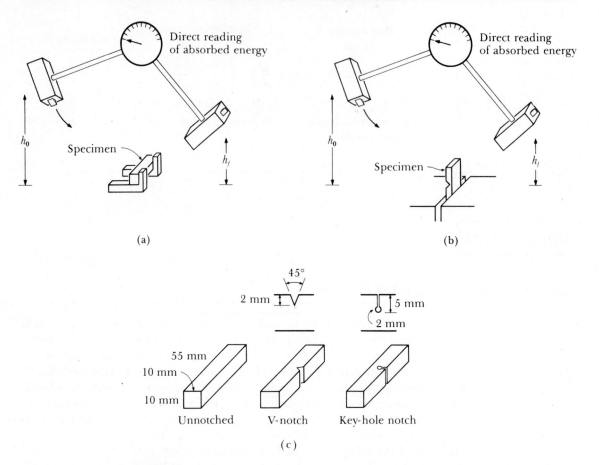

FIGURE 6-11   *The impact test. (a) The Charpy test, (b) the Izod test, and (c) dimensions of typical specimens.*

prior to failure. At low temperatures, the material is brittle, and little deformation at the point of fracture is observed. The *transition temperature* is the temperature at which the material changes from ductile to brittle failure.

A material that may be subjected to an impact blow during service must have a transition temperature *below* the temperature of the material's surround-

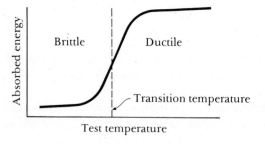

FIGURE 6-12   *Typical results from a series of impact tests.*

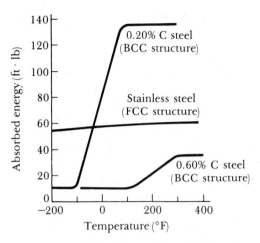

FIGURE 6-13 The Charpy V-notch properties for two plain-carbon steels (BCC structure) and an FCC stainless steel.

ings. For example, the transition temperature of a steel used for a carpenter's hammer should be below room temperature to prevent chipping of the steel.

Not all materials have a distinct transition temperature (Figure 6-13). BCC metals have transition temperatures but most FCC metals do not. FCC metals have high absorbed energies, with the energy decreasing gradually and slowly as the temperature decreases.

Notch Sensitivity. Notches caused by poor machining, fabrication, or design cause stresses to be concentrated, reducing the toughness of the material. The *notch sensitivity* of a material can be evaluated by comparing the absorbed energies of notched versus unnotched specimens. The absorbed energies are much lower in notched specimens if the material is notch sensitive, as in ductile cast iron (Figure 6-14). However, some materials, such as gray cast iron, are not notch sensitive.

Relationship to Stress-Strain Diagram. The energy required to break a material also corresponds to the area contained within the true stress–true strain diagram. Materials that have both high strength and high ductility have a good toughness (Figure 6-15). Ceramics and many composites, on the other hand, have poor toughness because they display virtually no ductility.

Use of Impact Properties. The absorbed energy and transition temperature are very sensitive to the loading conditions. For example, a higher rate at which energy is applied to the specimen will reduce the absorbed energy and increase the transition temperature. The size of the specimen also affects the results—smaller energies might be required to break thicker materials. Finally, the

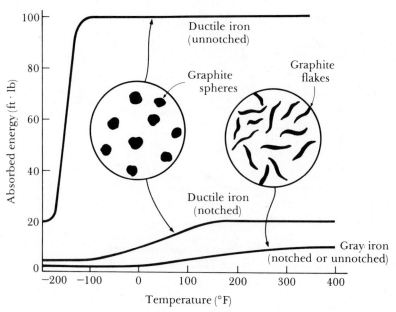

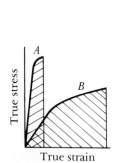

FIGURE 6-15
The area contained within the true stress–true strain curve is related to the impact energy. Although material **B** has a lower yield strength, it absorbs a greater energy than material **A**.

FIGURE 6-14 *The effect of internal and external notches on impact properties. Gray iron structures contain sharp graphite flakes that act as notches and produce low energies. Ductile iron structures contain spherical graphite nodules that do not act as notches. An external notch has a significant effect only on ductile iron.*

configuration of the notch may affect the behavior—a surface crack permits lower absorbed energies than does a V-notch. Because we often cannot predict or control all of these conditions, the impact test is best used for comparison and selection of materials rather than as a design criterion.

Fatigue Test

6–9 NATURE OF THE FATIGUE TEST

In many applications, a component is subjected to the repeated application of a stress below the yield strength of the material. This repeated stress may occur as a result of rotation, bending, or even vibration. Even though the stress is below the yield strength, the material may fail after a large number of applications of the stress. This mode of failure is known as *fatigue*.

A common method to measure the resistance to fatigue is the rotating cantilever beam test (Figure 6-16). One end of a machined, cylindrical specimen

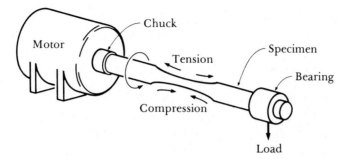

FIGURE 6-16 *The rotating cantilever beam fatigue tester.*

is mounted in a motor-driven chuck. A weight is suspended from the other end. The specimen initially has a tensile force acting on the top surface, while the bottom surface is compressed. After the specimen turns 90°, the locations that were originally in tension and compression have no stress acting on them. After a half revolution of 180°, the material that was originally in tension is now in compression. Thus, the stress at any one point goes through a complete cycle from zero stress to maximum tensile stress to zero stress to maximum compressive stress. The maximum stress acting on the specimen is given by

$$\sigma = 10.18 \frac{lF}{d^3} \tag{6-10}$$

where l is the length of the bar, F is the load, and d is the diameter.

After a sufficient number of cycles, the specimen may fail. Generally, a series of specimens are tested at different applied stresses and the stress is plotted versus the number of cycles to failure (Figure 6-17).

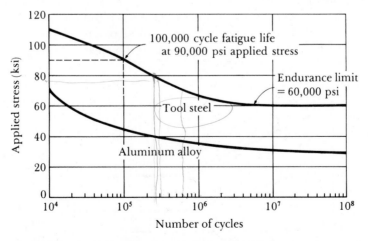

FIGURE 6-17 *The stress–number of cycles to failure curve for a tool steel and an aluminum alloy.*

6–10   RESULTS OF THE FATIGUE TEST

The fatigue test can tell us how long a part may survive or the maximum allowable loads that can be applied to prevent failure.

Fatigue Life.   The *fatigue life* tells us how long a component survives when a stress σ is repeatedly applied to the material. If we are designing a tool steel part that must undergo 100,000 cycles during its lifetime, the part must be designed so that the applied stress is lower than 90,000 psi (Figure 6-17).

Endurance Limit.   The *endurance limit*, which is the stress below which failure by fatigue never occurs, is our preferred design criterion. To prevent a tool steel part from failing, we must be sure that the applied stress is below 60,000 psi (Figure 6-17).

Fatigue Strength.   Some materials, including many aluminum alloys, have no true endurance limit. For these materials, we may specify a minimum fatigue life; then the *fatigue strength* is the stress below which fatigue does not occur within this time period. In many aluminum alloys, the fatigue strength is based on 500 million cycles.

Notch Sensitivity.   Fatigue cracks initiate at the surface of a stressed material, where the stresses are at a maximum. Any design or manufacturing defect at the surface concentrates stresses and encourages the formation of a fatigue crack. This susceptibility may be measured using a notched fatigue specimen (Figure 6-18). Sometimes highly polished surfaces are prepared in order to minimize the likelihood of a fatigue failure.

Endurance Ratio.   The fatigue resistance is related to the strength of the material at the surface. In many ferrous, or iron-base, alloys, the endurance limit is approximately one-half the tensile strength of the material. This ratio of endurance limit to tensile strength is the *endurance ratio*.

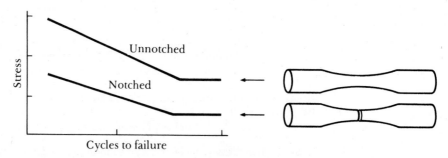

FIGURE 6-18   *The effect of a notch on the fatigue properties of a metal.*

$$\text{Endurance ratio} \; = \; \frac{\text{endurance limit}}{\text{tensile strength}} \approx 0.5 \tag{6-11}$$

If the tensile strength at the surface of the material increases, the resistance to fatigue also increases.

Temperature Effect.   Temperature influences the fatigue resistance. As the temperature of the material increases, the strength decreases and consequently both fatigue life and endurance limit decrease.

◇ | **Example 6-12**

A 650-lb force is applied to a tool steel bar rotating at 3000 cycles/min. The bar is 1 in. in diameter and 12 in. long. (a) Estimate the time before the bar fails and (b) calculate the diameter of the shaft that would prevent fatigue failure.

Answer:

(a) $\sigma \; = \; \dfrac{10.18 \, lF}{d^3} \; = \; \dfrac{(10.18)\,(12)\,(650)}{(1)^3} \; = \; 79{,}400 \, \text{psi}$

From Figure 6-17, the number of cycles to failure is 150,000. The time to failure is

$$t \; = \; \frac{250{,}000}{3000} \; = \; 83 \, \text{min}$$

(b) The endurance limit is 60,000 psi.

$$d^3 \; = \; \frac{(10.18)\,(12)\,(650)}{60{,}000} \; = \; 1.3234$$

$$d \; = \; 1.098 \, \text{in.}$$

Creep Test

6–11   NATURE OF THE CREEP TEST

If we apply a stress to a material at a high temperature, the material may stretch and eventually fail, even though the applied stress is less than the yield strength at that temperature. Plastic deformation at high temperatures is known as *creep*. Table 6-5 gives the approximate temperatures at which several metals begin to creep.

TABLE 6-5 Approximate temperatures at which creep becomes pronounced for selected metals and alloys

Metal	Temperature (°C)
Aluminum alloys	200
Titanium alloys	325
Low-alloy steels	375
High-temperature steels	550
Nickel and cobalt superalloys	650
Refractory metals (tungsten, molybdenum)	1000–1550

To determine the creep characteristics of a material, a constant stress is applied to a cylindrical specimen placed in a furnace (Figure 6-19). As soon as the stress is applied, the specimen stretches elastically a small amount ε_0 (Figure 6-20), depending on the applied stress and the modulus of elasticity of the material at the high temperature.

Dislocation Climb.   The high temperature permits dislocations in the metal to *climb*. In climb, atoms move either to or from the dislocation line by diffusion, causing the dislocation to move in a direction that is perpendicular, not parallel, to the slip plane (Figure 6-21). The dislocation can now escape from lattice

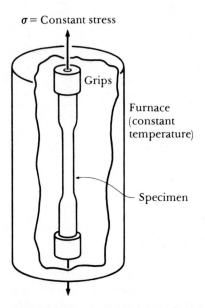

FIGURE 6-19 A specimen is placed in a furnace at an elevated temperature under a constant applied stress in the creep test.

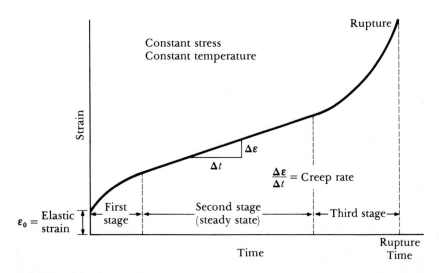

FIGURE 6-20   *A typical creep curve showing the strain produced as a function of time for a constant stress and temperature.*

imperfections that block the slip process. The dislocation, after climbing away from the imperfection, continues to slip and causes additional deformation of the specimen even at low applied stresses.

Creep Rate and Rupture Times.   During the creep test, the strain or elongation is measured as a function of time and plotted to give the creep curve (Figure 6-20). In the first stage of creep, many dislocations climb away from obstacles, slip, and contribute to deformation of the metal. Eventually, the rate at which dislocations climb away from obstacles equals the rate at which dislocations are blocked by other imperfections. This leads to second-stage, or steady-state, creep. The slope of the steady-state portion of the creep curve is the *creep rate*.

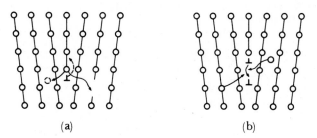

FIGURE 6-21   Dislocations can climb away from obstacles when atoms leave the dislocation line to create interstitials or to fill vacancies (a) or when atoms are attached to the dislocation line by creating vacancies or eliminating interstitials (b).

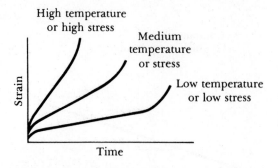

FIGURE 6-22 The effect of temperature or applied stress on the creep curve.

$$\text{Creep rate} \ = \ \frac{\Delta \text{ strain}}{\Delta \text{ time}} \qquad (6\text{-}12)$$

Eventually, during third-stage creep, necking begins, the stress increases, and the specimen deforms at an accelerated rate until failure occurs. The time required for failure to occur is the *rupture time*. Either a higher stress or a higher temperature reduces the rupture time and increases the creep rate (Figure 6-22).

◇ **Example 6-13**

The results of a creep test are given in Table 6-6. Calculate the creep rate in in./in./h.

Answer:

The data in Table 6-6 are plotted in Figure 6-23. From the slope of the steady-state portion of the curve

$$\text{Creep rate} \ = \ \frac{\Delta\varepsilon}{\Delta t} \ = \ \frac{0.021 - 0.009}{6000 - 1000} \ = \ \frac{0.012}{5000} \ = \ 2.4 \times 10^{-6}\,\text{in./in./h}$$

TABLE 6-6 Data from a creep test for Example 6-13

Strain (in./in.)	Time (h)
0.003	0
0.006	250
0.009	1000
0.012	2250
0.015	3500
0.018	4750
0.021	6000
0.024	7100
0.027	7500
0.030 (fracture)	7750

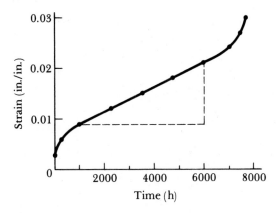

FIGURE 6-23 Graph of the data in Table 6-6 to produce a creep curve. The slope of the steady-state portion of the graph is the answer to Example 6-13.

6–12   USE OF CREEP DATA

Four of the many ways used to present the results from a series of creep tests are shown in Figure 6-24. The *stress-rupture curve* shown in Figure 6-24(a) permits us to estimate the expected lifetime of a component for a particular combination of stress and temperature. Figure 6-24(b) depicts the rupture time versus the reciprocal of temperature for a constant stress; this presentation of the data suggests an Arrhenius relationship for the rupture time and would permit an activation energy for the process to be calculated. The creep rate obtained for a particular combination of an applied stress and temperature can be estimated from Figure 6-24(c). The *Larson-Miller parameter*, illustrated in Figure 6-24(d), is used to consolidate the stress–temperature–rupture time relationship into a single curve.

◇   **Example 6-14**

Using the Larson-Miller parameter for ductile cast iron, as shown in Figure 6-24(d), determine the time required before the metal fails at an applied stress of 6000 psi and temperatures of 400°C and 600°C.

Answer:

The Larson-Miller parameter for 6000 psi is 34.3.

(a) At 400° C

$$34.3 = \frac{(400 + 273 \, \text{K})}{1000} (36 + 0.78 \ln t)$$

$$0.78 \ln t = (34.3) \left(\frac{1000}{673}\right) - 36 = 14.97$$

$$t = 2.2 \times 10^8 \, \text{h} = 25{,}100 \text{ years}$$

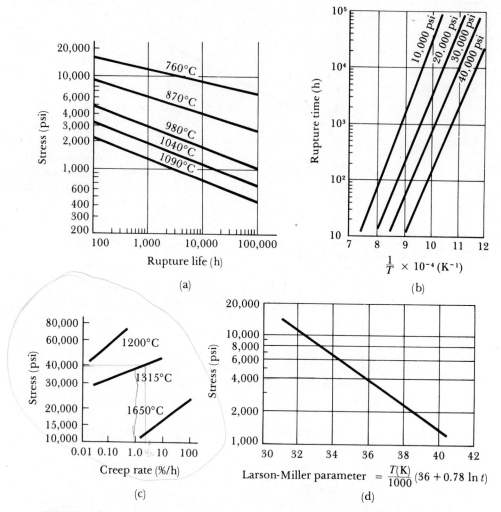

FIGURE 6-24   Results from a series of creep tests. (a) Stress-rupture curves for an iron-chromium-nickel alloy, (b) rupture time versus reciprocal temperature for a nickel heat-resistant alloy, (c) minimum creep rate curves for a tantalum alloy, and (d) Larson-Miller parameter for ductile cast iron.

(b) At $600°$ C

$$34.3 = \frac{(600 + 273\,\text{K})}{1000}(36 + 0.78 \ln t)$$

$$0.78 \ln t = (34.3)\left(\frac{1000}{873}\right) - 36 = 3.29$$

$$t = 67.9\,\text{h} = 2.8 \text{ days}$$

Hardness Test

6–13 NATURE OF THE HARDNESS TEST

The *hardness test* measures the resistance to penetration of the surface of a material by a hard object. A variety of hardness tests have been devised, but the most commonly used are the Rockwell test and the Brinell test (Figure 6-25).

In the *Brinell hardness test* a hard steel sphere, usually 10 mm in diameter, is forced into the surface of the material. The diameter of the impression left on the surface is measured and the Brinell hardness number (*BHN*) is calculated from the following equation.

$$BHN = \frac{F}{(\pi/2)D(D - \sqrt{D^2 - D_i^2})} \tag{6-13}$$

where F is the applied load in kilograms, D is the diameter of the indentor in millimeters, and D_i is the diameter of the impression in millimeters.

The *Rockwell hardness test* uses either a small diameter steel ball for soft materials or a diamond cone, or Brale, for harder materials. The depth of penetration of the indentor is automatically measured by the testing machine and converted to a Rockwell hardness number. Several variations of the Rockwell test are used, as shown in Table 6-7.

The Vickers and Knoop tests are *microhardness* tests; they form such small indentations that a microscope is required to obtain the measurement (Figure 6-26).

6–14 USE OF THE HARDNESS TEST

Hardness numbers are used primarily as a basis for comparison of materials, specifications for manufacturing and heat treatment, quality control, and correlation with other properties and behavior of materials. For example, Brinell

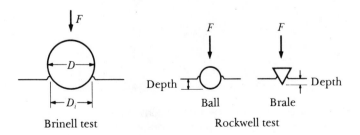

FIGURE 6-25 The Brinell and Rockwell hardness tests.

TABLE 6-7   Comparison of typical hardness tests

Test	Indentor	Load	Application
Brinell	10-mm ball	3000 kg	Cast iron and steel
Brinell	10-mm ball	500 kg	Nonferrous alloys
Rockwell A	Brale	60 kg	Very hard materials
Rockwell B	$\frac{1}{16}$-in. ball	100 kg	Brass, low-strength steel
Rockwell C	Brale	150 kg	High-strength steel
Rockwell D	Brale	100 kg	High-strength steel
Rockwell E	$\frac{1}{8}$-in. ball	100 kg	Very soft materials
Rockwell F	$\frac{1}{16}$-in. ball	60 kg	Aluminum, soft materials
Vickers	Diamond pyramid	10 kg	Hard materials
Knoop	Diamond pyramid	500 g	All materials

hardness is closely related to the tensile strength of steel by the relationship

$$\text{Tensile strength (psi)} \quad = \quad 500\,BHN \tag{6-14}$$

A Brinell hardness number can be obtained in just a few minutes with virtually no preparation of the specimen and without destroying the component, yet provides a close approximation for the tensile strength.

Hardness correlates well with wear resistance. A material used to crush or grind ore should be very hard to assure that the material is not eroded or abraded by the hard feed materials. Similarly, gear teeth in a transmission or

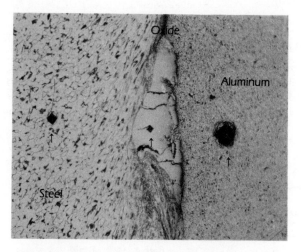

FIGURE 6-26 Microhardness impressions in an explosive bond joining aluminum to steel. The small size of the impression indicates that the inclusion trapped at the interface is harder than either aluminum or steel (× 200).

drive system of a vehicle should be hard so that the teeth do not wear out. Typically we find that polymer materials are exceptionally soft, metals have an intermediate hardness, and ceramics are exceptionally hard.

$\diamondsuit$ **Example 6-15**

A Brinell hardness test is performed on a steel using a 10-mm indentor with a load of 3000 kg. A 3.2-mm impression is measured on the surface of the steel. Calculate the *BHN*, the tensile strength, and the endurance limit of the steel.

Answers:

$$BHN = \frac{3000}{(\pi/2)(10)(10 - \sqrt{10^2 - 3.2^2})} = 363 \, \text{kg/mm}^2$$

$$\text{Tensile strength} = 500 \, BHN = (500)(363) = 181,600 \, \text{psi}$$

$$\text{Endurance limit} = 0.5 \text{ tensile strength} = (0.5)(181,600)$$

$$= 90,800 \text{ psi}$$

Fracture Mechanics

6–15   FRACTURE MECHANICS

Fracture mechanics is the discipline that studies the behavior of materials containing cracks or other small flaws. All materials contain some flaws; we wish to know the maximum stress that the material can withstand if it contains flaws of a certain size and geometry.

A typical fracture mechanics test may be performed by applying a tensile stress to a specimen prepared with a flaw of known size and geometry (Figure 6-27). If the specimen is thick enough, a "plain strain" condition is produced which gives the worst behavior of the material. The stress required to propagate a crack from the prepared flaw can be measured. The *stress intensity factor K* then can be calculated. In simple tests, the stress intensity factor is

$$K = f\sigma\sqrt{a\pi} \tag{6-15}$$

where f is a geometry factor for the specimen and flaw, σ is the applied stress, and a is the flaw size as defined in Figure 6-27. For thick plate, $f \approx 1$. If K is greater than or equal to a critical value K_{I_c}, the flaw grows and the material fails. The *critical fracture toughness* K_{I_c} is a property of the material (Table 6-8). A

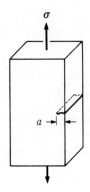

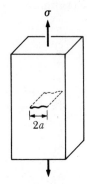

FIGURE 6-27
Schematic drawing of fracture toughness specimens with edge and internal flaws.

TABLE 6-8 The critical fracture toughness for selected materials

Material	Critical Fracture Toughness $(\text{psi} \cdot \text{in}^{1/2})$	Yield Strength (psi)
Al-Cu alloy	22,000	66,000
	33,000	47,000
Ti-6% Al-4% V	50,000	130,000
	90,000	125,000
Ni-Cr steel	45,800	238,000
	80,000	206,000
Al_2O_3	1,600	30,000
Si_3N_4	4,500	80,000
Transformation toughened ZrO_2	10,000	60,000
Si_3N_4-SiC composite	51,000	120,000

similar approach can be used to determine the ease with which a flaw grows in torsion, impact, fatigue, or other loading conditions.

The ability of the material to resist the growth of a crack depends on a large variety of factors, including the following.

1. Larger flaws reduce the permitted stress. Special manufacturing techniques have been devised to improve fracture toughness. For example, liquid aluminum is passed though a ceramic filter to remove impurity particles, whereas the AOD process (argon-oxygen decarburization) has been developed to produce steels containing fewer oxide inclusions.

2. Increasing the strength of a given metal tends to reduce fracture toughness, as shown for the titanium alloy in Table 6-8. This is often associated with the lower ductility that the stronger alloys possess.

3. Thicker materials have a lower stress intensity factor K than thin materials [Figure 6-28(a)]. However, the fracture toughness is less predictable for thin sections, due to the increased section size sensitivity as the thickness of the specimen approaches the size of the flaws.

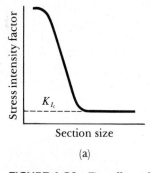

(a)

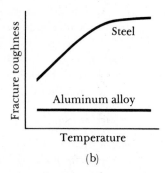

(b)

FIGURE 6-28 The effect of (a) section size, (b) temperature, and crystal structure on the stress intensity factor and fracture toughness of materials.

4. Increasing the temperature normally increases the fracture toughness of BCC and HCP metals. However, the fracture toughness of FCC metals is relatively unaffected by temperature [Figure 6-28(b)].

5. The ability of a material to deform is critical. In ductile metals, the material near the tip of the flaw can deform, helping to absorb energy and blunt further crack growth. Brittle materials such as glass cannot deform; consequently, the crack propagates with little energy required. We will point out in later chapters some of the methods used to improve the fracture toughness of brittle materials such as ceramics, thermosetting polymers, and composites.

6–16   IMPORTANCE OF FRACTURE MECHANICS

The fracture mechanics approach allows us to design and select materials while taking into account the inevitable presence of flaws. There are three variables to consider—the property of the material (K_{I_c}), the stress σ that the material must withstand, and the size of the flaw a.

We must be able to set an upper limit on the size of any flaw that is present by nondestructive testing. For example, ultrasonic testing or X-ray radiography may detect any flaw longer than 0.1 mm. This fixes the largest size a and gives us the worst condition that the material will face. If we know the magnitude of the applied stress, we can select a material that has a fracture toughness K_{I_c} large enough to prevent the flaw from growing. Or, if the material has already been specified, we can calculate the maximum permitted stress that can act on the material. Finally, if we know both the applied stress and the fracture toughness of the material, we can determine if our nondestructive testing capability is adequate.

◇ | **Example 6-16**

For a large plate, the geometry factor f is one. Suppose a steel casting alloy has a critical fracture toughness of $80{,}000\,\text{psi} \cdot \text{in}^{1/2}$. The steel will be exposed to a stress of $45{,}000\,\text{psi}$ during service. Calculate the minimum size of a crack at the surface that will grow. Repeat for an internal crack.

Answer:

$$K_{I_c} = f\sigma\sqrt{a\pi}$$
$$80{,}000 = (1)(45{,}000)\sqrt{a\pi}$$

For a surface crack: $a = \dfrac{1}{\pi}\left(\dfrac{80{,}000}{45{,}000}\right)^2 = 1\,\text{in.}$

For an internal crack: $2a = 2\,\text{in.}$

◇

SUMMARY

The mechanical behavior of materials is described by their mechanical properties, which are simply the results of idealized, simple tests. These tests are designed to represent different types of loading conditions. The tensile test describes the resistance of the material to a slowly applied stress; the results define the yield strength, ductility, and stiffness of the material. The fatigue test permits us to understand how a material performs when a cyclical stress is applied, the impact test indicates the shock resistance of the material, and the creep test provides information on the load-carrying ability of the material at high temperatures. The hardness test, besides providing a measure of the wear and abrasion resistance of the material, can be correlated to a number of other mechanical properties. Finally, fracture mechanics takes into account the presence of flaws in the material.

Although many other tests, including very specialized ones, are used to characterize mechanical behavior, the properties obtained from these six tests are most commonly found in handbooks. However, we should always recall that the properties listed in handbooks are average values from idealized tests and must be used with care.

GLOSSARY

Climb. Movement of a dislocation perpendicular to its slip plane by diffusion of atoms to or from the dislocation line.

Creep rate. The rate at which a material continues to stretch as a function of time when a stress is applied at a high temperature.

Creep test. Measures the resistance of a material to deformation and failure when subjected to a static load below the yield strength at an elevated temperature.

Double yield point. In certain materials, dislocations are initially pinned by lattice imperfections. A high stress is required to start plastic deformation, giving an upper yield point, while subsequent deformation occurs at a lower stress, giving a lower yield point.

Ductility. The ability of a material to be permanently deformed without breaking when a force is applied.

Elastic deformation. Deformation of the material that is recovered when the applied load is removed.

% Elongation. The total percent increase in the length of a specimen during a tensile test.

Endurance limit. The stress below which a material will not fail in a fatigue test.

Endurance ratio. The endurance limit divided by the tensile strength of the material. The ratio is about 0.5 for many ferrous metals.

Engineering strain. The amount that a material deforms per unit length in a tensile test.

Engineering stress. The applied load, or force, divided by the original cross-sectional area of the material.

Fatigue life. The number of cycles at a particular stress before a material fails by fatigue.

Fatigue strength. The stress required to cause failure by fatigue in 500 million cycles.

Fatigue test. Measures the resistance of a material to failure when a stress below the yield strength is repeatedly applied.

Fracture mechanics. The study of a material's ability to withstand a stress in the presence of a flaw.

Fracture toughness. The resistance of a material to failure in the presence of a flaw.

Hardness test. Measures the resistance of a material to penetration by a sharp object. Common tests include the Brinell test, Rockwell test, Knoop test, and Vickers test.

Hooke's law. The relationship between stress and strain in the elastic portion of the stress-strain curve.

Impact energy. The energy required to fracture a standard specimen when the load is suddenly applied.

Impact test. Measures the ability of a material to absorb a sudden application of a load without breaking. The Charpy test is a commonly used impact test.

Larson-Miller parameter. One of several parameters used to relate the stress, temperature, and rupture time in creep.

Modulus of elasticity. Young's modulus, or the slope of the stress-strain curve in the elastic region.

Modulus of rupture. The stress required to fracture a specimen in a bend test.

Necking. Local deformation of a tensile specimen. Necking begins at the tensile point.

Notch sensitivity. Measures the effect of a notch on the impact energy.

Offset yield strength. A yield strength obtained graphically that describes the stress that gives no more than a specified amount of plastic deformation.

Plastic deformation. Permanent deformation of the material when a load is applied, then removed.

% Reduction in area. The total percent decrease in the cross-sectional area of a specimen during the tensile test.

Rupture time. The time required for a specimen to fail by creep at a particular temperature and stress.

Stiffness. A qualitative measure of the elastic deformation produced in a material. A stiff material has a high modulus of elasticity.

Stress-rupture curve. A method of reporting the results of a series of creep tests by plotting the applied stress versus the rupture time.

Tensile strength. The stress that corresponds to the maximum load in a tensile test.

Tensile test. Measures the response of a material to a slowly applied uniaxial force. The yield strength, tensile strength, modulus of elasticity, and ductility are obtained.

Toughness. A qualitative measure of the impact properties of a material. A material that resists failure by impact is said to be tough.

Transition temperature. The temperature below which a material behaves in a brittle manner in an impact test.

True strain. The actual strain produced when a load is applied to a material.

True stress. The load divided by the actual area at that load in a tensile test.

Yield strength. The stress applied to a material that just causes permanent plastic deformation.

PRACTICE PROBLEMS

1. A 200-lb force is applied to a 0.1-in. diameter copper wire having a yield strength of 20,000 psi and a tensile strength of 40,000 psi. Determine (a) whether the wire will plastically deform and (b) whether the wire will experience necking.

2. Calculate the maximum force that a 0.1-in. thick and 2-in. wide nickel strip, having a yield strength of 45,000 psi and a tensile strength of 63,000 psi, can withstand with no plastic deformation.

3. A titanium bar 0.35 in. in diameter will stretch from a 6-in. length to a 6.02-in. length when a force of 5230 lb is exerted. Calculate the modulus of elasticity of the alloy.

4. The modulus of elasticity of nickel is 29.9×10^6 psi. Determine the length of the bar when a force of 1550 lb is applied to a 0.5 in. $\times$ 0.3 in. bar originally 36 in. long without causing plastic deformation.

5. We would like to plastically deform a 1-in. diameter bar of aluminum having a yield strength of 25,000 psi and a tensile strength of 40,000 psi. Can this be accomplished using a forging machine that can exert a maximum force of 15 tons?

6 A landing gear must be capable of supporting one-third of an airplane that weighs 200 tons. Just to be safe, we would like the landing gear to be able to support twice its maximum load. Determine the minimum cross-sectional area of the landing gear if it is made of a heat-treated steel having a yield strength of 125,000 psi and a tensile strength of 132,500 psi.

7. We would like to produce a copper plate that is 0.25 in. thick by a rolling process. The yield strength is 35,000 psi, the tensile strength is 55,000 psi, and the modulus of elasticity is 18×10^6 psi. Calculate the separation between the rolls that we should use, assuming that the rolls do not deflect.

8. We plan to stretch a steel bar until it has a final length of 72 in. What length should the bar be before the forming stress is removed? The yield strength is 62,000 psi, the tensile strength is 83,000 psi, and the modulus of elasticity is 30×10^6 psi.

9. A titanium bar 0.375 in. in diameter with a gage length of 2 in. is pulled in tension to failure. After failure, the gage length is 2.15 in. and the diameter is 0.352 in. Calculate the % elongation and % reduction in area.

10. The % elongation of a magnesium alloy is 12.5%. If a magnesium bar 24 in. long is pulled until it breaks, and if the deformation is uniform during this process, what is the final length of the bar? If necking occurred during this process, would the final length be longer or shorter than your calculation?

11. The following data were collected from a standard 0.505-in. diameter test specimen of magnesium.

Load (lb)		Gage Length (in.)
0		2.00000
1000		2.00154
2000		2.00308
3000		2.00462
4000		2.00615
5000		2.00769
5500		2.014
6000		2.050
6200	(maximum)	2.130
6000	(fracture)	2.255

After fracture, the gage length is 2.245 in. and the diameter is 0.466 in. Plot the data and calculate (a) the 0.2% offset yield strength, (b) the tensile strength, (c) the modulus of elasticity, (d) the % elongation, (e) the % reduction in area, (f) the stress at fracture, and (g) the true stress at fracture.

12. A standard 0.505-in. diameter tensile bar is machined from a copper-nickel alloy; the results of a tensile test are described in the following table.

Load (lb)		Gage Length (in.)
0		2.00000
1,000		2.00045
2,000		2.00091
3,000		2.00136
4,000		2.0020
6,000		2.020
8,000		2.052
10,000		2.112
11,000	(maximum)	2.280
9,000	(fracture)	2.750

After fracture, the gage length is 2.72 in. and the diameter is 0.362 in. Plot the data and calculate (a) the 0.2% offset yield strength, (b) the tensile strength, (c) the modulus of elasticity, (d) the % elongation, (e) the % reduction in area, (f) the stress at fracture, and (g) the true stress at fracture.

13. A 25-mm diameter tensile bar is prepared from a silver alloy and pulled in a tensile test with the following results.

Load (N)		Gage Length (mm)
0		50.0000
50,000		50.0613
100,000		50.1227
150,000		50.1848
175,000		50.50
200,000		51.35
225,000		52.90
231,000	(fracture)	53.40

After fracture, the gage length is 53.1 mm and the diameter is 24.25 mm. Plot the data and calculate (a) the 0.2% offset yield strength. (b) the tensile strength, (c) the modulus of elasticity, (d) the % elongation, (e) the % reduction in area, (f) the stress at fracture, and (g) the true stress at fracture.

14. A 15-mm diameter tensile bar of an aluminum alloy is pulled in a tensile test with the following results.

Load (N)		Gage Length (mm)
0		60.0000
10,000		60.0469
20,000		60.0938
30,000		60.1407
35,000		60.210
37,500		60.300
40,000		60.600
42,500		61.200
45,000	(maximum)	63.000
44,200	(fracture)	63.900

After fracture, the gage length is 63.66 mm and the diameter is 14.5 mm. Plot the data and calculate (a) the 0.2% offset yield strength, (b) the tensile strength, (c) the modulus of elasticity, (d) the % elongation, (e) the % reduction in area, (f) the stress at fracture, and (g) the true stress at fracture.

15. A three-point bend test is performed on a block of Al_2O_3 that is 5 in. long, 0.325 in. wide, and 0.25 in. tall and is resting on two supports 3 in. apart. A force of 360 lb is required to break the ceramic. Calculate the flexural strength in both psi and MPa.

16. A three-point bend test is performed on a block of silicon nitride that is 100 mm long, 8 mm wide, and 6 mm tall and is resting on two supports 80 mm apart. The modulus of rupture is 896 MPa. Calculate the force required to break the test block in newtons and in pounds.

17. Suppose a stainless steel is to be selected as a valve in a system designed to pump liquid helium at 4 K. A severe impact may be observed during opening and closing of the valve. Would you select a stainless steel that has an FCC or a BCC structure? Explain.

18. Fiberglass is made by introducing short glass fibers into a more ductile polymer matrix. Would you expect the fiberglass to be notch sensitive or insensitive in a series of impact tests? Explain.

19. A cylindrical tool steel specimen (Figure 6-17) which is 9 in. long and 0.5 in. in diameter is to be designed so that failure never occurs. What is the maximum load that can be applied?

20. A cylindrical aluminum test specimen (Figure 6-17) which is 8 in. long and 0.75 in. in diameter is to be designed so that failure does not occur within ten million cycles. What is the maximum load that can be applied?

21. A cylindrical tool steel test specimen (Figure 6-17) which is 10 in. long is to be subjected to a load of 1500 lb. What is the minimum diameter of the specimen if failure is never to occur?

22. A cylindrical aluminum test specimen (Figure 6-17) which is 250 mm in length is to be subjected to a load of 20,000 N. What is the minimum diameter of the specimen if failure within one million cycles is to be prevented?

23. A cylindrical tool steel test specimen (Figure 6-17) which is 6 in. long and 0.375 in. in diameter is to have a fatigue life of 500,000 cycles. What is the maximum allowable load?

24. A cylindrical aluminum test specimen (Figure 6-17) which is 7 in. long and 0.5 in. in diameter is rotating at 3000 rpm. If a load of 80 lb is applied to the specimen, how many hours of operation are expected before failure?

25. A cylindrical tool steel test specimen (Figure 6-17) which is 10 in. long and 1.5 in. in diameter is rotating at 60 rpm. If a load of 2650 lb is applied to the specimen, how many days of operation are expected before failure?

26. To avoid failure by fatigue, the maximum force that can be applied to the end of a rotating cylindrical steel specimen which is 5 in. long and 0.25 in. in diameter is 15 lb. Estimate the tensile strength of the steel.

27. Which of the following would you expect to show the temperature dependence of the creep rate ($\dot{\varepsilon}$): $\dot{\varepsilon} \sim T$; $\dot{\varepsilon} \sim T^2$; $\dot{\varepsilon} \sim 1/T$; or $\dot{\varepsilon} \sim \exp\left(-c/T\right)$? Explain your answer.

28. The following data are the results of a creep test performed on a material. The original gage length was 2.0 in. Calculate the creep rate in %/h.

Length of Specimen (in.)	Time (h)
2.020	50
2.038	200
2.046	500
2.060	1450
2.092	3450
2.122	5500
2.152	7500
2.180	8500
2.220 (rupture)	8950

29. A 2-in, diameter bar of an iron-chromium-nickel alloy is subjected to a load of 6000 lb. How many days will it survive at (a) 980° C, (b) 1040° C, and (c) 1090° C?

30. What is the maximum load that a 0.80 in. × 1.2 in. bar of an iron-chromium-nickel alloy can withstand at 870° C without failing within 10 years?

31. An iron-chromium-nickel alloy is to withstand a load of 2500 lb at 1090°C for 15 years. What is the minimum diameter of the bar that is necessary?

32. Develop an equation of the form $\sigma = ct^{-n}$ that will relate the applied stress σ to the rupture time for an iron-chromium-nickel alloy held at 760°C. Using this equation, calculate the maximum load that a 1-in. diameter bar of the alloy can withstand if it is to survive for 50 years.

33. An iron-chromium-nickel alloy is to operate for 10,000 h under a load of 3500 lb. What is the maximum operating temperature if the bar is 1.5 in. in diameter?

34. Suppose a stress of 20,000 psi is applied to the heat-resistant alloy shown in Figure 6-24(b). (a) Determine the equation for the Arrhenius relationship between the rupture time t_f and the operating temperature and (b) determine the rupture time if the alloy operates at 500°C.

35. Suppose the heat-resistant alloy shown in Figure 6-24(b) is to survive for a minimum of

20 years. Calculate the maximum temperature to which the alloy can be heated if the stress is 20,000 psi. (Use the equation developed in Problem 34.)

36. A 0.5-in. diameter bar of a tantalum alloy is originally 10 in. long. After operating at 1315°C for 50 h, its length must be less than 10.5 in. What is the maximum allowable load?

37. A 1.2-in. diameter bar of a tantalum alloy is originally 24 in. long. It operates at a load of 45,000 lb at 1315°C. Assuming it does not fail, what is the length after 8 h?

38. A ductile cast iron bar is to operate at a stress of 4000 psi for 2500 h without failing. What is the maximum allowable temperature?

39. A ductile cast iron bar is to operate at 600°C for 10,000 h. If the bar has a diameter of 0.75 in., what is the maximum allowable load?

40. A 3000-kg load is applied to a 10-mm diameter indentor, producing an impression having a diameter of 2.2 mm on a steel plate. Calculate the Brinell hardness number of the steel, then estimate the tensile strength and endurance limit of the steel.

41. A 500-kg load is applied to a 10-mm diameter indentor, producing an impression on a steel plate having a tensile strength of 75,000 psi. Estimate the diameter of the impression.

42. Estimate the Brinell hardness number of a steel having a tensile strength of 136,000 psi.

43. The fracture toughness of a material is 45,000 psi $\cdot$ in$^{1/2}$. What is the fracture toughness expressed in MPa $\cdot$ m$^{1/2}$?

44. A Ni–Cr steel with a yield strength of 238,000 psi (Table 6-8) contains internal flaws that may be as long as 0.001 in. What is the maximum allowable applied stress if these flaws are not to propagate? How does this compare with the yield strength?

45. An aluminum-copper alloy with a yield strength of 47,000 psi (Table 6-8) contains surface flaws that are 0.0003 in. deep. What is the maximum allowable applied stress if these flaws are not to propagate? How does this compare with the yield strength?

46. A ceramic has a fracture toughness of 2500 psi $\cdot$ in$^{1/2}$ and must experience an applied stress of 15,000 psi. Calculate the size of the maximum allowable (a) internal flaw and (b) surface flaw.

47. A polymer material has a fracture toughness of 1500 psi $\cdot$ in$^{1/2}$ and must experience an applied stress of 4500 psi. Calculate the size of the maximum allowable (a) internal flaw and (b) surface flaw.

47. A titanium alloy (Table 6-8) contains internal flaws that are 0.2 in. in length. Will these flaws propagate if a stress of one-half the yield strength is applied? Consider both a high-strength and a low-strength alloy.

49. A silicon nitride ceramic (Table 6-8) contains surface flaws that are 0.008 in. in length. Will these flaws propagate if a stress of one-half the yield strength is applied?

7

DEFORMATION, STRAIN HARDENING, AND ANNEALING

7–1 INTRODUCTION

In this chapter we will discuss three main topics—*cold working*, by which an alloy is simultaneously deformed and strengthened, *hot working*, by which an alloy is deformed at high temperatures without strengthening, and *annealing*, during which the effects of strengthening caused by cold working are eliminated or modified by heat treatment. The strengthening we obtain during cold working, which is brought about by increasing the number of dislocations, is called *strain hardening* or *work hardening*. By controlling these processes of deformation and heat treatment, we are able to process the material into a usable shape yet still improve and control the properties.

We should point out that the topics discussed in this chapter pertain almost exclusively to metals and alloys. Strain hardening, obtained by multiplication of dislocations, requires that the material have significant ductility. This is not the case for some of the other classes of materials, particularly ceramics. Deformation of thermoplastic polymers often produces a strengthening effect; however, the mechanism of deformation strengthening is quite different in polymers and will be discussed separately in Chapter 15.

Cold Working

7–2   RELATIONSHIP TO THE STRESS-STRAIN CURVE

A stress-strain curve is shown in Figure 7-1(a). As long as the stress does not exceed the yield strength σ_y, no permanent plastic deformation occurs and the elastic deformation is recovered. This is the condition we want to maintain after the finished part is put into service. However, when we wish to manufacture a part by deformation processing, the applied stress must exceed the yield strength, causing the metal to be permanently deformed into a useful shape.

If we apply a stress σ_1 that is greater than the yield strength [Figure 7-1(a)] we cause a permanent deformation, or strain ε_1, when the stress is removed. If we remove a sample from the metal that had been stressed to σ_1 and we retest that metal, we would obtain a different stress-strain curve [Figure 7-1(b)]. Our new test specimen would have a yield strength at σ_1 and would also have a higher tensile strength and a lower ductility. If we continue to apply a stress until we reach σ_2, then release the stress and again retest the metal, the new yield strength would be σ_2. Each time we apply a higher stress to the metal, the yield strength and tensile strength increase while the ductility decreases. We may eventually strengthen the metal until the yield, tensile, and breaking strengths are equal and there is no ductility [Figure 7-1(c)]. At this point the metal can be plastically deformed no further.

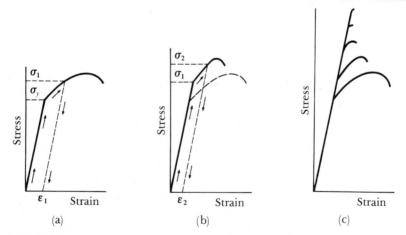

FIGURE 7-1   Development of strain hardening from the stress-strain diagram. (a) A specimen is stressed beyond the yield strength before the stress is removed. (b) Now the specimen has a higher yield strength and tensile strength, but lower ductility. (c) By repeating the procedure, the strength continues to increase and the ductility continues to decrease until the alloy becomes very brittle.

TABLE 7-1   Strain-hardening coefficients of typical metals and alloys

Metal	Crystal Structure	n	K (psi)
Titanium	HCP	0.05	175,000
Annealed alloy steel	BCC	0.15	93,000
Quenched and tempered medium-carbon steel	BCC	0.10	228,000
Molybdenum	BCC	0.13	105,000
Copper	FCC	0.54	46,000
Cu-30% Zn	FCC	0.50	130,000
Austenitic stainless steel	FCC	0.52	220,000

Adapted from G. Dieter, *Mechanical Metallurgy*, McGraw-Hill, 1961, and other sources.

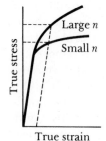

FIGURE 7-2 The true stress–true strain curves for metals with large and small strain-hardening coefficients. Larger degrees of strengthening are obtained for a given strain for the metal with the larger *n*.

By applying a stress that exceeds the original yield strength of the metal, we have *strain hardened* or *cold worked* the metal while simultaneously deforming the metal into a more useful shape.

Strain-Hardening Coefficient.   The response of the metal to cold working is given by the *strain-hardening coefficient n*, which is the slope of the plastic portion of the true stress–true strain curve in Figure 7-2 when a logarithmic scale is used.

$$\sigma_t = K\varepsilon_t^n \qquad (7\text{-}1)$$

or

$$\ln \sigma_t = \ln K + n \ln \varepsilon_t$$

The constant K is equal to the stress when $\varepsilon_t = 1$.

The strain-hardening coefficient is relatively low for HCP metals but is higher for BCC and particularly for FCC metals (Table 7-1). Metals with a low strain-hardening coefficient have a poor response to cold working.

7–3   DISLOCATION MULTIPLICATION

We obtain strengthening during deformation by increasing the number of dislocations in the metal. Before deformation a metal contains about 10^6 cm of dislocation line per cubic centimeter of metal. This is a relatively small number of dislocations.

When we apply a stress greater than the yield strength, dislocations begin to slip. Eventually, a dislocation moving on its slip plane encounters obstacles which pin the ends of the dislocation line. As we continue to apply the stress, the dislocation attempts to move by bowing in the center. The dislocation may move so far that a loop is produced (Figure 7-3). When the dislocation loop finally

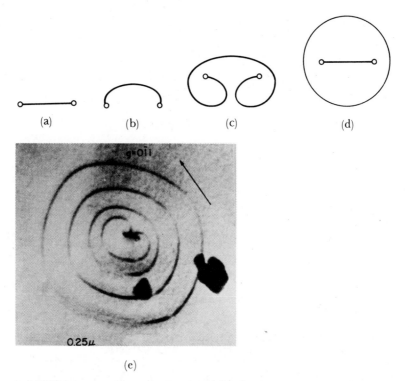

(a)   (b)   (c)   (d)

(e)

FIGURE 7-3 The Frank-Read source can generate dislocations. (a) A dislocation is pinned at its ends by lattice defects. (b) As the dislocation continues to move, the dislocation bows, eventually bending back on itself, (c). Finally the dislocation loop forms (d) and a new dislocation is created. (e) Electron micrograph of a Frank-Read source (× 30,000). From J. Brittain, "Climb Sources in Beta Prime-NiAl," *Metallurgical Transactions,* Vol. 6A, April 1975.

touches itself, a new dislocation is created. The original dislocation is still pinned and can create additional dislocation loops. This mechanism for generating dislocations is called a *Frank-Read source;* Figure 7-3(e) shows an electron micrograph of a Frank-Read source.

The number of dislocations may increase to about 10^{12} cm of dislocation line per cubic centimeter of metal. We know that the more dislocations we have, the more likely they are to interfere with one another, and the stronger the metal becomes.

Ceramic materials may contain dislocations and can even be strain hardened to a small degree. However, ceramics are normally so brittle that significant deformation and strengthening are not possible.

Certain polymers have an excellent response to strengthening by cold working. But the mechanism for strain hardening is quite different in polymers and involves alignment and possibly crystallization of the long, chainlike molecules. We will discuss this mechanism in more detail in Chapter 15.

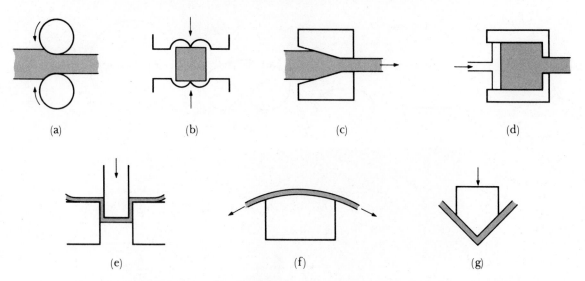

FIGURE 7-4 Schematic drawings of deformation processing techniques. (a) Rolling, (b) forging, (c) drawing, (d) extrusion, (e) deep drawing, (f) stretch forming, and (g) bending.

7-4 PROPERTIES VERSUS PERCENT COLD WORK

Many techniques are used to simultaneously shape and strengthen a metal by cold working (Figure 7-4). By controlling the amount of deformation we achieve using these processes, we control the amount of strain hardening. We normally measure the amount of deformation by defining the percent cold work.

$$\text{Percent cold work} = \frac{A_0 - A_f}{A_0} \times 100 \qquad (7\text{-}2)$$

where A_0 is the original cross-sectional area of the metal and A_f is the final cross-sectional area after deformation.

◇ | **Example 7-1**

Calculate the percent cold work when a 1 cm × 6 cm × 20 cm plate is deformed by rolling to a 0.5-cm plate (Figure 7-5).

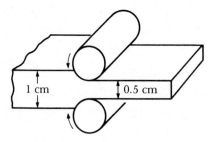

1 cm   0.5 cm

FIGURE 7-5 Diagram showing the rolling of a 1 cm plate to a 0.5 cm plate.

Answer:

The final dimensions of the plate will be 0.5 cm × 6 cm × 40 cm since, due to friction between the plate and the rolls, the width does not change during deformation.

$$\text{Percent cold work} \ = \ \frac{A_0 - A_f}{A_0} \times 100$$

$$= \ \frac{(1)\,(6) \ - \ (0.5)\,(6)}{(1)\,(6)} \times 100 \ = \ 50\%$$

If we measure the properties of the rolled plate, we find that the strength has increased but the ductility has decreased.

The effect of cold work on the mechanical properties of commercially pure copper is shown in Figure 7-6. As the percent cold work increases, both the yield and the tensile strength increase; however, the ductility decreases and approaches zero. The metal will break if more cold work is attempted. Therefore, there is a maximum amount of cold work or deformation that we can do to a metal.

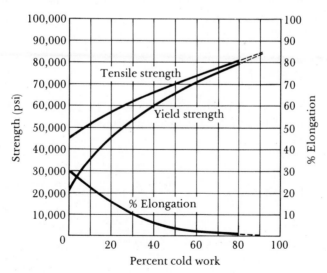

FIGURE 7-6   *The effect of cold work on the mechanical properties of copper.*

  Example 7-2

Determine if a 1 cm × 6 cm × 20 cm copper plate can be cold rolled to a 0.05 cm × 6 cm × 400 cm plate.

Answer:

The amount of cold work required is

$$\text{Percent cold work} = \frac{A_0 - A_f}{A_0} \times 100$$

$$= \frac{(1)(6) - (0.05)(6)}{(1)(6)} \times 100 = 95\%$$

The required cold work exceeds the maximum of 90% cold work for the alloy in Figure 7-6. Therefore, we cannot make the 0.05-cm thick material with a single rolling operation.

Use of the Cold-Working Relationship. We can use the relationship between cold work and mechanical properties to predict the properties of a metal or an alloy if we know the amount of cold work that is achieved during processing. We can then decide whether the component has adequate strength at critical locations.

When we wish to select a material for a component that requires certain minimum mechanical properties, we can design the deformation process. We can first determine the percent cold work that is necessary and then, using the final dimensions we desire, calculate the original metal dimensions from the cold work equation.

 Example 7-3

A copper rod 0.5 in. in diameter is to be reduced to 0.3 in. in diameter. Determine the expected mechanical properties.

Answer:

$$\text{Percent cold work} = \frac{A_0 - A_f}{A_0} \times 100 = \frac{(\pi/4)(d_0^2) - (\pi/4)(d_f^2)}{(\pi/4)(d_0^2)} \times 100$$

$$= \frac{(0.5)^2 - (0.3)^2}{(0.5)^2} \times 100$$

$$= 64\%$$

From Figure 7-6

$$\text{Yield strength} = 73{,}000\,\text{psi}$$
$$\text{Tensile strength} = 75{,}000\,\text{psi}$$
$$\%\ \text{Elongation} = 1.5\%$$

◇ | **Example 7-4**

Suppose we want to reduce a copper plate to a final thickness of 0.1 cm with at least 65,000 psi tensile strength, 60,000 psi yield strength, and 5% elongation. (a) How much cold work is required? (b) Calculate the original plate thickness before rolling.

Answer:

(a) From Figure 7-6, we need at least 35% cold work to produce a tensile strength of 65,000 psi and 40% cold work to produce a yield strength of 60,000 psi, but less than 45% cold work to meet the 5% elongation requirement. Any cold work between 40% and 45% will be satisfactory.

(b) From the cold work equation

$$\text{Percent cold work} \ = \ 40 \ = \ \frac{t_{min.} - 0.1}{t_{min.}} \times 100 \qquad t_{min.} \ = \ 0.167 \, \text{cm.}$$

$$\text{Percent cold work} \ = \ 45 \ = \ \frac{t_{max.} - 0.1}{t_{max.}} \times 100 \qquad t_{max.} \ = \ 0.182 \, \text{cm.}$$

7–5 MICROSTRUCTURE OF COLD-WORKED METALS

During deformation, a fibrous microstructure is produced as the grains within the metal become elongated (Figure 7-7).

Anisotropic Behavior. During deformation, the grains rotate as well as elongate, causing certain crystallographic directions and planes to become aligned. Consequently, preferred orientations, or textures, are developed which cause anisotropic behavior.

In processes such as wire drawing, a *fiber texture* is produced. In BCC metals, $\langle 110 \rangle$ directions line up with the axis of the wire [Figure 7-8(a)]. In FCC metals, $\langle 111 \rangle$ or $\langle 100 \rangle$ directions are aligned. Fortunately, this gives the highest strength along the axis of the wire, which is what we desire. A somewhat similar situation occurs when polymer materials are drawn into fibers; during drawing, the polymer chains change from a random orientation and line up side by side along the length of the fiber. As in metals, the strength is greatest along the axis of the polymer fiber.

In processes such as rolling, both a preferred direction and plane are produced [(Figure 7-8(b)], giving a *sheet texture*. In BCC metals the texture may be $\{100\} \langle 110 \rangle$ and in FCC metals the texture may be $\{112\} \langle 111 \rangle$ or $\{100\} \langle 112 \rangle$.

The properties of a rolled sheet or plate depend on the direction in which we apply the stress (Table 7-2). If the sheet is properly oriented to the applied

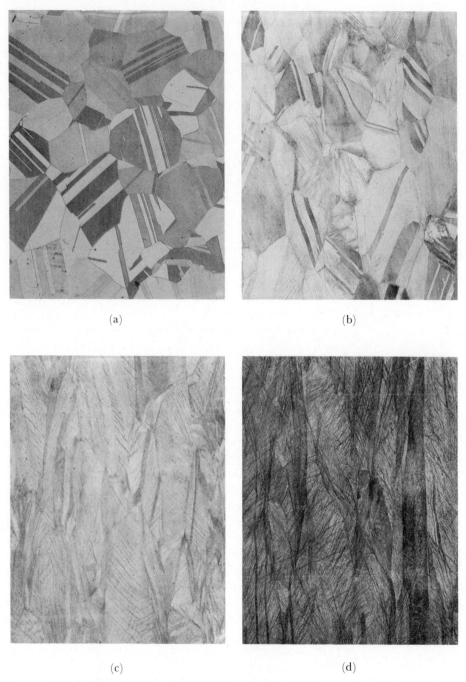

(a)

(b)

(c)

(d)

FIGURE 7-7 The fibrous grain structure of copper after cold working. (a) 0% cold work; (b) 31% cold work; (c) 67% cold work; (d) 82% cold work (× 200).

TABLE 7-2   Anisotropic properties produced by sheet textures during cold rolling.

Metal	Tensile Strength (ksi)			% Elongation		
	0°	45°	90°	0°	45°	90°
Deoxidized Cu						
0% Cold work	31	31	31	50	50	50
50% Cold work	53	53	55	6.0	5.2	5.7
90% Cold work	64	63	67	5.7	5.0	4.0
90% Cu-10% Zn						
0% Cold work	37	37	37	45	45	45
37% Cold work	59	60	63	5	4	3
56% Cold work	74	74	77	4	3	3
95% Cold work	82	87	95	2.7	2.7	3.2

Note: 0° = Parallel to rolling direction; 90° = perpendicular to rolling direction.

Adapted from *ASM Metals Handbook*, Vol. 1, 8th Ed., 1961.

stress during use, high strengths may be achieved. However, premature failure may occur if we apply a stress from a different direction.

Textures, as one would expect, become more pronounced as the amount of deformation increases.

Alignment and Deformation of Second Phases.   Any inclusions (foreign particles such as oxides) or second-phase grains that are present in the original structure are also aligned during deformation. Soft inclusions normally deform and elongate; hard inclusions may not deform but are still aligned in the direction of deformation. Elongated inclusions, called *stringers*, act as tiny internal notches and reduce the mechanical properties of the cold-worked metal (Figure 7-9). When the part is further processed or put into service, we must be sure that high tensile stresses do not act on the sharp inclusions, causing nucleation of cracks.

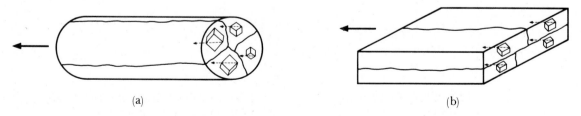

(a)   (b)

FIGURE 7-8   Development of anisotropic behavior in (a) drawn and (b) rolled products. Note the orientation of the unit cells in the grains.

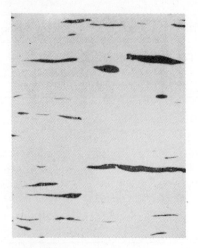

FIGURE 7-9   Elongated iron sulfide inclusions, or stringers, produced during the rolling of a steel (× 250). From *Metals Handbook*, Vol. 9, 9th Ed., American Society for Metals, 1985.

7–6   RESIDUAL STRESSES

Residual stresses, or stresses that are stored within the metal, develop during deformation. When a stress is applied to the metal, a small portion of that stress, perhaps about 10%, is stored internally within the structure as a tangled network of dislocations. The residual stresses increase the total energy of the structure.

The residual stresses are not uniform throughout the deformed metal. For example, high compressive residual stresses may be present at the surface of a rolled plate and high tensile stresses may be stored in the center (Figure 7-10). If we machine a small amount of metal from one surface of a cold-worked part,

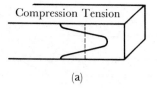

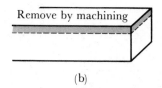

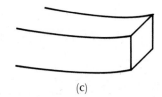

(a)   (b)   (c)

FIGURE 7-10.   (a) Compressive residual stresses at the surface are balanced by tensile stresses in the center of a cold-worked bar. (b) If one surface is machined, part of the compressive stresses in that surface are removed, upsetting the stress balance. (c) To restore the balance, the bar distorts.

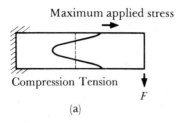
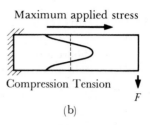

FIGURE 7-11   The compressive residual stresses can be harmful or beneficial. In (a), a bending force applies a tensile stress on the top of the beam. Since there are already tensile residual stresses at the top, the load-carrying characteristics are poor. In (b), the top contains compressive residual stresses. Now the load-carrying characteristics are very good.

we remove metal that contains only compressive residual stresses. To restore the balance, the plate must distort.

Residual stresses also affect the ability of the part to carry a load (Figure 7-11). If a tensile stress is applied to a material that already contains tensile residual stresses, the total stress acting on the part is the sum of the applied and residual stresses. However, if compressive stresses are stored at the surface of a metal part, an applied tensile stress must first balance the compressive residual stresses. Now the part may be capable of withstanding a larger than normal load.

◇ | **Example 7-5**

A cold-worked part has a yield strength of 20,000 psi. Calculate the tensile stress that can be applied without permanent deformation if (a) tensile residual stresses of 10,000 psi and (b) compressive residual stresses of 10,000 psi are present at the surface of the part.

Answer:

(a) If both the residual stress and the applied stress are tensile, the stresses are additive. The applied stress must be less than $20,000 - 10,000 = 10,000$ psi. The part can withstand only half of the expected load due to the presence of the residual stresses.

(b) In this case, the total applied stress that can be supported is $20,000 + 10,000 = 30,000$ psi. In this example, the compressive residual stresses are helpful.

Sometimes components that are subject to fatigue failure can be strengthened by *shot peening*. Bombarding the surface with steel shot propelled at a high velocity introduces compressive residual stresses at the surface; the compressive stresses significantly increase the resistance of the metal surface to fatigue failure.

7-7   CHARACTERISTICS OF COLD WORKING

There are a number of advantages and limitations of strengthening a metal by cold working or strain hardening.

1. We can simultaneously strengthen the metal while producing a desired final shape.

2. We can obtain excellent dimensional tolerances and surface finishes by the cold-working process.

3. The cold-working process is an inexpensive method for producing large numbers of small parts. However, for large parts, the amount of cold work is limited. If too much deformation is attempted, the metal may fail during processing. In addition, high forces, requiring large and expensive forming equipment, must be applied to exceed the yield strength in large parts.

4. Some metals, such as HCP magnesium, are rather brittle at room temperature. Only a small degree of cold working can be accomplished without causing the part to embrittle and fail.

5. Ductility, electrical conductivity, and corrosion resistance are impaired by the cold-working process. However, cold working reduces electrical conductivity less than many of the other strengthening processes, such as solid solution strengthening (Figure 7-12). This makes cold working a more satisfactory way to strengthen conductor materials, such as copper wires used for transmission of electrical power.

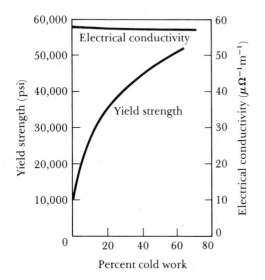
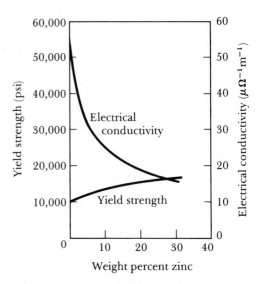

FIGURE 7-12 A comparison of strengthening copper by (a) cold working and (b) solid solution strengthening with zinc. Note that cold working produces a greater strengthening effect yet has little effect on electrical conductivity.

◇ **Example 7-6**

We wish to increase the yield strength of a copper wire from 10,000 psi to 15,000 psi. Compare the change in electrical conductivity if we achieve strengthening by cold working rather than by adding zinc.

Answer:

From Figure 7-12, we can obtain the required yield strength either with 2% cold work or by adding 19% Zn. For 2% cold work, the electrical conductivity is virtually unaffected, but 19% Zn reduces the electrical conductivity from $58 \times 10^6 \Omega^{-1} m^{-1}$ to $20 \times 10^6 \Omega^{-1} m^{-1}$.

 ◇

6. Residual stresses and anisotropic behavior may be introduced during cold working. These characteristics may be either harmful or beneficial, depending on how they can be controlled.

7. Some deformation processing techniques can be accomplished only if cold working occurs. For example, wire drawing requires that a rod be pulled through a die to produce a smaller cross-sectional area (Figure 7-13). For a given draw force F_d, a different stress is produced in the original and final wire. The stress on the initial wire must exceed the yield strength of the metal to cause deformation. The stress on the final wire must be less than its yield strength to prevent failure. This is accomplished only if the wire strain hardens during drawing.

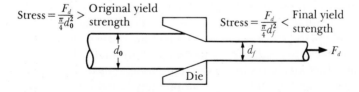

FIGURE 7-13 The wire-drawing process. The force F_d acts on both the original and final diameters. Thus, the stress produced in the final wire is greater than that in the original. If the wire did not strain harden during drawing, the final wire would break before the original wire is drawn through the die.

◇ **Example 7-7**

A copper rod 0.25 in. in diameter is to be drawn through a 0.20-in. diameter die in a wire-drawing process. What is the force required to deform the metal? Is the force sufficient to break the wire after it has been formed?

Answer:

$$\text{Percent cold work} = \frac{A_0 - A_f}{A_0} \times 100 = \frac{(0.25)^2 - (0.20)^2}{(0.25)^2} \times 100 = 36\%$$

The initial yield strength, from Figure 7-6, with 0% cold work is 22,000 psi. The final yield strength with 36% cold work is 58,000 psi. The force required to deform the initial wire is

$$F = \sigma_y A_0 = (22,000)\left(\frac{\pi}{4}\right)(0.25)^2 = 1080\,\text{lb}$$

The stress acting on the wire after passing through the die is

$$\sigma = \frac{F}{A_f} = \frac{1080}{(\pi/4)(0.20)^2} = 34,380\,\text{psi}$$

Since 34,380 psi < 58,000 psi, the wire will not break.

Annealing

7–8 THREE STAGES OF ANNEALING

Annealing is a heat treatment designed to eliminate the effects of cold working and to restore the cold-worked metal to the original soft, ductile condition. Several applications for annealing are employed. First, annealing may be used to completely eliminate the strain hardening achieved during cold working; the final part is soft and ductile but still has a good surface finish and dimensional accuracy. Second, after annealing, additional cold work could be done, since the ductility is restored. By combining repeated cycles of cold working and annealing, large total deformations may be achieved. Finally, annealing at a low temperature may be used to eliminate the residual stresses produced during cold working without affecting the mechanical properties of the finished part.

There are three stages in the annealing process. The effects of cold working and the three stages of annealing on the properties of brass are shown in Figure 7-14.

Recovery. The original cold-worked microstructure is composed of deformed grains containing a large number of tangled dislocations. When we first heat the metal, the additional thermal energy permits the dislocations to move and form the boundaries of a *polygonized* subgrain structure (Figure 7-15). The dislocation density, however, is virtually unchanged. This low-temperature treatment is called *recovery*.

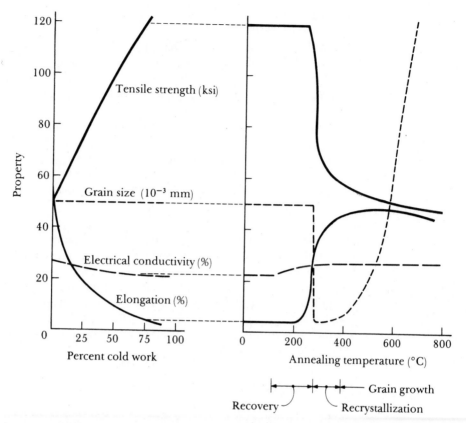

FIGURE 7-14   The effect of cold work and annealing on the properties of a Cu-35% Zn alloy.

Because the number of dislocations is not reduced during recovery, the mechanical properties of the metal are relatively unchanged. However, residual stresses are reduced or even eliminated when the dislocations are rearranged; recovery is often called a *stress relief anneal*. In addition, recovery restores high

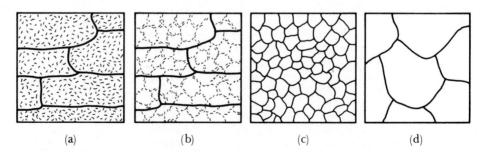

FIGURE 7-15   The effect of annealing temperature on the microstructure of cold-worked metals. (a) Cold worked, (b) after recovery, (c) after recrystallization, and (d) after grain growth.

electrical conductivity to the metal; combining strain hardening with recovery permits us to produce copper or aluminum wire for transmission of electrical power that is strong yet still has high conductivity. Finally, recovery often improves the resistance of the material to corrosion.

Recrystallization. *Recrystallization* occurs by the nucleation and growth of new grains containing few dislocations. When the metal is heated above the *recrystallization temperature*, approximately 0.4 times the absolute melting temperature of the metal, rapid recovery eliminates residual stresses and produces the polygonized dislocation structure. New grains then nucleate at the cell boundaries of the polygonized structure, eliminating most of the dislocations (Figure 7-15). Because the number of dislocations is greatly reduced, the recrystallized metal has a low strength but a high ductility.

Grain Growth. At still higher annealing temperatures, both recovery and recrystallization rapidly occur, producing the fine recrystallized grain structure. However, the energy associated with the large amount of grain boundary area makes the fine structure unstable at high temperatures. To reduce this energy, the grains begin to grow, with favored grains consuming the smaller grains (Figure 7-15). This phenomenon is called *grain growth* and was described in Chapter 5. Grain growth, illustrated for a copper–zinc alloy in Figure 7-16, is almost always undesirable.

◇ | **Example 7-8**

From the data in Table 7-3, determine the temperatures at which recovery, recrystallization, and grain growth begin in a Cu-12.5% Zn alloy.

TABLE 7-3 The effect of annealing temperature on the properties of a Cu-12.5% Zn alloy. (See Example 7-8.)

Annealing Temperature (°C)	Grain Size (mm)	Tensile Strength (psi)	% Elongation	Electrical Conductivity ($\times 10^6 \, \Omega^{-1} \, m^{-1}$)
25	0.100	80,000	5	16
100	0.100	80,000	5	16
150	0.100	80,000	5	17
200	0.100	80,000	5	19
250	0.100	80,000	5	20
300	0.005	75,000	9	20
350	0.008	55,000	30	21
400	0.012	48,000	40	21
500	0.018	40,000	48	21
600	0.025	39,000	48	22
700	0.050	38,000	47	22

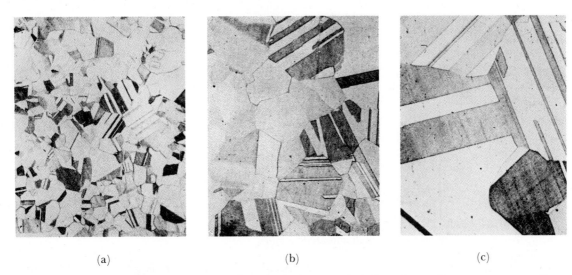

(a)   (b)   (c)

FIGURE 7-16   Photomicrographs showing the effect of annealing temperature on the grain size in brass. Twin boundaries can also be observed in the structures. (a) Annealed at 400°C, (b) annealed at 650°C, (c) annealed at 800°C (× 75). From R. Brick and A. Phillips, *The Structure and Properties of Alloys,* McGraw-Hill, 1949.

Answer:

The start of recovery is indicated by an increase in the electrical conductivity, or at a temperature between 150°C and 200°C.

The start of recrystallization is indicated by a decrease in grain size and tensile strength and an increase in % elongation. These changes begin between 300°C and 350°C.

Grain growth is indicated by the change in grain size. Grain growth occurs at an accelerated rate near 600°C.

7–9   CONTROL OF ANNEALING

There are three important factors that must be considered when selecting an annealing heat treatment—the recrystallization temperature, the size of the recrystallized grains, and the grain growth temperature.

Recrystallization Temperature.   The recrystallization temperature is affected by a variety of processing variables.

1. The recrystallization temperature decreases when the amount of cold work increases. Greater amounts of cold work make the metal less stable and

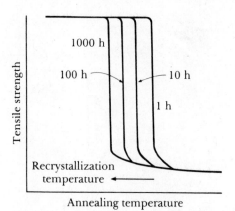

FIGURE 7-17 *The effect of annealing time on the recrystallization temperature.*

encourage nucleation of recrystallized grains. There is a minimum amount of cold work, about 30% to 40%, below which recrystallization will not occur.

2. A smaller original cold-worked grain size also reduces the recrystallization temperature by providing more sites—the former grain boundaries—at which new grains can nucleate.

3. Pure metals recrystallize at lower temperatures than solid solution-strengthened alloys. Often this is helpful. For example, alloys that are to be brazed or soldered may resist annealing and softening during the joining operation more effectively than would a pure metal.

4. Increasing the annealing time reduces the recrystallization temperature (Figure 7-17), since more time is available for nucleation and growth of the new recrystallized grains. However, temperature is far more important. Doubling the annealing time reduces the recrystallization temperature by only about 10°C.

5. Higher melting point alloys have a higher recrystallization temperature. Since recrystallization is a diffusion-controlled process, the recrystallization temperature is roughly proportional to $0.4 T_m$ (absolute). Typical recrystallization temperatures for selected metals are shown in Table 7-4.

Recrystallized Grain Size.   A number of factors also influence the size of the recrystallized grains. Reducing the annealing temperature, the time required to heat to the annealing temperature, or the annealing time will reduce the grain size by minimizing the opportunity for grain growth. Increasing the initial cold work also reduces the final grain size by providing a greater number of nucleation sites for new grains. Finally, the presence of a second phase in the microstructure helps prevent grain growth and keeps the recrystallized grain size small.

Grain Growth Temperature.   Heat treatment times can be reduced, often at considerable economic savings, by performing the heat treatment at higher temperatures. However, care must be taken to prevent grain growth. This

TABLE 7-4   Typical recrystallization temperatures for selected metals

Metal	Melting Temperature (°C)	Recrystallization Temperature (°C)
Sn	232	< Room temperature
Cd	321	< Room temperature
Pb	327	< Room temperature
Zn	420	< Room temperature
Al	660	150
Mg	650	200
Ag	962	200
Au	1064	200
Cu	1085	200
Fe	1538	450
Pt	1769	450
Ni	1453	600
Mo	2610	900
Ta	2996	1000
W	3410	1200

Adapted from R. Brick, A. Pense, and R. Gordon, *Structure and Properties of Engineering Materials*, McGraw-Hill, 1977.

applies to other types of heat treatments besides annealing. Occasionally, inclusions are deliberately introduced into an alloy in order to pin the grain boundaries and prevent grain growth, even at higher than normal temperatures.

7–10   INFLUENCE OF ANNEALING ON MATERIAL PROCESSING

The effects of recovery, recrystallization, and grain growth are important in the processing and eventual use of a metal or an alloy.

Deformation Processing.   By taking advantage of the annealing heat treatment, we can increase the amount of deformation we can accomplish. If we are required to reduce a 5-in. thick plate to a 0.05-in. thick sheet, we can do the maximum permissible cold work, anneal to restore the metal to its soft, ductile condition, then cold work again. We can repeat the cold work–anneal cycle repeatedly until we approach the proper thickness. The final cold-working step can be designed to produce the final dimensions and properties we require (Example 7-9).

Annealing also reduces the anisotropic behavior introduced by cold working; however, *annealing textures*, or preferred orientations, still persist. If an annealed sheet containing the annealing texture is to be deep drawn into a beverage can, certain segments of the sheet will deform more than other segments. This anisotropic deformation causes local thinning of the metal. On the other hand, the annealing texture is useful in silicon-iron sheet material used for electrical transformers. The alloy magnetizes most readily in the (001) directions, which fortunately correspond to the annealing texture in the alloy.

High-Temperature Service.   Neither strain hardening nor grain size strengthening would be appropriate for an alloy that is to be used at elevated temperatures, as in creep-resistant applications. Suppose we were to produce a strong, fine-grained structure by cold working the metal. When the metal is placed into service at a high temperature, recrystallization immediately causes a catastrophic decrease in strength. In addition, if the temperature were high enough, the strength would continue to decrease due to growth of the newly recrystallized grains. We would need to utilize different strengthening mechanisms for high-temperature applications.

Joining Processes.   When we join a cold-worked metal using a welding process, the metal adjacent to the weld may heat above the recrystallization and grain growth temperatures. This region is called the *heat-affected zone*. The structure and properties in the heat-affected zone of a weld are shown in Figure 7-18. The properties are catastrophically reduced by the heat of the welding process. Welding processes such as electron-beam welding or laser welding, which provide high rates of heat input for brief times, minimize the exposure of the metal to temperatures above recrystallization and minimize this type of damage.

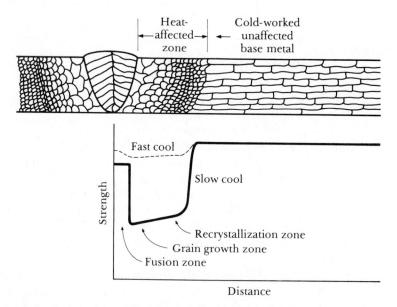

FIGURE 7-18 *The structure and properties surrounding a fusion weld in a cold-worked metal. Note the loss in strength due to recrystallization and grain growth in the heat-affected area.*

◇ | **Example 7-9**

Design the processing sequence required to reduce a 1 cm × 6 cm copper strip to a 0.1 cm × 6 cm strip having greater than 60,000 psi yield strength and 5% elongation.

Answer:

We determined in Example 7-4 that to obtain the required properties, a cold work of 40% to 45% is required. An intermediate thickness of 0.167 cm to 0.182 cm must be obtained before this final cold work. Let's check to see if we can cold work from 1 cm to the intermediate thickness.

$$\text{Percent cold work} = \frac{t_0 - t_f}{t_0} \times 100 = \frac{1 - 0.182}{1} \times 100 = 81.8\%$$

$$\text{Percent cold work} = \frac{t_0 - t_f}{t_0} \times 100 = \frac{1 - 0.167}{1} \times 100 = 83.3\%$$

A maximum of 90% cold work is permitted for the copper in Figure 7-6. Thus, our complete process is as follows.

(a) Cold work the 1-cm strip, which must originally be in the annealed condition, 81.8% to 83.3% to an intermediate thickness of 0.167 cm to 0.182 cm.

(b) Anneal above the recrystallization temperature to eliminate the strengthening caused by the large amount of cold work.

(c) Cold work 40% to 45% from 0.167 cm or 0.182 cm to the final dimension of 0.1 cm. This gives the correct final dimensions and properties.

Hot Working

7–11   CHARACTERISTICS OF THE HOT-WORKING PROCESS

We can deform a metal into a useful shape by hot working rather than cold working the metal. *Hot working* is defined as plastically deforming the metal at a temperature above the recrystallization temperature. During hot working, the metal is continually recrystallized (Figure 7-19). Hot working provides several advantages and introduces several limitations when compared with cold working.

Lack of Strengthening.   No strengthening occurs during deformation by hot working; consequently, the amount of plastic deformation is almost unlimited.

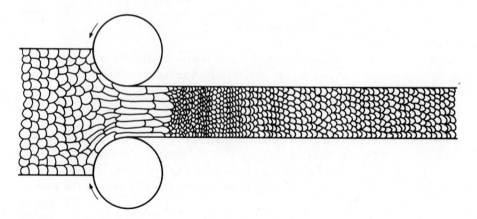

FIGURE 7-19   During hot working, the elongated, anisotropic grains immediately recrystallize. If the hot-working temperature is properly controlled, the final hot-worked grain size can be very fine.

A very thick plate can be reduced to a thin sheet in a continuous series of operations. The first steps in the process are carried out well above the recrystallization temperature to take advantage of the lower strength of the metal. The last step is performed just above the recrystallization temperature, using a large percent deformation, in order to produce the finest possible grain size. We could also use hot working to perform the bulk of the deformation process, followed by a final cold-working step to produce the final dimensions and strain hardening (Example 7-10).

Hot working is well suited for forming large parts, since the metal has a low yield strength and high ductility at elevated temperatures. In addition, HCP metals such as magnesium have more active slip systems at hot-working temperatures; the higher ductility permits larger deformations than are possible by cold working.

Elimination of Imperfections.   Some imperfections in the original metal may be eliminated or their effects minimized. Gas pores can be closed and welded shut during hot working—the internal lap formed when the pore is closed is eliminated by diffusion during the forming and cooling process. Composition differences in the metal can be reduced as hot working brings the surface and center of the plate closer together, thereby reducing diffusion distances.

Hot isostatic pressing (HIP), which is often used to consolidate metal or ceramic powders into a nearly finished state, can also be used to minimize imperfections. The pressure for deformation is introduced by application of a gas or fluid (Figure 7-20); porosity in castings, powder metallurgy parts, or ceramic components may be eliminated in this manner.

Anisotropic Behavior.   The final properties in hot-worked parts are not isotropic. The forming rolls or dies, which are normally at a lower temperature than the metal, cool the surface more rapidly than the center of the part. The

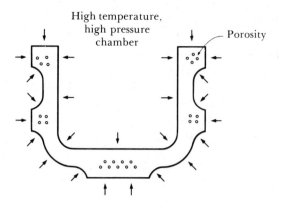

FIGURE 7-20 The hot isostatic pressing (HIP) process can be used to seal imperfections in a material.

surface then has a finer grain size than the center. In addition, a fibrous structure is produced because inclusions and second-phase particles are elongated in the working direction. Finally, a texture, or preferred orientation, similar to annealing textures may be produced.

Surface Finish and Dimensional Accuracy. The surface finish is usually poorer than that obtained by cold working. Oxygen may react with the metal at the surface, forming oxides, which are forced into the surface during forming. In some metals, such as tungsten and beryllium, hot working must be done in a protective atmosphere.

The dimensional accuracy is also more difficult to control during hot working. A greater elastic strain must be considered, since the modulus of elasticity is low at hot-working temperatures. In addition, the metal contracts as it cools from the hot-working temperature. The combination of elastic strain and thermal contraction requires that the part be made oversized during deformation; forming dies must be carefully designed and precise temperature control is necessary if accurate dimensions are to be obtained.

◇ | **Example 7-10**

Describe a process by which a 10 cm × 6 cm copper strip can be reduced to 0.1 cm × 6 cm, having a yield strength of 60,000 psi and 5% elongation.

Answer:

From the results obtained in the previous examples, the quickest way to produce the finished product is (a) hot work the strip from 10 cm to between 0.167 cm and 0.182 cm, then (b) cold work the strip from between 0.167 cm and 0.182 cm to the final 0.1 cm, using between 40% and 45% cold work.

◇

7–12   DEFORMATION BONDING PROCESSES

Both cold working and hot working are used to join or weld metals. *Deformation bonding* requires that (a) impurities on the joining surfaces be removed or broken up into discrete particles by the deformation process and (b) the pressure applied be sufficient to bring the atoms on each surface into intimate contact. By satisfying these two requirements, a large contact area is produced at the mating surfaces across which atom-to-atom attraction causes bonding to occur.

In *cold indentation welding* [Figure 7-21(a)], two thin sheets of metal are joined as the sheets are deformed between punches. The strength of the bond may be

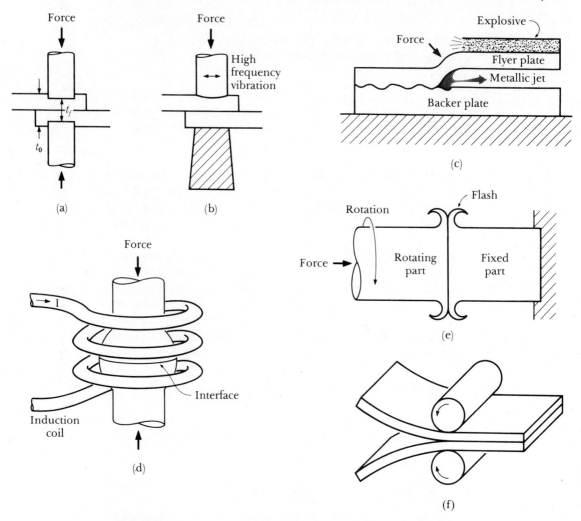

FIGURE 7-21   Schematic diagrams of typical deformation bonding processes. (a) Cold indentation welding, (b) ultrasonic welding, (c) explosive bonding, (d) hot pressure welding, (e) friction or inertia welding, and (f) roll bonding.

greater than the surrounding base material, even though the joint is reduced in thickness, because of cold working of the joint during welding. However, when too much deformation is attempted, the joint becomes too thin to support a large applied load.

Most deformation-bonding processes use high temperatures to assist in the bonding process. *Ultrasonic welding* of thin sheets of metal resembles the cold indentation process [Figure 7-21(b)]. However, the vibrating motion caused by an ultrasonic probe breaks up oxides with much less deformation and thinning of the joint than the cold indentation process. The vibrating action, often at a frequency of 20,000 Hz or more, may also raise the temperature of the joint to the recrystallization temperature. The metals joined in *explosive bonding* also begin at room temperature; the intense energy produced during explosive bonding strips away surface impurities and forces the surfaces together at high pressures [Figure 7-21(c)]. But the energy also increases the temperature at the bonded interface, often up to the melting temperature of the metal.

Hot pressure bonding and *friction* or *inertia* welding rely on localized heating and deformation for the joining process [Figure 7-21(d) and (e)]. An induction coil might be used to supply the heat for hot pressure bonding. Heat develops in friction or inertia welding due to the frictional forces between the rotating surfaces. Often *roll bonding* [Figure 7-21(f)] uses elevated temperatures to provide bonding of sheet or plate material. The manufacture of laminar composite materials, including U.S. coinage, is a typical application of roll bonding.

7–13 SUPERPLASTIC FORMING

Some alloys, when specially heat treated and processed, can be uniformly deformed an exceptional amount, perhaps as much as 1000%; this behavior is called *superplasticity*. These parts can be formed into very complicated shapes in one or only a few forming dies. Common superplastic alloys include Ti-6% Al-4% V and Zn-23% Al. Several conditions are required for an alloy to display superplastic behavior.

1. The metal must have a very fine grain structure, with grain diameters being less than about 0.005 mm.

2. The alloy must be deformed at a high temperature, often near 0.5 to 0.65 times the absolute melting point of the alloy.

3. A very slow rate of forming, or *strain rate*, must be employed. In addition, the stress required to deform the alloy must be very sensitive to the strain rate. If necking begins to occur, the necked region strains at a higher rate; the higher strain rate in turn strengthens the necked region, stops the necking, and the uniform deformation continues [Figure 7-22(a)].

4. The grain boundaries in the alloy should allow grains to easily slide over one another and rotate when a stress is applied [Figure 7-22(b)]. The proper temperature and a fine grain size are necessary for this to occur.

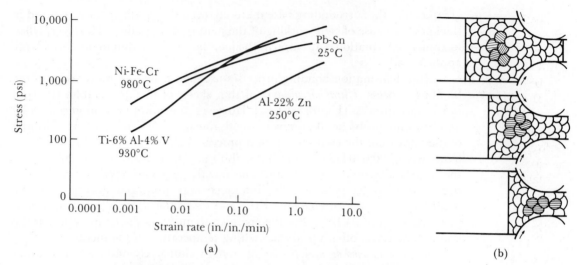

FIGURE 7-22 *Superplastic deformation. (a) The sensitivity of the metal's deformation to strain rate prevents necking and permits large deformations. (b) Grain boundary sliding rather than distortion of the grains may occur.*

SUMMARY

The properties of materials, and in particular of metals, can be controlled by combinations of deformation and simple heat treatments. When a metal is deformed by cold working, strain hardening occurs because additional dislocations are introduced into the structure. Very large increases in strength can be obtained in this manner. This process combines a materials processing technique with a materials strengthening technique.

However, because ductility is simultaneously reduced, the amount of strain hardening is limited. Furthermore, residual stresses, which often are detrimental, may be introduced. By annealing, we can eliminate all or a portion of the effects of strain hardening. A low-temperature recovery treatment causes stress relief without reducing strength; a higher temperature recrystallization treatment eliminates all of the effects of strain hardening. Excessive temperatures, however, cause grain growth.

Deformation and annealing can be combined in one step—hot working—to improve our materials processing ability. Large changes in shape are possible at high temperatures since the material is not strain hardened. Combining hot working and cold working may permit us to achieve both processing of the material into a useful shape and controlled improvement of properties.

GLOSSARY

Annealing. A heat treatment used to eliminate part or all of the effects of cold working.

Cold indentation welding. One method of joining metals by a cold-working process. The two surfaces are squeezed together between two punches, bringing the two surfaces into atom-to-atom contact by extensive deformation.

Cold working. Deformation of a metal below the recrystallization temperature. During cold working, the number of dislocations increases, causing the metal to be strengthened as its shape is changed.

Deformation bonding. A group of materials joining techniques by which the two surfaces are forced together at high pressures and often high temperatures. The pressure breaks up impurities at the surfaces and brings the materials into atom-to-atom contact, permitting bonding to occur.

Explosive bonding. A deformation bonding technique by which the high pressures and temperatures are produced by the detonation of a layer of explosive spread on one of the surfaces.

Fiber texture. A preferred orientation obtained in drawing processes by which grains are preferentially elongated in the drawing direction. Certain crystallographic directions in each grain also line up with the drawing direction, causing anisotropic behavior.

Frank-Read source. A pinned dislocation which, under an applied stress, produces additional dislocations. This mechanism is at least partly responsible for strain hardening.

Friction welding. A group of deformation bonding processes, which includes inertia welding, by which two surfaces are heated by friction caused as the parts rotate against one another and are finally joined when the heated surfaces are forced together under high pressures.

Heat-affected zone. The area adjacent to a weld that is heated during the welding process above some critical temperature at which a change in the structure, such as grain growth or recrystallization, occurs.

Hot pressure bonding. Producing a weld by forcing two heated metal surfaces into intimate contact at a high pressure.

Hot working. Deformation of a metal above the recrystallization temperature. During hot working only the shape of the metal changes; the strength remains relatively unchanged because no strain hardening occurs.

Polygonized structure. A subgrain structure produced in the early stages of annealing. The subgrain boundaries are a network of dislocations rearranged during heating.

Preferred orientation. An alignment of grains, inclusions, or other microstructural features in a particular direction or plane in a material as a result of its processing.

Recovery. A low-temperature annealing heat treatment designed to eliminate residual stresses introduced during deformation without reducing the strength of the cold-worked material.

Recrystallization. A medium-temperature annealing heat treatment designed to eliminate all of the effects of the strain hardening produced during cold working. Recrystallization must be accomplished above the recrystallization temperature.

Recrystallization temperature. The temperature above which the effects of strain hardening are eliminated during annealing. The recrystallization temperature is not a constant for a material but depends on the amount of cold work, the annealing time, and other factors.

Residual stresses. Stresses introduced in the material during processing which, rather than causing deformation of the material, remain stored in the structure. Later release of stresses as deformation can be a problem.

Sheet texture. A preferred orientation obtained in rolling processes. The grains are aligned so that a preferred crystallographic direction rotates parallel to the rolling direction and a preferred crystallographic plane rotates parallel to the sheet surface.

Shot peening. Introducing compressive

residual stresses at the surface of a part by bombarding that surface with steel shot. The residual stresses may improve the overall performance of the material.

Strain hardening. Strengthening of a material by increasing the number of dislocations by deformation, or cold working. Also known as work hardening.

Strain-hardening coefficient. The effect that strain has on the resulting strength of the material. A material with a high strain-hardening coefficient obtains high strength with only small amounts of deformation or strain.

Strain rate. The rate at which a material is deformed. A material may behave much differently if it is slowly pressed into a shape rather than smashed rapidly into a shape by an impact blow.

Stress relief anneal. The recovery stage of the annealing heat treatment, during which residual stresses are relieved without reducing the mechanical properties of the material.

Stringer. Inclusions that are deformed or aligned with the direction of deformation during hot or cold working.

Superplasticity. The ability of a material to deform uniformly by an exceptionally large amount. Careful control over temperature, grain size, and strain rate are required for a material to behave in a superplastic manner.

Ultrasonic welding. A special deformation welding technique in which the load that forces the two surfaces together is partly introduced by a very high frequency vibration. Bonding is achieved with very little total deformation.

PRACTICE PROBLEMS

1. A 0.505-in. diameter bar with a gage length of 2 in. is subjected to a tensile test. When a force of 12,000 lb is applied, the specimen has a diameter of 0.497 in., and a gage length of 2.007 in. When a force of 23,000 lb is applied, the diameter is 0.426 in. and the gage length is 2.354 in. (a) Determine the strain-hardening coefficient. (b) Would you expect the metal to have the FCC structure? Explain.

2. A copper-nickel alloy tensile bar, originally having a 0.505-in. diameter and a 2-in. gage length, has a strain-hardening coefficient of 0.48. The bar fails at an engineering stress of 98,000 psi; its gage length at the time of failure is 2.26 in. and its diameter is 0.475 in.; no necking occurred. Calculate the true stress on the bar when the engineering strain was 0.05 in./in.

3. The Frank-Read source in Figure 7-3(e) has created four dislocation loops from the original dislocation line. Estimate the total dislocation line present in the photograph and determine the percent increase in the length of dislocations produced by the deformation.

4. Suppose we begin with a copper plate 3 in. thick. Using Figure 7-6, calculate the total percent cold work if (a) we reduce the plate to 0.5 in. thick and (b) we reduce the plate first to 1 in., then later to 0.5 in.

5. Suppose we begin with an aluminum bar 3 in. thick. Using Figure 7-23, calculate the total percent cold work if (a) we reduce the bar to a 0.5-in. diameter and (b) we reduce the bar first to a 1-in. diameter, then later to a 0.5-in. diameter.

6. Calculate (a) the percent cold work and (b) the final properties if we reduce a 3105 aluminum plate from an original thickness of 2 in. to a final thickness of 0.25 in. (See Figure 7-23.)

7. Calculate (a) the percent cold work and (b) the final properties if we reduce a Cu-30% Zn brass bar from an original diameter of 1 in. to a final diameter of 0.28 in. (See Figure 7-24.)

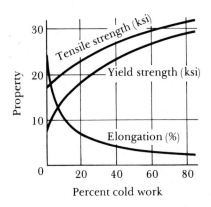

FIGURE 7-23 The effect of percent cold work on the properties of a 3105 aluminum alloy.

8. Calculate (a) the total percent cold work and (b) the final properties if we reduce a copper plate from 1.5 in. to 1.0 in. to 0.75 in. to 0.60 in. in three passes through a rolling mill. (See Figure 7-6.)

9. We wish to produce a 3105 aluminum rod having a final yield strength of at least 25,000 psi and a final diameter of 0.225 in. Using Figure 7-23, calculate the minimum original diameter required.

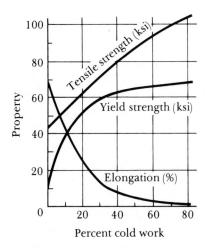

FIGURE 7-24 The effect of percent cold work on the properties of a Cu-30% Zn brass.

10. We wish to produce a Cu-30% Zn brass plate having a final % elongation of at least 20% and a final thickness of 0.375 in. Using Figure 7-24, calculate the maximum original thickness required.

11. We would like to produce a copper plate 0.25 in. thick having at least 10% elongation and a 45,000 psi yield strength. Is this possible? If so, calculate (a) the required percent cold work and (b) the original thickness of the plate. (See Figure 7-6.)

12. We would like to produce a Cu-30% Zn wire 0.10 in. in diameter with at least 20% elongation and a 60,000 psi tensile strength. Is this possible? If so, calculate (a) the required percent cold work and (b) the original diameter of the wire. (See Figure 7-24.)

13. We would like to produce a 3105 aluminum wire 0.05 in. in diameter with at least 10% elongation and a 25,000 psi tensile strength. Is this possible? If so, calculate (a) the required percent cold work and (b) the original diameter of the wire. (See Figure 7-23.)

14. We would like to produce a Cu-30% Zn wire 0.2 in. in diameter having at least 10% elongation and a 70,000 psi tensile strength. The original diameter of the bar is 4 in. The maximum allowable cold work is 75%. Describe the steps required, including percent cold work and intermediate thicknesses. (See Figure 7-24.)

15. We would like to produce a copper sheet 0.15 in. thick from an original plate that is 1.8 in. thick. The final required properties include at least a 50,000 psi yield strength and 5% elongation. The maximum allowable cold work per pass is 80%. Describe the steps required, including percent cold work and intermediate thicknesses. (See Figure 7-6.)

16. A 0.20-in. titanium wire is passed through a 0.18-in. diameter die in a wire-drawing process, producing a wire with a 60,000 psi yield strength and an 85,000 psi tensile strength. If the modulus of elasticity of the titanium is 16×10^6 psi, calculate the diameter of the final wire product.

17. A 0.060-in. magnesium wire with a yield strength of 25,000 psi is to be produced by a wire-drawing process. If the modulus of elasticity of the magnesium is 6.5×10^6 psi, calculate the necessary diameter of the opening in the die.

18. We plan to draw a 0.25-in. diameter Cu-30% Zn wire having a yield strength of 10,000 psi into a 0.20-in. diameter wire. (a) Using Figure 7-24, calculate the draw force, assuming no friction. (b) Will the draw force cause the drawn wire to break? (Prove by calculating the maximum force that the drawn wire can withstand.)

19. A 3105 aluminum wire 0.10 in. in diameter is to be made having a tensile strength of 25,000 psi. Using Figure 7-23, determine (a) the original diameter of the wire, (b) the required draw force, and (c) whether the as-drawn wire will survive the drawing process.

20. Successful wire drawing requires that strain hardening occur during deformation; what other process(es) in Figure 7-4 might also require strain hardening?

21. (a) From the data below, estimate the recovery, recrystallization, and grain growth temperatures. (b) Recommend a suitable temperature for a stress relief heat treatment. (c) Recommend a suitable temperature for a hot-working process. (d) Estimate the melting temperature of the alloy.

Annealing Temperature (°C)	Electrical Conductivity ($\times 10^5 \Omega^{-1} cm^{-1}$)	Yield Strength (psi)	Grain Size (in.)
200	0.85	120,000	0.004
400	0.86	120,000	0.004
600	1.08	120,000	0.004
800	1.24	120,000	0.004
1000	1.25	90,000	0.0015
1200	1.25	84,000	0.0015
1400	1.26	82,000	0.0025
1600	1.26	80,000	0.0065
2000	1.26	79,000	0.015

22. (a) From the data below, estimate the recovery, recrystallization, and grain growth temperatures. (b) Recommend a suitable temperature for a stress relief heat treatment. (c) Recommend a suitable temperature for a hot-working process. (d) Estimate the melting temperature of the alloy.

Annealing Temperature (°C)	Residual Stresses (psi)	Yield Strength (psi)	Grain Size (mm)
100	45,000	80,000	0.20
200	45,000	80,000	0.20
300	40,000	80,000	0.20
400	0	80,000	0.20
500	0	45,000	0.06
600	0	38,000	0.06
700	0	37,000	0.08
800	0	37,000	0.18

23. Aluminum is often added to liquid steel, causing deoxidation by producing tiny Al_2O_3 inclusions which are dispersed uniformly throughout the steel after solidification. What effects will the alumina particles have on the recrystallized grain size and the temperature at which grain growth will become a problem?

24. From the photomicrographs in Figure 7-16, estimate the ASTM grain size numbers and plot the grain size number versus the annealing temperature.

25. Using the data in Table 7-4, plot the recrystallization temperature versus the melting temperature of each metal, using absolute temperatures. Measure the slope and compare with the expected relationship between these two temperatures. Is our approximation a good one?

26. When copper is joined to nickel by a roll bonding process, a minimum of 85% deformation is required. Measure the thickness of a quarter, then estimate the minimum combined original thickness of the copper and nickel alloy sheets from which the quarter was produced.

27. Two sheets of annealed aluminum alloy 3105, each 0.10 in. thick, are to be joined by cold indentation welding. A total deformation of 67% is required at the joint to obtain bonding. (a) What is the final thickness of the joint? (b) What is the final yield strength of the material in the joint? (c) Estimate the force that each individual sheet could withstand and the force that the joint could withstand if the force was applied parallel to the surface of the sheets. (Refer to Figure 7-23.)

28. We plan to produce a cold indentation weld in annealed copper alloy (Figure 7-6) sheet material. Each sheet is 0.2 in. thick. The final thickness of the joint is 0.15 in. (a) Estimate the percent cold work done in the joining process. (b) Estimate the yield strength in the original material and in the final joint.

29. Suppose two sheets of aluminum alloy 3105 were to be joined by cold indentation welding. What effect, if any, would there be on the joining process if the aluminum sheet had been cold worked 50% prior to joining compared with the originally annealed sheet? (Refer to Figure 7-23.)

30. Annealing of a nickel alloy requires 1 h at 620°C. If you wanted to complete the annealing process in 15 min, what temperature would you recommend?

31. You would like to produce the following products. For each one, tell whether you would recommend hot working, cold working, or a combination of cold working and annealing; tell which, if any, of the processes shown in Figure 7-4 would be suitable; and explain your choice.

(a) Paper clips

(b) I-beams that will be welded to produce a portion of a bridge

(c) Copper tubing that will connect a water faucet to the main copper plumbing

(d) The steel tape in a tape measure

(e) A head for a carpenter's hammer formed from a round rod

32. What deformation bonding process would you recommend for each of the following? Explain your choices.

(a) Joining a 4-in. diameter steel shaft to a 4-in. diameter stainless steel shaft

(b) Sealing an electronic device in a protective box made from 0.02-in. thick stainless steel

(c) Joining a 1-in. thick steel plate to a 0.25-in. thick titanium plate, where the length and width of the plates are each 6 ft.

8

SOLIDIFICATION AND GRAIN SIZE STRENGTHENING

8–1 INTRODUCTION

In almost all metals and alloys, as well as in many semiconductors, composites, ceramics, and polymers, the material at one point in its processing is a liquid. The liquid then solidifies as it cools below the freezing temperature. The material may be used in the as-solidified condition or may be further processed by mechanical working or heat treatment. The structures produced during the solidification process affect the mechanical properties and influence the type of further processing needed to achieve the required properties. In particular, the grain size and shape may be controlled by solidification.

In this chapter we will introduce the fundamental principles of solidification, concentrating on the behavior of pure materials. In subsequent chapters we will see how solidification differs in alloys and multiple-phase materials.

8–2 NUCLEATION

During solidification, the atomic arrangement changes from at best a short-range order to a long-range order, or crystal structure. Solidification requires two steps—nucleation and growth. *Nucleation* occurs when a small piece of solid forms from the liquid. The solid must achieve a certain minimum critical size before it is stable. *Growth* of the solid occurs as atoms from the liquid are attached to the tiny solid until no liquid remains.

We expect a material to solidify when the liquid cools to just below the freezing temperature because the energy associated with the crystalline structure

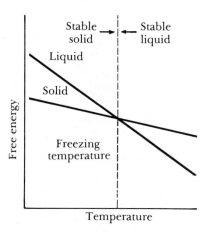

FIGURE 8-1   The volume free energy versus temperature for a pure metal. Below the freezing temperature, the solid has a lower free energy and is stable.

of the solid is then less than the energy of the liquid. As the temperature falls further below the freezing temperature, the energy difference becomes larger, making the solid even more stable (Figure 8-1). We might refer to this energy difference as the *volume free energy* ΔF_v.

However, in order for the solid to form, an interface must be created separating the solid from the liquid (Figure 8-2). A *surface free energy* σ is associated with this interface; the larger the surface, the greater the increase in surface energy. When the liquid cools to the freezing temperature, atoms in the liquid cluster together to produce a small region that resembles the solid material. This small solid particle is called an *embryo*. The total change in free energy produced when the embryo forms is the sum of the decrease in volume free energy and the increase in surface free energy.

$$\Delta F = \tfrac{4}{3}\pi r^3 \Delta F_v + 4\pi r^2 \sigma \qquad\qquad (8\text{-}1)$$

where $\tfrac{4}{3}\pi r^3$ is the volume of a spherical embryo of radius r, $4\pi r^2$ is the surface area of a spherical embryo, σ is the surface free energy, and ΔF_v is the volume free energy, which is a negative change.

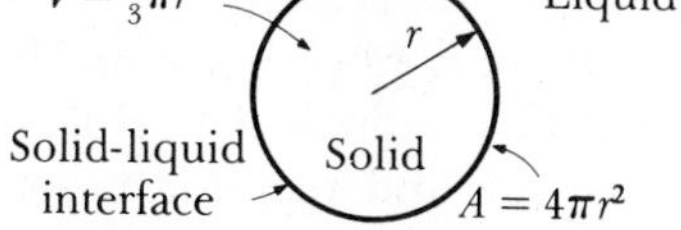

$V = \tfrac{4}{3}\pi r^3$ Liquid

Solid-liquid interface   Solid r $A = 4\pi r^2$

FIGURE 8-2   An interface is created when a solid forms from the liquid.

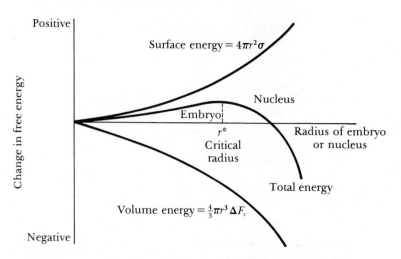

FIGURE 8-3 The total free energy of the solid-liquid system changes with the size of the solid. The solid is an embryo if its radius is less than the critical radius and is a nucleus if its radius is greater than the critical radius.

The total change in the free energy depends on the size of the embryo (Figure 8-3). If the embryo is very small, further growth of the embryo would cause the free energy to increase. Instead of growing, the embryo remelts and causes the free energy to decrease. Thus, the metal remains liquid. Since the liquid is present below the equilibrium freezing temperature, the liquid is undercooled. The *undercooling* is the equilibrium freezing temperature minus the actual temperature of the liquid. Nucleation has not occurred and growth cannot begin, even though the temperature is below the equilibrium freezing temperature!

If the embryo is large, the total energy decreases when the size of the embryo increases. The solid that now forms is stable, nucleation has occurred, and growth of the solid particle, which is now called a *nucleus*, begins.

Nucleation only occurs when enough atoms spontaneously cluster together to produce a solid with a radius greater than the *critical radius* r^*, corresponding to the maximum on the total free energy curve.

Homogeneous Nucleation. As the temperature of the liquid cools further below the equilibrium freezing temperature, there is a greater probability that atoms will cluster together to form an embryo larger than the critical radius. In addition, at larger undercoolings there is a larger volume free energy difference between the liquid and the solid; this reduces the critical size of the nucleus. *Homogeneous nucleation* occurs when the undercooling becomes large enough to permit the embryo to exceed the critical size.

We can estimate the size of the critical nucleus if we differentiate the total free energy equation. The differential with respect to r is zero when $r = r^*$, since the free energy curve is then at a maximum.

TABLE 8-1 Values for freezing temperature, latent heat of fusion, surface energy, and maximum undercooling for selected metals

Metal	Freezing Temperature (°C)	Latent Heat of Fusion (J/cm^3)	Surface Energy (ergs/cm^2)	Maximum Undercooling Observed (°C)
Ga	30	488	56	76
Bi	271	543	54	90
Pb	327	237	33	80
Ag	962	965	126	250
Cu	1085	1628	177	236
Ni	1453	2756	255	480
Fe	1538	1737	204	420

Adapted from B. Chalmers, *Principles of Solidification*, John Wiley & Sons, 1964.

$$\frac{d}{dr}(\Delta F) = \frac{d}{dr}(\tfrac{4}{3}\pi r^3 \Delta F_v + 4\pi r^2 \sigma) = 0$$

$$4\pi r^{*2} \Delta F_v + 8\pi r^* \sigma = 0$$

$$r^* = \frac{-2\sigma}{\Delta F_v} \qquad (8\text{-}2)$$

The volume free energy is given by the expression

$$\Delta F_v = \frac{-\Delta H_f \Delta T}{T_m} \qquad (8\text{-}3)$$

where ΔH_f is the latent heat of fusion of the metal, T_m is the equilibrium freezing temperature in K, and $\Delta T = T_m - T$ is the undercooling when the liquid temperature is T. The *latent heat of fusion* represents the heat that is given up during the liquid-solid transformation. By combining Equations (8-2) and (8-3)

$$r^* = \frac{2\sigma T_m}{\Delta H_f \Delta T} \qquad (8\text{-}4)$$

As the undercooling increases, the critical radius required for nucleation decreases. Table 8-1 presents values for σ and ΔH_f for selected metals. As an approximation, homogeneous nucleation occurs when

$$\Delta T = 0.2 T_m (\text{K}) \qquad (8\text{-}5)$$

◇ | **Example 8-1**

The freezing temperature of pure copper is 1085°C. Estimate the undercooling required for homogeneous nucleation.

Answer:

$$\Delta T = 0.2 T_m = (0.2)(1085 + 273) = 272°C$$

Undercoolings of this magnitude are never observed in the normal processing of molten copper.

 Example 8-2

Calculate the size of the critical radius and the number of atoms in the critical nucleus when solid copper forms by homogeneous nucleation.

Answer:

$$\Delta T = 0.2 T_m = 272°C, \quad T_m = 1358 \, K$$
$$\Delta H_f = 1628 \, J/cm^3 = 1628 \times 10^7 \, ergs/cm^3$$
$$\sigma = 177 \, ergs/cm^2$$
$$r^* = \frac{2\sigma T_m}{\Delta H_f \Delta T} = \frac{(2)(177)(1358)}{(1628 \times 10^7)(272)} = 10.85 \times 10^{-8} \, cm$$

The lattice parameter for FCC copper is $a_0 = 3.615 \, \text{Å} = 3.615 \times 10^{-8} \, cm$

$$V_{\text{unit cell}} = (a_0)^3 = (3.615 \times 10^{-8})^3 = 47.24 \times 10^{-24} \, cm^3$$
$$V_{r*} = \tfrac{4}{3}\pi r^3 = (\tfrac{4}{3}\pi)(10.85 \times 10^{-8})^3 = 5350 \times 10^{-24} \, cm^3$$

The number of unit cells in the critical nucleus is

$$\frac{5350 \times 10^{-24}}{47.24 \times 10^{-24}} = 113 \, \text{unit cells}$$

Since there are four atoms in each unit cell of FCC metals, the number of atoms in the critical nucleus must be

$$(4 \, atoms/cell)(113 \, cells/nucleus) = 452 \, atoms/nucleus$$

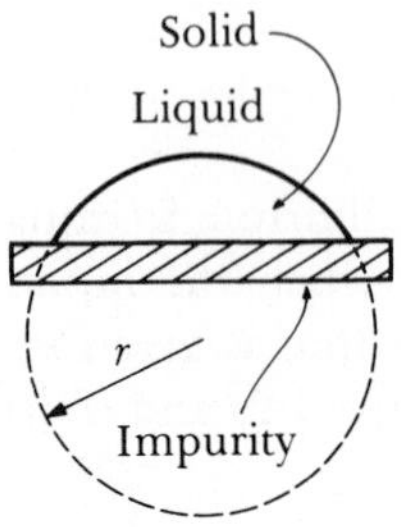

Solid
Liquid

r

Impurity

FIGURE 8-4 A solid forming on an impurity can assume the critical radius with a smaller increase in the surface energy. Thus, heterogeneous nucleation can occur with relatively low undercoolings.

Heterogeneous Nucleation. Except in unusual laboratory experiments, homogeneous nucleation never occurs in liquid metals. Instead impurities in contact with the liquid, either suspended in the liquid or on the walls of the container that holds the liquid, provide a surface on which the solid can form (Figure 8-4). Now, a radius of curvature greater than the critical radius is achieved with very little total surface between the solid and liquid. Only a few atoms must cluster together to produce a solid particle that has the required radius of curvature. Much less undercooling is required to achieve the critical size, so nucleation occurs more readily. Nucleation on impurity surfaces is known as *heterogeneous nucleation*.

Grain Size Strengthening by Nucleation.   Sometimes we may intentionally introduce impurity particles into the liquid. Such practices in metals are called *grain refinement* or *inoculation*. For example, a combination of 0.02% to 0.05% titanium and 0.01% to 0.03% boron is added to many liquid aluminum alloys. Solid titanium boride particles form and serve as effective sites for heterogeneous nucleation. The grain refining or inoculation procedure produces a large number of grains, each grain beginning to grow from one nucleus. The greater grain boundary surface area more effectively blocks slip, or movement of dislocations, and provides *grain size strengthening*.

Glasses.   In extreme cases of very rapid cooling, nucleation of the crystalline solid never occurs. Instead an unstable amorphous, or noncrystalline, solid forms. A short-range order of the atoms in this solid gives the structure a *glassy* appearance.

In many ceramic and polymer materials, nucleation of the solid crystalline structure is prevented at normal or even slow cooling rates. The ability to produce ceramic and polymer glasses by relatively simple and economical manufacturing processes gives us the transparent materials we need for so many uses.

However in metals, cooling rates of 10^6 °C/s or faster are required to suppress nucleation of the crystal structure. This rapid cooling rate can be obtained by directing tiny droplets of the molten metal onto a chilled copper surface; when the droplets hit the surface, they spread out as a thin film and cool rapidly. Originally, this process was called "splat" cooling. More recently, the production of metallic glasses, as well as other unique structures, by rapid cooling has been termed *rapid solidification processing*. The high cooling rates are obtained by producing metal powder particles by atomization or by forming continuous thin metallic ribbons, about 0.0015 in. in thickness.

Metallic glasses include complex iron-nickel-boron alloys containing chromium, phosphorus, cobalt, and other elements. The metal glasses may obtain strengths in excess of 500,000 psi while retaining fracture toughness of more than 10,000 psi · in$^{0.5}$. Excellent corrosion resistance, magnetic properties, and other physical properties make these materials attractive for certain electrical applications, including transformer cores.

8–3   GROWTH

Once solid nuclei have formed, growth occurs as atoms are attached to the solid surface. In pure metals, the nature of the growth of the solid during solidification depends on how heat is removed from the solid-liquid system. Two types of heat must be removed—the specific heat of the liquid and the latent heat of fusion. The *specific heat* is the heat required to change the temperature of a unit weight of the material by one degree. The specific heat must be removed first, either by radiation into the surrounding atmosphere or by conduction into the surrounding mold, until the liquid cools to the freezing temperature. The latent heat of

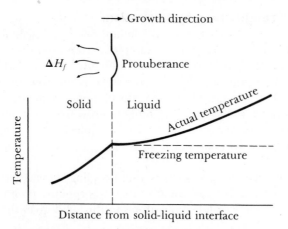

FIGURE 8-5   When the temperature of the liquid is above the freezing temperature, a protuberance on the solid-liquid interface will remelt, leading to maintenance of a planar interface. Latent heat is removed from the interface through the solid.

fusion, which represents the energy that is evolved as the disordered liquid structure transforms to a more stable crystal structure, must be removed from the solid-liquid interface before solidification is completed. The manner in which we remove the latent heat of fusion determines the growth mechanism and final structure.

Planar Growth.   Let's suppose that a well-inoculated liquid cools slowly, under equilibrium conditions. The temperature of the liquid metal is greater than the freezing temperature and the temperature of the solid is at or below the freezing temperature. The latent heat of fusion must be removed by conduction from the solid-liquid interface through the solid to the surroundings for solidification to continue. Any small protuberance that begins to grow on the interface is surrounded by liquid metal above the freezing temperature (Figure 8-5). The growth of the protuberance then stops until the remainder of the interface catches up. This growth mechanism, known as *planar growth*, occurs by the movement of a smooth solid-liquid interface into the liquid.

Dendritic Growth.   When nucleation is poor, the liquid undercools to a temperature below the freezing temperature before the solid forms (Figure 8-6). Under these conditions, a small solid protuberance called a *dendrite*, which forms at the interface, is encouraged to grow. As the solid dendrite grows, the latent heat of fusion is conducted into the undercooled liquid, raising the temperature of the liquid towards the freezing temperature. Secondary and tertiary dendrite arms can also form on the primary stalks to speed the evolution of the latent heat. Dendritic growth continues until the undercooled liquid warms to the freezing temperature. Any remaining liquid then solidifies by planar growth. The difference between planar and dendritic growth arises because of the

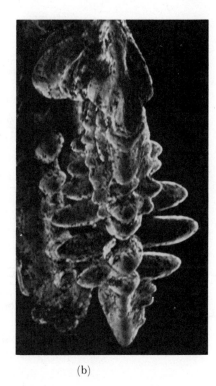

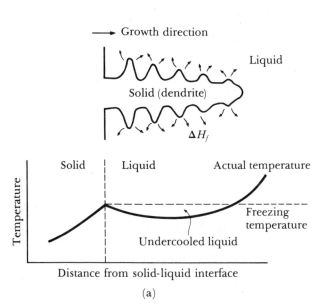

Growth direction

Liquid

Solid (dendrite)

ΔH_f

Solid | Liquid   Actual temperature

Freezing temperature

Undercooled liquid

Temperature

Distance from solid-liquid interface

(a)   (b)

FIGURE 8-6   (a) If the liquid is undercooled, a protuberance on the solid-liquid interface can rapidly grow as a dendrite. The latent heat of fusion is removed by raising the temperature of the liquid back to the freezing temperature. (b) Scanning electron micrograph of dendrites in steel (× 15).

different sinks for the latent heat. The container or mold must absorb the heat in planar growth, but the undercooled liquid absorbs the heat in dendritic growth.

In pure metals, dendritic growth normally represents only a small fraction of the total growth.

$$\text{Dendritic fraction} \ = \ f \ = \ \frac{c\Delta T}{\Delta H_f} \tag{8-6}$$

where c is the specific heat of the liquid. The numerator represents the heat that the undercooled liquid can absorb, and the latent heat in the denominator represents the total heat that must be given up during solidification. As the undercooling ΔT increases, more dendritic growth occurs.

◇ **Example 8-3**

Calculate the fraction of growth that occurs dendritically in copper which (a) nucleates homogeneously and (b) nucleates heterogeneously with 10°C undercooling.

Answer:

The latent heat of fusion for copper is $1628\,\text{J/cm}^3$, the specific heat is $4.4\,\text{J/cm}^3 \cdot {}^\circ\text{C}$, and, from Example 8-1, the undercooling for homogeneous nucleation is $272\,{}^\circ\text{C}$.

(a) For homogeneous nucleation

$$f = \frac{c\Delta T}{\Delta H_f} = \frac{(4.4)\,(272)}{1628} = 0.735$$

(b) For $10\,{}^\circ\text{C}$ undercooling

$$f = \frac{(4.4)\,(10)}{1628} = 0.027$$

8–4 SOLIDIFICATION TIME

The rate at which growth of the solid occurs during solidification depends on the cooling rate, or rate of heat extraction. A fast cooling rate produces rapid solidification or short solidification times. The time required for a simple casting to solidify completely can be calculated using *Chvorinov's rule*.

$$t_s = B\left(\frac{V}{A}\right)^2 \tag{8-7}$$

where t_s is the time required for the casting to solidify, V is the volume of the casting, A is the surface area of the casting in contact with the mold, and B is a *mold constant*. The mold constant depends on the properties and initial temperatures of both the metal and the mold. Almost always, a shorter solidification time produces a finer grain size and a stronger casting.

◇ | **Example 8-4**

Two castings are produced under identical conditions. One casting has the dimensions $2\,\text{cm} \times 8\,\text{cm} \times 16\,\text{cm}$, and the dimensions of the second casting are $3\,\text{cm} \times 6\,\text{cm} \times 8\,\text{cm}$. Which casting will be stronger?

Answer:

The casting that freezes in the shortest time should have the higher strength. From Chvorinov's rule

$$t_{s_1} = B\left[\frac{(2)\,(8)\,(16)}{(2)\,(2)\,(8) + (2)\,(2)\,(16) + (2)\,(8)\,(16)}\right]^2$$

$$= B\left(\frac{256}{352}\right)^2 = 0.53B$$

$$t_{s_2} = B\left[\frac{(3)(6)(8)}{(2)(3)(6) + (2)(3)(8) + (2)(6)(8)}\right]^2$$

$$= B\left(\frac{144}{180}\right)^2 = 0.64B$$

Because $t_1 < t_2$, the $2\,\text{cm} \times 8\,\text{cm} \times 16\,\text{cm}$ casting freezes faster and is stronger.

The solidification time also affects the size of the dendrites that grow. Normally, the dendrite size is characterized by measuring the distance between the secondary dendrite arms (Figure 8-7). The *secondary dendrite arm spacing*, or *SDAS*, is reduced when the casting freezes more rapidly. Because there is less time available to transfer heat, additional dendrite arms develop and grow to assist with the evolution of the latent heat. The finer, more extensive dendritic network serves as a more efficient conductor of the latent heat to the undercooled liquid. The *SDAS* is related to the solidification time by

$$SDAS = kt_s^n \tag{8-8}$$

where n and k are constants depending on the composition of the metal. This relationship is shown in Figure 8-8 for several alloys. Small secondary dendrite arm spacings are associated with higher strengths and improved ductility (Figure 8-9).

Rapid solidification processing is used to produce exceptionally fine secondary dendrite arm spacings; a common method is to produce very fine liquid droplets using special atomization processes. The tiny droplets freeze at a rate of about $10^4\,°\text{C/s}$. This cooling rate is not rapid enough to form a metallic glass but does produce the tiny dendritic structure shown in Figure 8-10. By carefully

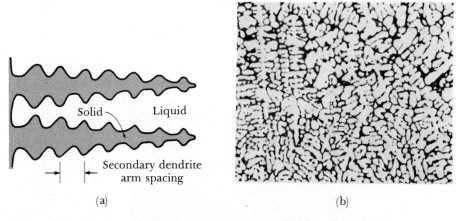

FIGURE 8-7 (a) The secondary dendrite arm spacing *SDAS*. (b) Dendrites in an aluminum alloy ($\times$ 50). From *Metals Handbook*, Vol. 9, 9th Ed., American Society for Metals, 1985.

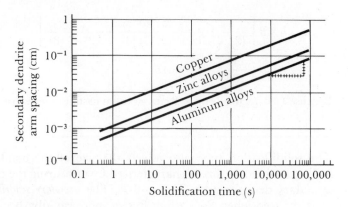

FIGURE 8-8   The effect of solidification time on the secondary dendrite arm spacings of copper, zinc, and aluminum.

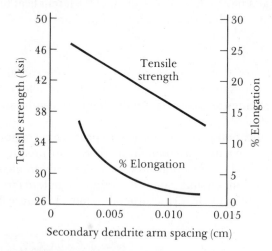

FIGURE 8-9   The effect of the secondary dendrite arm spacing on the properties of an aluminum casting alloy.

FIGURE 8-10   Tiny dendrites exposed within a titanium powder particle produced by rapid solidification processing (× 2200). From J. D. Ayers and K. Moore, "Formation of Metal Carbide Powder by Spark Machining of Reactive Metals," *Metallurgical Transactions*, Vol. 15A, June 1984, p. 1120.

consolidating the solid droplets by powder metallurgy processes, improved properties in the material can be obtained.

◇ **Example 8-5**

Determine the constants in the equation that describes the relationship between secondary dendrite arm spacing and solidification time for aluminum alloys (Figure 8-8).

Answer:

We can obtain the slope n on a log-log plot by measuring the slope on the graph. In Figure 8-8, five equal units are marked on the vertical scale and 12 equal units are marked on the horizontal scale. The slope is

$$n = \tfrac{5}{12} = 0.42$$

The constant k is the value of $SDAS$ when $t_s = 1$, since

$$\log SDAS = \log k + n \log t_s$$

If $t_s = 1$, $n \log t_s = 0$, and $SDAS = k$, from Figure 8-8

$$k = 8 \times 10^{-4} \, \text{cm}$$

◇ **Example 8-6**

Calculate the $SDAS$ and the tensile strength you expect to find at the center of a 1 in. × 8 in. × 12 in. aluminum casting. The mold constant in Chvorinov's rule for aluminum alloys is 45 min/in^2.

Answer:

$$V_{\text{casting}} = (1)(8)(12) = 96 \, \text{in}^3$$
$$A_{\text{casting}} = (2)(1)(8) + (2)(1)(12) + (2)(8)(12) = 232 \, \text{in}^2$$

From Chvorinov's rule

$$t_s = B\left(\frac{V}{A}\right)^2 = (45)\left(\frac{96}{232}\right)^2 = 7.7 \, \text{min} = 462 \, \text{s}$$

From Example 8-5

$$SDAS = (8 \times 10^{-4})t_s^{0.42} = (8 \times 10^{-4})(462)^{0.42}$$
$$= 105 \times 10^{-4} \, \text{cm} = 0.0105 \, \text{cm}$$

From Figure 8-9

$$\text{Tensile strength} = 39,000 \, \text{psi}$$

8–5   COOLING CURVES

We can summarize our discussion to this point by examining a cooling curve, or how the temperature of the metal changes with time (Figure 8-11). The liquid metal is poured into a mold at the *pouring temperature*. The difference between the pouring temperature and the freezing temperature is the *superheat*. The liquid metal cools as the specific heat of the liquid is extracted by the mold until the liquid reaches the freezing temperature. The slope of the cooling curve before solidification begins is the *cooling rate* $\Delta T/\Delta t$.

If effective heterogeneous nuclei are present in the liquid metal, solidification begins at the freezing temperature, as shown in Figure 8-11(a). A *thermal arrest*, or plateau, is produced because of the evolution of the latent heat of fusion. The latent heat keeps the remaining liquid at the freezing temperature until all of the liquid has solidified and no more heat can be evolved. Growth under these

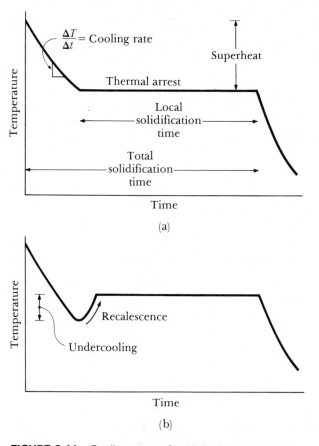

FIGURE 8-11   Cooling curves for (a) liquids that nucleate with no undercooling and (b) liquids that require large undercoolngs for nucleation.

conditions is planar. The *total solidification time* of the casting is the time required to remove both the specific heat of the superheated liquid and the latent heat of fusion. This is measured from the time of pouring until solidification is complete and is given by Chvorinov's rule. The *local solidification time* is the time required to remove only the latent heat of fusion at a particular location in the casting and is measured from when solidification begins until solidification is completed.

If undercooling develops due to poor nucleation, the cooling curve dips below the freezing temperature, as shown in Figure 8-11(b). After the solid finally nucleates, dendritic growth occurs. The latent heat, however, is absorbed by the undercooled liquid, raising the temperature of the liquid back to the freezing temperature. This phenomenon is known as *recalescence*. After the temperature of the remaining liquid is raised to the freezing temperature, a thermal arrest occurs until solidification is completed by planar growth.

8–6 CASTING OR INGOT STRUCTURE

Molten metals are poured into molds and permitted to solidify. Often the mold produces a finished shape, or casting. In other cases, the mold produces a simple shape, called an *ingot*, that requires extensive plastic deformation or machining before a finished product is created.

In either a casting or an ingot, a *macrostructure* is produced which can consist of as many as three parts (Figure 8-12).

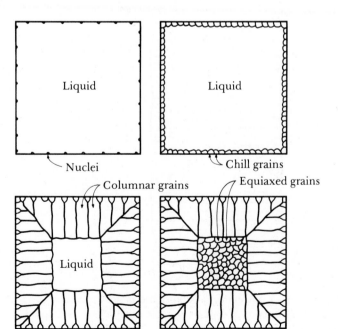

FIGURE 8-12

Development of the macrostructure of a casting during solidification.
(a) Nucleation begins.
(b) The chill zone forms.
(c) Preferred growth produces the columnar zone. (d) Additional nucleation creates the equiaxed zone.

Chill Zone.   The *chill zone* is a narrow band of randomly oriented grains at the surface of the casting. The metal at the mold wall is the first to cool to or below the freezing temperature. The mold wall also provides many surfaces at which heterogeneous nucleation may take place. Therefore, a large number of grains begin to nucleate and grow.

Columnar Zone.   The *columnar zone* contains elongated grains oriented in a particular crystallographic direction. As heat is removed from the casting by the mold material, the grains in the chill zone begin to grow in the direction opposite to the heat flow, or from the coldest towards the hottest areas of the casting. This usually means that the grains grow perpendicular to the mold wall.

Grains grow fastest in certain crystallographic directions. In metals with a cubic crystal structure, grains in the chill zone that have a $\langle 100 \rangle$ direction perpendicular to the mold wall grow faster than other less favorably oriented grains (Figure 8-13). Eventually, the grains in the columnar zone have $\langle 100 \rangle$ directions that are parallel to one another, giving the columnar zone anisotropic properties.

The formation of the columnar zone is influenced primarily by growth, rather than nucleation, phenomena. The grains may be composed of many dendrites if the liquid is originally undercooled. Or solidification may proceed by planar growth of the columnar grains if no undercooling has occurred.

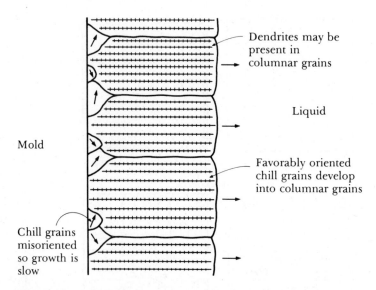

FIGURE 8-13 Competitive growth of the grains in the chill zone results in only those grains with favorable orientations developing into columnar grains.

Equiaxed Zone.   In most cases, a pure metal continues to grow in a columnar manner until all of the liquid has solidified. However, in alloys and in special circumstances in pure metals, an equiaxed zone forms in the center of the casting or ingot. The *equiaxed zone* contains new, randomly oriented grains, often caused by a low pouring temperature, alloying elements, or grain refining or inoculating agents. These grains grow as relatively round, or equiaxed, grains with a random orientation and stop the growth of the columnar grains. The formation of the equiaxed zone is a nucleation-controlled process and causes that portion of the casting to have isotropic behavior.

8–7 SOLIDIFICATION DEFECTS

Although there are a large number of potential defects that can be produced during solidification, two deserve special mention.

Shrinkage.   Almost all materials are more dense in the solid state than in the liquid state (Figure 8-14). During solidification, the material contracts, or shrinks, as much as 7% (Table 8-2).

If the shrinkage is unidirectional (Figure 8-15), only one dimension of the solid casting would be smaller than the dimensions of the mold. The mold could then be made oversized by the appropriate amount in order to compensate for the shrinkage.

However, in most situations, the bulk of the shrinkage occurs as *cavities*, if solidification begins at all surfaces of the casting, or as *pipes*, if one surface solidifies more slowly than the others. In either case, the casting is defective. A common technique for controlling cavity and pipe shrinkage is to place a *riser*, or an extra reservoir of metal, adjacent and connected to the casting, (Figure 8-16). As the casting solidifies and shrinks, liquid metal flows from the riser into the casting to fill the shrinkage void. We need only assure that the riser freezes

TABLE 8-2 Shrinkage during solidification for selected materials

Material	Shrinkage (%)
Al	7.0
Cu	5.1
Mg	4.0
Zn	3.7
Fe	3.4
Pb	2.7
Ga	+ 3.2 (expansion)
H_2O	+ 8.3 (expansion)

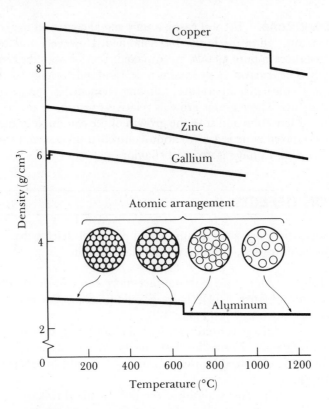

FIGURE 8-14   The density of selected metals versus temperature. Most metals have a higher density as solids than as liquids and thus contract during solidification. Note that gallium has the opposite behavior.

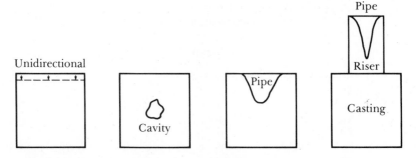

FIGURE 8-15   Several types of macroshrinkage can occur including (a) unidirectional, (b) cavity, and (c) pipe. (d) Risers can be used to help compensate for shrinkage.

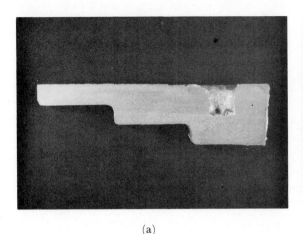
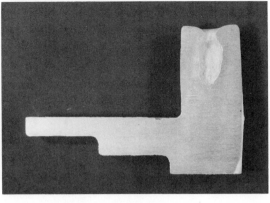

(a)   (b)

FIGURE 8-16 Sections through an aluminum casting. (a) Because no riser is used, concentrated shrinkage is present in the thick part of the casting. (b) Shrinkage is contained in the riser, thus producing a sound casting.

after the casting and that there is an internal liquid channel that connects the liquid in the riser to the last liquid to solidify in the casting. Chvorinov's rule can be used to help design the size of the riser.

◇ | **Example 8-7**

A cylindrical riser with a height equal to twice its diameter is to compensate for shrinkage in a $2\,\mathrm{cm} \times 8\,\mathrm{cm} \times 16\,\mathrm{cm}$ casting (Figure 8-17). Estimate the minimum size of the riser.

Answer:

We know that the riser must freeze after the casting.

$$t_{\text{riser}} > t_{\text{casting}} \quad \text{so} \quad B\left(\frac{V}{A}\right)^2_r > B\left(\frac{V}{A}\right)^2_c$$

$$\left(\frac{V}{A}\right)_r > \left(\frac{V}{A}\right)_c$$

$$V_c = (2)(8)(16) = 256$$

$$A_c = (2)(2)(8) + (2)(2)(16) + (2)(8)(16) = 352$$

$$V_r = \frac{\pi}{4} D^2 H = \frac{\pi}{4} D^2 (2D) = \frac{\pi}{2} D^3$$

$$A_r = 2\left(\frac{\pi}{4} D^2\right) + \pi D H = 2\left(\frac{\pi}{4} D^2\right) + \pi D (2D) = \tfrac{5}{2}\pi D^2$$

$$\frac{(\pi/2)(D)^3}{(5\pi/2)(D)^2} > \frac{256}{352}$$

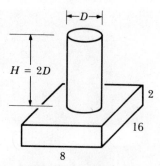

FIGURE 8-17   Geometry of casting for Example 8-7.

$$\frac{D}{5} > 0.727$$

$$D > 3.64\,\text{cm}$$

$$H > 7.28\,\text{cm}$$

$$V_r > 75.75\,\text{cm}^3$$

Although the volume of the riser is much smaller than that of the casting, the riser freezes more slowly due to its compact shape.

Interdendritic shrinkage is found when extensive dendritic growth occurs (Figure 8-18). Liquid metal may be unable to flow from a riser through the fine dendritic network to the solidifying metal. Consequently, small shrinkage pores are produced throughout the casting. This defect, also called *microshrinkage* or *shrinkage porosity*, is difficult to prevent by the use of risers. Fast cooling rates may reduce problems with interdendritic shrinkage; the dendrites may be shorter, permitting liquid to flow through the dendritic network to the solidifying solid interface. In addition, any shrinkage that remains may be finer and more uniformly distributed.

Gas Porosity.   Many metals dissolve a large quantity of gas when they are liquid. Aluminum, for example, dissolves hydrogen. However, when the aluminum solidifies, the solid metal retains in its structure only a small fraction of the hydrogen (Figure 8-19). The excess hydrogen forms bubbles that may be trapped in the solid metal, producing *gas porosity*. The porosity may be spread uniformly throughout the casting or may be trapped between dendrite arms.

The amount of gas that can be dissolved in the molten metal is given by Sievert's law.

$$\text{Percent of gas} \;=\; K\sqrt{p_{\text{gas}}} \tag{8-9}$$

where p_{gas} is the partial pressure of the gas in contact with the metal and K is a

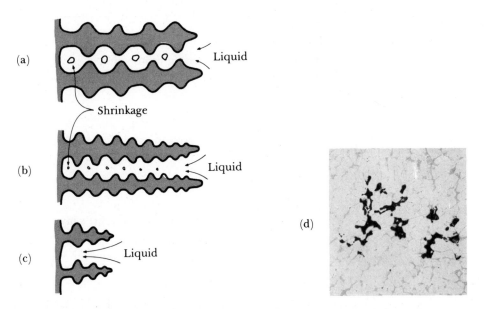

FIGURE 8-18 (a) Shrinkage can occur between the dendrite arms. Small secondary dendrite arm spacings result in smaller, more evenly distributed shrinkage porosity (b), while short primary arms can help avoid shrinkage (c). (d) Interdendritic shrinkage in an aluminum alloy (× 80).

constant which, for a particular metal-gas system, increases with increasing temperature. We can minimize gas porosity in castings by keeping the liquid temperature low, by adding materials to the liquid to combine with the gas and form a solid, or by assuring that the partial pressure of the gas remains low. The latter may be achieved by placing the molten metal in a vacuum chamber or bubbling an inert gas through the metal. Because p_{gas} is low in the vacuum or inert gas, the gas leaves the metal, enters the vacuum or inert gas, and is carried away.

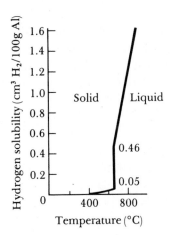

FIGURE 8-19 The solubility of hydrogen gas in aluminum. Although solid aluminum contains very little hydrogen, the amount of hydrogen that dissolves in liquid aluminum is large and increases rapidly with temperature.

Example 8-8

After melting at atmospheric pressure, molten copper contains 0.01% O. How much oxygen would remain if the molten copper were placed in a vacuum at 10^{-6} atm?

Answer:

The ratio of the partial pressure of oxygen before and during the vacuum treatment will be the same as the ratio of the total pressures. Thus

$$\frac{p_{\text{initial}}}{p_{\text{vacuum}}} = \frac{1}{10^{-6}} = 10^6$$

By forming a ratio, the constant K in Sievert's law cancels.

$$\frac{\% \ O_{\text{initial}}}{\% \ O_{\text{vacuum}}} = \frac{K\sqrt{p_{\text{initial}}}}{K\sqrt{p_{\text{vacuum}}}} = \sqrt{\frac{p_{\text{initial}}}{p_{\text{vacuum}}}} = \sqrt{10^6}$$

$$\% \ O_{\text{vacuum}} = \frac{\% \ O_{\text{initial}}}{\sqrt{10^6}} = (0.01)(10^{-3}) = 1 \times 10^{-5}\%$$

Oxygen is often dissolved in liquid steel during the steel-making process. During solidification, the dissolved oxygen combines with carbon, which is present as an alloying element, and carbon monoxide (CO) gas bubbles are trapped in the steel casting. The dissolved oxygen can be completely eliminated, however, if aluminum is added prior to solidification. The aluminum combines with the oxygen, producing solid alumina (Al_2O_3). In addition to eliminating gas porosity, the tiny Al_2O_3 inclusions are instrumental in pinning grain boundaries and thus preventing grain growth during later high-temperature heat treatments. Unfortunately, the completely deoxidized steel, known as *killed* or fine-grained steel, often displays a deep shrinkage pipe or cavity [Figure 8-20(a)].

Sometimes steels are only partly deoxidized. By adding a small amount of aluminum, a *rimmed* steel is produced in which enough CO is precipitated to just offset the solidification shrinkage [Figure 8-20(b)]. In addition to having less concentrated shrinkage, a rimmed steel helps produce smooth, attractive surfaces on steel sheet after subsequent processing.

 Example 8-9

Castings for steel pump housings used in the chemical industry are normally killed, whereas steel used to make sheet metal for automobile fenders is rimmed. Explain why the different deoxidation practices are used.

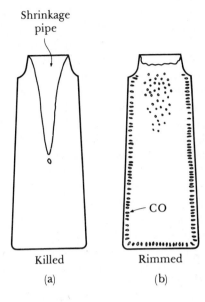

Shrinkage pipe

CO

Killed

(a)

Rimmed

(b)

FIGURE 8-20 The ingot structure of (a) a killed and (b) a rimmed steel after different degrees of deoxidation with aluminum. The killed steel, which is free of gas porosity, contains a deep pipe, and the rimmed steel, which contains distributed porosity, is relatively free of shrinkage.

Answer:

Steel castings cannot be treated to eliminate gas bubbles after solidification; therefore, gas bubbles must be prevented by killing the steel to remove all oxygen. In addition, the inclusions produced during deoxidation may prevent grain growth when the castings are heat treated.

Steel sheet, on the other hand, is formed from a casting by a series of plastic deformation steps in which the thickness of the steel is continually reduced. During deformation, the gas bubbles produced in the original casting are squeezed shut and eliminated by diffusion. The rimmed steel permits a smooth surface during the final deformation steps.

8–8 CONTROL OF CASTING STRUCTURE

As a general rule, we control solidification so that we produce a casting macrostructure containing a large number of small equiaxed grains. This permits the casting to have isotropic properties and improved strength due to grain size strengthening. In addition, we wish to make any dendrites as small as possible, which again improves the strength of the casting and refines both microshrinkage and gas porosity.

In order to obtain this desired structure, we must first assure that widespread nucleation occurs by using appropriate grain refining or inoculating agents. Second, we may encourage rapid solidification to assure that the secondary dendrite arm spacing within the grains is very small. The rate of solidification for any given metal can be influenced by the size of the casting, the mold material, and the casting process. Thick castings solidify more slowly than thin castings. Mold materials having a high density, thermal conductivity, and heat capacity produce more rapid solidification.

Figure 8-21 summarizes four of the dozens of casting processes. The processes are divided into several groups—sand molds, ceramic molds, and metal molds. The processes using metal molds tend to give the highest strength castings due to rapid solidification. Ceramic molds, because they are good insulators, give the slowest cooling and lowest strength castings.

Directional Solidification.   There are some applications for which a small equiaxed grain structure in the casting is not desired. Castings used for blades and vanes in turbine engines are an example (Figure 8-22). These castings are often made of cobalt or nickel superalloys by investment casting.

In conventionally cast parts, an equiaxed grain structure is produced. However, blades and vanes for turbine and jet engines fail along transverse grain boundaries. Better creep and fracture resistance are obtained using the directionally solidified (DS) technique. In the DS process, the mold is heated from one end and cooled from the other, producing a columnar microstructure with all of the grain boundaries running in the longitudinal direction of the part. No grain boundaries are present in the transverse direction [Figure 8-22(b)].

Still better properties are obtained by using a single crystal (SC) technique. Solidification of columnar grains again begins at a cold surface; however, due to the helical connection, only one columnar grain is able to grow to the main body of the casting [Figure 8-22(c)]. The single crystal casting has no grain boundaries at all and has its crystallographic planes and directions in an optimum orientation; in Figure 8-23, best resistance to fatigue in a nickel alloy is obtained for the [100] orientation.

Solidification of single crystals is also necessary in producing silicon semiconductor wafers, from which electronic devices such as integrated circuits are built.

8–9 SOLIDIFICATION AND METALS JOINING

Solidification is also important in the joining of metals by fusion welding. Some typical fusion-welding processes are illustrated in Figure 8-24. In the fusion-welding processes, a portion of the metals to be joined are melted, and in many instances additional molten filler metal is added. The pool of liquid metal is called the *fusion zone*. When the fusion zone subsequently solidifies, the original pieces of metal are joined together.

During solidification of the fusion zone, nucleation is not required. Heat is extracted most rapidly from the fusion zone through the original pieces of metal, which act as heat sinks. The liquid in the fusion zone first cools to the freezing temperature at the edges of the weld (Figure 8-25). But there are already solid

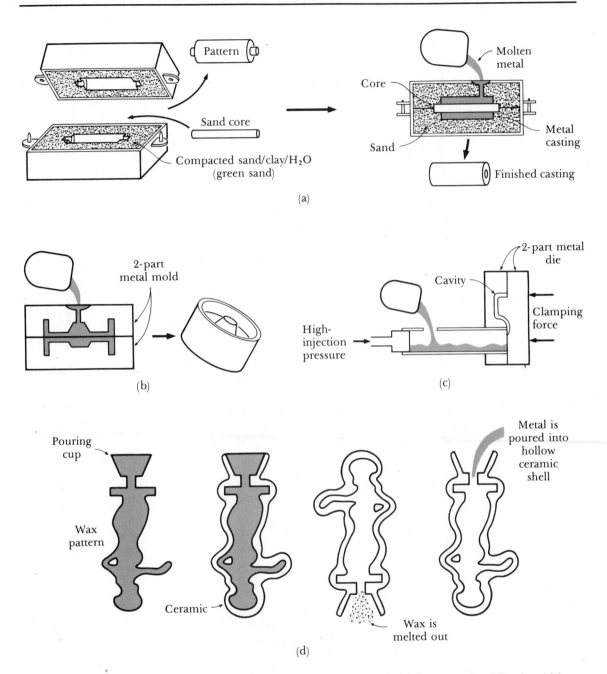

FIGURE 8-21 Four of the typical casting processes. (a) Green sand molding, in which clay-bonded sand is packed around a pattern. Sand cores can produce internal cavities in the casting. (b) The permanent mold process, in which metal is poured into an iron or steel mold. (c) Die casting, in which metal is injected at high pressures into a steel die. (d) Investment casting, in which a wax pattern is surrounded by a ceramic; after the wax is melted and drained, metal is poured into the mold.

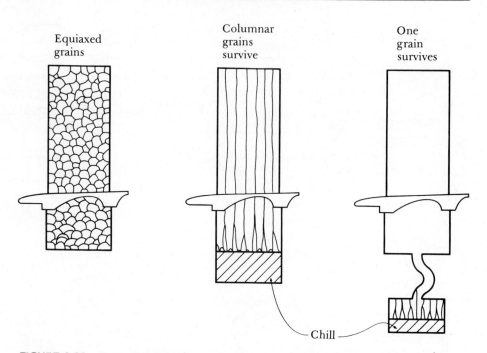

FIGURE 8-22   *Controlling grain structure in turbine blades. (a) Conventional equiaxed grains, (b) directionally solidified columnar grains, and (c) single crystal.*

grains of the original material at those locations! Consequently, the solid simply begins to grow from these grains, frequently in a columnar manner. Growth of the solid grains in the fusion zone from the preexisting grains is called *epitaxial growth.*

The structure and properties in the fusion zone depend on many of the same variables as in a metal casting. Addition of inoculating agents to the fusion zone reduces the grain size. Fast cooling rates or short solidification times promote a

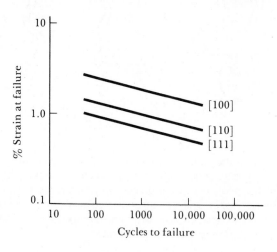

FIGURE 8-23   Effect of single crystal orientation on the fatigue properties of a nickel alloy. The percent strain is proportional to the applied stress.

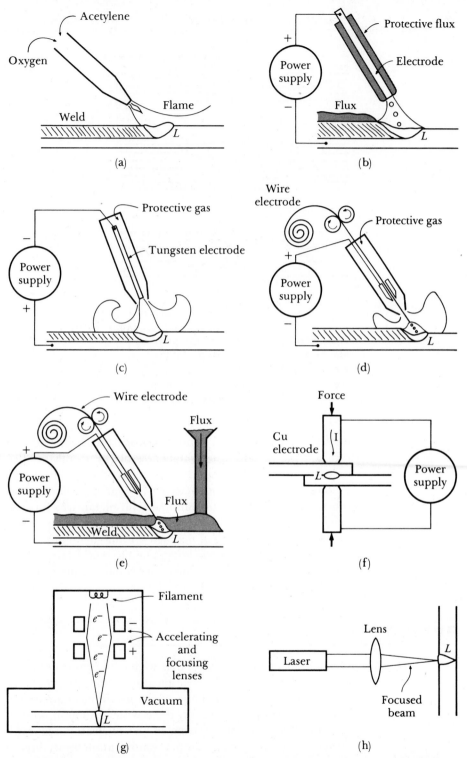

FIGURE 8-24 Typical fusion-welding processes. (a) Oxyacetylene welding, (b) shielded-metal arc welding, (c) gas-tungsten arc welding, (d) gas-metal arc welding, (e) submerged arc welding, (f) resistance welding, (g) electron-beam welding, and (h) laser welding.

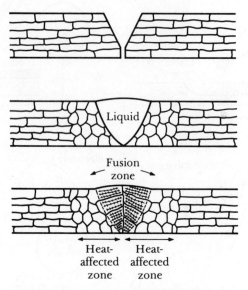

FIGURE 8-25 A schematic diagram of the fusion zone and the solidification of the weld during fusion welding. (a) Initial prepared joint. (b) Weld at the maximum temperature, with joint filled with filler metal. (c) Weld after solidification.

finer microstructure and improved properties. Factors that increase the cooling rate include increasing the thickness of the metal, smaller fusion zones, low original metal temperatures, and the type of welding process. Oxyacetylene welding, for example, uses relatively low intensity flames; consequently, welding times are long and the surrounding solid metal, which becomes very hot, is not an effective heat sink. Arc-welding processes provide a more intense heat source, thus minimizing heating of the surrounding metal and providing faster cooling. Resistance welding, laser welding, and electron-beam welding are exceptionally intense heat sources and produce very rapid cooling rates and potentially strong welds.

SUMMARY

One of the first opportunities that we have to control the mechanical properties of a material occurs during solidification of the liquid melt. During solidification we control the size and shape of the grains to improve overall properties, to obtain uniform properties, or, if we wish, to obtain anisotropic behavior. We exercise this control by assuring proper nucleation and growth through inoculation or grain refining, proper solidification time, and correct metal temperature—in other words, we must control the materials processing technique. We find that improved nucleation of grains and rapid cooling give smaller grains and thus provide grain size strengthening. Special structures and properties may be obtained by rapid solidification processing. Furthermore, by proper treatment of the molten material and correct casting procedures, we are able to prevent or control gas and shrinkage voids. In addition to providing an improved casting, these precautions also improve our ability to further process the material by deformation techniques and heat treatment.

GLOSSARY

Cavity shrinkage. A large void within a casting caused by the volume contraction that occurs during solidification.

Chill zone. A region of small, randomly oriented grains that forms at the surface of a casting due to heterogeneous nucleation.

Chvorinov's rule. The solidification time of a casting is directly proportional to the square of the volume to surface area ratio of the casting.

Columnar zone. A region of elongated grains having a preferred orientation that forms as a result of competitive growth during the solidification of a casting.

Cooling rate. The change in temperature for a given change in time. Rapid cooling rates normally give stronger castings.

Critical radius r^*. The minimum size that must be formed by atoms clustering together in the liquid before the solid particle is stable and begins to grow.

Dendrite. The treelike structure of the solid that grows when an undercooled liquid nucleates.

Dendritic growth. Rapid growth of a solid dendrite when an undercooled liquid nucleates and grows.

Directional solidification. Assuring that a casting grows from one direction only. This technique is used to cast high temperature-resistant turbine blades.

Embryo. A tiny particle of solid that forms from the liquid as atoms cluster together. The embryo is too small to grow.

Epitaxial growth. Growth of a liquid onto an existing solid material without the need for nucleation.

Equiaxed zone. A region of randomly oriented grains in the center of a casting produced as a result of widespread nucleation.

Fusion zone. The portion of a weld heated to produce all liquid during the welding process. Solidification of the fusion zone provides joining.

Gas porosity. Bubbles of gas trapped within a casting during solidification due to the lower solubility of the gas in the solid compared to the liquid.

Grain refinement. The addition of heterogeneous nuclei in a controlled manner to increase the number of grains in a casting.

Grain size strengthening. By reducing the size of the grains, and causing an increase in the amount of grain boundary area, a material may be strengthened.

Heterogeneous nucleation. Formation of a critically sized solid from the liquid on an impurity surface.

Homogeneous nucleation. Formation of a critically sized solid from the liquid by the clustering together of a large number of atoms at a high undercooling.

Ingot structure. The macrostructure of a casting, including the chill zone, columnar zone, and equiaxed zone.

Inoculation. The addition of heterogeneous nuclei in a controlled manner to increase the number of grains in a casting.

Interdendritic shrinkage. Small, frequently isolated pores between the dendrite arms formed by the shrinkage that accompanies solidification. Also known as microshrinkage or shrinkage porosity.

Latent heat of fusion ΔH_f. The heat evolved when a liquid solidifies. The latent heat of fusion is related to the energy difference between the solid and the liquid.

Local solidification time. The time required for a particular location in a casting to solidify once nucleation has begun.

Nucleus. A tiny particle of solid that forms from the liquid as atoms cluster together. Because these particles are large enough to be stable, nucleation has occurred and growth of the solid can begin.

Pipe shrinkage. A large conical-shaped void at the surface of a casting caused by the volume contraction that occurs during solidification.

Planar growth. The growth of a smooth solid-liquid interface during solidification when no undercooling of the liquid is present.

Pouring temperature. The temperature of the metal when it is poured into a mold during the casting process.

Rapid solidification processing. Producing unique material structures by promoting unusually high cooling rates during freezing.

Recalescence. The increase in the temperature of a solidifying liquid that is growing dendritically. The increase is caused by the transfer of the latent heat of fusion into the undercooled liquid.

Riser. An extra reservoir of liquid metal connected to a casting. If the riser freezes after the casting, the riser can provide liquid metal to compensate for shrinkage.

Secondary dendrite arm spacing (*SDAS*). The distance between the centers of two adjacent secondary dendrite arms.

Sievert's law. The amount of a gas that dissolves in a metal is proportional to the partial pressure of that gas in the surroundings.

Solidification. The transformation of a liquid to a solid material.

Specific heat. The heat required to change the temperature of a unit weight of the material one degree.

Superheat. The pouring temperature minus the freezing temperature.

Surface free energy σ. The increase in energy associated with the surface between a growing solid and a liquid.

Thermal arrest. A plateau on the cooling curve during the solidification of a material. The thermal arrest is due to the evolution of the latent heat of fusion during solidification.

Total solidification time. The time required for the casting to completely solidify after the casting has been poured.

Undercooling. The temperature to which the liquid metal must cool below the equilibrium freezing temperature before nucleation occurs.

Volume free energy ΔF_v. The change in free energy of a material when the material solidifies.

PRACTICE PROBLEMS

1. Plot the maximum observed undercooling versus the freezing temperature for the metals listed in Table 8-1. Do the data in the table confirm the relationship expressed in Equation 8-5?

2. Calculate the temperature and the number of degrees of undercooling at which lead (Pb) should nucleate homogeneously.

3. Calculate the size of the critical radius when lead nucleates homogeneously.

4. Calculate the number of atoms in the critical radius when lead nucleates homogeneously. The lattice parameter of FCC lead is about 4.9489 Å.

5. Calculate the temperature and the number of degrees of undercooling at which silver (Ag) should nucleate homogeneously.

6. Calculate the size of the critical radius when silver nucleates homogeneously.

7. Calculate the number of atoms in the critical radius when silver nucleates homogeneously. The lattice parameter of FCC silver is about 4.0862 Å.

8. Suppose that a nucleus of lead is formed homogeneously with an undercooling of only 10°C. How many atoms would have had to spontaneously group together in order for this to occur?

9. Suppose that a nucleus of silver is formed homogeneously with an undercooling of only 25°C. How many atoms would have had to spontaneously group together in order for this to occur?

10. Calculate and plot how the total free energy ΔF changes with the radius of a spherical nucleus of nickel at undercoolings of 25°C and 350°C.

11. Estimate the percent of solidification that will occur in a dendritic manner when nickel nucleates (a) with 10°C undercooling, (b) with 100°C undercooling, and (c) homogeneously. The specific heat of nickel is 4.1 J/cm^3 · °C.

12. For silver, plot the percent of solidification that will occur in a dendritic manner versus the undercooling produced prior to solidification. The specific heat of silver is 3.25 J/cm^3 · °C.

13. If 38% of solidification of iron occurs by dendritic growth, estimate the undercooling that must have been achieved prior to nucleation. The specific heat of the iron liquid is 5.78 J/cm^3 · °C.

14. If a 3-in. cube solidifies in 35 min, calculate (a) the mold constant and (b) the total

solidification time for a 1 in. × 7 in. × 15 in. plate.

15. If a 3 cm × 8 cm × 18 cm plate solidifies in 16 min, calculate (a) the mold constant and (b) the total solidification time for a 6 cm diameter sphere.

16. The following data are obtained from a series of copper castings. Determine the mold constant B and the exponent n in Chvorinov's rule for copper.

Size (in.)	Solidification Time (min)
0.25 × 0.25 × 6	0.2
1 × 2 × 8	2.7
3 diameter sphere	5.9
6 cube	18.0

17. During conventional casting of an aluminum alloy, Chvorinov's rule is given by $t_s = 45(V/A)^{1.6}$, using units of inches and minutes. Using the evaporative pattern casting process (see Chapter 1), a series of aluminum castings solidifies according to the following data. Determine the constants B and n in Chvorinov's rule and determine the effect of the Styrofoam pattern on solidification time.

Size (in.)	Solidification Time (min)
0.25 × 6 × 6	1.3
0.5 × 6 × 6	3.5
1 × 6 × 6	8.5
3 × 6 × 6	25.9

18. Figure 8-7(b) shows a photograph of an aluminum alloy. Estimate (a) the secondary dendrite arm spacing and (b) the local solidification time for that area of the casting.

19. Figure 8-26 shows a photograph of a copper alloy. Estimate (a) the secondary dendrite arm spacing and (b) the local solidification time for that area of the casting.

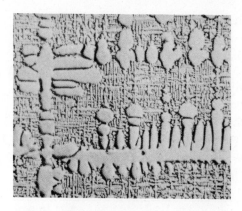

FIGURE 8-26 Dendrites in a copper-base alloy, for Problem 8-19 (× 30). From *Metals Handbook*, Vol. 9, 9th Ed., American Society for Metals, 1985.

20. Which aluminum casting will have the higher strength—a 1 in. × 8 in. × 8 in. plate or a 1.25 in. × 1.25 in. × 8 in. bar? Estimate the secondary dendrite arm spacing in the center of each casting, using the equation for solidification time given in Problem 8-17.

21. Which copper casting will have the higher strength—a cylindrical bar 1.5 in. in diameter and 8 in. tall or a 1 in. × 12 in. × 12 in. plate?

22. Calculate (a) the diameter of the sphere that would freeze in the same time as a 2.5 in. × 6 in. × 9 in. plate and (b) the ratio of the volume of the sphere to the volume of the plate.

23. We need a riser for an aluminum casting that takes at least 25 min to solidify. Suppose that the mold constant is 45 min/in². Which one of the following riser shapes will satisfy this requirement with the least riser volume or mass —a cube, a sphere, or a cylinder with an *H/D* ratio of 1.0?

24. Calculate the volume, diameter, and height of the cylindrical riser having an *H/D* ratio of 2 that would be required to prevent shrinkage in a steel casting having dimensions of 2 in. × 9 in. × 10 in.

25. Calculate the volume, diameter, and height of the cylindrical riser having an *H/D* ratio of 1.5 that would be required to prevent shrinkage in an aluminum casting having dimensions of 3 in. × 4 in. × 5 in.

26. A step-block casting is shown in Figure 8-27, along with an attached riser. Compare the solidification times of each casting section and riser and decide if the riser will prevent shrinkage.

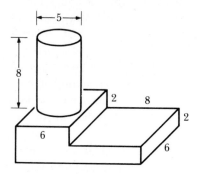

FIGURE 8-27 Step-block casting for Problem 8-26.

27. A step-block casting is shown in Figure 8-28, along with an attached riser. Compare the solidification times of each casting section and riser and decide if the riser will prevent shrinkage.

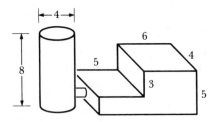

FIGURE 8-28 Step-block casting for Problem 8-27.

28. The following data were obtained from castings poured using a superalloy. Determine the constants *k* and *m* that relate secondary dendrite arm spacing to local solidification time.

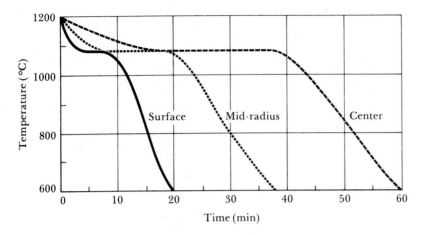

FIGURE 8-29 Cooling curves for Problem 8-31.

Solidification Time (s)	Dendrite Arm Spacing (mm)
300	0.020
1000	0.032
2500	0.046
5000	0.060

29. When aluminum powder particles are produced by rapid solidification processing, solidification times of 0.0005 s are observed. Estimate the secondary dendrite arm spacing of the metal powders and compare to the *SDAS* of an aluminum casting that is 1 in. × 10 in. × 10 in., using the expression for Chvorinov's rule given in Problem 8-17.

30. The solidification time of a typical aluminum weld is about 4 s. Compare the secondary dendrite arm spacing in the fusion zone of the weld to that in a 3 in. × 5 in. × 8 in. aluminum casting whose mold constant is 45 min/in² and whose exponent *n* is 1.5.

31. The cooling curves in Figure 8-29 were obtained from the surface, mid-radius, and center of a continuously cast copper bar. Determine the local solidification time and the *SDAS* at each location. Is the surface of a casting expected to be weaker or stronger than the center?

32. A cooling curve is shown in Figure 8-30. Determine (a) the pouring temperature, (b) the freezing temperature, (c) the superheat, (d) the cooling rate just before freezing begins, (e) the total solidification time, (f) the local solidification time, and (g) the probable identity of the material.

33. A cooling curve is shown in Figure 8-31. Determine (a) the pouring temperature, (b) the freezing temperature, (c) the superheat, (d) the cooling rate just before freezing begins, (e) the total solidification time, (f) the local solidification time, and (g) the undercooling.

34. Suppose a 5-in. cube of copper is allowed to solidify and all of the shrinkage occurs in the center of the casting as a spherical cavity. Estimate the diameter and volume of the shrinkage cavity.

35. The entire shrinkage in a 5-in. spherical casting is found as a spherical cavity with a diameter of 1.75 in. Estimate the percent volume change during solidification.

36. A 3 cm × 6 cm × 8 cm copper casting is produced which weighs 1220 g. Calculate the percent shrinkage that occurred during the solidification process. What would be the volume of the shrinkage cavity at the center of the casting?

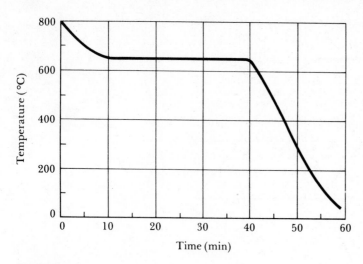

FIGURE 8-30   Cooling curve for Problem 8-32.

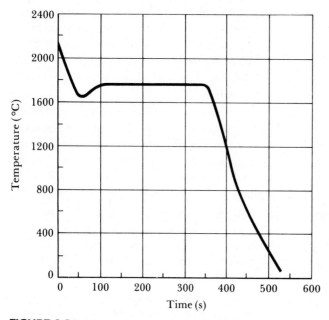

FIGURE 8-31   Cooling curve for Problem 8-33.

37. A 1 in. × 12 in. × 12 cm copper casting is produced which weighs 13 lb. Calculate the percent shrinkage that occurred during the solidification process. If the shrinkage is distributed as spherical pores, each having a diameter of 0.02 in., how many pores are present in the casting?

38. Suppose liquid aluminum completely fills a metal mold that is 3 in. × 3 in. × 36 in. and is controlled so that all of the shrinkage occurs along the length of the casting. What is the length of the aluminum casting immediately after solidification?

39. The density of liquid bismuth is 10.067 g/cm^3 and the density of solid bismuth is 9.5 g/cm^3. Determine (a) the percent volume change during freezing and (b) whether the casting shrinks or expands during freezing.

40. Suppose we produce an aluminum casting containing gas porosity. What forming process discussed in Chapter 7 would best be used to help close up the porosity without creating any major change in the dimensions of the casting?

41. Figure 8-10 shows the dendrites in a titanium powder particle that has been rapidly solidified. Assuming that the size of the titanium dendrites is related to solidification time by the same relationship as in aluminum, estimate the solidification time of the powder.

9

SOLIDIFICATION AND SOLID SOLUTION STRENGTHENING

9–1 INTRODUCTION

The mechanical properties of materials can be controlled by the addition of point defects, in particular substitutional and interstitial atoms. The point defects disturb the atomic arrangement in the lattice and interefere with the movement of dislocations, or slip. The point defects cause the material to be solid solution strengthened.

In addition, the introduction of point defects changes the composition of the material and influences the solidification behavior. We will examine this effect by introducing the equilibrium phase diagram. From the phase diagram we can predict how a material will solidify under both equilibrium and nonequilibrium conditions.

9-2 PHASES AND THE UNARY PHASE DIAGRAM

Pure materials have many engineering applications, but frequently, particularly when improved mechanical properties are required, alloys or mixtures of materials are used. There are two types of alloys—single-phase alloys and multiple-phase alloys. In this chapter we will examine the behavior of single-phase alloys. As a first step, let us define a phase and determine how the phase rule helps us to determine the state—solid, liquid, or gas—in which a pure material will exist.

Phase. A *phase* has the following characteristics: (a) a phase has the same structure or atomic arrangement throughout; (b) a phase has roughly the same

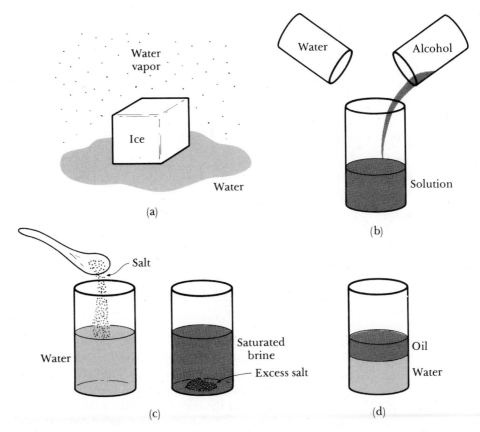

FIGURE 9-1   *Illustration of phases and solubility. (a) The three forms of water—gas, liquid, and solid—are each a phase. (b) Water and alcohol have unlimited solubility. (c) Salt and water have limited solubility. (d) Oil and water have virtually no solubility.*

composition and properties throughout; and (c) there is a definite interface between the phase and any surrounding or adjoining phases. For example, we could enclose a block of ice in a vacuum chamber [Figure 9-1(a)]. The ice would begin to melt and, in addition, some of the water might vaporize. Under these conditions we would have three phases coexisting—solid H_2O, liquid H_2O, and gaseous H_2O. Each of these forms of H_2O is a distinct phase; each has a unique atomic arrangement, unique properties, and a definite boundary between each form. In this case the phases have identical compositions, but that is not sufficient to permit us to call the entire system one phase.

Phase Rule.   The *Gibbs phase rule* describes the state of a material and has the general form

$$F = C - P + 2 \qquad (9\text{-}1)$$

In the phase rule, C is the number of components, usually elements or compounds, in the system; F is the number of degrees of freedom, or the number of

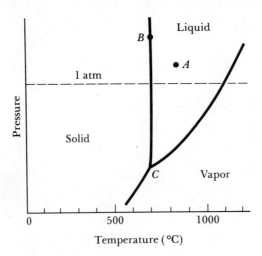

FIGURE 9-2 Schematic unary phase diagram for magnesium, showing the melting and boiling temperatures at one atmosphere pressure.

variables, such as temperature, pressure, or composition, that are allowed to change independently without changing the number of phases in equilibrium; and P is the number of phases present. The constant 2 in the equation implies that both temperature and pressure are allowed to change.

As an example of the use of the phase rule, let us consider the case of pure magnesium. Figure 9-2 shows a *unary*, or one-component, phase diagram in which the lines divide the liquid, solid, and vapor phases. In the unary phase diagram, there is only one component—in this case, magnesium. However, depending on the temperature and pressure, there may be one, two, or even three phases present at any one time; these phases are solid magnesium, liquid magnesium, and magnesium vapor. Note that at atmospheric pressure (one atmosphere), given by the dashed line, the intersection with the lines in the phase diagram give the usual melting and boiling temperatures for magnesium. At very low pressures the solid can sublime, or go directly to a vapor without melting when it is heated.

Suppose we have a pressure and temperature that puts us at point A in the phase diagram, which suggests that the magnesium should be all liquid. At this point, the number of components C is one (magnesium) and the number of phases is also one (liquid). The phase rule tells us that

$$F = C - P + 2 = 1 - 1 + 2 = 2$$

or there are two degrees of freedom. This means that, at least within limits, we can change either the pressure, the temperature, or both and still be in an all-liquid portion of the diagram. Or, put another way, we would have to fix both temperature and pressure to know precisely where we are in the diagram.

However, at point B we are at the boundary between the solid and liquid portions of the diagram. The number of components C is still one, but at point B the solid and liquid can coexist, or the number of phases P is two. From the phase rule.

$$F = C - P + 2 = 1 - 2 + 2 = 1$$

or there is only one degree of freedom. For example, if we change the temperature, the pressure must also be adjusted if we are to stay on the boundary where the liquid and solid coexist. We no longer have the freedom to change pressure and temperature independently if both solid and liquid are to be present. On the other hand, if we fix the pressure, the phase diagram tells us the temperature that we must have if solid and liquid are to coexist.

Finally, consider point C in the figure. At this point, solid, liquid, and vapor coexist. While the number of components is still one, there are three phases. The number of degrees of freedom is

$$F = C - P + 2 = 1 - 3 + 2 = 0$$

Now we have no degrees of freedom; all three phases can coexist only if both the temperature and the pressure are fixed. This point is known as the *triple point*.

◇ | **Example 9-1**

Although magnesium is a very lightweight metal, it is not often used in vehicles intended to enter the outer space environment. Why?

Answer:

In outer space, the pressure is very low. Even at cold temperatures, it is possible that the solid magnesium will begin to change to a vapor, causing metal loss that could damage the space vehicle.

9–3   SOLUBILITY AND SOLUTIONS

When we begin to combine different materials, as when we add alloying elements to a metal, we produce solutions. We are interested in how much of each material we can combine without producing an additional phase. In other words, we are interested in the solubility of one material in another.

Unlimited Solubility.   Suppose we begin with a glass of water and a glass of alcohol. The water is one phase and the alcohol is a second phase. If we pour the water into the alcohol and stir, only one phase is produced [Figure 9-1(b)]. The glass contains a solution of water and alcohol that has unique structure, properties, and composition. Water and alcohol are soluble in each other.

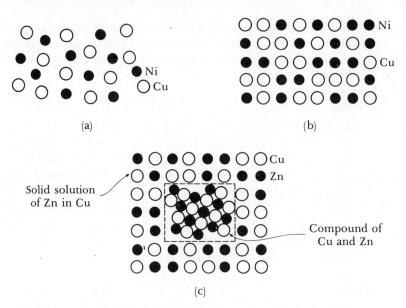

(a)

(b)

(c)

FIGURE 9-3 (a) Liquid copper and liquid nickel are completely soluble in each other. (b) Solid copper-nickel alloys display complete solid solubility, with copper and nickel atoms occupying random lattice sites. (c) In copper-zinc alloys containing more than 40% Zn, a second phase forms due to limited solubility of zinc in copper.

Furthermore, they display *unlimited solubility*—regardless of the ratio of water and alcohol, only one phase is produced by mixing them together.

Similarly, if we were to mix a container of liquid copper and a container of liquid nickel, only one liquid phase would be produced. The liquid alloy has the same composition, properties, and structure everywhere [Figure 9-3(a)]. Liquid nickel and copper also have unlimited solubility; regardless of the relative amounts of nickel and copper, only one liquid phase is produced.

If the liquid copper-nickel alloy solidifies and cools to room temperature, only one solid phase is produced. After solidification the copper and nickel atoms do not separate, but instead are randomly located at the lattice points of an FCC lattice. Within the solid phase, the structure, properties, and composition are uniform and no interface exists between the copper and nickel atoms. Therefore, copper and nickel also have unlimited solid solubility. The solid phase may be called a *solid solution* [Figure 9-3(b)].

Limited Solubility. When we add a small quantity of salt (one phase) to a glass of water (a second phase) and stir, the salt dissolves completely in the water [Figure 9-1(c)]. Only one phase—salty water or brine—is found. However, if we add too much salt to the water, the excess salt sinks to the bottom of the glass. Now we have two phases—water that is saturated with salt plus excess solid salt. We find that salt has a *limited solubility* in water.

If we add a small amount of liquid zinc to liquid copper, a single liquid solution is produced. When that copper-zinc solution cools and solidifies, a single solid solution having an FCC structure results, with copper and zinc atoms randomly located at the normal lattice points. However, if the liquid solution contains more than about 40% Zn, the excess zinc atoms combine with some of the copper atoms to form a CuZn compound [Figure 9-3(c)]. Two solid phases now coexist—a solid solution of copper saturated with about 40% Zn plus a CuZn compound. The solubility of zinc in copper is limited.

In the extreme case, there may be no solubility of one material in another. This is the case for oil and water [Figure 9-1(d)] or for copper-lead alloys.

9–4   CONDITIONS FOR UNLIMITED SOLID SOLUBILITY IN METALS

In order for an alloy system, such as copper-nickel, to have unlimited solid solubility, certain conditions must be satisfied. These condition are known as the Hume-Rothery rules and are as follows.

1. The atoms of the metals must be of similar size, with no more than a 15% difference in atomic radius. Otherwise the lattice strain produced by the differently sized atoms is too great to permit unlimited solubility.

2. The metals must have the same crystal structure; if not, there must be some point at which there is a transition from one phase to a second phase with a different structure.

3. The atoms of the metals must have the same valence; otherwise the valence electron difference may encourage the formation of compounds rather than solutions.

4. The atoms of the metals must have about the same electronegativity. If the electronegativies differ significantly, compounds again tend to form, as when sodium and chlorine combine to form sodium chloride.

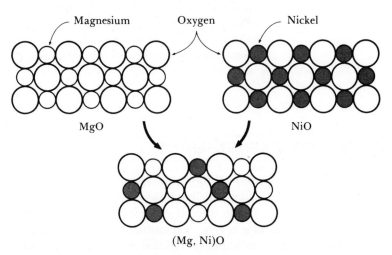

FIGURE 9-4 MgO and NiO have similar crystal structures, ionic radii, and valences; thus the two ceramic materials can form solid solutions.

Hume-Rothery's conditions must be met, but are not necessarily sufficient, in order for two metals to have unlimited solid solubility.

Similar behavior is observed between certain compounds, including ceramic materials. Figure 9-4 shows schematically the structure of MgO and NiO. But the Mg and Ni ions are similar in size and valence and consequently can replace one another in a sodium chloride-type of lattice, forming a complete series of solid solutions of the form (Mg, Ni)O.

◇ **Example 9-2**

Determine which of the following alloy systems might be expected to display unlimited solid solubility: Ag-Au, Al-Si, Ca-Al, K-Ba, Ag-Cu.

Answer:

From Hume-Rothery's conditions

Ag-Au: Both have a valence of 1, both are in the same column of the periodic table and have about the same electronegativity, and both are FCC. $r_{Ag} = 1.445 \times 10^{-8}$ cm, $r_{Au} = 1.442 \times 10^{-8}$ cm, $\Delta r/r = 0.2\% < 15\%$. Silver and gold have complete solid solubility.

Al-Si: Aluminum is FCC; silicon is DC. Aluminum and silicon have limited solid solubility.

Ca-Al: $r_{Ca} = 1.97 \times 10^{-8}$ cm, $r_{Al} = 1.43 \times 10^{-8}$ cm, $\Delta r/r = 37.8\% > 15\%$. Calcium and aluminum have limited solid solubility.

K-Ba: Potassium has a valence of 1 and barium has a valence of 2. They have limited solid solubility.

Ag-Cu: Both have a valence of 1, both are in the same column of the periodic table and have similar electronegativities, and both are FCC. $r_{Ag} = 1.445 \times 10^{-8}$ cm, $r_{Cu} = 1.28 \times 10^{-8}$ cm, $\Delta r/r = 12.9\% < 15\%$. Silver and copper satisfy all of Hume-Rothery's conditions, yet silver and copper do not display unlimited solid solubility!

9–5 SOLID SOLUTION STRENGTHENING

By producing solid solution alloys, we cause *solid solution strengthening*. In the copper-nickel system, we have intentionally introduced a solid substitutional atom (say nickel) into the original lattice (say copper). The copper-nickel alloy has a strength that is greater than that of pure copper. Similarly, by adding less than 40% Zn to copper, the zinc behaves as a substitutional atom which strengthens the copper-zinc alloy compared with pure copper.

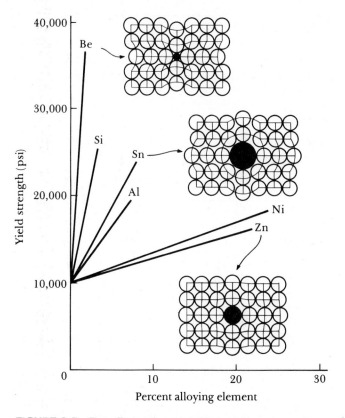

FIGURE 9-5 The effects of several alloying elements on the yield strength of copper. Nickel and zinc atoms are about the same size as copper atoms, but beryllium and tin atoms are much different from copper atoms. Both increasing atomic size difference and amount of alloying element increase solid solution strengthening.

Degree of Solid Solution Strengthening. The degree of solid solution strengthening depends on two factors. First, a large difference in atomic size between the original (or solvent) atom and the added (or solute) atom will increase the strengthening effect. A larger size difference produces a greater disruption of the initial lattice, making slip of dislocations more difficult (Figure 9-5).

Second, the greater the amount of alloying element added, the greater the strengthening effect (Figure 9-5). A Cu-20% Ni alloy is stronger than a Cu-10% Ni alloy. Of course, if too much of a large or small atom is added, the solubility limit may be exceeded and a different strengthening mechanism—*dispersion strengthening*—may be produced. This mechanism will be discussed in Chapter 10.

◇ | **Example 9-3**

From the atomic radii, show whether the size difference between copper atoms and alloying atoms accurately predicts the amount of strengthening found in Figure 9-5.

Answer:

The atomic radii and percent size difference are shown below.

Metal	Radius (Å)	$\dfrac{r - r_{Cu}}{r_{Cu}} \times 100$
Cu	1.278	
Zn	1.332	+ 4.2%
Al	1.432	+ 12.0%
Sn	1.509	+ 18.1%
Ni	1.243	− 2.7%
Si	1.176	− 8.0%
Be	1.14	− 10.8%

For atoms larger than copper, namely zinc, aluminum, and tin, increasing the size difference increases the strengthening effect. Likewise for smaller atoms, increasing the size difference increases the strengthening. Note that the larger atoms generally produce less strengthening than the smaller atoms.

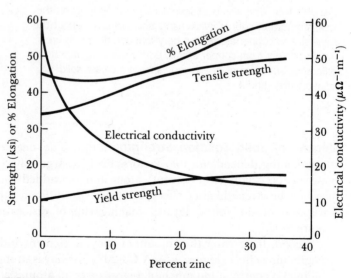

FIGURE 9-6   The effect of additions of zinc to copper on the properties of the solid solution-strengthened alloy. The increase in % elongation with increasing zinc content is not typical of solid solution strengthening.

Effect of Solid Solution Strengthening on Properties.   The results of the effects of solid solution strengthening on the properties of the material include the following (Figure 9-6).

1. The yield strength, tensile strength, and hardness of the alloy are greater than for the pure metals.

2. Almost always, the ductility of the alloy is less than that of the pure metal. Only rarely, as in copper-zinc alloys, does solid solution strengthening increase both strength and ductility.

3. Electrical conductivity of the alloy is much lower than that of the pure metal. Solid solution strengthening of copper or aluminum wires used for transmission of electrical power is not recommended due to this pronounced effect.

4. The resistance to creep, or loss of strength at elevated temperatures, is improved by solid solution strengthening. High temperatures do not cause a catastrophic change in the properties of solid solution-strengthened alloys.

9–6 ISOMORPHOUS PHASE DIAGRAMS

A *phase diagram* shows the phases and their compositions at any combination of temperature and alloy composition. When only two elements are present in the alloy, a *binary phase diagram* can be constructed. *Isomorphous* binary phase diagrams are found in a number of metallic and ceramic systems. In the isomorphous systems, which include the copper-nickel and NiO–MgO systems (Figure 9-7), only one solid phase forms; the two components in the system display complete solid solubility. There are several valuable pieces of information that we can obtain from these phase diagrams.

Liquidus and Solidus Temperatures.   The upper curve on the diagram represents the *liquidus* temperature for all of the copper-nickel alloys. We must heat a copper-nickel alloy above the liquidus to produce a completely liquid alloy that can then be cast into a useful shape. The liquid alloy begins to solidify when the temperature cools to the liquidus temperature.

The *solidus* temperature for the copper-nickel alloys is the lower curve. A copper-nickel alloy is not completely solid until the metal cools below the solidus temperature. If we use a copper-nickel alloy at high temperatures, we must be sure that the service temperature is below the solidus so that no melting occurs.

The copper-nickel alloys melt and freeze over a range of temperatures, between the liquidus and solidus. The temperature difference between the liquidus and solidus is the *freezing range* of the alloy. Within the freezing range, two phases coexist—a liquid and a solid. The solid is a solution of copper and nickel atoms; solid phases are typically designated by a lowercase Greek letter, such as α.

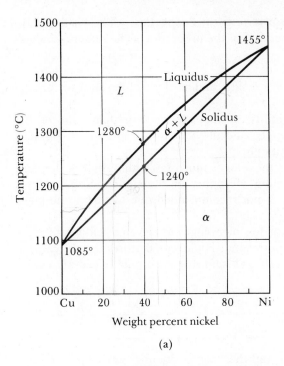

(a)

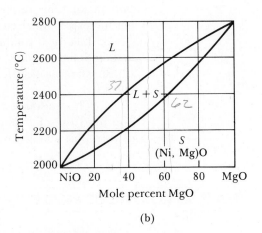

(b)

FIGURE 9-7 The equilibrium phase diagrams for the copper-nickel and NiO-MgO systems. The liquidus and solidus temperature are shown for a Cu-40% Ni alloy.

◇ **Example 9-4**

Determine the liquidus temperature, solidus temperature, and freezing range for the Cu-40% Ni alloy shown in Figure 9-7.

Answer:

$$T_{\text{liquidus}} = 1280°C$$
$$T_{\text{solidus}} = 1240°C$$
$$\text{Freezing range} = 1280 - 1240 = 40°C$$

Phases Present.   Often we are interested in which phases are present in an alloy at a particular temperature. If we plan to make a casting, we must be sure that the metal is initially all liquid; if we plan to heat treat an alloy component, we must be sure that no liquid forms during the process. The phase diagram can be treated as a road map; if we know the coordinates—temperature and alloy composition—we can determine the phases present.

Example 9-5

Determine the phases present in a Cu-40% Ni alloy at 1300°C, 1250°C, and 1200°C.

Answer:

From Figure 9-7 we find

     1300°C: Only liquid L is present

     1250°C: Both liquid L and solid α are present

     1200°C: Only solid α is present

Composition of Each Phase.   Each phase present in an alloy has a composition, expressed as the percentage of each element in the phase. Usually the composition is expressed in weight percent (wt%). When only one phase is present in the alloy, the composition of the phase equals the overall composition of the alloy. If the original composition of the alloy changes, then the composition of the phase must also change.

However, when two phases coexist, such as liquid and solid, the compositions of the two phases differ from one another and also differ from the original overall composition. In this case, if the original composition changes slightly, the composition of the two phases is unaffected, providing that the temperature remains constant.

This difference is explained by the Gibbs phase rule. In this case, unlike the example of pure magnesium described earlier, we will keep the pressure fixed at one atmosphere, which is the normal case for binary phase diagrams. Now the phase rule can be rewritten as

$$F = C - P + 1 \tag{9-2}$$

where again C is the number of components, P is the number of phases, and F is the number of degrees of freedom. In a binary system, the number of components C is two; the degrees of freedom that we have include changing the temperature and changing the composition of the phases present. We can apply this form of the phase rule to the Cu–Ni system, as shown in Example 9-6.

 Example 9-6

Determine the degrees of freedom in a Cu-40% Ni alloy at 1300°C, 1250°C, and 1200°C.

Answer:

At 1300° C, $P = 1$, since only one phase, liquid, is present. $C = 2$, since both copper and nickel atoms are present. Thus

    $F = 2 - 1 + 1 = 2$

We must fix both the temperature and the composition of the liquid phase to completely describe the state of the copper-nickel alloy in the liquid region.

At 1250° C, $P = 2$, since both liquid and solid are present. $C = 2$, since copper and nickel atoms are present. Now

$$F = 2 - 2 + 1 = 1$$

If we fix the temperature in the two-phase region, the compositions of the two phases are also fixed. Or, if the composition of one phase is fixed, the temperature and composition of the second phase are automatically fixed.

At 1200°C, $P = 1$, since only one phase, solid, is present. $C = 2$, since both copper and nickel atoms are present. Again

$$F = 2 - 1 + 1 = 2$$

and we must fix both temperature and composition to completely describe the solid.

Because there is only one degree of freedom in a two-phase region of a binary phase diagram, the compositions of the two phases are always fixed when we specify the temperature. This is true even if the overall composition of the alloy changes. This permits us to use a tie line to determine the composition of the two phases. A *tie line* is a horizontal line within a two-phase region drawn at the temperature of interest (Figure 9-8). Tie lines are not used in single-phase regions. In an isomorphous system, the tie line connects the liquidus and solidus

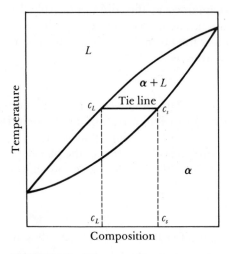

FIGURE 9-8 When an alloy is present in a two-phase region, a tie line at the temperature of interest fixes the composition of the two phases. This is a consequence of Gibbs phase rule, which provides only one degree of freedom.

points at the specified temperature. The ends of the tie line represent the compositions of the two phases in equilibrium.

◇ | **Example 9-7**

Determine the composition of each phase in a Cu-40% Ni alloy at 1300°C, 1270°C, 1250°C, and 1200°C. (See Figure 9-9.)

Answer:

A vertical line at 40% Ni represents the overall composition of the alloy.

1300°C: The only phase present is liquid. The liquid must contain 40% Ni, the overall composition of the alloy.

1270°C: Two phases are present. A horizontal line within the $\alpha + L$ field is drawn. The endpoint at the liquidus, which is in contact with the liquid region, is at 37% Ni. The endpoint at the solidus, which is in contact with the α region, is at 50% Ni. Therefore, the liquid contains 37% Ni and the solid contains 50% Ni.

1250°C: Again two phases are present. The tie line drawn at this temperature shows that the liquid contains 32% Ni and the solid contains 45% Ni.

1200°C: Only solid α is present, so the solid must contain 40% Ni.

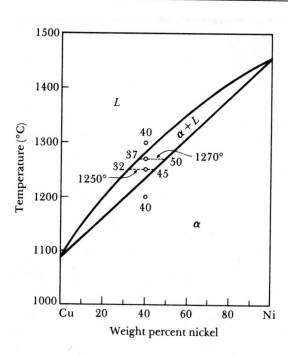

FIGURE 9-9 Tie lines and phase compositions for a Cu-40% Ni alloy at several temperatures.

In Example 9-7, we find that the solid α contains more nickel than the overall alloy and the liquid L contains more copper than the original alloy. Generally, the higher melting point element, in this case nickel, is concentrated in the first solid that forms.

Amount of Each Phase (the Lever Law).   Lastly, we are interested in the relative amounts of each phase present in the alloy. These amounts are normally expressed as weight percent (wt%).

In single-phase regions, the amount of the single phase is 100%. However, in two-phase regions we must calculate the amount of each phase. One technique is to perform a materials balance, as shown in Example 9-8.

◇ | **Example 9-8**

Calculate the amounts of α and L at 1250°C in the Cu-40% Ni alloy shown in Figure 9-10.

Answer:

Let's say that x = fraction of the alloy that is solid.

$$(\% \text{ Ni in } \alpha)(x) + (\% \text{ Ni in } L)(1 - x) = (\% \text{ Ni in alloy})$$

By multiplying and rearranging

$$x = \frac{(\% \text{ Ni in alloy}) - (\% \text{ Ni in } L)}{(\% \text{ Ni in } \alpha) - (\% \text{ Ni in } L)}$$

From the phase diagram at 1250°C

$$x = \frac{40 - 32}{45 - 32} = \frac{8}{13} = 0.62$$

If we convert from weight fraction to weight percent, the alloy at 1250°C contains 62% α and 38% L.

◇

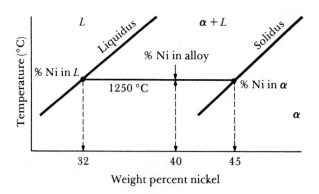

FIGURE 9-10 A tie line at 1250°C in the copper-nickel system that is used in Example 9-8 to find the amount of each phase.

To calculate the amounts of liquid and solid, we construct a lever on our tie line, with the fulcrum of our lever being the original composition of the alloy. The leg of the lever opposite to the composition of the phase whose amount we are calculating is divided by the total length of the lever to give the amount of that phase. In Example 9-8, note that the denominator represents the total length of the tie line and the numerator is the portion of the lever that is opposite the composition of the solid which we are trying to calculate.

The *lever law* in general can be written as

$$\text{Phase percent} = \frac{\text{opposite arm of lever}}{\text{total length of tie line}} \times 100 \qquad (9\text{-}3)$$

We can work the lever law in any two-phase region of a binary phase diagram. The lever law calculation is not used in single-phase regions because the answer is trivial—there is 100% of that phase present.

◇ **Example 9-9**

Determine the amount of each phase in the Cu-40% Ni alloy shown in Figure 9-9 at 1300°C, 1270°C, 1250°C, and 1200°C.

Answer:

1300°C: There is only one phase, so 100% L

$$1270°C: \%\, L = \frac{50 - 40}{50 - 37} \times 100 = 77\%$$

$$\%\, \alpha = \frac{40 - 37}{50 - 37} \times 100 = 23\%$$

$$1250°C: \%\, L = \frac{45 - 40}{45 - 32} \times 100 = 38\%$$

$$\%\, \alpha = \frac{40 - 32}{45 - 32} \times 100 = 62\%$$

1200°C: There is only one phase, so 100% α

◇ **Example 9-10**

Sometimes we wish to express composition as atomic percent (at%) rather than weight percent (wt%). Express the composition of a phase containing 40 wt% Ni-60 wt% Cu in atomic percent.

Answer:

Select as a base 100 g of the alloy, so there are 40 g of nickel and 60 g of copper in the base. From this data, the atomic weights of nickel and copper, and the Avogadro number N_A, we can calculate the atomic percent.

$$\frac{40\,\text{g Ni}}{58.71\,\dfrac{\text{g}}{\text{g} \cdot \text{mole}}} \times 6.02 \times 10^{23}\,\frac{\text{atoms}}{\text{g} \cdot \text{mole}} = 4.1 \times 10^{23}\,\text{Ni atoms}$$

$$\frac{60\,\text{g Cu}}{63.54\,\dfrac{\text{g}}{\text{g} \cdot \text{mole}}} \times 6.02 \times 10^{23}\,\frac{\text{atoms}}{\text{g} \cdot \text{mole}} = 5.7 \times 10^{23}\,\text{Cu atoms}$$

$$\text{at\% Ni} = \frac{4.1 \times 10^{23}}{4.1 \times 10^{23} + 5.7 \times 10^{23}} \times 100 = 42\%$$

$$\text{at\% Cu} = \frac{5.7 \times 10^{23}}{4.1 \times 10^{23} + 5.7 \times 10^{23}} \times 100 = 58\%$$

9–7 RELATIONSHIP BETWEEN STRENGTH AND THE PHASE DIAGRAM

We have previously mentioned that a copper-nickel alloy may be stronger than either pure copper or pure nickel due to solid solution strengthening. The change in mechanical properties of a series of copper-nickel alloys is shown in conjunction with the phase diagram in Figure 9-11.

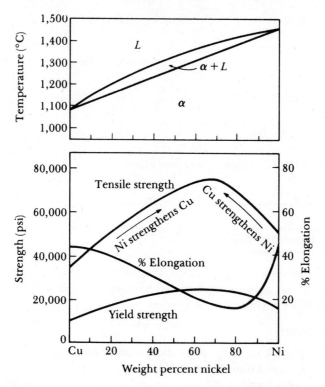

FIGURE 9-11 The mechanical properties of copper-nickel alloys. Copper is strengthened by up to 60% Ni and nickel is strengthened by up to 40% Cu.

The strength of the copper increases by solid solution strengthening until about 60% Ni is added. On the other hand, pure nickel is solid solution strengthened by the addition of copper until 40% Cu is added. The maximum strength is obtained for a Cu-60% Ni alloy, known as *Monel*. The maximum is closer to the pure nickel side of the phase diagram because pure nickel is stronger than pure copper.

9–8   SOLIDIFICATION OF A SOLID SOLUTION ALLOY

When an alloy such as Cu-40% Ni is melted and cooled, soldification requires that both nucleation and growth occur. Heterogeneous nucleation permits little or no undercooling, so solidification begins when the liquid reaches the liquidus temperature. The phase diagram (Figure 9-12), with a tie line drawn at the liquidus temperature, tells us that the first solid to form has a composition of Cu-52% Ni. Growth of the solid requires that the latent heat of fusion, which evolves as the liquid solidifies, be removed from the solid-liquid interface. In

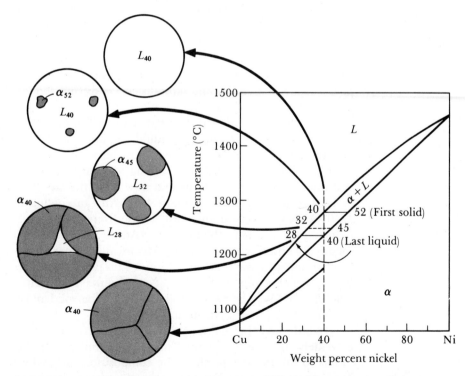

FIGURE 9-12 *The change in structure of a Cu-40% Ni alloy during equilibrium solidification. The nickel and copper atoms must diffuse during cooling in order to satisfy the phase diagram and produce a uniform, equilibrium structure.*

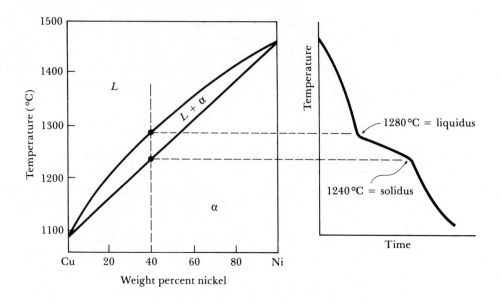

FIGURE 9-13 The cooling curve for an isomorphous alloy during solidification. The changes in slope of the cooling curve indicate the liquidus and solidus temperatures, in this case for a Cu-40% Ni alloy.

addition, diffusion must occur so that the compositions of the solid and liquid phases follow the solidus and liquidus curves during cooling. The latent heat of fusion is removed over a range of temperatures so that the cooling curve shows a change in slope, rather than a flat plateau (Figure 9-13).

At the start of freezing, the liquid contains Cu-40% Ni and the first solid contains Cu-52% Ni. Nickel atoms must have diffused to and concentrated at the first solid to form. But after cooling to 1250°C, solidification has advanced and the phase diagram tells us that now all of the liquid must contain 32% Ni and all of the solid must contain 45% Ni. On cooling from the liquidus to 1250°C, some nickel atoms must diffuse from the first solid to the new solid, reducing the nickel in the first solid. Additional nickel atoms diffuse from the solidifying liquid to the new solid. Meanwhile copper atoms have concentrated, by diffusion, into the remaining liquid. This process must continue until the last liquid, which contains Cu-28% Ni, solidifies and forms a solid containing Cu-40% Ni. Just below the solidus, all of the solid must contain a uniform concentration of 40% Ni throughout.

In order to achieve this equilibrium final structure, the cooling rate must be extremely low. Sufficient time must be permitted for the copper and nickel atoms to diffuse and produce the compositions given by the phase diagram. In most practical casting situations, the cooling rate is too rapid to permit equilibrium.

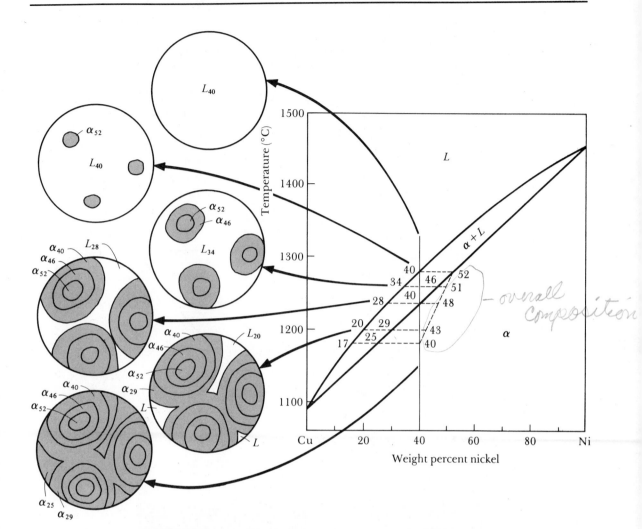

FIGURE 9-14 The change in structure of a Cu-40% Ni alloy during nonequilibrium solidification. Insufficient time for diffusion in the solid produces a segregated structure.

9–9 NONEQUILIBRIUM SOLIDIFICATION OF SOLID SOLUTION ALLOYS

When cooling is too rapid for atoms to diffuse and produce the equilibrium conditions, unusual structures are produced in the casting. Let's see what happens to our Cu-40% Ni alloy on rapid cooling.

Again the first solid, containing 52% Ni, forms on reaching the liquidus temperature (Figure 9-14). On cooling to 1260°C, the tie line tells us that the liquid contains 34% Ni and the solid which forms at that temperature contains

46% Ni. Since diffusion occurs rapidly in liquids, we expect the tie line to accurately predict the liquid composition. However, diffusion in solids is comparatively slow. The first solid that forms still has about 52% Ni, but the new solid contains only 46% Ni. We might find that the average composition of the solid is 51% Ni. This gives a different nonequilibrium solidus than that given by the phase diagram. As solidification continues, the nonequilibrium solidus line continues to separate from the equilibrium solidus.

When the temperature reaches 1240°C, the equilibrium solidus line, a significant amount of liquid remains. We could estimate the amount of liquid by performing a lever law calculation, where the ends of the lever are given by the

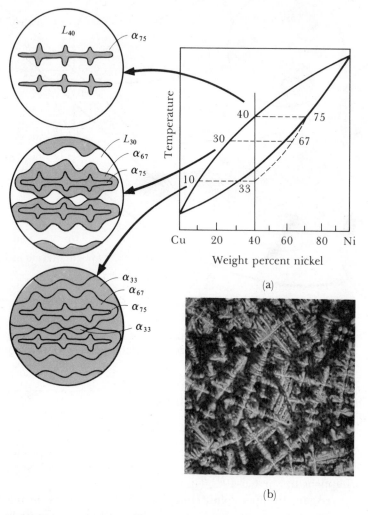

(a)

(b)

FIGURE 9-15 (a) Development of interdendritic segregation during solidification. (b) Photomicrograph of segregation in a copper-nickel alloy (×75).

liquidus point and the nonequilibrium solidus point. The liquid will not completely solidify until we cool to 1190°C, where the nonequilibrium solidus intersects the original composition of 40% Ni. At that temperature liquid containing 17% Ni solidifies, giving solid containing 25% Ni. The average composition of the solid is 40% Ni, but the composition is not uniform.

The actual location of the nonequilibrium solidus line and the final nonequilibrium solidus temperature depend on the cooling rate. Faster cooling rates cause greater departures from equilibrium.

◇ **Example 9-11**

Calculate the composition and amount of each phase in a Cu-40% Ni alloy that is present under the nonequilibrium conditions shown in Figure 9-14 at 1300°C, 1280°C, 1260°C, 1240°C, 1200°C, and 1150°C. Compare with the equilibrium compositions and amounts of each phase.

Answer:

Temperature	Equilibrium		Nonequilibrium	
1300°C	L: 40% Ni 100% L		L: 40% Ni 100% L	
1280°C	L: 40% Ni 100% L		L: 40% Ni 100% L	
	α: 52% Ni ~0% α		α: 52% Ni ~0% α	
1260°C	L: 34% Ni $\dfrac{46-40}{46-34} = 50\%\ L$		L: 34% Ni $\dfrac{51-40}{51-34} = 65\%\ L$	
	α: 46% Ni $\dfrac{40-34}{46-34} = 50\%\ \alpha$		α: 51% Ni $\dfrac{40-34}{51-34} = 35\%\ \alpha$	
1240°C	L: 28% Ni 0% L		L: 28% Ni $\dfrac{48-40}{48-28} = 40\%\ L$	
	α: 40% Ni 100% α		α: 48% Ni $\dfrac{40-28}{48-28} = 60\%\ \alpha$	
1200°C	α: 40% Ni 100% α		L: 20% Ni $\dfrac{43-40}{43-20} = 13\%\ L$	
			α: 43% Ni $\dfrac{40-20}{43-20} = 87\%\ \alpha$	
1150°C	α: 40% Ni 100% α		α: 40% Ni 100% α	

9–10 SEGREGATION

The nonuniform composition produced by nonequilibrium solidification is known as *segregation*. Figure 9-15 shows the development of *interdendritic segregation* or *microsegregation*, sometimes known as *coring*, which occurs over short distances between small dendrite arms. Dendritic growth is typical during the solidification of solid solution alloys, even when no thermal undercooling occurs.

Because of nonequilibrium solidification, the centers of the dendrites, which represent the first solid to freeze, are rich in the higher melting point element in the alloy. The regions between the dendrites are rich in the lower melting point element, since these regions represent the last liquid to freeze. Although we still have just one solid phase α, with an FCC structure, the composition and properties of α differ from one region to the next. We would expect the casting to have poorer properties as a result.

Hot Shortness. Microsegregation can cause *hot shortness*, or melting of the lower melting point interdendritic material at temperatures below the equilibrium solidus. When we heat the Cu-40% Ni alloy to 1225°C, below the equilibrium solidus but above the nonequilibrium solidus, the low-nickel regions between the dendrites melt.

Homogenization. We can reduce the interdendritic segregation and problems with hot shortness by using a *homogenization heat treatment*. If we heat the casting to a temperature below the nonequilibrium solidus and hold at that temperature for a long time, diffusion occurs (Figure 9-16). The nickel atoms in the centers of the dendrites diffuse to the interdendritic regions; copper atoms diffuse in the opposite direction. Since the diffusion distances are relatively short, only a few hours are required to eliminate most of the composition differences.

Macrosegregation. *Macrosegregation* occurs over a large distance, between the surface and the center of the casting (Figure 9-17). The surface, which freezes

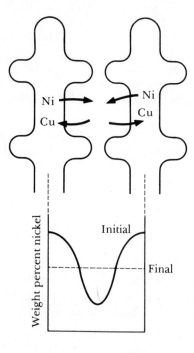

FIGURE 9-16 Microsegregation between dendrites can be reduced by a homogenization heat treatment. Counterdiffusion of nickel and copper atoms may eventually eliminate the composition gradients and produce a homogeneous composition.

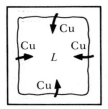

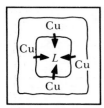

Ni > 40%

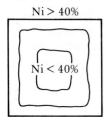

Ni < 40%

FIGURE 9-17
Macrosegregation occurs over a large distance in a casting and cannot be eliminated by homogenization in a practical period of time.

first, contains slightly more than the average amount of the higher melting point metal (nickel is an example). The center of the casting contains more of the lower melting point metal (say copper). Fortunately, the difference in composition is rather small and usually is not significant. We cannot eliminate macrosegregation by a homogenization treatment because the diffusion distances are too great. Macrosegregation can be reduced by *hot working*, which was discussed in Chapter 7.

Other Controls over Segregation. In addition to homogenization or hot working, we may be able to minimize segregation by other means. One of these methods is rapid solidification processing. If cooling is exceptionally rapid, the atoms will be "frozen" into random and uniformly distributed positions in the lattice. In this case, there is not even enough time for the atoms to begin to diffuse to produce the high concentration of high melting point components in the center of the first solid to form.

We may also find, when trying to produce pure materials, that impurities will be rejected into the last regions to freeze, resulting in segregation. This is a problem particularly when producing high-purity silicon used as the basis for semiconducting electronic devices. One way of improving the purity is to use a process known as *zone refining*. A rod of the material is moved at a very slow speed through a furnace with a very narrow high-temperature zone [Figure 9-18(a)]. The rod is melted over only a short length at any particular time. The impurities, which have a low solubility in the solid metal, collect in the molten zone. As the molten zone moves down the rod, the impurities are gradually carried to the end of the rod. After several passes, the impurities will have collected at the end, which is then removed from the rod.

The *Czochralski process* is also used to produce high-purity silicon: in this case, a single seed crystal is dipped into a bath of molten silicon and then is slowly withdrawn. Surface tension causes the molten silicon to adhere to and then solidify onto the seed crystal as it is withdrawn [Figure 9-18(b)]. Segregation problems, introduced due to convection currents in the molten silicon caused by gravity, limit the size of the silicon rods that can be grown. In a laboratory in outer space, where no gravity effects are present, convection currents and segregation can be minimized, and larger, purer rods of silicon can be grown.

9-11 CASTABILITY OF ALLOYS WITH A FREEZING RANGE

In alloys with a long freezing range, a pasty or mushy solid plus liquid region develops before solidification is complete. Liquid metal does not easily flow through this mushy region to compensate for soldification shrinkage: massive risers may not even be sufficient. Consequently, widespread interdendritic shrinkage is typically found. The pure metals and alloys with a short freezing range, on the other hand, tend to produce little or no mushy region; consequently, concentrated shrinkage voids which can be controlled by risers are typically found (Figure 9-19).

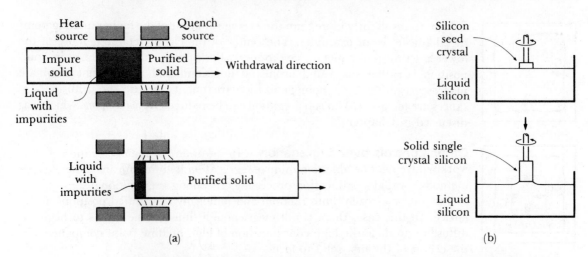

FIGURE 9-18 Producing pure single crystal semiconductor materials by (a) zone refining, where impurities segregate into a narrow molten zone and are carried to the end of the rod, and (b) growth of single crystals by slowly withdrawing a seed crystal from a molten bath of metal.

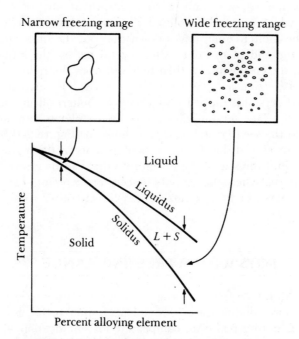

FIGURE 9-19 The effect of freezing range on shrinkage. Narrow freezing range alloys display concentrated shrinkage, while wide freezing range alloys contain dispersed shrinkage.

SUMMARY

Solid solution strengthening is the first of the techniques by which we strengthen a material by alloying. In this chapter we have shown that the strength of single-phase alloys is improved by the controlled addition of point defects. The response of the material to solid solution strengthening is determined by the type of element, in particular the atomic size difference and, at least up to a point, the amount of that element. Solid solution strengthening may provide a significant increase in the strength of the material which may be preserved at elevated temperatures, providing effective barriers against creep.

The amount of solid solution strengthening may be limited by the solubility of the alloying element. However, even in this case the original phase is strengthened by whatever amount of alloying element is soluble. In more complex multiple-phase materials, each phase is often strengthened by solid solution strengthening, even though the bulk of strengthening is by another mechanism.

Finally, solid solution strengthening complicates the solidification process of the material. The material now solidifies over a range of temperatures, providing an opportunity for composition differences to be produced from point to point in the material. This change in the solidification pattern, which is predicted by the phase diagram, introduces additional problems in controlling the solidification process and may even necessitate a homogenization heat treatment or restrict the temperatures at which a material can be used.

GLOSSARY

Binary phase diagram. A phase diagram in which there are only two components.

Czochralski process. A method used to produce single crystals of high-purity materials by a special solidification process.

Freezing range. The temperature difference between the liquidus and solidus temperatures.

Gibbs phase rule. This describes the number of degrees of freedom, or the number of variables that must be fixed to specify the temperature and composition of a phase.

Homogenization heat treatment. The heat treatment used to reduce the segregation caused during nonequilibrium solidification.

Hot shortness. Melting of the lower melting point nonequilibrium material that forms due to segregation, even though the temperature is below the equilibrium solidus temperature.

Hume-Rothery rules. The conditions that an alloy system must meet if the system is to display unlimited solid solubility. Hume-Rothery's rules are necessary but are not sufficient.

Isomorphous phase diagram. A phase diagram that displays unlimited solid solubility.

Lever law. A technique for determining the amount of each phase in a two-phase system.

Limited solubility. When only a maximum amount of a solute material can be dissolved in a solvent material.

Liquidus. The temperature at which the first solid begins to form during solidification.

Macrosegregation. Presence of composition differences in a material over large distances due to nonequilibrium solidification.

Microsegregation. Presence of concentration differences in a material over short distances due to nonequilibrium solidification. Also known as interdendritic segregation or coring.

Phase. A material having the same composition, structure, and properties everywhere under equilibrium conditions.

Phase diagram. A diagram showing the phases and the phase compositions at each combination of temperature and overall composition.

Segregation. Presence of nonequilibrium composition differences in a material, often caused by insufficient time for diffusion during solidification.

Solid solution. A solid phase that contains a mixture of more than one element, with the elements combining to give a uniform composition everywhere.

Solid solution strengthening. Increasing the strength of a material by introducing point defects into the structure in a deliberate and controlled manner.

Solidus. The temperature below which all liquid has completely solidified.

Solubility. The amount of one material that will completely dissolve in a second material without creating a second phase.

Tie line. A horizontal line drawn in a two-phase region of a phase diagram to assist in determining the compositions of the two phases.

Triple point. A pressure and temperature at which three phases are in equilibrium.

Unary phase diagram. A phase diagram in which there is only one component.

Unlimited solubility When the amount of one material that will dissolve in a second material without creating a second phase is unlimited.

Zone refining. A method which uses the segregation of impurities during solidification as a means for producing high-purity materials.

Practice Problems

1. The triple point for water occurs at 0.007 atm and 0.0075°C. Using this information and your knowledge of the behavior of water at atmospheric pressure, construct a schematic unary phase diagram.

2. The unary phase diagram for SiO_2 is shown in Figure 14-6. Locate the triple point where solid, liquid, and vapor coexist and give the temperature and the type of solid present. What do the other "triple" points indicate?

3. Which of the following systems would be expected to display unlimited solid solubility?

(a) Au–Ag (b) Ag–Na (c) Al–Cu (d) Al–Li

4. Which of the following systems would be expected to display unlimited solid solubility?

(a) Cs–K (b) Al–Au (c) U–W (d) Mo–Ta

5. Suppose 1 at % of the following elements are added to aluminum without exceeding the solubility limit. Which one would be expected to give the higher strength alloy? Will any alloying element have unlimited solid solubility in aluminum?

(a) Ge (b) Ag (c) Cu (d) Mg (e) Li (f) Zn

6. Determine the liquidus temperature, solidus temperature, and freezing range for a Cu–25% Ni alloy.

7. Determine the liquidus temperature, solidus temperature, and freezing range for a Cu-50% Ni alloy.

8. Determine the phases present, the compositions of each phase, and the amounts of each phase in wt % for a Cu-25% Ni alloy at 1150°C, 1200°C, and 1250°C.

9. Determine the phases present, the compositions of each phase, and the amounts of each phase in wt % for a Cu–50% Ni alloy at 1250°C, 1300°C, and 1350°C.

10. Figure 9-7(b) shows that NiO and MgO display complete solid solubility. What does this suggest concerning the crystal structures of NiO and MgO? What structure does each have? What does this suggest about the ionic radii and valences of the cations in each ceramic compound? Do NiO and MgO satisfy Hume-Rothery's conditions?

11. Determine the phases present, the compositions of each phase, and the amounts of each phase in mole % for a NiO–50% MgO mixture at 2200°C, 2400°C, and 2600°C.

12. A total of 100 g of NiO and MgO is heated to 2500°C, producing a structure containing 33 mole % liquid and 67 mole % solid. How many grams of NiO must be in the mixture?

13. The MgO–FeO system is shown in Figure 9-20. What crystal structure do you expect each ceramic compound to have? (You may need to review Chapter 3.) Based on ionic radii and valences of the cations in each ceramic compound, do FeO and MgO satisfy Hume-Rothery's conditions?

14. For a MgO-60% FeO mixture, determine (a) the liquidus temperature, (b) the solidus temperature, (c) the composition of each phase at 2200°C, and (d) the amount of each phase at 2200°C.

15. For a MgO–30% FeO mixture, determine (a) the liquidus temperature, (b) the solidus temperature, (c) the composition of each phase at 2400°C, and (d) the amount of each phase at 2400°C.

16. Plot how the freezing range depends on the % FeO in the MgO-FeO system. At what composition is the maximum freezing range obtained?

17. At 2000°C, a mixture of MgO and FeO contains 60% liquid. What is the composition of the original mixture?

18. A mixture of MgO and FeO is partly melted, giving a solid containing 60% FeO. Determine the temperature at which the mixture is held. Can you determine the composition of the original mixture? Explain and point out whether this result confirms the phase rule.

19. A MgO-40% FeO ceramic is heated until only 25% solid remains. By trial and error, estimate the temperature to which the ceramic was heated.

20. Figure 9-21 shows the Nb-W phase diagram. Verify that this system satisfies Hume-Rothery's rules.

21. For a Nb-70% W alloy, determine (a) the liquidus temperature, (b) the solidus temperature, (c) the composition of each phase at 3000°C, and (d) the amount of each phase at 3000°C.

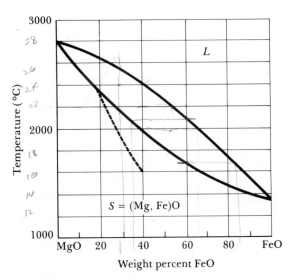

FIGURE 9-20 The equilibrium phase diagram for the MgO–FeO system.

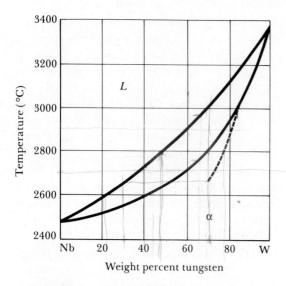

FIGURE 9-21 *The equilibrium phase diagram for the Nb-W system.*

22. Plot how the freezing range depends on the % W in the Nb-W system. At what composition is the maximum freezing range obtained? Would you expect shrinkage defects to be more concentrated in a Nb-70% W alloy or a Nb-30% W alloy?

23. When a Nb-W alloy is heated to 2800°C, only 33% solid remains. Determine the composition of the original alloy.

24. A Nb-60% W alloy is heated until 40% liquid is produced. By trial and error, determine the temperature that must have been obtained.

25. A Nb-30% W alloy is heated to 2600°C, producing a mixture of solid and liquid. Determine (a) the composition of the original alloy in atomic percent, (b) the composition of the solid and liquid phases in both wt % and at %, and (c) the amount of solid and liquid phases on a weight basis.

26. A MgO-40% FeO ceramic is heated to 2300°C, producing a mixture of solid and liquid. Determine (a) the composition of the original alloy in mole percent, (b) the composition of the solid and liquid phases in both

wt% and mol%, and (c) the amount of solid and liquid phases on a weight basis.

27. Suppose we prepare a Nb–W alloy by mixing together equal numbers of atoms of each, then heat the alloy to 2800°C. Calculate the composition of the alloy in wt%, then determine which phases are present.

28. Suppose we combine 50 cm³ of Nb with 30 cm³ of W and heat to 2800°C. Calculate the composition of the alloy in wt %, then determine which phases are present.

29. A MgO-60% FeO ceramic is allowed to cool from the liquid. Under equilibrium conditions, calculate the composition of (a) the first solid to form on freezing and (b) the last liquid to solidify.

30. A Nb-70% W alloy is allowed to cool from the liquid. Under equilibrium conditions, calculate the composition of (a) the first solid to form on freezing and (b) the last liquid to solidify.

31. Consider a Nb-70% W alloy which cools rapidly to produce the nonequilibrium solidus shown as a dashed line in Figure 9-21. (a)

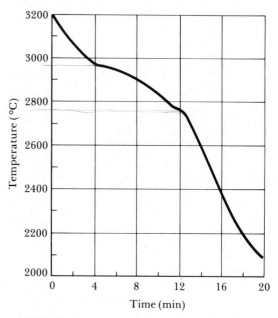

FIGURE 9-22 *Cooling curve of Nb-W alloy for Problem 9-33.*

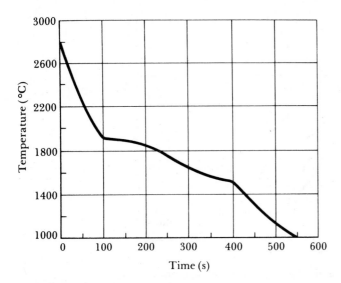

FIGURE 9-23 Cooling curve of MgO-FeO ceramic for Problem 9-34.

What is the composition of the first solid to form? (b) Determine the composition and amount of each phase present during equilibrium solidification at 2900°C, 2800°C, and 2700°C. (c) Determine the composition and amount of each phase present during nonequilibrium solidification at 2900°C, 2800°C, and 2700°C. (d) Under equilibrium conditions, determine the compositions of the last liquid to freeze and the last solid to form. (e) Under nonequilibrium conditions, determine the compositions of the last liquid to freeze and the last solid to form. (f) Compare the equilibrium and nonequilibrium solidus temperatures.

32. Consider a MgO-40% FeO ceramic which cools to produce the nonequilibrium solidus shown as a dashed line in Figure 9-20. (a) What is the composition of the first solid to form? (b) Determine the composition and amount of each phase present during equilibrium solidification at 2200°C, 2000°C, and 1800°C. (c) Determine the composition and amount of each phase present during nonequilibrium solidification at 2200°C, 2000°C, and 1800°C. (d) Under equilibrium conditions, determine the

compositions of the last liquid to freeze and the last solid to form. (e) Under nonequilibrium conditions, determine the compositions of the last liquid to freeze and the last solid to form. (f) Compare the equilibrium and nonequilibrium solidus temperatures.

33. Consider the cooling curve for the Nb–W alloy shown in Figure 9-22. Determine (a) the liquidus temperature, (b) the solidus temperature. (c) the freezing range, (d) the pouring temperature, (e) the superheat, (f) the local solidification time, (g) the total solidification time, and (h) the composition of the alloy.

34. Consider the cooling curve for the MgO–FeO ceramic shown in Figure 9-23. Determine (a) the liquidus temperature, (b) the solidus temperature, (c) the freezing range, (d) the pouring temperature, (e) the superheat, (f) the local solidification time, (g) the total solidification time, and (h) the composition of the ceramic.

35. Cooling curves are shown in Figure 9-24 for the bismuth-antimony system. Based on these curves, construct the Bi-Sb phase diagram.

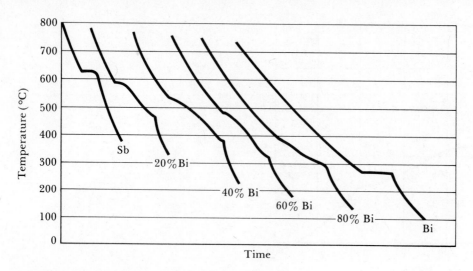

FIGURE 9-24   Cooling curves for several alloys in the Bi-Sb system (Problem 9-35).

10

SOLIDIFICATION AND DISPERSION STRENGTHENING

10–1 INTRODUCTION

When the solubility of a material is exceeded by adding too much of an alloying element, a second phase forms and a two-phase alloy is produced. The boundary between the two phases is a surface at which the atomic arrangement is not perfect. As a result, this boundary interferes with the slip of dislocations and strengthens the material. The general term for strengthening by the introduction of a second phase is *dispersion strengthening*.

In this chapter we will first discuss the fundamentals of dispersion strengthening to determine the structure we should aim to produce. Next we will examine the types of reactions that produce multiple-phase alloys. Finally, we will look in some detail at how we control dispersion strengthening by controlling the solidification process.

10–2 PRINCIPLES OF DISPERSION STRENGTHENING

More than one phase must be present in any dispersion-strengthened alloy. We call the continuous phase, which usually is present in larger amounts, the *matrix*. The second phase, usually present in smaller amounts, is called the *precipitate*. In some cases, two phases form simultaneously. We will define these structures differently, often calling the intimate mixture of phases a *microconstituent*.

There are some general considerations for determining how the characteristics of the matrix and precipitate affect the overall properties of the alloy (Figure 10-1).

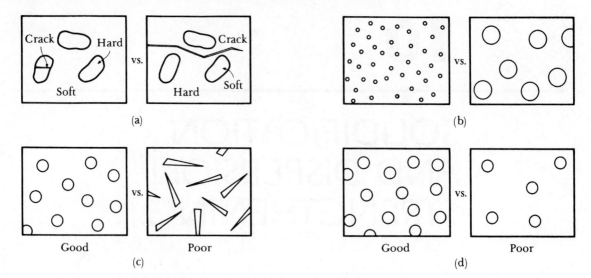

FIGURE 10-1 *Considerations for effective dispersion strengthening. (a) The precipitate should be hard and discontinuous, (b) the precipitate particles should be small and numerous, (c) the precipitate particles should be round rather than needlelike, and (d) larger amounts of precipitate increase strengthening.*

1. The matrix should be soft and ductile, while the precipitate should be hard and brittle. The precipitate acts as a very strong obstacle to slip of dislocations in the matrix. However, the matrix provides at least some ductility to the overall alloy.

2. The hard, brittle precipitate should be discontinuous, while the soft, ductile matrix should be continuous. If the precipitate were continuous, cracks could propagate through the entire structure. However, cracks in the discontinuous, brittle precipitate are arrested by the precipitate-matrix interface.

3. The precipitate particles should be small and numerous, increasing the likelihood that they interfere with the slip process.

4. The precipitate particles should be round, rather than needlelike or sharp-edged. The rounded shape is less likely to initiate a crack or to act as a notch.

5. Larger amounts of the precipitate increase the strength of the alloy.

10–3   INTERMETALLIC COMPOUNDS

Often dispersion-strengthened alloys contain an intermetallic compound. An *intermetallic compound* is made up of two or more elements, producing a new phase with its own composition, crystal structure, and properties. Intermetallic compounds are almost always very hard and brittle but can provide excellent dispersion strengthening of the softer matrix.

Stoichiometric intermetallic compounds have a fixed composition. Steels are strengthened by a stoichiometric intermetallic compound, Fe_3C, that has a fixed

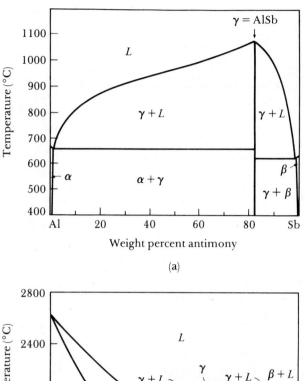

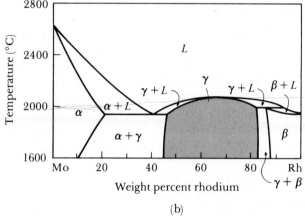

FIGURE 10-2   *The aluminum-antimony phase diagram includes a stoichiometric intermetallic compound* γ. *(b) The molybdenum-rhodium phase diagram includes a nonstoichiometric intermetallic compound* γ.

ratio of three iron atoms to one carbon atom. Stoichiometric intermetallic compounds are usually easy to detect in phase diagrams because they are represented by a vertical line [Figure 10-2(a)].

Nonstoichiometric intermetallic compounds can have a range of compositions. In the molybdenum-rhodium system, the γ phase is an intermetallic compound [Figure 10-2(b)]. Because the molybdenum-rhodium atom ratio is not rigidly fixed, the γ phase can contain from 45 wt % to 83 wt % Rh at 1600°C.

Usually intermetallic compounds are used to best advantage by dispersing them into a softer, more ductile matrix, as we will see in this and subsequent

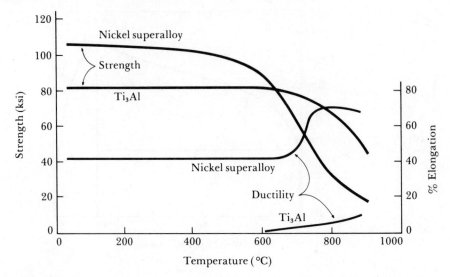

FIGURE 10-3 *The strength and ductility of the intermetallic compound Ti$_3$Al compared with that of a conventional nickel superalloy.*

chapters. However, there is considerable interest in using intermetallics by themselves, taking advantage of their high melting point, stiffness, and resistance to oxidation and creep. A variety of new materials based on intermetallic compounds are being developed, particularly for aircraft and engine components. These new materials, which include Ti$_3$Al and Ni$_3$Al, maintain their strength and even develop usable ductility at elevated temperatures (Figure 10-3). Special care is needed to improve the ductility of these normally brittle materials, and processing the intermetallics into useful shapes is still very difficult.

◇ | **Example 10-1**

A stoichiometric compound forms in the calcium-magnesium system at 45.3% Ca. What is the formula of the intermetallic compound?

Answer:

We must convert weight percent to atomic percent to obtain the atom ratio. The atomic mass of calcium is 40.08 g/g · mole and the atomic mass of magnesium is 24.31 g/g · mole. Assume a base of 100 g of the alloy.

$$\text{Ca atoms} = \frac{45.3 \text{ g Ca}}{40.08 \dfrac{\text{g}}{\text{g} \cdot \text{mole}}} \left(6.02 \times 10^{23} \frac{\text{atoms}}{\text{g} \cdot \text{mole}} \right) = 6.8 \times 10^{23}$$

$$\text{Mg atoms} = \frac{54.7 \text{ g Mg}}{24.31 \dfrac{\text{g}}{\text{g} \cdot \text{mole}}} \left(6.02 \times 10^{23} \frac{\text{atoms}}{\text{g} \cdot \text{mole}} \right) = 13.5 \times 10^{23}$$

$$\frac{\text{Mg atoms}}{\text{Ca atoms}} = \frac{13.5 \times 10^{23}}{6.8 \times 10^{23}} = 2$$

There are two magnesium atoms for each calcium atom, so the compound must be Mg_2Ca.

◇

10–4 PHASE DIAGRAMS CONTAINING THREE-PHASE REACTIONS

Many combinations of two elements produce more complicated phase diagrams than the isomorphous systems. These systems contain reactions that involve three separate phases, as defined in Figure 10-4.

Each of the reactions can be identified in a complex phase diagram by the following procedure.

1. Locate a horizontal line on the phase diagram. The horizontal line, which indicates the presence of a three-phase reaction, represents the temperature at which the reaction occurs under equilibrium conditions.

Eutectic	$L \rightarrow \alpha + \beta$	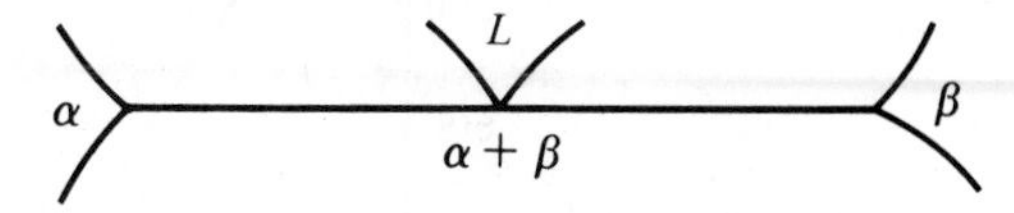
Peritectic	$\alpha + L \rightarrow \beta$	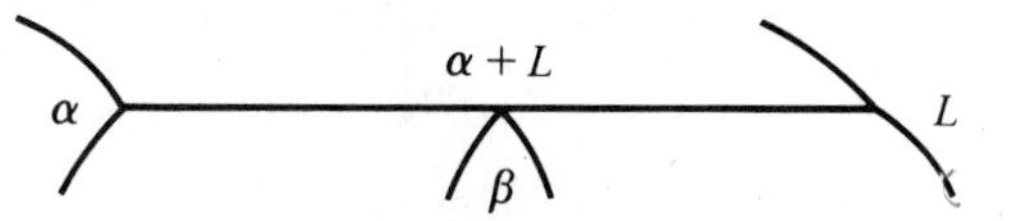
Monotectic	$L_1 \rightarrow L_2 + \alpha$	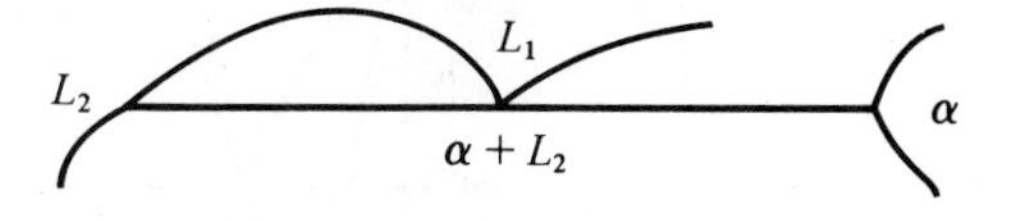
Eutectoid	$\gamma \rightarrow \alpha + \beta$	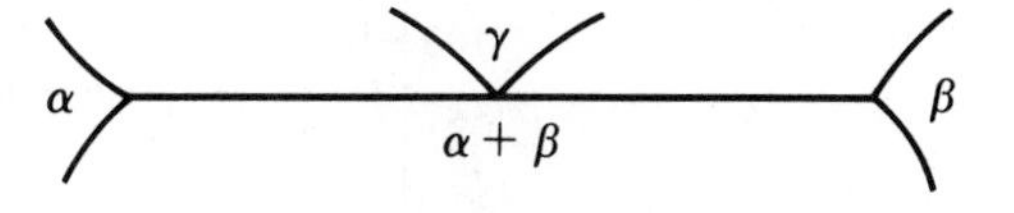
Peritectoid	$\alpha + \beta \rightarrow \gamma$	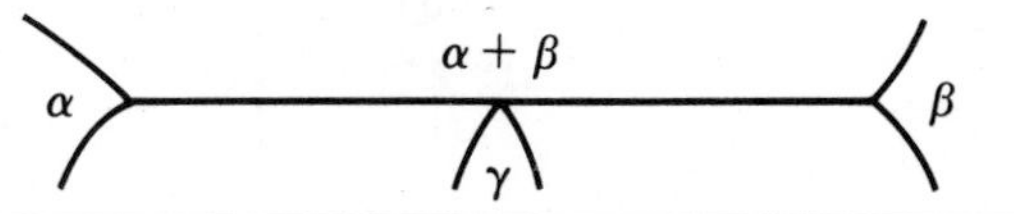

one liquid forms 2 solids

FIGURE 10-4 The five most important three-phase reactions in binary phase diagrams.

2. Locate three distinct points on the horizontal line—the two endpoints plus a third point, often near the center of the horizontal line. The center point represents the composition at which the three-phase reaction occurs.

3. Look immediately above the center point and identify the phase or phases present; look immediately below the center point and identify the phase or phases present. Then write in reaction form the phase(s) above the center point transforming to the phase(s) below the point. Compare this reaction with those in Figure 10-4 to identify the reaction.

◇   **Example 10-2**

Consider the phase diagram in Figure 10-5. Identify the three-phase reactions that occur.

Answer:

We find horizontal lines at 2000°C, 1400°C, 1100°C, 600°C, and 400°C.

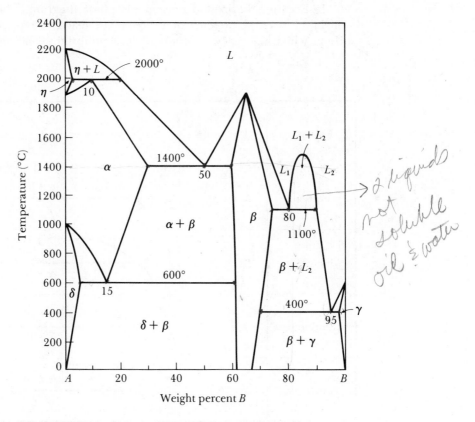

FIGURE 10-5 A hypothetical phase diagram for Example 10-2.

2000°C: The center point is at 10% B; $\eta + L$ are present above the point, α is present below. The reaction is

$$\eta + L \rightarrow \alpha \text{ or a } \textit{peritectic}$$

1400°C: This reaction occurs at 50% B.

$$L \rightarrow \alpha + \beta \text{ or a } \textit{eutectic}$$

1100°C: This reaction occurs at 80% B.

$$L_1 \rightarrow \beta + L_2 \text{ or a } \textit{monotectic}$$

600°C: This reaction occurs at 15% B.

$$\alpha \rightarrow \delta + \beta \text{ or a } \textit{eutectoid}$$

400°C: This reaction occurs at 95% B.

$$L_2 \rightarrow \beta + \gamma \text{ or a } \textit{eutectic}$$

The eutectic, peritectic, and monotectic reactions are a part of the solidification process. Alloys used for casting or soldering often take advantage of the low melting point of the eutectic reaction. Monotectic alloys improve the machining characteristics of the finished product, as in many brass and bronze alloys. Peritectic reactions are avoided if possible because nonequilibrium solidification and segregation frequently accompany the reaction.

The eutectoid and peritectoid reactions are completely solid-state reactions. The eutectoid forms the basis for the heat treatment of several alloy systems, including steel. The peritectoid reaction is extremely slow, producing undesirable, nonequilibrium structures in alloys.

All of these three-phase reactions occur at a fixed temperature and composition. The Gibbs phase rule for a three-phase reaction is

$$F = C - P + 1 = 2 - 3 + 1 = 0$$

since there are two components C in a binary phase diagram and three phases P are involved in the reaction. When the three phases are in equilibrium during the reaction, there are no degrees of freedom. The temperature and the composition of each phase involved in the three-phase reaction are fixed.

◇ **Example 10-3**

What three-phase reaction occurs at the aluminum-rich end of the Al-Cu phase diagram, (Figure 11-1)? According to the phase rule, the temperature and composition of each phase must be fixed in the three-phase reaction. What are the temperature of the reaction and the compositions of each phase in the reaction?

Answer:

The horizontal line at 548°C represents a three-phase reaction. Above the center point at 33.2% Cu, the alloy is all liquid. Below the center point, $\alpha + \theta$ form. Thus the reaction $L \rightarrow \alpha + \theta$ is a eutectic.

The temperature of the reaction is 548°C. The compositions of each phase are given by the three points on the line.

   α: 5.65% Cu θ: 52.5% Cu L: 33.2% Cu

 Example 10-4

What three-phase reaction occurs at 1700°C in the SiO_2–CaO ceramic phase diagram [Figure 14-22(c)]?

Answer:

The reaction occurs near 5% CaO. Above 1700°C, only liquid is present. Below 1700°C, we must be in a two-phase region containing almost pure SiO_2 and liquid. During the reaction, L_1 containing about 5% CaO transforms to SiO_2 plus L_2 containing about 30% CaO. Therefore, a monotectic reaction has occurred. The presence of the dome-shaped miscibility gap should have helped us come to this conclusion.

10–5 THE EUTECTIC PHASE DIAGRAM

The lead-tin system contains only a simple eutectic reaction (Figure 10-6). This alloy system is the basis for the most common alloys used for soldering. Let's examine four classes of alloys in this system.

Solid Solution Alloys. Alloys that contain 0% to 2% Sn behave exactly like the copper-nickel alloys; a single-phase solid solution α forms during solidification and remains after the alloy has cooled to room temperature (Figure 10-7). These alloys are strengthened by solid solution strengthening, by strain hardening, and by controlling the solidification process to refine the grain structure.

Alloys That Exceed the Solubility Limit. Alloys containing between 2% and 19% Sn also solidify to produce a single solid solution α. However, as the alloy continues to cool, a solid-state reaction occurs, permitting a second solid phase, β, to precipitate from the original α phase (Figure 10-8).

The α is a solid solution of tin in lead. Because Hume-Rothery's conditions are not satisfied, however, the solubility of tin in the α solid solution is limited.

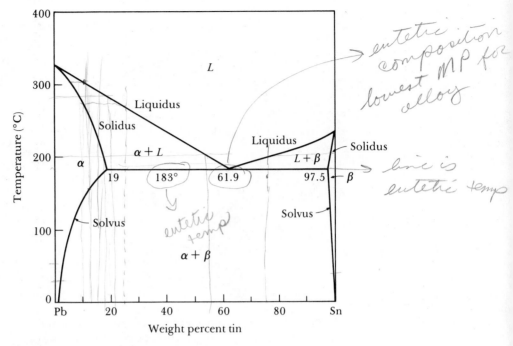

eutectic composition lowest MP for alloy

line is eutectic temp

eutectic temp

FIGURE 10-6   The lead-tin equilibrium phase diagram.

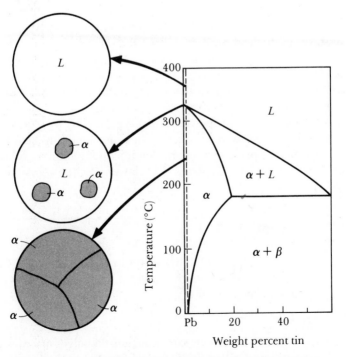

FIGURE 10-7   Solidification and microstructure of a Pb-1% Sn alloy. The alloy is a single-phase solid solution.

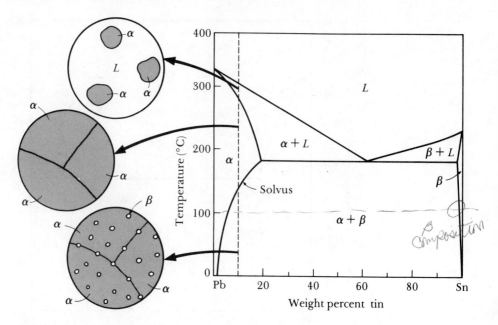

FIGURE 10-8   *Solidification, precipitation, and microstructure of a Pb-10% Sn alloy. Some dispersion strengthening occurs as the β solid precipitates.*

At 0°C, only 2% Sn can dissolve in α. As the temperature increases, more tin dissolves in the lead until, at 183°C, the solubility of tin in lead has increased to 19% Sn. This is the *maximum solubility* of tin in lead. The solubility of tin in solid lead at any temperature is given by the *solvus* line. Any alloy containing between 2% and 19% Sn cools past the solvus, the solubility limit is exceeded, and a small amount of β forms to satisfy the phase diagram.

We control the properties of this type of alloy by several techniques, including solid solution strengthening of the α portion of the structure, controlling the microstructure produced during solidification, and controlling the amount and characteristics of the β phase. This latter mechanism is one type of dispersion strengthening.

◇ | **Example 10-5**

Determine (a) the solubility of tin in solid lead at 100°C, (b) the maximum solubility of lead in solid tin, and (c) the amount of β that forms if a Pb-10% Sn alloy is cooled to 0°C.

Answer:

(a) The 100°C temperature intersects the solvus line at 5% Sn. The solubility of tin in lead at 100°C is therefore 5% Sn.

(b) The maximum solubility of lead in tin, which is found from the tin-rich side of the phase diagram, occurs at the eutectic temperature of 183°C and 2.5% Pb.

(c) At 0°C, the 10% Sn alloy is in an $\alpha + \beta$ region of the phase diagram. By drawing a tie line at 0°C and working the lever law, we find that

α: 2% Sn β: 100% Sn

$$\%\beta = \frac{10 - 2}{100 - 2} \times 100 = 8.2\%$$

Eutectic Alloys.   The alloy containing 61.9% Sn has the eutectic composition (Figure 10-9). Above 183°C the alloy is all liquid and therefore must contain 61.9% Sn. After the liquid cools to 183°C, the eutectic reaction begins.

$$L_{61.9\% \, Sn} \rightarrow \alpha_{19\% \, Sn} + \beta_{97.5\% \, Sn}$$

Two solid solutions—α and β—are formed during the eutectic reaction. The compositions of the two solid solutions are given by the ends of the eutectic line.

Diffusion must occur during the eutectic reaction, as liquid containing 61.9% Sn transforms to the lead-rich solid α and the tin-rich solid β. During solidification, growth of the eutectic requires both removal of the latent heat of fusion and redistribution of the two different atom species by diffusion. Since solidification occurs completely at 183°C, the cooling curve (Figure 10-10) is

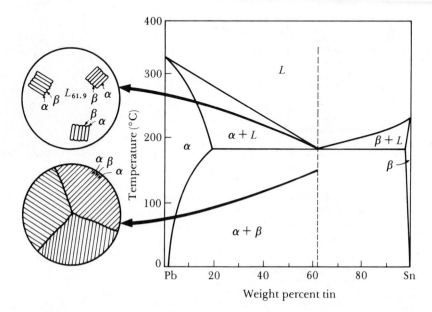

FIGURE 10-9   Solidification and microstructure of the eutectic alloy Pb-61.9% Sn.

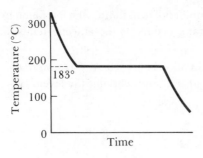

FIGURE 10-10 The cooling curve for a eutectic alloy is a simple thermal arrest, since eutectics freeze or melt at a single temperature.

similar to that of a pure metal; that is, a thermal arrest or plateau occurs at the eutectic temperature.

In order for atoms to be redistributed during eutectic solidification, a characteristic microstructure must develop. In the lead-tin system, the solid α and β phases grow from the liquid in a *lamellar*, or platelike, arrangement (Figure 10-11). The lamellar structure permits the lead and tin atoms to move through the liquid, in which diffusion is easy, without having to move an appreciable distance. This lamellar structure is characteristic of numerous other eutectic systems.

The product of the eutectic reaction is a unique and characteristic arrangement of the two solid phases called the *eutectic microconstituent*. In the Pb-61.9% Sn alloy, 100% of the eutectic microconstituent is formed, since all of the liquid goes through the reaction.

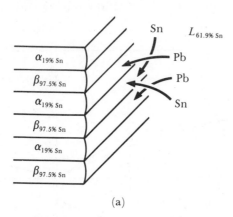

(a)

(b)

FIGURE 10-11 (a) Atom redistribution during lamellar growth of a lead-tin eutectic. Tin atoms from the liquid preferentially diffuse to the β plates and lead atoms diffuse to the α plates. (b) Photomicrograph of the lead-tin eutectic microconstituent ($\times$ 400).

◇ **Example 10-6**

Determine the amount and composition of each phase that is present in the eutectic microconstituent in a lead-tin alloy.

Answer:

The eutectic microconstituent contains 61.9% Sn. We work the lever law at a temperature just below the eutectic, say at 182°C, since that is the temperature at which the eutectic reaction is just completed. The fulcrum of our lever is 61.9% Sn. The ends of the tie line coincide approximately with the ends of the eutectic line.

$$\alpha: \text{Pb-19\% Sn} \qquad \% \ \alpha \ = \ \frac{97.5 \ - \ 61.9}{97.5 \ - \ 19} \times 100 \ = \ 45\%$$

$$\beta: \text{Pb-97.5\% Sn} \qquad \% \ \beta \ = \ \frac{61.9 \ - \ 19}{97.5 \ - \ 19} \times 100 \ = \ 55\%$$

opposite arm divided by total length

◇

Hypoeutectic and Hypereutectic Alloys.   As an alloy containing between 19% and 61.9% Sn cools, the liquid begins to solidify at the liquidus temperature. However, solidification is completed by going through the eutectic reaction (Figure 10-12). This solidification sequence occurs any time the vertical line corresponding to the original composition of the alloy crosses both the liquidus and the eutectic.

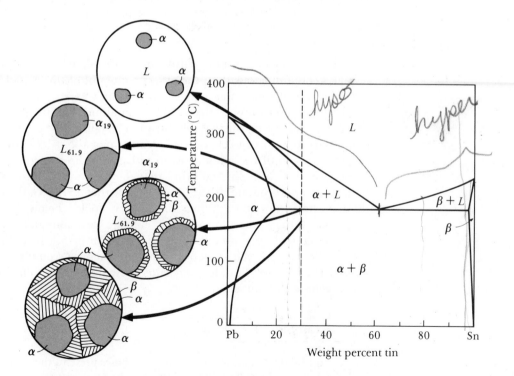

FIGURE 10-12   The solidification and microstructure of a hypoeutectic alloy (Pb-30% Sn).

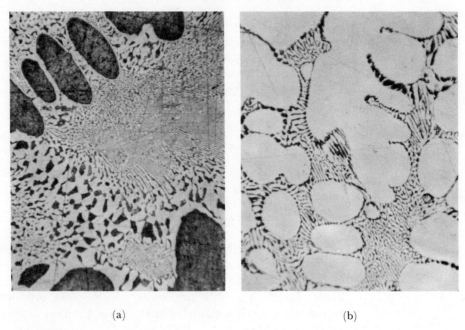

(a)   (b)

FIGURE 10-13 (a) A hypoeutectic lead-tin alloy. (b) A hypereutectic lead-tin alloy. The dark constituent is the lead-rich solid α, the light constituent is the tin-rich solid β, and the fine plate structure is the eutectic ($\times$ 400).

Alloys with compositions of 19% to 61.9% Sn are called *hypoeutectic* alloys, or alloys containing less than the eutectic amount of tin. An alloy to the right of the eutectic composition, between 61.9% and 97.5% Sn, is *hypereutectic*.

Let's consider a hypoeutectic alloy containing Pb-30% Sn and follow the changes in structure during solidification (Figure 10-12). On reaching the liquidus temperature of 260°C, solid α containing about 12% Sn nucleates. The solid α grows until the alloy cools to just above the eutectic temperature. At 184°C, we draw a tie line and find that the solid α contains 19% Sn and the remaining liquid contains 61.9% Sn. We note that at 184°C, the liquid contains the eutectic composition! When the alloy is cooled below 183°C, all of the remaining liquid goes through the eutectic reaction and transforms to a lamellar mixture of α and β. The microstructure shown in Figure 10-13(a) results. Notice that the eutectic microconstituent surrounds the solid α that formed between the liquidus and eutectic temperatures. The eutectic microconstituent is continuous.

◇ | **Example 10-7**

For a Pb-30% Sn alloy, determine the phases present, their amounts, and their compositions at 300°C, 200°C, 184°C, 182°C, and 0°C.

Answer:

Temperature (°C)	Composition		Amount
300	L	L: 30% Sn	$L = 100\%$
200	$\alpha + L$	L: 55% Sn	$L = \dfrac{30 - 18}{55 - 18} \times 100 = 32\%$
		α: 18% Sn	$\alpha = \dfrac{55 - 30}{55 - 18} \times 100 = 68\%$
184	$\alpha + L$	L: 61.9% Sn	$L = \dfrac{30 - 19}{61.9 - 19} \times 100 = 26\%$
		α: 19% Sn	$\alpha = \dfrac{61.9 - 30}{61.9 - 19} \times 100 = 74\%$
182	$\alpha + \beta$	α: 19% Sn	$\alpha = \dfrac{97.5 - 30}{97.5 - 19} \times 100 = 86\%$
		β: 97.5% Sn	$\beta = \dfrac{30 - 19}{97.5 - 19} \times 100 = 14\%$
0	$\alpha + \beta$	α: 2% Sn	$\alpha = \dfrac{100 - 30}{100 - 2} \times 100 = 71\%$
		β: 100% Sn	$\beta = \dfrac{30 - 2}{100 - 2} \times 100 = 29\%$

We call the solid α phase that formed when we cooled from the liquidus to the eutectic the *primary* or *proeutectic microconstituent*. This solid α did not take part in the eutectic reaction. Often we find that the amounts and compositions of the microconstituents are of more use to us than the amounts and compositions of the phases.

◇ **Example 10-8**

Determine the amounts and compositions of each microconstituent in a Pb-30% Sn alloy immediately after the eutectic reaction has been completed.

Answer:

The microconstituents are primary α and eutectic. We can determine their amounts and compositions if we look at how they form. The primary α is all of the solid α that forms before the alloy cools to the eutectic temperature; the eutectic microconstituent is all of the liquid that goes through the reaction. At a temperature just above the eutectic, say 184°C, the amounts and compositions of the two phases are

$$\alpha: 19\% \text{ Sn} \qquad \% \ \alpha \ = \ \frac{61.9 - 30}{61.9 - 19} \times 100 \ = \ 74\% \ = \ \% \text{ primary } \alpha$$

$$L: 61.9\% \text{ Sn} \qquad \% \ L \ = \ \frac{30 - 19}{61.9 - 19} \times 100 \ = \ 26\% \ = \ \% \text{ eutectic}$$

When the alloy cools below the eutectic to 182°C, nothing changes except that the liquid at 184°C transforms to eutectic. The solid α present at 184°C remains unchanged and is the primary microconstituent.

The cooling curve for the type of alloy in Example 10-8 is a composite of those for solid solution alloys and "straight" eutectic alloys (Figure 10-14). A change in slope occurs at the liquidus as primary α begins to form. Evolution of the latent heat of fusion slows the cooling rate as the solid α grows. When the alloy cools to the eutectic temperature, a thermal arrest is produced as the eutectic reaction proceeds at 183°C. The solidification sequence is similar in a hypereutectic alloy, giving the microstructure shown in Figure 10-13(b).

10–6 STRENGTH OF EUTECTIC ALLOYS

Each phase in the eutectic alloy is, to some degree, solid solution strengthened. In the lead-tin system, α, which is a solid solution of tin in lead, is stronger than pure lead. Some eutectic alloys can be strengthened by cold working. We also

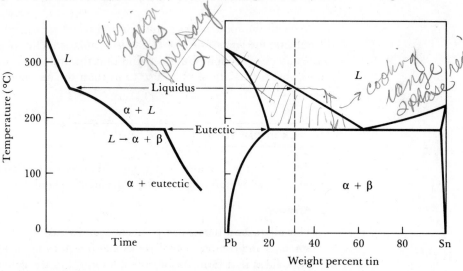

FIGURE 10-14 The cooling curve for a hypoeutectic Pb-30% Sn alloy.

control grain size by adding appropriate inoculants or grain refiners. However, we can best influence the properties by controlling the amount, size, shape, and distribution of the two solid phases in the eutectic.

Eutectic Grain Size.   *Eutectic grains* each nucleate and grow independently. Within each grain, the orientation of the lamellae in the eutectic microconstituent is identical. The orientation changes on crossing a eutectic grain boundary [Figure 10-18(a)]. We are able to refine the eutectic grain size, and consequently improve the strength of the eutectic alloy, by inoculation.

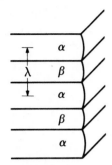

FIGURE 10-15
The interlamellar spacing in a eutectic microstructure.

Interlamellar Spacing.   The *interlamellar spacing* of a eutectic is the distance from the center of one α lamella to the center of the next α lamella (Figure 10-15). A small interlamellar spacing indicates that the individual lamellae are thin and consequently the amount of α-β interface area is large. A small interlamellar spacing therefore suggests that the strength of the eutectic is high.

The interlamellar spacing is determined primarily by the growth rate of the eutectic.

$$\lambda = cR^{-1/2} \qquad\qquad (10\text{-}1)$$

where R is the growth rate (cm/s) and c is a constant. The interlamellar spacing for the lead-tin eutectic is shown in Figure 10-16. We can increase the growth rate R, and consequently reduce the interlamellar spacing, by increasing the cooling rate or reducing the solidification time.

Amount of Eutectic.   By changing the composition of the alloy away from the eutectic composition, a primary microconstituent will form. We control the properties of the alloy by controlling the relative amounts of the primary microconstituent and the eutectic. In the lead-tin system, we find that as

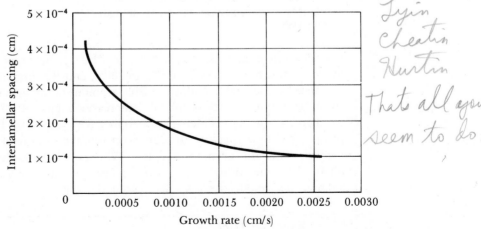

FIGURE 10-16   The effect of growth rate on the interlamellar spacing in the lead-tin eutectic.

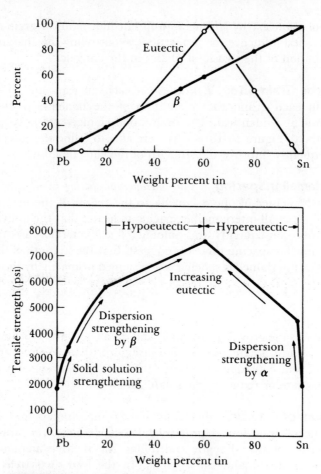

FIGURE 10-17 *The effect of composition and strengthening mechanism on the tensile strength of lead-tin alloys.*

the tin content increases from 19% to 61.9%, the amount of the eutectic microconstituent changes from 0% to 100% and the amount of the primary α is reduced accordingly. With increasing amounts of the stronger eutectic microconstituent, the strength of the alloy increases (Figure 10-17). Similarly, when we increase the lead added to tin from 2.5% to 38.1% Pb, the amount of primary β in the hypereutectic alloy decreases, the amount of the strong eutectic increases, and the strength increases. Whenever both individual phases have about the same strength, the eutectic alloy is expected to have the highest strength due to effective dispersion strengthening.

◇ | **Example 10-9**

Calculate the total % β and the % eutectic for the following lead-tin alloys: 10% Sn, 20% Sn, 50% Sn, 60% Sn, 80% Sn, 95% Sn.

Answer:

From the phase diagram (Figure 10-6) we want to calculate the total amount of β at room temperature, about 25°C, while we calculate the eutectic just above the eutectic temperature.

Alloy	% β	% Eutectic
Pb-10% Sn	$\dfrac{10-2}{100-2} \times 100 = 8.2\%$	0%
Pb-20% Sn	$\dfrac{20-2}{100-2} \times 100 = 18.4\%$	$\dfrac{20-19}{61.9-19} \times 100 = 2.3\%$
Pb-50% Sn	$\dfrac{50-2}{100-2} \times 100 = 49\%$	$\dfrac{50-19}{61.9-19} \times 100 = 72.3\%$
Pb-60% Sn	$\dfrac{60-2}{100-2} \times 100 = 59.2\%$	$\dfrac{60-19}{61.9-19} \times 100 = 95.6\%$
Pb-80% Sn	$\dfrac{80-2}{100-2} \times 100 = 79.6\%$	$\dfrac{97.5-80}{97.5-61.9} \times 100 = 49.2\%$
Pb-95% Sn	$\dfrac{95-2}{100-2} \times 100 = 94.9\%$	$\dfrac{97.5-95}{97.5-61.9} \times 100 = 7\%$

The % β and % eutectic are included in Figure 10-17 to show their relationship to the tensile strength of lead-tin alloys.

Shape of the Eutectic.   The shapes of the two phases in the eutectic microconstituent are influenced by the cooling rate, the presence of impurity elements, and the nature of the alloy (Figure 10-18).

In some alloys, the eutectic microconstituent is composed of thin, discontinuous plates of a brittle phase. The aluminum-silicon eutectic phase diagram (Figure 10-19) forms the basis for a number of important commercial alloys. Unfortunately, the silicon portion of the eutectic grows as thin, flat plates that appear needlelike in a photomicrograph [Figure 10-18(b)]. The silicon platelets concentrate stresses and reduce ductility and toughness.

The nature of the eutectic structure in aluminum-silicon alloys is altered by modification. *Modification* causes the silicon phase to grow as thin, interconnected rods between aluminum dendrites [Figure 10-18(c)], improving both tensile strength and % elongation (Figure 10-20). In two dimensions, the modified silicon appears to be composed of tiny, round particles. Rapidly cooled alloys, such as those in die casting, are naturally modified during solidification. At slower cooling rates, however, 0.02% Na or 0.01% Sr must be added to cause modification.

The shape of the primary phase is also important. Often the primary phase grows in a dendritic manner; decreasing the secondary dendrite arm spacing of

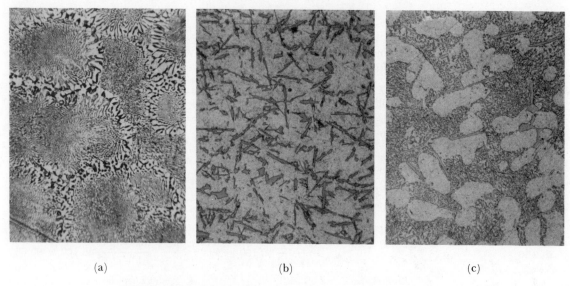

(a)   (b)   (c)

FIGURE 10-18   Typical eutectic microstructures. (a) Grains in the lead-tin eutectic (×300), (b) needlelike silicon plates in the aluminum-silicon eutectic (×100), and (c) rounded silicon rods in the modified aluminum-silicon eutectic (×100).

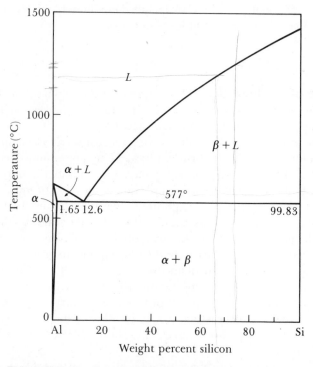

FIGURE 10-19   The aluminum-silicon phase diagram.

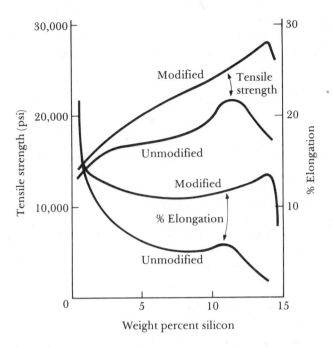

the primary phase may improve the properties of the alloy. Sometimes, however, the primary phase may take a different shape. In hypereutectic aluminum-silicon alloys, β is the primary phase. Because β is hard, the hypereutectic alloys are wear resistant; consequently, the hypereutectic aluminum-silicon alloys are considered for engine blocks. However, the primary silicon is normally very coarse [Figure 10-21(a)], causing poor castability, poor machinability, and

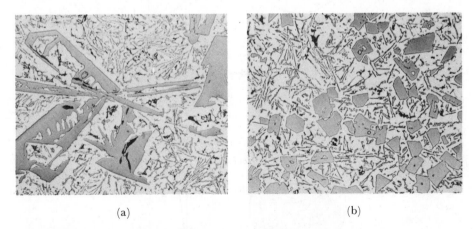

(a)   (b)

FIGURE 10-21 The effect of hardening with phosphorus on the microstructure of hypereutectic aluminum-silicon alloys. (a) Coarse primary silicon; (b) fine primary silicon as refined by phosphorus addition (× 75). From *Metals Handbook*, Vol. 7, 8th Ed., American Society for Metals, 1972.

gravity segregation (where the primary β floats to the surface of the casting during freezing). Addition of 0.05% P encourages nucleation of primary silicon, refines its size, and minimizes its deleterious qualities [Figure 10-21(b)].

◇ **Example 10-10**

Aluminum engine blocks for automobiles are manufactured using a hyper-eutectic Al-17% Si alloy. Estimate the amount of primary silicon and total amount of silicon phase present in the microstructure.

Answer:

From a tie line at 578°C in Figure 10-19

$$\% \text{ Primary } \beta = \frac{17 - 12.6}{99.83 - 12.6} \times 100 = 5.0\%$$

From a tie line at 576°C

$$\% \beta = \frac{17 - 1.65}{99.83 - 1.65} \times 100 = 15.6\%$$

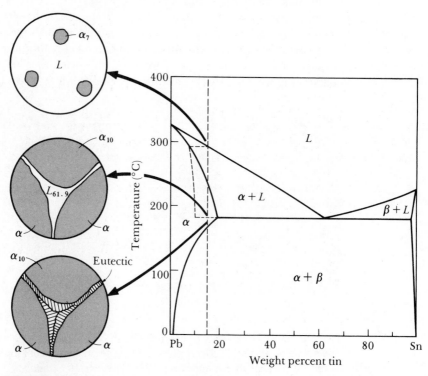

FIGURE 10-22 Nonequilibrium solidification and microstructure of a Pb-15% Sn alloy. A nonequilibrium eutectic microconstituent can form due to rapid solidification.

10–7   NONEQUILIBRIUM FREEZING IN THE EUTECTIC SYSTEM

Suppose we have an alloy, such as Pb-15% Sn, that ordinarily solidifies as a solid solution alloy. No eutectic solidification is expected in this alloy; the last liquid should freeze near 230°C, well above the eutectic. However, if the alloy cools too quickly, diffusion does not occur rapidly enough to permit the composition of the primary α to follow the solidus curve. A nonequilibrium solidus curve is produced (Figure 10-22). The primary α continues to grow until just above 183°C, when the remaining nonequilibrium liquid contains 61.9% Sn. This liquid then transforms to the eutectic microconstituent, surrounding the primary α.

◇ | **Example 10-11**

Calculate the amount of nonequilibrium eutectic that forms in a Pb-15% Sn alloy for the nonequilibrium solidus in Figure 10-22.

Answer:

If we draw a tie line just above the eutectic line that runs from the nonequilibrium solidus to the liquidus curve, we can perform the lever law calculations.

$$\% \text{ Primary } \alpha \; = \; \frac{61.9 - 15}{61.9 - 10} \times 100 \; = \; 90.4\%$$

$$\% \text{ Eutectic } \; = \; \frac{15 - 10}{61.9 - 10} \times 100 \; = \; 9.6\%$$

The presence of the nonequilibrium eutectic can make the alloy *hot short*, or melt at temperatures below the equilibrium melting point. When heat treating an alloy such as Pb-15% Sn, we must keep the maximum temperature below the eutectic temperature of 183°C to prevent hot shortness.

10–8   THE PERITECTIC REACTION

Peritectic reactions are found in a variety of alloys, including low-carbon steels (Figure 10-23). In the peritectic reaction, a liquid reacts with solid δ to form a second solid phase γ. After the reaction begins, the γ phase separates the two reacting phases. For the peritectic reaction to continue, atoms must diffuse through solid γ. But diffusion through a solid is much slower than diffusion through a liquid. As the peritectic reaction continues, the γ layer gets thicker and the reaction slows down even more. Unless the cooling rate is very slow, a segregated microstructure is produced.

In steels, segregation during the peritectic reaction is not a severe problem. Most steels eventually undergo significant plastic deformation and heat treatment, which reduce the segregation effects.

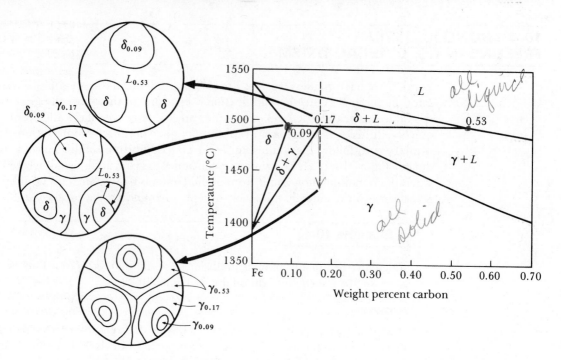

FIGURE 10-23   Solidification and microstructure that develop as a result of the peritectic reaction.

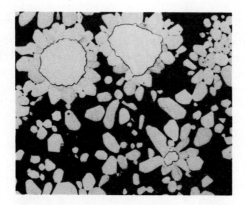

FIGURE 10-24   The peritectic reaction in a Cd-10% Cu alloy begins when rounded primary grains of Cd_8Cu_{15} react with the liquid (dark) to produce a halo of Cd_3Cu grains (× 100). From *Metals Handbook*, Vol. 9, 9th Ed., American Society for Metals, 1985.

Figure 10-24 illustrates the microstructure present during the peritectic reaction in a Cd-Cu alloy, where the reaction $L + Cd_8Cu_{15} \rightarrow Cd_3Cu$ occurs.

10–9   THE MONOTECTIC REACTION

We are aware that oil and water do not mix. They have very little solubility in one another and can be regarded as immiscible liquids. Certain alloy systems behave in the same manner (Figure 10-25). Liquid copper and liquid lead are completely soluble at high temperatures. However, alloys containing between 36% and 87% Pb separate into two liquids during cooling. The two liquids coexist in the *miscibility gap*, or dome, that is characteristic of all alloys that undergo the monotectic reaction.

During the solidification of a hypomonotectic copper-lead alloy, we find that solid α, almost pure copper, forms first. The liquid composition shifts to the monotectic composition, 36% Pb. Next the liquid transforms to more solid α plus

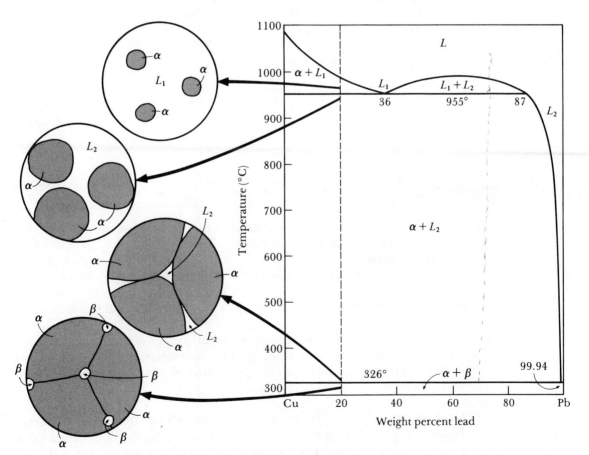

FIGURE 10-25   Solidification and microstructure of a hypomonotectic copper-lead alloy.

a second liquid containing 87% Pb. From the lever law, we find that only a relatively small amount of the second liquid is present. On further cooling, the second liquid eventually goes through a eutectic reaction, producing α and β, where β is nearly pure lead.

The microstructure of the copper-lead alloy contains spherical β particles randomly distributed throughout a copper matrix. Little dispersion strengthening occurs because the lead-rich β phase is very soft.

◇ Example 10-12

Calculate the amount of lead-rich liquid that forms immediately after the monotectic reaction in a Cu-5% Pb alloy. How much lead-rich solid β is present at room temperature?

Answer:

From a tie line at 954°C

$$L_2 = \frac{5 - 0}{87 - 0} \times 100 = 5.7\%$$

From a tie line at 25°C

$$\beta = \frac{5 - 0}{100 - 0} \times 100 = 5.0\%$$

10–10 TERNARY PHASE DIAGRAMS

Many alloy systems are based on three or even more elements. When three elements are present, we have a *ternary* alloy. To describe the changes in structure with temperature, we must draw a three-dimensional phase diagram. Figure 10-26 shows a hypothetical ternary phase diagram made up of elements *A*, *B*, and *C*. Note that two binary eutectics are present on the two visible faces of the diagram; a third binary eutectic between elements *B* and *C* is hidden on the back of the plot.

It is difficult to use the three-dimensional ternary plot; however, we can present the information from the diagrams in two dimensions by any of several methods, including the liquidus plot, the isothermal plot, and a vertical section called an isopleth.

Liquidus Plot.   We note in Figure 10-26 that the temperature at which freezing begins is shaded. We could transfer these temperatures for each composition onto a triangular diagram, as in Figure 10-27, and plot the liquidus temperatures as isothermal contours. This presentation is helpful in predicting the

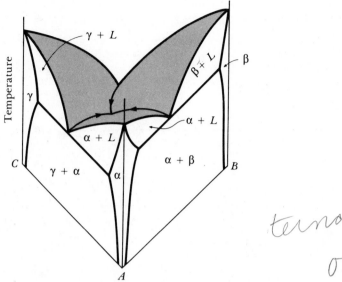

FIGURE 10-26 Hypothetical ternary phase diagram. Binary phase diagrams are present at the three faces.

freezing temperature of an alloy. The liquidus plot also gives the identity of the primary phase that will form during solidification for any given alloy composition.

◇   **Example 10-13**

Using the liquidus plot in Figure 10-27, determine the liquidus temperature and the primary phase that forms during solidification for the following alloys:

A-10% B-60% C A-20% B-20% C A-40% B-40% C

Answer:

The composition A-10% B-60% C is located at point x in the figure; by interpolating the isotherms in this region, the liquidus temperature is approximately 270°C. The primary phase, as indicated in the diagram, is γ.

The composition A-20% B-20% C is located at point y; the 300°C isotherm which intersects point y is the liquidus temperature. The primary phase that forms in this section of the diagram is α.

The composition A-40% B-40% C is located at point z; the liquidus at this point is 350°C, and the point is in the primary β region.

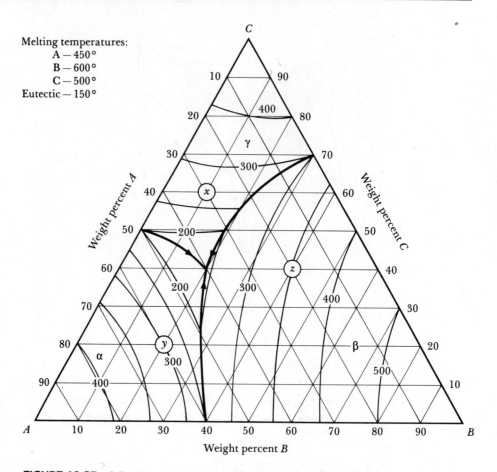

Melting temperatures:
  A — 450°
  B — 600°
  C — 500°
Eutectic — 150°

FIGURE 10-27 A liquidus plot for the hypothetical ternary phase diagram.

Isothermal Plot. The isothermal plot shows the phases present in any alloy at a particular temperature and is useful in predicting the phases and their amounts and compositions at that temperature. Figure 10-28 shows an isothermal plot from Figure 10-26 at room temperature. From this figure we could, for instance, show that an alloy containing A-10% B-10% C is composed of all α (point x); alloy A-50% B-10% C is a mixture of α and β (point y); and alloy A-30% B-30% C is a mixture of α, β, and γ (point z).

Isopleth Plot. Finally, we can present certain groups of alloys by vertical sections, also called *isopleths*. These sections represent a fixed composition of one of the elements, while the amounts of the other two elements are allowed to vary. These plots show how the phases and structures change when the temperature varies and when two of the elements present change their respective amounts. Figure 10-29 shows an isopleth through the hypothetical diagram in Figure 10-26 at a constant 40% C. An alloy containing A-40% B-40% C will begin to

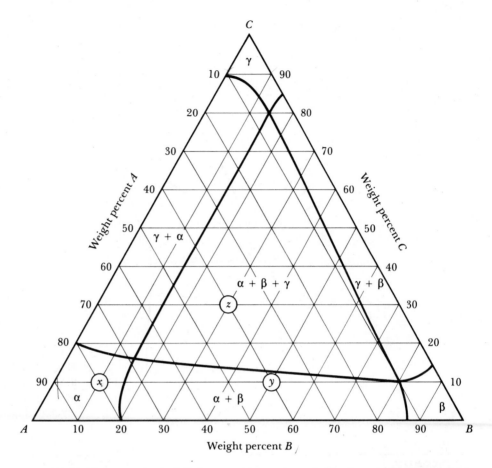

FIGURE 10-28 An isothermal plot at room temperature for the hypothetical ternary phase diagram.

freeze near 350°C, with primary β forming first. Near 275°C, γ will also begin to form. Finally, at 160°C, α forms and the last liquid freezes. The final microstructure contains α, β, and γ.

SUMMARY

By taking advantage of three-phase reactions and solubility limits, we control the properties of a material by dispersion strengthening. The eutectic reaction permits us to control dispersion strengthening by solidification techniques. As a result many common casting, brazing, and soldering alloys are based on eutectics. We are able to control the structure and properties by standard solidification techniques—inoculation and cooling rate or solidification time. Now we have the additional flexibility of controlling the amount, size, shape, and distribution of the two phases in the eutectic structure.

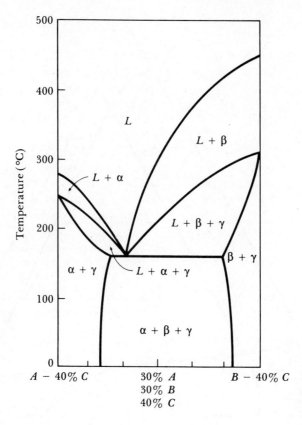

FIGURE 10-29 An isopleth through the hypothetical ternary phase diagram at a constant 40% C.

GLOSSARY

Dispersion strengthening. Increasing the strength of a material by mixing together more than one phase. By proper control of the size, shape, amount, and individual properties of the phases, excellent combinations of properties can be obtained.

Eutectic. A three-phase reaction in which one liquid phase solidifies to produce two solid phases.

Eutectoid. A three-phase reaction in which one solid phase transforms to two different solid phases.

Hyper-. A prefix indicating that the composition of an alloy is more than the composition at which a three phase reaction occurs.

Hypo-. A prefix indicating that the composition of an alloy is less than the composition at which a three-phase reaction occurs.

Interlamellar spacing. The distance between the center of a lamella or plate of one phase and the center of the adjoining lamella or plate of the same phase.

Isopleth. A vertical section through a ternary phase diagram showing the phases present at any temperature when the amount of one of the components is fixed.

Isothermal plot. A horizontal section through a ternary phase diagram showing the phases present at a particular temperature.

Lamella. A thin plate of a phase that forms during certain three-phase reactions, such as the eutectic or eutectoid.

Liquidus plot. A two-dimensional plot showing the temperature at which a three-component alloy system begins to solidify on cooling.

Matrix. Typically the first solid material to form during cooling of an alloy. Usually, the matrix is continuous and a second phase precipitates from it. However, in some complex alloys, the matrix is more difficult to define.

Microconstituent. A phase or mixture of phases in an alloy that has a distinct appearance. Frequently, we describe a microstructure in terms of the microconstituents rather than the actual phases.

Miscibility gap. A region in a phase diagram in which two phases, with essentially the same structure, do not mix, or have no solubility in one another. This is common in liquids, such as oil and water, but also is observed in solids.

Monotectic. A three-phase reaction in which one liquid transforms to a solid and a second liquid on cooling.

Nonstoichiometric intermetallic compound. A phase formed by the combination of two components into a compound having a structure and properties different from either component. The nonstoichiometric compound has a variable ratio of the components present in the compound.

Peritectic. A three-phase reaction in which a solid and a liquid combine to produce a second solid on cooling.

Peritectoid. A three-phase reaction in which two solids combine to form a third solid on cooling.

Precipitate. A solid phase that forms from the original matrix phase when the solubility limit is exceeded. In most cases, we try to control the formation of the precipitate to produce the optimum dispersion strengthening.

Primary microconstituent. The microconstituent which forms before the start of a three-phase reaction.

Solvus. A solubility line that separates a single solid phase region from a two solid phase region in the phase diagram.

Stoichiometric intermetallic compound. A phase formed by the combination of two components into a compound having a structure and properties different from either component. The stoichiometric intermetallic compound has a fixed ratio of the components present in the compound.

Ternary phase diagram. A phase diagram between three components showing the phases present and their compositions at various temperatures. This requires a three-dimensional plot.

PRACTICE PROBLEMS

1. A hypothetical phase diagram is shown in Figure 10-30. (a) What intermetallic compound is present? Is it stoichiometric or nonstoichiometric? (b) Identify the 4 solid solutions present in the system. (c) Identify the three-phase reactions by writing down the temperature, the reaction in equation form, the composition of each phase in the reaction, and the name of the reaction.

2. A hypothetical phase diagram is shown in Figure 10-31. (a) What intermetallic compound is present? Is it stoichiometric or nonstoichiometric? (b) Identify the 4 solid solutions present in the system. (c) Identify the three-phase reactions by writing down the temperature, the reaction in equation form, the composition of each phase in the reaction, and the name of the reaction.

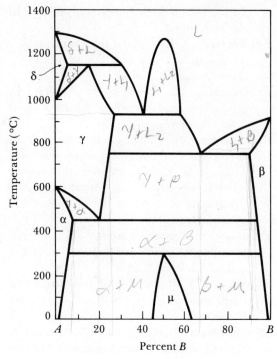

FIGURE 10-30 Hypothetical phase diagram for Problems 10-1, 10-9, 10-20, and 10-21.

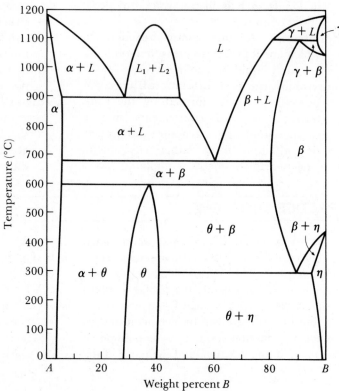

FIGURE 10-31 Hypothetical phase diagram for Problem 10-2.

3. A portion of the Cu-Al phase diagram is shown in Figure 13-8(d). (a) What intermetallic compounds are present in the system? Are they stoichiometric or nonstoichiometric compounds? (b) Identify the solid solution present in the system. (c) Identify the three-phase reactions by writing down the temperature, the reaction in equation form, and the name of the reaction.

4. The Co-Mo phase diagram is shown in Figure 10-32. (a) What intermetallic compounds are present? Are they stoichiometric or non-stoichiometric compounds? (b) Identify the solid solutions present in the system. (c) Identify the three-phase reactions by writing down the temperature, the reaction in equation form, the composition of each phase in the reaction, and the name of the reaction.

5. The SiO₂–MgO phase diagram for a ceramic system is shown in Figure 14-22(d). (a) What intermediate compounds are present? Is their composition fixed or variable? (b) Is there any solid solubility of one oxide in the other? Is this unusual for ceramic systems? (c) Identify the three-phase reactions by writing down the temperature, the reaction in equation form, and the name of the reaction.

6. A portion of the ZrO₂–CaO phase diagram for a ceramic system is shown in Figure 14-23(b). (a) Is there any solid solubility of CaO in ZrO₂? (b) Identify the three-phase reactions by writing down the temperature, the reaction in equation form, and the name of the reaction.

7. Consider a Pb-12% Sn alloy. During solidification, determine (a) the composition of the first solid to form, (b) the amounts and compositions of each phase at 290°C, (c) the liquidus, solidus, and solvus temperatures, (d) the amounts and compositions of each phase at 200°C, and (e) the amounts and compositions of each phase at 0°C. (f) Suppose, due to rapid cooling, that the composition of the last primary α was 6% Sn. Calculate the amount of nonequilibrium eutectic microconstituent.

8. Consider an Al-10% Mg alloy [Figure 13-1(b)]. During solidification, determine (a) the composition of the first solid to form, (b) the amounts and compositions of each phase at 550°C, (c) the liquidus, solidus, and solvus temperatures, (d) the amounts and compositions of each phase at 400°C, and (e) the amounts and compositions of each phase at 0°C. (f) Suppose, due to rapid cooling, that the composition of the last primary α was 7% Mg. Calculate the amount of nonequilibrium eutectic microconstituent.

9. Locate the eutectic reaction in the hypothetical phase diagram in Figure 10-30. Calculate the amount of each phase in the eutectic microconstituent.

10. Locate the eutectic in the Co-Mo phase diagram in Figure 10-32. Calculate the amount of each phase in the eutectic microconstituent. Assuming that the solid solutions are soft and

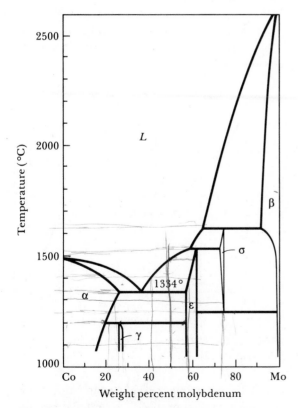

FIGURE 10-32 The Co-Mo equilibrium phase diagram (Problems 10-4, 10-10, and 10-14).

ductile, will the eutectic microconstituent be ductile or brittle?

11. Locate the eutectic reactions in the SiO_2–MgO phase diagram [Figure 14-22(d)]. Calculate the amount of each phase in each of the eutectic microconstituents. Would you expect the eutectic microconstituents to be ductile or brittle?

12. Consider a Pb-25% Sn alloy. Determine (a) if the alloy is hypoeutectic or hypereutectic, (b) the composition of the first solid to form during solidification, (c) the amounts and compositions of each phase at 184°C, (d) the amounts and compositions of each phase at 182°C, and (e) the amounts and compositions of each microconstituent at 182°C.

13. Consider a Pb-75% Sn alloy. Determine (a) if the alloy is hypoeutectic or hypereutectic, (b) the composition of the first solid to form during solidification, (c) the amounts and compositions of each phase at 184°C, (d) the amounts and compositions of each phase at 182°C, and (e) the amounts and compositions of each microconstituent at 182°C.

14. Consider a Co-50% Mo alloy (Figure 10-32). Determine (a) if the alloy is hypoeutectic or hypereutectic, (b) the composition of the first solid to form during solidification, (c) the amounts and compositions of each phase at 1400°C, (d) the amounts and compositions of each phase at 1300°C, and (e) the amounts and compositions of each phase at 1100°C.

15. Consider an Al-75% Si alloy. Determine (a) if the alloy is hypoeutectic or hypereutectic, (b) the composition of the first solid to form during solidification, (c) the amounts and compositions of each phase at 578°C, (d) the amounts and compositions of each phase at 576°C, and (e) the amounts and compositions of each microconstituent at 576°C.

16. A Pb-Sn alloy contains 35% primary α and 65% eutectic. What is the composition of the overall alloy?

17. A Pb-Sn alloy contains 26% α and 74% β at room temperature. What is the composition of the overall alloy? Is the alloy hypoeutectic or hypereutectic?

18. An Al-Si alloy contains 40% α and 60% β at 450°C. What is the composition of the overall alloy? Is the alloy hypoeutectic or hypereutectic?

19. An Al-Si alloy contains 18% primary β and 82% eutectic. What is the composition of the overall alloy?

20. In the hypothetical phase diagram in Figure 10-30, 30% γ and 70% β are found at 740°C. What is the overall composition of the alloy?

21. In the hypothetical phase diagram in Figure 10-30, 25% primary β and 75% eutectic are observed. What is the overall composition of the alloy?

22. A cooling curve for a Pb-Sn alloy is shown in Figure 10-33. Determine (a) the pouring temperature, (b) the superheat, (c) the liquidus temperature, (d) the eutectic temperature, (e) the freezing range, (f) the local solidification time, and (g) the total solidification time. From the cooling curve, determine the composition of the alloy.

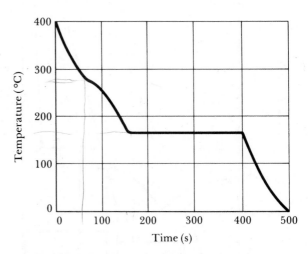

FIGURE 10-33 Cooling curve for Pb-Sn alloy (Problem 10-22).

23. A cooling curve for an Al-Si alloy is shown in Figure 10-34. Determine (a) the pouring temperature, (b) the superheat, (c) the liquidus temperature, (d) the eutectic tempera-

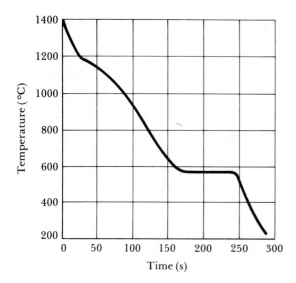

FIGURE 10-34 *Cooling curve for Al-Si alloy (Problem 10-23).*

ture, (e) the freezing range, (f) the local solidification time, and (g) the total solidification time. From the cooling curve, determine the composition of the alloy.

24. Construct a phase diagram from the following information: element *A* melts at 1200°C, and element *B* melts at 1000°C; element

B has a maximum solubility of 10% in element *A*, and element *A* has a maximum solubility of 20% in element *B*; the number of degrees of freedom from the phase rule is zero when the temperature is 800°C and there is 45% *B* present; at room temperature, 3% *A* is soluble in *B* and 0% *B* is soluble in *A*.

25. Cooling curves are obtained for a series of Cd-Zn alloys (Figure 10-35). Use the cooling curves to produce the Cd-Zn phase diagram. The maximum solubility of Zn in Cd is 2.6%, the maximum solubility of Cd in Zn is about 2%, and the solubilities at room temperature are about zero.

26. What fraction of the solidification of an Fe-0.35% C alloy occurs by the peritectic reaction?

27. What fraction of the solidification of an Fe-0.15% C alloy occurs by the peritectic reaction?

28. Calculate the amounts of *β* and liquid that must combine to produce 100% *γ* phase by the peritectic reaction in the Al-Li system (Figure 13-4).

29. An Al-Li alloy (Figure 13-4) contains a microstructure composed of 30% *γ* and 70% liquid just below the peritectic temperature.

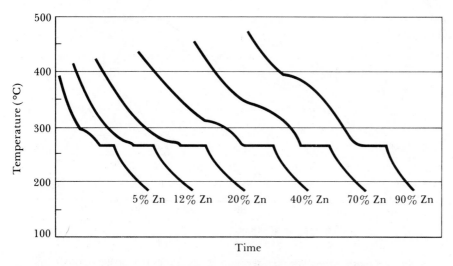

FIGURE 10-35 *Cooling curves for several Cd-Zn alloys (Problem 10-25).*

Determine the overall composition of the alloy. If this composition solidified under nonequilibrium conditions, would you expect to have more or less liquid present at this temperature?

30. Calculate the amounts and compositons of the two liquids present in a Cu-70% Pb alloy at 956°C.

31. A Cu-Pb alloy contains 75% α and 25% L_2 immediately after the monotectic reaction. Calculate (a) the composition of the overall alloy and (b) the amounts of L_1 and L_2 just before the monotectic reaction.

32. What is the solubility of lead in copper? Discuss this observation in terms of Hume-Rothery's conditions.

33. When a ternary eutectic reaction occurs, there are zero degrees of freedom. How many solid phases are produced as a result of the ternary eutectic? How many total phases are in equilibrium?

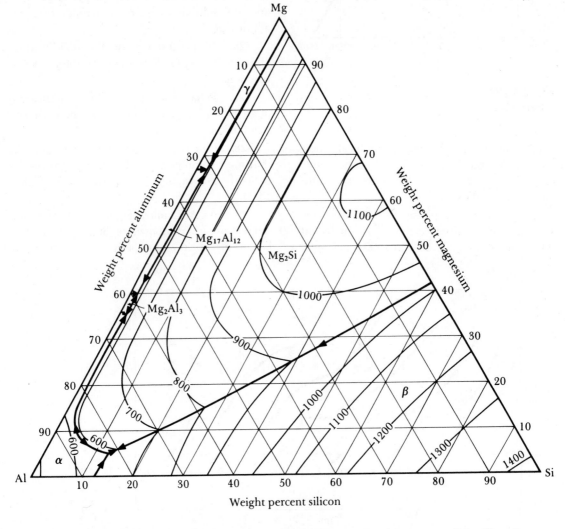

FIGURE 10-36 The liquidus plot for the aluminum-silicon-magnesium ternary phase diagram.

34. Consider the liquidus plot in Figure 10-27. For a constant 20% C, draw a graph showing how the liquidus temperatures changes from 80% A-0% B-20% C to 0% A-80% B-20% C.

35. Consider the liquidus plot for the Al-Si-Mg alloy in Figure 10-36. For a constant 20% Al, draw a graph showing how the liquidus temperature changes from 20% Al-80% Mg-0% Si to 20% Al-0% Mg-80% Si.

36. From the hypothetical ternary phase diagrams in Figures 10-27 and 10-28, determine for an A-40% B-20% C alloy (a) the liquidus temperature, (b) the primary phase, and (c) each phase present at room temperature.

37. From the hypothetical ternary phase diagrams in Figures 10-27 and 10-28, determine for an A-10% B-60% C alloy (a) the liquidus temperature, (b) the primary phase, and (c) each phase present at room temperature.

11

DISPERSION STRENGTHENING BY PHASE TRANSFORMATION AND HEAT TREATMENT

11–1 INTRODUCTION

In this chapter we will further discuss dispersion strengthening as we introduce a variety of solid-state transformation processes, including age hardening and the eutectoid reaction. We will also examine how nonequilibrium phase transformations, in particular the martensitic reaction, can provide strengthening. These dispersion-strengthening techniques require a heat treatment.

As we discuss these strengthening mechanisms, we must keep in mind the characteristics that produce the most desirable dispersion strengthening, as discussed in Chapter 10. The matrix should be relatively soft and ductile and the precipitate or second phase should be hard and brittle; the precipitate should be round and discontinuous; the precipitate particles should be small and numerous; and in general the more precipitate we have, the stronger will be the alloy.

11–2 NUCLEATION AND GROWTH IN SOLID-STATE REACTIONS

In order for a precipitate to form from a solid matrix, both nucleation and growth must occur. The total change in free energy required for nucleation of a spherical solid precipitate from the matrix is

$$\Delta F = \tfrac{4}{3}\pi r^3 \Delta F_v + 4\pi r^2 \sigma + \tfrac{4}{3}\pi r^3 \varepsilon \tag{11-1}$$

The first two terms include the volume free energy change and the surface energy change, just as in solidification [Equation (8-1)]. However, the third term takes into account the *strain energy* ε introduced when the precipitate forms in a solid, rigid matrix. The precipitate does not occupy the same volume that is displaced, so additional energy is required to permit the precipitate to be accommodated in the matrix.

As in solidification, nucleation occurs most easily on surfaces already present in the structure, thereby minimizing the surface energy term. Thus, the precipitate nucleates and grows most easily at the grain boundaries of the matrix or at other lattice defects. Increasing the number of lattice defects permits us to exercise some control over the number of nuclei produced.

Growth of the precipitate normally occurs by long-range diffusion and redistribution of the atoms to satisfy the phase diagram. These reactions proceed relatively slowly, since the atoms must diffuse in the solid, but occur more readily at high temperatures, where diffusion is more rapid. Thus, the growth rates are usually controlled primarily by controlling temperature. The relationship between growth and nucleation plays an important role in the phase transformation in solid-state reactions, just as in the solidification of materials.

11–3 ALLOYS STRENGTHENED BY EXCEEDING THE SOLUBILTIY LIMIT

In Chapter 10 we pointed out that lead-tin alloys containing about 2% to 19% Sn can be dispersion strengthened because the solubility of tin in lead is exceeded.

A similar situation occurs in aluminum-copper alloys. For example, the Al-4% Cu alloy, shown in Figure 11-1, is completely α, or an aluminum solid solution, above 500°C. On cooling below the solvus temperature, a second phase, θ, precipitates. The θ phase, which is a hard, brittle intermetallic compound $CuAl_2$, provides dispersion strengthening.

◇ **Example 11-1**

Calculate the amount of θ that forms at room temperature when an Al-4% Cu alloy slowly cools.

Answer:

From a tie line at 25°C

α: 0.02% Cu

θ: 53.5% Cu

$$\% \; \theta \; = \; \frac{4 \; - \; 0.02}{53.5 \; - \; 0.02} \; \times \; 100 \; = \; 7.4\%$$

Even this amount of θ is capable of providing effective dispersion strengthening if properly controlled.

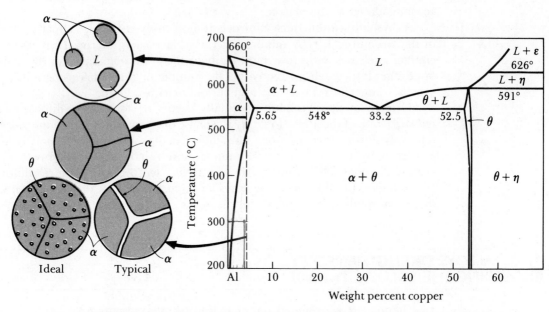

FIGURE 11-1 The aluminum-copper phase diagram and the microstructures that may develop during cooling of an Al = 4% Cu alloy.

◇ | **Example 11-2**

Calculate the amount of θ in the aluminum-copper eutectic microconstituent. Explain why most aluminum-copper alloys are designed to avoid the eutectic reaction.

Answer:

We obtain all eutectic when the alloy contains Al-33.2% Cu. Thus

$$\% \; \theta \; = \; \frac{33.2 \; - \; 5.65}{52.5 \; - \; 5.65} \; \times \; 100 \; = \; 58.8\%$$

Most of the eutectic is composed of the hard, brittle compound θ. The eutectic microconstituent will be brittle and, since the eutectic is continuous, the overall alloy will be brittle.

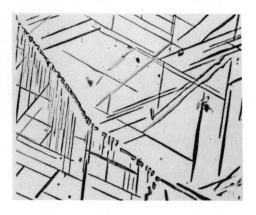

FIGURE 11-2 The Widmanstatten structure in a copper-titanium alloy ($\times$ 420). From *Metals Handbook*, Vol. 9, 9th Ed., American Society for Metals, 1985.

Unfortunately, we often are unable to control the precipitation of the second phase so that the requirements of good dispersion strengthening are satisfied. The second phase, such as θ in the aluminum-copper system, may not have a desirable size, shape, or distribution. Several factors influence the shape of the precipitate.

Widmanstatten Structure.   The second phase may grow so that certain planes and directions in the precipitate are parallel to preferred planes and directions in the matrix. This growth mechanism minimizes strain and surface energies and permits faster growth rates. Widmanstatten growth produces a characteristic appearance for the precipitate, such as plates, needles, rods, or even cubes. Particularly when the needlelike shape is produced (Figure 11-2), the Widmanstatten precipitate may embrittle the alloy. On the other hand, these structures may make it more difficult for cracks to propagate, therefore increasing the fracture toughness of certain alloys.

Interfacial Energy Relationships.   We expect the precipitate to have a spherical shape in order to minimize surface energy. However, the shape of the precipitate is also influenced by the *interfacial energy* associated with both the boundary between the matrix grains (γ_m) and the boundary between the matrix and the precipitate (γ_p). The interfacial surface energies fix a *dihedral angle θ* between the matrix-precipitate interface that in turn determines the shape of the precipitate [Figure 11-3(a)]. The relationship is

$$\gamma_m = 2\gamma_p \cos \frac{\theta}{2} \tag{11-2}$$

If the dihedral angle is small, the precipitate may be continuous. If the precipitate is also hard and brittle, the thin film that surrounds the matrix grains causes the alloy to be very brittle. On the other hand, discontinuous and even spherical precipitates form when the dihedral angle is large [Figure 11-3(b)].

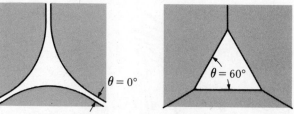

FIGURE 11-3(a)   The effect of surface energy and the dihedral angle on the shape of a precipitate.

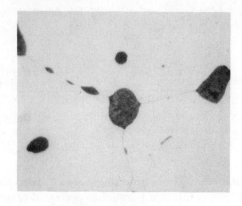

FIGURE 11-3(b)   A small precipitate of lead in copper (× 500).

◇ | **Example 11-3**

Calculate the ratio between γ_p and γ_m required to produce a continuous precipitate at grain boundaries.

Answer:

We obtain a continuous precipitate when $\theta = 0°$.

$$\frac{\gamma_m}{2\gamma_p} = \cos 0 = 1$$

$$\frac{\gamma_p}{\gamma_m} = \frac{1}{2}$$

◇ | **Example 11-4**

From the photomicrograph of the copper-lead alloy in Figure 11-3(b), determine the dihedral angle and calculate the interfacial energy between the copper matrix and the lead precipitate. The energy of copper grain boundaries is 646 ergs/cm^2.

Answer:

The lead precipitate is round, so $\theta = 180°$. From Equation (11-2)

$$\gamma_m = 2\gamma_p \cos\frac{\theta}{2} = 2\gamma_p \cos\frac{180}{2} = 2\gamma_p \cos 90 = 0$$

$$\gamma_p = \frac{646}{0} = \infty$$

The energy really isn't infinity. This curious result means that the energy of the copper-lead interface is so large that the lead will produce the smallest possible surface area, or will be spherical.

Cooling Rate. The rate at which the alloy cools past the solvus line determines the time available for diffusion and consequently affects the shape of the precipitate. Fast cooling rates help offset the effect of very low dihedral angles and permit a discontinuous rather than a continuous grain boundary precipitate to form. Figure 11-4 compares the microstructure of the Al-4% Cu alloy for two different cooling rates.

In our example of the Al-4% Cu alloy, slow cooling permits the hard, brittle θ phase to form as a thin, almost continuous, film at the α grain boundaries. The slow-cooled Al-4% Cu alloy does not have a desirable microstructure. Some improvement is obtained by increasing the rate of cooling as the alloy crosses the solvus line; however, optimum properties are still not obtained.

Coherent Precipitate. Even if we produce a uniform distribution of discontinuous θ precipitate, the precipitate may not significantly disrupt the

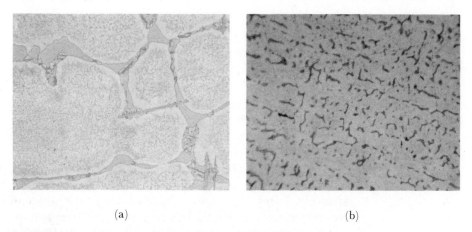

(a)   (b)

FIGURE 11-4 Photomicrographs of an Al-4% Cu alloy. Slow cooling (a) produces a more continuous, embrittling θ than fast cooling (b) ($\times$ 500).

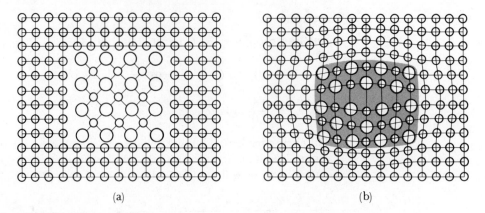

(a)   (b)

FIGURE 11-5 (a) A noncoherent precipitate has no relationship with the crystal structure of the surrounding matrix. (b) A coherent precipitate forms so that there is a definite relationship between the precipitate's and the matrix's crystal structure.

surrounding matrix structure. Consequently, the precipitate blocks slip only if it lies directly in the path of the dislocation [Figure 11-5(a)].

But when a *coherent precipitate* forms, the planes of atoms in the lattice of the precipitate are related to, or even continuous with, the planes in the lattice of the matrix [Figure 11-5(b)]. Now a widespread disruption of the matrix lattice is created and the movement of a dislocation is impeded even if the dislocation merely passes near the coherent precipitate. A special heat treatment, such as age hardening, may be required to produce the coherent precipitate.

11–4 AGE HARDENING OR PRECIPITATION HARDENING

Age hardening, or *precipitation hardening*, is designed to produce a uniform dispersion of a fine, hard coherent precipitate in a softer, more ductile matrix. The Al-4% Cu alloy is a classical example of an age hardenable alloy. There are three steps in the age-hardening heat treatment (Figure 11-6).

Step 1: Solution Treatment. The alloy is first heated to a temperature above the solvus temperature and held until a homogeneous solid solution α is produced. This step dissolves the θ precipitate and reduces any segregation present in the original alloy.

We could heat the alloy to just below the solidus temperature and increase the rate of homogenization. However, the presence of a nonequilibrium eutectic microconstituent may cause melting. Thus, the aluminum-copper alloy is solution treated between the solvus and eutectic temperatures, assuring that any eutectic microconstituent in the alloy does not melt. In the Al-4% Cu alloy, this treatment would be done between 500°C and 548°C.

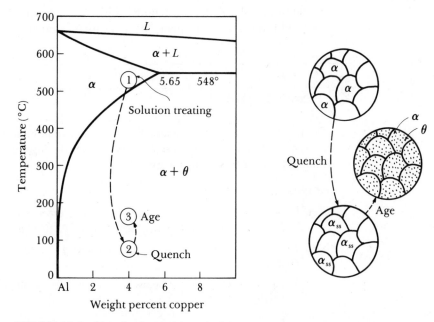

FIGURE 11-6 *The aluminum-rich end of the aluminum-copper phase diagram showing the three steps in the age-hardening heat treatment and the microstructures that are produced.*

Step 2: Quench. After solution treatment, the alloy, which contains only α in its structure, is rapidly cooled, or quenched. The atoms do not have time to diffuse to potential nucleation sites and permit the θ phase to form. After the quench the structure still contains only α. The α is a *supersaturated solid solution*, containing excess copper, and is not an equilibrium structure.

Step 3: Age. Finally, the supersaturated α is heated to a temperature below the solvus temperature. At this aging temperature, atoms are able to diffuse short distances. Because the supersaturated α is not stable, the extra copper atoms diffuse to numerous nucleation sites and a precipitate forms and grows. Eventually, if we hold the alloy for a sufficient time at the aging temperature, the equilibrium $\alpha + \theta$ structure is produced.

◇ | **Example 11-5**

Compare the composition of the α solid solution in the Al-4% Cu alloy at room temperature when the alloy cools under equilibrium conditions and when the alloy is quenched.

Answer:

From Figure 11-6, a tie line can be drawn at room temperature. The composition of the α determined from the tie line is about 0.02% Cu. However, the composition of the α after quenching is still 4% Cu. Since α contains more than the equilibrium copper content, the α is supersaturated with copper.

◇   **Example 11-6**

The magnesium-aluminum phase diagram is shown in Figure 13-1(b). Suppose a Mg-8% Al alloy is responsive to an age-hardening heat treatment. Recommend a heat treatment for the alloy.

Answer:

Step 1: Solution treat at a temperature between the solvus and the eutectic to avoid hot shortness. Thus, heat between 340°C and 437°C.

Step 2: Quench to room temperature fast enough to prevent the precipitate from forming.

Step 3: Age at a temperature below the solvus, or at some temperature below 340°C.

Nonequilibrium Precipitates during Aging.   During aging of aluminum-copper alloys, a continuous series of precipitates forms before the equilibrium θ is produced. At the start of aging, the copper atoms concentrate on {100} planes in the α matrix and produce very thin precipitates called *Guinier–Preston*, or GP-I, zones. As aging continues, more copper atoms diffuse to the precipitate and the GP-I zones thicken into thin disks, or GP-II zones. With continued diffusion, the precipitates develop a greater degree of order and are called θ'. Finally, the stable θ precipitate is produced.

The nonequilibrium precipitates—GP-I, GP-II, and θ'—are coherent precipitates. The strength of the alloy increases with aging time as these coherent phases grow in size during the initial stages of the heat treatment. When these coherent precipitates are present, the alloy is in the *aged* condition. Figure 11-7 shows the structure of an aged alloy.

When the stable noncoherent θ phase precipitates, the strength of the alloy decreases. Now the alloy is in the *overaged* condition. The θ still provides some dispersion strengthening, but with increasing time, the θ grains grow larger and less numerous and even the simple dispersion strengthening effect diminishes.

11–5   EFFECTS OF AGING TEMPERATURE AND TIME

The properties of an age hardenable alloy depend on both the temperature and the time for aging (Figure 11-8). At an aging temperature of 260°C, diffusion in the Al-4% Cu alloy is rapid and precipitates quickly form. As aging continues,

FIGURE 11-7 An electron micrograph of aged Al-15% Ag showing coherent γ' plates and round GP zones, $\times$ 40,000. (Courtesy J. B. Clark.)

we progress from tiny GP-I and GP-II zones to θ'. The strength reaches a maximum after less than 0.1 h exposure. Overaging occurs if the alloy is held for longer than 0.1 h.

At 190°C, which is a typical aging temperature for many of the aluminum alloys, longer times are required to produce the optimum strength. However, there are several benefits to using the lower temperature. First, the maximum

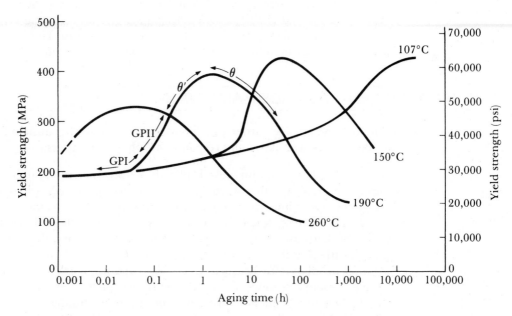

FIGURE 11-8 The effect of aging temperature and time on the yield strength of an Al-4% Cu alloy.

strength increases as the aging temperature decreases. Second, the strength maintains its maximum over a longer period of time. This broader peak permits the heat treater to make a small miscalculation in temperature or time but still produce the required properties. Third, the properties are more uniform. If the alloy is aged for only 10 min at 260°C, the surface of the part reaches the proper temperature and strengthens, but the center remains cool and ages only slightly.

◇ **Example 11-7**

The operator of a furnace left for his hour lunch break without removing the Al-4% Cu alloy from the aging furnace. Compare the effect on the yield strength of the extra hour of aging for aging temperatures of 190°C and 260°C.

Answer:

At 190°C, the peak strength of 400 MPa occurs at 6 h (Figure 11-8). After 7 h, the strength is essentially the same.

At 260°C, the peak strength of 340 MPa occurs at 0.06 h. However, after 1.06 h, the strength decreases to 250 MPa.

Thus, the higher aging temperature gives a lower peak strength and makes the strength more sensitive to aging time.

Aging at either 190°C or 260°C is called *artificial aging* because the alloy is heated to produce precipitation. Some solution-treated and quenched alloys age at room temperature; this is called *natural aging*. Natural aging requires long times, often several days, to reach the maximum strength. However, the peak strength is higher than that obtained in artificial aging and no overaging occurs.

11–6 REQUIREMENTS FOR AGE HARDENING

Not all alloys are age hardenable. Four conditions must be satisfied for an alloy to have a true age-hardening response during heat treatment.

1. The phase diagram must display decreasing solid solubility with decreasing temperature. In other words, the alloy must form a single phase on heating above the solvus line, then enter a two-phase region on cooling.
2. The matrix should be relatively soft and ductile and the precipitate should be hard and brittle. In most age hardenable alloys, the precipitate is a hard, brittle intermetallic compound.
3. The alloy must be quenchable. We cannot quench some alloys rapidly enough to suppress the formation of the second phase.
4. The precipitate that forms must be coherent with the matrix structure in order to develop the maximum strength and hardness. Furthermore, its size, shape, and distribution must be controlled.

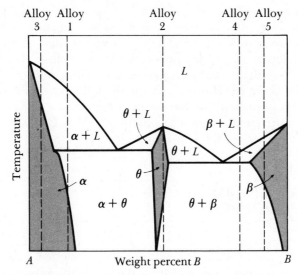

FIGURE 11-9   A hypothetical phase diagram for use in Example 11-8.

◇ | **Example 11-8**

Discuss the likelihood that each alloy shown in Figure 11-9 will display an age-hardening response.

Answer:

Alloy 1: The solvus displays increasing solid solubility as the temperature decreases; thus, an age-hardening response is impossible.

Alloy 2: The matrix is a hard, brittle intermetallic compound but the precipitate is more likely to be softer and more ductile. Thus, an age-hardening response is unlikely.

Alloy 3: This alloy is single phase up to the solidus temperature. No age hardening can occur.

Alloy 4: This is a two-phase alloy up to the melting or eutectic temperature. A slight aging response might be possible in this type of alloy, but the effect will be only small.

Alloy 5: We now have decreasing solid solubility with decreasing temperature, and the precipitate is a hard, brittle intermetallic compound. This alloy might be a potential candidate for age hardening.

In many alloys, the strengthening effect is limited because only a certain amount of precipitate can form. But rapid solidification processing may be able to provide an even better age-hardening response by extending the solubility of the alloying element in the matrix. By rapid freezing, more alloying element can be trapped as a supersaturated solid solution so that, during aging, a larger fraction of the hard, brittle dispersant may precipitate. For example, we normally add less than 5.65% copper to aluminum, since this is the maximum solubility of Cu in Al. However, by rapidly solidifying an Al-10% Cu alloy, we may be able to prevent the formation of the eutectic during freezing, produce a supersaturated solid solution also containing 10% Cu, and then, during aging, double the amount of the $CuAl_2$ precipitate.

11–7   USE OF AGE HARDENABLE ALLOYS AT HIGH TEMPERATURES

Based on our previous discussion, we would not select an age-hardened Al-4% Cu alloy for use at high temperatures. At service temperatures ranging from above room temperature to 500°C, the alloy overages and loses its strength rapidly. Above 500°C, the second phase redissolves in the matrix and we do not even obtain dispersion strengthening. In general, the aluminum age hardenable alloys are suited only for service near room temperature. However, some magnesium alloys may maintain their strength to about 250°C and certain nickel superalloys resist overaging at 1000°C.

We also have problems when welding age hardenable alloys (Figure 11-10). During welding the metal adjacent to the weld is heated. The heat-affected area contains two principle zones. The lower temperature zone near the unaffected base metal is exposed to temperatures just below the solvus and may overage. The higher temperature zone is solution treated, eliminating the effects of age hardening. If the solution-treated zone cools slowly, stable θ may form at the grain boundaries, embrittling the weld area. Very fast welding processes such as electron-beam welding, complete reheat treatment of the area after welding, or welding the alloy in the solution-treated condition improve the quality of the weld.

11–8   RESIDUAL STRESSES DURING QUENCHING

When an age hardenable alloy is quenched, the center of the part cools more slowly than the surface. The rapidly cooled surface contracts due to the coefficient of thermal expansion and contraction. The contracting surface applies a compressive stress to the center which, because the center is still hot, soft, and ductile, deforms. Later, the center cools and attempts to contract. But the contraction of the center is restrained by the cold, strong surface. The center is placed in tension while the surface is compressed. Consequently, a residual stress pattern is produced in the quenched part (Figure 11-11). The residual stresses cause distortion, warpage, or even cracking of the part.

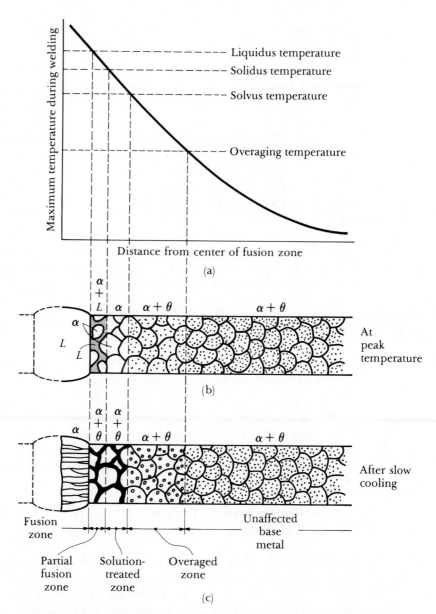

FIGURE 11-10 Microstructural changes that occur in age-hardened alloys during fusion welding. (a) Peak temperature that occurs during welding, (b) microstructure in the weld at the peak temperature, and (c) microstructure in the weld after slowly cooling to room temperature.

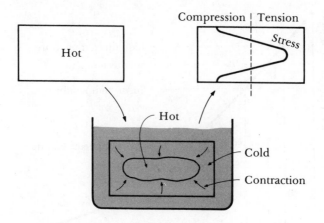

FIGURE 11-11 *The residual stress pattern produced in a solution-treated and quenched alloy during the age-hardening process. Note that the surface is in compression and the center is in tension.*

To help minimize problems due to residual stresses, age hardenable alloys are not quenched any more rapidly than necessary. Aluminum-base alloys are normally quenched in hot water, at about 80°C, rather than cold water.

11–9   THE EUTECTOID REACTION

In Chapter 10, we defined the eutectoid as a solid-state reaction in which one solid phase transforms to two other solid phases.

$$S_1 \rightarrow S_2 + S_3 \tag{11-3}$$

The formation of the two solid phases permits us to obtain dispersion strengthening. As an example of how we can use the eutectoid reaction to control the microstructure and properties of an alloy, let's examine the iron-carbon system, which is the basis for steels and cast irons.

The Iron-Cementite Phase Diagram.   Figure 11-12 shows the Fe-Fe$_3$C phase diagram. The following features should be noted.

Solid Solutions.   Iron goes through two allotropic transformations during heating or cooling. Immediately after solidification, iron forms a BCC structure called *δ-ferrite*. On further cooling, the iron transforms to an FCC structure called γ, or *austenite*. Finally, iron transforms back to the BCC structure at lower temperatures; this structure is called α, or *ferrite*. Both of the ferrites and the austenite are solid solutions of interstitial carbon atoms in iron (Figure 11-13). Because interstitial holes in the FCC lattice are somewhat larger than holes in the BCC lattice, a greater number of carbon atoms can be accommodated in FCC iron. Thus, the maximum solubility of carbon in austenite is 2.11% C, while the maximum solubility of carbon in BCC iron is much lower—0.0218% C in α and 0.09% C in δ. The solid solutions are relatively soft and ductile but are stronger than pure iron due to solid solution strengthening by the carbon.

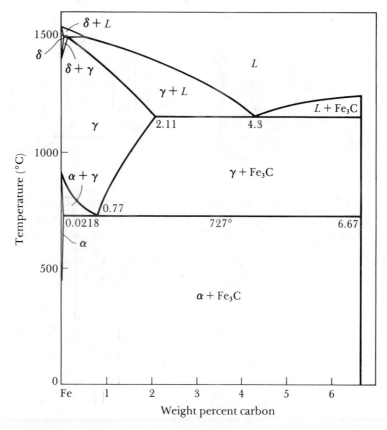

FIGURE 11-12 The Fe-Fe₃C phase diagram.

◇ | **Example 11-9**

Calculate the size of the interstitial sites for carbon atoms in δ, α, and γ. From these results, explain the difference in the maximum solubility of carbon in each phase. The atomic radii are shown in Table 11-1.

TABLE 11-1 Size of the atoms in steel, depending on the crystal structure

Atom	Crystal Structure	Radius (Å)
Fe	α	1.24
Fe	γ	1.29
Fe	δ	1.27
C		0.71

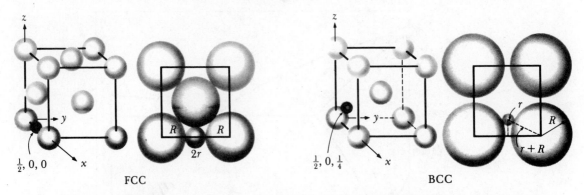

FCC   BCC

FIGURE 11-13   The unit cells of FCC and BCC iron, including the interstitial sites for carbon.

Answer:

The largest interstitial site in BCC iron has the coordinates $\frac{1}{2}$, 0, $\frac{1}{4}$. The ratio between the radius of the iron atom and the radius of the interstitial site can be calculated with the aid of Figure 11-13.

$$\left(r_{Fe} + r_{interstitial}\right)^2 = \left(\frac{a_0}{4}\right)^2 + \left(\frac{a_0}{2}\right)^2 = \left(\frac{5}{16}\right)a_0^2 = \left(\frac{5}{16}\right)\left(\frac{4r_{Fe}}{\sqrt{3}}\right)^2 = \frac{5r_{Fe}^2}{3}$$

$$r_{Fe} + r_{interstitial} = \frac{\sqrt{5}r_{Fe}}{\sqrt{3}}$$

$$\frac{r_{interstitial}}{r_{Fe}} = 0.291$$

The largest interstitial site in FCC iron has the coordinates $\frac{1}{2}$, 0, 0. From Figure 11-13.

$$2r_{Fe} + 2r_{interstitial} = a_0 = \frac{4r_{Fe}}{\sqrt{2}}$$

$$r_{Fe} + r_{interstitial} = \sqrt{2}r_{Fe}$$

$$\frac{r_{interstitial}}{r_{Fe}} = 0.414$$

Therefore

$$\alpha \text{ site} = (0.291)(1.24) = 0.36 \text{ Å}$$
$$\gamma \text{ site} = (0.414)(1.29) = 0.53 \text{ Å}$$
$$\delta \text{ site} = (0.291)(1.27) = 0.37 \text{ Å}$$

The interstitial sites are all smaller than the carbon atom, $r_C = 0.71$ Å, causing low solubility and good solid solution strengthening. The solubility is about 100 times greater in austenite than in ferrite because of the larger hole in FCC unit cells.

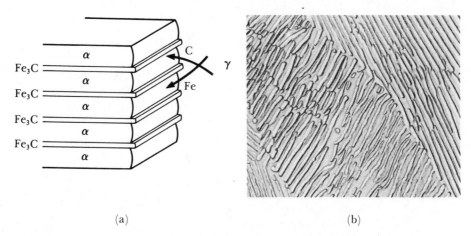

(a)   (b)

FIGURE 11-14 Growth and structure of pearlite. (a) Redistribution of carbon and iron. (b) Photomicrograph of the pearlite lamellae (× 2000). From *Metals Handbook*, Vol. 7, 8th Ed., American Society for Metals, 1972.

Intermetallic Compounds.   A stoichiometric intermetallic compound Fe_3C, or *cementite*, forms when the solubility of carbon in solid iron is exceeded. The Fe_3C contains 6.67% C, is extremely hard and brittle, and is present in all of the commercial steels. By properly controlling the amount, size, and shape of Fe_3C, we control the degree of dispersion strengthening and the properties of the steel.

Eutectoid Reaction.   If we heat an alloy containing the eutectoid composition of 0.77% C above 727°C, we produce a structure containing only austenite grains. When austenite cools to 727°C, the eutectoid reaction begins.

$$\gamma_{0.77\% \, C} \rightarrow \alpha_{0.0218\% \, C} + Fe_3C_{6.67\% \, C} \tag{11-4}$$

As in the eutectic reaction, the two phases that form have different compositions, so atoms must diffuse during the reaction (Figure 11-14). Most of the carbon in the austenite diffuses to the Fe_3C, but a greater percentage of iron atoms diffuse to α. This redistribution of atoms is easiest if the diffusion distances are short, which is the case when the α and Fe_3C grow as thin lamellae, or plates.

Pearlite.   The lamellar structure of α and Fe_3C that develops in the iron-carbon system is called *pearlite*. Pearlite is a microconstituent in steel. The lamellae in pearlite are much finer than the lamellae in the lead-tin eutectic because the iron and carbon atoms must diffuse through solid austenite rather than through liquid.

◇ | **Example 11-10**

Calculate the amounts of ferrite and cementite present in pearlite.

Answer:

Since pearlite must contain 0.77% C, then using the lever law

$$\% \; \alpha \; = \; \frac{6.67 \, - \, 0.77}{6.67 \, - \, 0.0218} \; \times \; 100 \; = \; 88\%$$

$$\% \; Fe_3C \; = \; \frac{0.77 \, - \, 0.0218}{6.67 \, - \, 0.0218} \; \times \; 100 \; = \; 12\%$$

◇

From Example 11-10, we find that most of the pearlite is composed of ferrite. In fact, if we examine the pearlite closely, we find that the Fe_3C lamellae are surrounded by α. The pearlite structure therefore produces effective dispersion strengthening—the continuous ferrite phase is relatively soft and ductile and the hard brittle cementite is dispersed.

Primary Microconstituents.   Hypoeutectoid steels contain less than 0.77% C, and hypereutectoid steels contain more than 0.77% C. Ferrite is the primary or proeutectoid microconstituent in hypoeutectoid alloys, and cementite is the primary or proeutectoid microconstituent in hypereutectoid alloys. If we heat a hypoeutectoid alloy containing 0.60% C above 750°C, only austenite remains in the microstructure. Figure 11-15 shows what happens when the austenite

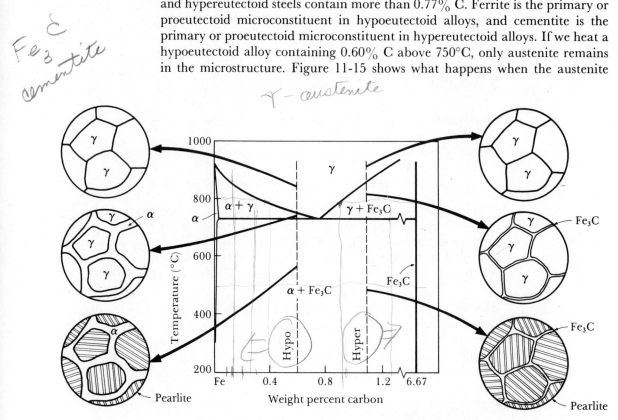

FIGURE 11-15   The evolution of the microstructure of hypoeutectoid and hypereutectoid steels during cooling in relationship to the phase diagram.

cools. Just below 750°C, ferrite precipitates and grows, usually at the austenite grain boundaries. Primary ferrite continues to grow until the temperature falls to 727°C. The remaining austenite at that temperature is now surrounded by ferrite and has changed in composition from 0.60% C to 0.77% C. Subsequent cooling to below 727°C causes all of the remaining austenite to transform to pearlite by the eutectoid reaction. The final structure contains two phases—ferrite and cementite—arranged as two microconstituents—primary ferrite and pearlite.

◇ **Example 11-11**

Calculate the amounts and compositions of phases and microconstituents in an Fe-0.60% C alloy at 726°C. → *one degree lower than eutectoid temp*

Answer:

Know for exam

The phases are ferrite and cementite. Using a tie line and working the lever law at 726°C, we find

$$\alpha: 0.0218\% \ C \qquad \% \ \alpha \ = \ \frac{6.67 - 0.60}{6.67 - 0.0218} \times 100 \ = \ 91.3\%$$

$$Fe_3C: 6.67\%C \qquad \% \ Fe_3C \ = \ \frac{0.60 - 0.0218}{6.67 - 0.0218} \times 100 \ = \ 8.7\%$$

The microconstituents are primary ferrite and pearlite. If we construct a tie line just above 727°C, we can calculate the amounts and compositions of ferrite and austenite just before the eutectoid reaction starts. All of the austenite at that temperature will transform to pearlite; all of the ferrite will remain as primary ferrite.

$$\text{Primary } \alpha: 0.0218\% \ C \quad \% \ \text{Primary } \alpha \ = \ \frac{0.77 - 0.60}{0.77 - 0.0218} \times 100 = 22.7\%$$

$\alpha + Fe_3{}^C$ ⟵  $\text{Pearlite: } 0.77\% \ C \qquad \% \ \text{Pearlite} \ = \ \dfrac{0.60 - 0.0218}{0.77 - 0.0218} \times 100 = 77.3\%$

◇

The final microstructure contains islands of pearlite surrounded by the primary ferrite [Figure 11-16(a)]. This permits the alloy to be strong, due to the dispersion-strengthened pearlite, yet ductile, due to the continuous primary ferrite.

In hypereutectoid alloys, however, the primary phase is Fe_3C, which again forms at the austenite grain boundaries. After the austenite cools through the eutectoid reaction, the steel contains hard, brittle cementite surrounding islands of pearlite [Figure 11-16(b)]. Now, because the hard, brittle microconstituent is continuous, the steel is also brittle. Fortunately, we can improve the microstructure and properties of the hypereutectoid steels by heat treatment.

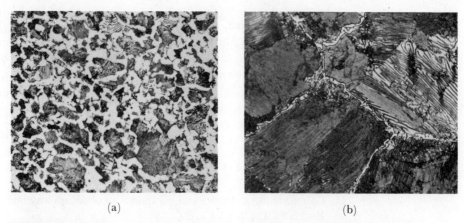

(a)   (b)

FIGURE 11-16   (a) A hypoeutectoid steel showing primary α (white) and pearlite
(× 400). (b) A hypereutectoid steel showing primary Fe₃C surrounding pearlite (× 800).
From *Metals Handbook*, Vol. 7, 8th Ed., American Society for Metals, 1972.

11–10   CONTROLLING THE EUTECTOID REACTION

We can control dispersion strengthening in the eutectoid alloys in much the
same way that we did in eutectic alloys.

Controlling the Amount of the Eutectoid.   By changing the composition of
the alloy, we change the amount of the hard second phase. As the carbon content
of a steel increases towards the eutectoid composition of 0.77% C, the amounts
of Fe_3C and pearlite increase, thus increasing the strength. However, this
strengthening effect eventually peaks and the properties level out or even
decrease when the carbon content is too high (Table 11-2).

TABLE 11-2   The effect of carbon on the strength of steels

	Slow Cooling			Fast Cooling		
Carbon (%)	Yield Strength (psi)	Tensile Strength (psi)	% Elongation	Yield Strength (psi)	Tensile Strength (psi)	% Elongation
0.15	41,250	56,000	37	47,000	61,500	37
0.20	42,750	57,200	36.5	50,250	64,000	36
0.30	49,500	67,250	31	50,000	75,500	32
0.40	51,250	75,250	30	54,250	85,500	28
0.50	53,000	92,250	24	62,000	108,500	20
0.60	54,000	90,750	23	61,000	112,500	18
0.80	54,500	89,250	25	76,000	146,500	11
0.95	55,000	95,250	13	72,500	147,000	9.5

After *Metals Progress Materials and Processing Databook, 1981.*

Example 11-12

Calculate the amounts of Fe_3C and pearlite in steels containing 0.2% C, 0.4% C, 0.8% C, and 1.2% C. Then plot the strength, % Fe_3C, and % pearlite versus the carbon content.

Answer:

Using the tie line and lever law in the iron-carbon diagram we obtain the following.

Carbon (%)	Fe_3C (%)	Pearlite (%)
0.20	$\dfrac{0.20 - 0.0218}{6.67 - 0.0218} \times 100 = 2.7\%$	$\dfrac{0.2 - 0.0218}{0.77 - 0.0218} \times 100 = 23.8\%$
0.40	$\dfrac{0.40 - 0.0218}{6.67 - 0.0218} \times 100 = 5.7\%$	$\dfrac{0.4 - 0.0218}{0.77 - 0.0218} \times 100 = 50.5\%$
0.80	$\dfrac{0.80 - 0.0218}{6.67 - 0.0218} \times 100 = 11.7\%$	$\dfrac{6.67 - 0.8}{6.67 - 0.77} \times 100 = 99.5\%$
1.20	$\dfrac{1.2 - 0.0218}{6.67 - 0.0218} \times 100 = 17.7\%$	$\dfrac{6.67 - 1.2}{6.67 - 0.77} \times 100 = 92.7\%$

These amounts and the strengths from Table 11-2 are plotted in Figure 11-17.

Controlling the Austenite Grain Size.   Under normal conditions, pearlite grows as grains or *colonies*. Within each colony the orientation of the lamellae is identical. The colonies nucleate most easily at the grain boundaries of the original austenite grains. We can increase the number of pearlite colonies by reducing the prior austenite grain size, usually by using low temperatures to produce the austenite or by deoxidizing the steel with aluminum. Typically, we can increase the strength of the alloy by reducing the size or increasing the number of colonies.

Controlling the Cooling Rate.   By increasing the cooling rate during the eutectoid reaction, we reduce the distance that the atoms are able to diffuse. Consequently, the lamellae produced during the reaction are finer or more closely spaced. By producing a finer pearlite, we increase the strength of the alloy. The strength of the alloy is closely related to the interlamellar spacing, or distance between the lamellae (Figure 11-18).

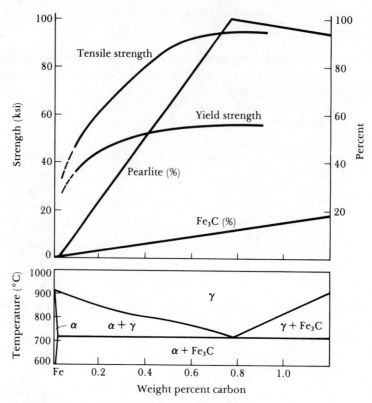

FIGURE 11-17   The strength, % Fe$_3$C, and % pearlite versus the
carbon content in slowly cooled steels.

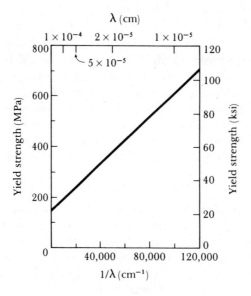

FIGURE 11-18   The effect of
interlamellar spacing of pearlite on the
yield strength of pearlite.

◇ | **Example 11-13**

Estimate the interlamellar spacing and strength of the pearlite shown in Figure 11-14.

Answer:

If we count the number of lamellar spacings in the upper right of Figure 11-14, remembering that interlamellar spacing is from one α plate to the next α plate, we find 14 spacings over a 2-cm distance. Due to magnification $\times 2000$, the distance is 0.001 cm. The interlamellar spacing λ is

$$\lambda = \frac{0.001}{14} = 7.14 \times 10^{-5} \, cm$$

From Figure 11-18, we can estimate the yield strength of the pearlite to be 200 MPa or 30,000 psi.

Controlling the Transformation Temperature. The solid-state eutectoid reaction is rather slow and the steel may cool below the equilibrium eutectoid temperature before the transformation begins. The transformation temperature affects the fineness of the structure (Figure 11-19), the time required for transformation, and even the arrangement of the two phases. This information is contained in the *time-temperature-transformation (TTT)* diagram (Figure 11-20). This diagram, also called the *isothermal transformation (I-T)* diagram or the *C*-curve, permits us to predict the structure, properties, and heat treatment required in steels.

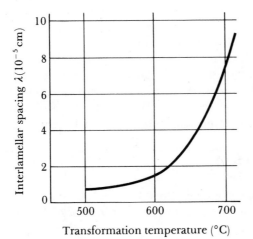

FIGURE 11-19 The effect of the austenite transformation temperature on the interlamellar spacing in pearlite.

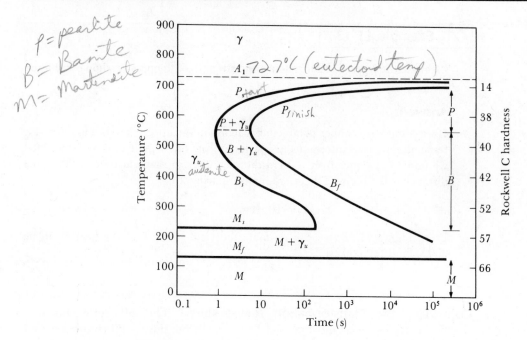

P = pearlite
B = Bainite
M = Martensite

FIGURE 11-20 The time-temperature-transformation (TTT) diagram for a eutectoid steel.

1. Nucleation and growth of pearlite: If we quench to just below the eutectoid temperature, the austenite is only slightly undercooled. Long times are required before stable nuclei for ferrite and cementite form; nucleation does not begin until near the pearlite start (P_s) time. After pearlite begins to grow, atoms diffuse rapidly and coarse pearlite is produced; the transformation is complete at the pearlite finish (P_f) time.

Austenite quenched to a lower temperature is more highly undercooled. Consequently, nucleation occurs more rapidly and the P_s is shorter. However, diffusion is also slower, so atoms diffuse only short distances and a finer pearlite is produced. Even though growth rates are slower, the overall time required for the transformation is reduced because of faster nucleation. Finer pearlite forms in shorter times as we reduce the isothermal transformation temperature to about 550°C, which is the *nose* or *knee* of the TTT curve (Figure 11-20).

◇ | **Example 11-14**

Describe the complete heat treatment and the microstructure after each step required to isothermally produce a hardness of R_c 32 in a eutectoid steel.

Answer:

Note that Rockwell C hardnesses are shown as a function of transformation temperature in the TTT diagram (Figure 11-20). A hardness of R_c 32 is obtained by transforming at 650°C, where the P_s time is 5 s and the P_f time is 50 s. The heat treatment and microstructures are as follows.

1. Austenitize above 727°C and hold for about 1 h. The steel contains 100% austenite.
2. Quench to 650°C and hold for at least 50 s. After 5 s, pearlite nucleates from the unstable austenite. Pearlite grows until, after 50 s, the microstructure contains 100% pearlite. The pearlite has a medium fineness. (Note that we returned to time zero when we quenched!)
3. Cool in air to room temperature. The microstructure remains all pearlite.

2. Nucleation and growth of bainite: At a temperature just below the nose of the TTT diagram, nucleation occurs rapidly but diffusion is slow. No transformation is detected until somewhat longer times, and total transformation times increase due to very slow growth.

In addition, we find a different microstructure. At low transformation temperatures, the lamellae in pearlite would have to be extremely thin and consequently the boundary area between the ferrite and Fe_3C lamellae would be very large. Because of the energy associated with the ferrite-cementite interface, the total energy of the steel would have to be very high. The steel can reduce its internal energy by permitting the cementite to precipitate as discrete, rounded particles in a ferrite matrix. This new microconstituent, or arrangement of ferrite and cementite, is called *bainite*. Transformation begins at a bainite start (B_s) time and ends at a bainite finish (B_f) time.

◇ **Example 11-15**

Excellent combinations of hardness, strength, and toughness are obtained from bainite. One heat treater austenitized a eutectoid steel at 750°C, quenched and held the steel at 250°C for 15 min, and finally permitted the steel to cool to room temperature. Did he produce the required bainitic structure?

Answer:

Let's examine the heat treatment using Figure 11-20. After heating at 750°C, the microstructure is 100% γ. After quenching to 250°C, unstable austenite remains for slightly more than 100 s, when fine bainite begins to grow. After 15 min, or 900 s, about 50% fine bainite has formed and the remainder of the steel still contains unstable austenite. As we will see later, the unstable austenite transforms to martensite when the steel is cooled to room temperature and the final structure is a mixture of bainite and hard, brittle martensite. The heat treatment was not successful. The heat treater should have held the steel at 250°C for at least 10^4 s, or about 3 h.

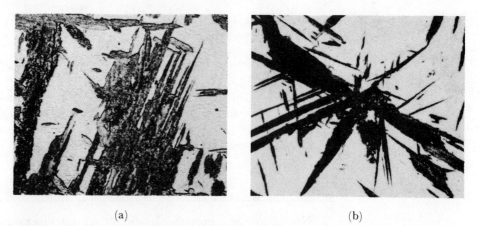

(a)   (b)

FIGURE 11-21   (a) Upper bainite (gray, feathery plates) (× 600). (b) Lower bainite (dark needles) (× 400). From *Metals Handbook*, Vol. 8, 8th Ed., American Society for Metals, 1973.

The times required for austenite to begin and finish its transformation to bainite increase and the bainite becomes finer as the transformation temperature continues to decrease. The bainite that forms just below the nose of the curve is called *coarse bainite, upper bainite,* or *feathery bainite.* The bainite that forms at lower temperatures is called *fine bainite, lower bainite,* or *acicular bainite.* Figure 11-21 shows typical microstructures of bainite.

Figure 11-22 shows the effect of transformation temperature on the properties of a eutectoid steel. As the temperature decreases, there is a general trend towards higher strength and lower ductility due to the finer microstructure that is produced.

11–11   THE MARTENSITIC REACTION

Martensite is a phase that forms as the result of a diffusionless solid-state transformation. Cobalt, for example, transforms from an FCC to an HCP crystal structure by a slight shift in the atom locations which alters the stacking sequence of close-packed planes. Because the reaction does not depend on diffusion, the martensite reaction is *athermal,* or the reaction depends only on the temperature, not the time. The martensite reaction often proceeds rapidly, at speeds approaching the velocity of sound in the material.

Martensite in Steels.   In steels with less than about 0.2% C, the FCC austenite transforms to a supersaturated BCC martensite structure. In higher carbon steels, the martensite reaction occurs as FCC austenite transforms to BCT (body centered tetragonal) martensite. The relationship between the FCC austenite and the BCT martensite [Figure 11-23(a)] shows that carbon atoms in the $\frac{1}{2}$, 0, 0

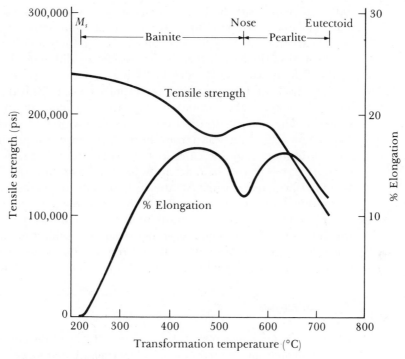

FIGURE 11-22   The effect of transformation temperature on the properties of a eutectoid steel.

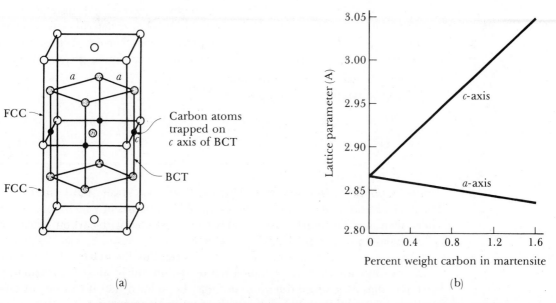

(a)   (b)

FIGURE 11-23   (a) The unit cell of BCT martensite is related to the FCC austenite unit cell. (b) As the percent of carbon increases, more interstitial sites are filled by the carbon atoms and the tetragonal structure of the martensite becomes more pronounced.

type of interstitial sites in the FCC cell can be trapped during the transformation to the body-centered structure, causing the tetragonal structure to be produced. As the carbon content of the steel increases, a greater number of carbon atoms are trapped in these sites, thereby increasing the difference between the a- and c-axes of the martensite [Figure 11-23(b)].

The steel must be quenched, or rapidly cooled, from the stable austenite region to prevent the formation of pearlite, bainite, or primary microconstituents. The martensite reaction begins in a eutectoid steel when austenite cools below 220°C, the martensite start (M_s) temperature (Figure 11-20). The amount of martensite increases as the temperature decreases. When the temperature passes below the martensite finish temperature (M_f), the steel should contain 100% martensite. At any intermediate temperature, the amount of martensite does not change as the time at that temperature increases.

The composition of martensite must be the same as that of the phase from which it forms. There is no long-range diffusion during the transformation that can change the composition. Thus in iron-carbon alloys, the initial austenite composition and the final martensite composition are the same.

◇ Example 11-16

A steel containing 0.40% C is heated to 740°C and then quenched. Determine the amount and composition of the martensite that forms.

Answer:

When the steel is heated to 740°C, a mixture of ferrite and austenite forms (Figure 11-12). We can use a tie line and the lever law to determine the amount and composition of austenite in the two-phase region, then equate the austenite to the martensite. Figure 12-2 shows the eutectoid region in greater detail.

$$\frac{\text{Austenite}}{\text{composition}} = \frac{\text{Martensite}}{\text{composition}} : \text{Fe-}0.68\% \text{ C}$$

$$\% \text{ Martensite} = \% \text{ Austenite} = \frac{0.40 - 0.021}{0.68 - 0.021} \times 100 = 58\%$$

Properties of Steel Martensite.   Martensite in steels is very hard and brittle. The BCT crystal structure has no close-packed slip planes in which dislocations can easily move. The martensite is highly supersaturated with carbon, since iron normally contains less than 0.0218% C at room temperature, and martensite contains the amount of carbon present in the steel. Finally, martensite has a fine grain size and an even finer substructure within the grains. Consequently, martensite has little or no ductility and may be so hard that it can be cut only with special tools. Because of this behavior, steel martensite by itself is not normally used. In the next section we will discuss how we can "temper" the martensite to produce more desirable properties.

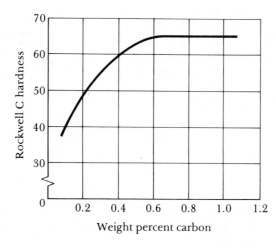

FIGURE 11-24 The effect of carbon content on the hardness of martensite in steels.

The structure and properties of the steel martensites depend significantly on the carbon content of the alloy (Figure 11-24). When the carbon content is low, the martensite grows in a "lath" shape, composed of bundles of flat, narrow plates that grow side by side [Figure 11-25(a)]. This martensite is not very hard. At a higher carbon content, plate martensite grows, in which flat, narrow plates grow individually rather than as bundles [Figure 11-25(b)]. The hardness is much greater in the higher carbon, plate martensite structure, partly due to the greater distortion, or large c/a ratio, of the crystal structure.

Martensite in Other Systems. The characteristics of the martensite reaction are different in other alloy systems. For example, martensite can form in iron-base alloys that contain little or no carbon by a transformation of the FCC crystal structure to a BCC crystal structure. In certain high-manganese steels

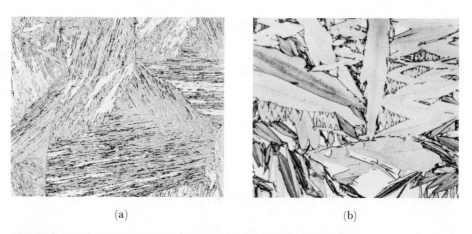

(a)   (b)

FIGURE 11-25 (a) Lath martensite in low-carbon steel ($\times$ 80). (b) Plate martensite in high-carbon steel ($\times$ 400). From *Metals Handbook*, Vol. 8, 8th Ed., American Society for Metals, 1973.

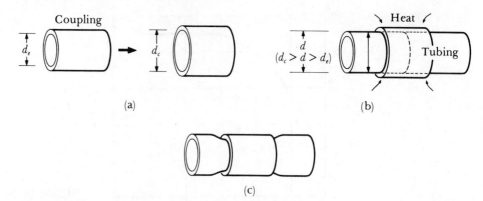

FIGURE 11-26 Use of memory alloys for coupling tubing. A memory alloy coupling is expanded (a) so it fits over the tubing (b). When the coupling is heated, it shrinks back to its original diameter (c)

and stainless steels, the FCC structure changes to an HCP crystal structure during the martensite transformation. In addition, the martensite reaction may occur during the transformation of many polymorphic ceramic materials.

The properties of martensite in other alloys are also different from the properties of steel martensite. In titanium alloys, the BCC titanium transforms to an HCP martensite structure during quenching. However, the titanium martensite is softer and weaker than the original structure.

Martensitic Alloys with a Memory. A unique property possessed by some alloys that undergo the martensitic reaction is the "memory" effect. A Ni-50% Ti alloy and several copper-base alloys can be given a sophisticated thermo-mechanical treatment to produce a martensitic structure. At the end of the treatment, the metal has been deformed to a predetermined shape. The metal can then be deformed into a second shape; but the metal changes back to the original shape when the temperature is increased. The metal remembers its predetermined shape! One commercial application for the memory effect in these martensitic alloys is in couplings for tubing (Figure 11-26). The coupling is set into a small diameter, then deformed into a larger diameter. The coupling, which is slipped over the tubing, contracts back to its predetermined shape after heating. A strong bond is thus produced between the tubes. Other applications include actuating levers, orthodontal braces, blood clot filters, engines, and perhaps eventually artificial hearts.

11–12 TEMPERING OF MARTENSITE

Martensite is not an equilibrium structure. When martensite in a steel is heated to some temperature below the eutectoid temperature, the stable α and Fe_3C precipitate. This process is called *tempering*. The decomposition of martensite in

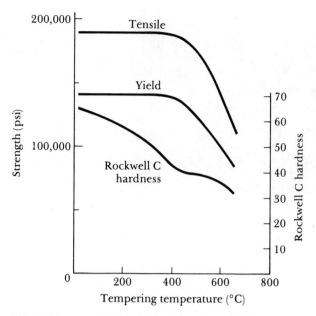

FIGURE 11-27 *Effect of tempering temperature on the properties of a eutectoid steel.*

steels causes the strength and hardness of the martensite to decrease, while the ductility and impact properties are improved (Figure 11-27).

At low tempering temperatures, the martensite may form two transition phases—a lower carbon martensite and a very fine nonequilibrium ε-carbide, or $Fe_{2.4}C$. The steel is still strong, brittle, and perhaps even harder than before tempering. At higher temperatures, the stable α and Fe_3C form and the steel becomes softer and more ductile. If the steel is tempered just below the eutectoid temperature, the Fe_3C becomes very coarse and the dispersion-strengthening effect is greatly reduced. By selecting the appropriate tempering temperature, a wide range of properties can be obtained. The product of the tempering process is a microconstituent called *tempered martensite* (Figure 11-28).

SUMMARY

Many types of solid-state transformations occur and can be controlled by proper heat treatments in materials. These heat treatments are designed to provide an optimum distribution of two or more phases in the microstructure. The resulting dispersion strengthening caused by the phases permits us to obtain a wide variety of structures and properties in materials.

In the most common of these transformations—exceeding the solubility limit, age hardening, control of the eutectoid, and the martensitic reaction—we strive to produce a final microstructure containing a uniform distribution of

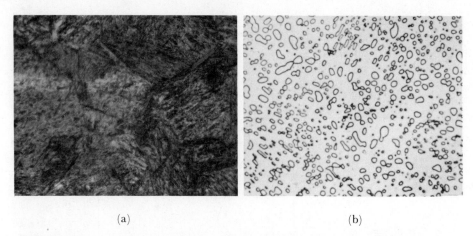

(a)   (b)

FIGURE 11-28   Tempered martensite in steel. (a) Tempered at a low temperature (× 500), (b) tempered at a high temperature for a long time (× 1000). Note the large Fe₃C particles at the higher temperature. From *Metals Handbook*, Vol. 9, 9th Ed., American Society for Metals, 1985.

many tiny, hard precipitate particles in a softer, more ductile matrix. By doing this, we are able to provide effective obstacles to the movement of dislocations, thus providing strength but still maintaining at least usable ductility and toughness.

Careful control of the heat treatment temperatures and times is essential in obtaining the proper microstructure. Phase diagrams assist in selecting the appropriate temperatures, but experimental data are needed to finally obtain the optimum combination of times, temperatures, and compositions.

Finally, since optimum properties are obtained through heat treatment, we must bear in mind that the structure and properties may change when the material is used at elevated temperatures. Overaging, overtempering, and loss of coherency may occur as a natural extension of the phenomena governing these transformations when the material is placed into service.

GLOSSARY

Age hardening. A special dispersion-strengthening heat treatment. By solution treatment, quenching, and aging, a coherent precipitate forms that provides a substantial strengthening effect. Also known as precipitation hardening.

Artificial aging. Reheating a solution-treated and quenched alloy to a temperature below the solvus in order to provide the thermal energy required for a precipitate to form.

Athermal transformation. When the amount of the transformation depends only on the temperature, not on the time.

Austenite. The name given to the FCC crystal structure of iron.

Bainite. A two-phase microconstituent, containing ferrite and cementite, that forms in steels that are isothermally transformed at relatively low temperatures.

Cementite. The hard, brittle intermetallic compound Fe_3C that when properly dispersed provides the strengthening in steels.

Coherent precipitate. A precipitate whose crystal structure and atomic arrangement have a continuous relationship with the matrix from which the precipitate formed. The coherent precipitate provides excellent disruption of the atomic arrangement in the matrix and provides excellent strengthening.

Dihedral angle. The angle that defines the shape of a precipitate particle in the matrix. The dihedral angle is determined by the relative surface energies.

Ferrite. The name given to the BCC crystal structure of iron.

Guinier-Preston zones. Tiny clusters of atoms that precipitate from the matrix in the early stages of the age-hardening process. Although the GP zones are coherent with the matrix, they are too small to provide optimum strengthening.

Interfacial energy. The energy associated with the boundary between two phases.

Isothermal transformation. When the amount of a transformation at a particular temperature depends on the time permitted for the transformation.

Martensite. A metastable phase formed in steel and other materials by a diffusionless, athermal transformation.

Natural aging. When a coherent precipitate forms from a solution-treated and quenched age hardenable alloy at room temperature, providing optimum strengthening.

Pearlite. A two-phase lamellar microconstituent, containing ferrite and cementite, that forms in steels that are cooled in a normal fashion or are isothermally transformed at relatively high temperatures.

Solution treatment. The first step in the age-hardening heat treatment. The alloy is heated above the solvus temperature to dissolve any second phase and to produce a homogeneous single-phase structure.

Strain energy. The energy required to permit a precipitate to fit into the surrounding matrix during nucleation and growth of the precipitate.

Supersaturated solid solution. The solid solution formed when a material is rapidly cooled from a high-temperature single-phase region to a low-temperature two-phase region without the second phase precipitating. Because the quenched phase contains more alloying element than the solubility limit, it is supersaturated in that element.

Tempering. A low-temperature heat treatment used to reduce the hardness of martensite by permitting the martensite to begin to decompose to the equilibrium phases.

Widmanstatten structure. The precipitation of a second phase from the matrix when there is a fixed crystallographic relationship between the precipitate and matrix crystal structures. Often needle-like or platelike structures form in the Widmanstatten structure.

PRACTICE PROBLEMS

1. Under equilibrium conditions, what is the maximum amount of θ that can form during aging of an Al-3.5% Cu alloy at 300°C? What is the maximum amount of θ that can form in the same alloy if natural aging occurs?

2. The θ particles that form in an Al-3% Cu alloy aged at 200°C have an average diameter of 500×10^{-8} cm and a density of $4.26\,g/cm^3$. Calculate the number of these particles per cubic centimeter of the alloy.

3. The Al-Li phase diagram is shown in Figure 13-4. Describe a heat treatment required to produce age hardening in an Al-2% Li alloy, including appropriate temperatures.

4. In rapid solidification processing, an Al-12% Cu alloy could be quenched from above the liquidus temperature to produce a supersaturated solid solution of α. Compare the percent θ produced when this alloy is aged at 250°C with the percent θ produced when a conventional Al-4% Cu alloy is aged at the same temperature.

5. In rapid solidification processing, an Al-8% Li alloy could be quenched from above the liquidus temperature to produce a supersaturated solid solution of α. Compare the percent β produced when this alloy is aged at 300°C with the percent β produced when a conventional Al-2% Li alloy is aged at the same temperature.

6. Calculate the amount of each phase and the amount of each microconstituent in an Fe-0.25% C alloy at 700°C.

7. Calculate the amount of each phase and the amount of each microconstituent in an Fe-1.25% C alloy at 700°C.

8. The microstructure of a steel contains 9% Fe₃C and 91% α at 500°C. What is the carbon content of the steel? Is the steel hypoeutectoid or hypereutectoid?

9. The microstructure of a steel contains 35% α and 65% γ at 800°C. What is the carbon content of the steel?

10. The microstructure of an iron-carbon alloy contains 25% Fe₃C and 75% γ at 800°C. What is the carbon content of the alloy?

11. The microstructure of a steel contains 18% Fe₃C and 82% α at room temperature. What is the carbon content of the steel? Is the steel hypoeutectoid or hypereutectoid?

12. The microstructure of an iron-carbon alloy contains 33% proeutectoid Fe₃C and 67% pearlite at 700°C. What is the carbon content of the alloy?

13. The microstructure of a steel contains 33% proeutectoid α and 67% pearlite at 700°C. What is the carbon content of the steel?

14. When a steel is heated, the austenite phase contains 0.4% C and constitutes 60% of the structure. Estimate the temperature and overall carbon content of the steel.

15. When an iron-carbon alloy is heated, the austenite phase contains 1.0% C and constitutes 80% of the structure. Estimate the temperature and overall carbon content of the alloy.

16. The density of α-Fe is 7.87 g/cm³, and the density of cementite is 7.66 g/cm³. If a cementite lamella in pearlite is 3×10^{-3} cm thick, calculate the thickness of the ferrite lamella.

17. Calculate the density of pearlite, using the densities of each phase given in Problem 16.

18. When we examine the microstructure of a steel, we find that the structure contains about 20 vol% primary α and 80 vol % pearlite. Using the densities of α (Problem 16) and pearlite (Problem 17), estimate the percent carbon in the steel.

19. When we examine the microstructure of a steel, we find that the structure contains about 15 vol % primary Fe₃C and 85 vol % pearlite. Using the densities of Fe₃C (Problem 16) and pearlite (Problem 17), estimate the percent carbon in the steel.

20. Determine the percent of monoclinic ZrO₂ present in the eutectoid product formed when tetragonal ZrO₂ transforms [Figure 14-23(b)].

21. Calculate the amount of each phase in the eutectoid reaction that occurs in the Co-Mo system (Figure 10-32) when σ cools. Based on this result, do you expect the eutectoid product to be ductile or brittle? Do you expect any alloy containing the eutectoid product to be ductile or brittle?

22. Calculate the amount of each phase in the eutectoid reaction that occurs in the Cu-Al system [Figure 13-8(d)] when β cools. Based on this result, do you expect the eutectoid product to be ductile or brittle? Do you expect any alloy containing the eutectoid product to be ductile or brittle?

23. Calculate the amount of eutectoid microconstituent when an Al-72% Zn alloy (Figure 13-13) cools under equilibrium conditions.

24. Calculate the amount of eutectoid microconstituent when a Cu-3% Be alloy [Figure 13-8(e)] cools under equilibrium conditions.

25. Suppose austenite in a eutectoid steel is transformed to pearlite at 650°C. Estimate (a) the interlamellar spacing and (b) the yield strength of the pearlite.

26. Suppose the strength of an isothermally transformed eutectoid steel is 65,000 psi (448 MPa). Estimate (a) the interlamellar spacing and (b) the transformation temperature.

27. A eutectoid steel is to be isothermally transformed to produce a hardness of $R_c 45$. Determine (a) the microconstituent that is produced and (b) the transformation temperature required.

28. A eutectoid steel is to be isothermally transformed to produce a hardness of $R_c 38$. Determine (a) the microconstituent that is produced and (b) the transformation temperature required.

29. We would like to produce a bainitic structure in a eutectoid steel having a hardness of $R_c 52$. Describe the complete heat treatment that would be required.

30. We would like to produce a pearlitic structure in a eutectoid steel having a hardness of $R_c 30$. Describe the complete heat treatment that would be required.

31. Describe the microstructure in a eutectoid steel that has been heated to 800°C for 1 h, quenched to 600°C and held for 100 s, and finally quenched to room temperature.

32. Describe the microstructure in a eutectoid steel that has been heated to 800°C for 1 h, quenched to 300°C and held for 1000 s, and finally quenched to room temperature.

33. Describe the microstructure in a eutectoid steel that has been heated to 800°C for 1 h, quenched to 400°C and held for 1000 s, and finally quenched to room temperature.

34. Describe the microstructure in a eutectoid steel that has been heated to 800°C for 1 h, quenched to 300°C and held for 60 s, and finally quenched to room temperature.

35. Describe the microstructure in a eutectoid steel that has been heated to 800°C for 1 h, quenched to room temperature and held for 1000 s, reheated to 300°C and held for 10,000 s, and finally quenched to room temperature.

36. A steel containing 0.2% C is held at 800°C for 1 h and is then quenched to room temperature. Calculate the composition and the amount of the martensite that forms during the quench. Refer to Figure 12-2.

37. A steel containing 1.4% C is held at 800°C for 1 h and is then quenched to room temperature. Calculate the composition and amount of the martensite that forms during the quench. Refer to Figure 12-2.

38. A steel microstructure contains 40% martensite containing 0.7% C. Determine (a) the temperature from which the steel was quenched and (b) the overall carbon content of the steel. Refer to Figure 12-2.

39. A steel microstructure contains 85% martensite containing 0.9% C. Determine (a) the temperature from which the steel was quenched and (b) the overall carbon content of the steel. Refer to Figure 12-2.

40. The lattice parameter of FCC austenite is about 3.6 Å. Calculate the volume change that occurs when a steel containing 0.5% C is quenched to produce martensite. Does the steel expand or contract during quenching?

41. A dilatometer is a device used to measure the volume change that occurs during a transformation. If a 3% increase in volume is measured when a steel is quenched from the austenite region, estimate by trial and error the carbon content of the steel. The lattice parameter of the original FCC austenite is 3.6 Å.

42. You would like to produce a quenched and tempered eutectoid steel having a yield strength of 100,000 psi. Describe the complete heat treatment, including approximate temperatures, that would be needed to produce this structure and strength.

43. You would like to produce a quenched and tempered eutectoid steel having a tensile strength of at least 150,000 psi but a hardness below R_c40. What range of tempering temperatures would be satisfactory?

44. In eutectic alloys, the eutectic microconstituent is generally the continuous one, but in the eutectoid structures, the primary microconstituent is normally continuous. Explain how this difference is a natural consequence of the reactions.

45. Describe how the memory metals might be useful as plates to be surgically placed around broken bones to provide more rapid healing.

46. List several everyday items for which the memory metals might be useful (eyeglass frames are one example).

Part III

ENGINEERING MATERIALS

Although there is a tremendous variety of engineering materials, the mechanical properties of each can be predicted and controlled by understanding the atomic bonding, atomic arrangement, and strengthening mechanisms discussed in the previous sections. This is particularly evident in Chapters 12 and 13, in which we examine the characteristics of specific metals and alloys. We will extensively utilize the ideas of solid solution strengthening, strain hardening, and dispersion strengthening for the ferrous and nonferrous alloys.

In ceramics and polymers, Chapters 14 and 15, the importance of atomic bonding and atomic arrangement will also be very evident. Although the strengthening mechanisms that apply to metals are less applicable to these materials, we will be able to draw many parallels between them. For example, deformation of certain polymers produces a fibrous microstructure that provides strengthening, just as in metals, but for different reasons. By producing copolymers, we in a sense provide solid solution strengthening. Phase diagrams play an important part in understanding the behavior of ceramics, although we often utilize the phase diagrams differently from the way we do in metals. We explain and control the mechanical properties of ceramics and polymers by mechanisms that do not involve dislocation movement.

Composite materials, Chapter 16, are even more difficult to categorize because of the many types and intended uses of the materials. The behavior of some of the composites can be explained in terms of dispersion strengthening. However, many composites are designed to provide special characteristics that go beyond the conventional methods for controlling the structure-property relationship.

12

FERROUS ALLOYS

12-1 INTRODUCTION

Metals and alloys are often divided into two groups—ferrous and nonferrous.

Ferrous alloys, which are based primarily on iron-carbon alloys, include plain-carbon steels, alloy and tool steels, stainless steels, and cast irons. These groups of ferrous alloys have a wide variety of characteristics and applications. All of the strengthening mechanisms we have discussed apply to at least some of these materials.

In this chapter, we will discuss in some detail how we use the eutectoid reaction to control the heat treatment and properties of steels. The importance of alloying elements for these heat treatments will be pointed out. Finally, we will examine two special classes of ferrous alloys—stainless steels and cast irons.

12-2 REVIEW OF THE FE-FE₃C PHASE DIAGRAM

The $Fe\text{-}Fe_3C$ phase diagram provides the basis for understanding the ferrous alloys. We examined the phase diagram in Chapter 11, but let's quickly review it before going on.

Solid Solutions. There are three solid solutions of importance—δ-ferrite, austenite (γ), and ferrite (α)—and one intermetallic compound—cementite or iron carbide (Fe_3C). In addition, a metastable phase—martensite—can form on rapid cooling.

Three-Phase Reactions. The three-phase reactions are

$$\text{Peritectic: } L_{0.53\% \, C} + \delta_{0.09\% \, C} \rightarrow \gamma_{0.17\% \, C}$$

$$\text{Eutectic: } L_{4.3\% \, C} \rightarrow \gamma_{2.11\% \, C} + Fe_3C_{6.67\% \, C}$$

$$\text{Eutectoid: } \gamma_{0.77\% \, C} \rightarrow \alpha_{0.0218\% \, C} + Fe_3C_{6.67\% \, C}$$

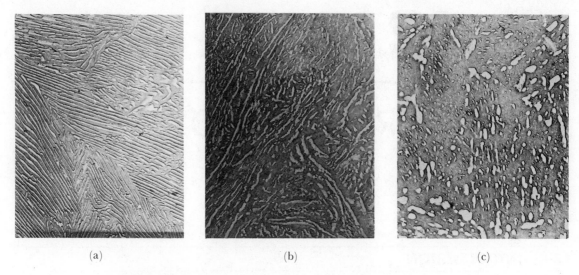

FIGURE 12-1   Electron micrographs of (a) pearlite, (b) bainite, and (c) tempered martensite, illustrating the differences in cementite size and shape among these three microconstituents (× 7500). Courtesy Association of Iron and Steel Engineers; from *The Making, Shaping, and Treating of Steel*, 10th Ed.

The dividing point between steels and cast irons is 2.11% C, or where the eutectic reaction becomes possible.

Microconstituents.   Several microconstituents may form, depending on how we control the eutectoid reaction. Pearlite is a lamellar mixture of ferrite and cementite. Bainite is a nonlamellar mixture of ferrite and cementite obtained by transformation of austenite at a large undercooling. Either primary ferrite or primary cementite may form, depending on the original composition of the alloy. Tempered martensite, a mixture of very fine cementite in ferrite, forms when martensite is reheated following its formation.

All of the heat treatments of a steel are directed towards producing the mixture of ferrite and cementite which gives the proper combination of properties. Figure 12-1 shows that the cementite becomes more rounded and finer as we progress from pearlite to bainite to tempered martensite. The next few sections describe some of the techniques by which we can exercise control over these structures.

We will concentrate on the eutectoid portion of the diagram (Figure 12-2). The solubility lines and the eutectoid isotherm are identified by A_3, A_{cm}, and A_1. The A_3 shows the temperature at which ferrite starts to form on cooling; the A_{cm} shows the temperature at which cementite starts to form; and the A_1 is the eutectoid temperature.

Designations.   The AISI (American Iron and Steel Institute) and SAE (Society of Automotive Engineers) designation systems (Table 12-1) use a four- or five-digit number, where the first two numbers refer to the major alloying

TABLE 12-1 Compositions of selected AISI-SAE steels

AISI-SAE Number	% C	% Mn	% Si	% Ni	% Cr	Others
1020	0.18–0.23	0.30–0.60				
1040	0.37–0.44	0.60–0.90				
1060	0.55–0.65	0.60–0.90				
1080	0.75–0.88	0.60–0.90				
1095	0.90–1.03	0.30–0.50				
1140	0.37–0.44	0.70–1.00				0.08–0.13% S
1340	0.38–0.43	1.60–1.90	0.15–0.30			
1541	0.36–0.44	1.35–1.65				
4140	0.38–0.43	0.75–1.00	0.15–0.30		0.80–1.10	0.15–0.25% Mo
4340	0.38–0.43	0.60–0.80	0.15–0.30	1.65–2.00	0.70–0.90	0.20–0.30% Mo
4620	0.17–0.22	0.45–0.65	0.15–0.30	1.65–2.00		0.20–0.30% Mo
4820	0.18–0.23	0.50–0.70	0.15–0.30	3.25–3.75		0.20–0.30% Mo
5120	0.17–0.22	0.70–0.90	0.15–0.30		0.70–0.90	
52100	0.98–1.10	0.25–0.45	0.15–0.30		1.30–1.60	
6150	0.48–0.53	0.70–0.90	0.15–0.30		0.80–1.10	0.15% min. V
8620	0.18–0.23	0.70–0.90	0.15–0.30	0.40–0.70	0.40–0.60	0.15–0.25% V
9260	0.56–0.64	0.75–1.00	1.80–2.20			

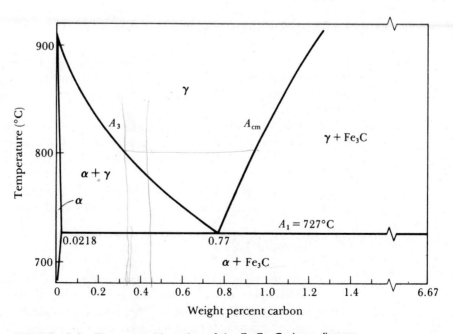

FIGURE 12-2 The eutectoid portion of the Fe-Fe₃C phase diagram.

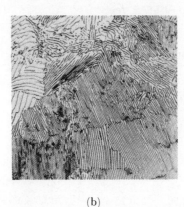

(a)   (b)   (c)

FIGURE 12-3 (a) Hypoeutectoid steel with primary ferrite (white) and pearlite (× 250). (b) Eutectoid steel containing only pearlite colonies (× 350). (c) Hypereutectoid steel with primary Fe$_3$C and pearlite (× 850). From *Metals Handbook*, Vols. 7 and 8, 8th Ed., American Society for Metals, 1972, 1973.

elements present and the last two or three numbers refer to the percent carbon. An AISI 1040 steel is a plain-carbon steel with 0.40% C. An SAE 10120 steel is a plain-carbon steel containing 1.20% C. An AISI 4340 steel is an alloy steel containing 0.40% C.

Structures. We described the microstructures of typical equilibrium-cooled steels in Chapter 11. These structures are reviewed in Figure 12-3. A hypoeutectoid steel, such as a 1050 steel, contains 100% γ after heating to equilibrium above the A_3 temperature. When the steel cools to between the A_3 and the A_1, primary α forms, normally outlining the austenite grain boundaries. When the steel cools to just above the A_1 temperature, a tie line tells us that the remaining austenite contains 0.77% C. This remaining austenite transforms by the eutectoid reaction to produce pearlite.

A 10120 steel is hypereutectoid and contains 100% γ above the A_{cm}. After cooling to between the A_{cm} and the A_1, primary Fe$_3$C forms at the austenite grain boundaries. The austenite present just above the A_1 temperature transforms to pearlite.

A 1077 steel goes through only the eutectoid reaction. Above the A_1 temperature, the structure is all austenite; below the A_1, the structure is all pearlite.

◇ | **Example 12-1**

Most automobile axles are forged from a 1050 steel. Preliminary examination of the steel prior to forging and heat treatment reveals that the microstructure contains about 60% pearlite and 40% primary α. Calculate the amount of each phase and microconstituent that you expect to find in a 1050 steel and determine if the steel is indeed a 1050 steel.

Answer:

The amounts of each phase and microconstituent in a 1050 steel can be calculated using the lever law. You may wish to review Example 11-11.

$$\alpha = \frac{6.67 - 0.5}{6.67 - 0.0218} \times 100 = 92.8\%$$

$$Fe_3C = \frac{0.5 - 0.0218}{6.67 - 0.0218} \times 100 = 7.2\%$$

$$Pearlite = \frac{0.5 - 0.0218}{0.77 - 0.0218} \times 100 = 63.9\%$$

$$Primary\ \alpha = \frac{0.77 - 0.5}{0.77 - 0.0218} \times 100 = 36.1\%$$

The calculated amount of pearlite—63.9%—compares closely with the amount estimated from the microstructure—60%. The steel is probably a 1050 steel.

◇ **Example 12-2**

An as-received bar stock is observed to contain about 95% pearlite and 5% primary Fe$_3$C. Calculate the carbon content of the steel and determine the grade, or AISI-SAE number, of the steel.

Answer:

Let's calculate the percent carbon using the lever law.

$$Pearlite = \frac{6.67 - \%\ C}{6.67 - 0.77} \times 100 = 95$$

$$6.67 - \%\ C = 0.95\ (6.67 - 0.77)$$

$$\%\ C = 1.065$$

We notice from Table 12-1 that a range of carbon is permitted in AISI-SAE steels. Thus, the grade of our steel is closest to a 10110 steel, with approximately 1.10% C.

◇ **Example 12-3**

A steel to be used as a spring is estimated to contain 10% Fe$_3$C and 90% ferrite. Can we consider this steel to be a eutectoid steel?

Answer:

Let's calculate the percent carbon using the lever law.

$$\alpha = \frac{6.67 - \% \ C}{6.67 - 0.0218} \times 100 = 90$$

$$6.67 - \% \ C = 0.9 \ (6.67 - 0.0218)$$

$$\% \ C = 0.69$$

A 1080 steel, which is approximately a eutectoid steel, should, from Table 12-1, contain 0.75% to 0.88% C. Since our steel contains only 0.69% C, it should not be considered a eutectoid steel.

12–3   SIMPLE HEAT TREATMENTS

Four simple heat treatments are commonly used for steels. These heat treatments (Figure 12-4) are used to accomplish one of three purposes.

Process Anneal—Eliminating Cold Work.   The ferrite in steels with less than 0.25% C is strengthened by cold working. The recrystallization heat treatment used to eliminate the effect of cold working is called a *process anneal*. The process anneal is done 80°C to 170°C below the A_1 temperature.

Annealing and Normalizing—Controlling Dispersion Strengthening. Plain-carbon steels are dispersion strengthened by controlling the amount, size, shape, and distribution of Fe_3C. As the carbon increases, more Fe_3C is present and, up to a point, the strength of the steel increases.

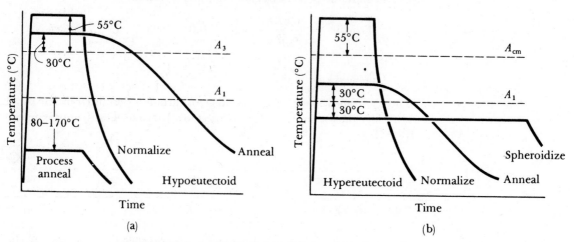

FIGURE 12-4   Schematic summary of the simple heat treatments for (a) hypoeutectoid steels and (b) hypereutectoid steels.

We can refine the Fe_3C by controlling the cooling rate as the austenite transforms to pearlite. If we permit very slow cooling, the pearlite is coarse—this heat treatment is called *annealing* or a full anneal. Faster cooling produces fine pearlite—this heat treatment is *normalizing*.

1. Hypoeutectoid steels are annealed by heating the steel about 30°C above the A_3 temperature to produce homogeneous austenite. This is the *austenitizing treatment*. Then the steel is furnace cooled. By permitting both the furnace and the steel to cool together, slow cooling rates are produced. Because lots of time is available for diffusion, primary ferrite and pearlite are coarse and the steel has a low strength and good ductility.

2. Hypereutectoid steels are annealed by first heating to 30°C above the A_1. The steel is not heated above the A_{cm} to produce all austenite because, on slow cooling, Fe_3C would form as a continuous film on the austenite grain boundaries and cause embrittlement. Austenitizing just above the A_1 permits the Fe_3C to become rounded. After austenitizing, the steel is furnace cooled to produce discontinuous Fe_3C and coarse pearlite.

3. Steels are normalized by heating to 55°C above either the A_3 or A_{cm}, depending on the composition of the steel. After austenitizing, the steel is removed from the furnace and air cooled. Air cooling gives faster cooling rates and finer pearlite. The hypereutectoid steel can be normalized above the A_{cm} because, due to the faster cooling rate, the Fe_3C has less opportunity to form as a continuous film at the austenite grain boundaries.

Figure 12-5 shows the typical properties obtained by annealing and normalizing.

Spheroidizing—Improving Machinability.   High-carbon steels, which contain a large amount of Fe_3C, have poor machining characteristics. During the spheroidizing treatment, which requires long times at about 30°C below the A_1, the Fe_3C changes into large, spherical particles in order to reduce boundary area. The microstructure, known as *spheroidite*, now has a continuous matrix of soft, machinable ferrite (Figure 12-6). After machining, the steel is given a more sophisticated heat treatment to produce the required properties. A similar structure would occur if martensite were tempered just below the A_1 for long times, as shown in Figure 11-28(b).

◇   **Example 12-4**

Recommend temperatures for the process annealing, annealing, normalizing, and spheroidizing of 1020, 1077, and 10120 steels.

Answer:

If we consult Figure 12-2, we can determine the critical A_1, A_3, or A_{cm} temperatures for each steel. We can then specify the heat treatment based on these temperatures.

	1020	1077	10120
Critical temperatures	$A_1 = 727°C$ $A_3 = 830°C$	$A_1 = 727°C$	$A_1 = 727°C$ $A_{cm} = 895°C$
Process annealing	$727 - (80 \text{ to } 170) =$ $557°C \text{ to } 647°C$	Not done	Not done
Annealing	$830 + 30 =$ $860°C$	$727 + 30 =$ $757°C$	$727 + 30 =$ $757°C$
Normalizing	$830 + 55 =$ $885°C$	$727 + 55 =$ $782°C$	$895 + 55 =$ $950°C$
Spheroidizing	Not done	$727 - 30 =$ $697°C$	$727 - 30 =$ $697°C$

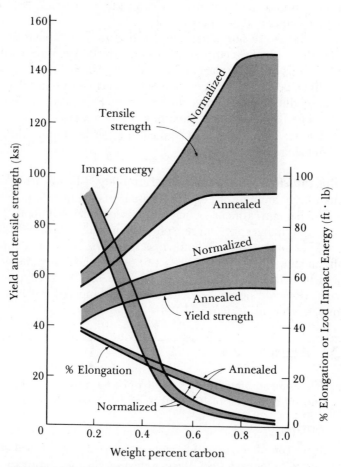

FIGURE 12-5 The effect of carbon and heat treatment on the properties of plain-carbon steels.

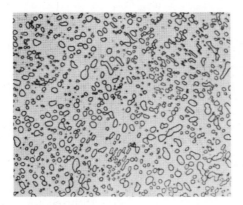

FIGURE 12-6   The microstructure of spheroidite with Fe_3C particles dispersed in a ferrite matrix (× 850). From *Metals Handbook*, Vol. 7, 8th Ed., American Society for Metals, 1972.

12–4   ISOTHERMAL HEAT TREATMENTS AND DISPERSION STRENGTHENING

The effect of transformation temperature on the properties of a 1080 (eutectoid) steel was discussed in Chapter 11. As the isothermal transformation temperature decreases, pearlite become progressively finer before bainite begins to form instead. At very low temperatures, martensite is obtained. As the microstructure of a 1080 steel becomes finer, better dispersion strengthening is obtained. Also, the fine, rounded microstructure of bainite produces higher strengths and hardnesses than pearlitic structures while maintaining usable ductility and toughness.

Austempering and Isothermal Annealing.   The isothermal transformation heat treatment used to produce bainite is called *austempering*, and simply involves austenitizing the steel, quenching to some temperature below the nose of the TTT curve, and holding at that temperature until all of the austenite transforms to bainite (Figure 12-7).

Annealing and normalizing are usually used to control the fineness of pearlite. However, pearlite formed by an *isothermal anneal* (Figure 12-7) may give more uniform properties, since the cooling rates and microstructure obtained during annealing and normalizing vary across the cross section of the steel.

Isothermal Transformations in Hypoeutectoid and Hypereutectoid Steels.   In either a hypoeutectoid or a hypereutectoid steel, the TTT diagram must reflect the possible formation of a primary phase. The isothermal transformation diagrams for a 1050 and a 10110 steel are shown in Figure 12-8. The

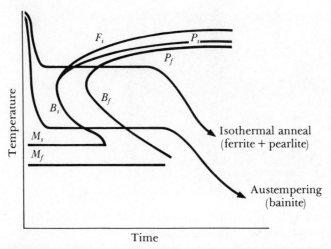

FIGURE 12-7 *The austempering and isothermal anneal heat treatments.*

most remarkable change is the presence of a "wing" which begins at the nose of the curve and becomes asymptotic to the A_3 or A_{cm} temperature. The wing represents the ferrite start (F_s) time in hypoeutectoid steels or the cementite start (C_s) time in hypereutectoid steels.

When a 1050 steel is austenitized, quenched, and held at a temperature between the A_1 and the A_3, primary ferrite nucleates and grows; eventually an equilibrium amount of ferrite and austenite result. Similarly, primary cementite will nucleate and grow to its equilibrium amount in a 10110 steel held between the A_{cm} and A_1 temperatures.

If an austenitized 1050 steel is quenched to a temperature between the nose and the A_1 temperatures, primary ferrite again nucleates and grows until reaching the equilibrium amount. The remainder of the austenite then transforms to pearlite. A similar situation, producing primary cementite and pearlite, is found for the hypereutectoid steel.

If we quench below the nose of the curve, only bainite forms, regardless of the carbon content of the steel. Bainite, unlike pearlite, does not have a fixed composition.

◇ | **Example 12-5**

A 1050 steel is isothermally heat treated to give a hardness of $R_c 23$. Describe the heat treatment and the amount of each microconstituent after each step of the heat treatment.

Answer:

From Figure 12-2, the A_3 temperature is 755°C. The desired hardness is obtained by transforming the steel at 600°C, where $F_s = 1.0\,\mathrm{s}$, $P_s = 1.5\,\mathrm{s}$, and $P_f = 5\,\mathrm{s}$.

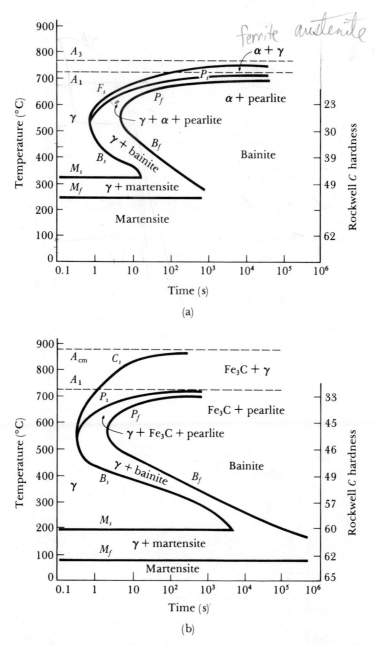

FIGURE 12-8   The TTT diagrams for (a) a 1050 and (b) a 10110 steel.

1. Austenitize at $755 + (30 \text{ to } 55) = 785°C$ to $810°C$ and hold for perhaps 1 h. The microstructure is 100% γ.

2. Quench and hold at $600°C$ for at least 5 s. Primary ferrite begins to precipitate from the unstable austenite after about 1.0 s. After 1.5 s pearlite begins to grow, and the austenite is completely transformed to

ferrite and pearlite after 5 s. From the lever law

$$\text{Primary } \alpha = \frac{0.77 - 0.5}{0.77 - 0.0218} \times 100 = 36\%$$

$$\text{Pearlite} = \frac{0.5 - 0.0218}{0.77 - 0.0218} \times 100 = 64\%$$

3. Cool in air to room temperature. The structure will still contain the equilibrium amounts of primary ferrite and pearlite.

Interrupting the Isothermal Transformation.   Complicated microstructures are produced by interrupting the isothermal heat treatment. For example, we could austenitize the 1050 steel (Figure 12-9) at 800°C, quench to 650°C and hold for 10 s (permitting some ferrite and pearlite to form), then quench to 350°C and hold for 1 h (3600 s). Whatever unstable austenite remained before quenching to 350°C transforms to bainite. The final structure is ferrite, pearlite, and bainite.

We could complicate the treatment further by interrupting the treatment at 350°C after 1 min (60 s) and quenching. Any austenite remaining after 1 min at 350°C forms martensite. The final structure now contains ferrite, pearlite,

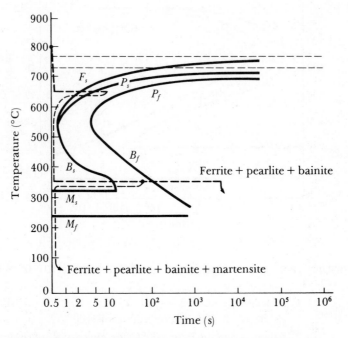

FIGURE 12-9 *Producing complicated structures by interrupting the isothermal heat treatment of a 1050 steel.*

FIGURE 12-10 Dark feathers of bainite surrounded by light martensite, obtained by interrupting the isothermal transformation process (× 1500). From *Metals Handbook*, Vol. 9, 9th Ed., American Society for Metals, 1985.

bainite, and martensite. Note that each time we change the temperature we start at zero time!

Figure 12-10 shows the structure obtained by interrupting the transformation to bainite of a 0.5% C steel by quenching the remaining austenite to martensite.

Because such complicated mixtures of microconstituents produce unpredictable properties, these structures are seldom produced intentionally.

◇ **Example 12-6**

A 1050 steel is held at 800°C for 1 h, quenched to 700°C and held for 50 s, quenched to 400°C and held for 20 s, and finally quenched to room temperature. What is the final microstructure of the steel? Use the TTT diagram (Figure 12-8) for a 1050 steel.

Answer:

1. After 1 h at 800°C, 100% austenite forms.
2. Ferrite begins to form after 20 s at 700°C but, after 50 s, the steel contains only ferrite and unstable austenite.
3. Immediately after quenching to 400°C, the steel is still only ferrite and austenite. Bainite begins to form after 3 s and, after 20 s, the steel contains ferrite, bainite, and still some unstable austenite.
4. After quenching to room temperature, the remaining austenite crosses the M_s and M_f temperatures and transforms to martensite. The final structure is ferrite, bainite and martensite.

12–5   QUENCH AND TEMPER HEAT TREATMENTS

We can obtain an even finer dispersion of Fe_3C if we first quench the austenite to produce martensite, then temper. During tempering an intimate mixture of ferrite and cementite forms from the martensite, as discussed in Chapter 11. By controlling the quench and temper heat treatment, we also control the final properties of the steel (Figure 12-11). Several factors affect the martensite reaction and the execution of the quench and temper heat treatment in plain-carbon steels.

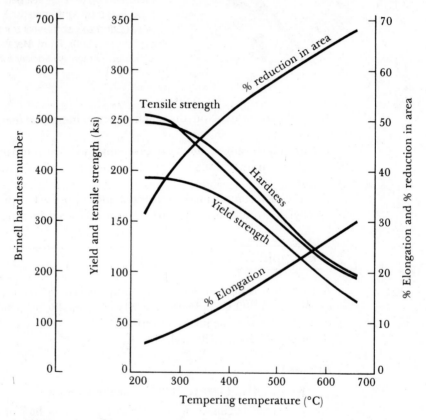

FIGURE 12-11   The effect of tempering temperature on the mechanical properties of a 1050 steel.

◇ | **Example 12-7**

Select a quench and temper heat treatment that will produce a yield strength of 145,000 psi and an elongation greater than 15% in a 1050 steel.

Answer:

From Figure 12-11, we find that the yield strength exceeds 145,000 psi if the steel is tempered below 460°C, while the elongation exceeds 15% if tempering is done above 425°C. One possible heat treatment is as follows.

1. Austenitize above the A_3 temperature of 755°C for 1 h. An appropriate temperature may be $755 + 55 = 810$°C.
2. Quench rapidly to room temperature. Since the M_f is about 250°C, martensite will form.
3. Temper by heating the steel to 440°C. Normally, 1 h will be sufficient if the steel is not too thick.
4. Cool to room temperature.

Effect of Carbon on the M_s and M_f.   The martensite start and finish temperatures are reduced when the carbon content increases (Figure 12-12). High-carbon steels must be refrigerated to produce all martensite.

Retained Austenite.   As the martensite plates form during quenching, they surround and isolate small pools of austenite (Figure 12-13). For the remaining austenite to transform, the surrounding martensite must deform, but the strong

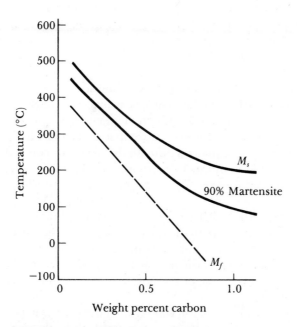

FIGURE 12-12 The effect of carbon on the M_s and M_f temperatures in plain-carbon steels.

Austenite

Martensite

(a)

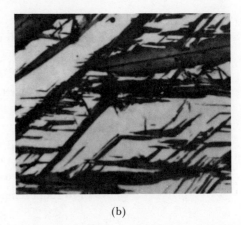

(b)

FIGURE 12-13 Formation of retained austenite in steel during quenching. The pools of austenite cannot transform due to the restraint of the strong surrounding martensite. (a) Schematic diagram. (b) Dark martensite needles surround retained austenite in high-carbon steel (× 1000). From *Metals Handbook*, Vol. 8, 8th Ed., American Society for Metals, 1973.

martensite resists the transformation. Either the existing martensite cracks or the austenite remains trapped in the structure as *retained austenite*.

Retained austenite can be a serious problem. Martensite softens and becomes more ductile during tempering. After tempering, the retained austenite cools below the M_s and M_f temperatures and transforms to martensite, since the surrounding tempered martensite can deform. But now the steel contains more hard, brittle martensite! A second tempering step may be needed to eliminate the martensite formed from the retained austenite.

Residual Stresses and Cracking. Residual stresses are also produced because of the volume change. The hard surface is placed in tension, while the center is compressed. If the residual stresses are high enough, quench cracks form at the surface (Figure 12-14). However, if we first cool to just above the M_s and hold until the temperature equalizes through the steel, subsequent quenching permits all of the steel to transform to martensite at about the same time. This heat treatment is called *marquenching* or *martempering* (Figure 12-15).

Quench Rate. In the TTT diagram, we assumed that we could cool from the austenitizing temperature to the transformation temperature instantly. Because this is not true, undesired microconstituents may form during quenching. For example, pearlite may form as the steel cools past the nose of the curve, particularly since the time of the nose is less than one second in plain-carbon steels.

The rate at which the steel cools during quenching depends on two primary factors. First, the surface always cools faster than the center of the steel. In addition, as the size of the part increases, the cooling rate at any location is

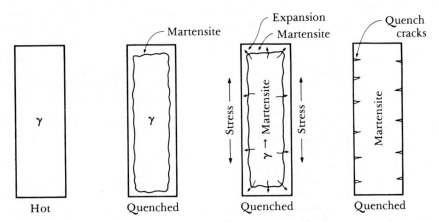

FIGURE 12-14 Formation of quench cracks due to residual stresses produced during quenching. The figure illustrates the development of stresses as the austenite transforms to martensite during cooling.

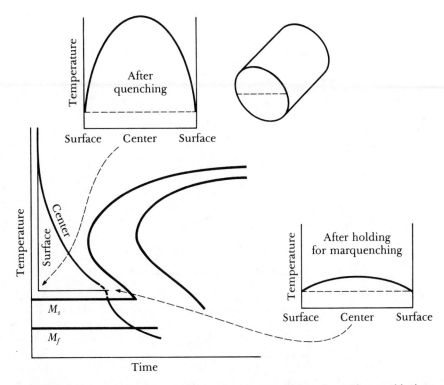

FIGURE 12-15 The marquenching heat treatment, designed to reduce residual stresses and quench cracking.

TABLE 12-2   The H coefficient, or severity of the quench, for several quenching media

Medium	H Coefficient	Cooling Rate in 1-in. Bar (°C/s)
Oil (no agitation)	0.25–0.30	16–20
Oil (violent agitation)	0.80–1.10	40–50
H_2O (no agitation)	0.90–1.00	44–47
H_2O (violent agitation)	4.0	190
Brine (no agitation)	2.0	90
Brine (violent agitation)	5.0	230

From *The Making, Shaping, and Treating of Steel*, 9th Ed., United States Steel, 1971.

slower. Second, the cooling rate depends on the temperature and heat transfer characteristics of the quenching medium (Table 12-2). Quenching in oil, for example, produces a lower H coefficient, or slower cooling rate, than quenching in water or brine.

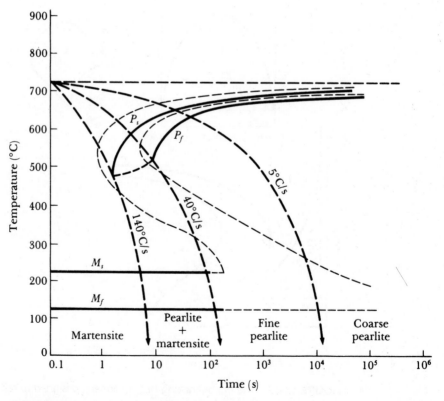

FIGURE 12-16   The CCT diagram (solid lines) for a 1080 steel compared with the TTT diagram (dashed lines).

Continuous Cooling Transformation Diagrams.   We can develop a *continuous cooling transformation (CCT)* diagram by determining the microstructures produced in a steel at various rates of cooling. The CCT curve for a 1080 steel is shown in Figure 12-16. The CCT diagram differs from the TTT diagram (Figure 11-20) in that transformations begin at slightly longer times and no bainite region is observed.

If we cool a 1080 steel at 5°C/s, the CCT diagram tells us that we obtain coarse pearlite; we have annealed the steel. Cooling at 35°C/s gives fine pearlite, or a normalizing heat treatment. Cooling at 100°C/s permits pearlite to start forming, but the reaction is incomplete and the remaining austenite changes to martensite. We obtain 100% martensite and thus are able to perform a quench and temper heat treatment, only if we cool faster than 140°C/s.

Other steels have more complicated CCT diagrams. Figure 12-17 shows the CCT diagram for a slightly alloyed 1020 steel. Cooling rates between 10°C/s and 20°C/s give a combination of ferrite, pearlite, bainite, and martensite.

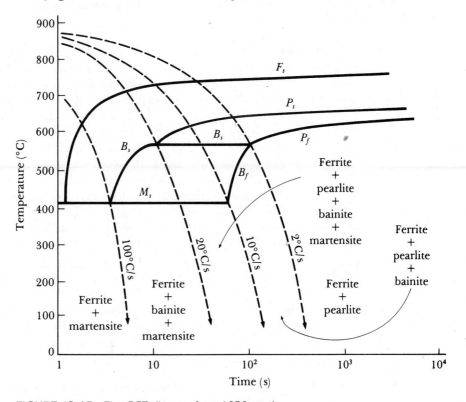

FIGURE 12-17   *The CCT diagram for a 1020 steel.*

◇ | **Example 12-8**

A 1020 steel cools at 8°C/s when quenched in oil and at 50°C/s when quenched in water. What microstructure is produced by each of the heat treatments?

Answer:

From Figure 12-17, an 8°C/s cooling curve crosses the F_s, P_s, B_s, and B_f lines. The structure is ferrite, pearlite and bainite.

At 50°C/s, the cooling curve crosses the F_s, B_s, and M_s lines. The structure is ferrite, bainite, and martensite. There may also be a small amount of retained austenite.

12–6   EFFECT OF ALLOYING ELEMENTS

Alloying elements are added to steels to (a) provide solid solution strengthening of ferrite, (b) cause the precipitation of alloy carbides rather than Fe_3C, (c) improve corrosion resistance and other special characteristics of the steel, and (d) improve the hardenability. Improving hardenability is most important in alloy and tool steels.

Hardenability.   In plain-carbon steels, the nose of the TTT and CCT curves occurs at very short times; hence, very fast cooling rates are required to produce all martensite. In thin sections of steel, the rapid quench produces distortion and cracking. In thick steels, we are unable to produce martensite. All common alloying elements in steel shift the TTT and CCT diagrams to longer times, permitting us to obtain all martensite even in thick sections at slow cooling rates. Figure 12-18 shows the TTT and CCT curves for a 4340 steel.

Hardenability refers to the ease with which martensite forms. Plain-carbon steels have low hardenability—only very high cooling rates produce all martensite. Alloyed steels have high hardenability—even cooling in air may produce martensite. Hardenability does *not* refer to the hardness of the steel. A low-carbon, high-alloy steel may easily form martensite but, because of the low carbon content, the martensite is not hard.

◇ **Example 12-9**

Using the TTT and CCT diagrams, compare the hardenabilities of 4340 and 1050 steels by determining (a) the times required for isothermal transformation to occur at 650°C and (b) the cooling rates required to produce martensite.

Answer:

The transformation times of the two steels, determined from Figures 12-8 and 12-18, are listed below.

	1050	4340
F_s	2 s	200 s
P_s	5 s	3,000 s
P_f	20 s	10,000 s

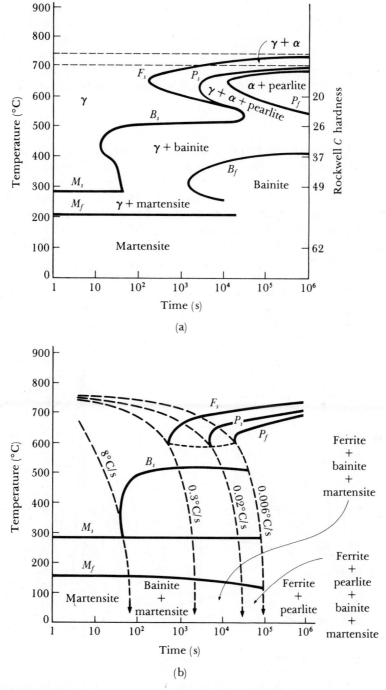

FIGURE 12-18 (a) TTT and (b) CCT curves for a 4340 steel.

From the CCT diagram for the 4340 steel, a cooling rate greater than 8°C/s is required to produce all martensite. A CCT curve for the 1050 steel is not available; however, if we examine the CCT diagrams for the 1020 and 1080 steels, we find that the 8°C/s cooling rate gives all pearlite in the 1080 steel and a mixture of ferrite, pearlite, and bainite in the 1020 steel. The alloying elements in the 4340 steel substantially increase the hardenability.

Effects on the Phase Diagram.   When alloying elements are added to steel, the binary Fe-Fe$_3$C phase diagram is altered (Figure 12-19). First, alloying elements reduce the carbon content at which the eutectoid reaction occurs. Thus, a plain-carbon steel containing 0.6% C is hypoeutectoid, whereas an alloy steel containing 0.6% C and 2% Mo is hypereutectoid. The alloy steel may contain primary cementite.

Some alloying elements, such as nickel, reduce the A_1, A_3, and A_{cm} temperatures. A tempering or service temperature that is satisfactory for plain-carbon steels may exceed the A_1 of the nickel alloy steel, permitting austenite to form. However, most alloying elements increase the transformation temperatures. When these steels are austenitized at the usual 30°C to 55°C above the A_3 or A_{cm} lines for plain-carbon steels, the temperature may still be below the

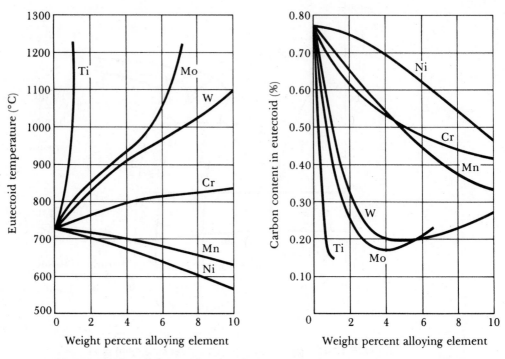

FIGURE 12-19   The effect of selected alloying elements on the eutectoid temperature and on the percent carbon in the eutectoid.

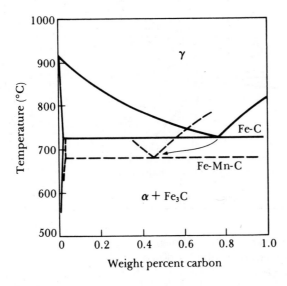

FIGURE 12-20 The effect of 6% manganese on the eutectoid portion of the Fe-Fe₃C phase diagram.

actual transformation temperatures of the alloy steel. Thus, 100% austenite does not form and the heat treatment is not successful. Figure 12-20 shows the effect of 6% manganese on the phase diagram.

◇ | **Example 12-10**

A 1060 steel can be given a subcritical anneal at 700°C; as a consequence of the heat treatment, the steel is softened and any effects of prior cold work are eliminated. Suppose a steel containing 0.6% C and 8% Ni is given such a heat treatment. Describe the structure and discuss the appropriateness of the heat treatment.

Answer:

In the nickel steel, the amount of carbon at the eutectoid composition is 0.54% C. The A_1 temperature for the nickel steel is 600°C. Consequently, when the nickel steel is held at 700°C, austenite will form. The steel is now hypereutectoid rather than hypoeutectoid. After cooling from the heat-treating temperature, primary cementite can form at the grain boundaries and embrittle the steel.

◇

Martensite Start and Finish Temperatures.   The alloying elements reduce the M_s and M_f temperatures. Thus, alloy steels may have to be quenched to lower than normal temperatures to produce martensite. One expression for the M_s temperature is

$$M_s = 539 - 36\% \text{ C} - 39\% \text{ Mn} - 39\% \text{ Cr} - 19\% \text{ Ni} \qquad (12\text{-}1)$$

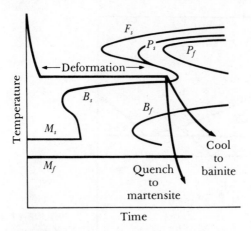

FIGURE 12-21   When alloying elements introduce a bay region into the TTT diagram, the steel can be ausformed.

Shape of the TTT Diagram.   Alloying elements may introduce a "bay" region into the TTT diagram, as in the case of the 4340 steel (Figure 12-18). The bay region is used as the basis for a thermomechanical heat treatment known as *ausforming*. A steel can be austenitized, quenched to the bay region, plastically deformed, and finally quenched to produce martensite (Figure 12-21).

Tempering.   Alloying elements reduce the rate of tempering compared with a plain-carbon steel (Figure 12-22). This effect may permit the alloy steels to operate more successfully at higher temperatures than plain-carbon steels.

12–7   HARDENABILITY CURVES

CCT diagrams are unavailable for many steels; furthermore, accurate cooling rates are difficult to determine. Instead, a *Jominy test* (Figure 12-23) is used to compare hardenabilities of steels. A steel bar 4 in. long and 1 in. in diameter is austenitized, placed into a fixture, and sprayed at one end with water. This procedure produces a range of cooling rates—very fast at the quenched end, almost air cooling at the opposite end. After the test, hardness measurements are made along the test specimen and plotted to produce a *hardenability curve* (Figure 12-24). The distance from the quenched end is the *Jominy distance* and is related to the cooling rate (Table 12-3).

Virtually any steel transforms to martensite at the quenched end. Thus, the hardness at zero Jominy distance is determined solely by the carbon content of the steel. At larger Jominy distances, the cooling rates are slower and there is a greater likelihood that bainite or pearlite will form instead of martensite. An alloy steel with a high hardenability maintains a rather flat hardenability curve;

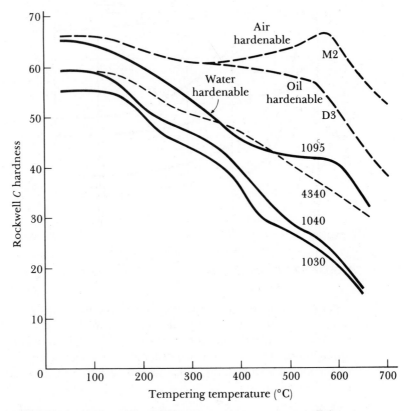

FIGURE 12-22   The effect of alloying elements on the tempering curves of steels. M2 and D3 refer to tool steels.

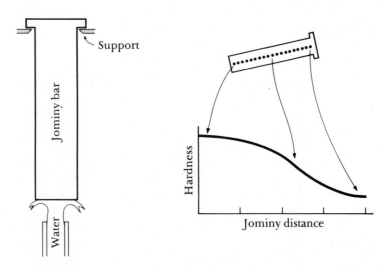

FIGURE 12-23   The Jominy test for determining the hardenability of a steel.

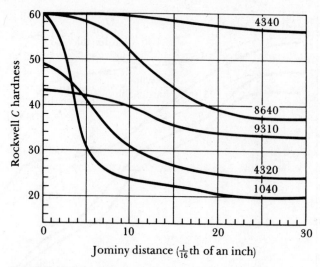

FIGURE 12-24   The hardenability curves for several steels.

a plain-carbon steel has a curve that drops off quickly. The hardenability is determined primarily by the alloy content in the steel.

We can use hardenability curves in selecting or replacing steels in practical applications. The fact that two different steels cool at the same rate if quenched under identical conditions helps in this selection process. A simple example is described below.

TABLE 12-3   The relationship between cooling rate and Jominy distance

Jominy Distance (in.)	Cooling Rate (°C/s)
$\frac{1}{16}$	315
$\frac{2}{16}$	110
$\frac{3}{16}$	50
$\frac{4}{16}$	36
$\frac{5}{16}$	28
$\frac{6}{16}$	22
$\frac{7}{16}$	17
$\frac{8}{16}$	15
$\frac{10}{16}$	10
$\frac{12}{16}$	8
$\frac{16}{16}$	5
$\frac{20}{16}$	3
$\frac{24}{16}$	2.8
$\frac{28}{16}$	2.5
$\frac{36}{16}$	2.2

◇ **Example 12-11**

A gear made from 9310 steel is found to wear at an excessive rate. The hardness at a critical location is R_c40. We decide that a hardness of at least R_c50 is needed at that location to resist wear. Which steel(s) in Figure 12-24 would be most appropriate?

Answer:

A hardness of R_c40 in a 9310 steel corresponds to a Jominy distance of $\frac{10}{16}$ in. The other steels cool at this same rate and therefore have the following hardnesses at the critical location.

1040 R_c24
4320 R_c31
8640 R_c52
4340 R_c60

Both the 8640 and 4340 steels are appropriate. The 4320 steel has too low a carbon content ever to reach R_c50; the 1040 has enough carbon, but the hardenability is too low.

In another simple technique, we utilize the severity of the quench and the Grossman chart (Figure 12-25) to determine the hardness at the center of a round bar. The bar diameter and H coefficient, or severity of the quench in Table 12-2, give the Jominy distance at the center of the bar. We can then determine the hardness from the hardenability curve of the steel. (See Example 12-12.)

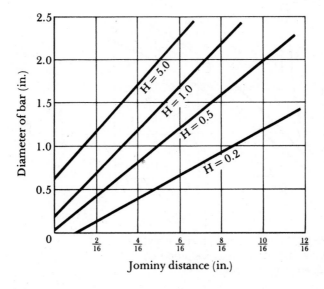

FIGURE 12-25 The Grossman chart used to determine the hardenability of a steel from the bar diameter and severity of the quench.

◇ | **Example 12-12**

An 8640 steel bar 1 in. in diameter must have a minimum hardness of $R_c 58$ throughout the cross section. Which of the following quenching media will be satisfactory: still oil, still water, or agitated brine?

Answer:

First, find the H coefficient from Table 12-2. Then, using the 1-in. diameter, determine the Jominy distance that corresponds to each quench medium from Figure 12-25. Finally, determine the hardness for each Jominy distance from the hardenability curve in Figure 12-24.

Medium	H Coefficient	Jominy Distance (in.)	R_c Hardness
Still oil	0.30	$\frac{7}{16}$	56
Still water	1.00	$\frac{3}{16}$	59
Agitated brine	5.00	$\frac{1}{16}$	60

The minimum $R_c 58$ hardness is obtained by quenching in either still water or agitated brine.

Finally, a simplified technique can be demonstrated using the data in Figure 12-26, which shows the Jominy distance obtained at selected locations in a quenched steel plate.

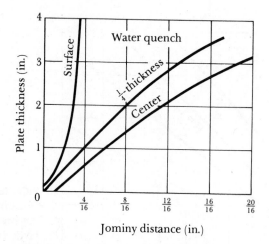

FIGURE 12-26 *Effect of section size and location on the Jominy distance produced when quenching steel plate.*

◇ | **Example 12-13**

Determine the hardness profile across the cross section of a 4320 quenched plate 2 in. thick.

Answer:

From Figures 12-24 and 12-26 we can determine the hardness across the plate, which can then be plotted (Figure 12-27).

Location	Jominy Distance (in.)	R_c Hardness
Surface	$\frac{3}{16}$	45
$\frac{1}{4}$-thickness	$\frac{8}{16}$	34
Center	$\frac{11}{16}$	30

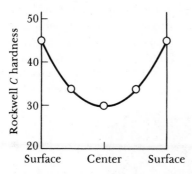

FIGURE 12-27   Hardness profile across the cross section of a 4320 quenched plate 2 in. thick.

12–8   SPECIAL STEELS

There are many special categories of steels, including tool steels, high strength–low alloy steels, microalloyed steels, dual phase steels, and maraging steels.

Tool steels are usually high-carbon steels that obtain high hardnesses by a quench and temper heat treatment. Their applications include cutting tools in machining operations, dies for die casting, forming dies, and others in which a combination of high strength, hardness, toughness, and temperature resistance is needed. A wide variety of tool steels is available, as shown in Table 12-4.

Alloying elements improve the hardenability and high-temperature stability of the tool steels (Figure 12-22). The water hardenable steels such as 1095 must be quenched rapidly to produce martensite and also soften rapidly even at

TABLE 12-4   Compositions of selected tool steels

Steel	% C	% Mn	% W	% Mo	% Cr	% V
W1 (water hardening)	0.6–1.4					
O1 (oil hardening)	0.9	1.0	0.5		0.5	
A2 (air hardening)	1.0			1.0	5.0	
S1 (shock resisting)	0.5		2.5	0.5	1.5	
D1 (chromium cold work)	1.0			1.0	12.0	
H11 (chromium hot work)	0.35			1.5	5.0	0.4
H20 (tungsten hot work)	0.35		9.0		2.0	
H41 (molybdenum hot work)	0.65		1.5	8.0	4.0	1.0
T1 (tungsten high speed)	0.75		18.0		4.0	1.0
M1 (molybdenum high speed)	0.85		1.5	8.5	4.0	1.0

relatively low temperatures; oil hardenable steels form martensite more easily, temper more slowly, but still soften at high temperatures. The air hardenable and special tool steels may harden to martensite while cooling in air; in addition, these steels may not soften until near the A_1 temperature. In fact, the highly alloyed tool steels may pass through a *secondary hardening peak* near 500°C as the normal cementite dissolves and hard alloy carbides precipitate. The alloy carbides are particularly stable, resist growth or spheroidization, and are important in establishing the high temperature resistance of these steels.

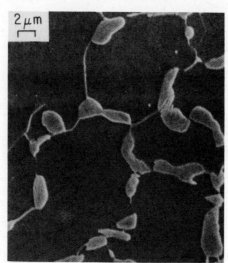

FIGURE 12-28 Microstructure of a dual phase steel, showing islands of light martensite in a ferrite matrix (× 2500). From G. Speich, "Physical Metallurgy of Dual-Phase Steels," *Fundamentals of Dual-Phase Steels*, The Metallurgical Society of AIME, 1981.

High strength–low alloy (HSLA) steels and *microalloyed steels* are low-carbon steels containing small amounts of alloying elements. The HSLA steels are specified on the basis of yield strength, with grades corresponding to 40,000–80,000 psi; the steels contain the least amount of alloying element that still provides the proper yield strength without heat treatment. In microalloyed steels, careful processing, permitting the precipitation of carbides and nitrides of Cb, V, Ti, or Zr, provides dispersion strengthening and helps produce a fine grain size.

Dual phase steels contain a uniform distribution of ferrite and martensite, with the dispersed martensite providing yield strengths of 60,000–145,000 psi. These low-carbon steels do not contain enough alloying elements to have good hardenability using normal quenching processes. But when the steel is heated into the ferrite plus austenite portion of the phase diagram, the austenite phase becomes enriched in carbon, which provides the needed hardenability. During quenching, only the austenite portion transforms to martensite (Figure 12-28).

Maraging steels are low-carbon, highly alloyed steels that derive their strength from a combination of solid solution strengthening and the precipitation hardening of a low-carbon martensite. The steels are austenitized and quenched to produce a soft martensite that contains less than 0.3% C. When the martensite is aged at about 500°C, a variety of intermetallic compounds, including Ni_3Ti, Fe_2Mo, and Ni_3Mo, may precipitate.

12–9 SURFACE TREATMENTS

Low-carbon steels have low strength and hardness but good ductility and toughness, whereas high-carbon steels have the opposite behavior. We can, by proper heat treatment, produce a structure that is hard and strong at the surface, so that excellent wear and fatigue resistance are obtained, but at the same time gives a soft, ductile, tough core which provides good resistance to impact failure.

Selectively Heating the Surface. We could begin by rapidly heating the surface of a medium-carbon steel above the A_3 temperature—the center remains below the A_1. After the steel is quenched, the center is still a soft mixture of ferrite and pearlite, while the surface is martensite (Figure 12-29). The depth of the martensite layer is the *case depth*. Tempering produces the desired hardness at the surface. We can provide local heating of the surface by using a gas flame, an induction coil, a laser beam, or an electron beam. We can, if we wish, harden only selected areas of the surface that are most subject to failure by fatigue or wear.

Carburizing and Nitriding. For even better toughness, we start with a low-carbon steel, which usually contains alloying elements for improved hardenability. Carbon is diffused into the surface of the steel at a temperature above

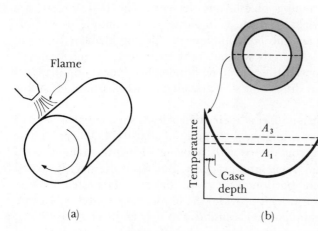

FIGURE 12-29 (a) Surface hardening by localized heating. (b) Only the surface heats above the A_1 temperature and is quenched to martensite. (c) The photograph shows the case depth in a surface-hardened 1.5-in. diameter 1050 steel automobile axle.

the A_3 (Figure 12-30). A high carbon content is produced at the surface due to rapid diffusion and the high solubility of carbon in austenite. When the steel is then quenched and tempered, the surface becomes a high-carbon tempered martensite, while the center remains soft and ductile. The thickness of the hardened surface, again called the *case depth*, depends on the amount of carbon present at the surface of the steel, the temperature, and the time (Figure 12-31). The case depth is much smaller in carburized steels than in flame- or induction-hardened steels.

    Pack carburizing requires that the steel be surrounded by charcoal, which burns to produce an atmosphere of carbon monoxide from which carbon atoms

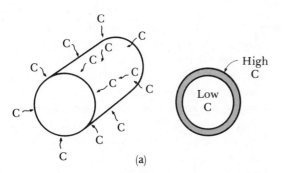
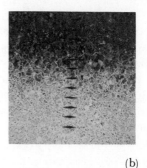

Case depth ≈ 0.03 in.

FIGURE 12-30 (a) Carburizing of a low-carbon steel to produce a high-carbon, wear-resistant surface. (b) The carburized case in a 1010 steel. The microhardness indentations show that the high-carbon case is harder than the low-carbon interior ($\times 25$).

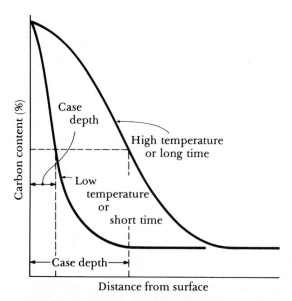

FIGURE 12-31   The case depth depends on the temperature and time of carburizing.

diffuse into the steel. In *gas carburizing*, the steel is placed in a sealed furnace containing carbon monoxide. *Liquid carburizing* requires that the steel be placed in a bath of molten cyanide, which contains carbon.

Nitrogen provides a hardening effect similar to that of carbon. In *cyaniding*, the steel is immersed in a liquid cyanide bath which permits both carbon and nitrogen to diffuse into the steel. In *carbonitriding*, a gas containing carbon monoxide and ammonia is generated and both carbon and nitrogen diffuse into the steel. Finally, only nitrogen diffuses into the surface from a gas in *nitriding*. Nitriding is carried out below the A_1 temperature.

In each of these processes, compressive residual stresses are introduced at the surface, providing excellent fatigue resistance in addition to the good combination of hardness, strength, and toughness.

12–10   WELDABILITY OF STEEL

During welding, the metal nearest the weld heats above the A_1 temperature and austenite forms (Figure 12-32). During cooling, the austenite in this heat-affected zone transforms to a new structure, dependent on the cooling rate and the CCT diagram for the steel. Plain low-carbon steels have such a low hardenability that normal cooling rates seldom produce martensite. However, an alloy steel may have to be preheated to slow down the cooling rate or postheated to temper any martensite that forms.

A steel that is originally quenched and tempered has two problems during welding. First, the portion of the heat-affected zone that heats above the A_1 may

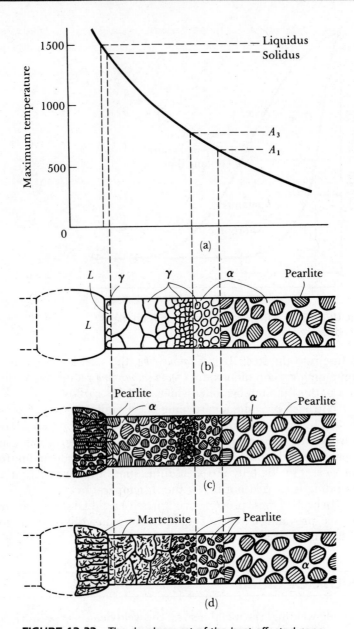

FIGURE 12-32 The development of the heat-affected zone in a weld. (a) The maximum temperature at any point, (b) the structure at the maximum temperature, (c) the structure after cooling in a steel of low hardenability, and (d) the structure after cooling in a steel of high hardenability.

form martensite after cooling. Second, a portion of the heat-affected zone below the A_1 may overtemper. Normally, we should not weld a steel in the quenched and tempered condition.

◇ | **Example 12–14**

Compare the structures in the heat-affected zones of welds in 1080 and 4340 steels if the cooling rate in the heat-affected zone is 5°C/s.

Answer:

From the CCT diagrams, Figures 12-16 and 12-18, the cooling rate in the weld produces the following structures.

1080: 100% pearlite

4340: Bainite and martensite

The high hardenability of the alloy steel reduces the weldability, permitting martensite to form and embrittle the weld.

12–11   STAINLESS STEELS

Stainless steels are selected for their excellent resistance to corrosion. All true stainless steels contain a minimum of about 12% Cr, which permits a thin, protective surface layer of chromium oxide to form when the steel is exposed to oxygen.

There are four categories of stainless steels based on crystal structure and strengthening mechanism. Examples and characteristics of each type are included in Table 12-5.

The Iron-Chromium-Carbon Phase Diagram.   Figure 12-33 shows the iron-chromium phase diagram. The chromium produces a *gamma loop*. As the amount of chromium increases, the temperature range in which austenite is stable decreases until the austenite completely disappears. Thus with high chromium contents, the ferritic, or α, structure is present at all temperatures. Chromium is a *ferrite stabilizing element*.

Figure 12-34 illustrates the effect of chromium on the iron-carbon phase diagram. Chromium causes the austenite region to shrink while the ferrite region increases in size. For high-chromium, low-carbon compositions, ferrite is present as a single phase up to the solidus temperature.

Ferritic Stainless Steels.   Ferritic stainless steels contain up to 30% Cr and less than 0.12% C. Because of the BCC structure, the ferritic stainless steels have good strengths and moderate ductilities derived from solid solution strengthen-

TABLE 12-5 Compositions and properties of selected stainless steels

Steel	% C	% Cr	% Ni	Others	Tensile Strength (psi)	Yield Strength (psi)	% Elongation	Condition
Austenitic								
201	0.15	16–18	3.5–5.5	5.5–7.5% Mn	95,000	45,000	40	Annealed
304	0.08	18–20	8.0–10.5		75,000	30,000	30	Annealed
					185,000	140,000	9	Cold worked
304L	0.03	18–20	8–12		75,000	30,000	30	Annealed
316	0.08	16–18	10–14	2–3% Mo	75,000	30,000	30	Annealed
321	0.08	17–19	9–12	Ti (5 × % C)	85,000	35,000	55	Annealed
347	0.08	17–19	9–13	Nb (10 × % C)	90,000	35,000	50	Annealed
Ferritic								
430	0.12	16–18			65,000	30,000	22	Annealed
442	0.12	18–23			75,000	40,000	20	Annealed
Martensitic								
416	0.15	12–14		0.60% Mo	180,000	140,000	18	Quenched
431	0.20	15–17	1.25–2.50		200,000	150,000	16	and
440C	0.95–1.2	16–18		0.75% Mo	285,000	275,000	2	tempered
Precipitation hardening								
17-4	0.07	16–18	3–5	0.15–0.45% Nb	190,000	170,000	10	Age
17-7	0.09	16–18	6.5–7.8	0.75–1.25% Al	240,000	230,000	6	hardened

Adapted from *ASM Metals Handbook*, Vol. 3, 9th Ed., 1980.

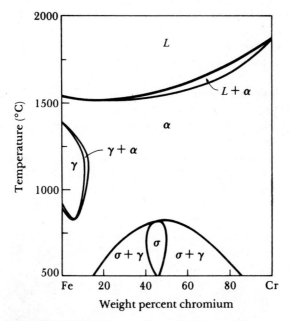

FIGURE 12-33   The iron-chromium phase diagram.

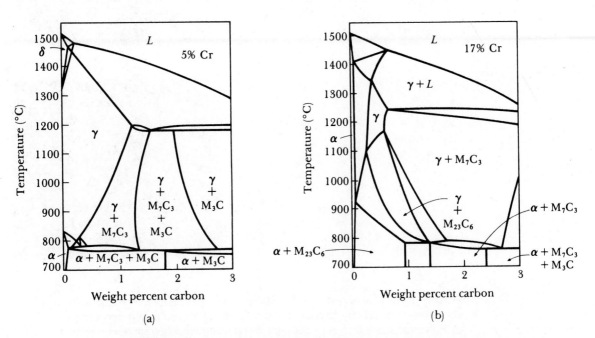

FIGURE 12-34   The effect of chromium on the iron-carbon phase diagram, shown as isopleths at (a) 5% Cr and (b) 17% Cr.

ing and strain hardening. When the carbon or chromium contents are high, precipitation of carbide particles provides dispersion strengthening but also embrittles the alloy. Ferritic stainless steels have excellent corrosion resistance and moderate formability and are relatively inexpensive.

Martensitic Stainless Steels.   From Figure 12-34, we find that a 17% Cr-0.5% alloy heated to 1200°C produces 100% austenite, which transforms to martensite on quenching. The martensite is then tempered to produce high strengths and hardnesses.

The chromium content is usually less than 17% Cr; otherwise the austenite field becomes so small that very stringent control over both austenitizing temperature and carbon content is required. Lower chromium contents also permit

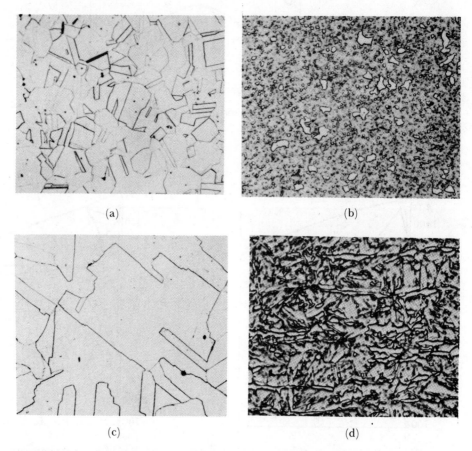

(a)

(b)

(c)

(d)

FIGURE 12-35 (a) Annealed ferritic stainless steel (× 85). (b) Martensitic stainless steel containing large primary carbides and small carbides formed during tempering (× 350). (c) Austenitic stainless steel (× 500). (d) Precipitation-hardened stainless steel with ferrite plates in a martensite matrix (× 1000). From *Metals Handbook,* Vols. 7 and 8, 8th ed., American Society for Metals, 1972, 1973.

the carbon content to vary from about 0.1% to 1.0%, allowing martensites of different hardnesses to be produced.

Since the chromium gives the steel high hardenability, an air or oil quench permits martensite to form. The martensitic stainless steels have tempering curves similar to those of high-alloy tool steels. Little softening occurs until a tempering temperature near 500°C is reached. A secondary hardening peak may be observed if alloy carbides form. Figure 12-35 includes the structure of a typical martensitic stainless steel.

The low chromium content and the presence of two phases cause tempered martensitic stainless steels to have less corrosion resistance than the other stainless steels. However, the combination of hardness, strength, and corrosion resistance makes the alloys attractive for many applications, including high-quality knives, ball bearings, and valves.

Austenitic Stainless Steels.   Nickel, which is an *austenite stabilizing element,* increases the size of the austenite field while nearly eliminating ferrite from the iron-chromium-carbon alloys (Figure 12-36). If the carbon content is below about 0.03%, the carbides do not form and the steel is virtually all austenite at room temperature (Figure 12-35).

The FCC austenitic stainless steels have excellent ductility, formability, and corrosion resistance. Strength is obtained by extensive solid solution strengthen-

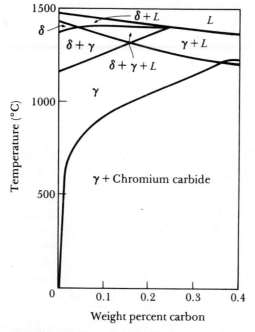

FIGURE 12-36 *A section of the iron-chromium-nickel-carbon phase diagram at a constant 18% Cr-8% Ni.*

ing, and the austenitic stainless steels may be cold worked to higher strengths than the ferritic stainless steels. The steels have excellent low-temperature impact properties, since they have no transition temperature. Furthermore, the austenitic stainless steels are not ferromagnetic. Unfortunately, the high nickel and chromium contents make the alloys expensive.

◇ | **Example 12–15**

Describe a simple test to separate high-nickel stainless steel from low-nickel stainless steel.

Answer:

The high-nickel stainless steels are austenitic, whereas the low-nickel stainless steels are probably ferritic or martensitic. An ordinary magnet will be attracted to the low-nickel ferritic and martensitic steels but will not be attracted to the high-nickel austenitic steel.

◇ | **Example 12-16**

Which of the stainless steels would you select for a pump used to transport liquid helium at 4 K?

Answer:

Because of the extremely low temperature, a material with good low-temperature properties is necessary. The austenitic stainless steels might serve best, since they do not have a ductile-brittle transition temperature and thus have better low-temperature toughness.

Precipitation Hardening (PH) Stainless Steels. The composition of the precipitation hardening, or PH, stainless steels is similar to that of the austenitic stainless steels except for the presence of aluminum, niobium, or tantalum. The PH stainless steels derive their properties from solid solution strengthening, strain hardening, age hardening, and the martensitic reaction. High mechanical properties are obtained even with low carbon contents.

A heat treatment for a typical 17-7 PH stainless steel is shown in Figure. 12-37. After fabrication of the steel in the annealed condition, three steps are required.

1. Conditioning. This step, done at 760°C to 955°C, prepares the austenite for subsequent transformation to martensite.

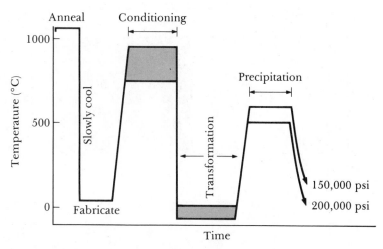

FIGURE 12-37 *Schematic diagram showing the steps in the heat treatment of a PH stainless steel.*

2. Quenching and transformation. The steel is cooled to 15°C or below to permit austenite to transform to martensite.

3. Precipitation. The steel is reheated to 500°C to 600°C, permitting Ni_3Al and other precipitates to form from the martensite. Higher strengths are obtained with lower aging temperatures.

Cast Irons

12–12 SOLIDIFICATION OF CAST IRONS

Cast irons are iron-carbon-silicon alloys, typically containing 2% to 4% C and 0.5% to 3% Si, that pass through the eutectic reaction during solidification.

The microstructures of the five important types of cast irons are shown schematically in Figure 12-38. *Gray cast iron* contains small, interconnected graphite flakes that cause low strength and ductility. *White cast iron* is a hard, brittle, unmachinable alloy containing massive amounts of Fe_3C. *Malleable cast iron* is produced by the heat treatment of white iron, causing irregular but rounded clumps of graphite to precipitate. This graphite form permits good strength, ductility, and toughness in the iron. *Ductile* or *nodular cast iron* contains spheroidal graphite particles obtained during solidification by the addition of small amounts of magnesium to the molten iron. Properties are similar to those of malleable iron. *Compacted graphite cast iron* contains rounded but interconnected

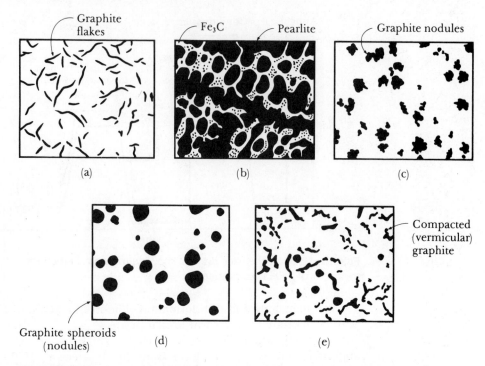

FIGURE 12-38 *Schematic drawings of the five types of cast iron: (a) gray iron, (b) white iron, (c) malleable iron, (d) ductile iron, and (e) compacted graphite iron.*

(vermicular) graphite also produced during solidification by the addition of magnesium. The structure and properties are intermediate between gray and ductile irons.

To understand the origin of these cast irons, we must examine the phase diagram, solidification, and phase transformations of the alloys.

The Iron-Carbon-Silicon Phase Diagram.   Figure 12-39 shows the Fe–C phase diagram. The dashed lines refer to the metastable relationship between iron and cementite, which we used when studying the structure and treatment of steels. The solid lines show the equilibrium reactions between iron and graphite. Although the iron-graphite lines represent the stable reactions, nucleation of graphite in iron-carbon alloys is so difficult that we almost always obtain more than the 6°C undercooling which allows the metastable reactions to occur.

However, when we add silicon to the iron, the temperature difference between the stable and metastable eutectic reactions increases (Figure 12-40). There is a larger temperature interval, during which the stable iron-graphite eutectic reaction can nucleate and grow during cooling; thus silicon is a *graphite stabilizing element.*

Silicon also reduces the amount of carbon in the eutectic. A portion of the silicon can be treated as carbon, permitting us to define a *carbon equivalent* (CE).

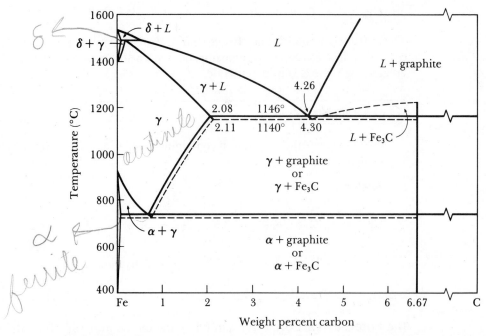

FIGURE 12-39 The iron-carbon phase diagram showing the relationship between the stable iron-graphite equilibria (solid lines) and the metastable iron-cementite reactions (dashed lines).

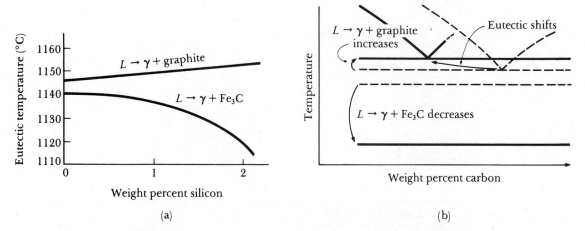

FIGURE 12-40 The effect of silicon on (a) the eutectic temperature and (b) the phase diagram. Increasing silicon increases the difference between the two eutectic temperatures and reduces the percent carbon in the eutectic.

$$\mathrm{CE} \;=\; \% \; \mathrm{C} + \tfrac{1}{3}\% \; \mathrm{Si} \qquad\qquad (12\text{-}2)$$

The eutectic composition is always near 4.3% CE. Any cast iron with a carbon equivalent less than 4.3% is hypoeutectic; any cast iron with a carbon equivalent greater than 4.3% is hypereutectic.

◇ | **Example 12-17**

Determine whether a cast iron containing 3.6% C and 2.4% Si is hypoeutectic or hypereutectic.

Answer:

$$\mathrm{CE} \;=\; \% \; \mathrm{C} + \frac{\% \; \mathrm{Si}}{3} \;=\; 3.6 + \frac{2.4}{3} \;=\; 4.4$$

Since the CE is greater than 4.3, the iron is hypereutectic.

The Eutectic Reaction.   We can relate the temperature at which the eutectic reaction occurs to the type of iron that forms: if solidification occurs between the two eutectic temperatures, a graphite-containing cast iron, such as gray, ductile, or compacted graphite iron, is produced. If solidification occurs below the lower eutectic temperature, a white cast iron forms. Some of the conditions that favor each of the two eutectic reactions are shown in Table 12-6.

TABLE 12-6   *Conditions favoring each of the eutectic reactions in cast irons*

Formation of Gray Iron, Ductile Iron, Compacted Graphite Iron $L \rightarrow \gamma + \mathrm{Gr}$	Formation of White Iron $L \rightarrow \gamma + \mathrm{Fe_3C}$
High CE	Low CE
High silicon	Low silicon
Cu, Ni	Cr, Mo, V, Bi, Te, others
Slow cooling	Fast cooling
Thick castings	Thin castings
Inoculation	No inoculation

 | **Example 12-18**

Estimate the amount of graphite and the amount of $\mathrm{Fe_3C}$ in the stable and metastable eutectic microconstituents.

Answer:

If we assume a pure iron-carbon alloy, the lever laws are

$$\text{Graphite} \; = \; \frac{4.26 \, - \, 2.08}{100 \, - \, 2.08} \; \times \; 100 \; = \; 2.2\%$$

$$\text{Fe}_3\text{C} \; = \; \frac{4.30 \, - \, 2.11}{6.67 \, - \, 2.11} \; \times \; 100 \; = \; 48.0\%$$

When Fe_3C forms, it becomes the continuous phase in the eutectic microconstituent, making the white iron hard and brittle.

In some cases, both eutectic reactions may occur. In *mottled iron*, solidification begins above the lower eutectic temperature but is completed below that temperature. The mixture of white and gray cast iron [Figure 12-41(a)] is undesirable under any conditions. *Chilled iron* is produced when the surface of the casting cools rapidly enough to cause white cast iron to form while the center of the casting, which cools more slowly, forms gray cast iron. This iron [Figure 12-41(b)] is sometimes useful for inexpensive wear-resistant components.

The depth to which white cast iron forms can serve as a simple method for checking the carbon equivalent of a cast iron. By pouring specially designed castings which are subsequently broken, the *chill depth*, or depth of white iron below the surface, can be measured and related to the carbon equivalent (Figure 12-42).

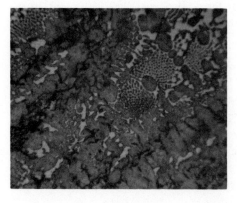

Chill depth ≈ 0.25 in.

(a)   (b)

FIGURE 12-41   (a) Mottled iron containing graphite flakes, Fe_3C (white), and pearlite (gray) ($\times 200$). (b) Fracture surface of chilled iron, showing white iron at the rapidly cooled end of the casting.

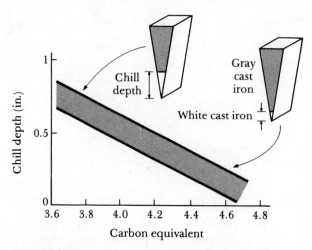

FIGURE 12-42 The relationship between chill depth and carbon equivalent. The wedge-shaped casting produces rapid freezing at the tip, giving white iron, and slower cooling at the thick end, giving gray iron.

◇ | **Example 12-19**

Inexpensive scissors with a hard cutting edge are sometimes made from gray iron. Describe how the appropriate hardness is obtained.

Answer:

The cross section of the blade of the scissors is a wedge. If the carbon equivalent is correctly adjusted, the tip of the wedge cools rapidly enough to produce white iron, while the remainder of the blade forms gray iron. The white iron, at the cutting surface, is hard and can be sharpened.

12–13 THE MATRIX STRUCTURE IN CAST IRONS

The matrix structure and properties of each type of cast iron are determined by how the austenite transforms during the eutectoid reaction. Under equilibrium conditions, and if some graphite is already present as a result of solidification, austenite transforms to graphite and ferrite.

$$\gamma \rightarrow \alpha + \text{graphite}$$

Carbon atoms diffuse from the austenite to existing graphite particles produced during solidification, leaving behind the low-carbon ferrite. The stable transformation to ferrite and graphite is encouraged if the cooling rate is slow

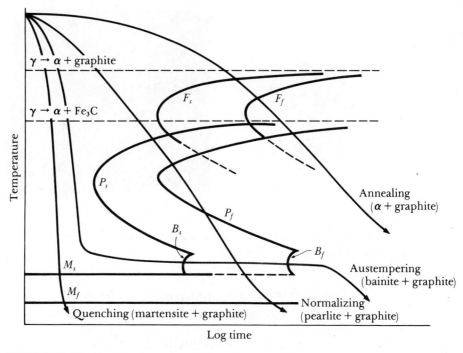

FIGURE 12-43 *The transformation diagram for austenite in a cast iron.*

(permitting time for diffusion) and if the graphite particles are fine and closely spaced (giving shorter diffusion distances).

The transformation diagram in Figure 12-43 describes how the austenite might transform during heat treatment. *Annealing*, which requires furnace cooling after austenitizing, produces a slow cooling rate and a ferritic matrix. *Normalizing*, or air cooling, gives faster cooling and a pearlitic matrix. The cast irons can also be quenched and tempered, isothermally transformed to produce bainite (austempered), or surface hardened. Austempered ductile iron, with strengths of over 200,000 psi, have recently become of great interest for a variety of applications, including gears.

The matrix structure can also be controlled by the composition of the iron. Ferrite is encouraged by higher silicon contents. However, most alloying elements, even 0.05% Sn or 0.50% Cu, favor the formation of pearlite.

12–14 CHARACTERISTICS AND PRODUCTION OF THE CAST IRONS

In order to produce each of the five important types of cast iron, we must carefully control the eutectic solidification, often by adding modifiers to encourage proper eutectic growth, as well as the eutectoid transformation or heat treatment of the iron.

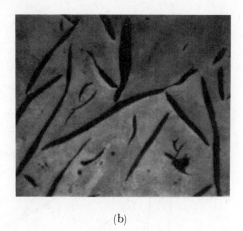

(a)   (b)

FIGURE 12-44   (a) Sketch and (b) photomicrograph of the flake graphite in gray cast iron (× 100).

Gray Cast Iron.   Gray iron is the most common of the cast irons. Solidification produces interconnected graphite flakes, which resemble a number of potato chips glued together at a single location (Figure 12-44). The point at which the flakes are connected is the original graphite nucleus. The gray cast iron contains many such clusters, or *eutectic cells*, of graphite flakes, with each cell representing one nucleation event. Inoculation, produced by adding small amounts of iron-silicon alloys, or rapid cooling rates help produce finer graphite flakes with a smaller eutectic cell size, thus improving strength. The graphite flakes, which resemble tiny cracks within the cast iron structure, concentrate stresses so that the gray iron possesses a low tensile strength and behaves in a brittle manner, with elongations of only 1% or less.

The gray irons are often specified by a class number of 20 to 80 (Table 12-7). A class 20 gray iron has a nominal tensile strength of 20,000 psi. However, in thick castings, coarse graphite flakes and a ferrite matrix produce tensile strengths as low as 12,000 psi (Figure 12-45). For very thin castings, fine graphite and pearlite give tensile strengths near 40,000 psi. If cooling is too fast, carbides may form.

At lower carbon equivalents, the nominal tensile strength of the cast iron is higher, since a smaller amount of graphite forms during solidification. Even higher strengths can be obtained by alloying or heat treatment.

In spite of its low tensile strength and ductility, gray iron has a number of attractive properties. The flakes do not act as stress raisers under compressive loading, so with proper design gray iron can carry large loads. The machinability of gray iron is excellent, since the graphite flakes act as chip breakers. Resistance to sliding wear is good; the porous graphite flakes absorb and hold lubricant and, because graphite is soft and slippery, may even provide self-lubrication. Very good thermal conductivity is obtained due to the interconnected graphite flakes. Vibration damping characteristics are exceptional, particularly when the

TABLE 12-7 Typical compositions and properties of gray irons

Class	Nominal Tensile Strength (psi)	Tensile Strength Range (psi)	% C	% Si
20	20,000	12,000–40,000	3.4–3.7	2.3–2.8
25	25,000		3.3–3.5	2.0–2.3
30	30,000	20,000–46,000	3.2–3.5	2.0–2.3
35	35,000		3.1–3.4	1.9–2.2
40	40,000	28,000–54,000	3.0–3.3	1.8–2.1
60	60,000	44,000–66,000	Max. 3.8% CE	

Adapted from *Iron Castings Handbook*, ed. C. Walton and T. Opar, Iron Castings Society, 1981.

flakes are coarse. Because of this characteristic, gray iron is used for engine blocks and machine tool bases. Gray iron does not make a very good bell, however.

White Cast Iron. Low carbon equivalent white irons containing about 2.5% C and 1.5% Si are an intermediate product in the manufacture of malleable iron, as will be discussed in the next section. A group of highly alloyed white irons are used for their hardness and resistance to abrasive wear. Elements such as chromium, nickel, and molybdenum are added so that, in addition to the alloy carbides formed during solidification, martensite can be produced during subsequent heat treatment. White cast irons are shown in Figure 12-46.

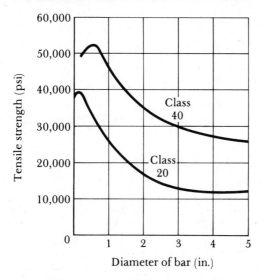

FIGURE 12-45 The effect of cooling rate or casting size on the tensile properties of two gray cast irons.

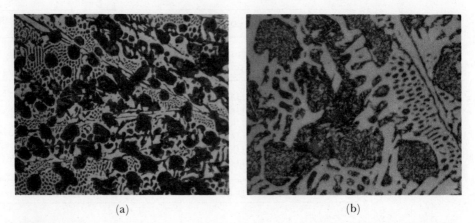

(a)   (b)

FIGURE 12-46 (a) Normal white iron containing Fe_3C (white) and pearlite ($\times 100$). (b) Martensitic white iron with Fe_3C (white) and a matrix of martensite needles and retained austenite ($\times 250$).

Malleable Cast Iron. Malleable iron is produced by heat treating unalloyed white iron. During the malleabilizing heat treatment, the cementite formed during solidification is decomposed and graphite clumps, or nodules, are produced. The nodules, or temper carbon, often resemble popcorn. The rounded graphite shape permits the malleable iron to have a good combination of strength and ductility.

The production of malleable iron requires several steps (Figure 12-47).

1. A white cast iron must be produced that decomposes readily during heat treatment. A carbon equivalent of about 3% keeps graphite flakes from forming during solidification yet permits Fe_3C to decompose in short times.

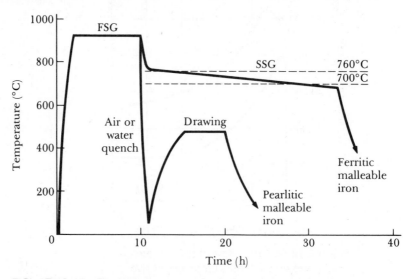

FIGURE 12-47 The heat treatments for ferritic and pearlitic malleable irons.

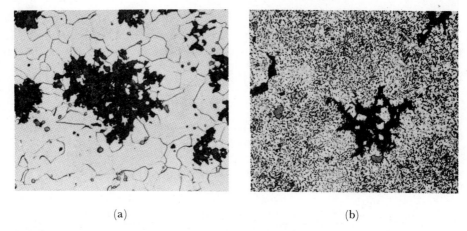

(a)   (b)

FIGURE 12-48   (a) Ferritic malleable iron with graphite nodules and small MnS inclusions in a ferrite matrix (× 200). (b) Pearlitic malleable iron drawn to produce a tempered martensite matrix (× 500). From *Metals Handbook*, Vols. 7 and 8, 8th Ed., American Society for Metals, 1972, 1973.

2. Nucleation of the graphite nodules occurs as the white cast iron is slowly heated to the malleabilizing temperature.

3. *First stage graphitization* (FSG) at about 925°C decomposes the cementite to the stable austenite and graphite phases. The carbon in Fe_3C diffuses to the graphite nuclei produced during heating, leaving behind an austenite matrix.

4. The austenite decomposes during subsequent cooling from the FSG temperature. Two types of malleable cast irons can be produced—ferritic and pearlitic (Figure 12-48). To make *ferritic malleable iron*, the casting is cooled at 5°C/h to 15°C/h through the eutectoid temperature range, about 760°C to 700°C, to cause *second stage graphitization* (SSG). The austenite transforms to ferrite, with the excess carbon diffusing to the existing graphite nodules. Ferritic malleable iron has exceptional toughness compared with other irons because its low silicon content reduces the transition temperature below room temperature.

Pearlitic malleable iron is obtained when the austenite is cooled in air or oil. The iron is called pearlitic malleable whether the matrix contains pearlite from air cooling or martensite from the oil quench. In either case, the matrix is hard and brittle. To improve the ductility, the pearlitic malleable iron is drawn at a temperature below the eutectoid range. *Drawing* tempers the martensite or spheroidizes the pearlite, thus reducing the amount of *combined carbon*, or cementite. As the amount of combined carbon decreases, the strength of the pearlitic malleable iron decreases and ductility and toughness increase (Table 12-8).

◇ | **Example 12-20**

Much longer heat treatment times are required to make malleable iron when the percent chromium exceeds 0.1%. Explain.

Answer:

The chromium is an excellent carbide stabilizer (Table 12-6). Consequently, the decomposition of the cementite is slow because the cementite is relatively stable.

TABLE 12-8 Grades of malleable cast iron

Grade	Minimum Tensile Strength (psi)	Minimum Yield Strength (psi)	Minimum % Elongation	Drawing Temperature (°C)	Matrix Structure
32510	50,000	32,500	10		Ferrite
35018	53,000	35,000	18		
50005	70,000	50,000	5	750	
60004	80,000	60,000	4	730	Spheroidized
70003	85,000	70,000	3	710	pearlite
					or
80002	95,000	80,000	2	670	tempered
90001	105,000	90,000	1	< 600	martensite

Adapted from data in *Iron Castings Handbook*, ed. C. Walton and T. Opar, Iron Castings Society, 1981.

Ductile or Nodular Cast Iron. Ductile iron is produced by treating a relatively high carbon equivalent liquid iron with magnesium or cerium, causing spheroidal graphite to grow during solidification. Several steps are required to produce this iron (Figure 12-49).

1. *Desulfurization.* Sulfur causes graphite to grow in the flake rather than the spheroidal form. Low-sulfur irons are obtained by melting low-sulfur, high-quality charge materials; by melting in furnaces that remove sulfur from iron

FIGURE 12-49 Schematic diagram of the treatment of ductile iron.

during melting; or by mixing the iron with a desulfurizing agent such as calcium carbide.

2. *Nodulizing*. Magnesium, added in the nodulizing step, removes any sulfur and oxygen still in the liquid metal and provides a residual of 0.03% Mg which causes growth of spheroidal graphite. The magnesium is added near 1500°C. Unfortunately, magnesium vaporizes near 1150°C. Many nodulizing alloys contain magnesium diluted with ferrosilicon to reduce the violence of the reaction and permit higher magnesium recoveries. Special techniques also are required to keep the lightweight nodulizing alloy below the surface of the metal.

◇ | **Example 12-21**

A 1000-lb ladle of low-sulfur molten iron is treated with 16.7 lb of a ferrosilicon nodulizing alloy containing 9% Mg. The final analysis shows that 0.035% Mg is dissolved in the iron. What fraction of the magnesium added was vaporized or oxidized?

Answer:

The amount of magnesium added is

$$\text{lb Mg} = (0.09)(16.7) = 1.5$$

The percent magnesium added to the iron is

$$\frac{1.5}{1000} \times 100 = 0.15\%$$

The percent loss of magnesium is

$$\frac{0.15 - 0.035}{0.15} \times 100 = 77\%$$

This loss is typical during nodulizing. The recovery of the magnesium is only $100 - 77 = 23\%$.

◇

Fading, or gradual, nonviolent vaporization or oxidation of the magnesium, must also be controlled. If the iron is not poured within a few minutes after nodulizing, the iron reverts to gray iron.

3. *Inoculation*. Magnesium is an effective carbide stabilizer; consequently nodulizing causes white iron to form during solidification. Although any graphite that nucleates is conditioned by the magnesium to grow as spheres, practically no graphite forms! Consequently, we must inoculate the iron after nodulizing. Inoculants are usually ferrosilicon alloys containing 50% to 85% Si and small amounts of calcium, aluminum, strontium, or barium which produce nucleation sites for graphite. The inoculant effect also fades with time.

Compared with gray iron, ductile cast iron has excellent strength, ductility, and toughness. Ductility and strength are also higher than in malleable irons,

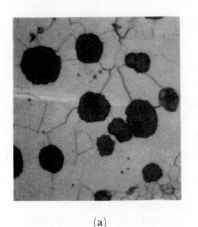
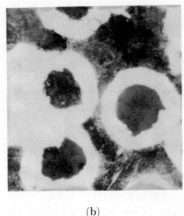
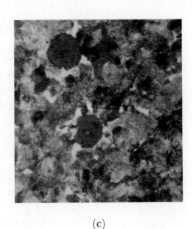

(a)   (b)   (c)

FIGURE 12-50   (a) Annealed ductile iron with a ferrite matrix (× 250). (b) As-cast ductile iron with a matrix of ferrite (white) and pearlite (× 250). (c) Normalized ductile iron with a pearlite matrix (× 250).

but due to the higher silicon content in ductile iron, the toughness may be slightly lower. Typical structures of ductile iron are shown in Figure 12-50, and properties are described in Table 12-9.

Compacted Graphite Cast Iron.   Compacted graphite cast iron is obtained by undernodulizing the molten metal, producing a residual of about 0.015% Mg. The graphite shape is intermediate between flakes and spheres, with numerous rounded rods of graphite that are interconnected to the nucleus of the eutectic cell (Figure 12-51). This graphite, sometimes called *vermicular* graphite, also forms when ductile iron fades.

The compacted graphite permits the cast iron to have strengths and ductilities that exceed that of gray cast iron but allows the iron to retain good thermal conductivity and vibration damping characteristics (Table 12-10).

TABLE 12-9   Grades of ductile cast iron

Grade	Tensile Strength (psi)	Yield Strength (psi)	% Elongation	Matrix Structure	Heat Treatment
60-40-18	60,000	40,000	18	Ferrite	Annealed
65-45-12	65,000	45,000	12	Mostly ferrite	As-cast
80-55-06	80,000	60,000	6	Mostly pearlite	As-cast
100-70-03	100,000	70,000	3	Pearlite	Normalized
120-90-02	120,000	90,000	2	Tempered martensite	Quenched and tempered

From *Iron Castings Handbook*, ed. C. Walton and T. Opar, Iron Castings Society, 1981.

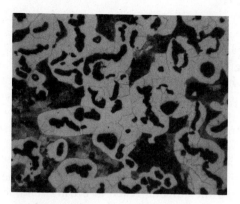

FIGURE 12-51 The structure of compacted graphite cast iron, with a matrix of ferrite (white) and pearlite (gray) (×250).

Furthermore, the ability to produce castings with smaller risers, which is typical of gray iron, is retained.

The treatment for the compacted graphite iron is similar to that for ductile iron. Normally, low-sulfur base metals are required, the magnesium is introduced in dilute form in special alloys containing titanium, the metal must be poured shortly after treatment to avoid fading, and the iron must be inoculated to assure that graphite rather than cementite forms.

TABLE 12-10 Comparison of compacted graphite cast iron with gray and ductile cast iron

Cast Iron	Tensile Strength (psi)	Yield Strength (psi)	% Elongation	Relative Damping Capacity
Gray iron	20,000–60,000		< 1.5	1
Compacted graphite iron				0.6
Grade A (< 10% pearlite)	40,000	28,000	5	
Grade B (20–80% pearlite)	50,000	40,000	1	
Grade C (> 80% pearlite	65,000	55,000	1	
Ductile iron	60,000–120,000	40,000–90,000	2–18	0.3

SUMMARY

At first glance, steels appear to be an exceptionally varied and complicated group of alloys. However, the properties of steels are determined primarily by the amount, size, shape, and distribution of the cementite, which in turn are controlled by heat treatment. Spheroidizing, which produces large, spheroidal cementite, gives the softest, most ductile steel. Annealing gives a coarse pearlitic structure containing lamellar cementite. Normalizing produces a finer pearlite and improved strength. Austempering (giving bainite) or quench and temper heat treatments (giving tempered martensite) produce exceptionally fine dispersions of cementite. The most important function of alloying elements is to improve our ability to perform these heat treatments or to help make the structures that are produced more stable at high temperatures.

Stainless steels, which must contain a minimum of 12% Cr, are selected for their excellent corrosion resistance. By proper control of alloying elements and heat treatment, a variety of structures and properties can be produced.

Cast irons undergo the eutectic reaction during solidification. Depending on the reaction, either cementite and austenite or graphite and austenite form. The shape of the graphite is controlled by the addition of modifiers, such as magnesium, and inoculants, producing spheroidal, compacted, or flake graphite cast irons. Cementite is decomposed by heat treatment to produce malleable iron. Furthermore, the decomposition of austenite to ferrite, pearlite, bainite, or martensite can be achieved by controlling alloying elements, cooling rate, or heat treatment.

GLOSSARY

Annealing (cast iron). A heat treatment used to produce a ferrite matrix in a cast iron by austenitizing, then furnace cooling.

Annealing (steel). A heat treatment used to produce a soft, coarse pearlite in a steel by austenitizing, then furnace cooling.

Ausforming. A thermomechanical heat treatment in which austenite is plastically deformed below the A_1 temperature, then permitted to transform to bainite or martensite.

Austempering. The isothermal heat treatment by which austenite transforms to bainite.

Austenitizing. Heating a steel or cast iron to a temperature where homogeneous austenite can form. Austenitizing is the first step in most of the heat treatments for steel and cast irons.

Carbon equivalent. Carbon plus one-third of the silicon in a cast iron.

Carburizing. A group of surface-hardening techniques by which carbon diffuses into a steel. Some related processes by which nitrogen is introduced include carbonitriding, cyaniding, and nitriding.

Case depth. The depth below the surface of a steel to which hardening occurs by surface hardening and carburizing processes.

Cast iron. Ferrous alloys containing sufficient carbon so that the eutectic reaction occurs during solidification.

Chill depth. The distance from the surface in which the cast iron freezes according to the cementite reaction rather than the graphite reaction.

Chilled iron. Cast iron designed so that the surface is hard, wear-resistant white iron while the center is softer, tougher gray or ductile iron.

Compacted graphite cast iron. A cast iron treated with small amounts of magnesium and titanium to cause graphite to grow during solidification as an interconnected, coral-shaped precipitate, giving properties midway between gray and ductile iron.

Drawing. Reheating a malleable iron in order to reduce the amount of carbon combined as cementite by spheroidizing pearlite, tempering martensite, or graphitizing both.

Dual phase steels. Special steels treated to produce martensite dispersed in a ferrite matrix.

Ductile cast iron. Cast iron treated with magnesium or cerium to cause graphite to precipitate during solidification as spheres, permitting excellent strength and ductility. Also known as nodular cast iron.

Eutectic cell. A cluster of graphite flakes produced during solidification which are all interconnected to a common nucleus.

Fading. The loss of the nodulizing or inoculating effect in cast irons as a function of time, permitting undesirable changes in microstructure and properties.

Ferrous alloys. Metal alloys based primarily on iron, including steel, stainless steel, and cast iron.

First stage graphitization. The first step in the heat treatment of a malleable iron, during which the massive carbides formed during solidification are decomposed to graphite and austenite.

Gamma loop. A number of alloying elements in iron, including chromium, reduce the temperature range over which austenite is stable, thus producing a loop in the phase diagram.

Gray cast iron. Cast iron which, during solidification, permits graphite flakes to grow, causing low strength and poor ductility.

Hardenability. The ease with which a steel can be quenched to form martensite. Steels with high hardenability form martensite even on slow cooling.

Hardenability curves. Graphs showing the effect of cooling rate on the hardness of a steel.

Inoculation. The addition of an agent to the molten cast iron that provides nucleation sites at which graphite precipitates during solidification.

Isothermal annealing. Heat treatment of a steel by austenitizing, cooling rapidly to a temperature between the A_1 and the nose of the TTT curve, and holding until the austenite transforms to pearlite.

Jominy distance. The distance from the quenched end of a Jominy bar. The Jominy distance is related to the cooling rate.

Jominy test. The test used to evaluate hardenability. An austenitized steel bar is quenched at one end only, thus producing a range of cooling rates along the bar.

Malleable cast iron. Cast iron obtained by a lengthy heat treatment during which cementite decomposes to produce rounded clumps of graphite. Good strength and ductility are obtained as a result of this structure.

Marquenching. Quenching austenite to a temperature just above the M_s and holding until the temperature is equalized throughout the steel before further cooling to produce martensite. This process reduces residual stresses and quench cracking. Also known as *martempering*.

Mottled iron. Cast iron produced when both cementite and graphite precipitate during solidification. Mottled iron is not desirable under any conditions.

Nodulizing. The addition of magnesium to molten cast iron to cause the graphite to precipitate as spheres rather than as flakes during solidification.

Normalizing. A simple heat treatment obtained by austenitizing and air cooling to produce a fine pearlitic structure.

Process anneal. A low-temperature heat treatment used to eliminate all or part of the effect of cold working in steels.

Retained austenite. Austenite that is unable to transform into martensite during quenching because of the volume expansion associated with the reaction.

Second stage graphitization. The second step in the heat treatment of malleable irons that are to have a ferritic matrix. The iron is cooled slowly from the first stage graphitization temperature so that austenite transforms to ferrite and graphite rather than pearlite.

Spheroidite. A microconstituent containing coarse spheroidal cementite particles in a matrix of ferrite, permitting excellent machining characteristics in high-carbon steels.

Stainless steels. A group of ferrous alloys that contain at least 12% Cr, providing extraordinary corrosion resistance.

Tempered martensite. The mixture of ferrite and cementite formed when martensite is tempered.

Tool steels. A group of high-carbon steels which provide combinations of high hardness, toughness, or resistance to elevated temperatures.

Vermicular graphite. The rounded, interconnected graphite that forms during the solidification of cast iron. This is the intended shape in compacted graphite iron but is a defective shape in ductile iron.

White cast iron. Cast iron which produces massive amounts of cementite rather than graphite during solidification. The white irons are very hard but brittle.

PRACTICE PROBLEMS

1. Calculate the amounts of ferrite, cementite, primary ferrite, and pearlite in a 1030 steel.

2. Calculate the amounts of ferrite, cementite, primary cementite, and pearlite in a 10140 steel.

3. A plain-carbon steel contains 14% pearlite and 86% primary ferrite. What is the probable AISI-SAE number for the steel?

4. A plain-carbon steel contains 92% ferrite and 8% cementite. What is the probable AISI-SAE number for the steel?

5. A spheroidized steel contains 15% cementite in a ferrite matrix. Estimate the AISI-SAE number for the steel.

6. Complete the table below.

	1050 Steel	10120 Steel
A_1 temperature		
A_3 or A_{cm} temperature		
Full annealing temperature		
Normalizing temperature		
Process annealing temperature		
Spheroidizing temperature		

7. A 1030 steel is to be hot rolled, with subsequent air cooling giving a normalized structure. What should be the last hot rolling temperature?

8. During spheroidizing of a 10100 steel, a uniform distribution of spherical cementite particles, each 0.002 cm in diameter, is produced. Estimate (a) the number of these particles per cubic centimeter of the steel and (b) the number of these particles per pound of steel. The density of ferrite is $7.87 \, \text{g/cm}^3$ and that of Fe_3C is $7.66 \, \text{g/cm}^3$.

9. Suppose a 1080 steel is available in either the pearlitic or spheroidized condition. In the pearlite, the cementite platelets are 0.00002 cm thick, while the ferrite platelets are 0.00007 cm thick. In the spheroidite, the cementite spheres are 0.003 cm in diameter. Estimate the total interface area between the ferrite and cementite in a cubic centimeter of each steel. Densities of each phase are given in Problem 12-8.

10. Describe the final microstructure obtained in a 1050 steel after each of the following heat treatments.

(a) Heat at 820°C, quench to 700°C and hold for 5 s, quench to room temperature.

(b) Heat at 820°C, quench to 400°C and hold for 1000 s, quench to room temperature.

(c) Heat at 820°C, quench to 600°C and hold for 100 s, quench to room temperature.

(d) Heat at 820°C, quench to 350°C and hold for 50 s, quench to room temperature.

(e) Heat at 820°C, quench to 600°C and hold for 100 s, quench to 400°C and hold for 100 s, cool slowly to room temperature.

11. Describe the final microstructure obtained in a 10110 steel after each of the following heat treatments.

(a) Heat to 900°C, quench to 700°C and hold for 5 s, quench to 300°C and hold for 100 s, quench to room temperature.

(b) Heat to 900°C, quench to 600°C and hold for 1 s, quench to room temperature.

(c) Heat to 900°C, quench to 400°C and hold for 10,000 s, cool to room temperature.

(d) Heat to 900°C, quench to 700°C and hold for 5 s, quench to 400°C and hold for 1000 s, cool to room temperature.

(e) Heat to 900°C, quench to 250°C and hold for 100 s, quench to room temperature, heat to 400°C and hold for 3600 s, cool to room temperature.

(f) Heat to 900°C, quench to 600°C and hold for 100 s, quench to room temperature.

12. Recommend an appropriate isothermal annealing heat treatment for a 1050 steel which will give a Rockwell C hardness of 25.

13. Recommend an appropriate austempering heat treatment for a 10110 steel which will give a Rockwell C hardness of 49.

14. After austempering, a 1050 steel has a hardness of R_c40. Estimate the temperature at which the austenite was transformed.

15. After an isothermal transformation heat treatment, a 10110 steel has a hardness of R_c57. Estimate the temperature at which the austenite was transformed.

16. A 1080 steel (Figure 11-20) is aus-

tempered to produce a hardness of R_c52. Estimate (a) the transformation temperature and (b) the minimum transformation time required.

17. Estimate the minimum times required to isothermally anneal (a) 1050, (b) 1080, and (c) 10110 steels at 600°C.

18. A 1050 steel is to be quenched and tempered to produce a minimum Brinell hardness of 300 and a minimum % elongation of 15%. Is this possible? If so, what tempering temperature would you recommend?

19. A 1050 steel is to be quenched and tempered to produce a minimum yield strength of 150,000 psi and a minimum % reduction in area of 55%. Is this possible? If so, what tempering temperature would you recommend?

20. In one case, a 1030 steel is austenitized at 850°C and quenched. In a second case, a 1030 steel is austenitized at 750°C and quenched. Calculate (a) the composition and (b) the amount of martensite formed in each example.

21. A 1020 steel is improperly austenitized and quenched, giving a structure composed of 30% martensite and 70% ferrite. Estimate (a) the carbon content of martensite and (b) the austenitizing temperature. What austenitizing temperature should have been used?

22. A 1050 steel develops surface cracks during quenching. Describe a heat treatment, including appropriate temperatures and times, that might help minimize this problem. What is this treatment called?

23. What difficulties would you expect if you attempted a marquenching heat treatment using a 1050 steel?

24. Suppose a new polymer quenching medium has an H coefficient of 0.5. Would this quenchant be more likely to produce quench cracks than a typical water quench?

25. We would like to perform a marquenching heat treatment for a 1050 steel. What is the minimum quenching temperature that we could use?

26. When a 1080 steel is quenched in a particular medium, a mixture of pearlite and martensite forms. What microstructure would be produced in an identical treatment for a 1020 steel?

27. When a 1020 steel is quenched in a particular medium, a mixture of ferrite, pearlite, and bainite is produced. What microstructure would form in an identical treatment for a 1080 steel?

28. A steel containing 8% Mn and 0.5% C is heated to 700°C. (a) Is this a hypoeutectoid or a hypereutectoid steel? (b) What phases will be present in the steel at 700°C? (c) Would this treatment be an appropriate spheroidizing treatment?

29. A steel containing 4% Mo and 0.4% C is heated to 850°C, which is well above the A_3 temperature for a plain carbon steel. (a) Is this a hypoeutectoid or a hypereutectoid steel? (b) What phases will be present in the steel at 850°C? (c) Would this be an appropriate austenitizing treatment if the steel is to be quenched to produce all martensite?

30. Using average values for the alloying elements, estimate the M_s temperature for the following steels: (a) 1140, (b) 4620, (c) 52100, (d) 8620, and (e) 6150.

31. For a 4340 steel, estimate (a) the maximum allowable time for a martempering heat treatment and (b) the maximum allowable time for an ausforming process. At what temperature would each of these treatments be conducted?

32. For a 4340 steel, describe the microstructure obtained for the following heat treatments.

(a) Heat at 800°C, quench to 300°C and hold for 100 s, quench to room temperature.

(b) Heat at 800°C, quench to 650°C and hold for 10^5 s, cool to room temperature.

(c) Heat at 800°C, quench to 700°C and hold for 10^4 s, quench to room temperature.

(d) Heat at 800°C, quench to 700°C and hold for 10^4 s, quench to 400°C and hold for 10^3 s, quench to room temperature.

(e) Heat at 800°C, quench to 600°C and hold for 100 s, quench to room temperature.

(f) Heat at 800°C, quench to 300°C and hold for 10^4 s, quench to room temperature.

33. A quenched and tempered steel is to be exposed to 500°C for brief periods of time. It is expected to maintain a hardness of $R_c 40$ after a number of these exposures. Which of the following steels would be acceptable: 1030, 1040, 1095, 4340?

34. A 4320 steel part quenched to give a hardness of $R_c 30$ is found to wear at an excessive rate. Estimate the hardness of the part if the steel were 1040, 9310, or 8640. Which of these alternatives might be better choices than the 4320 steel?

35. An 8640 steel part is to have a hardness between $R_c 45$ and $R_c 50$. What range of cooling rates must be obtained in order to satisfy this requirement?

36. Consider that the H coefficient of oil is 1.0, that of unagitated brine is 2.0, and that of agitated brine is 5.0. Estimate the cooling rate in the center of a 1-in. diameter bar produced by each of these quenching media.

37. A 2-in. diameter bar of 4320 steel is to have a hardness of $R_c 30$. What is the required severity of the quench (H coefficient)? What type of quenching medium would you recommend to produce the desired hardness with the least chance of cracking?

38. A 1-in. diameter bar of 8640 steel is to have a minimum hardness of $R_c 45$. Will (a) agitated brine, (b) still water, (c) still oil be a satisfactory quenching medium?

39. What is the maximum diameter of a 4320 steel bar quenched in unagitated brine that will develop a hardness of $R_c 35$?

40. An 8640 steel is cooled at 8°C/s. Estimate the hardness of the quenched steel.

41. A 4320 steel plate 1.5 in. thick is quenched in water. Plot the hardness profile across the cross section of the plate after quenching.

42. A 4320 steel plate 2 in. thick is water quenched. By plotting the hardness profile across the cross section of the plate, determine the depth of the plate which is hardened to at least $R_c 40$.

43. An 8640 steel plate is to be produced by water quenching, with the center of the plate having a minimum hardness of $R_c 50$. What is the maximum plate thickness that can undergo this treatment?

44. A 2-in. plate of 1040 steel is to have a minimum surface hardness of $R_c 50$ and a maximum center hardness of $R_c 25$. Is this possible using a water quench?

45. The center of a 2-in. diameter bar of 9310 steel has a hardness of $R_c 40$. Determine the hardness at the center of a 1-in. diameter bar of 4320 quenched in the same medium.

46. A 1010 steel is to be gas carburized using an atmosphere that produces 1.2% C at the surface of the steel. The case depth is defined as the distance below the surface that contains at least 0.5% C. If carburizing is done at $950°C$, determine the time required to produce a case depth of 0.01 in. (See Chapter 5 for review.)

47. A dual phase steel containing 0.15% C is to be produced by originally austenitizing at $750°C$. Determine the amounts of ferrite and martensite in the final structure. What are the carbon content and approximate R_c hardness of the martensite?

48. Occasionally, when an austenitic stainless steel is welded, the weld deposit may be slightly magnetic. Based on the Fe–Cr–Ni–C phase diagram, what phase would you expect is causing the magnetic behavior? Why might this phase have formed? What could you do to restore the nonmagnetic behavior?

49. Determine the carbon equivalent for a gray iron containing 3.2% C and 2.7% Si. Is the iron hypoeutectic or hypereutectic? What is the primary phase that will form during solidification?

50. A gray iron intended for particularly good vibration damping should have a carbon equivalent of at least 4.4%. If the iron contains 3.8% C, (a) what is the minimum amount of silicon that must be added and (b) what is the primary phase that will form during solidification?

51. What is the difference in microstructure produced when a 1040 steel is annealed versus when a gray cast iron is annealed?

52. Explain why the thermal conductivity of ductile or malleable cast iron is less than that of gray cast iron.

53. A gray iron contains 3.5% C and 2.4% Si. (a) Determine the carbon equivalent. (b) What is the class number for the iron? (c) What is the nominal tensile strength of the iron? (d) Estimate the tensile strength of this iron in a 3-in. diameter bar.

54. A class 40 gray iron casting is found to have a tensile strength of 30,000 psi. What is the expected diameter of the bar? Would you expect this casting to have more or less pearlite than normally would be expected?

55. Suppose you could double the number of graphite nodules produced when a white iron is heated to the malleabilizing temperature. What effect would this have on the time required for first stage graphitization and on the cooling rate required for second stage graphitization?

56. When the thickness of a ductile iron casting increases, the number of graphite nodules normally decreases. (a) What effect will this have on the amount of ferrite that is present in the matrix? (b) Suppose you observed the opposite effect of thickness on the amount of ferrite. How would you explain this difference?

57. Suppose a ductile iron containing 0.1% S is nodulized by adding 0.05% Mg. What fraction of the added magnesium will be lost by combining with sulfur to produce MgS? Will there be enough Mg left to permit graphite nodules to form?

58. We would like to produce a 65-45-12 grade of ductile cast iron without heat treating. (a) What is the major phase that should be present in the matrix? (b) Would increasing the number of graphite nodules produced during solidification help or hinder our efforts to produce this matrix? Explain.

59. We would like to produce a quenched and tempered ductile cast iron. (a) Would you expect the hardenability of the cast iron to be better or worse than for a plain carbon steel? Explain. (b) Would you expect the hardenability of the ductile iron to be better or worse than for a malleable cast iron? Explain.

60. A 32510 malleable cast iron contains large primary cementite after the heat treatment is completed. Is this undesirable structure a result of poor FSG or poor SSG treatment? Explain.

61. A 35018 malleable cast iron contains a mixture of ferrite and pearlite in the matrix after heat treatment. Is this undesirable structure a result of poor FSG or poor SSG treatment? Explain.

62. An Fe-3.6% C cast iron is allowed to solidify. Calculate (a) the amount of cementite in both weight percent and volume percent that forms if white iron is produced, (b) the amount of graphite in both weight percent and volume percent that forms if gray iron is produced, and (c) the change in volume during solidification for both gray and white cast irons. (The densities of the phases involved are $7.0 \, \text{g/cm}^3$ for the liquid, $7.69 \, \text{g/cm}^3$ for austenite, $7.66 \, \text{g/cm}^3$ for cementite, and $1.5 \, \text{g/cm}^3$ for graphite.)

63. Calculate the percent carbon in an iron-carbon alloy that will give no volume change when the liquid transforms to austenite plus graphite during solidification. (Use the data in Problem 12-62.)

64. Figure 12-52 illustrates the cooling curves at several distances from the surface of a gray cast iron. Based on these curves, estimate the chill depth in the casting.

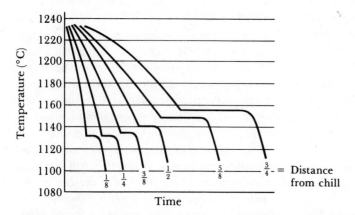

FIGURE 12-52 Cooling curves and thermal arrest temperatures used to determine the chill depth for Problem 12-64.

13

NONFERROUS ALLOYS

13-1 INTRODUCTION

Nonferrous alloys are based on metals other than iron. In this chapter, we will look at a cross section of the more important nonferrous engineering alloys, pointing out how the strengthening mechanisms introduced in Part II of the text are applied to specific types of alloys.

13-2 ALUMINUM ALLOYS

Aluminum is a lightweight metal, with a density of $2.70 \, \text{g/cm}^3$ or one-third the density of steel. Although aluminum alloys have relatively low tensile properties compared with steel, their strength-to-weight ratio, as defined below, is excellent.

$$\text{Strength-to-weight ratio} = \frac{\text{tensile strength}}{\text{density}} \qquad (13\text{-}1)$$

Aluminum is often used when weight is an important factor, as in aircraft and automotive applications.

Aluminum also responds readily to strengthening mechanisms. Table 13-1 compares the strength of pure annealed aluminum with that of alloys strengthened by various techniques. The alloys may be 30 times stronger than pure aluminum.

On the other hand, aluminum often does not display an endurance limit in fatigue, so failure eventually occurs even at rather low stresses. Because of its low melting temperature, aluminum does not perform well at elevated temperatures. Finally, aluminum alloys have a low hardness, leading to poor wear resistance.

◇ | **Example 13-1**

A steel cable $\frac{1}{2}$ in. in diameter has a yield strength of 45,000 psi. The density of steel is about 7.87 g/cm³. Based on the data in Table 13-4, determine (a) the maximum load that the steel cable can support, (b) the diameter of a cold-worked aluminum-manganese alloy (3003-H18) required to support the same load as the steel, and (c) the weight per foot of the steel cable versus the aluminum alloy cable.

Answer:

(a) Load $= F = \sigma_y A = 45,000 \left(\dfrac{\pi}{4}\right) \left(\dfrac{1}{2}\right)^2 = 8836 \, \text{lb}$

(b) The yield strength of the alloy is 27,000 psi. Thus

$$A = \frac{\pi}{4} d^2 = \frac{F}{\sigma_y} = \frac{8836}{27,000} = 0.327$$

$$d = 0.65 \, \text{in.}$$

(c) Density of steel $= \rho = 7.87 \, \text{g/cm}^3 = 0.2828 \, \text{lb/in}^3$

Density of aluminum $= \rho = 2.70 \, \text{g/cm}^3 = 0.097 \, \text{lb/in}^3$

Weight of steel $= Al\rho = \dfrac{\pi}{4}\left(\dfrac{1}{2}\right)^2 (12)(0.2828) = 0.666 \, \text{lb/ft}$

Weight of aluminum $= Al\rho = \dfrac{\pi}{4}(0.65)^2 (12)(0.097) = 0.386 \, \text{lb/ft}$

Although the yield strength of the aluminum is lower than that of the steel and the cable must be larger in diameter, the cable weighs only about half as much as the steel.

Designation.   Aluminum alloys can be subdivided into two major groups, wrought and casting alloys, based on their method of fabrication. *Wrought alloys*, which are shaped by plastic deformation, have compositions and microstructures significantly different from casting alloys, reflecting the different requirements of the manufacturing process. Within each major group we can divide the alloys into two subgroups—heat treatable and nonheat treatable alloys. Heat treatable alloys are age hardened, whereas nonheat treatable alloys are strengthened by solid solution strengthening, strain hardening, or dispersion strengthening.

Aluminum alloys are designated by the numbering system in Table 13-2. The first number specifies the principle alloying elements, and the remaining numbers refer to the specific composition of the alloy. Representative phase diagrams are shown in Figures 10-19, 11-1, 13-1, and 13-4.

TABLE 13-1 The effect of strengthening mechanisms in aluminum and aluminum alloys

Material	Tensile Strength (psi)	Yield Strength (psi)	% Elongation	Yield Strength (alloy) / Yield Strength (pure)
Pure annealed Al (99.999% Al)	6,500	2,500	60	
Commercially pure Al (annealed, 99% Al)	13,000	5,000	45	2.0
Solid solution strengthened (1.2% Mn)	16,000	6,000	35	2.4
75% cold worked pure Al	24,000	22,000	15	8.8
Dispersion strengthened (5% Mg)	42,000	22,000	35	8.8
Age hardened (5.6% Zn-2.5% Mg)	83,000	73,000	11	29.2

Adapted from data in *Metals Handbook*, Vol. 2, 9th Ed., American Society for Metals, 1979.

The degree of strengthening is given by the *temper designation* T or H, depending on whether the alloy is heat treated or strain hardened (Table 13-3). Other designations indicate whether the alloy is annealed (O), solution treated (W), or used in the as-fabricated condition (F). The numbers following the T or H indicate the amount of strain hardening, the exact type of heat treatment, or other special aspects of the processing of the alloy.

TABLE 13-2 Designation system for aluminum alloys

Wrought alloys

1xxx	Commercially pure Al (> 99% Al)	Not aged
2xxx	Al-Cu	Age hardenable
3xxx	Al-Mn	Not aged
4xxx	Al-Si and Al-Mg-Si	Age hardenable if magnesium is present
5xxx	Al-Mg	Not aged
6xxx	Al-Mg-Si	Age hardenable
7xxx	Al-Mg-Zn	Age hardenable

Casting alloys

1xx.x	Commercially pure Al	Not aged
2xx.x	Al-Cu	Age hardenable
3xx.x	Al-Si-Cu or Al-Mg-Si	Some are age hardenable
4xx.x	Al-Si	Not aged
5xx.x	Al-Mg	Not aged
7xx.x	Al-Mg-Zn	Age hardenable
8xx.x	Al-Sn	Age hardenable

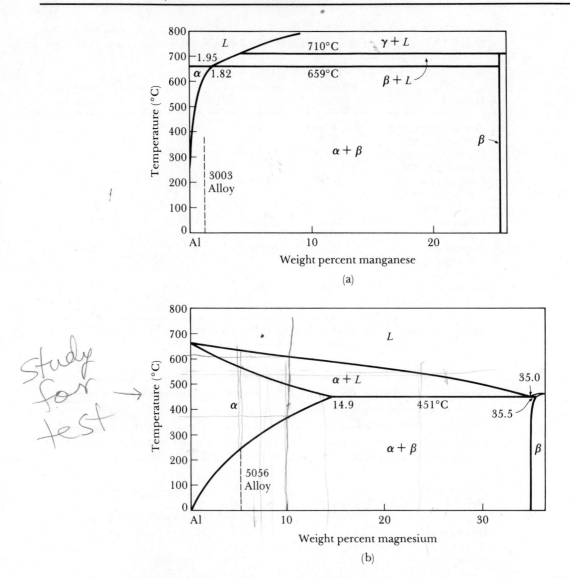

FIGURE 13-1   Portions of the phase diagrams for (a) aluminum-manganese and (b) aluminum-magnesium.

Wrought Alloys.   The 1xxx, 3xxx, 5xxx, and most of the 4xxx alloys are not age hardenable. Compositions and properties of typical alloys are shown in Table 13-4.

The 1xxx and 3xxx alloys are single-phase alloys except for the presence of small amounts of inclusions or intermetallic compounds (Figure 13-2). The properties of these alloys are controlled by strain hardening, solid solution strengthening, and grain size control. However, because the solubilities of the alloying elements in aluminum are small, the degree of solid solution strengthening

TABLE 13-3   Temper designations for aluminum alloys

F   As-fabricated (hot worked, forged, cast, etc.)
O   Annealed (in the softest possible condition)
H   Cold worked
    H1x—cold worked only. (x refers to the amount of cold work and strengthening.)
        H12—cold work that gives a tensile strength midway between the O and H14 tempers.
        H14—cold work that gives a tensile strength midway between the O and H18 tempers.
        H16—cold work that gives a tensile strength midway between the H14 and H18 tempers.
        H18—cold work that gives about 75% reduction.
        H19—cold work that gives a tensile strength greater than 2000 psi of that obtained by the H18 temper.
    H2x—cold worked and partly annealed.
    H3x—cold worked and stabilized at a low temperature to prevent age hardening of the structure.
W   Solution treated
T   Age hardened
    T1—cooled from the fabrication temperature and naturally aged.
    T2—cooled from the fabrication temperature, cold worked, and naturally aged.
    T3—solution treated, cold worked, and naturally aged.
    T4—solution treated and naturally aged.
    T5—cooled from the fabrication temperature and artificially aged.
    T6—solution treated and artificially aged.
    T7—solution treated and stabilized by overaging.
    T8—solution treated, cold worked, and artificially aged.
    T9—solution treated, artificially aged, and cold worked.
    T10—cooled from the fabrication temperature, cold worked, and artificially aged.

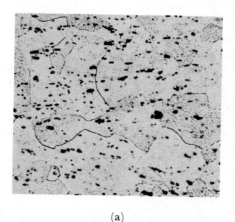 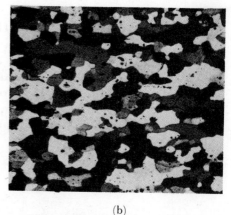

(a)   (b)

FIGURE 13-2   (a) FeAl$_3$ inclusions in annealed 1100 aluminum ($\times$ 350); (b) Mg$_2$Si precipitates in annealed 5457 aluminum alloy ($\times$ 75). From *Metals Handbook*, Vol. 7, 8th Ed., American Society for Metals, 1972.

TABLE 13-4   Properties of typical wrought aluminum alloys

Alloy		Tensile Strength (psi)	Yield Strength (psi)	% Elongation	Applications
Nonheat treatable					
1100-O	> 99% Al	13,000	5,000	40	Electrical components, foil,
1100-H18		24,000	22,000	10	corrosion resistance
3003-O	1.2% Mn	16,000	6,000	35	Beverage cans,
3003-H18		29,000	27,000	7	architectural uses
4043-O	5.2% Si	21,000	10,000	22	Filler metal for welding
4043-H18		41,000	39,000	1	
5056-O	5% Mg	42,000	22,000	35	Containers, marine
5056-H18		60,000	50,000	15	components
Heat treatable					
2024-O	4.4% Cu	27,000	11,000	20	
2024-T4		68,000	47,000	20	
4032-T6	12% Si-1% Mg	55,000	46,000	9	Transportation, aircraft,
6061-O	1% Mg-0.6% Si	18,000	8,000	27	aerospace, and other
6061-T6		45,000	40,000	15	high-strength applications
7075-O	5.6% Zn-2.5% Mg	38,000	15,000	17	
7075-T6		83,000	73,000	11	

From data in *Metals Handbook*, Vol. 2, 9th Ed., American Society for Metals, 1979.

is limited. For example, annealed 3003 alloy has a yield strength of only 6000 psi, compared with 5000 psi for commercially pure aluminum.

The 5xxx alloys contain two phases at room temperature—α, a solid solution of magnesium in aluminum, and Mg_2Al_3, a hard, brittle intermetallic compound. The aluminum-magnesium alloys are strengthened by a fine dispersion of Mg_2Al_3 as well as by strain hardening, solid solution strengthening, and grain size control. However, because the Mg_2Al_3 precipitate is not coherent, age-hardening treatments are not possible.

The 4xxx series alloys also contain two phases, α and nearly pure silicon, β. Alloys which contain both silicon and magnesium can be age hardened by permitting Mg_2Si to precipitate.

The 2xxx, 6xxx, and 7xxx alloys are age hardenable alloys. In each alloy, several coherent precipitates form before the final equilibrium phase is produced.

$$2xxx: \quad \alpha_{ss} \rightarrow \alpha + GP\text{-}I \rightarrow \alpha + GP\text{-}II \rightarrow \alpha + \theta' \rightarrow \alpha + \theta$$

$$6xxx: \quad \alpha_{ss} \rightarrow \alpha + GP \text{ zones} \rightarrow \alpha + \beta'(Mg_2Si) \rightarrow \alpha + \beta(Mg_2Si)$$

$$7xxx: \quad \alpha_{ss} \rightarrow \alpha + GP \text{ zones} \rightarrow \alpha + \eta'(MgZn_2) \rightarrow \alpha + \eta(MgZn_2)$$

Although excellent strengths and strength-to-weight ratios are obtained for these alloys, the amount of precipitate that can form is limited. In addition, they cannot be used at temperatures above approximately 175°C in the aged condition.

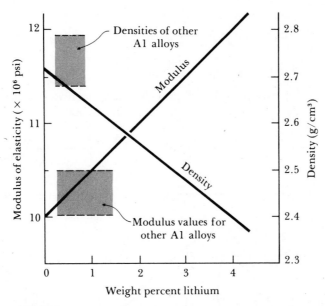

FIGURE 13-3   *The effect of lithium on the stiffness and density of aluminum alloys.*

A new class of alloys based on aluminum and lithium is being developed, particularly for the aerospace industry. Lithium has a density of $0.534\,g/cm^3$, so when it is added to aluminum, the overall density of the Al-Li alloy may be as much as 10% less than that of conventional aluminum alloys (Figure 13-3). The modulus of elasticity is increased, and the strength can equal or exceed that of conventional alloys.

The high strength of the Al-Li alloys is due to an age-hardening response during heat treatment. By conventional treatments, alloys containing 1% to 2% Li can be heat treated (Figure 13-4). Additional Li, up to 4%, can be introduced by rapid solidification processing, further enhancing both the light weight and maximum strength. Some difficulties are encountered with poor fracture toughness in these particular alloys; in addition, processing of the alloy must be done carefully in order to avoid the possibility of an explosion if the liquid alloy were to come into contact with water.

◇ **Example 13-2**

Determine the maximum amount of the precipitate that forms in aluminum alloy 5056 under equilibrium conditions at room temperature.

Answer:

From Table 13-4, the 5056 alloy contains 5% Mg. The phase diagram in Figure 13-1(b) permits us to calculate the amount of Mg_2Al_3 (β) precipitate.

Composition of α: 0% Mg

Composition of β: 35% Mg

$$\% \ \beta \ = \ \frac{5 - 0}{35 - 0} \ \times \ 100 \ = \ 14.3\%$$

The dispersed β precipitate provides a large increase in yield strength compared with the pure aluminum—22,000 psi versus 5000 psi.

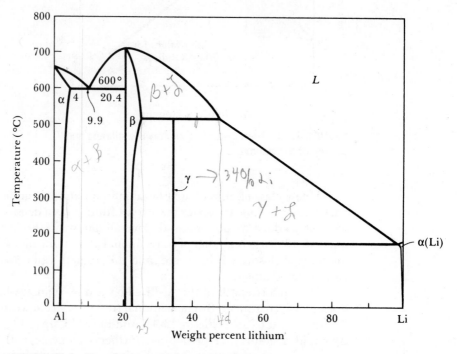

FIGURE 13-4   The aluminum-lithium phase diagram.

◇ | **Example 13-3**

Compare the strength-to-weight ratio of commercially pure aluminum with that of an age-hardened 7xxx series aluminum alloy.

Answer:

The density for both is about 2.70 g/cm³ or 0.097 lb/in³. The tensile strength for commercially pure aluminum is 13,000 psi, and the tensile strength for age-hardened 7075 Al-Mg-Zn alloy is 83,000 psi.

$$\text{Strength-to-weight ratio (Al)} \;=\; \frac{13,000}{0.097} \;=\; 134,000 \,\text{in.}$$

$$\text{Strength-to-weight ratio (7075)} \;=\; \frac{83,000}{0.097} \;=\; 856,000 \,\text{in.}$$

Casting Alloys.   Many of the common aluminum casting alloys shown in Table 13-5 contain enough silicon to cause the eutectic reaction, giving the alloys low melting points, good fluidity, and good castability. *Fluidity* is the ability of the liquid metal to flow through a mold without prematurely solidifying, and *castability* refers to the ease with which a good casting can be made from the alloy.

The properties of the aluminum-silicon alloys are controlled by solid solution strengthening of the α aluminum matrix, dispersion strengthening by the β phase, and solidification, which controls the primary grain size and shape as well as the nature of the eutectic microconstituent. Fast cooling obtained in die casting or permanent mold casting normally increases strength by refining grain size and the eutectic microconstituent (Figure 13-5). Grain refinement using boron and titanium additions, modification using sodium or strontium to change the eutectic structure, or hardening with phosphorus to refine the primary silicon are all done in certain alloys to improve the microstructure and

TABLE 13-5   Properties of typical aluminum casting alloys

Alloy		Tensile Strength (psi)	Yield Strength (psi)	% Elongation	Casting Process
201-T6	4.5% Cu	70,000	63,000	7	Sand
319-F	6% Si-3.5% Cu	27,000	18,000	2	Sand
		34,000	19,000	2.5	Permanent mold
356-T6	7% Si-0.3% Mg	33,000	24,000	3.5	Sand
		38,000	27,000	5	Permanent mold
380-F	8.5% Si-3.5% Cu	46,000	23,000	3.5	Permanent mold
384-F	11.2% Si-4.5% Cu-0.6% Mg	48,000	24,000	2.5	Permanent mold
390-F	17% Si-4.5% Cu-0.6% Mg	41,000	35,000	1	Die casting
443-F	5.2% Si	19,000	8,000	8	Sand
		23,000	9,000	10	Permanent mold
		33,000	16,000	9	Die casting
413-F	12% Si	43,000	21,000	2.5	Die casting
518-F	8% Mg	45,000	28,000	7	Sand
713-T5	7.5% Zn-0.7% Cu-0.35% Mg	30,000	22,000	4	Sand
850-T5	6.2% Sn-1% Ni-1% Cu	23,000	11,000	10	Sand

After data in *Metals Handbook*, Vol. 2, 9th Ed., American Society for Metals, 1979.

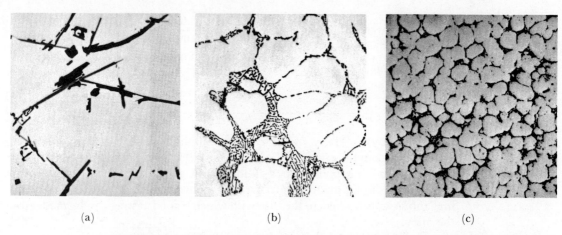

(a)   (b)   (c)

FIGURE 13-5 (a) Sand-cast 443 aluminum alloy containing coarse silicon and inclusions; (b) permanent-mold 443 alloy containing fine dendrite cells and fine silicon due to faster cooling; (c) die-cast 443 alloy with still a finer microstructure (× 350). From *Metals Handbook*, Vol. 7, 8th Ed., American Society for Metals, 1972.

thus the degree of dispersion strengthening. Many alloys also contain copper, magnesium, or zinc, permitting an age-hardening reaction by precipitation of $CuAl_2$, Mg_2Si, or $MgZn_2$.

◇ | **Example 13-4**

Aluminum casting alloys such as 201-T6 and 518-F have poor fluidity and shrinkage characteristics compared with the 413-F alloy. Explain why this might be the case, using the phase diagram.

Answer:

The 201 alloy contains 4.5% Cu, the 518 alloy contains 8% Mg, and the 413 alloy contains 12% Si. From the phase diagrams in Figures 10-19, 11-1, and 13-1, neither the aluminum-copper nor the aluminum-magnesium alloy passes through the eutectic reaction, but the 12% Si alloy is nearly a straight eutectic. We expect much better fluidity in alloys that contain the eutectic reaction.

In addition, shrinkage problems are accentuated for long freezing range alloys. From the phase diagrams

Freezing range of alloy 295 = 650 − 575 = 75°C

Freezing range of alloy 413 = 0°C

Freezing range of alloy 518 = 620 − 530 = 90°C

The short freezing range of the 413 alloy gives concentrated shrinkage that can be prevented by risers. The long freezing range of the other alloys gives microshrinkage that is difficult to control.

13–3   MAGNESIUM ALLOYS

Magnesium is lighter than aluminum, with a density of $1.74\,\text{g/cm}^3$, and melts at a lower temperature. Although magnesium alloys are not as strong as aluminum alloys, their strength-to-weight ratios are comparable. Consequently, magnesium alloys are used in aerospace applications, high-speed machinery, and transportation and materials handling equipment.

However, magnesium has a low modulus of elasticity ($6.5 \times 10^6\,\text{psi}$) and poor resistance to fatigue, creep, and wear. Magnesium also poses a hazard during casting and machining, since it combines easily with oxygen and burns. Finally, the response of magnesium to strengthening mechanisms is relatively poor (Example 13-5).

$\Diamond$ | **Example 13-5**

From the data in Table 13-7, (a) estimate the ratio by which the yield strength of magnesium can be increased by alloying and heat treatment and (b) compare the yield strength-to-weight ratio of high-strength magnesium alloys with that of high-strength aluminum alloys.

Answer:

(a) The yield strength of pure annealed magnesium is 13,000 psi and that of age-hardened AZ80A-T5 alloy is 40,000 psi.

$$\text{Ratio} = \frac{40,000}{13,000} = 3$$

When compared with a ratio of 30 in aluminum, the response of magnesium to strengthening is rather poor.

(b) The strength-to-weight ratio of AZ80A-T5 compared with that of 7075-T6 aluminum alloy (Table 13-4) is

$$\text{AZ80A-T5:} \quad \frac{40,000\,\text{psi}}{1.74\,\text{g/cm}^3} = \frac{40,000\,\text{psi}}{0.063\,\text{lb/in}^3} = 6.3 \times 10^5\,\text{in.}$$

$$\text{7075-T6:} \quad \frac{73,000\,\text{psi}}{2.70\,\text{g/cm}^3} = \frac{73,000\,\text{psi}}{0.097\,\text{lb/in}^3} = 7.5 \times 10^5\,\text{in.}$$

Designation.   Magnesium alloys can be either wrought or cast, and both heat treatable and nonheat treatable grades are available. Some binary phase diagrams involving magnesium alloys are shown in Figure 13-6.

The alloys are normally designated by a combination of letters and numbers (Table 13-6). The letters refer to the two major alloying elements present in the alloy, and the numbers refer to the approximate amounts of the major alloying elements. The temper designations, such as T and H, are the same as for the aluminum alloys.

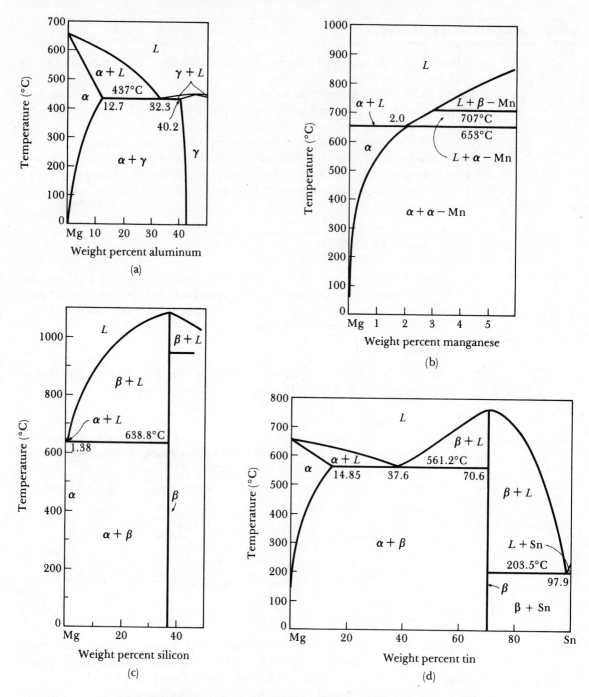

FIGURE 13-6 Binary phase diagrams for the (a) magnesium-aluminum, (b) magnesium-manganese, (c) magnesium-silicon, and (d) magnesium-tin systems.

TABLE 13-6   *Designation system for magnesium alloys*

A. Two letters indicate the major alloying elements.

A—Al	Q—Ag
E—Ce	S—Si
H—Th	T—Sn
K—Zr	Z—Zn
M—Mn	

B. Two (and sometimes three) numbers indicate the approximate amounts of alloying elements, rounded off to the nearest percent.
C. A final letter gives variations to the normal alloy.

◇ | **Example 13-6**

What would be the designation for a magnesium alloy containing 2.8% Th and 1.2% Mn?

Answer:

From Table 13-6, thorium is designated by H and manganese by M. To the nearest whole number, the alloy contains about 3% Th and 1% Mn. Thus, the alloy is designated HM31.

Structure and Properties.   Magnesium, which has the HCP structure, is less ductile than aluminum. However, the alloys do have some ductility because alloying increases the number of active slip planes. Some deformation and strain hardening can be accomplished at room temperature, and the alloys can be readily deformed at elevated temperatures. However, strain hardening produces a relatively small effect in pure magnesium due to the low strain-hardening coefficient.

As in aluminum alloys, the solubility of alloying elements in magnesium is limited. Alloys may be strengthened by either dispersion strengthening or age hardening. Some age-hardened magnesium alloys, such as those containing zirconium, thorium, silver, or cerium, have good resistance to overaging at temperatures as high as 300°C. Alloys containing up to 9% Li are being developed for even lighter weight. Properties of some typical magnesium alloys are listed in Table 13-7.

13–4   BERYLLIUM

Beryllium is lighter than aluminum, with a density of $1.848\,\text{g/cm}^3$, yet is stiffer than steel, with a modulus of elasticity of $42 \times 10^6\,\text{psi}$. Beryllium alloys, which have a yield strength of 30,000 to 50,000 psi, have high strength-to-weight ratios

TABLE 13-7 Properties of typical magnesium alloys

Alloy		Tensile Strength (psi)	Yield Strength (psi)	% Elongation
Casting alloys				
AM100-T6	10% Al-0.1% Mn	40,000	22,000	1
AZ81A-T4	7.6% Al-0.7% Zn	40,000	12,000	15
ZK61A-T6	6% Zn-0.7% Zr	45,000	28,000	10
AM60A-F	6% Al-0.13% Mn	32,000	19,000	6
ZE41A-T5	4.2% Zn-1.2% Ce	30,000	20,000	3.5
Wrought alloys				
AZ10A-F	1.2% Al-0.4% Zn	35,000	21,000	10
AZ80A-T5	8.5% Al-0.5% Zn	55,000	40,000	7
ZK40A-T5	4% Zn-0.45% Zr	40,000	37,000	4
HK31A-H24	3% Th-0.6% Zr	38,000	30,000	8
Pure Mg				
Annealed		23,000	13,000	3–15
Cold worked		26,000	17,000	2–10

After data in *ASM Metals Handbook*, Vol. 2, 9th Ed., 1979.

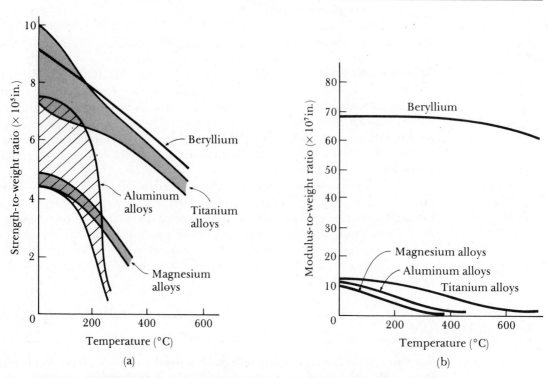

FIGURE 13-7 A comparison of the strength-to-weight (a) and modulus of elasticity-to-weight (b) ratios of beryllium and other nonferrous alloys.

and maintain both strength and stiffness to high temperatures (Figure 13-7). A number of aerospace and nuclear applications indicate that beryllium is potentially an excellent engineering material.

Unfortunately, beryllium is expensive, brittle, reactive, and toxic. It has the HCP structure and has limited ductility at room temperature. When exposed to the atmosphere at elevated temperatures, beryllium rapidly oxidizes to BeO, which is toxic. Consequently, sophisticated manufacturing techniques, such as vacuum casting, vacuum forging, and powder metallurgy, must be used.

13–5   COPPER ALLOYS

There are a myriad of copper-base alloys that take advantage of all of the strengthening mechanisms that we have discussed. Typical properties of alloys displaying these strengthening mechanisms are shown in Table 13-8.

The copper-base alloys are heavier than iron. Although the yield strength of some alloys is high, the strength-to-weight ratio is typically less than that of aluminum or magnesium alloys. The alloys have better resistance to fatigue,

TABLE 13-8   Properties of typical copper alloys obtained by different strengthening mechanisms

Material	Alloy and Temper Designation	Tensile Strength (psi)	Yield Strength (psi)	% Elongation	Strengthening Mechanism
Pure Cu, annealed		30,300	4,800	60	
Commercially pure Cu, annealed to coarse grain size	10100–OS050	32,000	10,000	55	
Commercially pure Cu, annealed to fine grain size	10100–OS025	34,000	11,000	55	Grain size
Commercially pure Cu, cold worked	10100–H10	57,000	53,000	4	Strain hardening
Annealed Cu-35% Zn	27000–OS050	47,000	15,000	62	
As-fabricated Cu-30% Ni	71500–M20	55,000	20,000	45	Solid solution
Annealed Cu-10% Sn	52400–OS035	66,000	28,000	68	
Annealed nickel silver	77000–OS035	60,000	27,000	40	
Cold-worked Cu-35% Zn	27000–H10	98,000	63,000	3	Solid solution +
Cold-worked Cu-30% Ni	71500–H80	84,000	79,000	3	Strain hardening
Age-hardened Cu-2% Be	17200–TF00	190,000	175,000	4	Age hardening
Quenched and tempered Cu-Al	95500–TQ50	110,000	60,000	5	Martensitic reaction
Cast manganese bronze	86500–F	71,000	28,000	30	Eutectoid reaction

Data from *Metals Handbook*, Vol. 2, 9th Ed., American Society for Metals, 1979.

TABLE 13-9 Designation system for copper-base alloys

Wrought alloys

100xx–159xx	Commercially pure Cu
160xx–199xx	Nearly pure Cu but age hardenable due to Cd, Be, or Cr
2xxxx	Cu-Zn (brass)
3xxxx	Cu-Zn-Pb (leaded brass)
4xxxx	Cu-Zn-Sn (tin bronze)
5xxxx	Cu-Sn and Cu-Sn-Pb (phosphor bronze)
6xxxx	Cu-Al (aluminum bronze), Cu-Si (silicon bronze), Cu-Zn-Mn (manganese bronze)
7xxxx	Cu-Ni (cupronickel), Cu-Ni-Zn (nickel silver)

Cast alloys

800xx–811xx	Commercially pure Cu
813xx–828xx	95–99% Cu
833xx–899xx	Cu-Zn alloys containing Sn, Pb, Mn, or Si
9xxxx	Other Cu alloys, including tin bronze, aluminum bronze, cupronickel, and nickel silver

creep, and wear than the lightweight aluminum and magnesium alloys. Many of the alloys also have excellent ductility, corrosion resistance, and electrical and thermal conductivity.

Copper alloys are also unique in that they may be selected to produce an appropriate decorative color. Pure copper is red. However, zinc additions produce a yellow color, and nickel produces a silver color.

The designation system for copper-base alloys and their tempers are shown in Tables 13-9 and 13-10. The temper designation system differs from that of aluminum and magnesium alloys.

Commercially Pure Copper. Coppers containing less than 1% impurities are used for electrical applications. Small amounts of cadmium or silver improve high-temperature hardness, and tellurium or sulfur improve machinability. Some coppers are dispersion strengthened by small amounts of Al_2O_3, which improve the hardness of the copper without significantly impairing conductivity. Any of these coppers can be strengthened by strain hardening, producing large increases in strength with relatively small decreases in conductivity.

◇ Example 13-7

You wish to select a conductive material for the contacts of a switch or relay. Which of the coppers would you choose to maximize the life of the device?

Answer:

We want a material that has a high hardness yet good conductivity. The hard ceramic particles of Al_2O_3 provide wear resistance yet do not enter into the lattice of the copper and so do not interfere with the electrical conductivity.

TABLE 13-10   Temper designation for copper alloys

Hxx—cold worked. (xx indicates the degree of cold work.)

	Percent Reduction in Thickness or Diameter
H01 $\frac{1}{4}$ hard	10.9
H02 $\frac{1}{2}$ hard	20.7
H03 $\frac{3}{4}$ hard	29.4
H04 hard	37.1
H06 extra hard	50.1
H08 spring hard	60.5
H10 extra spring	68.6
H12 special spring	75.1
H14 super spring	80.3

HRxx—cold worked and stress relieved. (xx refers to initial percent cold work.)
Mxx—as-manufactured. (xx refers to the type of manufacturing process.)
Oxx—annealed. (xx refers to annealing method.)
OSxxx—annealed to produce a particular grain size. (xxx refers to the grain diameter in 10^{-3} mm. Thus, OS025 gives a grain diameter of 0.025 mm.)
TB00—solution treated.
TDxx—solution treated and cold worked. (xx refers to percent cold work.)
TF00—age hardened.
THxx—cold worked and aged. (xx refers to degree of cold work.)
TLxx—aged and cold worked. (xx refers to degree of cold work.)
TQxx—quenched and tempered. (xx gives details of heat treatment.)

Strain Hardening.   The single-phase copper alloys are strengthened significantly by cold working. Examples of this effect are shown in Table 13-8. The FCC copper has excellent ductility and a high strain-hardening coefficient.

Solid Solution-Strengthened Alloys.   A number of copper-base alloys contain large quantities of alloying elements yet remain single phase. Important binary phase diagrams are shown in Figure 13-8.

The copper-zinc, or *brass*, alloys with less than 40% Zn form single-phase solid solutions of zinc in copper. The mechanical properties, even elongation, increase as the zinc content increases. These alloys can be cold formed into rather complicated yet corrosion-resistant components. *Manganese bronze* is a particularly high-strength alloy containing manganese as well as zinc for solid solution strengthening. A copper-nickel-zinc ternary alloy called *nickel silver* really has no silver present although it does have a silver color. The formability of these alloys approaches that of the copper-zinc alloys.

Tin bronzes, often called *phosphor bronzes*, may contain up to 10% Sn and remain single phase. The phase diagram predicts that the alloy will contain the Cu_3Sn (ε) compound. However, the kinetics of the reaction are so slow that the precipitate may not form.

Alloys containing less than about 9% Al or less than 3% Si are also single phase. These *aluminum bronzes* and *silicon bronzes* have good forming characteristics and are often selected for their good strength and excellent toughness.

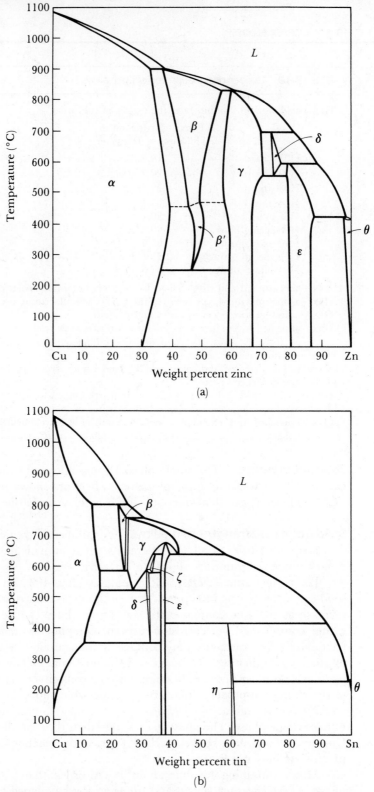

FIGURE 13-8 Binary phase diagrams for the (a) copper-zinc, (b) copper-tin, (c) copper-silicon, (d) copper-aluminum, and (e) copper-beryllium systems.

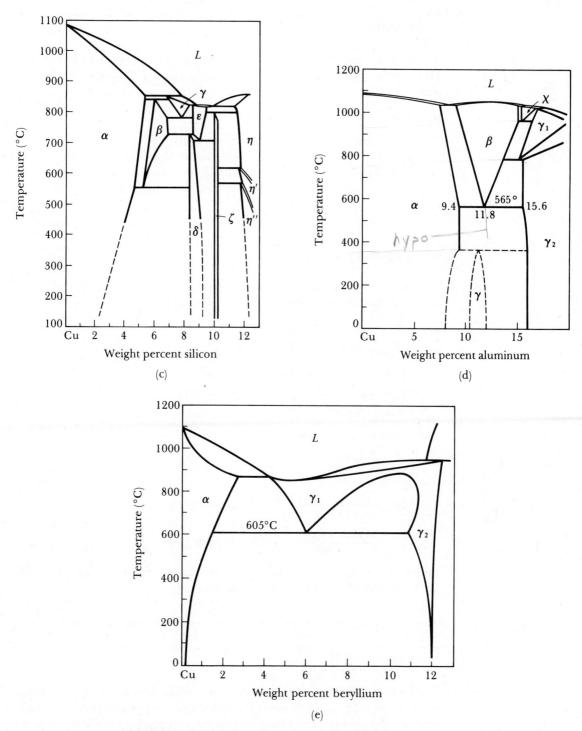

FIGURE 13-8 (Continued).

Example 13-8

We say that copper can contain up to 40% Zn or 9% Al and still be single phase. How do we explain this statement in view of the phase diagrams in Figure 13-8?

Answer:

According to the phase diagrams, the solubility of zinc in copper at room temperature is about 30% and the solubility of aluminum in copper is about 8%. The equilibrium precipitates do not form, however, because of the slow rates of diffusion at these temperatures, just as in the copper-tin alloys.

 Example 13-9

Compare the percentage increase in the yield strength of commercially pure annealed aluminum, magnesium, and copper by strain hardening.

Answer:

From Tables 13-4, 13-7, and 13-8, we find that

    Al increases from 5000 psi to 22,000 psi

    Mg increases from 13,000 psi to 17,000 psi

    Cu increases from 10,000 psi to 53,000 psi

The percentage increases are

$$\text{Al:} \quad \frac{22,000 - 5000}{5000} \times 100 = 340\%$$

$$\text{Mg:} \quad \frac{17,000 - 13,000}{13,000} \times 100 = 31\%$$

$$\text{Cu:} \quad \frac{53,000 - 10,000}{10,000} \times 100 = 430\%$$

Strain hardening is an effective way to strengthen copper and aluminum but produces little effect in HCP magnesium.

 Example 13-10

Suppose the effect of increasing the percent alloying element in copper produces a linear increase in yield strength. Determine the equations for the yield strength of copper as a function of zinc, nickel, and tin. Are the results consistent with the atomic radii of these elements?

Answer:

From Table 13-8, we find the yield strengths of pure copper and some typical alloys.

Pure copper: 10,000 psi

Cu-35% Zn: 15,000 psi

$$\text{Yield strength} = 10,000 + \frac{15,000 - 10,000}{35} \,(\% \text{ Zn})$$

$$= 10,000 + 143 \,(\% \text{ Zn})$$

Cu-30% Ni: 20,000 psi

$$\text{Yield strength} = 10,000 + \frac{20,000 - 10,000}{30} \,(\% \text{ Ni})$$

$$= 10,000 + 333 \,(\% \text{ Ni})$$

Cu-10% Sn: 28,000 psi

$$\text{Yield strength} = 10,000 + \frac{28,000 - 10,000}{10} \,(\% \text{ Sn})$$

$$= 10,000 + 1800 \,(\% \text{ Sn})$$

The atomic radii of the elements are $r_{Cu} = 1.278\,\text{Å}$, $r_{Zn} = 1.332\,\text{Å}$, $r_{Ni} = 1.243\,\text{Å}$, and $r_{Sn} = 1.405\,\text{Å}$. Copper, nickel, and zinc have about the same atomic radii, but tin atoms are much larger. Thus, addition of tin produces large lattice distortions and gives a larger solid solution-strengthening effect.

Age Hardenable Alloys.   A number of copper-base alloys display an age-hardening response, including zirconium copper, chromium copper, and beryllium copper. The copper-beryllium alloys are used for their high strength, their high stiffness (making them useful as springs), and their nonsparking qualities (making them useful for tools to be used near flammable gases and fluids).

Phase Transformations.   Aluminum bronzes that contain over 9% Al can form at least some β phase on heating above 565°C, the eutectoid temperature [Figure 13-8(d)]. On subsequent cooling, the eutectoid reaction produces a lamellar structure, or pearlite, that contains a brittle γ_2 compound. The low temperature peritectoid reaction, $\alpha + \gamma_2 \rightarrow \gamma$, normally does not occur. The eutectoid product is relatively weak and brittle. However, we can rapidly quench the β to produce martensite, or β', which has a high strength and low ductility. When β' is subsequently tempered, a combination of high strength, good ductility, and excellent toughness is obtained as fine platelets of α precipitate from the β' (Figure 13-9).

FIGURE 13-9 The microstructure of a quenched and tempered aluminum bronze containing alpha plates in a beta matrix (× 150). From *Metals Handbook,* Vol. 7, 8th Ed., American Society for Metals, 1972.

◇ | **Example 13-11**

Consider an aluminum bronze alloy containing 10% Al [Figure 13-8(d)]. (a) Calculate the amount of brittle γ_2 that will form on equilibrium cooling past the eutectoid temperature and (b) determine the phases you would expect to have during each step of the quench and temper heat treatment.

Answer:

(a) The amount of γ_2 that forms after the eutectoid reaction is

$$\gamma_2 = \frac{10 - 9.4}{15.6 - 9.4} \times 100 = 9.7\%$$

(b) The steps in the quench and temperature heat treatment are to hold at 900°C, quench, and temper between 400°C and 650°C.

The alloy is 100% β above 850°C. At 900°C, the alloy contains all β with a composition of 10% Al.

After quenching, the β transforms to β' martensite, which also contains 10% Al. The β' is supersaturated in copper.

During tempering, the copper supersaturation is removed by precipitation of α in a matrix of β. The phase diagram shows that α and β are stable above 565°C and therefore the tempering may lead to the equilibrium phases. However, if we temper at 400°C, the phase diagram shows that γ_2 and α should be present. Normally, the equilibrium γ_2 is not produced in the usual tempering times.

Leaded Copper Alloys. Virtually any of the wrought alloys may contain up to 4.5% Pb. The lead forms a monotectic reaction with the copper and produces tiny lead spheres as the last liquid to solidify. The lead improves the machining characteristics by improving chip-forming tendencies.

Even larger amounts of lead are used for copper casting alloys. The lead helps provide lubrication and embeddability, by which hard particles or grit are embedded in the soft lead spheres, and therefore helps to minimize wear. The large amounts of lead cannot be introduced to wrought alloys because the lead melts and embrittles the alloy during hot working.

13–6 NICKEL AND COBALT

Nickel and cobalt alloys are used for corrosion protection and for high-temperature resistance, taking advantage of their high melting points and high strengths. Nickel is FCC and has good formability; cobalt is an allotropic metal, with the FCC structure above 417°C and the HCP structure at lower temperatures. Special cobalt alloys are used for exceptional wear resistance and, because of resistance to human body fluids, for prosthetic devices. Typical alloys and their applications are listed in Table 13-11.

Nickel and Monel. Nickel and its alloys have excellent corrosion resistance and forming characteristics.

When copper is added to nickel, the maximum strength is obtained near 60% Ni. A number of alloys, called *Monels*, with approximately this composition are used for their strength and corrosion resistance in salt water and at elevated temperatures. Some of the Monels contain small amounts of aluminum and titanium. These alloys show an age-hardening response by the precipitation of γ', a coherent Ni_3Al or Ni_3Ti precipitate which nearly doubles the tensile properties. The precipitates resist overaging at temperatures up to 425°C (Figure 13-10).

Superalloys. Superalloys contain large amounts of alloying elements intended to produce a combination of high strength at elevated temperatures, resistance to creep at temperatures up to 1000°C, and resistance to corrosion. Yet these excellent high-temperature properties are obtained even though the melting temperature of the alloys is about the same as that for steels.

The three categories of superalloys—nickel base, iron-nickel base, and cobalt base—often have exotic names, befitting their use. Typical applications include vanes and blades for turbine and jet engines, heat exchangers, chemical reaction vessel components, and heat-treating equipment. Figure 13-11 shows the effect of temperature on the stress-rupture properties.

To obtain high strengths and creep resistance, the alloying elements must produce a strong, stable microstructure at high temperatures. Solid solution strengthening, dispersion strengthening, and precipitation hardening are generally employed.

TABLE 13-11 Compositions, properties, and applications for selected nickel and cobalt alloys

Material	Tensile Strength (psi)	Yield Strength (psi)	% Elongation	Strengthening Mechanism	Applications
Pure Ni (99.9% Ni)	50,000 95,000	16,000 90,000	45 4	Annealed Cold worked	Corrosion resistance
Ni-Cu alloys					
Monel 400 (Ni-31.5% Cu)	78,000	39,000	37	Annealed	Valves, pumps, heat exchangers
Monel K-500 (Ni-29.5% Cu-2.7% Al-0.6% Ti)	150,000	110,000	30	Aged	Shafts, springs, impellers
Ni superalloys					
Inconel 600 (Ni-15.5% Cr-8% Fe)	90,000	29,000	49	Carbides	Heat treatment equipment
Hastelloy B-2 (Ni-28% Mo)	130,000	60,000	61	Carbides	Corrosion resistance
Hastelloy G (Ni-20% Cr-20% Fe-7% Mo + Nb, Ta)	100,000	47,000	50	Aged	Chemical processing
MAR-M246 (Ni-10% Co-9% Cr-10% W + Ti, Al, Ta)	140,000	125,000	5	Aged	Jet engines
DS-Ni (Ni-2% ThO$_2$)	71,000	48,000	14	Dispersion	Gas turbines
Fe-Ni superalloys					
Incoloy 800 (Ni-46% Fe-21% Cr)	89,000	41,000	37	Carbides	Heat exchangers
Co superalloys					
Haynes 25 (50% Co-20% Cr-15% W-10% Ni)	135,000	65,000	60	Carbides	Jet engines
Stellite 6B (60% Co-30% Cr-4.5% W)	177,000	103,000	4	Carbides	Abrasive wear resistance

Data from *Metals Handbook*, Vol. 3, 9th Ed., American Society for Metals, 1980.

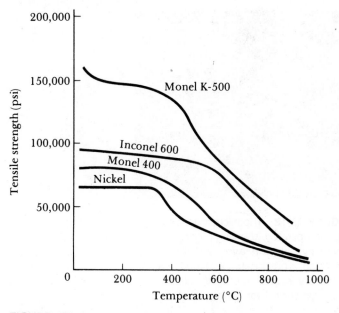

FIGURE 13-10 The effect of temperature on the tensile strength of several nickel-base alloys.

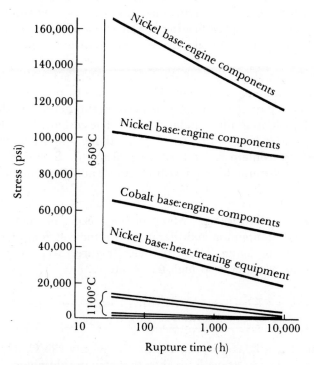

FIGURE 13-11 The stress-rupture behavior of selected superalloys at 650°C and 1100°C.

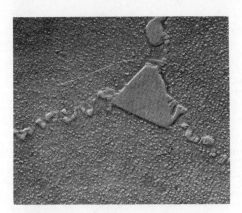

FIGURE 13-12 The microstructure of a typical superalloy, showing a grain boundary carbide network and the dispersed γ' precipitate ($\times$ 15,000). From *Metals Handbook*, Vol. 7, 8th Ed., American Society for Metals, 1972.

Solid Solution Strengthening. Large additions of chromium, molybdenum, and tungsten and smaller additions of tantalum, zirconium, niobium, and boron provide solid solution strengthening. Because no catastrophic metallurgical softening process occurs on heating, the effects of solid solution strengthening are stable and consequently make the alloy resistant to creep.

Carbide Dispersion. All of the alloys contain a small amount of carbon which, by combining with other alloying elements, produces a network of fine, stable carbide particles. The carbide network interferes with dislocation movement and prevents grain boundary sliding. The carbides include TiC, BC, ZrC, TaC, Cr_7C_3, $Cr_{23}C_6$, Mo_6C, and W_6C, although often the carbides are more complex and contain several alloying elements. Stellite 6B, a cobalt-base superalloy, has unusually good wear resistance at high temperatures due to these carbides.

Precipitation Hardening. Many of the nickel and nickel-iron superalloys which contain aluminum and titanium form the coherent precipitate γ' (Ni_3Al or Ni_3Ti) during aging. The γ' particles (Figure 13-12) increase strength and resistance to creep, even at high temperatures.

13–7 ZINC ALLOYS

Pure zinc is nearly as heavy as steel, melts at only 420°C, has the HCP crystal structure, and has a strength less than that of many aluminum alloys. With these properties, we might expect that zinc would seldom be used. Yet many appli-

TABLE 13-12 Properties of selected zinc alloys

Alloy	Tensile Strength (psi)	Yield Strength (psi)	% Elongation	Processing
Casting alloys				
Zn-4% Al	41,000		10	Die casting
Zn-12% Al	45,000	30,000	2	Sand casting
Zn-27% Al	61,000	53,000	5	Sand casting
Wrought alloys				
Zn-0.08% Pb	20,000		60	Hot work
	23,000		45	Cold work
Zn-1% Cu	29,000		40	Hot work
	36,000		30	Cold work
Zn-22% Al	58,000	51,000	11	Superplastic

Data from *Metals Handbook*, Vol. 2, 9th Ed., American Society for Metals, 1979.

cations are found for both wrought and cast zinc. Table 13-12 summarizes the properties of several zinc alloys.

Because zinc recrystallizes and creeps near room temperature, it has excellent ductility. However, strain hardening is negligible. Wrought zinc is used for batteries, photoengraving plates, and roofing components, such as gutters. A special wrought alloy, Zn-22% Al, displays superplastic behavior and can be formed into complex panels and cabinets.

Zinc die castings are produced using an alloy that contains about 4.5% Al. This eutectic alloy (Figure 13-13) has a low melting temperature and excellent fluidity, permitting the metal to fill thin, complicated molds. The castings are easily cleaned, then coated or plated to produce a decorative effect. Zinc alloys containing 12% to 27% Al have also been developed which have sufficient strength to compete with many copper alloys and cast irons. Figure 13-14 shows one of these alloys.

One problem with zinc alloys is related to the eutectoid reaction that occurs after solidification. Particularly in die casting, the eutectoid reaction is not completed during cooling. However, after some period of time, even at room temperature, additional transformation will occur, causing dimensional changes. A low-temperature annealing process may be performed to stabilize the structure prior to final machining.

13–8 TITANIUM ALLOYS

Titanium is a relatively lightweight metal that provides excellent corrosion resistance, high strength-to-weight ratio, and good high-temperature properties. Strengths up to 200,000 psi coupled with a density of 4.505 g/cm^3 provide the

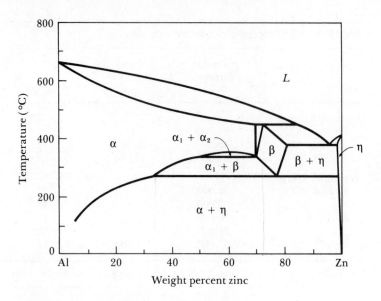

FIGURE 13-13   The aluminum-zinc phase diagram.

excellent mechanical properties, while an adherent, protective TiO_2 film provides excellent resistance to corrosion and contamination below 535°C. Above 535°C, the oxide film breaks down and small atoms such as carbon, oxygen, nitrogen, and hydrogen embrittle the titanium.

Titanium is allotropic, with the HCP crystal structure (α) at low temperatures and a BCC structure (β) above 882°C. Alloying elements provide solid solution strengthening and change the allotropic transformation temperature. The alloying elements can be divided into four groups, as summarized in Figure 13-15. Additions such as tin and zirconium provide solid solution strengthening

FIGURE 13-14   Microstructure of a Zn-27% Al alloy, with primary aluminum-rich dendrites (white) surrounded by a eutectoid product and interdendritic regions containing a eutectic product and precipitate particles (× 500). From *American Foundrymen's Society Transactions*, Vol. 88, 1980, p. 190.

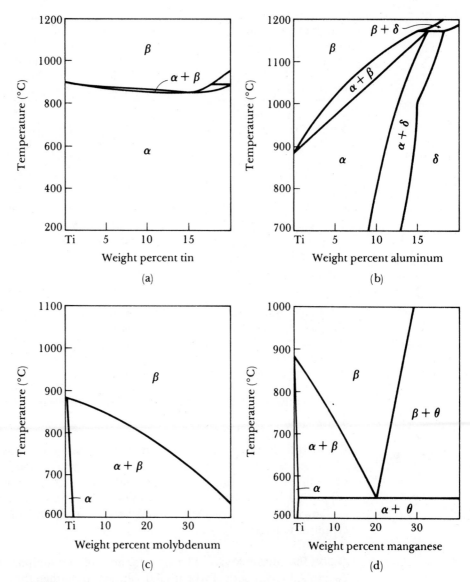

FIGURE 13-15 Portions of the phase diagrams for (a) titanium-tin, (b) titanium-aluminum, (c) titanium-molybdenum, and (d) titanium-manganese.

without affecting the transformation temperature. Aluminum, oxygen, hydrogen, and other alpha stabilizing elements increase the temperature at which α transforms to β. Beta stabilizers such as vanadium, tantalum, molybdenum, and niobium lower the transformation temperature, even causing β to be stable at room temperature. Finally, manganese, chromium, and iron produce a eutectoid reaction, reducing the temperature at which the α-β transformation occurs and producing a two-phase structure at room temperature. There are

TABLE 13-13   Properties of selected titanium alloys

Material	Tensile Strength (psi)	Yield Strength (psi)	% Elongation
Commercially pure Ti			
99.5% Ti	35,000	25,000	24
99.0% Ti	80,000	70,000	15
Alpha Ti alloys			
5% Al-2.5% Sn	125,000	113,000	15
Beta Ti alloys			
13% V-11% Cr-3% Al	187,000	176,000	5
Near alpha Ti alloys			
8% Al-1% Mo-1% V	140,000	120,000	14
6% Al-4% Zr-2% Sn-2% Mo	146,000	144,000	3
Alpha-beta Ti alloys			
8% Mn	140,000	125,000	15
6% Al-4% V	150,000	140,000	8
7% Al-4% Mo	170,000	150,000	10
6% Al-6% V-2% Sn	160,000	150,000	12

Data from *Metals Handbook*, Vol. 3, 9th Ed., American Society for Metals, 1980.

several categories for titanium and its alloys, which are summarized in Table 13-13.

Commercially Pure Titanium.   Unalloyed titanium is used for its superior corrosion resistance. Impurities, such as oxygen, dramatically increase the strength of the titanium. Commercially pure titanium is relatively weak (Figure 13-16) but has the best corrosion resistance. Applications include heat exchangers, piping, reactors, pumps, and valves for the chemical and petrochemical industries.

Alpha Titanium Alloys.   The most common of the all-alpha alloys contains 5% Al and 2.5% Sn, both of which solid solution strengthen the alpha. These alloys have good corrosion and oxidation resistance, maintain their strength well at high temperatures, have good weldability, and normally have good ductility and formability in spite of their HCP structure. The alpha alloys are annealed at high temperatures in the β region and then are cooled. Rapid cooling gives a fine acicular or needlelike α grain structure (Figure 13-17), whereas furnace cooling gives a more platelike structure.

Beta Titanium Alloys.   Although large additions of vanadium or molybdenum produce an entirely β structure at room temperature, none of the so-called beta alloys are actually alloyed to that extent. Instead, they are rich in β stabilizers,

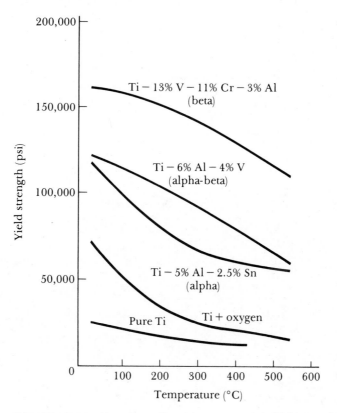

FIGURE 13-16 The effect of temperature on the yield strength of selected titanium alloys.

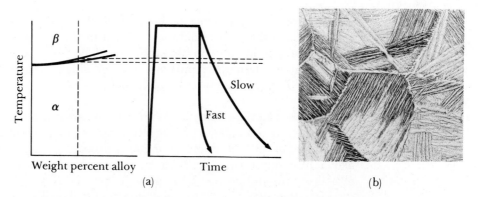

FIGURE 13-17 (a) Annealing and (b) microstructure of rapidly cooled alpha titanium (× 100). Both the grain boundary precipitate and the Widmanstatten plates are alpha. From *Metals Handbook*, Vol. 7, 8th Ed., American Society for Metals, 1972.

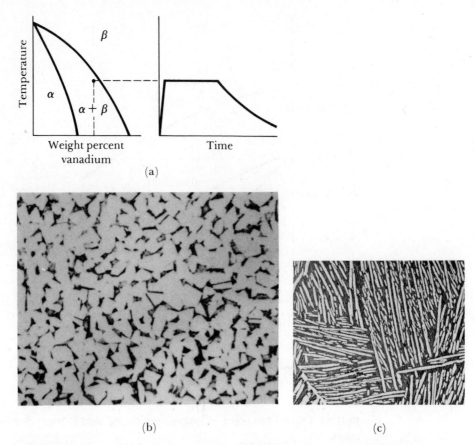

FIGURE 13-18   Annealing of an alpha-beta titanium alloy. (a) Annealing is done just below the α-β transus temperature, (b) slow cooling gives equiaxed α grains ($\times$ 250), and (c) rapid cooling gives needlelike α grains ($\times$ 2500). From *Metals Handbook*, Vol. 7, 8th Ed., American Society for Metals, 1972.

so that rapid cooling produces a metastable structure composed of all β. In the annealed condition, where only β is present in the microstructure, strength is derived from solid solution strengthening. The alloys can also be aged to produce higher strengths. Applications include high-strength fasteners, beams, and other fittings for aerospace applications.

Alpha-Beta Titanium Alloys.   With proper balancing of the α and β stabilizers, a mixture of α and β is produced at room temperature. Annealing provides a combination of high ductility, uniform properties, and good strength. The alloy is heated just below the β-transus temperature, permitting a small amount of α to remain and prevent grain growth (Figure 13-18). Slow cooling permits the formation of equiaxed α grains surrounding small islands of the β phase, while fast cooling, particularly from above the α-β transformation temperature, produces a needlelike alpha phase (Figure 13-18). When the equiaxed α struc-

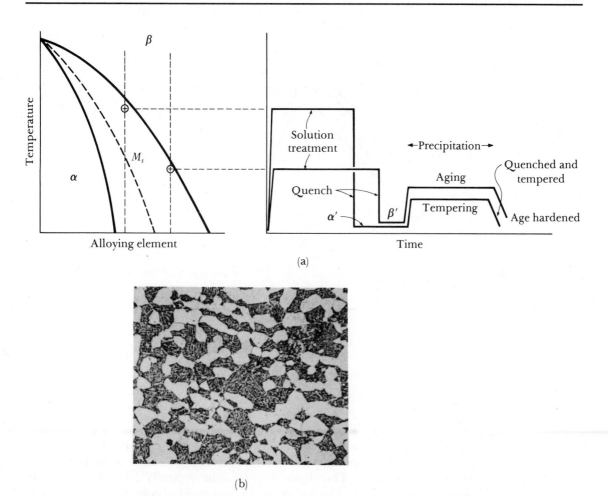

FIGURE 13-19   (a) Heat treatment and (b) microstructure of the alpha-beta titanium alloys. The structure contains primary α (large white grains) and a dark β matrix with needles of α formed during aging ($\times$ 250). From *Metals Handbook*, Vol. 7, 8th Ed., American Society for Metals, 1972.

ture is formed, cracks nucleate only with difficulty; when the acicular structure is produced, cracks nucleate more easily, but, because the cracks tend to follow the boundaries between the α and β, growth is more difficult. The "near alpha" alloys, which contain only a small amount of β, are usually used in the annealed condition.

More highly alloyed alpha-beta alloys can be heat treated to high strengths. The alloy is solution treated near the β-transus temperature (Figure 13-19). Next, the alloy is rapidly cooled to form a metastable supersaturated solid solution β' or titanium martensite α'. The alloy is then aged or tempered near 500°C. During aging, finely dispersed α and β precipitate from β' or α', increasing the strength of the alloy (Figure 13-20). Note that α' is a soft phase, just the opposite to the case of steel martensite.

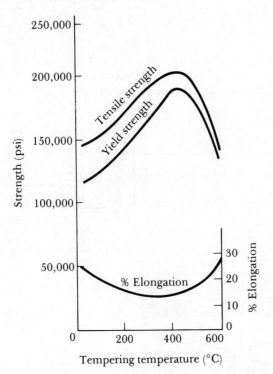

FIGURE 13-20 The effect of tempering on the properties of an α-β titanium alloy.

Normally, titanium martensite forms in the less highly alloyed alpha-beta alloys, whereas the supersaturated beta is more easily retained in alloys closer to the entirely β alloys. The titanium martensite typically has an acicular or needlelike appearance. During aging, the α precipitates in a Widmanstatten structure which improves the tensile properties as well as the fracture toughness of the alloy. Components for airframes, rockets, jet engines, and landing gear are typical applications for the heat-treated alpha-beta alloys.

◇ | **Example 13-12**

A Ti-8% Mn alloy is heat treated by holding at 705°C, water quenching, and reheating to 480°C. Describe the microstructure after (a) holding at 705°C and (b) reheating to 480°C.

Answer:

(a) Let's perform a lever law calculation at 705°C, which is just below the β-transus temperature at 780°C, using Figure 13-15(d).

$$\beta = \frac{8 - 0.5}{12 - 0.5} \times 100 = 65.2\%$$

$$\alpha = \frac{12 - 8}{12 - 0.5} \times 100 = 34.8\%$$

The alloy is solution treated below 780°C in order to retain some α in the microstructure, which in turn prevents grain growth of β. After quenching, the microstructure will remain the same.

(b) After aging at 480°C, the alloy is below the eutectoid and the stable phases are α and θ. If the aging step is long enough, the θ will precipitate from β and increase the strength.

 Example 13-13

Compare the approximate yield strength-to-weight ratio of the 13% V-11% Cr-3% Al beta titanium alloy with that of the 7075-T6 aluminum alloy.

Answer:

From Tables 13-4 and 13-13, we find that the yield strengths are 176,000 psi for the titanium alloy and 73,000 psi for the aluminum alloy. The ratios are

$$\text{Ti:} \quad \frac{176,000 \, \text{psi}}{4.505 \, \text{g/cm}^3} = \frac{176,000 \, \text{psi}}{0.163 \, \text{lb/in}^3} = 10.8 \times 10^5 \, \text{in.}$$

$$\text{Al:} \quad \frac{73,000 \, \text{psi}}{2.70 \, \text{g/cm}^3} = \frac{73,000 \, \text{psi}}{0.096 \, \text{lb/in}^3} = 7.6 \times 10^5 \, \text{in.}$$

The high strength and moderate density of the titanium give it an excellent strength-to-weight ratio.

Processing of Titanium Alloys. Titanium alloys are processed into useful shapes by casting, forming, and joining techniques. However, care must be taken to prevent contamination when the temperature of the alloy exceeds 535°C.

The alloys for titanium castings are melted in a vacuum furnace and poured into molds constructed from a ceramic material or graphite. Special precautions are also required to minimize contamination during welding. For example, gas-tungsten arc welding produces welds with the same resistance to corrosion as the base material. However, the welded area must be protected with an inert gas until the weld cools to below 535°C.

Some alloys, including the Ti-6% Al-4% V alpha-beta alloy, are superplastic and can be deformed as much as 1000%. During the slow process of deformation, the alloy is protected by an argon atmosphere. The superplastic forming can also be coupled with simultaneous diffusion bonding to produce complicated aircraft parts.

13–9   REFRACTORY METALS

The refractory metals—tungsten, molybdenum, tantalum, and niobium (or columbium)—have exceptionally high melting temperatures and consequently have the potential for high-temperature service. These metals find a variety of applications in aerospace and electronic components. Some important mechanical and physical properties of the refractory metals are given in Table 13-14.

Oxidation.   The refractory metals begin to oxidize between 200°C and 425°C and are rapidly contaminated or embrittled. Consequently, special precautions are required during casting, hot working, welding, or powder metallurgy. The metals must also be protected during service at elevated temperatures. For example, the tungsten filament in a light bulb is protected by a vacuum.

For some applications, the metals may be coated with a silicide or aluminide coating. The coating must (a) have a high melting temperature, (b) be compatible with the refractory metal, (c) provide a diffusion barrier to prevent contaminants from reaching the underlying metal, and (d) have a coefficient of thermal expansion similar to that of the refractory metal. Coatings are available that protect the metal to about 1650°C.

Forming Characteristics.   The refractory metals, which have the BCC crystal structure, display a ductile-to-brittle transition temperature. Because the transition temperatures for niobium and tantalum are below room temperature, these two metals can readily be formed. However, annealed molybdenum and tungsten normally have a transition temperature above room temperature, causing them to be brittle. Fortunately, if these metals are hot worked to produce a fibrous microstructure, the transition temperature is lowered and the forming characteristics are significantly improved (Figure 13-21).

Alloys.   Large increases in both room temperature and high-temperature mechanical properties are obtained by alloying. However, a limited number of alloys are available for each metal and the alloying elements are quite exotic.

TABLE 13-14   Properties of refractory metals

| Metal | Melting Temperature (°C) | Density (g/cm³) | Room Temperature | | | $T = 1000°C$ | |
			Tensile Strength (psi)	Yield Strength (psi)	% Elongation	Tensile Strength (psi)	Yield Strength (psi)
Nb	2470	8.66	45,000	20,000	25	17,000	8,000
Mo	2610	10.22	120,000	80,000	10	50,000	30,000
Ta	2996	16.6	50,000	35,000	35	27,000	24,000
W	3410	19.25	300,000	220,000	3	66,000	15,000

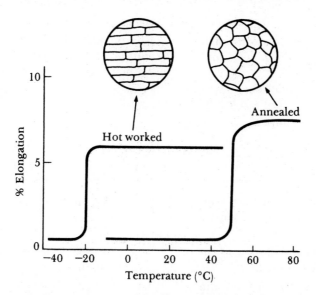

FIGURE 13-21 *The effect of temperature on the ductility of tungsten. Deformation reduces the transition temperature to below room temperature.*

These alloys typically are solid solution strengthened; in fact, tungsten and molybdenum form a complete series of solid solutions, much like copper and nickel. Some alloys, such as W-2% ThO_2, are dispersion strengthened by oxide particles during their manufacture by powder metallurgy processes.

SUMMARY

The nonferrous alloys take advantage of all of the strengthening mechanisms in materials. Depending on the characteristics of the base material, a tremendous range of properties can be obtained, including strong, lightweight materials such as aluminum, beryllium, and magnesium and high-temperature materials such as titanium, nickel, cobalt, and refractory metals.

GLOSSARY

Castability. The ease with which a metal can be poured into a mold to make a casting without producing defects or requiring unusual or expensive techniques to prevent casting problems.

Fluidity. The ability of liquid metal to fill up a mold cavity without prematurely freezing.

Nonferrous alloy. An alloy based on some metal other than iron.

Refractory metals. Metals having a melting temperature above that of titanium.

Superalloys. A group of nickel, iron-nickel, and cobalt alloys that have exceptional heat resistance, creep resistance, and corrosion resistance.

Tempers. A shorthand notation using letters and numbers to describe the processing of an alloy. H tempers refer to cold-worked alloys; T tempers refer to age-hardening treatments.

Wrought alloys. Alloys that are shaped by a deformation process.

PRACTICE PROBLEMS

1. Calculate the modulus of elasticity-to-density ratio for an Al-4% Li alloy and compare with the ratio for pure aluminum.

2. Explain why aluminum alloys containing more than 15% Mg are not used.

3. Determine the formula for the β phase in Al-Mn alloys.

4. Determine the formulas for the β and γ intermetallic compounds in the Al-Li system. Which of these is a stoichiometric intermetallic compound?

5. Calculate the amount of each phase in the Al-Li eutectic.

6. Estimate the secondary dendrite arm spacing for each structure in Figure 13-5 and, from Figure 8-8, estimate the solidification time obtained by each of the three casting processes. Do you expect higher strengths for die casting, permanent mold casting, or sand casting?

7. Would you expect a 2024-T9 aluminum alloy to be stronger or weaker than a 2024-T6 alloy? Explain.

8. Estimate the tensile strength expected for an 1100-H12 aluminum alloy.

9. Estimate the tensile strength expected for a 5056-H16 aluminum alloy.

10. Suppose you prepare an Al-8% Li alloy by rapid solidification processing to produce a single phase at room temperature. How much precipitate will form during aging?

11. How much Mg_2Al_3 (β) forms in a 518-T7 aluminum alloy?

12. Determine the designation for a magnesium alloy containing 4.3% Al and 11.9% Zn that has been hot rolled, cooled, and allowed to naturally age.

13. An HK31A-H24 magnesium alloy has a tensile strength of 38,000 psi. Would you expect that the alloy could have an H28 designation? Explain.

14. Suppose a 12 in. long round bar is to support a load of 500 lb without any permanent deformation. Calculate the minimum diameter of the bar if it is made of (a) ZK40A-T5 magnesium and (b) 390-F aluminum. Calculate the weight of the bar in each case.

15. A 24 in. long wire that is 0.1 in. in diameter is expected to elongate no more than 0.03 in. under load. What is the maximum force that can be applied if the wire is made of (a) aluminum and (b) beryllium?

16. A 27000-H10 copper alloy plate 0.25 in. thick is to be produced. Calculate the original thickness of the plate before rolling. Should the deformation be done by hot or cold working? Explain.

17. Figure 7-16 shows micrographs of a Cu-Zn alloy that has been annealed at different temperatures. Determine the "xxx" in the OSxxx designation obtained for each of the annealing temperatures.

18. Estimate the minimum tensile and yield strengths for a Cu-30% Zn alloy having the H10 temper. (See Figure 7-24.)

19. A 10100-OS050 copper alloy is cold extruded from a 0.45 in. diameter to a 0.23 in.

diameter. What is the approximate temper of the final bar?

20. Consider a Cu-20% Sn alloy. Describe the sequence by which the alloy freezes and cools to room temperature. Is it likely that the equilibrium $\alpha + \varepsilon$ structure will be produced?

21. We would like to produce a TQ temper in an aluminum bronze containing 13% Al. How much γ_2 precipitate will form during tempering at 450°C?

22. Suppose we would like to plastically deform a Cu-10% Pb alloy. Based on Figure 10-25, estimate the maximum temperature we should use during deformation. What would happen if we used a higher temperature?

23. Would you expect the fracture toughness of quenched and tempered aluminum bronze to be high or low? Explain. (You might compare this structure with some of those produced in $\alpha + \beta$ titanium alloys.)

24. Based on the micrograph in Figure 13-12, would you expect the γ' precipitate or the carbides to produce a greater strengthening effect in superalloys? Explain your answer.

25. Under equilibrium freezing, no eutectic microconstituent is expected in a Zn-27% Al alloy. However, some eutectic is generally found in the last regions to solidify. Using Figure 13-13, show how this nonequilibrium eutectic will form.

26. Under nonequilibrium cooling conditions, the centers of the dendrites that form in a Zn-27% Al alloy contain no eutectoid product, while a lamellar, pearlitic eutectoid structure is found near the edges of the dendrites. Using Figure 13-13, explain this behavior.

27. What is the eutectoid reaction that causes growth in the Zn-4% Al die casting alloys? What is the maximum temperature you would select to stabilize the die castings prior to finish machining?

28. When liquid Zn-27% Al alloy is very slowly cooled to permit solidification, shrinkage is found in the bottom half of the casting. In most other alloys, shrinkage is more commonly found

in the top half of the casting. Explain why the Zn-27% Al alloy is different.

29. By what mechanism does tin provide strengthening to titanium alloys?

30. When steel is joined using arc welding, only the liquid fusion zone must be protected by a gas or flux. However, when titanium is welded, both the front and back sides of the welded metal must be protected. Why must these precautions be taken when joining titanium?

31. Select an appropriate annealing temperature for a Ti-10% Mo alloy.

32. The Ti-V phase diagram is shown in Figure 13-22. For a Ti-10% V alloy, (a) estimate the solution-treating temperature that will control grain growth by retaining 10% α in the microstructure, (b) determine the phase(s) present after quenching to room temperature from the solution-treating temperature, (c) calculate the amount of each phase after reheating to produce equilibrium at 400°C, and (d) describe the microstructure after reheating at 400°C for 1 h. Is this an age-hardening treatment or a quench and temper treatment?

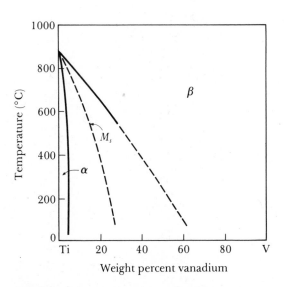

FIGURE 13-22 The titanium-vanadium phase diagram.

33. For a Ti-30% V alloy, (a) estimate the solution-treating temperature that will control grain growth by retaining 15% α in the microstructure, (b) determine the phase(s) present after quenching to room temperature from the solution-treating temperature, (c) calculate the amount of each phase after reheating to produce equilibrium at 400°C, and (d) describe the microstructure after reheating at 400°C for 1 h. Is this an age-hardening treatment or a quench and temper treatment?

34. What happens when the protective coating on a tungsten part expands more than the tungsten? What happens when the protective coating on a tungsten part expands less than the tungsten?

35. Compare the % elongation of the aluminum and copper alloys with that of the magnesium and titanium alloys. Which group generally tends to have the higher ductility? Why is this expected?

36. Determine the ratio between the yield strengths of the strongest aluminum, magnesium, copper, titanium, and nickel-copper alloys with the yield strengths of the pure metals. Compare the alloy systems and rank them in order of their response to strengthening mechanisms.

37. Determine the strength-to-weight ratios (using the yield strength) of the strongest aluminum, magnesium, copper, titanium, tungsten, and Monel alloys. Use the densities of the pure metals, in lb/in³, to form your ratios. Rank the alloy systems according to their strength-to-weight ratios.

38. Explain why aluminum and magnesium alloys are commonly used in aerospace and transportation applications, whereas copper and beryllium alloys are not.

39. Based on the phase diagrams, estimate the solubilities of nickel, zinc, silicon, aluminum, tin, and beryllium in copper at room temperature. Are these solubilities expected in view of Hume–Rothery's rules for solid solubility? Explain.

14

CERAMIC MATERIALS

14–1 INTRODUCTION

Ceramic materials, which are joined by ionic or covalent bonds, are complex compounds and solutions containing both metallic and nonmetallic elements. Ceramics typically are hard, brittle, high melting point materials with low electrical and thermal conductivity, good chemical and thermal stability, and high compressive strengths.

Ceramic materials have a wide range of applications, ranging from pottery, brick, tile, cooking ware, and soil pipe to glass, refractories, magnets, electrical devices, and abrasives. The tiles that protect the space shuttle are silica, a ceramic material. The newly discovered superconductive materials are also ceramics. In this chapter we will examine the structure of both crystalline and glassy ceramic materials, then summarize their processing, mechanical properties, and applications. In later chapters, the electrical, magnetic, thermal, and optical properties of ceramics will be discussed and contrasted with those of other materials.

14–2 REVIEW OF CRYSTALLINE CERAMIC STRUCTURES

In Chapter 3, a number of special structures produced as a result of ionic or covalent bonding restrictions were introduced. In ionically bonded ceramic materials, the arrangement of ions in the unit cells must satisfy ionic radius ratio requirements; this requires cations to fit into certain interstitial sites that provide the proper coordination. In addition, the ion arrangement within the unit cell must assure that a proper charge balance is obtained. A number of special unit cells, including the cesium chloride, sodium chloride, zinc blende, and fluorite structures, were discussed. These are summarized in Figure 14-1. In these crystalline unit cells, both short-range and long-range order of the ions are achieved.

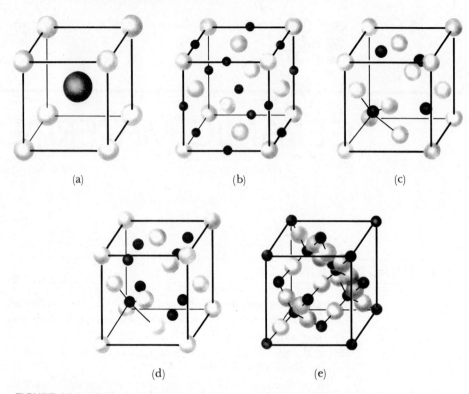

FIGURE 14-1 Typical crystal structures for ceramic compounds. (a) Cesium chloride, (b) sodium chloride, (c) zinc blende, (d) fluorite, and (e) cristobalite.

In addition, some ceramic materials are at least partly bonded by covalent bonding. An example of this is the cristobalite form of SiO_2, or silica, which is an exceptionally important raw material for ceramics. Again, the arrangement of the atoms in the unit cell must provide the proper coordination, maintain a charge balance, and in addition assure that the directionality of the covalent bonds is not violated. The cristobalite structure is also reviewed in Figure 14-1. In silica, or SiO_2, covalent bonding demands that the silicon atoms have four nearest neighbors—four oxygen atoms—thus creating a tetrahedral structure. The silicon-oxygen tetrahedra are the basic building blocks for silica, more complicated crystalline structures, and even glassy silicate structures.

Ionically bonded metal oxides are an important group of traditional ceramic materials. These compounds include MgO and CaO, both having the sodium chloride structure. They have relatively high melting temperatures (2800°C for MgO, 2570°C for CaO) and are often used in high-temperature refractory structures. A number of sulfides, chlorides, fluorides, and bromides also have the structures reviewed in Figure 14-1.

Perovskite Structure. Most of the ceramic materials have much more complicated crystal structures, however. The perovskite unit cell, which represents

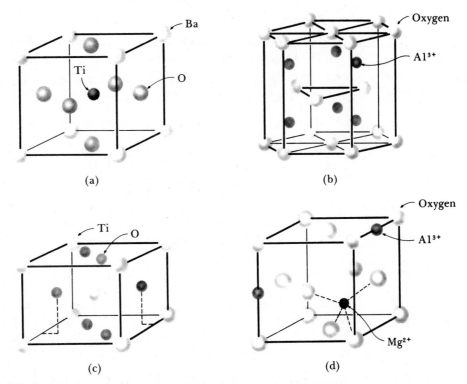

FIGURE 14-2   *Complex ceramic crystal structures. (a) Perovskite, (b) a portion of the corundum cell (two-thirds of the Al^{3+} sites are actually occupied), (c) rutile, and (d) spinel. Only one-eighth of the spinel unit cell is shown.*

one example of this increased complexity [Figure 14-2(a)], is found in several important electrical ceramics, such as $BaTiO_3$. In this structure, three different types of ions must be present. If barium ions are located at the corners of a cube, oxygen ions can be located at face-centered sites and titanium ions can be placed in body-centered sites. Distortion of the unit cell can produce an electrical signal, permitting some titanates to serve as transducers.

Corundum Structure.   One of the forms of alumina, Al_2O_3, has the corundum crystal structure, which is similar to a hexagonal close-packed structure; however, 12 aluminum ions and 18 oxygen ions are associated with each unit cell [Figure 14-2(b)]. Alumina is a common refractory and abrasive material.

◇ | **Example 14-1**

Corundum, or Al_2O_3, has a hexagonal unit cell [Figure 14-2(b)]. The lattice parameters for alumina are $a_0 = 4.75\,\text{Å}$ and $c_0 = 12.99\,\text{Å}$, and the density is about $3.98\,\text{g/cm}^3$. How many Al_2O_3 groups, Al^{3+} ions, and O^{2-} ions are present in a hexagonal prism having these dimensions?

Answer:

The molecular weight of alumina is $2(26.98) + 3(16) = 101.96 \, \text{g/g} \cdot \text{mole}$. The volume of the hexagonal prism is

$$V = a_0^2 c_0 \cos 30 = (4.75)^2(12.99) \cos 30 = 253.82 \, \text{Å}^3$$
$$= 253.82 \times 10^{-24} \, \text{cm}^3/\text{prism}$$

If x is the number of Al_2O_3 groups in the prism, then

$$3.98 = \frac{101.96x}{(253.82 \times 10^{-24})(6.02 \times 10^{23})}$$
$$x = \frac{(3.98)(253.82 \times 10^{-24})(6.02 \times 10^{23})}{101.96} = 6$$

Thus, a hexagonal prism having the dimensions given in the problem will contain 6 Al_2O_3 groups, with 12 aluminum ions and 18 oxygen ions.

Rutile Structure. The TiO_2, or rutile, crystal structure is a body-centered tetragonal structure [Figure 14-2(c)] with titanium ions at the corners and center of the cell and oxygen ions located at nonregular interstitial sites, causing distortion of the cell from cubic to tetragonal.

Spinel Structure. The spinel structure [Figure 14-2(d)], typical of $MgAl_2O_4$, has a cubic structure which can be viewed as consisting of eight smaller cubes. In each of the smaller cubes, the oxygen ions are located in normal face-centered cubic positions. Within the smaller cubes are four octahedral interstitial sites and eight tetrahedral interstitial sites; the Mg and both Al ions fit into three of these 12 available sites. In normal spinels, the divalent ions (like Mg) fit into tetrahedral sites and the trivalent ions (like Al) fit into octahedral sites. In inverse spinels, the divalent ion and half of the trivalent ions are located at octahedral sites. Many important electrical and magnetic ceramics, including Fe_3O_4 and other ferrites, have this structure.

A number of ceramic materials have recently become of critical importance as possible substitutes for metallic alloys for high-temperature applications and as portions of composite materials. These include many nitrides, carbides, and borides in which bonding is a combination of the metallic and ionic bonds. Tungsten carbide, WC, has a hexagonal structure, with tungsten atoms at the corners of the cell and a carbon atom at an interstitial location. One form of tungsten oxide, WO_2, has the rutile structure. Hafnium carbide (HfC) and hafnium nitride (HfN) have the sodium chloride structure. One of the forms of SiC has the zinc blende structure.

Graphite. Graphite, a crystalline form of carbon, is sometimes considered a ceramic material, even though carbon is an element rather than a combination of a metal and a nonmetal. Graphite has a hexagonal, layered structure (Figure 14-3).

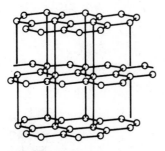

FIGURE 14-3 The hexagonal (but not HCP) structure of graphite.

◇ | **Example 14-2**

The perovskite structure of $TiCaO_3$ is shown in Figure 14-2. Does the Ti^{4+} ion in the center of the cell "rattle" around in its interstitial site? Calculate the size of the unit cell of $TiCaO_3$.

Answer:

Let's assume that the titanium ion is not present and determine the size of the interstitial hole. The ions will touch along a face diagonal in this case. The lattice parameter is

$$a_0 = \frac{2r_{Ca} + 2r_O}{\sqrt{2}} = \frac{2(0.99) + 2(1.32)}{\sqrt{2}} = 3.27\,Å$$

The size of the interstitial hole is

$$2r_{hole} = a_0 - 2r_O = 3.27 - 2(1.32) = 0.63$$
$$r_{hole} = 0.315\,Å$$

But the size of the hole, 0.315 Å, is less than that of the titanium ion, 0.68 Å. The titanium ion must push the surrounding ions apart.

Consequently, when the titanium ion is in place, the ions touch between the oxygen and titanium ions, and the actual lattice parameter is

$$a_0 = 2r_{Ti} + 2r_O = 2(0.68) + 2(1.32) = 4.00\,Å$$

14–3   SILICATE STRUCTURES

The silicate structures are based on the silica tetrahedron. Silica tetrahedra, SiO_4^{4-}, behave as ionic groups; the oxygen ions at the corners of the tetrahedra are attached to other ions, or one or more of the oxygen ions can be shared by two tetrahedral groups to satisfy the charge balance. Figure 14-4 summarizes these structures.

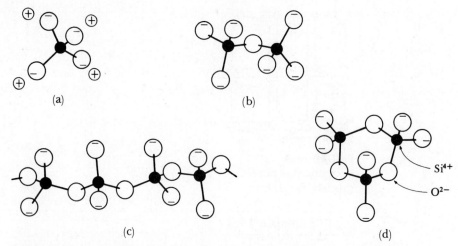

FIGURE 14-4 Arrangement of silica tetrahedra. (a) Orthosilicate island, (b) pyrosilicate island, (c) chain, and (d) ring. Positive ions are attracted to the silicate groups.

Silicate Compounds.   When two Mg^{2+} ions are available to combine with one tetrahedron, a compound Mg_2SiO_4, or forsterite, is produced. The two Mg^{2+} ions satisfy the charge requirements and balance the SiO_4^{4-} ions. The Mg_2SiO_4 groups in turn produce a three-dimensional crystalline structure. Similarly, Fe^{2+} ions can combine with silica tetrahedra to produce Fe_2SiO_4. Mg_2SiO_4 and Fe_2SiO_4 form a series of solid solutions known as *olivines* or *orthosilicates*.

Two silicate tetrahedra can combine by sharing one corner to produce a double tetrahedron, or a $Si_2O_7^{6-}$ ion. This ionic group can then combine with other ions to produce *pyrosilicate*, or double tetrahedron, compounds.

Ring and Chain Structures.   When two corners of the tetrahedron are shared with other tetrahedral groups, rings and chains with the formula $(SiO_3)_n^{2n-}$ form, where n gives the number of SiO_3^{2-} groups in the ring or chain. A large number of ceramic materials have this *metasilicate* structure. Wollastonite $(CaSiO_3)$ is built from Si_3O_9 rings; beryl $(Be_3Al_2Si_6O_{18})$ contains larger Si_6O_{18} rings; and enstatite $(MgSiO_3)$ has a chain structure.

Sheet Structures.   When the O:Si ratio gives the formula Si_2O_5, the tetrahedra combine to give sheet structures, including clay (Figure 14-5) and mica. Kaolinite, a common clay, is composed of a silicate sheet ionically bonded to a sheet composed of $AlO(OH)_2$, producing thin platelets of clay with the formula $Al_2Si_2O_5(OH)_4$. Montmorillonite, or $Al_2(Si_2O_5)_2(OH)_2$, contains two silicate sheets sandwiched around a central $AlO(OH)_2$ layer. The platelets are bonded to one another by weak Van der Waals bonds.

Silica.   Finally, when all four corners of the tetrahedra are shared with other silica tetrahedra, silica (SiO_2) is produced, as in cristobalite (Figure 14-1). Silica can exist in several allotropic forms. As the temperature increases, silica

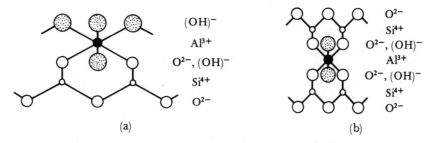

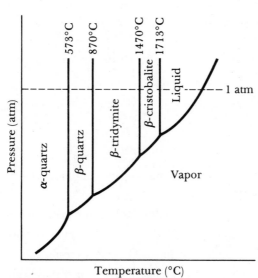

FIGURE 14-5   The silicate sheet structure that forms the basis for clays. (a) Kaolinite clay and (b) montmorillonite clay.

changes from α-quartz to β-quartz to β-tridymite to β-cristobalite to liquid. The pressure-temperature equilibrium diagram in Figure 14-6 shows the stable forms of silica.

The transformation from α-quartz to β-quartz is a *displacive transformation*. This transformation is identical to the martensite reaction; the quartz rapidly changes crystal structure by a slight distortion of the lattice involving second or further nearest neighbors (Figure 14-7). Similar transformations occur between different forms of tridymite and cristobalite. The higher temperature structure in the displacive type of transformation normally has a more open structure, a lower density, a higher heat capacity, and a more symmetrical crystal structure.

The transformation between β-quartz and β-tridymite and between β-tridymite and β-cristobalite are reconstructive rather than displacive. A *reconstructive transformation* requires that bonds between the atoms be broken and reestablished to produce a major change in crystal structure. Reconstructive

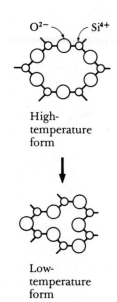

FIGURE 14-7
The displacive transformation in quartz.

FIGURE 14-6   The pressure-temperature phase diagram for SiO_2.

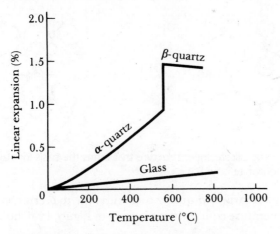

FIGURE 14-8 *The expansion of quartz. In addition to the regular, almost linear, expansion, a large, abrupt expansion accompanies a displacive transformation. However, glasses expand uniformly.*

transformations require a greater amount of energy, involve nucleation and growth, and occur more slowly than the displacive reaction.

When displacive transformations occur, there is an abrupt change in the dimensions of the ceramic crystal. These changes are shown in Figure 14-8 for quartz. High stresses and even cracking accompany these large volume changes in silica.

◇ | **Example 14-3**

Determine the type of silicate structure for each of the following complex ceramics.

$$CaO \cdot MnO \cdot 2SiO_2 \qquad Na_2O \cdot 2SiO_2 \qquad Sc_2O_3 \cdot 2SiO_2$$
$$3FeO \cdot Al_2O_3 \cdot 3SiO_2$$

Answer:

If we rearrange the chemical formulas of the ceramics, we can isolate the ratios of Si and O in the structures.

$$CaO \cdot MnO \cdot 2SiO_2 = CaMn(SiO_3)_2 \quad \text{or metasilicate}$$
$$Na_2O \cdot 2SiO_2 = Na_2(Si_2O_5) \quad \text{or sheet structure}$$
$$Sc_2O_3 \cdot 2SiO_2 = Sc_2(Si_2O_7) \quad \text{or pyrosilicate}$$
$$3FeO \cdot Al_2O_3 \cdot 3SiO_2 = Fe_3Al_2(SiO_4)_3 \quad \text{or orthosilicate}$$

14–4   IMPERFECTIONS IN CRYSTALLINE CERAMIC STRUCTURES

As in metals, the structures of ceramic materials contain a variety of imperfections.

Point Defects. Substitutional and interstitial solid solutions form in ceramic materials. The NiO–MgO system (Figure 9-7) and MgO–FeO system (Figure 9-20) display a complete series of substitutional solid solutions. Likewise, the orthosilicates, $(Mg,Fe)_2SiO_4$, display a complete range of solubility, with the Mg^{2+} and Fe^{2+} ions replacing one another completely in the silicate structure.

◇ | **Example 14-4**

Show that Mg_2SiO_4 and Fe_2SiO_4 should have complete solubility in one another.

Answer:

Both magnesium and iron have a valence of two and the crystal structures of the compound are identical, so we need to check the ionic radii. From Appendix B, $r_{Mg} = 0.66\,\text{Å}$ and $r_{Fe} = 0.74\,\text{Å}$.

$$\frac{\Delta r}{r_{Fe}} = \frac{0.74 - 0.66}{0.74} \times 100 = 10.8\%$$

Thus, the ionic radii of magnesium and iron are within 15% of one another and complete solid solubility might be expected.

Note that metallic Fe and Mg are not soluble—their atomic radii are very different, as are their crystal structures and electronegativities.

Often, the solid solubility of one phase in another is limited. Interstitial solid solutions are less likely in ceramics than in metals because the normal interstitial sites are already filled. For example, MgO (with the sodium chloride structure) has all of the octahedral sites filled and CaF_2 (the fluorite structure) has all of the tetrahedral sites filled.

Maintaining a balanced charge distribution is difficult when solid solution ions are introduced. However, charge deficiencies or excesses can be accommodated in ceramic materials in several ways. For example, if an Al^{3+} ion in the center of a montmorillonite clay platelet is replaced by a Mg^{2+} ion, the clay platelet has an extra negative charge. To equalize the charge, a positively charged ion, such as sodium or calcium, is adsorbed onto the surface of the clay platelet (Figure 14-9).

A second way of accommodating the unbalanced charge is to create vacancies. Normally, FeO contains equal numbers of Fe^{2+} and O^{2-} ions in the

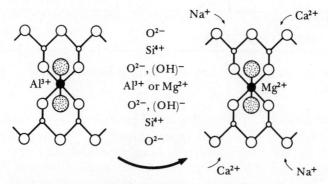

FIGURE 14-9 Replacement of an Al^{3+} ion by a Mg^{2+} ion in a montmorillonite clay platelet produces a charge imbalance which permits cations, such as sodium or calcium, to be attracted to the clay.

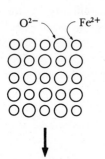

FIGURE 14-10 Formation of vacancies in FeO when ions with a different valence are substituted into the structure. To maintain equal charge, vacancies must be created.

sodium chloride structure. However, two Fe^{3+} ions can substitute for three Fe^{2+} ions, creating a vacancy where an iron ion normally would be (Figure 14-10).

◇ | **Example 14-5**

What wt % oxygen must be present in FeO to prevent any vacancies? What fraction of the iron sites are vacancies if FeO contains 25 wt % O?

Answer:

If no vacancies are present, the FeO should contain 50 at % Fe and 50 at % O.

$$wt\% \; O = \frac{(50 \, at\%)(16 \, g/g \cdot mole)}{(50 \, at\%)(16 \, g/g \cdot mole) + (50 \, at\%)(55.847 \, g/g \cdot mole)} \times 100$$

$$wt\% \; O = 22.3\%$$

If there is 25 wt % O present, then the atomic percent is

$$at\% \; O = \frac{\dfrac{25}{16 \, g/g \cdot mole}}{\dfrac{25}{16 \, g/g \cdot mole} + \dfrac{75}{55.847 \, g/g \cdot mole}} \times 100 = 53.8\%$$

$$at\% \; Fe = 100 - 53.8 = 46.2\%$$

Suppose we consider 100 oxygen ions. The number of Fe^{2+} and Fe^{3+} ions is

$$\frac{x}{46.2} = \frac{100}{53.8}$$

$$x = 86$$

The number of vacancies is $100 - 86 = 14$, so the fraction of the iron sites which are vacant is $\frac{14}{100} = 0.14$.

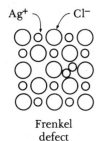

Frenkel
defect

Schottky
defect

FIGURE 14-11
The Frenkel and
Schottky defects
in ceramic crystal
structures.

We could also substitute more than one type of ion. For example, we can introduce a Li^+ ion and an Fe^{3+} ion to substitute for two Mg^{2+} ions in MgO. In this mechanism, vacancies do not need to be created.

Vacancies may also be present as Frenkel defects or Schottky defects (Figure 14-11). The Frenkel defect occurs when an ion leaves its normal position and a vacancy remains. The Schottky defect is a pair of vacancies—a cation vacancy and an anion vacancy.

Line Defects or Dislocations. Dislocations are observed in some ceramic materials, including LiF, sapphire (Al_2O_3), and MgO crystals. Even at high temperatures, however, the ceramic generally fails in a brittle manner before any appreciable slip and plastic deformation occur.

Surface Defects. Although stacking faults, small angle grain boundaries, and twin boundaries are observed in some ceramics, grain boundaries and the surface of the ceramic particles are normally much more important (Figure 14-12).

Typically, ceramics with a small grain size have improved mechanical properties compared with coarse-grained ceramics. Finer grain sizes help reduce stresses at grain boundaries due to anisotropic expansion and contraction. Normally, a fine grain size is produced by beginning with finer ceramic raw materials.

Particle surfaces, which represent planes of broken, unsatisfied covalent or ionic bonds, are reactive. Gaseous molecules, for example, may be adsorbed onto the surface to reduce the surface energy. In clay deposits, foreign ions may be attracted to the platelet surface (Figure 14-13), altering the composition of the clay.

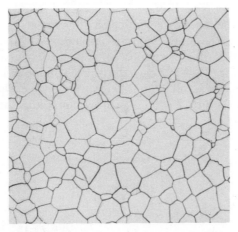

FIGURE 14-12 Grain structure in PLZT, a lead-lanthanum-zirconium titanate used as a ceramic sensor material (× 600). Courtesy of G. Haertling.

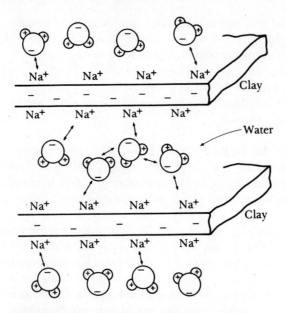

FIGURE 14-13 *The particle surface is important in the behavior and use of clays, adsorbing other ions and molecules and permitting the moist clay to bind coarser materials into ceramic bodies.*

14–5 GLASSES AND OTHER NONCRYSTALLINE CERAMIC MATERIALS

The most important of the noncrystalline ceramic materials are glasses. A glass is a solid material that has hardened and become rigid without crystallizing. A glass in some ways resembles an undercooled liquid. However, below the glass transition temperature (Figure 14-14), the rate of volume contraction on cooling is reduced and the material can be considered a glass rather than an undercooled liquid. The glassy structures are produced by joining together silica tetrahedra

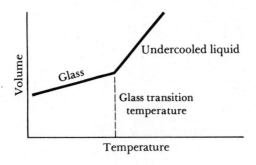

FIGURE 14-14 *The volume of silica versus temperature. The change in slope occurs at the glass transition temperature and shows the transition from an undercooled liquid to a glass.*

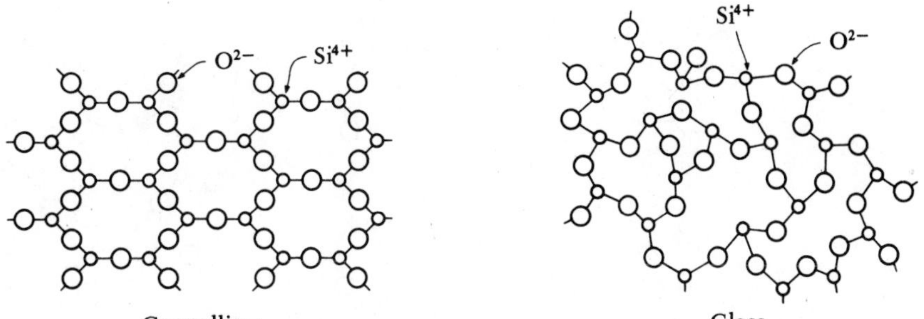

FIGURE 14-15 *Crystalline versus glassy silicate structures. Both structures have a short-range order, but only the crystalline structure has a long-range order.*

or other ionic groups to produce a solid but noncrystalline framework structure (Figure 14-15).

We can also find noncrystalline structures in exceptionally fine powders, as in gels or colloids. In these materials, particle sizes may be 100 Å or less. These amorphous materials, which include some cements and adhesives, are produced by condensation of vapors, electrodeposition, or chemical reactions.

Silicate Glasses.   The silicate glasses are the most widely used. Fused silica, formed from pure SiO_2, has a high melting point and the dimensional changes during heating and cooling are small (Figure 14-8). Generally, however, the silicate glasses contain additional oxides that act as glass formers, intermediates, or modifiers (Table 14-1). Oxides such as silica that have a high bond strength behave as *glass formers*. An *intermediate* oxide, such as lead or aluminum oxide, has

TABLE 14-1   Division of the oxides into glass formers, intermediates, and modifiers, and a comparison of the bond strengths

Metal in Oxide	Bond Strength (kcal/mol)	Metal in Oxide	Bond Strength (kcal/mol)	Metal in Oxide	Bond Strength (kcal/mol)
Glass formers		**Intermediates**		**Modifiers**	
B	119	Ti	73	La	58
Si	106	Zn	72	Y	50
Ge	108	Pb	73	Sn	46
P	99	Al	60	Ga	45
V	99	Be	63	In	43
As	73	Zr	61	Pb	39
Sb	76	Cd	60	Mg	37
Zr	81			Ca	32
				Ba	33
				Sr	32
				Na	20
				K	13
				Cs	10

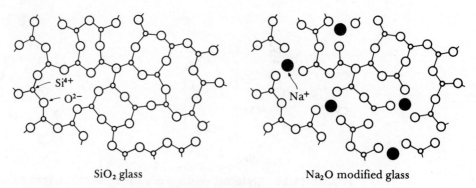

SiO$_2$ glass   Na$_2$O modified glass

FIGURE 14-16   The effect of Na$_2$O on the silica glass network. Soda is a modifier, disrupting the glassy network and reducing the ability to form a glass.

a lower bond strength; the intermediate oxide does not form a glass by itself but is incorporated into the network structure of the glass formers. A final group of oxides, the *modifiers*, have a low bond energy. Modifiers eventually cause the glass to devitrify, or crystallize.

Modified Silicate Glasses.   Modifiers break up the silica network if the oxygen to silicon ratio increases significantly. When adding Na$_2$O, for example, the sodium ions enter holes within the network rather than becoming part of the network. However, the oxygen ion that enters with the Na$_2$O does become part of the network (Figure 14-16). When this happens, there aren't enough silicon ions to combine with the extra oxygen ions and keep the network intact. Eventually, a high O : Si ratio causes the remaining silica tetrahedra to form chains, rings, or compounds and the silica no longer transforms to a glass. When the O : Si ratio is above about 2.5, silica glasses are difficult to form and above a ratio of three, a glass forms only when special precautions are taken, such as use of rapid cooling rates.

◇ | **Example 14-6**

How much Na$_2$O can be added to SiO$_2$ before the O : Si ratio exceeds 2.5 and the glass-forming tendencies are impaired?

Answer:

Let f_{Na_2O} be the mole fraction of Na$_2$O added to the glass. Then

$$\frac{\left(1\,\dfrac{\text{O ion}}{\text{Na}_2\text{O}}\right)(f_{Na_2O}) + \left(2\,\dfrac{\text{O ions}}{\text{SiO}_2}\right)(1 - f_{Na_2O})}{\left(1\,\dfrac{\text{Si ion}}{\text{SiO}_2}\right)(1 - f_{Na_2O})} = 2.5$$

$$\frac{f_{Na_2O} + 2(1 - f_{Na_2O})}{1 - f_{Na_2O}} = 2.5$$

$$f_{Na_2O} = 0.33 \qquad f_{SiO_2} = 0.67$$

$$\text{wt \% Na}_2\text{O} = \frac{(0.33)\,(\text{mol wt Na}_2\text{O})}{(0.33)\,(\text{mol wt Na}_2\text{O}) + (0.67)\,(\text{mol wt SiO}_2)} \times 100$$

$$= \frac{(0.33)\,(61.98)}{(0.33)\,(61.98) + (0.67)\,(60.08)} \times 100$$

$$= 34\%$$

$$\text{wt \% SiO}_2 = 66\%$$

◇

Nonsilicate Glasses. Glasses produced from BeF_2, GeO_2, aluminum phosphate, and boron phosphate also have a tetrahedral short-range structure. However, borate, B_2O_3, glass is formed by combining triangular units rather than tetrahedra. Some glasses are formed by combining silica and borate. Pyrex, for example, contains appreciable quantities of B_2O_3 in silica.

◇ **Example 14-7**

Determine the $O:Si$ ratio when 20 wt % B_2O_3 is added to SiO_2.

Answer:

The molecular weights of B_2O_3 and SiO_2 are 69.62 and 60.08 g/g · mole, respectively. We can calculate the mole fraction of B_2O_3.

$$f_{B_2O_3} = \frac{\dfrac{\text{wt \% B}_2\text{O}_3}{\text{mol wt B}_2\text{O}_3}}{\dfrac{\text{wt \% B}_2\text{O}_3}{\text{mol wt B}_2\text{O}_3} + \dfrac{\text{wt \% SiO}_2}{\text{mol wt SiO}_2}} = \frac{\dfrac{20}{69.62}}{\dfrac{20}{69.62} + \dfrac{80}{60.08}} = 0.177$$

$$\frac{O}{Si} = \frac{\left(3\,\dfrac{\text{O ions}}{B_2O_3}\right)(f_{B_2O_3}) + \left(2\,\dfrac{\text{O ions}}{SiO_2}\right)(1 - f_{B_2O_3})}{\left(1\,\dfrac{\text{Si ion}}{SiO_2}\right)(1 - f_{B_2O_3})}$$

$$\frac{O}{Si} = \frac{3(0.177) + 2(1 - 0.177)}{1(1 - 0.177)} = 2.65$$

A lead oxide glass forms if at least some silica or borate is present in PbO. Although lead oxide is not a glass former itself, it enters the glassy network of SiO_2 so effectively as an intermediate that some glasses may contain as much as 90% PbO.

The formation and stability of glasses can be determined using transformation diagrams similar to CCT and TTT diagrams in steels. When a liquid ceramic is cooled too slowly, a transformation line will be crossed and nucleation of a crystal will begin. Modification of the glass, much like addition of alloying elements to a steel, will permit glass to form on slower cooling or will delay or

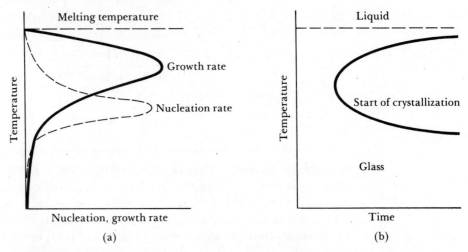

FIGURE 14-17 (a) The effect of temperature on the rate of nucleation and growth of crystals from a glass, and (b) the time required for a glass to begin to crystallize at various temperatures.

prevent devitrification. Figure 14-17(a) shows that when a glass is held at different temperatures, the rate of nucleation and growth of the solid ceramic crystals varies. At high temperatures, growth is rapid, but because few nuclei are available, long times are required for devitrification to begin. At low temperatures, many nuclei are available but growth is very slow, leading again to long total times. Figure 14-17(b) shows that, as a result of the nucleation and growth requirements, an intermediate temperature will give the shortest crystallization time. The rate of nucleation is generally more rapid at a temperature somewhat below that of the maximum crystal growth rate.

14–6   DEFORMATION AND FAILURE

Ceramic materials, both crystalline and noncrystalline, are generally considered to be very brittle, particularly at low temperatures. This brittle behavior is due to the inability of dislocations to easily move, a consequence of the large Burgers vector, the presence of relatively few slip systems, and the necessity to break strong ionic bonds and then force ions past oppositely charged ions during the slip process. Because slip and therefore plastic deformation are not likely to occur in ceramic materials, any crack that begins to form is not blunted by deformation of the material ahead of that crack. The crack instead continues to propagate.

Griffith Crack Theory.   Any crack or imperfection at the surface of a ceramic material not only produces brittle behavior but also limits the ability of the material to withstand a tensile stress. This is because any crack tends to concentrate and magnify the applied stress. The Griffith crack theory helps to

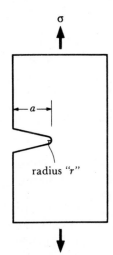

FIGURE 14-18
Schematic diagram of the Griffith crack in a ceramic.

explain this effect. Figure 14-18 shows a crack of length a at the surface of a brittle material. The radius r of the crack is also shown. When an applied tensile stress σ is applied, the actual stress acting at the crack tip is given by the Griffith equation

$$\sigma_{max} \cong 2\sigma\sqrt{a/r} \tag{14-1}$$

For very thin cracks (small r) or long cracks (large a), a significant concentration of the applied stress occurs, causing the crack to continue growing. Unless something is present in the structure to stop the crack growth, the stress concentration may continue to increase as the length of the crack increases. This same approach explains why the ceramic materials tend to have very low fracture toughness.

When a compressive stress is applied, the brittle behavior no longer imposes such a severe limit on the strength of the ceramic. Because compressive stresses try to close rather than open the crack, ceramics often have very good compressive strengths.

◇ **Example 14-8**

An advanced ceramic, sialon, has a tensile strength of 60,000 psi. What is the maximum tensile stress that can be applied to a thin crack 0.01 in. deep having a tip radius of 100 Å if fracture is not to occur?

Answer:

If the maximum stress in Equation 14-1 is greater than the tensile strength of 60,000 psi, the crack will propagate. Therefore

$$r = 100\,\text{Å} = 39.37 \times 10^{-8}\,\text{in.}$$
$$60,000 = 2\sigma\sqrt{0.01/39.37 \times 10^{-8}}$$
$$\sigma = 189\,\text{psi}$$

Toughening Methods. One of the important areas of research in ceramics is the development of tougher materials, leading to higher tensile and fracture strengths. Several methods have been devised to help solve this problem. One technique is to deliberately introduce compressive residual stresses at the surface of the material, where cracks are most likely to initiate. Production of *tempered glass* has long been used to obtain tough glasses; this is accomplished by rapidly cooling the hot glass. The surface of the glass cools first and contracts; later, the center cools and attempts to contract but is restrained from doing so by the rigid, strong surface. This produces high tensile stresses in the center but counterbalancing compressive stresses at the surface (Figure 14-19).

Another traditional method for improving toughness of ceramics is to surround the brittle ceramic particles with a softer, tougher matrix material.

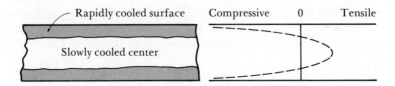

FIGURE 14-19 Tempered glass is cooled rapidly to produce compressive residual stresses at the surface.

This is done in producing *cermet* cutting tools and abrasives, which are really composite materials [Figure 14-20(a)]. As an example, very hard tungsten carbide (WC) particles are embedded in a matrix of cobalt metal. The composite part retains the high hardness and cutting ability of WC, but the softer, more ductile Co can deform and absorb energy. For high-temperature use, the hard ceramic often is embedded in a glassy matrix. As we will see shortly, glassy materials can deform and better absorb energy at elevated temperatures.

Another approach developed more recently is to create ceramic-ceramic composites by introducing ceramic fibers or agglomerates into the ceramic

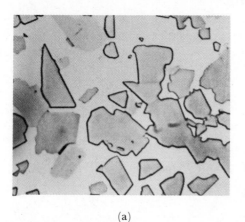

FIGURE 14-20 Microstructures of toughened ceramics. (a) A tungsten carbide–cobalt matrix cermet (× 1500), (b) a glass-ceramic composite reinforced with SiC fibers (× 100), and (c) transformation-toughened ZrO_2, containing plates of the tetragonal phase in a monoclinic matrix (× 15,000). From *Metals Handbook*, Vol. 9, 9th Ed., American Society for Metals, 1985.

(a)

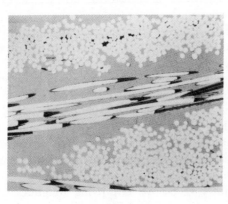

(b)

(c)

matrix. When a crack tries to propagate in the matrix, it encounters the interface between the matrix and the ceramic fiber, which helps block the propagation of the crack [Figure 14-20(b)].

Certain ceramic materials can be *transformation toughened*. In zirconia, for instance, the energy of a crack can be absorbed by a metastable phase present in the original structure. This absorption of the crack energy, which effectively blunts the crack growth, permits the metastable phase to transform into the more stable form and simultaneously helps to close the crack [Figure 14-20(c)]. Glasses can be made tougher by the same method; in one such approach, large, unstable tetragonal zirconia crystals are produced in a glass ceramic. The crystals undergo a spontaneous martensitic transformation to monoclinic crystals when a crack passing nearby stresses the unstable phase. Monoclinic crystals, larger and less symmetrical than tetragonal ones, essentially close the crack due to an expansion of 3% to 5% by volume, making it harder for the crack to grow. Such approaches may more than double the fracture toughness of the ceramic.

The processing of the ceramic is also critical in improving toughness. Improved processing techniques that produce exceptionally fine-grained, high-purity, completely dense ceramics also improve strength and toughness. Some of these techniques will be discussed in a later section. Another processing approach is to deliberately introduce many tiny microcracks which are too small themselves to propagate on their own; however, these tiny microcracks can help blunt other, larger cracks that may try to grow.

Viscous Flow of Glass. In a glass, deformation can occur by isotropic viscous flow if the temperature is sufficiently high. Groups of atoms, such as silicate islands, rings, or chains, move past one another in response to the stress, permitting deformation. The resistance to an applied stress is offered by the attraction between these groups of atoms. This resistance is related to the viscosity η of the glass, which in turn is dependent on temperature.

$$\eta = \eta_0 \exp\left(\frac{E_\eta}{RT}\right) \tag{14-2}$$

As the temperature increases, the viscosity decreases, viscous flow is easier, and the glass is more easily deformed. The activation energy E_η is related to the ease with which the atom groups move past one another. The addition of modifiers, such as Na_2O, breaks up the network structure, permits the atom groups to move more easily, reduces E_η, and reduces the viscosity and strength of the glass (Figure 14-21).

The viscosity of the glass is related to several critical processing temperatures. The *melting range*, in which the glass is fluid, occurs when the viscosity is very low, about 50 to 500 poise. The viscosities in the *working range* vary from 10^4 to 10^7 poise; in the working range, which occurs at temperatures just below the melting point, the glass can be deformed into useful shapes. At still lower temperatures, the *annealing point* occurs; here the viscosity is approximately 10^{13}

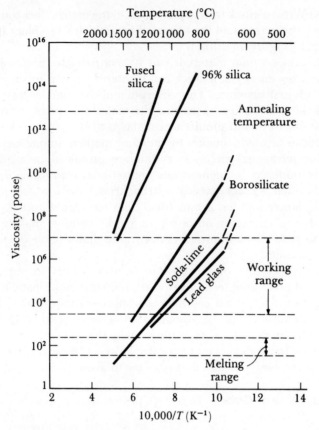

FIGURE 14-21 The effect of temperature and composition on the viscosity of glass.

poise and there may be just enough mobility of the glassy chains so that residual stresses can be reduced. At even lower temperatures, below the *strain point*, the glass is completely rigid.

◇ | **Example 14-9**

Estimate the activation energy for viscous flow in fused silica and soda-lime glass.

Answer:

From Figure 14-21 we can determine the slope of the curves for the two glasses. For SiO_2, two points are

$$\eta = 10^{10} \text{ at } \frac{1000}{T} = 0.6 \text{ or } T = 1667 \text{ K}$$

$$\eta = 10^{13} \text{ at } \frac{1000}{T} = 0.7 \text{ or } T = 1429 \text{ K}$$

Since

$$\ln \eta = \ln \eta_0 + \frac{E_\eta}{RT}$$

$$\ln 10^{10} = 23.025 = \ln \eta_0 + \frac{E_\eta}{1.987(1667)} = \ln \eta_0 + 0.0003 E_\eta$$

$$\ln 10^{13} = 29.933 = \ln \eta_0 + \frac{E_\eta}{1.987(1429)} = \ln \eta_0 + 0.00035 E_\eta$$

Subtracting,

$$23.025 - 29.933 = 0.0003 E_\eta - 0.00035 E_\eta$$

$$E_\eta = \frac{6.908}{0.00005} = 138,000 \, \text{cal/mol}$$

For soda-lime glass, two points are

$$\eta = 10^2 \text{ at } \frac{1000}{T} = 0.58 \quad \text{or} \quad T = 1724 \, \text{K}$$

$$\eta = 10^6 \text{ at } \frac{1000}{T} = 0.95 \quad \text{or} \quad T = 1053 \, \text{K}$$

$$\ln 10^2 = 4.605 = \ln \eta_0 + \frac{E_\eta}{1.987(1724)} = \ln \eta_0 + 0.00029 E_\eta$$

$$\ln 10^6 = 13.816 = \ln \eta_0 + \frac{E_\eta}{1.987(1053)} = \ln \eta_0 + 0.00048 E_\eta$$

$$E_\eta = \frac{9.211}{0.00019} = 48,500 \, \text{cal/mol}$$

◇

Creep in Ceramics. Because many ceramics are designed for high-temperature applications, creep resistance is an important property. The creep rate of the glasses is related closely to the viscosity and hence will increase exponentially with increasing temperature.

Creep rates of crystalline ceramics are not so dependent on dislocation climb as in metals; instead, grain boundary sliding, coalescence of microcracks, and porosity become more important. The presence of a noncrystalline phase, such as a glass, at the grain boundaries is particularly damaging. If solid crystals can be caused to precipitate in these glassy phases, the creep rate may be significantly reduced.

14–7 PHASE DIAGRAMS IN CERAMIC MATERIALS

The equilibrium binary phase diagrams for ceramic materials include solid solutions, miscibility gaps, and three-phase reactions. Tie lines and lever law calculations determine equilibrium compositions and amounts of phases. Some of the important phase diagrams are described in this section.

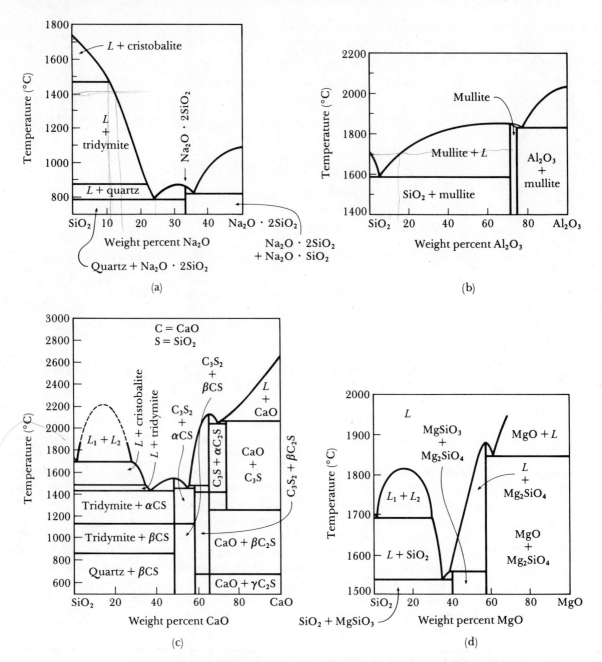

FIGURE 14-22 Binary ceramic phase diagrams with silica. (a) SiO_2–Na_2O, (b) SiO_2–Al_2O_3, (c) SiO_2–CaO, and (d) SiO_2–MgO.

SiO₂–Na₂O.   Additions of other oxides to silica act as fluxes and reduce the liquidus temperature of the ceramic. The most powerful is soda, or Na_2O, which reduces the melting temperature from $1720°C$ to as low as $800°C$ at the eutectic composition [Figure 14-22(a)]. The soda is often added to silica in glass-making processes.

◇ | **Example 14-10**

Glasses may be produced by melting silica and soda, then permitting the mixture to cool. What are the minimum glass-making temperatures for fused silica and for a window glass containing 15% Na_2O?

Answer:

From the silica-soda phase diagram, the melting point for fused silica, or pure SiO_2, is $1720°C$, and the liquidus temperature for SiO_2-15% Na_2O is $1240°C$. The silica-soda glass can be manufactured and shaped at a temperature nearly $500°C$ below that for fused silica.

◇

SiO₂–Al₂O₃.   The silica-alumina binary system shown in Figure 14-22(b) is the basis for many of the common clays and refractories. A eutectic occurs near 5% Al_2O_3, producing a low melting point material that is normally avoided. Even when rather high alumina contents are present, a substantial amount of liquid is present above $1595°C$.

A very important compound, mullite, is present in this system. Ceramics that contain substantial amounts of mullite have good resistance to high temperatures.

◇ | **Example 14-11**

We find that a SiO_2-Al_2O_3 fireclay brick can perform satisfactorily at $1700°C$ if no more than 20% liquid surrounds the mullite present in the microstructure. What is the minimum percent Al_2O_3 that must be in the refractory?

Answer:

Locate the liquid and mullite portion of the SiO_2–Al_2O_3 phase diagram. At $1700°C$, the compositions of the two phases are

L: 16% Al_2O_3 Mullite: 72% Al_2O_3

We can use the lever law to calculate the amount of alumina that gives 20% L.

$$\% \ L \ = \ \frac{72 - x}{72 - 16} \times 100 \ = \ 20$$

$$72 - x = (0.20)(72 - 16)$$
$$x = 60.8\% \ Al_2O_3$$

If the refractory contains between 60.8% and 72% Al_2O_3, less than 20% L forms at 1700°C.

SiO₂–CaO. The CaO reduces the melting temperature to as low as 1436°C [Figure 14-22(c)]. This *fluxing action* is used to advantage in the steel-making process. A stiff, high silica slag forms when iron is produced from iron ore. The addition of limestone, or $CaCO_3$, provides CaO that reduces the melting point and makes the slag less viscous.

SiO₂–MgO. The MgO lowers the melting point of SiO_2 a small amount [Figure 14-22(d)]. Some refractories for containing molten metal may be made from high MgO ceramics, taking advantage of the high melting temperatures of MgO and forsterite (Mg_2SiO_4).

CaO–Al₂O₃. These oxides are excellent fluxes for one another [Figure 14-23(a)], producing a low melting point liquid phase that rapidly attacks the remaining solid. If these two materials come into contact at high temperatures, they liquefy and fail rapidly.

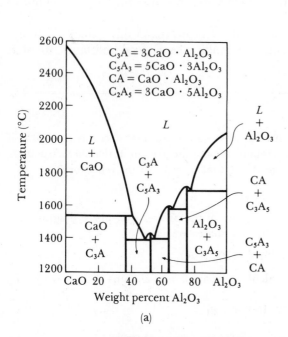

(a)

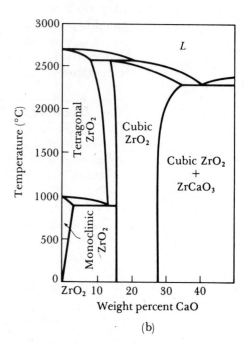

(b)

FIGURE 14-23 Phase diagrams for (a) $CaO–Al_2O_3$ and (b) $ZrO_2–CaO$.

ZrO$_2$–CaO. The high melting temperature of ZrO$_2$ is used for special refractories. However, pure ZrO$_2$ transforms from a tetragonal to a monoclinic structure on cooling, often causing cracks to form. Addition of CaO produces a cubic form of zirconia, which is stable at all temperatures, and improves the crack resistance [Figure 14-23(b)]. Other oxides, such as MgO, perform the same function as CaO.

◇ **Example 14-12**

Describe in detail the equilibrium solidification and cooling of a SiO$_2$-40% Al$_2$O$_3$ ceramic.

Answer:

The important temperatures for the 40% Al$_2$O$_3$ ceramic are the liquidus temperature (1810°C) and the eutectic temperature (1595°C).

$T > 1810°C$: The ceramic contains 100% L of composition 40% Al$_2$O$_3$.

$T = 1810°C$: Solid mullite begins to crystallize from the melt. The mullite contains 72% Al$_2$O$_3$.

$1595°C < T < 1810°C$: Mullite continues to grow and the composition of the liquid shifts towards 5% Al$_2$O$_3$. Just above the eutectic temperature, the amounts of each phase are

$$\text{Mullite} = \frac{40-5}{72-5} \times 100 = 52\%$$

$$\text{Liquid} = \frac{72-40}{72-5} \times 100 = 48\%$$

$T < 1595°C$: When the eutectic solidifies, cristobalite (containing 0% Al$_2$O$_3$) and more mullite form. The amounts of each phase and microconstituent are

$$\text{Mullite} = \frac{40-0}{72-0} \times 100 = 56\%$$

$$\text{Cristobalite} = \frac{72-40}{72-0} \times 100 = 44\%$$

$$\text{Primary mullite} = \frac{40-5}{72-5} \times 100 = 52\%$$

$$\text{Eutectic} = \frac{72-40}{72-5} \times 100 = 48\%$$

On subsequent cooling, the cristobalite may transform to tridymite or quartz.

CS = CaO · SiO$_2$ (wollastonite)
C$_3$S$_2$ = 3CaO · 2SiO$_2$
C$_2$S = 2CaO · SiO$_2$
C$_3$S = 3CaO · SiO$_2$
A$_3$S$_2$ = 3Al$_2$O$_3$ · 2SiO$_2$ (mullite)
C$_3$A = 3CaO · Al$_2$O$_3$
C$_{12}$A$_7$ = 12CaO · 7Al$_2$O$_3$
CA = CaO · Al$_2$O$_3$
CA$_2$ = CaO · 2Al$_2$O$_3$
CA$_6$ = CaO · 6Al$_2$O$_3$
CAS$_2$ = CaO · Al$_2$O$_3$ · 2SiO$_2$ (anorthite)
C$_2$AS = 2CaO · Al$_2$O$_3$ · SiO$_2$ (gehlenite)

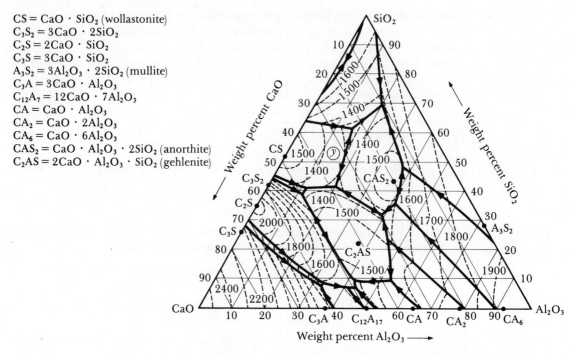

FIGURE 14-24   The liquidus plot for the CaO–SiO$_2$–Al$_2$O$_3$ ternary phase diagram.

CaO–SiO$_2$–Al$_2$O$_3$.   This ternary system includes the ceramics used as fluxes and slags for making steel, for cements, and for many enamels and glazes. Figure 14-24 shows the liquidus plot for the ternary system. Liquidus temperatures as low as 1170°C can be produced when the ceramic contains 15% Al$_2$O$_3$, 23% CaO, and 62% SiO$_2$.

SiO$_2$–CaO–Na$_2$O.   This ternary system is the basis for the soda-lime glasses, the most common of the commercial sheet and plate glasses. The silica-rich end of the phase diagram is shown in Figure 14-25. In practice, the solidification sequence does not follow the liquidus plot; instead, the modifying effect of the soda causes the liquid to cool as a glass.

◇ | **Example 14-13**

Compare the liquidus temperatures of soda-lime glasses of the following compositions:

Glass *A*:   74% SiO$_2$-13% CaO-13% Na$_2$O
Glass *B*:   74% SiO$_2$-6% CaO-20% Na$_2$O
Glass *C*:   80% SiO$_2$-7% CaO-13% Na$_2$O

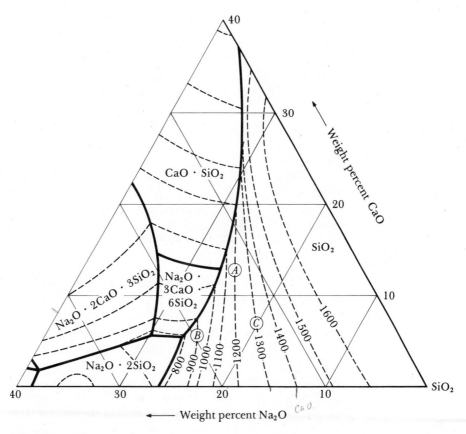

FIGURE 14-25 The liquidus plot for the SiO_2–CaO–Na_2O phase diagram.

Answer:

The locations of the three glasses are marked in Figure 14-25. The liquidus temperatures are

Glass A: Liquidus = 1200°C

Glass B: Liquidus = 900°C

Glass C: Liquidus = 1300°C

14–8 PROCESSING OF CERAMICS

Glasses are manufactured into useful articles by first producing a liquid, then cooling and shaping the liquid at a temperature where viscous flow is possible.

Crystalline ceramic materials are manufactured into useful articles by preparing a shape, or compact, composed of the raw materials in a fine powder

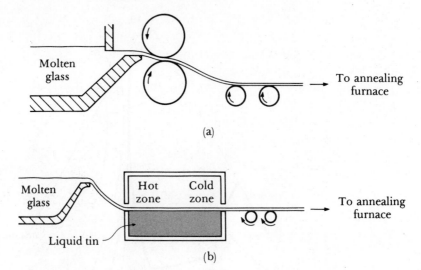

FIGURE 14-26 *Techniques for manufacturing sheet and plate glass. (a) Rolling, and (b) floating the glass on molten tin.*

form. The powders are then bonded using a variety of mechanisms, including chemical reaction, partial or complete vitrification, and sintering.

Forming Techniques for Glass.   Glass is formed at a high temperature, with the viscosity controlled so the glass can be shaped without breaking. Sheet and plate glass are produced when the glass is in the molten state, or melting range (Figure 14-21). Techniques include rolling the molten glass through water-cooled rolls or floating the molten glass over a pool of molten tin (Figure 14-26). The molten tin process produces an exceptionally smooth surface on the glass.

Some glass shapes, including large optical lenses, are produced by casting the molten glass into a mold.

Shapes such as containers or light bulbs can be formed by pressing, drawing, or blowing the glass into molds (Figure 14-27). The glass is heated to the working range so that the glass is formable but not "runny."

◇ | **Example 14-14**

Compare typical glass-forming temperatures for plate glass and for glass bottles made from pure silica versus soda-lime glasses.

Answer:

From Figure 14-21, we find that soda-lime glasses can be worked into bottles in a temperature range corresponding to

$$0.72 < \frac{1000}{T} < 1.0$$

or

$$T = \frac{1000}{0.72} = 1389\,\text{K} = 1116°\text{C} \quad \text{to} \quad T = \frac{1000}{1.0} = 1000\,\text{K} = 727°\text{C}$$

For making plate glass from soda-lime glass, the melting temperature range corresponds to

$$0.55 < \frac{1000}{T} < 0.62$$

so

$$T = \frac{1000}{0.55} = 1818\,\text{K} = 1545°\text{C} \quad \text{to} \quad T = \frac{1000}{0.62} = 1613\,\text{K} = 1340°\text{C}$$

By extrapolating the curve in Figure 14-21, fused silica might be worked at about

$$\frac{1000}{T} = 0.5 \quad \text{or} \quad T = \frac{1000}{0.5} = 2000\,\text{K} = 1727°\text{C}$$

However, exceptionally high temperatures would be needed to produce plate glass from pure SiO_2. The advantage of the soda and lime is evident.

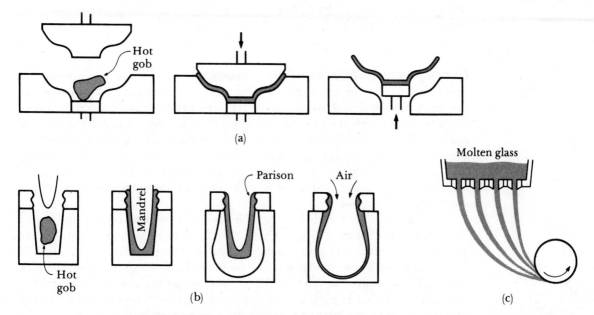

FIGURE 14-27 Techniques for forming glass products. (a) Pressing, (b) press and blow process, and (c) drawing of fibers.

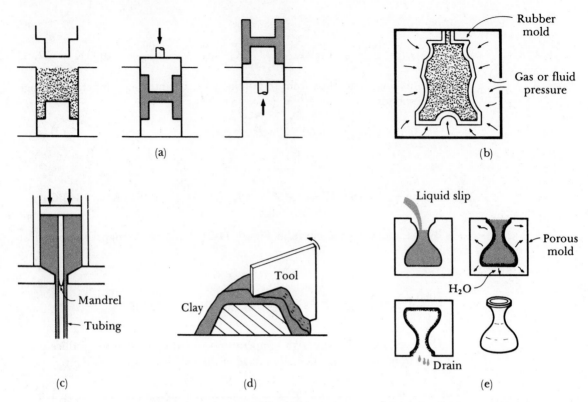

FIGURE 14-28 *Processes for shaping crystalline ceramics. (a) Pressing, (b) isostatic pressing, (c) extrusion, (d) jiggering, and (e) slip casting.*

Forming Techniques for Crystalline Ceramics.   Most crystalline ceramics are produced by a combination of a forming technique followed by high-temperature processing designed to develop the required structure and properties. These processes begin with the preparation of powders having a carefully controlled particle size. For traditional ceramics, the powders are usually prepared by grinding naturally occurring minerals; however, the more advanced ceramics, such as carbides, borides, and nitrides, require much more complex methods, including vapor-phase, laser, and plasma synthesis, for producing the proper chemistry of the powders.

The powders are blended, often mixed with water or an organic lubricant, and then formed into a shape (Figure 14-28). Dry or semidry mixtures are pressed to produce green compacts of suitable strength using mechanical or hydraulic compaction machines. For more uniform compaction of complex shapes, isostatic pressing may be done; the powders are placed into a rubber mold and subjected to high pressures through a gas or liquid medium. Higher moisture contents permit the powders to be more plastic or formable. Hydroplastic forming processes, including extrusion, jiggering, and hand working, can

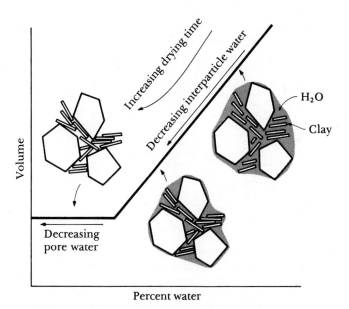

FIGURE 14-29 *The change in the volume of a ceramic body as the moisture is removed during drying. Dimensional changes cease after the interparticle water is gone.*

be done to these plastic mixes. Ceramic slurries containing large amounts of organic plasticizers can be injected into molds.

Even higher moisture contents permit the formation of a *slip*, or pourable slurry, containing the ceramic powder. The slip is poured into a porous mold. The water in the slip is drawn into the mold, leaving behind a soft solid which contains a low moisture content. When enough water has been drawn from the slip to produce a desired thickness of solid, the remaining liquid slip is poured from the mold. This leaves behind a hollow shell.

After cold forming, the ceramic bodies still are weak, contain water or other lubricants, and are porous. Subsequent drying and firing are then required.

Drying of Powder Compacts. After the ceramic bodies are formed, the excess moisture is removed. During drying, large dimensional changes occur (Figure 14-29). Most of the shrinkage occurs during the initial stages of drying as the interparticle water evaporates. The temperature and humidity are carefully controlled during drying to minimize stresses, distortion, or cracking.

Firing of Powder Compacts. During firing, the rigidity and strength of the ceramic increase. Firing, or sintering, causes additional shrinkage of the ceramic body as the pore size between the particles is reduced.

We must control four features of the final sintered microstructure—grain size, pore size, pore shape, and the amount of glass. We do this primarily by controlling the sintering temperature, the initial particle size, and fluxes.

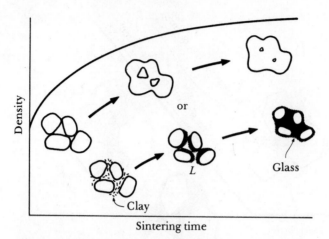

FIGURE 14-30 During sintering, diffusion produces bridges between the particles and eventually causes the pores to be filled in. Glass formation may also provide bonding.

During sintering, ions first diffuse along grain boundaries and surfaces to the points of contact between particles, providing bridging and connection of the individual grains (Figure 14-30). Further grain boundary diffusion closes the pores and increases the density, while the pores become more rounded. Finer initial particle sizes and higher temperatures accelerate the rate of pore shrinkage.

When the pores become so small that they no longer pin the grain boundaries, grain growth occurs. Further shrinkage of the pores requires volume diffusion, which is very slow. A second dispersed phase, which prevents grain growth, may permit fuller densification in shorter times.

Often during firing, *vitrification*, or melting, occurs. Fluxes or impurities in the ceramic body react with the remainder of the solids to produce a liquid phase at the grain surfaces. The liquid helps to eliminate porosity and changes to a glass after cooling. The presence of a glassy phase that serves as a binder is called the *ceramic bond*.

◇ | **Example 14-15**

A clay body is produced by mixing 100 lb of $CaO \cdot SiO_2$ with 50 lb of kaolinite clay, $Al_2O_3 \cdot 2SiO_2 \cdot 2H_2O$. The mixture is dried and fired at 1300°C until equilibrium is achieved. Determine the liquidus temperature.

Answer:

First we must determine the final composition of the ceramic after firing. We will need the molecular weights of the oxides to do this.

$$\text{Mol wt CaO} = 40.08 + 16 = 56.08 \text{ g/g} \cdot \text{mole}$$

$$\text{Mol wt SiO}_2 = 28.08 + 2(16) = 60.08 \text{ g/g} \cdot \text{mole}$$

$$\text{Mol wt Al}_2\text{O}_3 = 2(26.98) + 3(16) = 101.96 \text{ g/g} \cdot \text{mole}$$

$$\text{Mol wt H}_2\text{O} = 2(1) + 16 = 18.00 \text{ g/g} \cdot \text{mole}$$

In 100 lb of $CaO \cdot SiO_2$, the weight of each oxide is

$$CaO = \frac{56.08}{56.08 + 60.08} \times 100 = 48.3 \text{ lb}$$

$$SiO_2 = 100 - 48.3 = 51.7 \text{ lb}$$

In 50 lb of kaolinite, the weight of each oxide is

$$SiO_2 = \frac{2(60.08)}{2(60.08) + 101.96 + 2(18)} \times 50 = 23.3 \text{ lb}$$

$$Al_2O_3 = \frac{101.96}{2(60.08) + 101.96 + 2(18)} \times 50 = 19.7 \text{ lb}$$

$$H_2O = 50 - 23.3 - 19.7 = 7.0 \text{ lb}$$

After drying and firing, the water is driven off, so the total final weight of the ceramic is

$$\text{Total weight} = 100 + 50 - 7 = 143 \text{ lb}$$

The composition of the fired ceramic is

$$CaO = \frac{48.3}{143} \times 100 = 33.8\%$$

$$SiO_2 = \frac{51.7 + 23.3}{143} \times 100 = 52.4\%$$

$$Al_2O_3 = \frac{19.7}{143} \times 100 = 13.8\%$$

This composition is shown in Figure 14-24 as point y. The liquidus temperature is about 1340°C.

Hot Pressing and Reaction Bonding.   In some cases, particularly for the advanced ceramics, forming is done at high temperatures by hot pressing or hot isostatic pressing (HIP). In the HIP process, the powders are typically sealed in metal or glass containers prior to heating and pressing. This permits use of less lubricant and provides for at least some simultaneous sintering, resulting in parts that have less porosity and therefore higher mechanical properties. Hot-formed ceramics typically have better fracture toughness than ceramics produced by pressing and then sintering.

Some ceramics, such as Si_3N_4, can be produced by *reaction bonding*. Silicon is formed into a shape and then reacted with nitrogen at a high temperature to form the nitride. Reinforcement fibers such as SiC can be mixed with the silicon prior to reaction bonding to produce a composite. Although reaction bonding can be done at lower temperatures than hot pressing, lower densities and mechanical properties are obtained. For example, hot-pressed Si_3N_4 can have a flexural strength of 125,000 psi, compared with 30,000 psi for reaction bonding.

Sol Gel Processing.   The sol gel process can be used to produce and consolidate exceptionally fine, pure ceramic powders. A liquid colloidal solution is prepared which contains dissolved metallic ions. Hydrolysis reactions form an organometallic solution, or *sol*, composed of polymerlike chains containing the metallic ions and oxygen. Condensation of amorphous oxide particles occurs from the solution, producing a rigid gel. The gel is dried and then fired to provide sintering and densification of a final ceramic part; the sintering temperatures are low due to the highly reactive powders. Higher firing temperatures may permit the production of glasses and glass-ceramics. The sol gel process can be used to produce UO_2 for nuclear reactor fuels, perovskite structures such as barium titanate for electronic devices, alumina for high-strength structural applications, and a wide variety of other ceramics.

Porosity.   The pores in the fired ceramic may be either interconnected or closed. The *apparent porosity* measures the interconnected porosity and determines the permeability, or the ease with which gases and fluids seep through the ceramic component. The apparent porosity is determined by weighing the dry ceramic (W_d), then reweighing the ceramic both when it is suspended in water (W_s) and after it is removed from the water (W_w). Then

$$\text{Apparent porosity} = \frac{W_w - W_d}{W_w - W_s} \times 100 \tag{14-3}$$

The *true porosity* includes both interconnected and closed pores. The true porosity, which better correlates with the properties of the ceramic, is

$$\text{True porosity} = \frac{\rho - B}{\rho} \times 100 \tag{14-4}$$

where

$$B = \frac{W_d}{W_w - W_s} \tag{14-5}$$

B is the *bulk density* and ρ is the true density or specific gravity of the ceramic. The bulk density is the weight of the ceramic divided by its volume.

$\diamondsuit$ | **Example 14-16**

Silicon carbide particles are compacted and fired at a high temperature to produce a strong ceramic shape. The specific gravity of SiC is 3.2 g/cm³. The ceramic shape subsequently is weighed when dry (360 g), after soaking in water

(385 g), and suspended in water (224 g). Calculate the apparent porosity, the true porosity, and the fraction of the pore volume that is closed.

Answer:

$$\text{Apparent porosity} = \frac{W_w - W_d}{W_w - W_s} \times 100 = \frac{385 - 360}{385 - 224} \times 100 = 15.5\%$$

$$\text{Bulk density} = B = \frac{W_d}{W_w - W_s} = \frac{360}{385 - 224} = 2.24$$

$$\text{True porosity} = \frac{\rho - B}{\rho} \times 100 = \frac{3.2 - 2.24}{3.2} \times 100 = 30\%$$

The closed pore percentage is the true porosity minus the apparent porosity, or $30 - 15.5 = 14.5\%$. Thus

$$\text{Fraction closed pores} = \frac{14.5}{30} = 0.483$$

Cementation. By *cementation*, the ceramic raw materials are joined using a binder that does not require firing or sintering. A liquid resin, such as sodium silicate, aluminum phosphate, or Portland cement, coats the ceramic particles and provides bridges (Figure 14-31). A chemical reaction produces a solid that joins the particles together. Some typical cementation reactions are shown in Table 14-2. The locations of some typical cements in the CaO-SiO_2-Al_2O_3 phase diagram are shown in Figure 14-32.

Because cemented ceramic materials often have a high porosity and permeability, they may be used as ceramic filters. The binder systems are also used

FIGURE 14-31 A photograph of silica sand grains bonded with sodium silicate through the cementation mechanism (×60).

TABLE 14-2 Typical cementation reactions in ceramic systems

Plaster of paris

$$CaSO_4 \cdot \tfrac{1}{2}H_2O + \tfrac{3}{2}H_2O \rightarrow CaSO_4 \cdot 2H_2O$$

Calcined plaster of paris — Solid crystals

Calcium aluminate cement

$$3CaO \cdot Al_2O_3 + 6H_2O \rightarrow Ca_3Al_2(OH)_{12}$$

Tricalcium aluminate — Solid gel

Aluminum phosphate cement

$$Al_2O_3 + 2H_3PO_4 \rightarrow 2AlPO_4 + 3H_2O$$

Alumina — Phosphoric acid — Aluminum phosphate

Sodium silicate cement

$$xNa_2O \cdot ySiO_2 \cdot zH_2O + CO_2 \rightarrow glass$$

Liquid sodium silicate

Portland cement

$$3CaO \cdot Al_2O_3 + 6H_2O \rightarrow Ca_3Al_2(OH)_{12} + heat$$
$$2CaO \cdot SiO_2 + xH_2O \rightarrow Ca_2SiO_4 \cdot xH_2O + heat$$
$$3CaO \cdot SiO_2 + (x + 1)H_2O \rightarrow Ca_2SiO_4 \cdot xH_2O + Ca(OH)_2 + heat$$

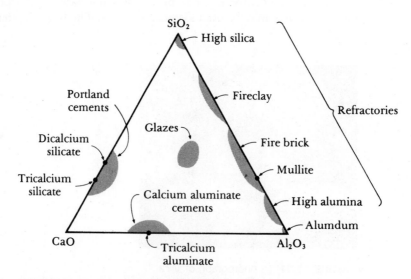

FIGURE 14-32 The approximate locations of typical cements, glazes, and refractories in the CaO–SiO$_2$–Al$_2$O$_3$ phase diagram.

to make molds for metal castings. The binder produces strong, rigid bonds between sand (silica) grains, yet permits mold gases to escape through the permeable mold rather than be trapped as gas defects in the casting.

Heat Treatment. Some ceramics are heat treated. Annealing, or heating the ceramic to high temperatures (Figure 14-21), reduces stresses. Large glass castings, for example, are often annealed and slowly cooled to prevent cracking. Glasses are also heat treated to cause *devitrification*, or precipitation of a crystalline phase from the glass.

Tempered glass is produced by quenching the surface of plate glass with air, causing the surface layers to cool and contract. When the center cools, its contraction is restrained by the already rigid surface, which is placed in compression. Prestressed glass is capable of withstanding much higher tensile stresses and impact blows than untempered glass.

Single Crystals. Single crystals of ceramic materials, which are useful for a variety of applications, including lasers and semiconductors, can also be produced. The Czochralski method, by which a single crystal is drawn from a molten bath, is one method sometimes used. A *hydrothermal* process can be used to produce single crystal silica; silica is dissolved in a water solution at high temperatures and pressures. The silica then precipitates out of solution as single crystals when the temperature is reduced.

14-9 APPLICATIONS AND PROPERTIES OF CERAMICS

There are a large variety of ceramic materials and applications for these materials.

Clay Products. Many ceramics are based primarily on clay to which a coarser material, such as quartz, and a flux material, such as feldspar, are added (Figure 14-33). Feldspars are a group of minerals including $(K,Na)_2O \cdot Al_2O_3 \cdot 6SiO_2$. The materials are mixed with water and the product is formed, dried, and fired. High clay contents improve the forming characteristics, permitting more complicated ceramic bodies to be produced. High feldspar contents reduce the liquidus temperature and consequently the firing temperature. To a certain degree, the silica is a "filler" material.

Brick and tile are pressed or extruded into shape, dried, and fired to produce a ceramic bond. Higher firing temperatures or finer original particle sizes produce more vitrification, less porosity, and higher density. The higher density improves mechanical properties but reduces the insulating qualities of the brick or tile.

Earthenware are porous clay bodies fired at relatively low temperatures; little vitrification occurs, the porosity is very high and interconnected, and

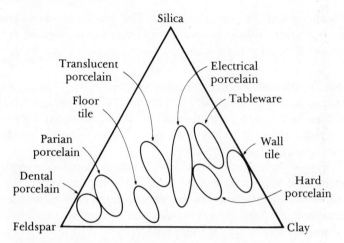

FIGURE 14-33   *The location of typical clay products in the silica-clay-feldspar phase diagram.*

the earthenware ceramics leak. These products must be covered with an impermeable glaze.

Higher firing temperatures, giving greater vitrification and less porosity, produce stoneware. The stoneware, which is used for drainage and sewer pipe, contains only 2% to 4% closed pores.

China and porcelain require even higher firing temperatures to cause complete vitrification and virtually no porosity.

Refractories.   Refractory materials must withstand high stresses at high temperatures. Most of the high melting point pure ceramic materials qualify as refractory materials; however, the pure oxide refractories are expensive and difficult to form into useful shapes. Instead, typical refractories are composed of coarse oxide particles, or *grog*, bonded by a finer refractory material. The finer material melts during firing, providing bonding. Typical refractory bricks contain about 20% to 25% apparent porosity, improving the thermal insulation.

The oxide refractory materials can be classed into three types—acidic, basic, and neutral (Table 14-3).

Common acidic refractories include fireclays, or silica-alumina ceramics. Pure silica is a good refractory material and is used to contain molten metal. In some applications, the silica may be bonded with small amounts of boron oxide, which melts and produces the ceramic bond.

When 3% to 8% alumina is added to silica, the ceramic material has a very low melting temperature and is not suited for refractory applications. Increasing the alumina content, perhaps by using more kaolinite clay, improves the refractoriness of the fireclay.

The fireclays are typically vitreous. In the high-alumina fireclays, however, substantial amounts of mullite form, permitting the refractories to have a

TABLE 14-3 Compositions of typical refractories

Refractory	SiO_2	Al_2O_3	MgO	Fe_2O_3	Cr_2O_3
Acidic					
Silica	95–97				
Superduty fire brick	51–53	43–44			
High-alumina fire brick	10–45	50–80			
Basic					
Magnesite			83–93	2–7	
Olivine	43		57		
Neutral					
Chromite	3–13	12–30	10–20	12–25	30–50
Chromite-magnesite	2–8	20–24	30–39	9–12	30–50

From *Ceramic Data Book*, Cahners Publishing Co., 1982.

good combination of high temperature resistance along with high hardness and mechanical properties. Figure 14-32 shows the locations of some typical refractories in the phase diagram.

Basic refractories include periclase (pure MgO), magnesite (MgO rich), dolomite (MgO plus CaO), and olivine (Mg_2SiO_4). The basic refractories are more expensive than the acidic refractories. However, in steel making and certain other high-temperature applications, the basic refractories must be used to provide compatibility with the metal.

Neutral refractories include chromite and chromite-magnesite. These refractories might be used to separate acidic and basic refractories, since the acidic and basic refractories attack one another.

Other refractory materials include zirconia (ZrO_2), zircon ($ZrO_2 \cdot SiO_2$), and a variety of nitrides, carbides, borides, and graphite. Most of the carbides, such as TiC and ZrC, do not resist oxidation well and their high-temperature applications are best suited to reducing conditions. However, silicon carbide is an exception—when SiC is oxidized at high temperatures, a thin layer of SiO_2 forms at the surface, protecting the SiC from further oxidation to about 1500°C. Nitrides and borides also have high melting temperatures and are less susceptible to oxidation. Some of the oxides and nitrides are possible materials for use in jet engines. Graphite is unique in that its strength increases as the temperature increases.

◇ | **Example 14-17**

How much kaolinite clay must be added to 100 g of quartz to produce a SiO_2-30% Al_2O_3 fireclay brick after firing?

Answer:

Kaolinite is $2SiO_2 \cdot Al_2O_3 \cdot 2H_2O$. Let's first assume that the H_2O has been driven off and find the amount of $2SiO_2 \cdot Al_2O_3$ required. The molecular weights of the oxides were obtained in Example 14-15. Suppose we let $x = $ the grams of $2SiO_2 \cdot Al_2O_3$. The weights of the oxides in the final refractory are

$$Al_2O_3 = (x) \frac{101.96}{101.96 + 2(60.08)} = 0.459x$$

$$SiO_2 = 100 + (x) \frac{2(60.08)}{101.96 + 2(60.08)} = 100 + 0.541x$$

Thus

$$30\% \ Al_2O_3 = \frac{0.459x}{0.459x + 100 + 0.541x} \times 100$$

$$(0.3)(x + 100) = 0.459x$$

$$x = \frac{30}{0.159} = 188.7 \ g \ 2SiO_2 \cdot Al_2O_3$$

The amount of clay, which includes water, is

$$Kaolinite = (188.7) \frac{222.12 + 2(18)}{222.12} = 219.3 \ g$$

◇

Electrical and Magnetic Ceramics.   Ceramics display a variety of useful electrical and magnetic properties. Some ceramics, including SiC, can serve as resistors and heating elements for furnaces. Other ceramics have semiconducting behavior and are used for thermistors and rectifiers.

Another group of ceramics, including barium titanate, display excellent dielectric, piezoelectric, and ferroelectric behavior. In particular, the piezo-electric properties of barium titanate make it an attractive material for capacitors and transducers.

Many of the clay bodies have excellent electrical insulation characteristics. Generally, electrical insulators should have a very low amount of porosity; thus, clay bodies that have completely vitrified, such as porcelain or glass, are used for electrical insulators for high voltages. However, when high temperature resistance at high frequencies is needed, as in automotive spark plugs, crystalline alumina performs better.

A new family of superconductors has been developed. Based on $YB_2Cu_3O_7$ and similar materials, superconductivity has been obtained at temperatures above the temperature of liquid nitrogen ($77°C$), rather than near absolute zero, as in previous materials.

Glasses.   Most of the commercial glasses are based on silica, with modifiers such as soda added to break down the network structure and reduce the

TABLE 14-4   Compositions of typical glasses

Glass	SiO_2	Al_2O_3	CaO	Na_2O	B_2O_3	MgO	PbO	Others
Fused silica	99							
Vycor	96				4			
Pyrex	81	2		4	12			
Glass jars	74	1	5	15		4		
Window glass	72	1	10	14		2		
Plate glass	73	1	13	13				
Light bulbs	74	1	5	16		4		
Fibers	54	14	16		10	4		
Thermometer	73	6		10	10			
Lead glass	67			6			17	10% K_2O
Optical flint	50			1			19	13% BaO, 8% K_2O, ZnO
Optical crown	70			8	10			2% BaO, 8% K_2O

After data collected by L. H. van Vlack, *Physical Ceramics for Engineers*, Addison-Wesley, 1964.

melting point. Calcium oxide is added to counteract the higher solubility of the glass in water, also caused by the soda. The most common commercial glass is soda-lime glass, containing approximately 75% SiO_2, 15% Na_2O, and 10% CaO. Although inexpensive, the soda-lime glasses have poor resistance to chemical attack and thermal shock. Table 14-4 compares the compositions of several typical glasses.

Borosilicate glasses, containing about 15% B_2O_3, have excellent chemical and dimensional stability; their uses include laboratory glassware (Pyrex) and perhaps the disposal of high-level radioactive nuclear waste. Aluminosilicate glass, with 20% Al_2O_3 and 12% MgO, and high-silica glasses, with 3% B_2O_3, are excellent for high-temperature resistance and for protection against heat or thermal shock. Fused silica, or virtually pure SiO_2, has the best resistance to temperature, thermal shock, and chemical attack, although it is also the most expensive.

Improved optical qualities compared with those of soda-lime glasses are obtained when the glass contains about 80% PbO. Particularly high quality fused silica is used for fiber-optic systems. Special optical qualities can also be obtained, including sensitivity to light. Photochromic glass, which is darkened by the ultraviolet portion of sunlight, is used for sunglasses. Photosensitive glass darkens permanently when exposed to ultraviolet light; if only selected portions of the glass are exposed and then immersed in hydrofluoric acid, etchings can be produced. Polychromatic glasses are sensitive to all light, not just ultraviolet radiation.

Glass-Ceramics.   Glass-ceramics are partially crystalline, partially glassy ceramics produced in such a way that virtually no pores remain in the microstructure. This is accomplished by initially forming the part from liquid glass to which special nucleating agents, such as TiO_2, have been added. The glass is then heat treated at a lower temperature to encourage the nucleation of exceptionally small and numerous crystals. These crystals are then allowed to grow until perhaps as much as 90% of the part has crystallized. As a consequence of this special structure, good mechanical strength and toughness can be achieved with a very low coefficient of thermal expansion and very good corrosion resistance at high temperatures. Some of these materials can even be machined using the same procedures used for metals, which is unusual considering the high hardness of ceramics.

Perhaps the most important glass-ceramic is based on the Li_2O–Al_2O_3–SiO_2 system. These materials are used for cooking utensils and ceramic tops for stoves. Other glass-ceramics are used in communication and computer applications.

Enamels and Glazes.   Glazes are added to the surface to make the ceramics impermeable, to provide protection and decoration, or for special purposes. Enamels are applied to metals. The enamels and glazes are clay products that vitrify easily during firing. A common composition is $CaO \cdot Al_2O_3 \cdot 2SiO_2$ (Figure 14-32).

Special colors can be produced in glazes and enamels by the addition of other minerals. Zirconium silicate gives a white glaze, cobalt oxide makes the glaze blue, chromium oxide produces green, lead oxide gives a yellow color, and a red glaze may be produced by adding a mixture of selenium and cadmium sulfides.

One of the problems encountered with a glaze or enamel is surface cracking, or crazing, which occurs when the glaze has a coefficient of thermal expansion different from that of the underlying ceramic material.

Fibers.   Fibers are produced from ceramic materials for several uses—as a reinforcement in composite materials, for weaving into fabrics, or for use in fiber-optic systems. Borosilicate glass fibers are the most commonly produced fibers and provide strength and stiffness in fiberglass. However, fibers can be produced from a variety of other ceramics, including alumina, silicon carbide, and boron carbide.

A special type of fibrous material is the silica tile used to provide the thermal protection system for the space shuttle. Silica fibers are bonded with a silica powder to produce an exceptionally lightweight tile with densities as low as $0.144 \, g/cm^3$; the tile is coated with special high-emissivity glazes to permit protection up to 1300°C.

Slags and Fluxes.   When metals are melted or refined, oxides are produced that float to the surface of the molten metal as a *slag*. The slags help purify or refine the metal, provided that the composition of the slag is properly adjusted. Consequently, ceramic fluxes are added to the slag for several reasons.

1. To reduce the melting temperature so the slag is more fluid.
2. To assure that a good refining action is produced; for example, CaO and CaF_2 remove sulfur from molten steel and cast iron.

3. To assure that the refractories containing the molten metal are not attacked. If acidic refractories are used to contain metal, acid oxides such as SiO_2 or Al_2O_3 may be added to assure that the liquid slag is also acid and compatible with the refractory.

◇ **Example 14-18**

Suppose a high-alumina refractory containing SiO_2-72% Al_2O_3 (mullite) is used to contain liquid cast iron at 1500°C. Large amounts of CaO are added to the iron in an attempt to lower the sulfur content of the iron. What effect will the CaO have on the refractory?

Answer:

From Figure 14-24, we see that CaO additions reduce the liquidus temperature of the refractory, which normally is 1850°C. A straight line drawn on the phase diagram from mullite to CaO shows that the liquidus temperature falls to about 1790°C when 10% CaO dissolves in the refractory, to 1650°C for 20% CaO, to 1550°C for 30% CaO, and to 1480°C for 50% CaO. The fluxing action of the basic CaO can eventually cause the acid SiO_2-Al_2O_3 refractory to fail.

14-10 ADVANCED CERAMIC MATERIALS

Advanced ceramics, also called technical, engineering, or structural ceramics, include carbides, borides, and nitrides as well as some oxides, and are used in relatively pure form. Often these materials are selected for both mechanical and physical properties (Table 14-5).

Alumina (Al_2O_3) is a commonly used crucible material used to contain molten metal or to operate at high temperatures where good strength is required. One classical example is for use as insulators in spark plugs. Some unique applications are also being found in dental and medical use, including restoration of teeth, bone filler, or orthopedic implants. Specially doped alumina can serve as a laser.

Boron carbide (B_4C) is somewhat unusual because it is very hard yet unusually lightweight. It finds uses in applications requiring excellent abrasion resistance and as a portion of bulletproof armor plate, although it has rather poor properties at high temperatures.

Silicon carbide (SiC) provides outstanding oxidation resistance at temperatures even above the melting point of steel. SiC often is used as a coating for metals, carbon-carbon composites, and other ceramics to provide protection at these extreme temperatures. SiC is also used as a particulate and fibrous reinforcement in both metal matrix and ceramic matrix composites (Figure 14-34).

TABLE 14-5   Mechanical properties of selected advanced ceramics

Material	Density (g/cm^3)	Tensile Strength (psi)	Flexural Strength (psi)	Compressive Strength (psi)	Young's Modulus (psi)	Fracture Toughness (psi · in$^{1/2}$)
Al$_2$O$_3$	3.98	30,000	80,000	400,000	56 × 10^6	5
SiC (sintered)	3.1	25,000	80,000	560,000	60 × 10^6	4
Si$_3$N$_4$ (reaction bonded)	2.5	20,000	35,000	150,000	30 × 10^6	3
Si$_3$N$_4$ (hot pressed)	3.2	80,000	130,000	500,000	45 × 10^6	5
Sialon	3.24	60,000	140,000	500,000	45 × 10^6	9
ZrO$_2$ (partially stabilized)	5.8	65,000	100,000	270,000	30 × 10^6	10
ZrO$_2$ (transformation toughened)	5.8	50,000	115,000	250,000	29 × 10^6	11

Silicon nitride (Si$_3$N$_4$) has similar properties to SiC, although its oxidation resistance and high-temperature strength are somewhat lower. Both silicon nitride and silicon carbide are likely candidates for components for automotive and gas turbine engines, permitting higher operating temperatures and better fuel efficiencies with less weight than traditional metals and alloys.

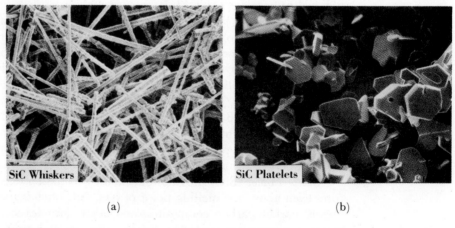

(a)   (b)

FIGURE 14-34 Silicon carbide reinforcement materials. (a) SiC whiskers and (b) SiC single crystal platelets. Courtesy of American Matrix, Inc.

Sialon ($Si_3Al_3O_3N_5$) is a relatively new advanced ceramic in which aluminum and oxygen are partially substituted for silicon and nitrogen in silicon nitride. The sialon crystals are typically embedded in a glassy phase based on Y_2O_3. The glassy phase is then allowed to devitrify by a heat treatment to improve the creep resistance. The result is a ceramic that is relatively lightweight with a low coefficient of thermal expansion, good fracture toughness, and a higher strength than many of the other common advanced ceramics. Sialon may find applications in engine components and other applications involving both high temperatures and wear conditions.

Zirconia (ZrO_2) is a particularly interesting advanced ceramic because of the unique way in which its properties can be controlled. Pure zirconia is polymorphic, with a tetragonal crystal structure transforming to a monoclinic structure by a martensitic reaction during cooling. There is a significant volume change of about 3% that accompanies this transformation and may lead to cracking of the zirconia parts.

By the addition of sufficient amounts of MgO, CaO, or Y_2O_3 to zirconia, a cubic crystal structure is produced that is stable at all temperatures, as shown in the phase diagram (Figure 14-23). This helps to prevent cracking because the phase transformation no longer occurs. This form of zirconia is called stabilized zirconia.

Urania (UO_2) is widely used as a nuclear reactor fuel. This material has exceptional dimensional stability because its crystal structure can accommodate the products of the fission process.

SUMMARY

The structure and properties of crystalline ceramic materials can be interpreted in terms of their complex crystal structures and phase diagrams. Because of their brittle behavior, they are often manufactured into useful components by pressing moist aggregates or powders into shapes, followed by drying and firing. This permits the particles to sinter and become solid. The crystalline ceramics typically have high melting temperatures and high hardness and are suitable for many high-temperature or corrosion-resistant applications.

The glassy ceramics have a nonequilibrium structure that, because of the effect on viscosity, permits us to shape the material into useful components by forming processes.

Modification of the structure of the ceramic materials cannot be interpreted in terms of slip, as we found in metals and alloys. Instead, additions to ceramics change the melting temperatures, influence the amount of glass that forms during sintering, or affect the physical properties of the ceramic. However, new processing techniques, including incorporation of ceramics into composite materials, are creating tough ceramics that may successfully compete with alternative materials.

GLOSSARY

Apparent porosity. The percent of a ceramic body that is composed of interconnected porosity.

Bulk density. The mass of a ceramic body per unit volume, including closed and interconnected porosity.

Cementation. Bonding ceramic raw materials into a useful product using binders that form a glass or gel without firing at high temperatures.

Ceramic bond. Bonding ceramic materials by permitting a glassy product to form at high firing temperatures.

Devitrification. The precipitation of a crystalline product from the glassy product, usually at high temperatures.

Displacive transformation. A change in crystal structure with changes in temperature or pressure without the necessity for significant amounts of diffusion.

Firing. Heating a ceramic body at a high temperature to cause a ceramic bond to form.

Flux. Additions to ceramic raw materials that reduce the melting temperature.

Glass-ceramics. Ceramic shapes formed in the glassy state and later allowed to crystallize during heat treatment to achieve improved strength and toughness.

Glass formers. Oxides with a high bond strength which easily produce a glass during processing.

Glass transition temperature. The temperature at which an undercooled liquid becomes a glass.

Griffith crack theory. Describes the influence of small cracks on the properties of brittle materials.

Grog. Coarse oxide particles that are bonded by finer minerals to produce refractory products.

Intermediates. Oxides that, when added to a glass, help to extend the glassy network, although the oxides normally do not form a glass themselves.

Metasilicates. A group of silicate structures having a ring or chain structure.

Modifiers. Oxides that, when added to a glass, disrupt the glassy network, eventually causing crystallization.

Orthosilicates. A group of silicate structures based on a single silicate tetrahedral unit. Also known as olivines.

Pyrosilicates. A group of silicate structures based on a pair of silicate tetrahedral units.

Reaction bonding. A ceramic processing technique by which a shape is made using one material which is later converted to a ceramic material by reaction with a gas.

Reconstructive transformation. Transformation from one crystal structure to another by a process requiring diffusion and rearrangement of atoms.

Refractories. A group of ceramic materials capable of withstanding high temperatures for prolonged periods of time.

Slag. An oxide product formed from a molten metal during melting or refining.

Slip. A liquid slurry that is poured into a mold. When the slurry begins to harden at the mold surface, the remaining liquid slurry is decanted, leaving behind a hollow ceramic casting.

Tempered glass. Glass that is prestressed during cooling to improve its strength.

Transformation toughening. Improving the toughness of ceramic materials by taking advantage of volume changes that accompany a polymorphic transformation induced by a crack.

True porosity. The percent of a ceramic body that is composed of both closed and interconnected porosity.

Viscous flow. Deformation of a glassy material at high temperatures.

Vitrification. Melting or formation of a glass.

PRACTICE PROBLEMS

1. Quartz (SiO_2) has a hexagonal crystal structure with lattice parameters of $a_0 = 4.913$ Å and $c_0 = 5.405$ Å and a density of 2.65 g/cm^3. Determine the number of silica (SiO_2) groups in quartz.

2. Determine (a) the direction in the unit cell along which ions are in contact in $CaTiO_3$, which has the perovskite structure, (b) the lattice parameter for the unit cell, and (c) the density of the unit cell.

3. The TiO_2 structure is shown in Figure 14-2(c). The lattice parameters for the tetragonal structure are $a_0 = 4.59$ Å and $c_0 = 2.96$ Å. (a) Determine the numbers of each ion in the unit cell, (b) calculate the packing factor for the unit cell, and (c) determine whether the oxygen and titanium ions touch along the [110] direction.

4. Tungsten carbide (WC) has a hexagonal structure with lattice parameters of $a_0 = 2.91$ Å and $c_0 = 2.84$ Å. If one W atom is located at each corner of the cell, estimate the density of the ceramic.

5. In TiO_2, oxygen ions are located at the following coordinates.

```
0.3056,  0.3056,  0
0.6944,  0.6944,  0
0.8056,  0.1944,  0.5
0.1944,  0.8056,  0.5
```

(a) Based on the radius ratios, what should be the coordination number for the titanium ions? (b) Based on the coordinates listed above and Figure 14-2(c), is the titanium ion at 0.5, 0.5, 0.5 equidistant from the number of oxygen ions predicted by this coordination number? (Use the lattice parameters from Problem 14-3 and calculate the actual distance between each oxygen ion and the central titanium ion.)

6. Determine whether the following are orthosilicate, pyrosilicate, metasilicate, or sheet types of ceramics.

$FeO \cdot SiO_2$ $CaO \cdot Al_2O_3 \cdot 2SiO_2$
$3MgO \cdot 2SiO_2$ $2MgO \cdot SiO_2$

$MgO \cdot CaO \cdot SiO_2$
$Al_2O_3 \cdot 2SiO_2$

7. Asbestos is a hydrated magnesium silicate that often has been used as a fibrous material in insulation and brake linings. Does this suggest that asbestos is an orthosilicate, pyrosilicate, metasilicate, or sheet type of material?

8. Is wollastonite ($CaSiO_3$) expected to have an orthosilicate, pyrosilicate, metasilicate, or sheet type of structure?

9. If 5% of the Al^{3+} ions in montmorillonite ($Al_2O_3 \cdot 4SiO_2 \cdot H_2O$) are replaced by Mg^{2+} ions, how many grams of K^+ ions will be attracted to 100 g of the clay?

10. Suppose 5% of the Fe^{2+} ions are removed when Fe^{3+} ions are introduced into FeO, which has the sodium chloride structure. Calculate (a) the number of vacancies per cubic centimeter, (b) the atomic percent oxygen in the FeO, and (c) the weight percent oxygen in the FeO.

11. Suppose 100 g of Mg_2SiO_4 are combined with 150 g of Fe_2SiO_4 to produce an olivine solid solution. Calculate the weight percent of Mg, Fe, Si, and O ions in the solid solution.

12. Suppose 25 mol % Na_2O is added to SiO_2. Calculate the O : Si ratio and determine whether this material will provide good glass-forming tendencies.

13. Suppose 25 wt % Na_2O is added to SiO_2. Calculate the O : Si ratio and determine whether this material will provide good glass-forming tendencies.

14. A glass composed of 70 mol % SiO_2, 15 mol % CaO, and 15 mol % Na_2O is produced. Determine the O : Si ratio of this glass. Does the material have good glass-forming tendencies?

15. How many grams of PbO can be added to 1 kg of SiO_2 before the O : Si ratio exceeds 2.5 and glass-forming tendencies are poor? Compare this with the number of grams of MgO that can be tolerated.

16. Alumina (Al_2O_3) has a tensile strength of about 30,000 psi. Calculate the minimum radius of a flaw 0.75 in. deep in order to prevent an applied stress of 12,000 psi from causing fracture.

17. Hot-pressed Si_3N_4 has a tensile strength of about 80,000 psi. If a crack 0.125 in. deep with a radius of 0.001 in. is present at the surface, what is the maximum allowable applied stress?

18. Partially stabilized ZrO_2 has a tensile strength of about 65,000 psi. A microcrack at the surface is 0.01 mm deep with a radius of 1000 Å. Is this crack likely to propagate when an applied stress of 3000 psi is applied? Show by suitable calculations.

19. From Figure 14-21, estimate the activation energy for viscous flow in a borosilicate glass.

20. A glass is found to have an annealing temperature of 700°C and enters the working range at or above 1200°C. Determine (a) the activation energy for viscous flow and (b) the approximate melting temperature for the glass.

21. Suppose you combine 5 mol of SiO_2 with 1 mol of Na_2O. Determine (a) the weight percent Na_2O in the ceramic, (b) the liquidus temperature of the ceramic, and (c) the amount of solid SiO_2 if the ceramic is held at 900°C long enough for the equilibrium phase to precipitate. (d) Explain in terms of the structure why this ceramic should have a lower activation energy for viscous flow than does pure SiO_2.

22. Calculate (a) the amount of mullite that must be combined with SiO_2 to produce an equilibrium structure of 30% liquid–70% mullite at 1800°C and (b) the weight percent Al_2O_3 present in this ceramic.

23. Suppose, when preparing forsterite (Mg_2SiO_4), that we add 10% more SiO_2 than is required. Determine (a) the weight percent MgO in the overall ceramic, (b) the percent forsterite

in the ceramic at 1600°C, and (c) the liquidus temperature for the ceramic.

24. How many moles of CaO must be added to 10 mol of ZrO_2 to assure that the zirconia is stabilized, or retains the same crystal structure until melting begins?

25. Determine the liquidus temperature of a ceramic prepared from 70% Al_2O_3, 20% SiO_2, and 10% CaO.

26. Suppose iron ore contains 35% SiO_2, with the remainder being iron oxide (FeO). How many grams of CaO per kilogram of iron ore should be added to a blast furnace operating at 1600°C to produce a completely liquid CaO–SiO_2 slag, assuming all of the iron oxide is reduced to produce liquid steel?

27. A fireclay containing 50% SiO_2, 40% Al_2O_3, and 10% CaO is melted and allowed to cool under equilibrium conditions. Determine the liquidus temperature.

28. A ceramic glaze is prepared containing 25% SiO_2, 45% Al_2O_3, and 30% CaO. (a) Determine the liquidus temperature of the glaze. (b) Based on the liquidus temperature, is this a reasonable composition for a glaze? (c) Do you expect the glaze to contain the equilibrium amounts of solids at room temperature? Explain.

29. Suppose 50 kg of $Al_2O_3 \cdot 4SiO_2 \cdot H_2O$ are combined with 100 kg of $2CaO \cdot Al_2O_3 \cdot SiO_2$ to produce a clay body. The ceramic is dried and fired at 1600°C. (a) Determine the weight and composition of the clay body after firing. (b) Did any melting occur during firing? Explain.

30. Suppose we combine 20 lb of mullite ($3Al_2O_3 \cdot 2SiO_2$), 30 lb of kaolinite clay ($Al_2O_3 \cdot 2SiO_2 \cdot 2H_2O$), and 15 lb of silica. The ceramic is dried and fired at 1500°C. (a) Determine the weight and composition of the clay body after firing. (b) Did any melting occur during firing? Explain. (c) What solid phases are present at room temperature for equilibrium cooling?

31. Suppose we combine 30 lb of $Na_2O \cdot SiO_2$ with 10 lb of $3CaO \cdot 2SiO_2$. (a) Determine the composition of the material after melting. (b)

What is the minimum temperature to which we would have to heat this material to produce all liquid, from which a glass could then be produced?

32. We would like to produce 100 kg of a soda-lime glass containing 80% SiO_2 with a liquidus temperature of 1400°C. How many kg of Na_2O and how many kg of CaO must be added?

33. After weighing out 250 g of Al_2O_3 particles, with a density of 3.9 g/cm^3, a ceramic body is produced by firing. When suspended in water, the ceramic part weighs 160 g. After removal from the water, the part weighs 295 g. Calculate (a) the apparent porosity, (b) the true porosity, and (c) the fraction of the pore volume that is closed.

34. A reaction bonded Si_3N_4 part weighs 500 g and occupies a volume of 200 cm^3. (a) What is the bulk density of the ceramic? (b) The actual density of Si_3N_4 is 3.2 g/cm^3. What is the true porosity of the ceramic part? (c) When the ceramic is weighed after soaking in water, the part weighs 525 g. What fraction of the pores are closed?

35. How much water is required to completely hydrate 90 lb of tricalcium aluminate cement?

36. Suppose 2 kg of calcined plaster of paris are mixed with 500 g of water. (a) How much extra water was added beyond that required for cementation? (b) What ratio of water to plaster of paris would be required to just provide the cementation reaction?

37. Suppose 30 kg of feldspar, 20 kg of montmorillonite clay, and 10 kg of silica are combined. According to Figure 14-33, for what type of application would this material be suited?

15

POLYMERS

15–1 INTRODUCTION

Polymers, which include such diverse materials as plastics, rubbers, and adhesives, are giant organic molecules having molecular weights of 10,000 to 1,000,000 g/g · mole. *Polymerization* is the process by which small molecules are joined to create these giant molecules. As the polymer molecules grow in size, the melting or softening point increases and the polymer becomes stronger and more rigid.

Polymers are lightweight, corrosion-resistant materials that typically have a low strength and stiffness but are not suitable for use at high temperatures. However, most polymers are relatively inexpensive and can be readily formed into a variety of shapes, ranging from plastic bags to mechanical gears to bathtubs. Polymers also can have a variety of useful physical properties. Some, for instance, are transparent and therefore can replace glass in a number of applications. Although most polymers are electrical insulators, special conductive polymers are now being developed. Some polymers have a very low coefficient of friction, others have the ability to convert light to electricity, and still others provide the basis for nonstick cookware. Polymers are used in an amazing number of applications, including toys, home appliances, structural and decorative items, coatings, paints, adhesives, automobile tires, foams, packaging, and many others.

15–2 POLYMER CATEGORIES AND STRUCTURES

Polymers can be classified into groups in several ways—by the method by which the molecules are synthesized, by their molecular structure, or by their mechanical and thermal behavior (Table 15-1).

Polymerization Mechanism. *Addition polymers* are produced by covalently joining the individual molecules, producing chains that may be thousands of

---

TABLE 15-1   Summary of the classification methods for polymers

Behavior	General Structure	Joining Mechanism
Thermoplastic	Flexible linear chains	Addition or condensation
Thermosetting	Rigid three-dimensional network	Usually condensation, but sometimes addition may produce the framework
Elastomers	Linear cross-linked chains	Usually addition produces both the chains and cross-links

---

molecules long without changing the chemistry of the reactants. *Condensation polymers* are produced when two or more types of molecules are joined by a chemical reaction that releases a by-product, such as water.

Polymer Structure.   *Linear polymers* form long chains containing thousands of molecules. These chains may be formed by either the addition or condensation reaction. Although the individual chains may be very strong, only weak secondary bonds hold one chain to another. *Network polymers* are three-dimensional framework structures produced by rigidly cross-linking chains with strong covalent bonds, with either an addition or condensation reaction.

Polymer Behavior.   The most commonly used method to describe polymers is by their behavior when heated. *Thermoplastic* polymers, as the name suggests, typically behave in a plastic, ductile manner. These polymers can be formed at elevated temperatures, cooled, and then reheated and reformed into a different shape, all without changing the final basic structure or properties of the polymer. The thermoplastic polymers, which have a linear structure, are easily recycled. *Thermosetting* polymers are network polymers often formed by a condensation reaction. The thermosetting polymers are generally stronger than the thermoplastics. However, these polymers cannot be reprocessed after they have been formed because part of the molecule—the by-product of the condensation reaction—has left the material. *Elastomers*, or rubbers, have an intermediate behavior but, more importantly, have the ability to elastically deform by enormous amounts without being permanently changed in shape.

In general, polymerization of all three categories begins with the production of long chains in which the atoms are strongly joined by covalent bonding. In thermoplastics, the chains are held together only by weak Van der Waals bonds; in thermosets, cross-linking agents rigidly bond the chains together by covalent bonds, producing improved strength; in elastomers, some cross-linking is accomplished, thereby helping to provide the unique elastic properties.

Representing Structures.   All of the polymers have a complex three-dimensional structure that is very difficult to describe pictorially. Figure 15-1 shows four ways we could represent a segment of polyethylene, the simplest of the linear addition thermoplastic polymers. The two-dimensional model in

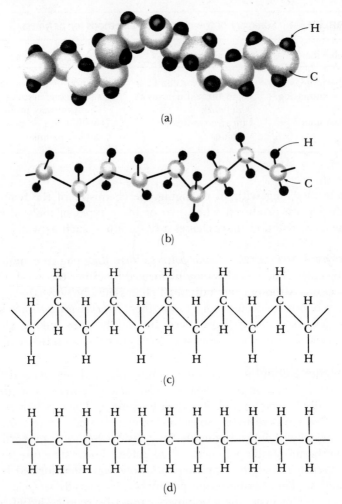

FIGURE 15-1 Four ways to represent the structure of polyethylene. (a) Solid three-dimensional model, (b) three-dimensional "space" model, (c) two-dimensional model showing the kinked nature of the polymer chain, and (d) simple two-dimensional model.

Figure 15-1(d) includes the essential elements of the polymer structure and, because of its simplicity, it is the representation we will use.

We will also encounter ring structures, such as the benzene ring found in styrene and phenol molecules. Rather than showing all of the atoms in the benzene ring, we may use a hexagon (Figure 15-2). When we encounter more complex rings, we will include all of the atoms.

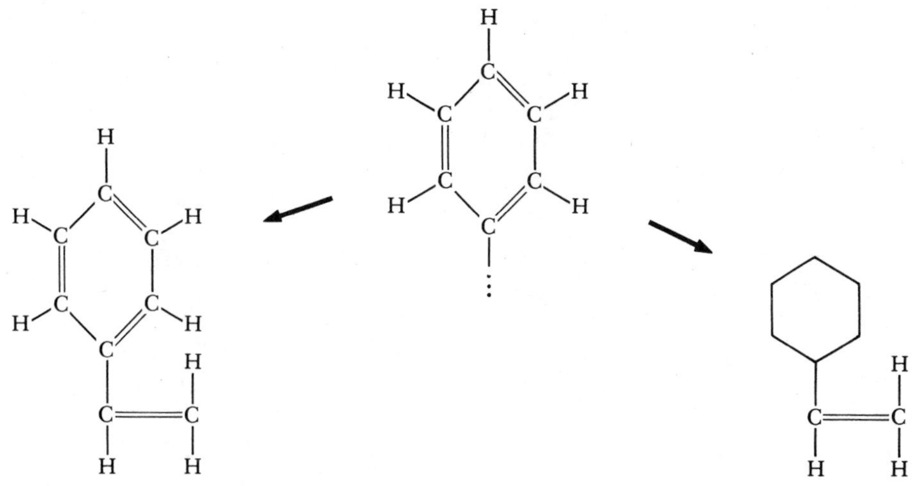

FIGURE 15-2 Two ways to represent the benzene ring. In this case, the benzene ring is shown attached to a pair of carbon atoms, producing styrene.

15–3   CHAIN FORMATION BY THE ADDITION MECHANISM

The formation of the most common polymer—polyethylene—from ethylene molecules is an example of addition polymerization. Ethylene, a gas, has the formula C_2H_4. The two carbon atoms are joined by double covalent bonds. Each carbon atom shares two of its electrons with the second carbon atom, while two hydrogen atoms are bonded to each of the carbon atoms (Figure 15-3). The ethylene molecule is called a *monomer*.

In the presence of heat, pressure, or a catalyst, the double bond between the carbon atoms is broken and replaced with a single covalent bond. Each carbon atom now has only seven electrons in its outer *sp* energy level, and the ethylene molecule is now called a *mer*. In order to satisfy the bonding requirements, the mer combines with additional ethylene mers, thereby assuring that each carbon atom again shares four covalent bonds. As the process continues and more ethylene mers are added, a chain composed of a backbone of carbon atoms is produced. The long molecular chains, or *polymers*, are called polyethylene.

Unsaturated Bonds.   Addition polymerization occurs because the original molecule contains a double covalent bond between the carbon atoms. The double bond is an *unsaturated bond*; by changing to a single bond, the carbon atoms are still joined but additional mers can be added to produce the chain.

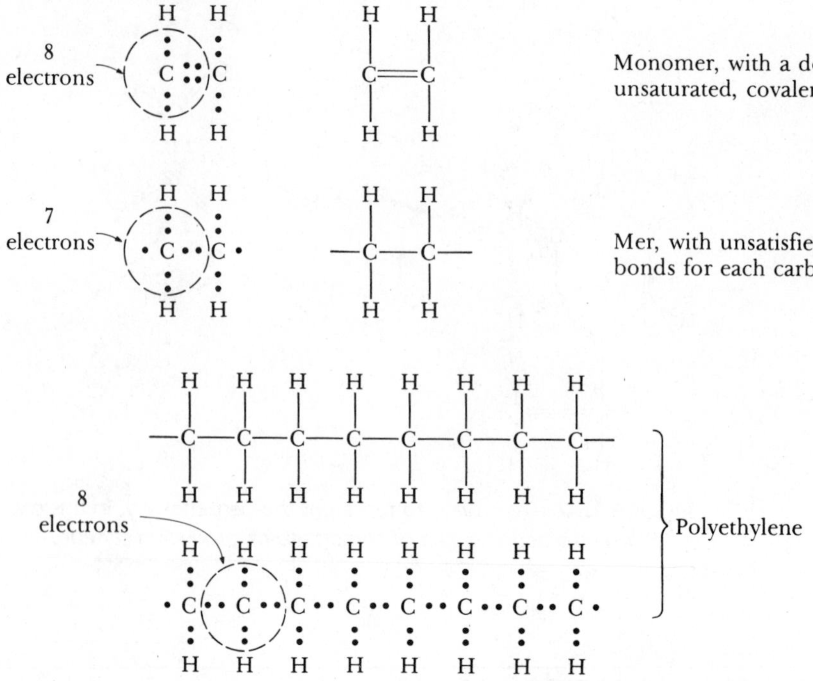

FIGURE 15-3 The addition reaction for producing polyethylene from ethylene molecules. The unsaturated double bond in the monomer is broken to produce a mer, which can then attract additional mers to either end to produce a chain.

◇ | **Example 15-1**

The monomer for propylene is shown in Table 15-6. Show how the monomers join to produce polypropylene.

Answer:

Polypropylene forms when the double covalent bond in the monomer is replaced by a single bond between the carbon atoms. The structure is then

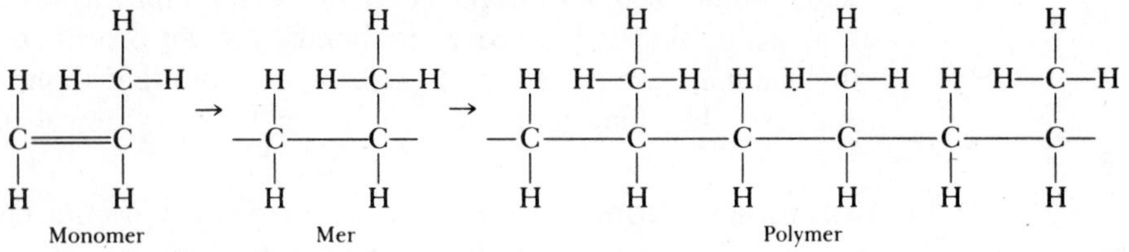

Tetrahedral Structure of Carbon. The structure of addition polymer chains is based on the nature of the covalent bonding in carbon. Carbon, like silicon,

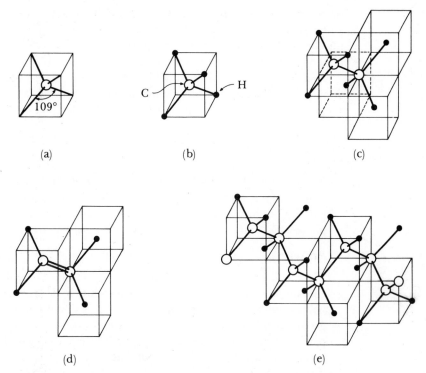

FIGURE 15-4 The tetrahedral structure of carbon can be combined in a variety of ways to produce solid crystals, nonpolymerizable gas molecules, and polymers. (a) Carbon tetrahedron, (b) methane, with no unsaturated bonds, (c) ethane, with no unsaturated bonds, (d) ethylene, with an unsaturated bond, and (e) polyethylene.

has a valence of four. The carbon atom shares its valence electrons with four surrounding atoms, producing a tetrahedral structure in which the four covalent bonds have a fixed angular relationship of 109° (Figure 15-4). In diamond, all of the atoms in the tetrahedron are carbon and the special FCC structure, diamond cubic, is produced.

However, in organic molecules, some of the positions in the tetrahedron are occupied by hydrogen, chlorine, fluorine, or even groups of atoms. Since the hydrogen atom has only one electron to share, the tetrahedron cannot be further extended. The structure in Figure 15-4(b) shows an organic molecule— methane—that could not undergo a simple addition polymerization process, since all four bonds are satisfied by hydrogen atoms.

The initial carbon atom could be joined with one covalent bond to a second carbon atom, with all of the other bonds involving hydrogen, as in ethane [Figure 15-4(c)]. But again, polymerization cannot occur.

However, in ethylene, the carbon atoms are joined by an unsaturated double bond, while the other sites are occupied by hydrogen atoms [Figure 15-4(d)]. During polymerization the double bond is broken and each carbon atom in the molecule attracts a new mer, eventually giving polyethylene [Figure 15-4(e)].

Functionality.   The *functionality* is the number of sites at which new molecules can be attached to the mer. In ethylene, there are two locations—each carbon atom—at which molecules can be attached. Thus, ethylene is bifunctional and only chains will form.

If there are three sites at which molecules can be attached, the mer is trifunctional and a three-dimensional network can form. Normally, trifunctional mers produce stronger polymers than bifunctional mers.

◇ | **Example 15-2**

Phenol molecules have the structure show below. The phenol molecules can be joined to one another when a hydrogen atom is removed from the ring and participates in a condensation reaction. What is the maximum functionality of phenol? Do you expect that a chain or a network structure will be produced?

Answer:

The phenol molecule will give up a hydrogen atom from any of the five corners containing only hydrogen atoms. The sixth hydrogen in the OH group is bonded too tightly to the ring. Thus, the functionality is five. However, due to geometrical limitations—if all five sites underwent a condensation reaction, the participating molecules would be crowded together—the maximum functionality is actually only three. Since phenol is at least trifunctional, a network structure is produced.

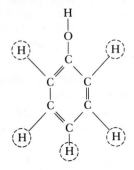

Initiation of the Addition Mechanism.   To begin the addition polymerization process, an initiator, such as hydrogen peroxide, H_2O_2, is added to ethylene (Figure 15-5). The covalent bonds between the oxygen atoms in the peroxide and between the carbon atoms in the ethylene are broken and an OH group is attached to one end of the ethylene mer. By terminating one end of the mer, the OH group acts as the nucleus for a chain.

Growth of the Addition Chain.   Once the chain is initiated, the reaction proceeds spontaneously. Bonding energies between atoms are shown in Table

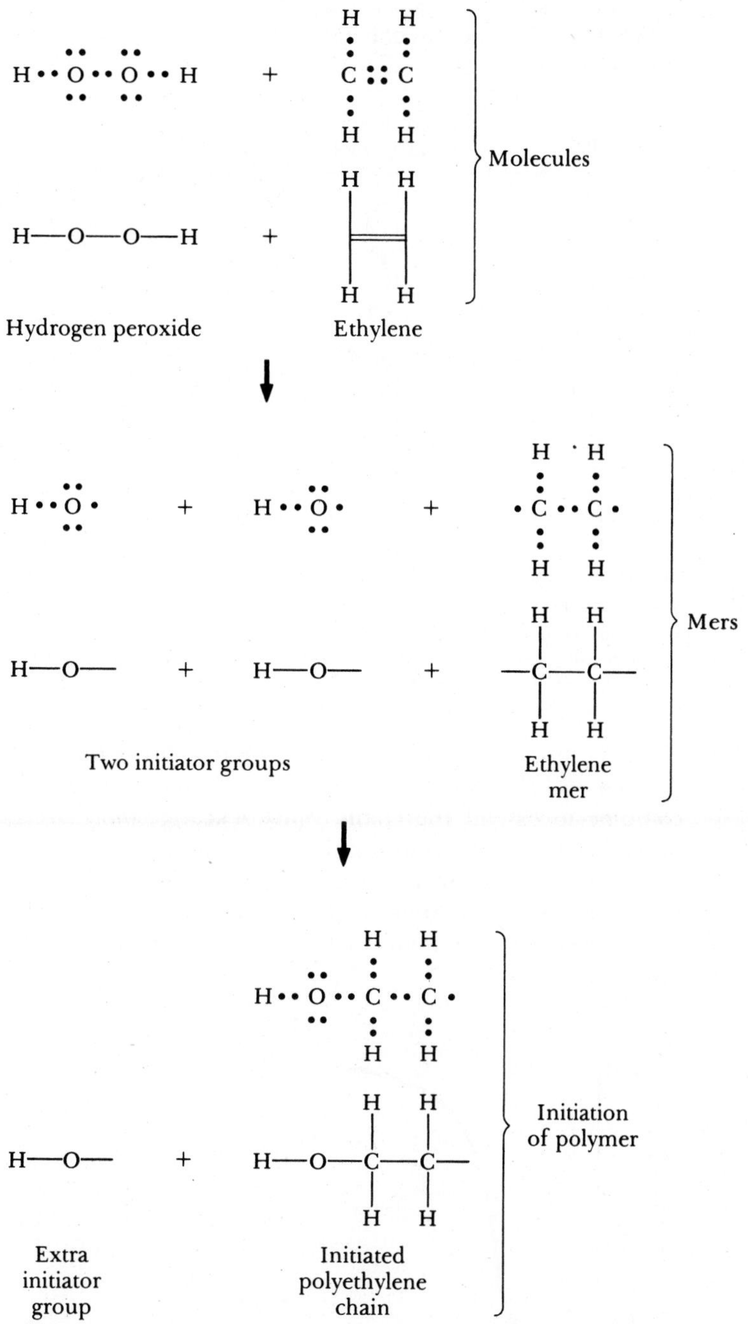

FIGURE 15-5   A polyethylene chain is initiated when hydrogen peroxide splits into two OH groups and a mer is formed. One of the OH groups is attached to the mer, initiating the chain.

TABLE 15-2   The strength of bonds in polymers

Type of Bond	Bond Strength (kcal/g $\cdot$ mole)
C—C	88
C=C	172
C≡C	230
C—H	104
C—Cl	86
C—N	73
C—O	86
C=O	127
N—H	110
O—H	119
H—H	104
N—O	73
O—O	51
C—F	103
O—S	132

Data from *Handbook of Chemistry and Physics*, 56th Ed., CRC Press, 1975.

15-2. We must introduce 172 kcal/g $\cdot$ mole to break the double carbon bond, but $2 \times 88 = 176$ kcal/g $\cdot$ mole are released when the two single covalent bonds combine to extend the chain. Since 4 kcal/g $\cdot$ mole are evolved, this energy difference favors the continuation of the polymerization process.

The kinetics of growth are depicted in Figure 15-6. Growth is initially slow but speeds up dramatically after initiation. Since energy is released during polymerization, the temperature may rise, increasing the growth rate still further.

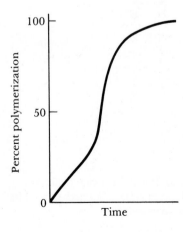

FIGURE 15-6   The rate of growth of the chains and the overall rate of polymerization are initially slow, but then continue at a high rate. When polymerization is nearly complete, the rate slows down again.

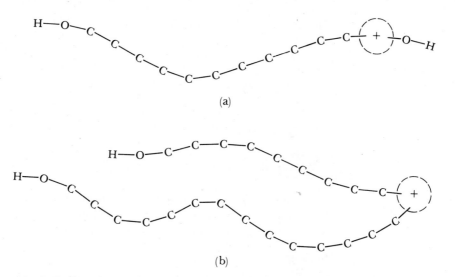

(a)

(b)

FIGURE 15-7   Termination of the addition chains can occur when (a) OH groups are attached to the ends of the chains or (b) two chains combine.

When polymerization is nearly complete, the remaining unattached mers must diffuse a long distance before reaching an active end of a chain. Consequently, the growth rate again decreases.

Termination of the Addition Chain.   The chains may be terminated by two mechanisms (Figure 15-7). First, the ends of two growing chains may join to produce a single large chain. Second, the active end of the chain may attract an initiator group, OH, which now terminates the chain. We can control the length of the chain by controlling the amount of initiator—if small amounts of initiator are added, less is available to terminate the chains and the chains are longer.

◇ | **Example 15-3**

If no heat is lost to the surroundings, estimate the increase in temperature that accompanies the polymerization of ethylene. The specific heat, or energy required to raise the temperature of one g of polyethylene by $1°C$, is 0.55 cal/g $\cdot$ $°C$.

Answer:

There are 4 kcal/g $\cdot$ mole or 4000 cal/g $\cdot$ mole evolved during polymerization. The molecular weight of the ethylene mer (C_2H_4) is

$$
\begin{aligned}
\text{Mol wt} &= (2\,\text{C atoms})(M_C) + (4\,\text{H atoms})(M_H) \\
&= (2\,\text{C atoms})(12\,\text{g/g}\cdot\text{mole}) + (4\,\text{H atoms})(1\,\text{g/g}\cdot\text{mole}) \\
&= 28\,\text{g/g}\cdot\text{mole}
\end{aligned}
$$

Then the temperature rise is

$$\Delta T = \frac{4000 \text{ cal/g} \cdot \text{mole}}{(0.55 \text{ cal/g} \cdot {}^\circ\text{C})(28 \text{ g/g} \cdot \text{mole})}$$

$$= 260^\circ\text{C}$$

 Example 15-4

One gram of hydrogen peroxide is added to 10,000 g of ethylene to serve as the initiator and terminator. Calculate the average molecular weight of the polymer if all of the hydrogen peroxide is consumed.

Answer:

Each hydrogen peroxide molecule will initiate and terminate one polymer chain, since the hydrogen peroxide changes into two OH groups. The molecular weight of hydrogen peroxide is

$$\text{Mol wt } H_2O_2 = 2(M_H) + 2(M_O)$$

$$= 2(1) + 2(16) = 34 \text{ g/g} \cdot \text{mole}$$

$$\text{Number of } H_2O_2 \text{ molecules} = \frac{1}{34} (6.02 \times 10^{23})$$

$$= 1.77 \times 10^{22}$$

For ethylene

$$\text{Mol wt } C_2H_4 = 2(12) + 4(1) = 28 \text{ g/g} \cdot \text{mole}$$

$$\text{Number of } C_2H_4 \text{ molecules} = \frac{10,000}{28} (6.02 \times 10^{23})$$

$$= 2.15 \times 10^{26} \text{ mers of } C_2H_4$$

$$\text{Number of mers/chain} = \frac{2.15 \times 10^{26}}{1.77 \times 10^{22}} = 12,147$$

Therefore, the molecular weight of the polymer is

$$\text{Mol wt polymer} = 12,147(28) = 340,116 \text{ g/g} \cdot \text{mole}$$

Chain Shape. During growth, the polymer chains twist and turn due to the tetrahedral nature of the covalent bond. Figure 15-8 illustrates two possible geometries in which a chain might grow.

The third atom in Figure 15-8(a) can be located at any position on the circle and still preserve the directionality of the covalent bond. A straight chain, as in

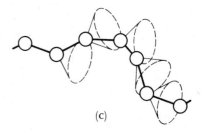

FIGURE 15-8 *The angular relationship between the bonds in the carbon chain can be satisfied when the third carbon atom is placed anywhere on the circle in (a). Depending on how the atoms are placed, the chain may be either straight (b) or highly twisted (c).*

Figure 15-8(b), could be produced, but it is more likely that the chain will be highly twisted, as in Figure 15-8(c).

The chains twist and turn in response to external factors, such as temperature or the availability and location of the next mer to be attached to the chain. Eventually, the chains become intertwined with other chains growing simultaneously. The appearance of the polymer chains may resemble that of a bucket of earthworms or a plate of spaghetti.

15–4   CHAIN FORMATION BY THE CONDENSATION MECHANISM

Linear polymers are also formed by condensation reactions, giving structures and properties that resemble those of linear addition polymers. Normally, heat, pressure, or a catalyst causes the reaction to proceed. The polymerization of dimethyl terephthalate and ethylene glycol to produce polyester is an important example (Figure 15-9).

During polymerization, the OH group of ethylene glycol combines with the CH_3 group of dimethyl terephthalate. A by-product, methyl alcohol, is driven off and the two monomers combine to produce a larger molecule. The monomers in this example are bifunctional. Consequently, polymerization and growth occur at both ends of our new molecule by the same reaction. Eventually, a long polymer chain—a polyester—is produced.

The length of the chain depends on the ease with which mers can diffuse to the ends and undergo the condensation reaction. Termination occurs when no more monomers reach the end of the chain to continue the reaction; in the condensation polymers, the length of the chain is no longer controlled by the amount of the initiator added.

◇ | **Example 15-5**

The linear polymer 6,6-nylon is formed when adipic acid and hexamethylene diamine combine by a condensation reaction. Show how this reaction occurs and determine the by-product that forms.

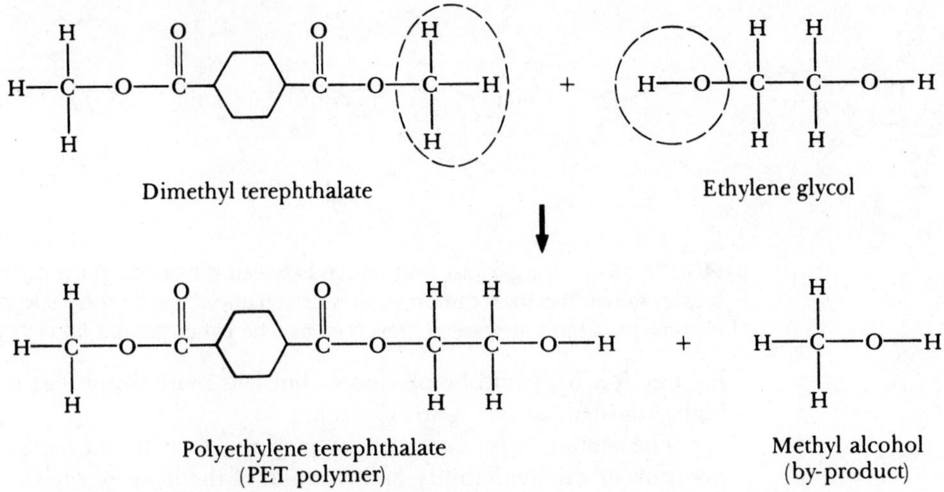

Dimethyl terephthalate   Ethylene glycol

Polyethylene terephthalate
(PET polymer)

Methyl alcohol
(by-product)

FIGURE 15-9 The condensation reaction for polyethylene terephthalate (PET), a common polyester. A CH_3 group and an OH group are removed from the monomers, producing methyl alcohol as a by-product.

Answer:

The molecular structures of the monomers are shown below. The linear nylon chain is produced when a hydrogen atom from the hexamethylene diamine combines with an OH group from adipic acid to form a water molecule.

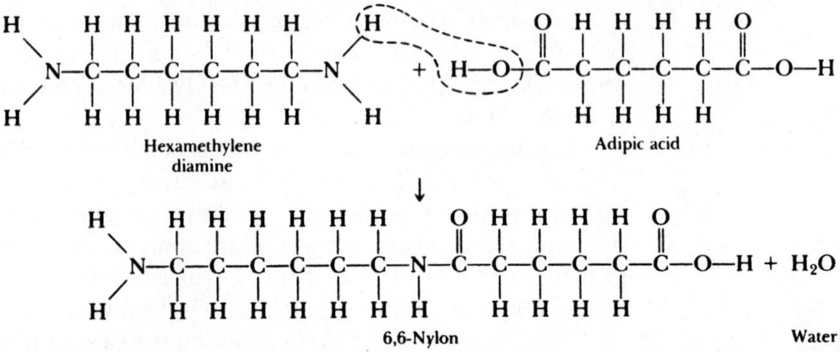

Hexamethylene
diamine   Adipic acid

6,6-Nylon   Water

Note that the reaction can continue at both ends of the new molecule; consequently, long chains may form. This polymer is called 6,6-nylon because both monomers contain six carbon atoms.

◇   **Example 15-6**

How much energy is consumed or evolved when 6,6-nylon is formed? Will the reaction be spontaneous?

Answer:

6,6-Nylon forms when a N—H bond and an O—C bond are broken on the hexamethylene diamine and adipic acid monomers. (See Example 15-5). A C—N bond forms to join the two monomers into a chain, and an O—H bond is formed to produce the by-product water. The change in energy is

$$\Delta \text{ Energy } = \text{ energy consumed } - \text{ energy evolved}$$
$$= (E_{N-H} + E_{O-C}) - (E_{C-N} + E_{O-H})$$
$$= (110 + 86) - (73 + 119) = +4 \text{ kcal/g} \cdot \text{mole}$$

Since energy is consumed, the reaction is not spontaneous.

◇ **Example 15-7**

Suppose 1000 g of dimethyl terephthalate are combined with ethylene glycol to produce PET. Determine how many grams of ethylene glycol are required and the total weight of the polymer after polymerization is completed.

Answer:

We need equal numbers of moles of the two monomers.

$$\text{Mol wt dimethyl terephthalate } = (10 \text{ C atoms})(12) + (4 \text{ O atoms})(16)$$
$$+ (10 \text{ H atoms})(1)$$
$$= 194 \text{ g/g} \cdot \text{mole}$$

$$\begin{array}{c}\text{Number of dimethyl} \\ \text{terephthalate molecules}\end{array} = \frac{1000 \text{ g}}{194} (6.02 \times 10^{23})$$
$$= 31.03 \times 10^{23}$$

We need the same number of molecules of ethylene glycol.

$$\text{Mol wt ethylene glycol } = (2 \text{ C atoms})(12) + (2 \text{ O atoms})(16)$$
$$+ (6 \text{ H atoms})(1)$$
$$= 62 \text{ g/g} \cdot \text{mole}$$

$$\text{Grams of ethylene glycol } = \frac{31.03 \times 10^{23}}{6.02 \times 10^{23}} (62) = 319.6 \text{ g}$$

For each of the monomers, one methyl alcohol will be evolved.

$$\text{Mol wt methyl alcohol } = (1 \text{ C atoms})(12) + (1 \text{ O atom})(16)$$
$$+ (4 \text{ H atoms})(1)$$
$$= 32 \text{ g/g} \cdot \text{mole}$$

$$\text{Grams of methyl alcohol } = \frac{(31.03 \times 10^{23})(2)}{(6.02 \times 10^{23})} (32)$$
$$= 329.9 \text{ g}$$

The (2) is in the above equation because there are 31.03×10^{23} mol of dimethyl terephthalate and an equal number of molecules of ethylene glycol.

$$\text{Total weight of polymer} \ = \ 1000 + 319.6 - 329.9 = 989.7\,\text{g}$$

15–5   DEGREE OF POLYMERIZATION

The degree of polymerization describes the average length to which a chain grows. If the polymer contains only one type of monomer, the degree of polymerization is the average number of molecules or mers that are present in the chain. We can also define the degree of polymerization as

$$\frac{\text{Degree of}}{\text{polymerization}} \ = \ \frac{\text{molecular weight of polymer}}{\text{molecular weight of mer}} \qquad (15\text{-}1)$$

When the chain is composed of more than one type of mer, we can define the average molecular weight of the mer as

$$\bar{M} \ = \ \Sigma f_i M_i \qquad\qquad (15\text{-}2)$$

where f_i is the molecular fraction of mers having molecular weight M_i.

If the polymer chain is formed by condensation, the molecular weight of the by-product must be subtracted from that of the mer

$$\bar{M} \ = \ \Sigma(f_i M_i - M_{\text{by-product}}) \qquad (15\text{-}3)$$

◇   **Example 15-8**

Calculate the degree of polymerization if polyethylene has a molecular weight of 100,000 g/g · mole.

Answer:

The molecular weight of the ethylene mer is

$$\text{Mol wt mer} \ = \ 2(12) + 4(1) = 28\,\text{g/g} \cdot \text{mole}$$
$$\text{Degree of polymerization} \ = \ \frac{100{,}000}{28} \ = \ 3571$$

◇   **Example 15-9**

Calculate the degree of polymerization if 6,6-nylon has a molecular weight of 120,000 g/g · mole.

Answer:

The nylon forms when hexamethylene diamine and adipic acid combine and release a molecule of water. When a long chain forms, there is on average one water molecule released for each reacting molecule. The molecular weights of the molecules are

$$
\begin{aligned}
\text{Mol wt diamine} &= (6\,\text{C atoms})(12) + (2\,\text{N atoms})(14) \\
&\quad + (16\,\text{H atoms})(1) \\
&= 116\,\text{g/g}\cdot\text{mole} \\
\text{Mol wt adipic acid} &= (6\,\text{C atoms})(12) + (4\,\text{O atoms})(16) \\
&\quad + (10\,\text{H atoms})(1) \\
&= 146\,\text{g/g}\cdot\text{mole} \\
\text{Mol wt water} &= (2\,\text{H atoms})(1) + (1\,\text{O atom})(16) \\
&= 18\,\text{g/g}\cdot\text{mole} \\
\text{Av. mol wt mer} &= 0.5(116) + 0.5(146) - 18 \\
&= 113\,\text{g/g}\cdot\text{mole} \\
\text{Degree of polymerization} &= \frac{120{,}000}{113} = 1062
\end{aligned}
$$

The degree of polymerization refers to the total number of monomers in the chain. Half of the 1062 mers are hexamethylene diamine and the other half are adipic acid.

Actually, the chains in the linear polymers are not all of the same length and thus each chain may have a different molecular weight. Two methods are used to calculate an average molecular weight for linear polymers.

The *weight average molecular weight* is obtained by dividing the chains into size ranges and determining the fraction of chains having molecular weights within that range (Table 15-3). The weight average molecular weight $\bar{M}_w$ is

$$
\bar{M}_w = \frac{\Sigma f_i M_i}{\Sigma f_i} \tag{15-4}
$$

where M_i is the mean molecular weight of each range and f_i is the weight fraction of the polymer having chains within that range.

The *number average molecular weight* $\bar{M}_n$ is based on the number fraction, rather than weight fraction, of the chains within each size range (Table 15-4). It is always smaller than the weight average molecular weight.

$$
\bar{M}_n = \frac{\Sigma x_i M_i}{\Sigma x_i} \tag{15-5}
$$

where M_i is again the mean molecular weight of each size range but x_i is the

TABLE 15-3 Data for the weight average molecular weight for the polymer in Example 15-10

Molecular Weight Range (g/g · mole)	Mean M_i	f_i	$f_i M_i$
0–5,000	2,500	0.01	25
5,000–10,000	7,500	0.05	375
10,000–15,000	12,500	0.07	875
15,000–20,000	17,500	0.23	4025
20,000–25,000	22,500	0.28	6300
25,000–30,000	27,500	0.22	6050
30,000–35,000	32,500	0.10	3250
35,000–40,000	37,500	0.03	1125
40,000–45,000	42,500	0.01	425
		$\Sigma = 1.00$	$\Sigma = 22{,}450$

TABLE 15-4 Data for the number average molecular weight for the polymer in Example 15-10

Molecular Weight Range (g/g · mole)	Mean M_i	x_i	$x_i M_i$
0–5,000	2,500	0.02	50
5,000–10,000	7,500	0.08	600
10,000–15,000	12,500	0.11	1375
15,000–20,000	17,500	0.19	3325
20,000–25,000	22,500	0.23	5175
25,000–30,000	27,500	0.25	6875
30,000–35,000	32,500	0.08	2600
35,000–40,000	37,500	0.03	1125
40,000–45,000	42,500	0.01	425
		$\Sigma = 1.00$	$\Sigma = 21{,}550$

fraction of the total number of chains within each range. Either $\bar{M}_w$ or $\bar{M}_n$ can be used to calculate the degree of polymerization.

◇ | **Example 15-10**

The weight fraction f_i and the number fraction x_i showing the distribution of chains of differing molecular weight of a polymer are shown in Tables 15-3 and 15-4. Calculate the weight and number average molecular weights from the data.

Answer:

$$\bar{M}_w = \frac{\Sigma f_i M_i}{\Sigma f_i} = \frac{22{,}450}{1} = 22{,}450 \, \text{g/g} \cdot \text{mole}$$

$$\bar{M}_n = \frac{\Sigma x_i M_i}{\Sigma x_i} = \frac{21,550}{1} = 21,550 \text{ g/g} \cdot \text{mole}$$

15-6   DEFORMATION OF THERMOPLASTIC POLYMERS

Before examining the different types of thermoplastic polymers and their behavior, we should first take note of how they deform when an external stress is applied. A stress-strain curve for a thermoplastic polymer might have the general form shown in Figure 15-10. This somewhat unusual shape is related to the way the polymer chains move relative to one another under load. Several differences may be noted compared with a typical ductile metal such as aluminum, in particular the decrease in stress immediately after yielding begins.

Elastic Behavior.   Elastic deformation in these polymers is due to two mechanisms. First, the covalent bonds within the chain stretch and distort, allowing the chains to elongate elastically. As soon as the stress is removed, this distortion can be recovered almost instantly; as a result, the initial portion of the stress-strain curve may be linear and a modulus of elasticity can be calculated from the diagram. This is similar to elasticity in metals and ceramics, which also deform elastically by stretching of metallic, ionic, or covalent bonds.

But in addition, entire segments of the polymer chains may be distorted; when the stress is removed, the segments move back to their original positions

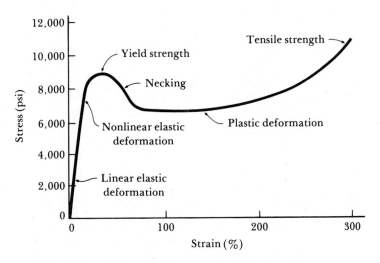

FIGURE 15-10   The stress-strain curve for nylon 6,6, a typical thermoplastic polymer.

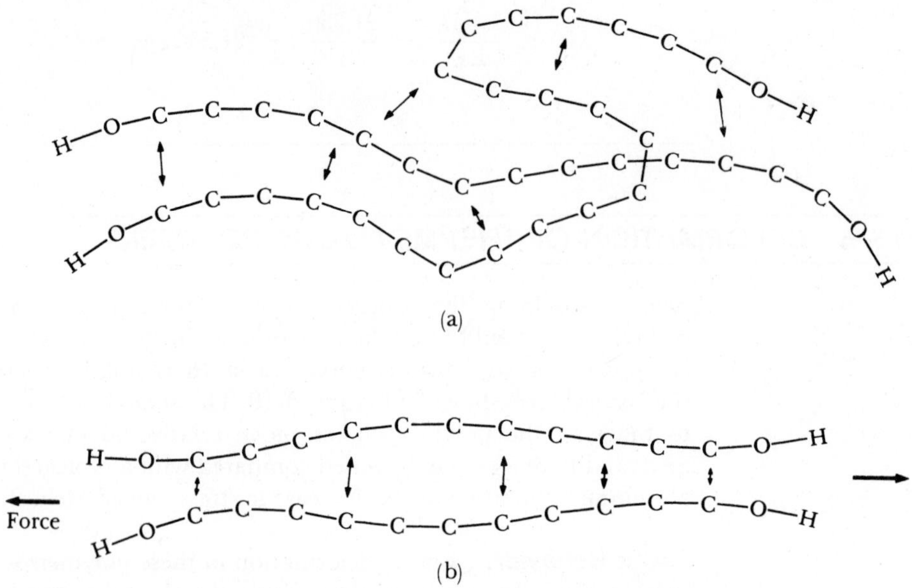

FIGURE 15-11 (a) The individual chains are held together loosely by Van der Waals bonds and mechanical interlocking. (b) When the polymer is stretched, the chains are straightened out and eventually begin to slide past one another.

only over a period of time, often hours or even months. This time-dependent elastic behavior may contribute to some nonlinear elastic behavior.

Plastic Behavior.   These polymers also experience plastic deformation when the stress exceeds the yield strength. Unlike the behavior of metals, however, plastic deformation is not a consequence of dislocation movement. Instead, viscous flow, or sliding of the chains past one another under load, causes permanent deformation. The drop in the stress beyond the yield point can be explained by this phenomenon. Initially, the chains may be highly tangled and intertwined. When the stress is sufficiently high, the chains begin to untangle and straighten. As the chains straighten, necking also occurs, permitting continued sliding of the chains at a lesser stress. However, eventually the chains become almost parallel and closer together; stronger Van der Waals bonding between the more closely aligned chains requires higher stresses to continue the deformation process (Figure 15-11).

Again, the ability of the stress to cause chain slippage is related to time. If the stress is applied slowly, the chains will be able to slide more easily past one another; if the stress is applied rapidly, sliding may be minimized and the polymer may tend to behave in a brittle manner.

Viscoelasticity.   The time dependency of elastic and plastic deformation of the thermoplastics is related to the *viscoelastic* behavior of the materials. The stress required to deform the material a certain percentage depends on both the total

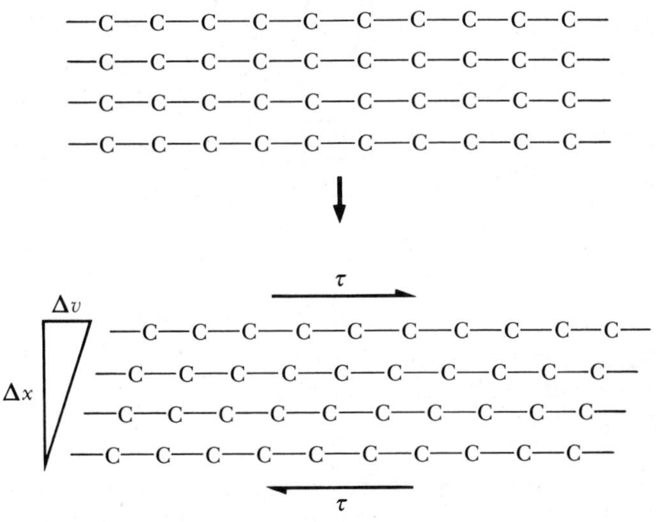

$$-C-C-C-C-C-C-C-C-C-C-$$
$$-C-C-C-C-C-C-C-C-C-C-$$
$$-C-C-C-C-C-C-C-C-C-C-$$
$$-C-C-C-C-C-C-C-C-C-C-$$

$\downarrow$

$\xrightarrow{\quad \tau \quad}$

Δv

Δx

$$-C-C-C-C-C-C-C-C-C-C-$$
$$-C-C-C-C-C-C-C-C-C-C-$$
$$-C-C-C-C-C-C-C-C-C-C-$$
$$-C-C-C-C-C-C-C-C-C-C-$$

$\xleftarrow{\quad \tau \quad}$

FIGURE 15-12 A shear stress τ causes the polymer chains to slide over one another by viscous flow. The velocity gradient, $\Delta v/\Delta x$, produces a displacement of the chains that depends on the viscosity η of the polymer.

percentage of deformation (or strain) and the rate of deformation (or strain rate). We would expect that higher total deformations would require higher applied stresses, since the chains would become more aligned and rigidly bonded; in addition, we expect that more rapid rates of deformation would require higher stresses, since less time is then allowed for slippage of the chains.

One way of describing the ease with which permanent deformation occurs is by the viscosity of the polymer. The viscosity η, as described in Figure 15-12, is defined by

$$\eta = \frac{\tau}{\Delta v/\Delta x} \tag{15-6}$$

where τ is the shear stress causing adjacent chains to slide and $\Delta v/\Delta x$ is the velocity gradient, which is related to how rapidly the chains are displaced relative to one another. If the viscosity is high, a large stress is required to cause a given displacement. Therefore, polymers with a high viscosity have less viscous deformation.

The temperature effect on viscosity is identical to that in glasses, as discussed in Chapter 14

$$\eta = \eta_0 \exp \frac{E_\eta}{RT} \tag{15-7}$$

where η_0 is a constant and E_η is the activation energy, which is related to the ease with which the chains slide past one another. As the temperature increases, the polymer is less viscous and deforms more easily.

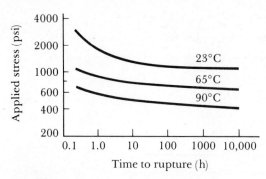

FIGURE 15-13 The effect of temperature on the stress-rupture behavior of high-density polyethylene.

Creep and Impact Properties. The viscoelastic behavior of the thermoplastic polymers can help explain the impact and creep properties, both of which are time-related phenomena. When very high rates of strain are applied, as in an impact test, there is not enough time for the chains to move and cause plastic deformation. As a result, the thermoplastics behave in a rather brittle manner and have poor impact values and poor fracture toughness.

On the other hand, when the stress is applied over a long period of time, substantial viscous flow occurs, even at relatively low temperatures. Consequently, the thermoplastic polymer continually strains or creeps. A higher temperature or stress increases the amount of creep. In fact, stress-rupture curves can be obtained just as in metals (Figure 15-13).

Another method for measuring the temperature dependence of polymers is the *heat deflection temperature*, which is the temperature at which a given deformation of a beam will occur under a standard load. High heat deflection temperatures indicate good resistance to creep.

The viscoelastic behavior is very helpful to us in producing components from thermoplastic polymers. The ease with which sliding of the chains occurs enables us to pour the polymer as we would a liquid in casting processes or easily form the polymer into a useful shape by deformation processing techniques. However, the viscoelastic behavior also is undesirable in that continued deformation, or creep, may occur even at room temperature. In the next sections, we will examine how we can better control the deformation behavior of the thermoplastics, still permitting us easily to form the material yet produce a rigid, dimensionally stable component.

15–7   EFFECT OF TEMPERATURE ON THERMOPLASTICS

Temperature influences the mechanical behavior of thermoplastics through its effect on the strength of the Van der Waals bonding between the chains. At high temperatures, the bonds are very weak and viscous flow occurs easily at little or

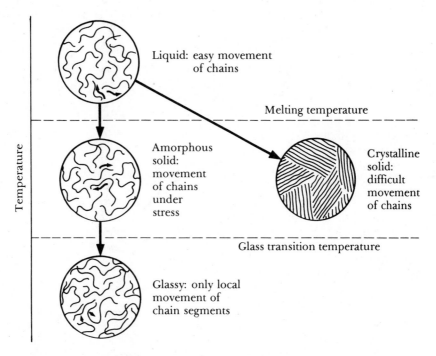

FIGURE 15-14   The effect of temperature on the structure and behavior of thermoplastic polymers.

no applied stress. As the temperature decreases, viscoelastic behavior becomes pronounced as the stronger bonds make chain sliding more difficult and complex. At very low temperatures, the polymer may be so viscous that no chain sliding occurs and the polymer behaves as a rigid solid. Several types of behavior can therefore be observed, depending on the temperature and chain structure. Figures 15-14 and 15-15 summarize these effects.

Degradation Temperature.   At very high temperatures, the covalent bonds between the atoms in the linear chain may be destroyed—the polymer burns or chars. This temperature T_d, the *degradation temperature*, limits the usefulness of the polymer and represents the upper temperature at which the polymer can be formed into a useful shape.

Liquid Polymers (the Viscous State).   At or above the *melting temperature* T_m, the viscosity is very low. Although the chains may be twisted and intertwined, they may move without an external force and, if a force is applied, the polymer flows with virtually no elastic strain occurring. The strength and modulus of elasticity are nearly zero and the polymer is suitable for casting and many other forming processes. The melting points of some typical polymers are shown in Table 15-5.

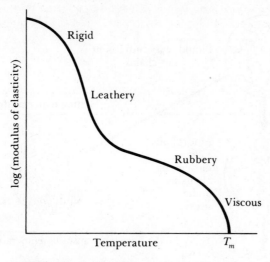

FIGURE 15-15 The effect of temperature on the modulus of elasticity for an amorphous thermoplastic polymer.

TABLE 15–5 Melting and glass transition temperatures for selected thermoplastics and elastomers

Polymer	$T_m(°C)$	$T_g(°C)$
Addition polymers		
Low-density (LD) polyethylene	115	− 120
High-density (HD) polyethylene	137	− 120
Polyvinyl chloride	175–212	87
Polypropylene	168–176	− 16
Polystyrene	240	85–125
Polyacrylonitrile	320	107
Teflon	327	
Polychlorotrifluoroethylene	220	
Polymethyl methacrylate (acrylic)		90–105
ABS		88–125
Condensation polymers		
Acetal	181	− 85
6,6-Nylon	265	50
Cellulose acetate	230	
Polycarbonate	230	145
Polyester	255	75
Elastomers		
Silicone		− 123
Polybutadiene	120	− 90
Polychloroprene	80	− 50
Polyisoprene	30	− 73

Amorphous Polymers (the Plastic State).   Just below the melting temperature, the polymer chains are still twisted and intertwined, giving an amorphous structure. The polymer develops some strength and elasticity and, if no load is applied, can retain its shape. Initially, the polymer behaves in a *rubbery* manner; when a stress is applied, both elastic and plastic deformation can occur easily. When the stress is removed, the elastic portion of the deformation is almost instantly recovered, but the polymer shape has been permanently deformed by chain movement. Large, permanent elongations can be achieved, and thus the polymer can easily be formed into useful shapes by molding, extrusion, or other forming methods.

As the temperature falls further, the polymer becomes stiffer and stronger and behaves in a *leathery* manner. Now the viscosity is higher, so the chain segments that contribute to elastic deformation no longer move rapidly. As a result, recovery of the total elastic deformation becomes time dependent after removal of an applied stress.

Glassy Polymers (the Rigid State).   As the temperature of the polymer continues to decrease, the viscosity becomes so high that only very localized movement of small molecular groups on the chain is possible. Elastic deformation is almost limited to stretching of the bonds, which requires high stresses and leads to a high modulus of elasticity. In addition, plastic deformation is very limited because the chains no longer slide over one another. Below the *glass transition temperature* T_g, the linear polymer becomes hard and brittle and behaves much the same as a ceramic glass. The stress-strain curve will show little plastic yielding prior to fracture, and other properties, such as the density or specific volume, change at a different rate (Figure 15-16).

The glassy polymer cannot be formed into useful shapes and may be too brittle for many applications. On the other hand, the glassy polymers are

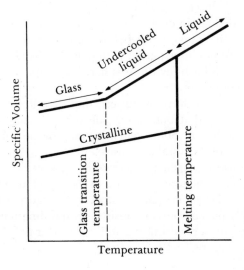

FIGURE 15-16 The relationship between the specific volume and the temperature of the polymer shows the melting and glass transition temperatures.

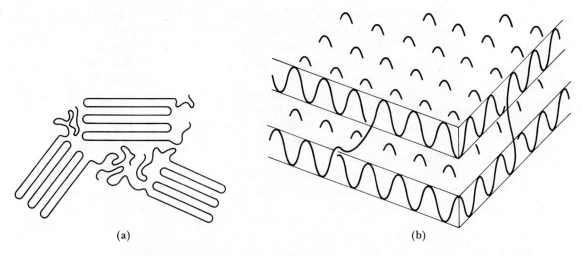

(a)   (b)

FIGURE 15-17   The folded chain model for crystallinity in polymers, shown in (a) two dimensions and (b) three dimensions.

significantly stronger and more rigid than the amorphous structure and also are more dimensionally stable, or more creep resistant.

The glass transition temperature is typically about 0.5 to 0.75 times the absolute melting temperature T_m, with simple addition chains having a lower T_g than the more complicated chains. Glass transition temperatures for a number of thermoplastics are included in Table 15-5.

Crystalline Polymers.   Many thermoplastics partially crystallize when cooled below the melting temperature, with the chains becoming closely aligned for appreciable distances. One model describing the shape of the chains is shown in Figure 15-17 for both two and three dimensions. In this *folded chain* model, the chains loop back on themselves, with each loop being approximately 100 carbon atoms long. The folded chain extends in three dimensions before moving off in a different orientation. This leaves behind small crystalline regions with a diameter of about 0.01 cm interspersed with patches of amorphous regions. The crystals can take various forms, with the spherulitic shape shown in Figure 15-18 being particularly common.

The crystalline regions have a unit cell that describes the regular packing of the chains. The crystal structure for polyethylene, shown in Figure 3-34, describes one such unit cell. Some polymers may be polymorphic; one of the nylons has three different crystal structures.

During crystallization, a sharp decrease in the specific volume (or increase in the density) occurs as the tangled chains in the liquid are rearranged into the more orderly, close-packed crystalline structure (Figure 15-16). The higher density, more closely packed polymer resists chain sliding more effectively than the amorphous structures, resulting in a stiffer, stronger polymer, particularly at high temperatures (Figure 15-19).

FIGURE 15-18   Photograph of spherulitic crystals in an amorphous matrix of nylon (× 200). From R. Brick, A. Pense, and R. Gordon, *Structure and Properties of Engineering Materials*, 4th Ed, McGraw-Hill, 1977.

Several factors influence crystallization. Crystallization is easiest for simple addition polymers, such as polyethylene, in which no bulky molecules or atom groups attached to the carbon chain are present to interfere with the close packing of the chains. Slow cooling during solidification, which permits more time for the chains to become aligned, encourages crystallization. Controlled crystallization can be achieved by first quenching to produce an amorphous structure, then annealing below the melting temperature to allow tiny crystals to nucleate and grow. Finally, slow deformation of the polymer between the melting and glass transition temperatures may promote crystallization by straightening the chains and bringing them into a parallel structure. Figure 15-20 shows one model for the crystalline structure obtained when the polymer is drawn into a fiber.

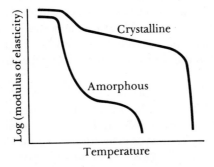

FIGURE 15-19   The effect of crystallinity on the modulus of elasticity of thermoplastic polymers.

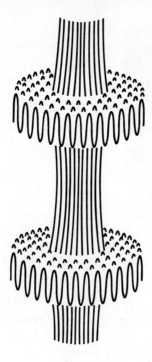

FIGURE 15-20 A model for crystallinity of a drawn polymer fiber. Note the folded chains in the areas in which the chains are not yet completely aligned.

Even in the simplest of polymers, such as polyethylene, crystallization is never complete. With increasing degrees of crystallinity, however, mechanical properties, including resistance to creep, are improved.

15–8 CONTROLLING THE STRUCTURE AND PROPERTIES OF THERMOPLASTICS

Now that we are familiar with the effects of deformation and temperature, we are ready to examine the many ways in which we can modify and control the properties of thermoplastic polymers. These methods can be grouped into three main categories—controlling the length of the individual chains, controlling the strength of the bonds within the chains, and controlling the strength of the bonds between the chains.

Degree of Polymerization.   As the linear chains increase in length, the polymer has a higher melting temperature and improved mechanical properties, including resistance to creep. We can use the ethylene monomer to illustrate this point. Ethylene is a gas. As the degree of polymerization and the molecular weight increase, the polymerized ethylene becomes a liquid, then eventually thickens into a wax. When the degree of polymerization is 200 to 500, the material is solid, has a high melting temperature, and is considered a polymer.

However, once the molecular weight of the polymer is greater than about 15,000 to 40,000 g/g · mole, the properties increase only slightly while the liquid polymer becomes more viscous and more difficult to cast or form.

Type of Monomer. The type of monomer from which the polymer is produced will influence the bonding between chains (which affects the ease with which the chains slide past one another) as well as the stiffness of the individual chains.

We can examine the effect of the monomer on bonding between chains by comparing monomers containing just two carbon atoms in the backbone.

where R can be one or more types of atoms or groups of atoms. In polyethylene, all four Rs are hydrogen atoms. The linear chains composed of these symmetrical monomers have relatively little resistance to sliding of the chains under a stress.

Vinyl compounds have one of the hydrogen atoms replaced with a different atom or atom group; several examples are included in Table 15-6. By adding a chlorine atom, we produce a vinyl chloride molecule, which can be polymerized to produce polyvinyl chloride, PVC. Adding a methyl group, CH_3, will produce polypropylene, PP. In polystyrene, PS, a benzene ring, C_6H_5, replaces the hydrogen atom, while polyacrylonitrile, PAN, has an extra carbon and nitrogen atom attached.

When two of the hydrogen atoms are replaced, a series of *vinylidene compounds* are formed; polyvinylidene chloride (more recognizable by the trade name Saran Wrap) and polymethyl methacrylate, PMMA (which includes Lucite and Plexiglas), are examples.

The effects of adding other atoms or atom groups to the carbon backbone in place of hydrogen atoms are illustrated by the typical properties given in Table 15-6. The larger atoms, such as chlorine, or groups of atoms, such as methyl or benzene groups, interfere with the sliding of the chains when a stress is applied; the higher viscosity helps to increase the strength and stiffness of the polymer. In addition to making it more difficult for the chains to slide, some of the more polar atoms or atom groups may provide stronger Van der Waals bonds between chains; the chlorine atom in PVC and the carbon-nitrogen group in PAN perform this double strengthening function.

In polytetrafluoroethylene (more commonly known as Teflon), all four hydrogen atoms are replaced by fluorine, the monomer is again symmetrical, and the strength of the polymer is not much greater than that of polyethylene. However, the C—F bond permits Teflon to have a high melting point with the added benefit of low-friction, nonstick characteristics that make the polymer useful for bearings and cookware.

TABLE 15-6 The mers and properties of selected thermoplastics produced by addition polymerization

Polymer	Structure	Tensile Strength (psi)	% Elongation	Modulus of Elasticity (ksi)	Density (g/cm³)	Izod Impact (ft·lb/in.)	Heat Deflection Temperature at 66 psi (°C)	Applications
Polyethylene (PE) low density (LD) high density (HD)		1,200–3,000 3,000–5,500	50–800 15–130	15–40 60–180	0.92 0.96	1–9 0.4–4	42 85	Packing films, wire insulation, squeeze bottles, tubing, household items
Polyvinyl chloride (PVC)		5,000–9,000	2–100	300–600	1.40			Pipe, valves, fittings, floor tile, wire insulation, vinyl automobile roofs
Polypropylene (PP)		4,000–6,000	10–700	160–220	0.90	0.4–1.0	115	Tanks, carpet fibers, rope, packaging
Polystyrene (PS)		3,200–8,000	1–60	380–450	1.06	0.35–0.45	82	Packaging and insulation foams, lighting panels, appliance and furniture components, egg cartons
Polyvinylidene chloride (PVPC)		3,500–5,000	160–240	50–80	1.15	0.4–1.0	60	Packaging, pipes, draperies

Polymer	Structure							Uses
Polyacrylonitrile (PAN)	H H, H–C–C≡N, C	9,000	3–4	510–580	1.15	1.5–4.8	78	Textile fibers, precursor for carbon fibers, food containers
Polymethyl methacrylate (PMMA) (acrylic-Plexiglas)	H H–C–H, H–C–C=O, H O–C–H, H	6,000–12,000	2–5	350–450	1.22	0.3–0.6	93	Windows, windshields, coatings, hard contact lenses, internally lighted signs
Polychlorotrifluoroethylene	F Cl, C–C, F F	4,500–6,000	80–250	150–300	2.15	2.5–2.7	125	Valve components, gaskets, tubing, electrical insulation
Polytetrafluoroethylene (Teflon) (PTFE)	F F, C–C, F F	2,000–7,000	100–400	60–80	2.17	3	120	Seals, valves, nonstick coatings

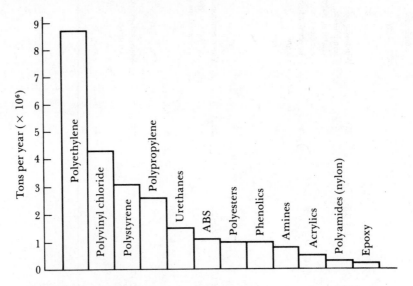

FIGURE 15-21 Consumption of various polymers in the United States in 1985. All others totaled less than 2.7 million tons.

As illustrated in Figure 15-21, the simple polymers such as polyethylene, polyvinyl chloride, polystyrene, and polypropylene make up the greatest percentage of polymers by use. However, a large number of polymers useful for special applications are formed from monomers that are much more complex, often by the condensation mechanism. These monomers may contain oxygen, nitrogen, or sulfur atoms or even benzene rings; the benzene rings incorporated into the chain are called *aromatic* groups. Table 15-7 shows several examples. The acetals contain a backbone of alternating carbon and oxygen atoms. Polyamides, which are characterized by a N—C=O atom group, include nitrogen atoms in the chain. Polyesters contain an O—C=O group and introduce oxygen and an

$$\overset{\displaystyle O}{\underset{\displaystyle \|}{}}$$

aromatic group into the chain. When the O—C—O group introduces two oxygen atoms into the chain, a polycarbonate is formed; in addition, aromatic groups may be introduced. Cellulose, polysulfones, polyimides, polyamide-imides, and aromatic polyamides also typify these complex polymer chains.

When these complex chains are formed, bonding within the chains is significantly increased. The chains are much more difficult to twist and turn, both elastically and plastically. Therefore, these polymers tend to have higher strengths, higher stiffnesses, and higher melting points than the simpler addition polymers. In some cases, significant improvements in impact properties, as illustrated in Table 15-7, can be gained from these complex chains, with polycarbonates being particularly remarkable.

Some of the complex thermoplastic chains may become so stiff that they act as rigid rods, even when heated above the melting point. These materials are

TABLE 15-7 The repeating units and properties for typical thermoplastics having complicated chain structures

Polymer	Structure	Tensile Strength (psi)	% Elongation	Modulus of Elasticity (ksi)	Density (g/cm³)	Izod Impact (ft · lb/in.)	Heat Deflection Temperature at 66 psi (°C)	Applications
Polyoxymethylene (acetal) (POM)		9,500–12,000	25–75	520	1.42	1.2–2.3	165	Plumbing fixtures, pens, bearings, gears, fan blades
Polyamide (nylon) (PA)		11,000–12,000	60–300	400–500	1.14	0.8–2.1	245	Bearings, gears, fibers, rope, automotive components, electrical components
Polyester (PET)		8,000–10,500	50–300	400–600	1.36	0.25–0.65	38	Fibers, photographic film, recording tape, boil-in-bag containers, beverage containers
Polycarbonate (PC)		9,000–11,000	110–130	300–400	1.2	14–16	138	Electrical and appliance housings, automotive components, football helmets, returnable bottles

TABLE 15-7 The repeating units and properties for typical thermoplastics having complicated chain structures

Polymer	Structure	Tensile Strength (psi)	% Elongation	Modulus of Elasticity (ksi)	Density (g/cm³)	Izod Impact (ft · lb/in.)	Heat Deflection Temperature at 66 psi (°C)	Applications
Cellulose		2,000–8,000	5–50	200–250	1.30	0.4	67	Textiles (rayon), cellophane packaging, Ping-Pong balls, adhesives and coatings, microfilm, safety goggles
Polyimide (PI)		11,000–17,000	8–10	300	1.39	1.5	320	Adhesives, circuit boards, fibers for space shuttle
Polyphenylene oxide (PPO)		7,000–9,600	35–60	350	1.06–1.38	6–7	107	Automotive, business machine, and electrical components
Polyetheretherketone (PEEK)		10,200	50–150	550	1.31	1.6	160	High-temperature electrical insulation and coatings

TABLE 15-7 (Continued)

Polymer							Applications
Polyphenylene sulfide (PPS)	9,500	1–2	480	1.3	< 0.5	135	Coatings, fluid-handling components, electronic components, hair dryer components
Polyether sulfone (PES)	12,200	30–80	350	1.37	1.6	200	Electrical components, coffeemakers, hair dryers, microwave oven components
	9,000	50–65	315	1.21	5	180	Traffic signals, microwave cookware
Polyetherimide (PEI)	15,200	60	430	1.27	1	210	Electrical, automotive, and jet engine components
Polyamide-imide (PAI)	17,000–27,000	15	600–730	1.39	2.5–4	267	Electronic components, aerospace and automotive applications

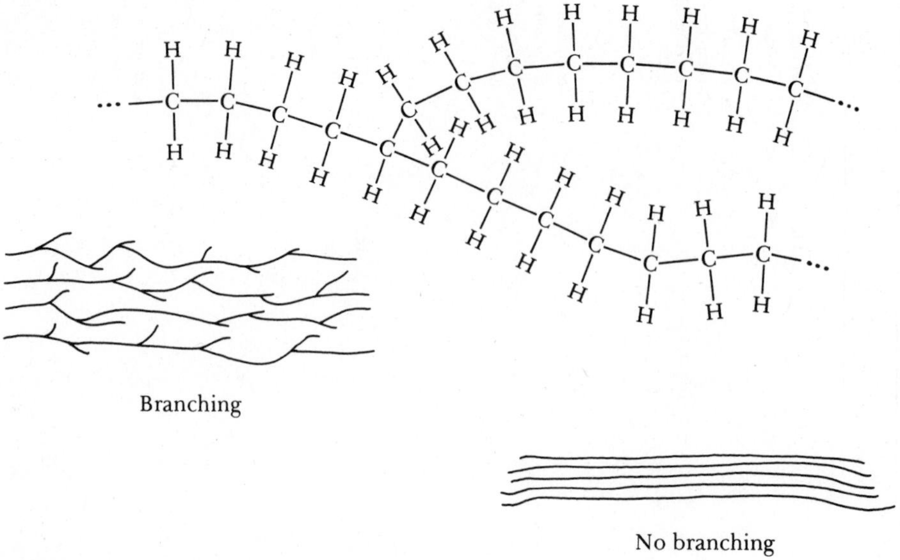

Branching

No branching

FIGURE 15-22 Branching can occur in linear polymers. Branching makes crystallization more difficult.

sometimes characterized as *liquid-crystalline* polymers. Some aromatic polyesters and aromatic polyamides (*aramids*) are examples of liquid-crystalline polymers and are easily made into high-strength fibers, as we will see in Chapter 16.

Branching. Branching occurs when an atom attached to the main linear chain is removed and replaced by another linear chain (Figure 15-22). This can occur several times per 100 carbon atoms in the polymer backbone. Branching prevents dense packing and crystallization of the chains, therefore reducing the density, stiffness, and strength of the polymer. Low-density (LD) polyethylene, which has many branches, is weaker than high-density (HD) polyethylene, which has virtually no branching (Table 15-6).

Copolymers. Linear addition chains composed of two or more types of molecules are called *copolymers*. For example, vinyl chloride and vinyl acetate can copolymerize (Figure 15-23). Both have unsaturated carbon bonds, permitting the mers to join without the formation of a by-product. Any ratio of the two molecules can be combined to give a range of properties.

   Another important copolymer is ABS, composed of acrylonitrile, butadiene (a synthetic rubber), and styrene (Figure 15-24). The styrene and acrylonitrile form a linear copolymer (SAN) that serves as a matrix. Styrene and butadiene also form a linear copolymer, BS rubber, that acts as the filler material. The combination of the two copolymers gives ABS an excellent combination of strength, rigidity, and toughness.

   As shown in Figure 15-25, the arrangement of the monomers can take several forms, including alternating copolymers, random copolymers, block copolymers, and grafted copolymers.

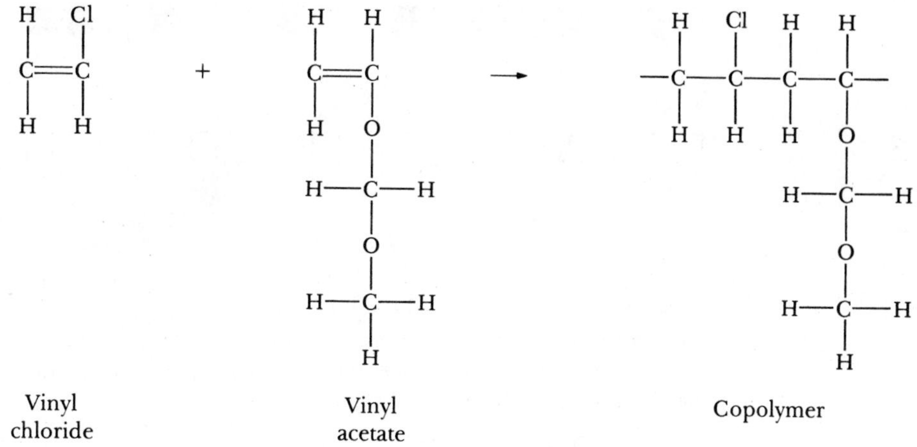

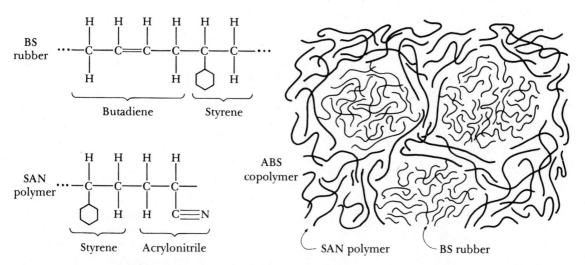

Vinyl
chloride

Vinyl
acetate

Copolymer

FIGURE 15-23 Copolymerization of vinyl chloride and vinyl acetate.

Structures of Nonsymmetrical Polymer Chains. When a polymer is formed from nonsymmetrical mers, the structure is determined by the locations of the nonsymmetrical atoms or atom groups. This is called *tacticity*. In the *syndiotactic* arrangement, the atoms or atom groups alternately occupy positions on opposite sides of the linear chain. The atoms are all on the same side of the chain in *isotactic* polymers, whereas the arrangement of the atoms is random in *atactic* polymers (Figure 15-26). The atactic structure, which is the least regular and least predictable, tends to give the poorest packing, lowest density, lowest strength and stiffness, and worst resistance to heat or chemical attack. These structures are more likely to have an amorphous structure with a relatively high glass

BS
rubber

Butadiene   Styrene

SAN
polymer

Styrene   Acrylonitrile

ABS
copolymer

SAN polymer   BS rubber

FIGURE 15-24 Copolymerization produces the polymer ABS, which is really made up of two copolymers, SAN and BS, grafted together.

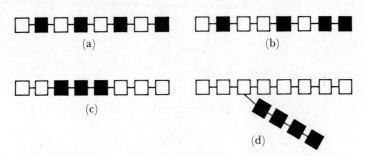

FIGURE 15-25 Four types of copolymers. (a) Alternating monomers, (b) random monomers, (c) block copolymers, and (d) grafted copolymers. Open blocks represent one type of mer; solid blocks represent a second type of mer.

transition temperature. Consequently, the syndiotactic or isotactic structures are normally preferred.

A different effect occurs in isoprene, or natural rubber (Figure 15-27). When isoprene is polymerized, one of the two double bonds is broken and the second is rearranged. In the *trans* form of isoprene, the hydrogen atom and the methyl group at the center of the mer are located on opposite sides of the chain. This arrangement leads to relatively straight chains, producing a very soft thermoplastic polymer called gutta percha. However, when both the hydrogen atom and the methyl group are on the same side of the chain, the monomer bends, or arcs, leading to a kinked chain that eventually forms a coil. This *cis*

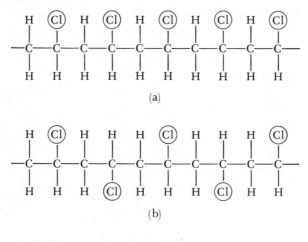

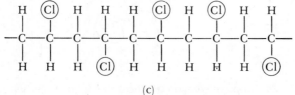

FIGURE 15-26 Three possible arrangements of nonsymmetrical mers. (a) Isotactic, (b) syndiotactic, and (c) atactic.

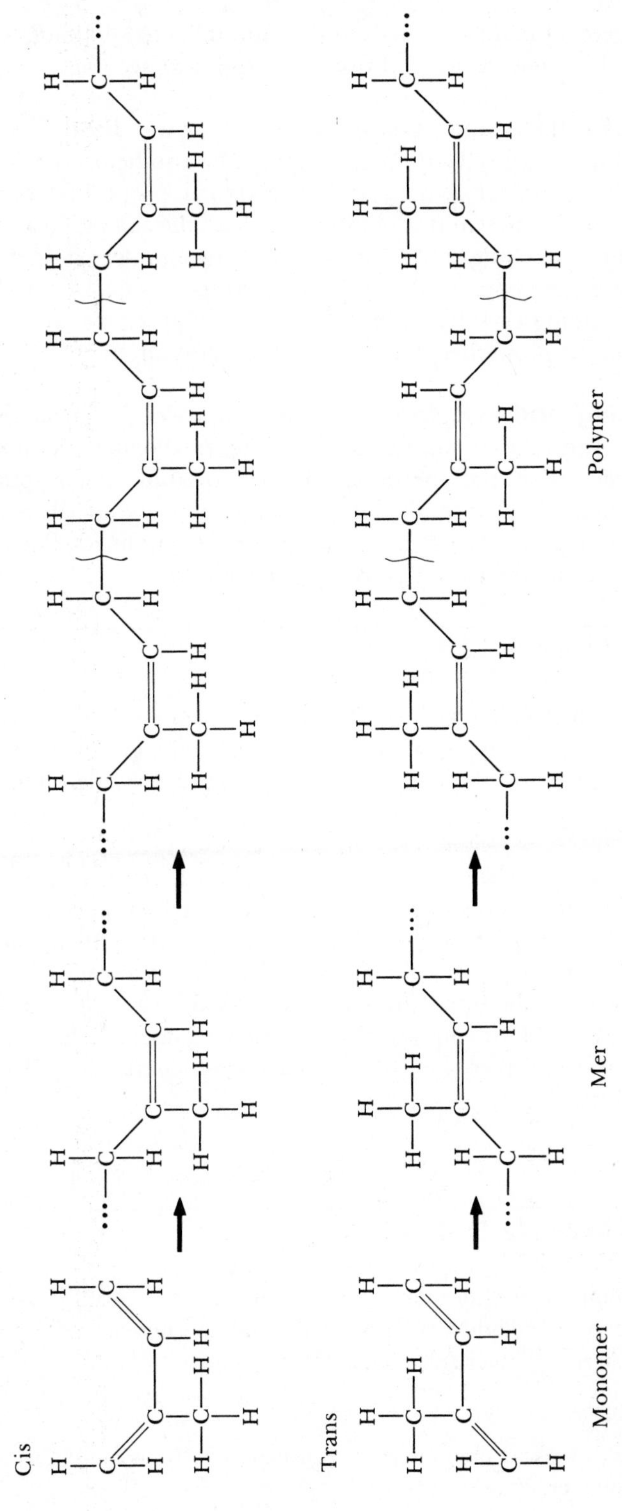

Cis

Trans

Monomer

Mer

Polymer

FIGURE 15-27 The *cis* and *trans* structures of isoprene.

arrangement makes it possible, after suitable cross-linking, for isoprene to serve as an elastomer, as we will discuss in the next section.

Crystallization and Deformation. As we have discussed previously, encouraging crystallization of the polymer also helps to improve density, resistance to chemical attack, and mechanical properties, even at higher temperatures, due to stronger bonding between the chains. In addition, deformation straightens and aligns the chains, producing a preferred orientation and increasing the degree of crystallinity. This leads to the equivalent of strain hardening. By combining crystallization and deformation we can also produce fibers with mechanical properties that exceed those of many metals and ceramics.

Blending and Alloying. We can improve the mechanical properties of many of the thermoplastics by blending or alloying. By mixing an immiscible elastomer with the thermoplastic, we produce a two-phase polymer. The elastomer, added as small globs, does not enter the structure as a copolymer but instead helps to absorb energy and improve toughness. Polycarbonates used for aircraft canopies are toughened in this manner.

◇ | **Example 15-11**

An amorphous polymer is pulled in a tensile test. After a sufficient stress is applied, necking is observed to begin on the gage length. However, the neck disappears as the stress continues to increase. Explain this behavior.

Answer:

Normally, when necking begins, the smaller cross-sectional area increases the stress at the neck and necking is accelerated. However, during the tensile test, the chains in the amorphous structure are straightened out and the polymer becomes more crystalline (Figure 15-28). When necking begins, the chains at the neck align and the polymer is locally strengthened sufficiently to resist further deformation at that location. Consequently, the remainder of the polymer, rather than the necked region, continues to deform until the neck disappears.

◇ | **Example 15-12**

Compare the mechanical properties of LD polyethylene, HD polyethylene, polyvinyl chloride, polypropylene, and polystyrene and explain their differences in terms of their structures.

Answer:

Let's look at the maximum tensile strength and modulus of elasticity for each polymer.

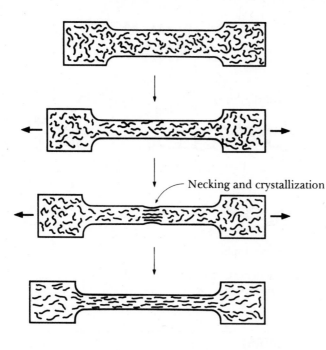

— Necking and crystallization

FIGURE 15-28 Necks are not stable in amorphous polymers because local alignment strengthens the necked region and reduces its rate of deformation.

Polymer	Tensile Strength (psi)	Modulus of Elasticity (ksi)	Structure
LD polyethylene	3000	40	Highly branched, amorphous structure with symmetrical mers
HD polyethylene	5500	180	Amorphous structure with symmetrical mers but little branching
Polypropylene	6000	220	Amorphous structure with small methyl side groups
Polystyrene	8000	450	Amorphous structure with benzene side groups
Polyvinyl chloride	9000	600	Amorphous structure with large chlorine atoms as side groups

We can conclude that

(a) Branching, which reduces the density and close packing of chains, reduces the mechanical properties of polyethylene.

(b) Adding atoms or atom groups other than hydrogen to the chain increases strength and stiffness. The methyl group in polypropylene provides some improvement, the benzene ring of styrene provides higher properties, and the chlorine atom in polyvinyl chloride provides a large increase in properties.

◇

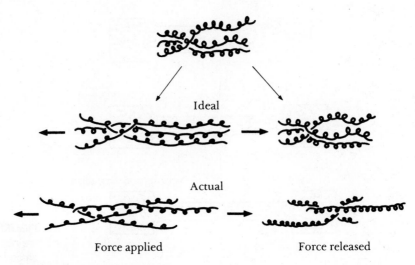

Ideal

Actual

Force applied   Force released

FIGURE 15-29 Schematic diagram showing the recoil of an ideal elastomer versus the actual behavior.

15–9 ELASTOMERS (RUBBERS)

A number of natural and synthetic linear polymers, called *elastomers*, display a large amount of elastic deformation when a force is applied. All of this deformation may be completely recovered when the stress is removed (Figure 15-29). A typical example is an elastic band.

In a typical elastomer monomer, two unsaturated bonds are present. Polymerization involves using one of these unsaturated bonds to produce a linear chain which is coiled due to the *cis* arrangment of the bonds. However, there is still one unsaturated bond left in the chain for each monomer. If a stress is applied to the polymer at this point, the elastomer will behave in a viscoelastic manner; the chains will uncoil and bonds will stretch, producing large amounts of elastic deformation, but the chains will also slide past one another, producing some nonrecoverable plastic deformation. The polymer behaves as a thermoplastic rather than an elastomer. To produce a usable elastomer, we must somehow prevent the viscous deformation.

Cross-Linking. We prevent viscous deformation while retaining large elastic deformation by cross-linking, using the remaining unsaturated bonds in the chains. By breaking these double bonds, we can insert other atoms or atom groups which rigidly connect adjacent chains by strong covalent bonds (Figure 15-30). *Vulcanization*, which joins the coiled chains with sulfur atoms, is a common method for cross-linking many of the elastomers.

The stress-strain curve for an elastomer is shown in Figure 15-31. Virtually all of the curve represents elastic deformation; thus the elastomers display a nonlinear elastic behavior. Initially the modulus of elasticity is relatively low,

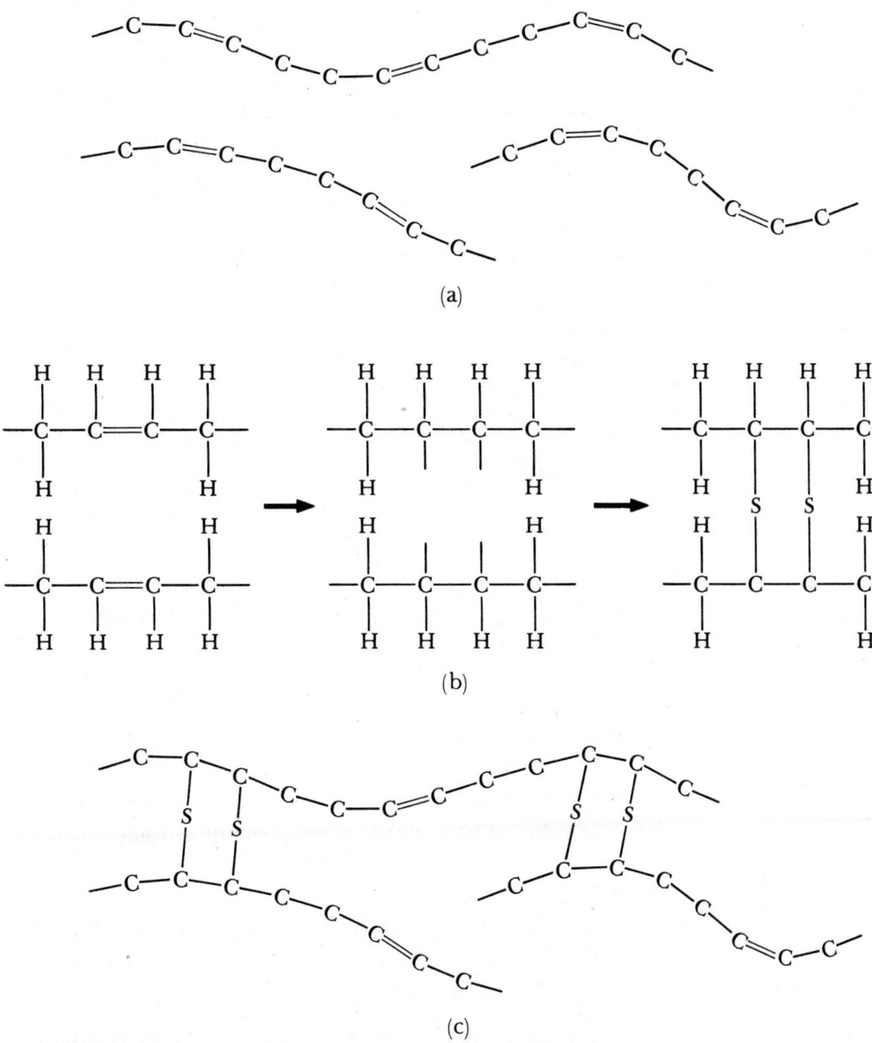

(a)

(b)

(c)

FIGURE 15-30   The individual elastomer chains (a) are joined by sulfur atoms (b) to produce the cross-linked rubber (c).

since most of the elastic deformation is due to uncoiling of the chains. However, after most of the uncoiling is accomplished, further elastic deformation occurs by stretching of the bonds, leading to a higher modulus of elasticity. Because of this nonlinear behavior, no modulus of elasticity is given for the typical elastomers listed in Table 15-8.

The elasticity of the rubber is determined by the number of cross-links, or the amount of sulfur added to the material. Low sulfur additions leave the rubber soft and flexible. Increasing the sulfur content restricts the uncoiling of the chains and the rubber becomes harder, more rigid, and brittle. Up to 30% to 40% sulfur can be added to provide cross-linking in elastomers.

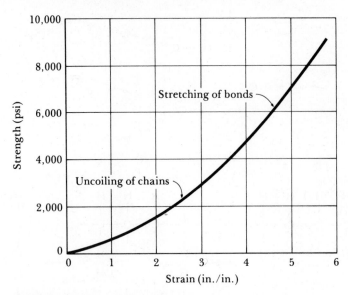

FIGURE 15-31 The stress-strain curve for an elastomer.
Virtually all of the deformation is elastic; therefore, the modulus
of elasticity varies as the strain changes.

◇ | **Example 15-13**

If 20% of the possible cross-linking sites in isoprene are used in vulcanization, calculate the wt% S that is present in the rubber.

Answer:

On inspection of the cross-linking process shown in Figure 15-30, we find that, on average, a maximum of one sulfur atom per monomer would be required to produce complete cross-linking.

$$\text{Mol wt isoprene} = (5\,\text{C atoms})(12) + (8\,\text{H atoms})(1) = 68\,\text{g/g}\cdot\text{mole}$$

$$\text{Mol wt sulfur} = 32\,\text{g/g}\cdot\text{mole}$$

$$\% \text{S} = \frac{(0.2)(1\,\text{S atom})(M_\text{S})}{(0.2)(1\,\text{S atom})(M_\text{S}) + (1\,\text{mer})(M_\text{isoprene})} \times 100$$

$$= \frac{(0.2)(32)}{(0.2)(32) + 68} \times 100 = 8.6\%$$

Polyisoprene is natural rubber and has been duplicated synthetically. A common synthetic rubber is the copolymer BS (or BSR), butadiene-styrene rubber (Figure 15-24). Several other common elastomers are listed in Table 15-8.

TABLE 15-8 The repeating units and properties of selected elastomers

Polymer	Structure	Tensile Strength (psi)	% Elongation	Density (g/cm³)	Applications
Polyisoprene	(structure)	3000	800	0.93	Tires
Polybutadiene	(structure)	3500		0.94	Industrial tires, vibration mounts
Polybutylene	(structure)	4000	350	0.92	Piping, insulation, coatings
Polychloroprene (Neoprene)	(structure)	3500	800	1.24	Hoses, cable sheathing

TABLE 15-8 (Continued)

Polymer	Structure	Tensile Strength (psi)	% Elongation	Density (g/cm³)	Applications
Butadiene-styrene (BS or SBR rubber)		600–3000	600–2000	1.0	Tires
Butadiene-acrylonitrile		700	400	1.0	Gaskets, fuel hoses
Silicone		350–1000	100–700	1.5	Gaskets, seals

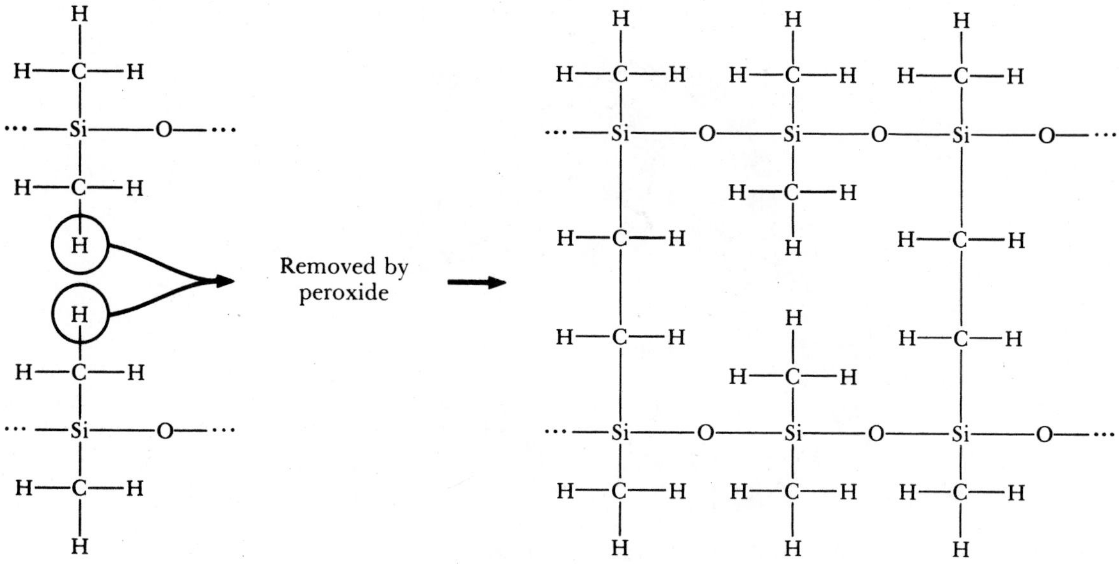

FIGURE 15-32 *Cross-linking in silicone rubbers.*

In some elastomers, a different cross-linking mechanism may be employed. Silicone rubbers are based on a linear chain composed of silicon and oxygen atoms. Because the chain contains no unsaturated bonds, cross-linking is obtained by introducing a peroxide, which removes hydrogen atoms from the CH_3 groups and permits carbon atoms from adjoining chains to be bonded. The silicone rubbers (Figure 15-32) provide high temperature resistance, permitting use of the rubber at temperatures as high as 315°C.

Thermoplastic Elastomers. Thermoplastic elastomers are a special group of polymers which do not rely on cross-linking to produce the large amount of elastic deformation. There are four main types of thermoplastic elastomers (TPEs)—styrene-butadiene block copolymers, olefinic copolymers, urethanes, and polyester block copolymers.

As an example, Figure 15-33 shows the structure of the SB thermoplastic elastomer, which is a block copolymer with butadiene monomers in the middle of the chain and styrene monomers at the ends. The styrene ends of adjoining chains are attracted to one another by strong polar bonds, producing domains, which tightly hold the chains together. In between the domains are rubbery areas containing butadiene monomers. Elastic deformation occurs by recoverable movement of the chains; however, sliding of the chains at normal temperatures is prevented by the styrene domains.

The SB block copolymers differ from the BS rubber discussed earlier in that cross-linking of the butadiene monomers is not necessary and in fact may be undesirable. The SB block copolymer, when heated, will deform in a viscous manner as would any other thermoplastic, making fabrication very easy, while

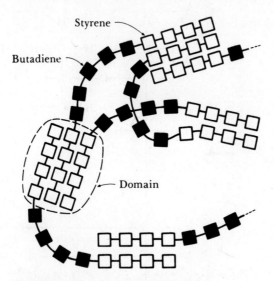

FIGURE 15-33 *The structure of the SB copolymer in a thermoplastic elastomer. The strong polar attractions between the styrene portions of the polymer provide elastic behavior without cross-linking of the butadiene.*

reverting to its elastomer characteristics at lower temperatures. Cross-linking of the butadiene would restrict the viscoelastic behavior and the ability to form the polymer.

Stress Relaxation.   When an elastomer is subjected to a constant strain rather than a constant stress, viscous flow causes the stress acting on the polymer to decrease over time. When a stress is applied to an elastomer, its chains uncoil to produce a corresponding strain. Over time, some viscous flow occurs even in vulcanized rubber as the chains are rearranged. This permits the stretched chains to recoil. Since the dimensions are fixed due to the constant strain, the force or stress acting on the polymer is reduced, or relaxed.

The rate at which stress relaxation occurs depends on the *relaxation time* λ, which is a property of the polymer. The stress after time t is given by

$$\sigma = \sigma_0 \exp\left(-\frac{t}{\lambda}\right) \tag{15-8}$$

where σ_0 is the original stress. The relaxation time depends on the viscosity and therefore is affected by the temperature.

$$\lambda = \lambda_0 \exp\left(\frac{E_\eta}{RT}\right) \tag{15-9}$$

where E_η is the activation energy for viscous flow and λ_0 is a constant.

◇ **Example 15-14**

A stress of 1000 psi is required to stretch polyisoprene around a set of books. After six weeks, the stress acting on the rubber is only 980 psi. How much stress will remain after one year?

Answer:

The stress has been reduced by stress relaxation. The strain is still unchanged. Thus,

$$\sigma = \sigma_0 \exp\left(-\frac{t}{\lambda}\right)$$

$$980 = 1000 \exp\left(-\frac{6}{\lambda}\right)$$

$$-\frac{6}{\lambda} = \ln\left(\frac{980}{1000}\right) = \ln(0.98) = -0.0202$$

$$\lambda = \frac{6}{0.0202} = 297 \text{ weeks}$$

Thus in one year (52 weeks),

$$\sigma = 1000 \exp\left(-\frac{52}{297}\right) = 1000 \exp(-0.175) = 1000(0.839)$$

$$= 839 \text{ psi}$$

15–10 THERMOSETTING POLYMERS

Thermosetting polymers typically are formed by first producing linear chains, then cross-linking the chains to produce a three-dimensional framework structure. A condensation reaction could be involved in forming the original chains or in cross-linking the chains. Often the thermosetting polymer materials are obtained in a two-part liquid resin. When the two parts are mixed, the cross-linking reaction begins. In other cases, heat and pressure are used to initiate the cross-linking action.

The functional groups for a number of thermosetting polymers are summarized in Table 15-9.

Phenolics. The condensation reaction joining phenol and formaldehyde molecules provides the basis for the phenolic resins (Figure 15-34). The oxygen atom in the formaldehyde molecule reacts with a hydrogen atom on each of two phenol molecules and water is evolved as the by-product. The two phenol molecules are then joined by the remaining carbon atom in the formaldehyde.

TABLE 15-9 Functional groups for several common thermosetting polymers

Polymer	Structure	Tensile Strength (psi)	% Elongation	Modulus of Elasticity (ksi)	Density (g/cm³)	Typical Applications
Phenolics		5,000–9,000	0–2	400–1300	1.27	Adhesives, coatings, laminates
Amines	Melamine Urea	5,000–10,000	0–1	1000–1600	1.50	Adhesives, cookware, electrical moldings
Polyesters		6,000–13,000	0–3	300–650	1.28	Electrical moldings, decorative laminates, matrix in fiberglass

	Structure					Applications
Epoxies		4,000–15,000	0–6	400–500	1.25	Adhesives, electrical moldings
Urethanes		5,000–10,000	3–6		1.30	Fibers, coatings, foams, insulation
Furan		3,000–4,500	0	1580	1.75	Binders for molding sands
Silicone		3,000–4,000	0	1200	1.55	Adhesives, gaskets, sealants

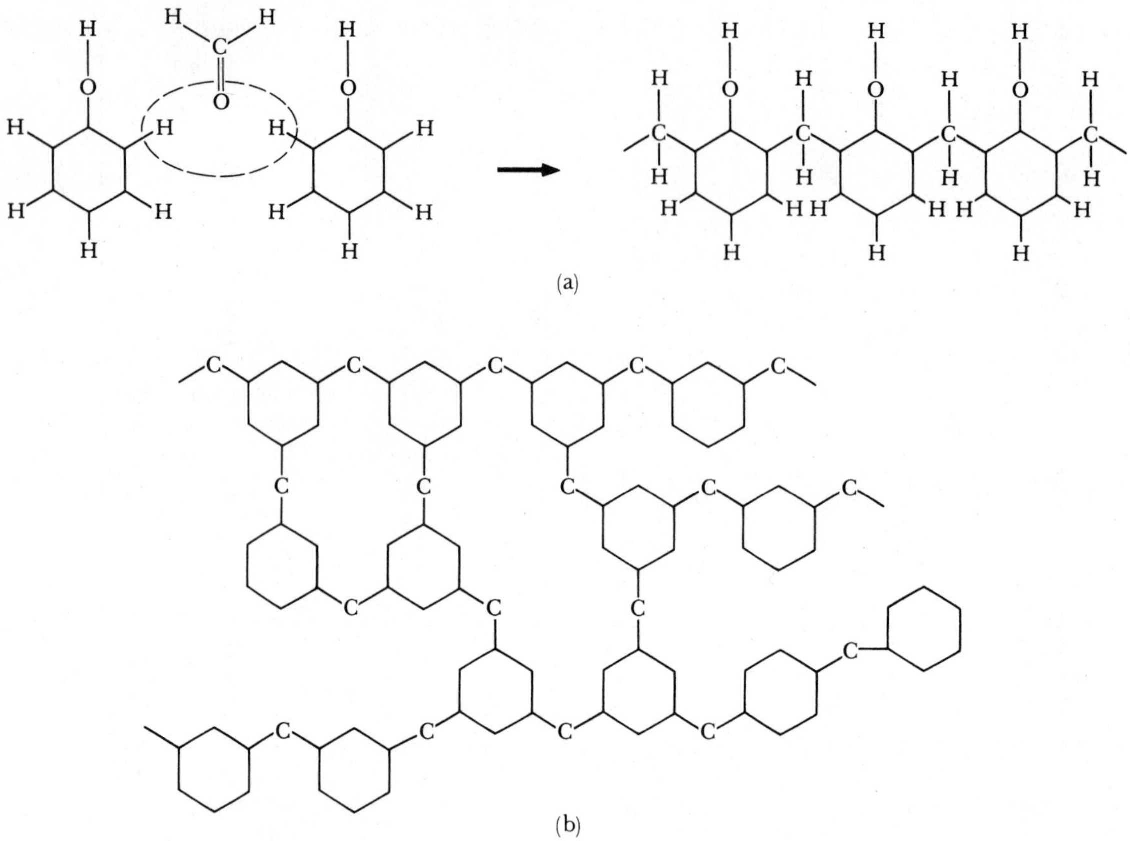

(a)

(b)

FIGURE 15-34 *Structure of a resole phenolic. In (a), two phenol rings are joined by a condensation reaction through a formaldehyde molecule. Eventually, a linear chain forms. In (b), excess formaldehyde serves as the cross-linking agent, producing a network, thermosetting polymer.*

Since phenol is trifunctional, the same reaction can occur at other locations on each of the rings, binding each phenol molecule to several others.

Initially, a linear chain is produced. Two important resins are produced from this reaction. The *resole* resins, which contain an excess of formaldehyde, are linear chains with little or no cross-linking. When the resole resins are subsequently heated, the excess formaldehyde provides the cross-linking required to produce the three-dimensional structure.

When phenol and formaldehyde are combined with excess phenol, a *novalac* chain is produced. The novalac resin can then be blended with hexamethylene-tetramine $(C_2N_4H_{12})$ or other agents which, when heated, provide cross-linking.

The phenolic resins are commonly used in adhesives, coatings, laminates, and even as binders for foundry sands and cores.

Amines. The amino resins, produced by combining urea-formaldehyde or melamine-formaldehyde, are similar to the phenolics. The urea or melamine molecules are joined by a formaldehyde link to produce linear chains. Excess formaldehyde can provide the cross-linking needed to give strong, rigid polymers suitable for adhesives, laminates, and molding materials for cookware and electrical applications.

◇ | **Example 15–15**

Describe the formation of an amino polymer using urea and formaldehyde.

Answer:

The molecular structures for the monomers of urea and formaldehyde are shown below. A linear chain can be formed when hydrogen atoms from the urea combine with the oxygen atom from the formaldehyde.

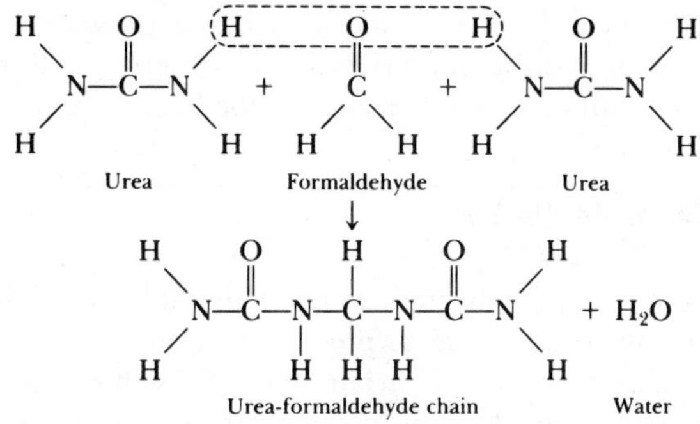

Urea Formaldehyde Urea

Urea-formaldehyde chain Water

This chain can be extended indefinitely as the condensation reaction continues. If additional formaldehyde is present, other hydrogen atoms may also react with the formaldehyde to cross-link the chains into a framework structure.

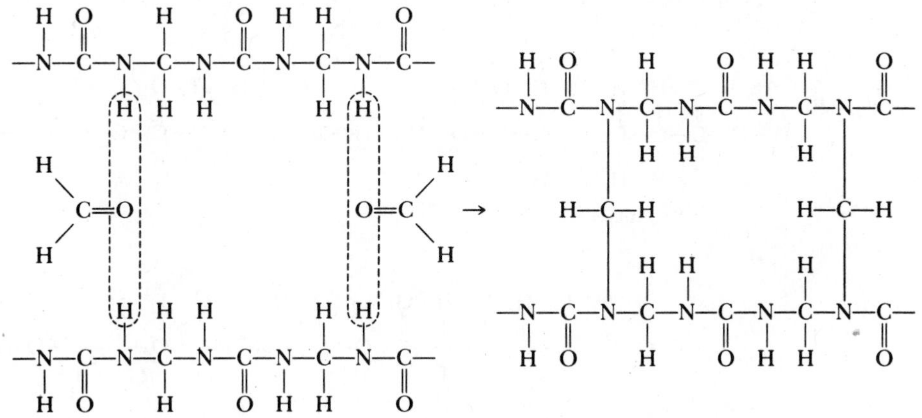

Amino polymer

The functionality of the urea monomer may be as great as four.

◇

Urethanes. Urethanes, which normally contain a $\text{N}-\overset{\overset{\displaystyle O}{\|}}{\text{C}}-\text{O}$ group, are produced when isocyanate molecules combine with other organic molecules to produce a linear polymer by an addition mechanism. The chains are then cross-linked. Depending on the degree of cross-linking, the urethanes may behave as thermosetting polymers, thermoplastic polymers, or elastomers. These polymers find application as fibers, coatings, and foams for furniture, mattresses, and insulation.

Polyesters. Polyesters form chains from acid and alcohol molecules by a condensation reaction, giving water as a by-product. The acid component of the chain contains an unsaturated carbon bond. A vinyl molecule, such as styrene, cross-links the chains by an addition reaction. Depending on the cross-linking, the polyesters can be either thermosetting or thermoplastic. Polyesters are used as molding or casting materials for a variety of electrical applications and decorative laminates and as a matrix for fiber-reinforced composites.

◇ | **Example 15-16**

A thermosetting polyester can be produced by combining adipic acid, ethylene glycol, and maleic acid to produce a linear condensation polymer chain containing an unsaturated carbon bond. Show how the chain develops and how styrene can provide cross-linking of the chains into a framework structure.

Answer:

The monomers are shown below. First, a chain forms.

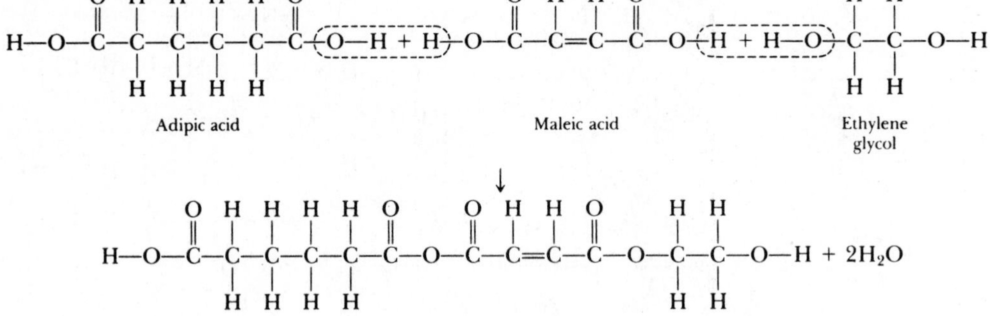

The linear chain contains an unsaturated carbon bond which, when styrene is added, is broken. The chains can then be joined by an addition mechanism.

```
   O  H  H  H  H  O     O  H  H  O        H  H
   ‖  |  |  |  |  ‖     ‖  |  |  ‖        |  |
—O—C——C——C——C——C——C——O——C——C══C——C——O————C——C——O—
      |  |  |  |           |  |           |  |
      H  H  H  H           H  H           H  H

                              +
                         H——C——H
                              ‖
                         H——C——⬡

                              +

      H  H  H  H                       H  H
      |  |  |  |                       |  |
—O——C——C——C——C——C——O——C——C══C——C——O————C——C——O—
    ‖  |  |  |  ‖     ‖  |  |  ‖        |  |
    O  H  H  H  O     O  H  H  O        H  H

   O  H  H  H  H  O     O  H  ↑  O        H  H
   ‖  |  |  |  |  ‖     ‖  |  |  ‖        |  |
—O—C——C——C——C——C——C——O——C——C——C——C——O————C——C——O—
      |  |  |  |           |           |  |
      H  H  H  H           H           H  H

                         H——C——H
                              |
                         H——C——⬡
                              |

      H  H  H  H           H           H  H
      |  |  |  |           |           |  |
—O——C——C——C——C——C——O——C——C——C——C——O————C——C——O—
    ‖  |  |  |  ‖     ‖  |  ↓  ‖        |  |
    O  H  H  H  O     O  H     O        H  H
```

◇

Epoxies. Complex monomers that contain a tight $C—O—C$ ring are polymerized by an addition mechanism into linear chains, which then join into
framework structures. During polymerization, the $C—O—C$ rings are broken
(called *ring scission*) and the bonds are rearranged to join the molecules (Figure
15-35). Epoxies are often used as adhesives, as rigid molded parts for electrical
applications, and as a matrix for composites.

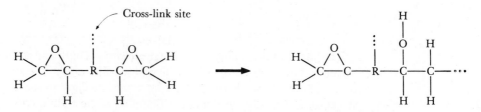

FIGURE 15-35   Ring scission is responsible for polymerization in epoxies.

Polyimides. Some of the polyimides can be cross-linked to produce a thermo-setting polymer. One special group—the bismaleimides (BMI)—can develop relatively good temperature resistance.

Interpenetrating Polymer Networks. Some special polymer materials can be produced when linear thermoplastic chains are intertwined through a thermo-setting framework. For example, nylon, acetal, or polypropylene chains can penetrate into a cross-linked silicone thermoset. In more advanced systems, two interpenetrating thermosetting framework structures can be produced. This structure helps provide superior mechanical properties.

Deformation of Thermosetting Polymers. Because of the rigid three-dimensional framework structure, groups of atoms cannot readily move in thermosetting polymers. In a tensile test, thermosetting polymers should have the same behavior as a brittle metal or ceramic.

Most of the thermosetting polymers have high strength, low ductility, a high modulus of elasticity, and poor impact properties compared with other polymers. As the temperature increases, the properties decrease due to greater stretching of bonds and degradation of the polymer.

◇ **Example 15-17**

What conclusions can we make concerning the mechanical properties of polymers when we compare linear addition thermoplastics, linear condensation thermo-plastics, and thermosetting polymers?

Answer:

Let's look at the range of maximum properties from Tables 15-6, 15-7, and 15-9.

Polymer	Tensile Strength (psi)	% Elongation	Modulus of Elasticity (ksi)
Linear addition thermoplastics	3000–12,000	5–800	40–600
Linear condensation thermoplastics	8000–17,000	10–300	250–600
Thermosetting polymers	4000–15,000	0–6	500–1600

The linear addition polymers have the lowest strength and stiffness but the highest ductility. The thermosets have the highest strength and stiffness but are brittle. Most linear condensation thermoplastics have intermediate properties; their molecular structure is normally more complex than that of the addition polymers, but they are not cross-linked as are the thermosets.

◇

15–11 ADDITIVES TO POLYMERS

Most polymers contain additives which impart special characteristics to the material.

Pigments. Pigments are used to produce colors in plastics and paints. The pigment must withstand the temperatures and pressures during processing of the polymer, must be compatible with the polymer, and must be stable.

Stabilizers. Stabilizers prevent deterioration of the polymer due to environmental effects. Antioxidants are added to ABS, polyethylene, and polystyrene. Heat stabilizers are required in processing polyvinyl chloride; otherwise, hydrogen and chlorine atoms may be removed as hydrochloric acid, causing the polymer to be embrittled. Stabilizers also prevent deterioration due to ultraviolet radiation.

Antistatic Agents. Most polymers, because they are poor conductors, build up a charge of static electricity. Antistatic agents attract moisture from the air to the polymer surface, improving the surface conductivity of the polymer and reducing the likelihood of a spark or discharge.

Flame Retardants. Most polymers, because they are organic materials, are flammable. Additives that contain chlorine, bromine, phosphorus, or metallic salts reduce the likelihood that combustion will occur or spread.

Plasticizers. Plasticizers are low molecular weight molecules or chains which, by reducing the glass transition temperature, improve the properties and forming characteristics of the polymer. Plasticizers are particularly important for polyvinyl chloride, which has a glass transition temperature well above room temperature. However, adding too much plasticizer may cause the polymer to become a liquid.

Fillers. Fillers are added for a variety of purposes. Perhaps the best known example is the addition of carbon black to rubber, which improves strength and wear resistance of tires. Some fillers, such as short fibers or flakes of inorganic materials, improve the mechanical properties of the polymer. Others, called *extenders*, permit a large volume of a polymer material to be produced with relatively little actual resin. Calcium carbonate, silica, and clay are frequently used extenders.

Reinforcement. The strength and rigidity of polymers are improved by introducing glass, polymer, or carbon filaments. For example, fiberglass consists of short filaments of glass in a polymer matrix.

15–12   FORMING OF POLYMERS

There are four major methods for producing polymer shapes—molding, extrusion, manufacture of films, and manufacture of fibers. The techniques used to form the polymers depend to a large extent on the nature of the polymer—in particular, whether it is thermoplastic or thermosetting. Typical processes are shown in Figures 15-36 and 15-37.

The greatest variety of techniques are used to form the thermoplastic polymers. The polymer is heated to near or above the melting temperature so that the polymer is plastic or liquid. The polymer is then cast or injected into a mold or forced into or through a die to produce the required shape. Thermoplastic elastomers can be formed in much the same manner.

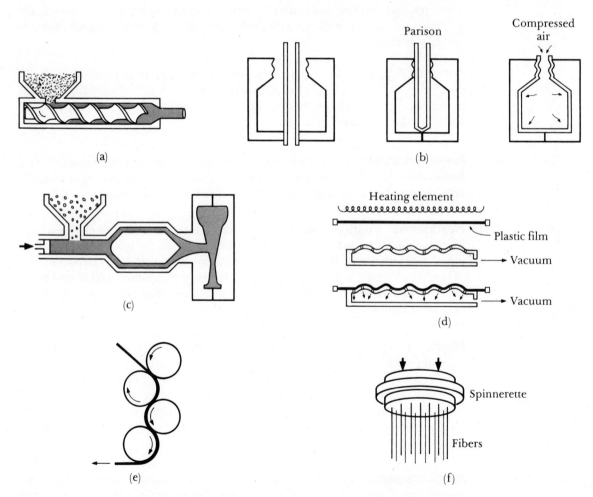

FIGURE 15-36   Typical forming processes for thermoplastic polymers. (a) Extrusion, (b) blow molding, (c) injection molding, (d) thermoforming, (e) calendaring, and (f) spinning.

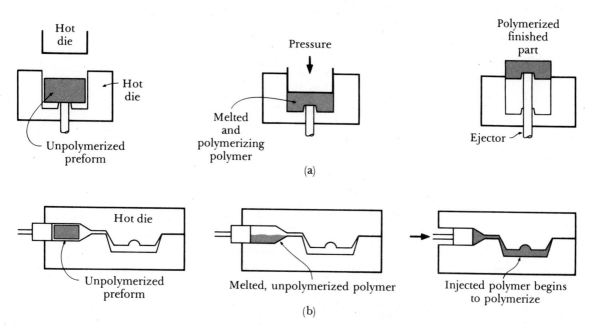

FIGURE 15-37 Typical forming processes for thermosetting polymers. (a) Compression molding and (b) transfer molding.

Fewer forming techniques are used for the thermosetting polymers because once polymerization has occurred and the framework structure is established, the thermosetting polymers are no longer capable of being formed. After vulcanization, elastomers also can be formed no further.

Extrusion. A screw mechanism forces the heated thermoplastic through a die opening to produce solid shapes, films, sheets, tubes, pipes, and even plastic bags. One special extrusion process for producing films is illustrated in Figure 15-38. Extrusion can also be used to coat wires and cables with either thermoplastics or elastomers.

Blow molding. A hot glob of plastic polymer, called a *parison*, is introduced into a die and, by gas pressure, expanded against the walls of the die. This process is used to produce plastic bottles, containers, and many other hollow shapes.

Injection Molding. Thermoplastics heated above the melting temperature can be forced into a closed die to produce a molding. This process is similar to the die casting of molten metals. A plunger or a special screw mechanism applies the pressure to force the hot polymer into the die. A wide variety of products, ranging from cups, combs, and gears to garbage cans, can be produced in this manner.

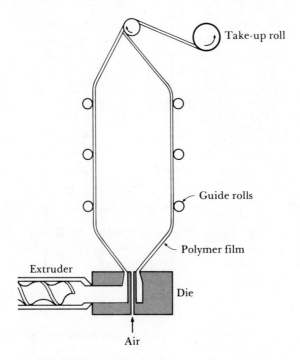

FIGURE 15-38 One technique by which polymer films can be produced. The film is extruded in the form of a bag, which is separated by air pressure until the polymer cools.

Thermoforming.   Thermoplastic polymer sheets heated into the plastic region can be formed over a die to produce such diverse products as egg cartons and decorative panels. The forming can be done using matching dies, a vacuum, or air pressure. Often, special polymer compositions, called sheet molding compounds (SMCs), are required for thermoforming processes.

Calendaring.   In a calendar, molten plastic is poured into a set of rolls with a small opening. The rolls, which may be embossed with a pattern, squeeze out a thin sheet of the polymer, usually polyvinyl chloride. Typical products include floor tile, shower curtains, and thin films.

Rotational Molding.   Molten polymer can be poured into a rotating mold, allowing centrifugal action to produce a thin molded shape, such as a camper top.

Drawing and Rolling.   These processes produce fibers or change the shape of extrusions. In addition to producing the final dimensions, these processes cause crystallization and a preferred orientation of the chains in thermoplastic polymers.

Spinning.   Filaments, fibers, and yarns may be produced by spinning, which is really an extrusion process. The molten thermoplastic polymer is forced through a die containing many tiny holes. The die, called a *spinnerette*, can rotate and produce a yarn. For some materials, including nylon, the fiber may subsequently be stretched to align the chains parallel to the axis of the fiber; this increases the strength of the fibers.

Foams.   Foamed products can be produced in polystyrene, urethanes, polymethyl methacrylate, and a variety of other polymers. The polymer is produced in the form of tiny beads, often containing a foaming agent which will decompose to nitrogen, carbon dioxide, or another gas when heated. During this preexpansion process, the bead may increase in diameter by as many as 50 times and will become hollow. The preexpanded beads can then be injected into a die, with the individual beads fusing together, to form exceptionally lightweight products with densities of perhaps only $0.015 \, \text{g/cm}^3$. Styrofoam cups, packaging, and insulation are some of the myriad applications for these foamed products.

Compression Molding.   Thermosetting polymers can be formed by placing the solid material into a heated die. Application of a high pressure and temperature causes the polymer to melt, fill the die, and immediately begin to harden.

Transfer Molding.   A double chamber is used in transfer molding of thermosetting polymers. The polymer is heated under pressure in one chamber; after melting, the polymer is injected into the adjoining die cavity. This process combines elements of both compression and injection molding and permits some of the advantages of injection molding to be used for thermosetting polymers.

Reaction Injection Molding (RIM).   Thermosetting polymers normally are difficult to injection mold. However, in reaction injection molding, the liquid resins are injected first into a mixer and then directly into a heated mold to produce a shape. The forming and curing occur simultaneously in the mold. In reinforced reaction injection molding (RRIM), a reinforcing material consisting of particles or short fibers is introduced with the liquid resins to produce a composite material.

Casting.   Many polymers can be cast into molds and permitted to solidify. The molds may be plate glass for producing individual thick plastic sheets or moving stainless steel belts for continuous casting of thinner sheets.

SUMMARY

Polymers fit into three main categories—thermoplastics, thermosets, and elastomers—depending on the structure and properties that are achieved. The properties of thermoplastics, which include good ductility but low strength, are controlled by preventing the linear chains from sliding, much like our control of slip in metals, and by stiffening the individual chains. The properties of elastomers, which include exceptional elastic elongations, are controlled by cross-linking. The structure of the thermosets is typified by a network that provides high strengths but poor ductility. The importance of structure and processing on the properties is particularly evident in polymer materials.

GLOSSARY

Addition polymers. Polymer chains built up by adding monomers together without creating a by-product.

Aramids. Polyamide polymers containing aromatic groups of atoms in the linear chain.

Branching. When a separate polymer chain is attached to another chain.

Cis. A monomer form in which the unsaturated bonds are geometrically located on the same side of the molecule.

Condensation polymer. Polymer chains built up by a chemical reaction between two or more molecules, producing a by-product.

Copolymer. An addition polymer produced by joining more than one type of monomer.

Cross-linking. Attaching chains of polymers together to produce a three-dimensional network polymer.

Degradation temperature. The temperature above which a polymer burns, chars, or decomposes.

Degree of polymerization. The number of monomers in a polymer.

Elastomers. Polymers possessing a highly kinked and partly cross-linked chain structure permitting the polymer to have exceptional elastic deformation.

Extenders. Additives or fillers to polymers that provide bulk at a low cost.

Functionality. The number of sites on a monomer at which polymerization can occur.

Glass transition temperature. The temperature below which the amorphous polymer assumes a rigid glassy structure.

Heat deflection temperature. The temperature at which a polymer will deform a given amount under a standard load.

Interpenetrating polymer networks. Polymer structures produced by intertwining two separate polymer structures or networks.

Isomer. A molecule that has the same composition as, but a structure different from, a second molecule.

Liquid-crystalline polymers. Exceptionally stiff polymer chains that act as rigid rods, even above their melting point.

Mer. The molecule from which a polymer is produced after the double covalent bond has been broken.

Monomer. The molecule from which a polymer is produced.

Parison. A hot glob of soft or molten polymer that is blown or formed into a useful shape.

Plasticizer. An additive that reduces the glass transition temperature, thus improving the formability of a polymer.

Reinforcement. Additives to polymers designed to provide significant improvement in strength. Fibers are typical reinforcements.

Spinnerette. An extrusion die containing many small openings through which the hot or molten polymer is forced to produce filaments. Rotation of the spinnerette twists the filaments into a yarn.

Stress relaxation. A reduction of the stress acting on a material over a period of time at a constant strain due to viscoelastic deformation.

Tacticity. Describes the location in the polymer chain of atoms or atom groups in nonsymmetrical mers.

Thermoplastic elastomers. Polymers that behave as thermoplastics at high temperatures but as elastomers at lower temperatures.

Thermoplastic polymers. Polymers that can be reheated and remelted numerous times, since no by-product forms during processing.

Thermosetting polymers. Polymers that polymerize at high temperatures, releasing a by-product and thus restricting their recyclability.

Trans. A monomer form in which the unsaturated bonds are geometrically located on opposite sides of the molecule.

Unsaturated bond. The double or even triple covalent bond joining two atoms together in an organic molecule. When a single covalent bond replaces the unsaturated bond, polymerization can occur.

Viscoelasticity. The deformation of a polymer by viscous flow of the chains or segments of the chains when a stress is applied.

Vulcanization. Cross-linking elastomer chains by introducing sulfur atoms at elevated temperatures and pressures.

PRACTICE PROBLEMS

1. The formula for formaldehyde is HCHO. (a) Draw the structure of the formaldehyde molecule and mer. (b) Does formaldehyde polymerize to produce an acetal polymer by an addition reaction or a condensation reaction? Try to produce a polymer by each method. (c) Draw the acetal structure produced from formaldehyde.

2. Show that the angle between the covalent bonds in the carbon tetrahedron is approximately 109°.

3. The distance between the centers of two adjacent carbon atoms in linear polymers is approximately 1.5 Å. Determine the length of a polyethylene chain containing 10,000 mers.

4. Suppose the distance between the centers of two adjacent carbon atoms in polyvinyl chloride is 1.5 Å. Determine the molecular weight of a polymer chain that is 2500 Å in length.

5. The molecular weight of a polypropylene polymer is 200,000 g/g · mole. Determine the degree of polymerization of the polymer.

6. The degree of polymerization in polyethylene is 7500. Calculate the number of polyethylene chains in 1 g of the polymer.

7. Calculate (a) the weight and (b) the volume of polystyrene when we produce 10^{20} chains of polystyrene having a degree of polymerization of 10,000.

8. Suppose a polypropylene rope weighs 0.25 lb per foot. If the degree of polymerization of the polymer is 5000, calculate the number of polypropylene chains in a 10-ft length of rope.

9. Calculate (a) the grams of H_2O_2 initiator that must be added to 1000 g of methyl methacrylate to produce a degree of polymerization of 5000 and (b) the total number of chains produced. (See Table 15-6.)

10. If 10 g of hydrogen peroxide (H_2O_2) is added to 2000 g of acrylonitrile, calculate (a) the degree of polymerization of the polyacrylonitrile and (b) the total number of chains that are formed.

11. Suppose 1 kg of vinyl chloride is to be polymerized. (a) How many g · mole of vinyl chloride are required? (b) How much energy is released during polymerization? (c) If the specific heat of polyvinyl chloride is 0.24 cal/g · °C, calculate the temperature change during polymerization.

12. Suppose 1 kg of phenylene oxide is polymerized. The monomer for polyphenylene oxide is shown below. (a) Describe the polymerization of phenylene oxide, including the by-product formed. (b) How many individual condensation steps are required? (c) How many grams of the by-product are produced?

13. The polymerization of nylon 6,6 is shown in Example 15-5. Calculate (a) the energy involved when 500 g of monomer are polymerized and (b) the change in temperature when polymerization occurs. The specific heat of nylon is about 0.4 cal/g · °C.

14. Suppose a cellulose polymer as described in Table 15-7 is produced. (a) Sketch the structure of the cellulose monomer, assuming that water is produced as the by-product. (b) If 2 kg of the cellulose monomer are polymerized, calculate the weight of the polymer and the weight of the evolved water.

15. A copolymer is produced by combining 1 kg of ethylene and 2 kg of propylene with 5 g of hydrogen peroxide. Calculate (a) the degree of polymerization and (b) the molecular weight of the polymer.

16. Chlorotrifluoroethylene and ethylene are combined to give 5 kg of a copolymer with a degree of polymerization of 2500 and a molecular weight of 128,600 g/g · mole. Determine (a) the amount in grams of each monomer used and (b) the amount in grams of hydrogen peroxide initiator required.

17. Would you expect that addition polymerization would occur faster or slower than condensation polymerization? Explain.

18. From the data in the table, determine the weight average molecular weight and the number average molecular weight of the polymer.

Molecular Weight Range (g/g · mole)	f_i	x_i
0–5,000	0.02	0.04
5,000–10,000	0.08	0.09
10,000–15,000	0.16	0.18
15,000–20,000	0.26	0.29
20,000–25,000	0.27	0.24
25,000–30,000	0.14	0.12
30,000–35,000	0.05	0.03
35,000–40,000	0.02	0.01

19. Figure 15-39 shows the distribution of molecular weight of polymer chains, by both weight fraction and number fraction. Estimate the weight average molecular weight and the number average molecular weight of the polymer.

20. Explain why you would prefer that the number average molecular weight of a polymer be as close as possible to the weight average molecular weight.

21. Using Table 15-5, construct a graph to determine the relationship between the glass transition temperatures and the melting temperatures of the addition thermoplastic polymers.

22. The viscosity of polymers is typically 10^{11} poise at the glass transition temperature and 10^4 poise 35°C above the glass transition temperature. (a) Estimate the activation energy for viscous flow in polyacrylonitrile. (b) Estimate the viscosity at the melting temperature of PAN.

23. Silly Putty is a polymer material used as a child's toy. This material will bounce like a rubber ball but will flow like a liquid when left undisturbed. Explain this behavior.

24. You can change the grain size of crystalline polymers. What would be the effect of annealing a crystalline polymer at a temperature below the melting temperature? What would be the effect of allowing a liquid polymer to crystallize at different temperatures below the melting temperature?

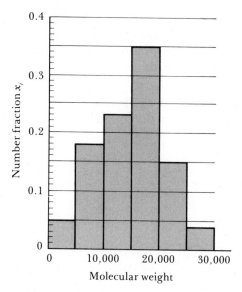

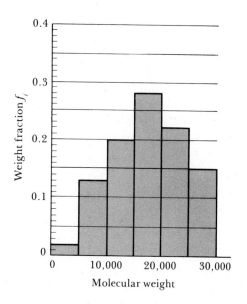

FIGURE 15-39 *Data for Problem 15-19.*

25. From Figure 3-34, calculate the theoretical density of crystalline polyethylene and compare with the actual density. What does this tell you about the percent crystallinity of the polymer?

26. Describe the relative tendencies of the following polymers to crystallize:

(a) Branched polyethylene versus linear polyethylene

(b) Polyethylene versus polyethylene-polypropylene copolymer

(c) Isotactic polypropylene versus atactic polypropylene

27. Suppose a thermoplastic polymer can be produced in sheet form either by rolling or by continuous casting. In which case would you expect to obtain the higher strength? Explain.

28. A strip of polyethylene 10 in. long and having a degree of polymerization of 1000 is pulled in tension until all of the polyethylene chains are straightened, without distorting the covalent bonds and without allowing any sliding of the chains. If the length of the strip is now 75 in., estimate the original average distance between the ends of the chains.

29. How much sulfur is required to provide 75% cross-linking of 5 kg of polybutadiene?

30. How many grams of polychloroprene will be completely vulcanized by 10 g of sulfur?

31. What fraction of the possible cross-linking sites are used if 50 g of sulfur are added to 5 kg of polyisoprene?

32. A butadiene-styrene elastomer is produced. We find that 30 g of sulfur will completely cross-link 400 g of the elastomer. What fraction of the mers in the elastomer are styrene?

33. We would like to produce 10 kg of butadiene-acrylonitrile elastomer. Calculate and plot the amount of sulfur required to provide 50% cross-linking versus the weight percent acrylonitrile present in the rubber.

34. You want to produce a complex component from an elastomer. Should you vulcanize the rubber before or after the forming operation? Explain.

35. Figure 15-31 shows the stress-strain curve for an elastomer. From the curve, calculate and plot the modulus of elasticity versus strain and explain the results.

36. A stress of 1500 psi is applied to an elastomer. After 3 weeks, the stress is observed to be 1400 psi. What period of time is required before the stress falls to 1200 psi?

37. An elastomer is stretched using a stress of 750 psi. At 27°C, the stress is reduced to 700 psi after 1 year. At 75°C, the stress falls to 700 psi after 1 week. Determine the activation energy for viscous flow in the elastomer.

38. The degree of polymerization normally is not used to characterize network polymers such as phenolics. Explain.

39. Suppose we wish to produce a linear amine chain by combining 5 kg of urea with formaldehyde. Calculate (a) the amount of formaldehyde required to produce the chains, (b) the amount of water evolved, and (c) the final weight of the polymer. (d) How much additional formaldehyde is needed to completely cross-link the linear chains into a network?

40. Explain why a thermosetting network polyester cannot be produced using only adipic acid and ethylene glycol (Example 15-16).

41. A thermosetting polyester is produced by combining 100 g of adipic acid, 150 g of maleic acid, and 50 g of ethylene glycol. Calculate the amount of styrene required to completely cross-link the chains.

42. A polyester is produced by combining 500 g of maleic acid and 250 g of ethylene glycol under conditions that give a degree of polymerization of 700. Then 30 g of styrene are added. Calculate the fraction of the possible cross-linking sites that are used.

43. 100 g of phenol are polymerized to produce a phenolic thermosetting polymer. What is the total amount of formaldehyde required to produce the chains and to satisfy one-third of the possible cross-linking sites?

44. Many paints are polymeric materials. Explain why plasticizers are added to paints. What must happen to the plasticizers after the paint is applied?

45. The density of the Styrofoam used to make a cup is about 3 lb per cubic foot. An average-sized cup weighs about 2.5 g. What volume of solid polystyrene is required to produce 1000 cups?

46. Compare the tensile strength-to-weight ratios of the following materials, using data from tables in Chapters 13, 14, and 15. How do the polymers compare with other materials?

poly-	poly-	poly-
propylene	acrylonitrile	etherimide
2024-T4	aged Cu-2% Be	AM100-T6
aluminum		magnesium
Al_2O_3	SiC	hot-pressed
		Si_3N_4

47. Compare the modulus of elasticity-to-density ratio for the following materials, using Table 6-3, Table 14-5, and data in this chapter. Are the properties of the lightweight polymers comparable with those of other materials?

low-density polyethylene	polyacrylonitrile
polyphenylene sulfide	Al_2O_3
hot-pressed Si_3N_4	aluminum alloys
titanium alloys	steels

16

COMPOSITE MATERIALS

16–1 INTRODUCTION

Composites are produced when two materials are joined to give a combination of properties that cannot be attained in the original materials. Composite materials may be selected to give unusual combinations of stiffness, strength, weight, high-temperature performance, corrosion resistance, hardness, or conductivity. Composites can be metal-metal, metal-ceramic, metal-polymer, ceramic-polymer, ceramic-ceramic, or polymer-polymer. Metal-ceramic composites, for example, include cemented carbide cutting tools, silicon carbide fiber-reinforced titanium, and enameled steel.

Composites can be placed into three categories—particulate, fiber, and laminar—based on the shapes of the materials (Figure 16-1). Concrete, a mixture of cement and gravel, is a particulate composite; fiberglass, containing glass fibers embedded in a polymer, is a fiber-reinforced composite; and plywood, having alternating layers of wood veneer, is a laminar composite. If the reinforcing particles are uniformly distributed, particulate composites have isotropic properties; fiber composites may be either isotropic or anisotropic; laminar composites always have anisotropic behavior.

16–2 DISPERSION-STRENGTHENED COMPOSITES

By stretching slightly our definition of a composite, we can consider a special group of dispersion-strengthened materials containing particles 100 Å to 2500 Å in diameter as particulate composites. These *dispersoids*, usually some type of metallic oxide, are introduced into the matrix by means other than traditional phase transformations. Even though the small particles are not coherent with the matrix, they block the movement of dislocations and produce a pronounced strengthening effect. Only small amounts of the dispersed material are required.

At room temperature, the dispersion-strengthened composites may be weaker than traditional age-hardened alloys, which contain a coherent pre-

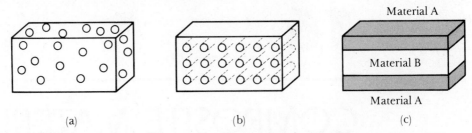

FIGURE 16-1 Comparison of the three types of composite materials. (a) Particulate composite, (b) fiber-reinforced composite, and (c) laminar composite.

cipitate. However, because the composites do not catastrophically soften by overaging, overtempering, grain growth, or coarsening of the dispersed phase, the strength of the composite decreases only gradually with increasing temperature, as illustrated by the SAP material in Figure 16-2. Furthermore, creep resistance is superior to that of metals and alloys (Figure 16-3).

Rapid solidification processing techniques which are being developed to produce certain titanium alloys also take advantage of this type of dispersion strengthening. In these materials, erbium and other rare earth elements combine with oxygen to produce tiny but stable oxide dispersoids.

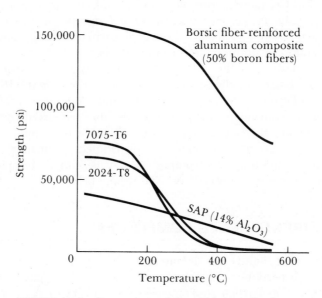

FIGURE 16-2 Comparison of the yield strength of dispersion-strengthened sintered aluminum powder (SAP) composite with that of two conventional two-phase high-strength aluminum alloys. The composite has benefits above about 300°C. A fiber-reinforced aluminum composite is shown for comparison.

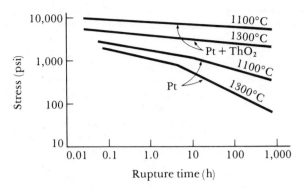

FIGURE 16-3 Dispersion-strengthened platinum, containing 12.5% ThO_2, has much higher creep resistance than pure platinum.

Considerations in Selecting the Dispersant. The properties of the dispersion-strengthened composites are optimized if we consider the following guidelines.

1. The dispersed phase, typically a hard, stable oxide, must be an effective obstacle to slip.

2. The dispersed material must have an optimum size, shape, distribution, and amount.

3. The dispersed material must have a low solubility in the matrix material. Furthermore, no chemical reactions should occur between the dispersant and the matrix. Alumina does not readily dissolve in aluminum; thus alumina is an effective dispersant for aluminum alloys. However, copper oxide will dissolve in copper at high temperatures; the $Cu-Cu_2O$ system would not be effective.

4. Good bonding must be achieved between the dispersed material and the matrix. A small solubility of the dispersant in the matrix may help produce a good, firm bond.

Examples of Dispersion-Strengthened Composites. Table 16-1 lists some materials of interest. Perhaps the classic example is the sintered aluminum powder, or SAP, composite. The SAP material has an aluminum matrix strengthened by up to 14% Al_2O_3. The composite can be formed by a powder metallurgy process; aluminum and alumina powders are blended, compacted at high pressures, and sintered. In a second technique, the aluminum powder is treated to give a continuous oxide film on each particle. When the powder is compacted, the oxide film fractures into tiny particles which are surrounded by the aluminum metal during sintering.

Another important group of dispersion-strengthened composites includes thoria-dispersed metals (Figure 16-4). Even 1% to 2% ThO_2, or thoria, significantly strengthens nickel, tungsten, and superalloys. TD-nickel is produced in several ways, including *internal oxidation*. The thorium is present in nickel as

TABLE 16-1   Examples and applications of selected dispersion-strengthened composites

System	Applications
Ag-CdO	Electrical contact materials
Al-Al$_2$O$_3$	Possible use in nuclear reactors
Be-BeO	Aerospace and nuclear reactors
Co-ThO$_2$, Y$_2$O$_3$	Possible creep-resistant magnetic materials
Ni-20% Cr-ThO$_2$	Turbine engine components
Pb-PbO	Battery grids
Pt-ThO$_2$	Filaments, electrical components
W-ThO$_2$, ZrO$_2$	Filaments, heaters

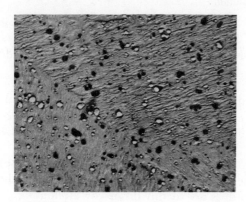

FIGURE 16-4   Electron micrograph of TD-nickel. The dispersed ThO$_2$ particles have a diameter of 3000 Å or less (× 2000). From *Oxide Dispersion Strengthening*, p. 714, Gordon and Breach, 1968. © AIME.

an alloying element; after a powder metallurgy compact is made, oxygen is allowed to diffuse into the metal, react with the thorium, and produce thoria. Yttria (Y$_2$O$_3$), alumina (Al$_2$O$_3$), and other oxides perform the same function.

◇ | **Example 16-1**

Suppose 2 wt % ThO$_2$ is added to nickel. Each ThO$_2$ particle has a diameter of 1000 Å. How many particles are present in each cubic centimeter?

Answer:

The densities of ThO$_2$ and nickel are 9.69 g/cm^3 and 8.9 g/cm^3, respectively. The volume fraction is

$$f_{\text{ThO}_2} = \frac{\dfrac{2}{9.69}}{\dfrac{2}{9.69} + \dfrac{98}{8.9}} = 0.0184 \text{ cm}^3 \text{ ThO}_2 \text{ per cm}^3$$

The volume of each ThO_2 sphere is

$$V_{ThO_2} = \tfrac{4}{3}\pi r^3 = \tfrac{4}{3}\pi (0.5 \times 10^{-5}\,cm)^3 = 0.52 \times 10^{-15}\,cm^3$$

$$\text{Number of}\ ThO_2 = \frac{0.0184}{0.52 \times 10^{-15}} = 35.4 \times 10^{12}\ particles/cm^3$$

16–3   TRUE PARTICULATE COMPOSITES

The true particulate composites contain large amounts of coarse particles that do not effectively block slip. The particulate composites are designed to produce unusual combinations of properties rather than to improve strength. These composites include many combinations of metals, ceramics, and polymers. A number of these systems will be discussed to illustrate the wide range of materials, manufacturing processes, and applications that are used. First, however, let us briefly examine one method sometimes used for predicting certain composite properties.

Rule of Mixtures.   Certain properties of a particulate composite depend only on the relative amounts and properties of the individual constituents. The *rule of mixtures* can accurately predict these properties. The density of a particulate composite, for example, is

$$\rho_c = \Sigma f_i \rho_i = f_1 \rho_1 + f_2 \rho_2 + \cdots + f_n \rho_n \tag{16-1}$$

where ρ_c is the density of the composite, $\rho_1, \rho_2, \ldots, \rho_n$ are the densities of each constituent in the composite, and $f_1, f_2, \ldots, f_n$ are the volume fractions of each constituent. Unfortunately, properties such as hardness and strength cannot be predicted by the rule of mixtures.

Cemented Carbides.   Cemented carbides, or *cermets*, contain hard ceramic particles dispersed in a metallic matrix. Tungsten carbide inserts used for cutting tools in machining operations are typical of this group. Tungsten carbide, WC, is extremely hard and can cut quenched and tempered steels. The carbide is also very stiff, so close tolerances can be held during machining, and it has a very high melting temperature, so high temperatures generated during rapid machining can be tolerated. Unfortunately, tools constructed from tungsten carbide are extremely brittle.

To improve toughness, tungsten carbide particles are combined with cobalt powder and pressed into powder compacts. The compacts are heated above the melting temperature of the cobalt. The liquid cobalt surrounds each of the solid tungsten carbide particles (Figure 16-5). After solidification, the cobalt serves as the binder for tungsten carbide and provides good impact resistance. As the tungsten carbide particles at the cutting surface become dull, they either fracture

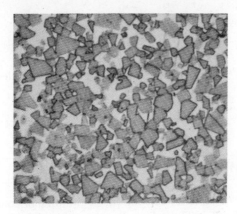

FIGURE 16-5 Microstructure of tungsten carbide—20% cobalt-cemented carbide (× 1300). From *Metals Handbook*, Vol. 7, 8th Ed., American Society for Metals, 1972.

TABLE 16-2 Grades and composites for typical cemented carbide cutting tools

Grade	% WC	% Co	% TaC	% TiC	Applications
C-1	90	6–12	0–3		
C-2	93	5–6	0–2		Stainless steels,
C-3	93	4–6	0–2		cast irons,
C-4	97	3–4			nonferrous alloys
C-5	73	8–10	10–12	6–8	
C-6	74	5–9	10–12	7–12	
C-7	76	4–6	5–10	10–14	Steels
C-8	75	4–6	4–8	10–20	

or pull out of the cobalt matrix and expose new, sharp-edged particles that continue to provide good cutting. For finish machining, the amount of cobalt binder is intentionally reduced so the particles pull out easily and the tool remains sharp. For rough grinding, more cobalt is added to improve toughness. Table 16-2 shows several classes of cemented carbides.

◇ | **Example 16-2**

A cemented carbide cutting tool used for machining contains 75 wt % WC, 15 wt % TiC, 5 wt % TaC, and 5 wt % Co. Estimate the density of the composite.

Answer:

First, we must convert the weight percentages to volume fraction. The densities of the components of the composite are

$$\rho_{WC} = 15.77 \, \text{g/cm}^3 \quad \rho_{TiC} = 4.94 \, \text{g/cm}^3$$
$$\rho_{TaC} = 14.5 \, \text{g/cm}^3 \quad \rho_{Co} = 8.90 \, \text{g/cm}^3$$

$$f_{WC} = \frac{\dfrac{75}{15.77}}{\dfrac{75}{15.77} + \dfrac{15}{4.94} + \dfrac{5}{14.5} + \dfrac{5}{8.9}} = \frac{4.76}{8.70} = 0.547$$

$$f_{TiC} = \frac{\dfrac{15}{4.94}}{8.70} = 0.349$$

$$f_{TaC} = \frac{\dfrac{5}{14.5}}{8.70} = 0.040$$

$$f_{Co} = \frac{\dfrac{5}{8.90}}{8.70} = 0.064$$

From the rule of mixtures, the density of the composite is

$$
\begin{aligned}
\rho_c = \Sigma f_i \rho_i &= (0.547)(15.77) + (0.349)(4.94) + (0.040)(14.5) \\
&\quad + (0.064)(8.9) \\
&= 11.50 \, \text{g/cm}^3
\end{aligned}
$$

◇

Abrasives.   Grinding and cutting wheels are formed from alumina (Al_2O_3), silicon carbide (SiC), cubic boron nitride (BN), and diamond. To provide toughness, the abrasive particles are bonded by a glass or polymer matrix. As the hard particles wear, they fracture or pull out of the matrix, exposing new cutting surfaces.

Electrical Contacts.   Materials used for electrical contacts in switches and relays must have a good combination of wear resistance and electrical conductivity. Otherwise, the contacts erode, causing poor contact and arcing. Particulate composites, such as tungsten-reinforced silver, provide materials having the proper combination of hardness and conductivity.

A tungsten powder compact is made using conventional powder metallurgical processes (Figure 16-6), producing high interconnected porosity. Liquid silver is then vacuum infiltrated to fill the interconnected voids. Both the silver and the tungsten are continuous. Thus, the pure silver efficiently conducts current while the hard tungsten provides wear resistance.

◇   **Example 16-3**

A silver-tungsten composite for an electrical contact is produced by first making a porous tungsten powder metallurgy compact, then infiltrating pure silver into the pores. The density of the tungsten compact before infiltration is $14.5 \, \text{g/cm}^3$.

Calculate the volume fraction of porosity and the final weight percent of silver in the compact after infiltration.

Answer:

The densities of pure tungsten and pure silver are $19.3 \, \text{g/cm}^3$ and $10.49 \, \text{g/cm}^3$. We can assume that the density of a pore is zero, so from the rule of mixtures

$$\rho_c = f_W \rho_W + f_{\text{pore}} \rho_{\text{pore}}$$
$$14.5 = f_W (19.3) + f_{\text{pore}} (0)$$
$$f_W = 0.75$$
$$f_{\text{pore}} = 1 - 0.75 = 0.25$$

After infiltration, the volume fraction of silver will equal the volume fraction of pores.

$$f_{\text{Ag}} = f_{\text{pore}} = 0.25$$

$$\text{wt} \% \, \text{Ag} = \frac{(0.25)(10.49)}{(0.25)(10.49) + (0.75)(19.3)} \times 100 = 15.3\%$$

◇

Polymers.   Many engineering polymers which contain fillers and extenders are particulate composites. A classic example is carbon black in vulcanized rubber. Carbon black consists of tiny carbon spheroids only 50 Å to 5000 Å in diameter. The carbon black improves the strength, stiffness, hardness, wear resistance, and heat resistance of the rubber.

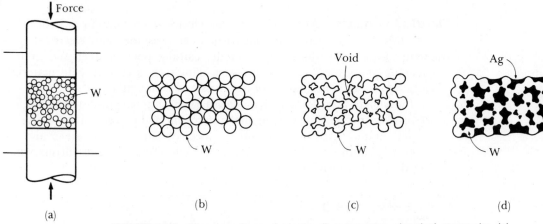

FIGURE 16-6 The steps in producing a silver-tungsten electrical composite. (a) Tungsten powders are pressed, (b) a low-density compact is produced, (c) sintering joins the tungsten powders, and (d) liquid silver is infiltrated into the pores between the particles.

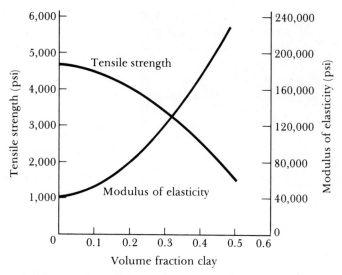

FIGURE 16-7   *The effect of clay on the properties of polyethylene.*

Extenders, such as calcium carbonate, solid glass spheres, and various clays, are added so that a smaller amount of the more expensive polymer is required. The extenders stiffen the polymer, increase the hardness and wear resistance, increase thermal conductivity, and improve resistance to creep; however, strength and ductility normally decrease (Figure 16-7). Introducing hollow glass spheres may impart the same changes in properties while significantly reducing the weight of the composite.

Other special properties can be obtained. Polyethylene may contain metallic powders, such as lead, to improve absorption of fission products in nuclear application, or bronze, to improve the electrical conductivity of the polymer and permit the polymer to be electroplated.

Foundry Molds and Cores.   Molds and cores used to make metal castings frequently consist of silica sand grains bonded by a matrix of either an organic or an inorganic resin. The sand grains are refractory, insulating materials that do not react with the molten metal. Common binders include phenolic resins, urethane resins, furan resins, and sodium silicate. The resins coat the individual sand grains and provide bridging (Figure 16-8). The voids between the sand grains permit gases to escape from the mold rather than being trapped in the metal casting.

Compocasting.   One unusual technique for producing particulate-reinforced castings is based on the thixotropic behavior of partly liquid-partly solid melts. A liquid alloy is allowed to cool until about 40% solids have formed; during solidification the solid-liquid mixture is vigorously stirred to break up the dendritic structure (Figure 16-9). The resulting solid-liquid slurry has *thixotropic*

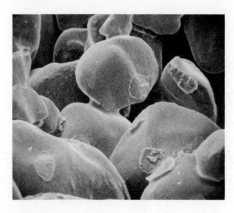

FIGURE 16-8 A scanning electron microscope view of the bridging of a thermosetting polymer that binds sand grains in a shell mold or core for the foundry industry. Note some fractured bonds (× 60).

behavior—the slurry behaves as a solid when no stress is applied, but flows like a liquid when pressure is exerted.

If a particulate material is introduced to the molten metal during cooling and stirring, a uniform dispersion is produced. The thixotropic slurry containing the particulate is injected into a die under pressure; this process is termed *compocasting*. A variety of materials, including Al_2O_3, SiC, TiC, and glass beads, have been incorporated into aluminum and magnesium alloys by this technique.

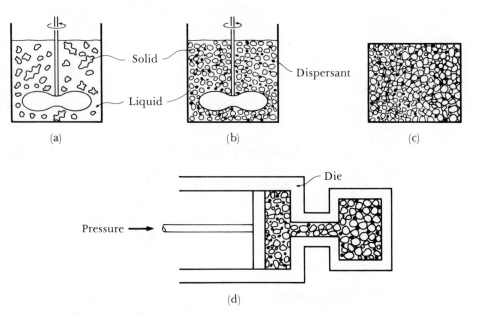

FIGURE 16-9 In compocasting, (a) a solidifying alloy is stirred to break up the dendritic network, (b) a reinforcement is introduced into the slurry, (c) when no force is applied, the solid-liquid does not flow, and (d) high pressures cause the solid-liquid mixture to flow into a die.

◇ | **Example 16-4**

We wish to compocast an aluminum alloy containing 9% Cu and 25% Al_2O_3. We find that compocasting is best accomplished when the alloy is die cast at a temperature that gives 60% liquid and 40% solid. From the phase diagram for aluminum-copper (Figure 11-1), estimate the appropriate casting temperature.

Answer:

To determine the temperature at which 60% liquid is present in the $\alpha + L$ region, let's perform the lever law calculation at several temperatures. The liquidus temperature is 630°C and the eutectic temperature is 548°C.

$$\text{At } 600°C: \quad L = \frac{9 - 3}{21 - 3} \times 100 = 33\%$$

$$\text{At } 620°C: \quad L = \frac{9 - 2}{14 - 2} \times 100 = 58\%$$

$$\text{At } 625°C: \quad L = \frac{9 - 1.5}{11 - 1.5} \times 100 = 79\%$$

We obtain 60% L near 620°C.

16–4 FIBER-REINFORCED COMPOSITES

The fiber-reinforced composites improve strength, fatigue resistance, stiffness, and strength-to-weight ratio by incorporating strong, stiff, brittle fibers into a softer, more ductile matrix. The matrix material transmits the force to the fibers and provides ductility and toughness, while the fibers carry most of the applied force. Unlike dispersion-strengthened composites, the strength of the composite is increased at both room temperature and elevated temperatures (Figure 16-2).

A tremendous variety of reinforcing materials are employed. For centuries, straw has been used to strengthen mud bricks. Steel reinforcing bar is introduced into concrete structures. Glass fibers in a polymer matrix produce fiberglass for transportation and aerospace applications. Fibers made of boron, carbon, and polymers provide exceptional reinforcement. Even tiny single crystals of ceramic materials, called *whiskers*, are being developed.

The reinforcing materials also are arranged in a variety of orientations (Figure 16-10). Short, randomly oriented glass fibers are usually present in fiberglass. Unidirectional arrangements of continuous fibers may be used to deliberately produce anisotropic properties. Fibers can be woven into fabrics or produced in the form of tapes. Alternating layers of tapes can be changed in orientation.

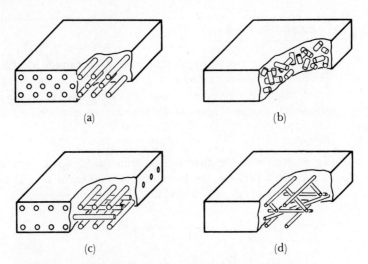

(a)

(b)

(c)

(d)

FIGURE 16-10 Various morphologies of fiber-reinforced composites. (a) Continuous unidirectional fibers, (b) randomly oriented discontinuous fibers, (c) orthogonal fibers, and (d) multiple-ply fibers.

16–5 PREDICTING PROPERTIES OF FIBER-REINFORCED COMPOSITES

The rule of mixtures always predicts the density of fiber-reinforced composites.

$$\rho_c = f_m \rho_m + f_f \rho_f \tag{16-2}$$

where the subscripts m and f refer to the matrix and the fiber. In addition, the rule of mixtures accurately predicts the electrical and thermal conductivity of fiber-reinforced materials along the fiber direction if the fibers are continuous and unidirectional.

$$K_c = f_m K_m + f_f K_f \tag{16-3}$$

$$\sigma_c = f_m \sigma_m + f_f \sigma_f \tag{16-4}$$

where k is the thermal conductivity and σ is the electrical conductivity. Unfortunately, with the exception of density, the properties are much more difficult to predict if the fibers are not continuous and unidirectionally aligned.

◇ | **Example 16-5**

The density of a composite made from boron fibers in an epoxy matrix is $1.8\,\text{g/cm}^3$. The density of boron is $2.36\,\text{g/cm}^3$ and that of epoxy is $1.38\,\text{g/cm}^3$. Calculate the volume fraction of boron fibers in the composite.

Answer:

If f_B is the volume fraction of boron, then $1 - f_B$ is the volume fraction of epoxy.

$$\rho_c = f_B\rho_B + f_E\rho_E = f_B\rho_B + (1 - f_B)\rho_E$$

$$1.8 = f_B(2.36) + (1 - f_B)(1.38)$$

$$1.8 = 2.36f_B + 1.38 - 1.38f_B$$

$$f_B = \frac{1.80 - 1.38}{2.36 - 1.38} = 0.429$$

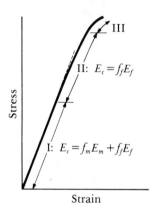

FIGURE 16-11 The stress-strain curve for a fiber-reinforced composite. At low stresses, the modulus of elasticity is given by the rule of mixtures. At higher stresses, the matrix deforms and the rule of mixtures is no longer obeyed.

Modulus of Elasticity. When a load is applied parallel to continuous, unidirectional fibers, the rule of mixtures accurately predicts the modulus of elasticity.

$$E_c = f_m E_m + f_f E_f \tag{16-5}$$

However, when the applied stress is very large, the matrix begins to deform and the stress-strain curve is no longer linear (Figure 16-11). Since the matrix now contributes little to the stiffness of the composite, the modulus can be approximated by

$$E_c = f_f E_f \tag{16-6}$$

When the load is applied perpendicular to the fibers, each component acts independently of the other. The modulus of the composite is now

$$\frac{1}{E_c} = \frac{f_m}{E_m} + \frac{f_f}{E_f} \tag{16-7}$$

◇ **Example 16-6**

Derive the rule of mixtures for the modulus of elasticity of a fiber-reinforced composite when a stress is applied along the axis of the fibers.

Answer:

The total force acting on the composite is the sum of the forces carried by each constituent.

$$F_c = F_m + F_f$$

Since $F = \sigma A$

$$\sigma_c A_c = \sigma_m A_m + \sigma_f A_f$$

$$\sigma_c = \sigma_m \left(\frac{A_m}{A_c}\right) + \sigma_f \left(\frac{A_f}{A_c}\right)$$

If the fibers have a uniform cross section, the area fraction equals the volume fraction f.

$$\sigma_c = \sigma_m f_m + \sigma_f f_f$$

From Hooke's law, $\sigma = \varepsilon E$. Therefore

$$E_c \varepsilon_c = E_m \varepsilon_m f_m + E_f \varepsilon_f f_f$$

If the fibers are rigidly bonded to the matrix, both the fibers and the matrix must stretch equal amounts, so

$$\varepsilon_c = \varepsilon_m = \varepsilon_f$$
$$E_c = f_m E_m + f_f E_f$$

◇   **Example 16-7**

Derive the equation for the modulus of elasticity of a fiber-reinforced composite when a stress is applied perpendicular to the axis of the fiber.

Answer:

In this example, the strains are no longer equal; instead, the weighted sum of the strains in each component equals the total strain in the composite, while the stresses in each component are equal.

$$\varepsilon_c = f_m \varepsilon_m + f_f \varepsilon_f$$

$$\frac{\sigma_c}{E_c} = f_m \left(\frac{\sigma_m}{E_m}\right) + f_f \left(\frac{\sigma_f}{E_f}\right)$$

Since $\sigma_c = \sigma_m = \sigma_f$

$$\frac{1}{E_c} = \frac{f_m}{E_m} + \frac{f_f}{E_f}$$

Strength.   The strength of a composite depends on the bonding between the fibers and the matrix and is limited by deformation of the matrix. Consequently, the strength is almost always less than that predicted by the rule of mixtures.

Other properties, such as ductility, impact properties, fatigue properties, and creep properties, are even more difficult to predict than the tensile properties.

$\diamondsuit$ | **Example 16-8**

Borsic-reinforced aluminum containing 40 vol % fibers is an important high-temperature, lightweight composite material. Estimate the density, modulus of elasticity, and strength parallel to the fiber axis. Also estimate the modulus of elasticity perpendicular to the fibers.

Answer:

The properties of the individual components are shown below.

Material	Density (g/cm^3)	Modulus of Elasticity (psi)	Tensile Strength (psi)
Fibers	2.36	55,000,000	400,000
Aluminum	2.70	10,000,000	5,000

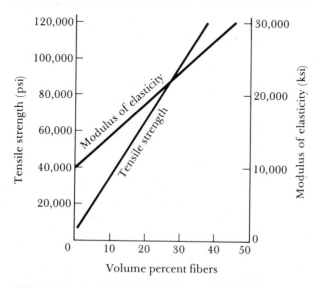

FIGURE 16-12   *The influence of volume percent Borsic fibers on the properties of Borsic-reinforced aluminum parallel to the fibers.*

From the rule of mixtures

$$\rho_c = (0.6)(2.7) + (0.4)(2.36) = 2.56\,\text{g/cm}^3$$
$$E_c = (0.6)(10 \times 10^6) + (0.4)(55 \times 10^6) = 28 \times 10^6\,\text{psi}$$
$$\text{TS}_c = (0.6)(5,000) + (0.4)(400,000) = 163,000\,\text{psi}$$

Perpendicular to the fibers

$$\frac{1}{E_c} = \frac{0.6}{10 \times 10^6} + \frac{0.4}{55 \times 10^6} = 0.06727 \times 10^{-6}$$
$$E_c = 14.9 \times 10^6\,\text{psi}$$

The actual modulus and strength parallel to the fibers are shown in Figure 16-12. The calculated modulus of elasticity, 28×10^6 psi, is exactly the same as the measured modulus. However, the estimated strength, 163,000 psi, is substantially higher than the actual strength, about 130,000 psi. We also note that the modulus of elasticity is very anisotropic.

 Example 16-9

Glass fibers in nylon provide reinforcement. If the nylon contains 30 vol % E glass, what fraction of the applied stress is carried by the glass fibers?

Answer:

The modulus of elasticity for each component of the composite is

$$E_{\text{glass}} = 10.5 \times 10^6\,\text{psi} \qquad E_{\text{nylon}} = 0.4 \times 10^6\,\text{psi}$$

Both the nylon and the glass fibers have equal strain if bonding is good, so

$$\varepsilon_c = \varepsilon_m = \varepsilon_f$$

$$\varepsilon_m = \frac{\sigma_m}{E_m} = \varepsilon_f = \frac{\sigma_f}{E_f}$$

$$\frac{\sigma_f}{\sigma_m} = \frac{E_f}{E_m} = \frac{10.5 \times 10^6}{0.4 \times 10^6} = 26.25$$

$$\text{Fraction} = \frac{\sigma_f}{\sigma_f + \sigma_m} = \frac{1}{1 + \dfrac{\sigma_m}{\sigma_f}} = \frac{1}{1 + \dfrac{1}{26.25}} = 0.96$$

Almost all of the load is carried by the glass fibers.

16–6 CONTROLLING THE CHARACTERISTICS OF FIBER-REINFORCED COMPOSITES

A large number of factors must be considered when designing a fiber-reinforced composite.

Aspect Ratio. Fibers can be short, long, or even continuous. Their dimensions are often characterized by the aspect ratio l/d, where l is the fiber length and d is the diameter.

In general, we find that the properties of the composite are improved when the aspect ratio is large. Brittleness of the fibers is often caused by surface imperfections; by making the diameter as small as possible, the fiber has less surface area and consequently fewer of these flaws that might initiate brittle fracture. We also prefer long fibers; because the ends of each fiber carry less of the load than the remainder of the fiber, it is more difficult for the short fibers to carry the imposed loads. For example, nylon reinforced with carbon fibers with an aspect ratio of 30 has a tensile strength of 16,000 psi; longer fibers with an aspect ratio of 800 produce a strength of 35,000 psi.

Continuous fibers are often difficult to produce and introduce into the matrix, while short fibers can easily be incorporated into the matrix but produce relatively poor reinforcement. In many fiber-reinforced systems, discontinuous fibers with an aspect ratio greater than some critical value are used to provide an acceptable compromise between processing ease and properties.

Volume Fraction of Fibers. A greater volume fraction of fibers increases the strength and stiffness of the composite (Figure 16-12). However, an upper limit of about 80% fibers occurs, when the fibers can no longer be completely surrounded by the matrix.

Orientation of Fibers. Randomly oriented fibers with a small aspect ratio can give isotropic behavior, but the optimum mechanical properties may not be achieved. When a composite is created using unidirectionally aligned fibers, optimum stiffness and strength can be obtained if the applied load is parallel to the fibers, but these properties may be very poor if the load is perpendicular to the fibers (Figure 16-13). Unidirectionally oriented fibers give very anisotropic behavior and, in more complicated stress conditions in which large loads are applied from several directions, may not perform adequately.

Fortunately, one of the unique characteristics of fiber-reinforced composites is that their properties can be tailored to meet different types of loading conditions. This can be done by using multiple plies of continuous fibers, with the orientation of the fibers differing in each ply, or by producing fabrics in which the fibers have different orientations caused by the different weaves used (Figure 16-14). If all of the load is in one direction, all of the fibers should be oriented parallel to that load. If the composite is loaded from two perpendicular directions, an orthogonal pattern of fibers can be formed. Fibers can be added in still

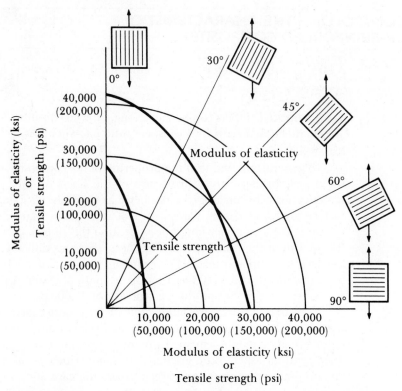

FIGURE 16-13 *The effect of fiber orientation with respect to the applied stress for fiber-reinforced composites for a boron-reinforced titanium alloy.*

other directions, say at a 30° or 45° angle to the usual fiber direction, to meet additional high stresses. Although maximum strength and stiffness decrease somewhat compared with use of only unidirectionally aligned fibers, the properties become more uniform and the composite is better able to withstand the applied loads, particularly in the two-dimensional plane containing the fibers.

Fabrics can also be woven to provide reinforcement in three directions. In even the simplest of fabric weaves, the fibers in each individual layer of fabric have some small degree of orientation in a third direction. Even better three-dimensional reinforcement can occur when the fabric layers are knitted or stitched together; more complicated three-dimensional weaves, such as that in Figure 16-14(c), can also be used.

Properties of the Fiber. The fiber material should be strong, stiff, and lightweight. If the composite is to be used at elevated temperatures, the fiber should also have a high melting temperature. The strength, stiffness, and weight are often referred to by the *specific strength* and *specific modulus* of the material.

$$\text{Specific strength} = \frac{\sigma}{\rho} \tag{16-8}$$

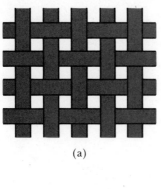

(a)

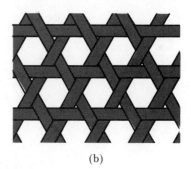

(b)

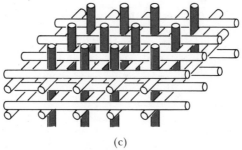

(c)

FIGURE 16-14 Three of the many weave patterns for fibers, including (a) biaxial weave, (b) triaxial weave, and (c) three-dimensional weave.

$$\text{Specific modulus} \ = \ \frac{E}{\rho} \tag{16-9}$$

where σ is the yield strength, ρ is the density, and E is the modulus of elasticity. Of course, we prefer fibers with a high specific strength and specific modulus. Some typical data for fibers are shown in Table 16-3 and Figure 16-15. The highest specific modulus is usually found in materials having a low atomic number and covalent bonding, such as carbon and boron. These two elements also have a high strength and melting temperature.

Aramid fibers, of which Kevlar and Nomex are the best known examples, are aromatic polyamide polymers strengthened by a backbone containing benzene rings (Figure 16-16) and are examples of liquid-crystalline polymers in that the polymer chains are rodlike and very stiff. Specially prepared polyethylene fibers are also available. Both the aramid and polyethylene fibers have excellent strength and stiffness but are limited to low-temperature use (Figure 16-17). Polyethylene fibers, because of their lower density, have superior specific strength and specific modulus.

Ceramic fibers and whiskers, including alumina, glass, and silicon carbide, have excellent strength and stiffness. Glass fibers, which are the most commonly used, include pure silica, S glass (SiO_2-25% Al_2O_3-10% MgO), and E glass (SiO_2-18% CaO-15% Al_2O_3). Although they are considerably denser than

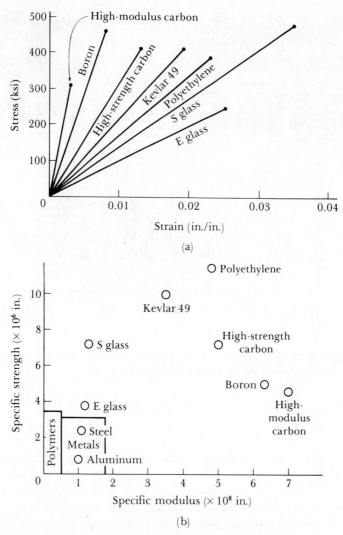

FIGURE 16-15 *Properties of fibers. (a) The stress-strain curve for individual fibers and (b) comparison of the specific strength and specific modulus of fibers versus metals and polymers.*

the polymer fibers, the ceramics can be used at much higher temperatures. Beryllium and tungsten, although metallically bonded, have a high modulus that makes them attractive fiber materials for certain applications.

◇   **Example 16-10**

Compare the strength of Kevlar fibers with that of carbon fibers at 30°C and 300°C.

TABLE 16-3 Properties of selected fiber-reinforcing materials

Material	Density (g/cm^3)	Tensile Strength (ksi)	Modulus of Elasticity $(\times 10^6 \, psi)$	Melting Temperature $(°C)$	Specific Modulus $(\times 10^7 \, in.)$	Specific Strength $(\times 10^6 \, in.)$
E glass	2.55	500	10.5	< 1725	11.4	5.6
S glass	2.50	650	12.6	< 1725	14.0	7.2
SiO_2	2.19	850	10.5	1728	13.3	10.8
Al_2O_3	3.95	300	55	2015	38.8	2.1
ZrO_2	4.84	300	50	2677	28.6	1.7
Carbon HS (high strength)	1.50	820	40	3700	74.2	7.4
Carbon HM (high modulus)	1.50	270	77	3700	143	5.0
BN	1.90	200	13	2730	18.8	2.9
Boron	2.36	500	55	2030	64.7	4.7
B_4C	2.36	330	70	2450	82.4	3.9
SiC	4.09	300	70	2700	47.3	2.0
TiB_2	4.48	15	74	2980	45.6	0.1
Be	1.83	185	44	1277	77.5	2.8
W	19.4	580	59	3410	8.5	0.8
Polyethylene	0.97	375	17	147	48.8	10.8
Nylon	1.14	120	0.4	249	1.0	2.9
Kevlar	1.44	650	18	500	34.7	10.1
Al_2O_3 whiskers	3.96	3000	62	1982	43.4	21.0
BeO whiskers	2.85	1900	50	2550	48.5	18.5
B_4C whiskers	2.52	2000	70	2450	76.9	22.1
SiC whiskers	3.18	3000	70	2700	60.8	26.2
Si_3N_4 whiskers	3.18	2000	55		47.8	17.5
Graphite whiskers	1.66	3000	102	3700	170	50.2
Cr whiskers	7.2	1290	35	1890	13.4	4.9

Adapted partially from L. J. Broutman, "Mechanical Properties of Fiber Reinforced Plastics," *Composite Engineering Laminates*, ed. G. H. Dietz, The M.I.T. Press, 1969.

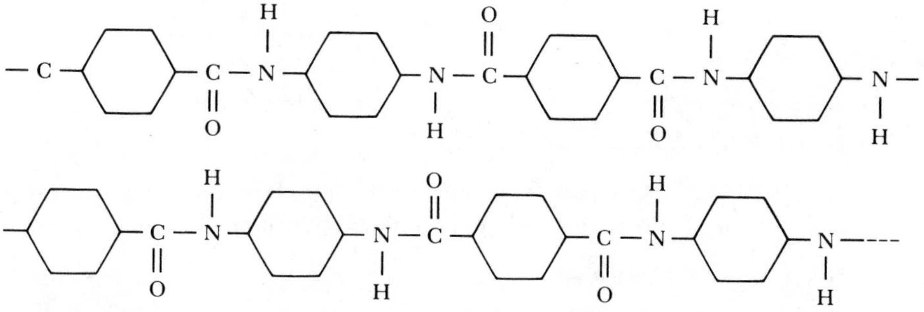

FIGURE 16-16 The structure of Kevlar. The fibers are joined by secondary bonds between oxygen and hydrogen atoms on adjoining chains.

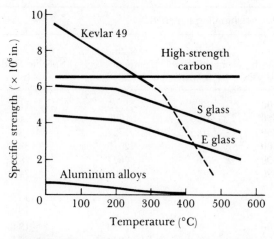

FIGURE 16-17 *The effect of temperature on the strength of several fibers used for fiber-reinforced composites.*

Answer:

From Figure 16-17, we find the specific strength of each fiber at the two temperatures of interest.

Material	Specific Strength ($\times 10^6$ in.)	Density (lb/in^3)	Strength (psi)
Kevlar at 30°C	9.5	0.052	494,000
Kevlar at 300°C	6.0	0.052	312,000
Carbon at 30°C	6.5	0.054	351,000
Carbon at 300°C	6.5	0.054	351,000

The carbon fibers, although weaker at low temperatures, maintain their strength to high temperatures.

◇ Example 16-11

Compare the specific modulus and specific strength of a 1040 annealed steel and a 2024-T4 aluminum alloy with those of boron.

Answer:

The yield strengths for the aluminum alloy and steel are found in Tables 11-2 and 13-4.

Material	Yield Strength (psi)	Modulus of Elasticity ($\times 10^6$ psi)	Density (lb/in^3)
2024-T4 aluminim	47,000	10	0.097
1040 steel	51,250	30	0.284
Boron	400,000	55	0.085

From these data, we can calculate the specific strength and specific modulus for each material.

$$\text{Specific strength (2024-T4 aluminum)} = \frac{47,000}{0.097} = 485,000 \text{ in.}$$

$$\text{Specific strength (1040 steel)} = \frac{51,250}{0.284} = 180,000 \text{ in.}$$

$$\text{Specific strength (boron fiber)} = \frac{400,000}{0.085} = 4,700,000 \text{ in.}$$

$$\text{Specific modulus (2024-T4 aluminum)} = \frac{10 \times 10^6}{0.097} = 1.03 \times 10^8 \text{ in.}$$

$$\text{Specific modulus (1040 steel)} = \frac{30 \times 10^6}{0.284} = 1.06 \times 10^8 \text{ in.}$$

$$\text{Specific modulus (boron fiber)} = \frac{55 \times 10^6}{0.085} = 6.47 \times 10^8 \text{ in.}$$

Matrix Properties.   Compared with the fibers, matrix materials are usually tough and ductile. Their purpose is to support the fibers, allowing the load to be transmitted to the fibers and preventing cracks in broken fibers from propagating throughout the entire composite. The matrix preferably should also be strong so that it contributes to the overall strength of the composite. Finally, the melting temperature of the matrix usually limits the high-temperature use of the composite.

Most polymer materials, both thermoplastics and thermosets, are available in short glass fiber–reinforced grades, with the fibers improving the strength and stiffness of the polymers. These composites are formed into useful shapes by the usual polymer processes discussed in Chapter 15. Polymer matrices, particularly polyesters and epoxies, are also used in more advanced fiber-reinforced composites. For example, these thermosetting polymers are the usual reinforcements in fiberglass. Thermosetting aromatic polyimides are used for somewhat higher temperature applications. Unfortunately, these composites are useful only at relatively low temperatures—a maximum of about 80°C for polyesters to 315°C for polyimides.

Some metal matrix composites, including boron fiber–reinforced aluminum, have been used. The metal matrix permits the composite to operate at higher temperatures, but producing the composite is normally more difficult and expensive than for the polymer matrix materials. One particularly important application for metal matrix composites is the next generation of jet engine parts, which may be fiber-reinforced superalloys.

Amazingly, hard, brittle ceramics may also be used as a matrix in composites. The ceramic matrix composites have very good properties at elevated temperatures and are lighter in weight than the high-temperature metal matrix composites. In a later section, we will look at how we can develop some degree of toughness in these materials.

Bonding and Failure. Particularly in polymer and metal matrix composites, good bonding must be obtained between the various constituents. The fibers must be firmly bonded to the matrix material if the load is to be properly transmitted from the matrix to the fibers. In addition, the fibers may pull out of the matrix during loading, reducing the strength and fracture resistance of the composite, if bonding is poor. In some cases, special coatings may be used to improve bonding. Glass fibers may be coated with a silane "keying" agent (called *sizing*) to improve bonding and moisture resistance in fiberglass composites. Carbon fibers are similarly coated with an organic material to improve bonding. Boron fibers have been coated with silicon carbide or boron nitride to improve bonding with an aluminum matrix; in fact, these fibers have been called Borsic fibers to reflect the presence of the silicon carbide (SiC) coating.

Another property that must be considered when combining fibers into a matrix is the similarity between the coefficients of thermal expansion for the two materials. If the fiber expands and contracts at a rate much different from that of the matrix, bonding can be disrupted and premature failure can occur.

In many composites, individual plies or layers of fabric are joined together. Bonding between these layers must also be good or another problem—*delamination*—may occur. The layers may tear apart under load and cause failure. Using composites in which a three-dimensional weave has been used to orient the fibers will also help prevent delamination.

16–7 MANUFACTURING FIBERS AND COMPOSITES

Producing a fiber-reinforced composite involves several steps, including producing the fibers, arranging the fibers into bundles or fabrics, and introducing the fibers into the matrix.

Making the Fiber. Metallic fibers, glass fibers, and many polymer fibers (including nylon, aramid, and polyacrylonitrile) can be formed by various drawing processes, as described in Chapter 7 (wire drawing of metal) and Chapter 15 (using the spinnerette for glass fibers). Fibers may be as fine as 0.001 cm in diameter.

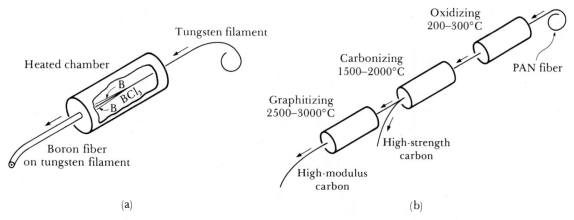

FIGURE 16-18 *Methods for producing (a) boron and (b) carbon fibers.*

Boron and carbon are too brittle and reactive to be worked by conventional drawing processes. Boron fiber is produced by *chemical vapor deposition* (CVD) [Figure 16-18(a)]. A very fine, 0.0002 cm diameter heated tungsten filament is used as a substrate, passing through a seal into a heated chamber. Vaporized boron compounds, such as BCl_3, are introduced into the chamber, decompose, and permit boron to precipitate onto the tungsten wire. The final fiber may be 0.0004 to 0.0032 cm in diameter.

Carbon fibers about 0.0001 cm in diameter are made by carbonizing, or pyrolizing, an organic filament, which is more easily drawn or spun into thin, continuous lengths [Figure 16-18(b)]. The organic filament, known as a *precursor*, is often rayon (a cellulosic polymer), polyacrylonitrile (**PAN**), or pitch (various aromatic organic compounds). High temperatures decompose the organic polymer, driving off all of the elements but carbon. As the carbonizing temperature increases from 1000°C to 3000°C, the tensile strength decreases while the modulus of elasticity increases (Figure 16-19). Drawing the carbon filaments at critical times during carbonizing may produce desirable preferred orientations in the final carbon filament.

Whiskers, which are single crystals of exceptional fineness, are discontinuous, with aspect ratios of 20 to 1000. Because the whiskers contain no mobile dislocations, slip cannot occur and the whiskers have exceptionally high strengths. The technology for producing whiskers is often very complex.

Arranging the Fibers.   The exceptionally fine filaments are normally bundled together as rovings, yarns, or tows. *Yarns* and *rovings* may contain 1000 to 10,000 filaments, while *tows* may contain several hundred thousand filaments (Figure 16-20).

Often, fibers are chopped into short lengths of 1 cm or less. These fibers, also called *staples*, are easily incorporated into the matrix. They are normally present in composites in a random orientation, which, depending on the application of the composite, may be either an advantage or disadvantage.

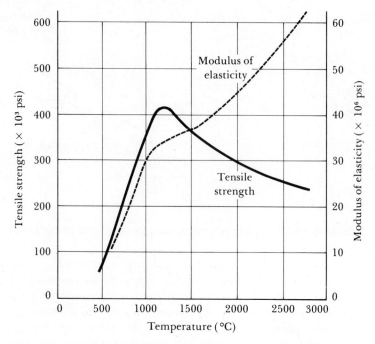

FIGURE 16-19 *The effect of heat treatment temperature on the strength and modulus of elasticity of carbon fibers.*

More continuous fibers for polymer matrix composites can be processed into mats or fabrics. *Mats* contain nonwoven, randomly oriented fibers loosely held together by a polymer resin. The fibers can also be woven into two-dimensional or even three-dimensional fabrics, as described previously. The fabrics may then be impregnated with a polymer resin. In either mats or fabrics, the resins at this point in the processing are usually thermosetting polymers that have not yet been completely polymerized; these mats or fabrics are called *prepregs*.

When unidirectionally aligned fibers are to be introduced into a polymer matrix, *tapes* may be produced. Individual fibers can be unwound from spools

FIGURE 16-20 A scanning electron micrograph of a carbon tow, containing many individual carbon filaments (× 200).

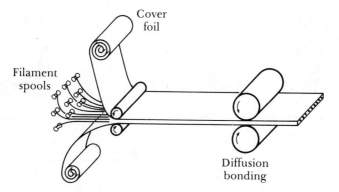

FIGURE 16-21 *Production of fiber tapes by encasing fibers between metal cover sheets by diffusion bonding.*

onto a mandrel, which determines the spacing of the individual fibers, and prepregged with a polymer resin. These tapes, only one fiber diameter thick, may be up to 48 in. wide. Figure 16-21 illustrates that tapes can also be produced by covering the fibers with upper and lower layers of metal foil that are then joined by diffusion bonding.

When unidirectionally aligned fibers are to be introduced into a metal matrix, fibers or fiber tapes may be precoated with the metal. This can be done by passing the fiber through a molten metal bath, electrodepositing metal onto the fiber, spraying the fiber with metal droplets melted in a plasma or laser torch, or vapor depositing metal onto the fibers.

Producing the Composite. A variety of methods for producing composite parts are used, depending on the application and materials. Short fiber-reinforced composites are normally formed by mixing the fibers with a liquid or plastic matrix, then using relatively conventional techniques, such as injection molding for polymer-base composites or casting (including compocasting) for metal matrix composites. Polymer matrix composites can also be produced by a spray-up method, in which short fibers mixed with a resin are sprayed against a form and cured.

Special techniques, however, have been devised for producing composites using continuous fibers, either in unidirectionally aligned, mat, or fabric form (Figure 16-22). In *hand lay-up* techniques, the tapes, mats, or fabrics are placed against a form, saturated with a polymer resin, rolled to assure good contact and freedom from porosity, and finally cured. Fiberglass car and truck bodies might be made in this manner, which is generally slow and labor intensive.

Tapes and fabrics can also be placed in a die and formed by *bag molding* processes. High-pressure gases or a vacuum are introduced to force the individual plies together so that good bonding is achieved during curing. Large polymer matrix components for the skins of military aircraft have been produced by these techniques. In *matched die molding*, short fibers or mats are placed into a two-part die; when the die is closed, the composite shape is formed.

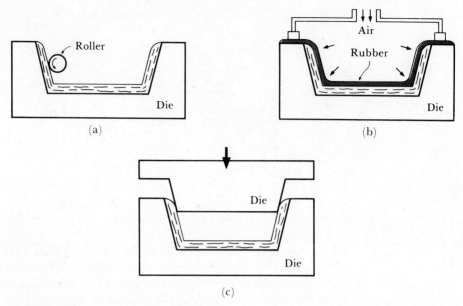

(a)

(b)

(c)

FIGURE 16-22   Producing composite shapes in dies by (a) hand lay-up, (b) pressure bag molding, and (c) matched die molding.

Filament winding is used to produce products such as pressure tanks and rocket motor castings (Figure 16-23). One or more continuous fibers are wrapped around a form or mandrel to gradually build up a solid shape that may even be several feet in thickness. The filament can be dipped in the polymer matrix resin prior to winding, or the resin can be impregnated around the fiber during or after winding. Curing completes the production of the composite part.

Pultrusion and Pulforming.   *Pultrusion* is used to extrude either a thermoplastic or a thermosetting polymer matrix around fibers to form a simple-shaped

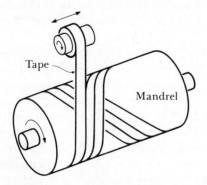

FIGURE 16-23   Producing composite shapes by filament winding.

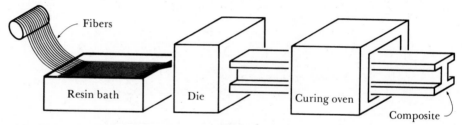

FIGURE 16-24 *Producing composite shapes by pultrusion.*

product with a constant cross section, such as round, rectangular, pipe, plate, or sheet shapes (Figure 16-24). Fibers or mats are drawn from spools, passed through a polymer resin bath for impregnation, and gathered together to produce a particular shape before entering a heated die for curing. Curing of the resin is normally accomplished almost immediately, so a continuous product is produced. The pultruded stock can subsequently be formed into somewhat more complicated shapes, such as fishing poles, golf club shanks, and ski poles. In *pulforming*, more complicated shapes, including springs, can be produced.

Metal matrix composites with continuous fibers are more difficult to produce than are the polymer matrix composites. Casting processes that force liquid around the fibers using capillary rise, pressure casting, vacuum infiltration, or continuous casting are illustrated in Figure 16-25. Various solid-state compaction processes can also be used. Figure 16-26 illustrates how several tapes can be placed into a closed die and deformed into shape; interdiffusion between the individual tapes can produce a solid form. Powder metallurgy techniques can also be used; metal powder can be poured around the fiber, followed by compaction and sintering at high temperatures.

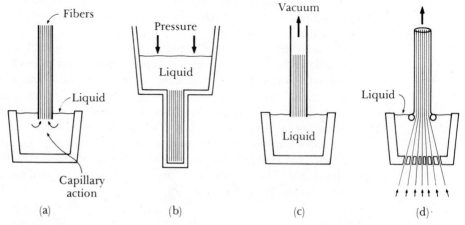

FIGURE 16-25 *Casting techniques for producing composite materials. (a) Capillary rise, (b) pressure casting, (c) vacuum infiltration, and (d) continuous casting.*

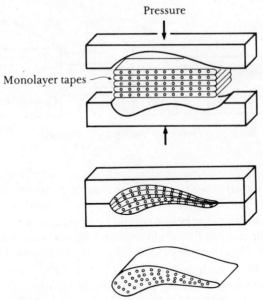

FIGURE 16-26   Closed die deformation and bonding of a composite composed of several tapes.

16–8   FIBER-REINFORCED SYSTEMS AND APPLICATIONS

Before completing our discussion of fiber-reinforced composites, let's look at the behavior and applications of some of the most common of these materials. Figure 16-27 compares the specific modulus and specific strength of several composites with those of metals and polymers. Note that the values in this figure are lower than those in Figure 16-15, since we are now looking at the composite, not just the fiber.

Some fiber-reinforced composites have been used for a long period of time. Reinforced concrete, in which concrete is reinforced by steel rods which improve the strength and prevent the structure from collapsing if the concrete fails, is a common construction material. Reinforcement of automobile and truck tires with nylon, aramid, or steel fibers improves the strength and life of the fiber-reinforced rubber. Fiberglass containing short fibers has long been used to produce components for automotive, marine, and appliance products. The world record in pole-vaulting increased dramatically when fiberglass poles were developed. The improvement in strength and stiffness of three polymers by glass fibers is shown in Figure 16-28.

Advanced Composites.   The term *advanced composites* is often used when the composite is intended to provide service in very critical applications, as in the aerospace industry. The advanced composites normally are polymer matrix

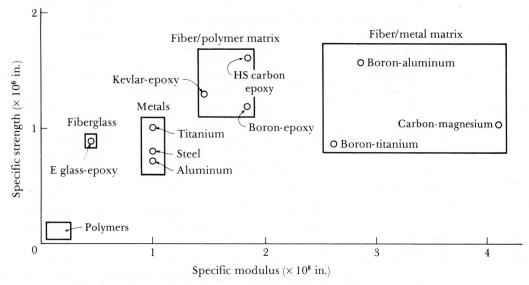

FIGURE 16-27   A comparison of the specific modulus and specific strength of several composite materials with metals and polymers.

composites reinforced with high-strength polymer, metal, or ceramic fibers. Some typical applications for these composites are listed in Table 16-4. In all cases, exceptionally good combinations of strength, stiffness, and light weight are required. In addition, improvements are generally observed in fatigue and creep resistance (Figure 16-29).

Advanced composites are used extensively in both structural and skin applications in modern aircraft, taking advantage of the high specific strength.

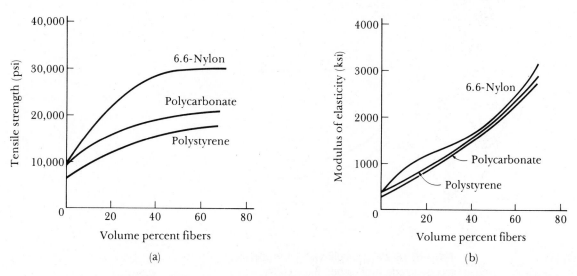

FIGURE 16-28   The effect of randomly oriented glass fibers in composites with several polymer matrices.

TABLE 16-4   Examples of fiber-reinforced materials and applications

Material	Applications
Borsic aluminum	Fan blades in engines, other aircraft and aerospace applications
Kevlar-epoxy	Aircraft, aerospace (including space shuttle), boat hulls,
Kevlar-polyester	sporting goods (including tennis rackets, golf club shafts, fishing rods), flak jackets
Graphite-polymer	Aerospace, automotive, sporting goods
Glass-polymer	Lightweight automotive applications, water and marine applications, corrosion-resistant applications, sporting goods equipment, aircraft and aerospace components

Carbon and aramid fibers in a polymer matrix are the most common examples, although the recently developed polyethylene fibers may soon gain wide acceptance. Carbon fibers are used extensively where particularly good stiffness is required; aramid, and to an even greater extent, polyethylene, fibers are better suited to high-strength applications in which toughness and damage resistance are more important. Compared with carbon, the polymer fibers have better ductility, about the same strength, and lower density. Unfortunately, the polymer fibers lose their strength at relatively low temperatures, as do all of the polymer matrices (Figure 16-30).

The advanced composites are also frequently used for sporting goods. Tennis rackets, golf clubs, skis, ski poles, and fishing poles often contain carbon or aramid fibers because the higher stiffness provides better performance. In the case of golf clubs, carbon fibers allow less weight in the shank and therefore more weight in the head. Fabric reinforced with polyethylene fibers is used for light weight sails for racing yachts.

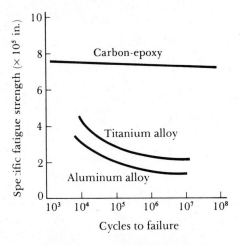

FIGURE 16-29 Comparison of the specific fatigue strength of carbon fiber–epoxy composites with that of titanium and aluminum alloys.

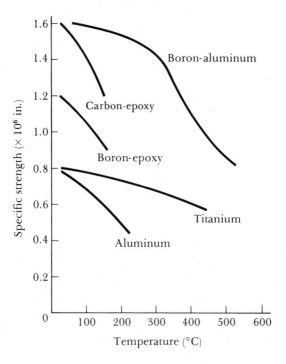

FIGURE 16-30 *The specific strength versus temperature for several composites and metals.*

A unique application for aramid fiber composites is in armor. Tough Kevlar composites provide better ballistic protection than do other materials, making them suitable for lightweight, flexible bulletproof clothing.

Hybrid composites are also becoming important. These composites are composed of two or more types of fibers. For instance, Kevlar fibers may be mixed with carbon fibers to improve the toughness of a stiff composite, or Kevlar might be mixed with glass fibers to improve stiffness. Particularly good tailoring of the composite to meet specific applications can be achieved by controlling the amounts and orientations of each fiber.

Tough composites can also be produced if careful attention is paid to the choice of materials and processing techniques. Better fracture toughness in the usually rather brittle composites can be obtained by using long fibers, amorphous (such as PEEK and PPS) rather than crystalline or cross-linked matrices, thermoplastic elastomer matrices, or interpenetrating network polymers.

Metal Matrix Composites.   These materials, which are strengthened by metal or ceramic fibers, provide better high-temperature resistance than the normal advanced composites due to the higher melting temperatures of the metal matrix. Aluminum reinforced with Borsic fibers has been used extensively in aerospace applications. Unfortunately, the metal matrix composites are much heavier than the advanced polymer matrix composites.

Metal matrix composites may find important applications in components for rocket or aircraft engines. Superalloys reinforced with metal fibers, such as

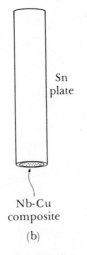
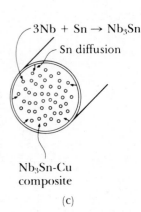

FIGURE 16-31 *The manufacture of composite superconductor wires. (a) Niobium wire is surrounded with copper during forming. (b) Tin is plated onto Nb-Cu composite wire. (c) Tin diffuses to niobium to produce the Nb$_3$Sn-Cu composite.*

tungsten, or ceramic fibers, such as SiC or B$_4$N, may permit these materials to maintain their strength at higher temperatures, permitting jet engines to operate more efficiently.

A unique application for metal matrix composites is in the superconducting wire required for fusion reactors. The intermetallic compound Nb$_3$Sn has good superconducting properties but is very brittle. To produce Nb$_3$Sn wire, pure niobium wire is surrounded by copper as the two metals are formed into a wire composite (Figure 16-31). The niobium-copper composite wire is then coated with tin. The tin diffuses through the copper and reacts with the niobium to produce the intermetallic compound.

Ceramic-Ceramic Composites. Composites containing ceramic fibers in a ceramic matrix are also finding applications. Two important uses will be discussed to illustrate the unique properties that can be obtained with these materials.

Carbon-carbon composites are used for extraordinary temperature resistance in aerospace applications. Carbon-carbon composites can operate at temperatures of up to 3000°C and, in fact, are stronger at high temperatures than at low temperatures (Figure 16-32). Carbon-carbon composites are made by forming a polyacrylonitrile or carbon fiber fabric into a mold, then impregnating the fabric with an organic resin, such as a phenolic. The part is then pyrolyzed to convert the phenolic resin to carbon. The composite, which is still soft and porous, may be impregnated and pyrolyzed several more times, continually increasing the density, strength, and stiffness. Finally the part is coated with silicon carbide to protect the carbon-carbon composite from oxidation. Strengths of 300,000 psi and stiffnesses of 50×10^6 psi can be obtained.

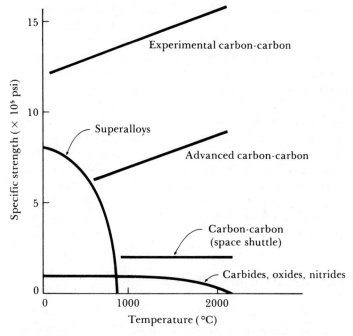

FIGURE 16-32   A comparison of the specific strength of various carbon-carbon composites with that of other high-temperature materials versus temperature.

TABLE 16-5   Effect of SiC reinforcement fibers on the properties of selected ceramic materials

Material	Flexural Strength (psi)	Fracture Toughness $(psi \cdot in^{1/2})$
Al_2O_3	80,000	5
Al_2O_3/SiC	115,000	8
SiC	72,000	4
SiC/SiC	110,000	23
ZrO_2	30,000	5
ZrO_2/SiC	65,000	20
Si_3N_4	68,000	4
Si_3N_4/SiC	115,000	51
Glass	9,000	1
Glass/SiC	120,000	17
Glass ceramic	30,000	2
Glass ceramic/SiC	120,000	16

Carbon-carbon composites have been used as nose cones and leading edges of high-performance aerospace vehicles such as the space shuttle and as brake discs on racing cars and commercial jet aircraft.

Ceramic fiber–ceramic matrix composites are being developed to provide improved strength and toughness (Table 16-5). Ceramics are brittle, with very poor fracture toughness. Once a crack begins, propagation of the crack occurs rapidly and catastrophic failure will occur. However, a crack moving through the matrix of a composite will encounter a fiber. If the bonding is poor between the matrix and fiber, the crack will have to propagate around the fiber in order to continue the fracture process. In addition, poor bonding will allow the fiber to begin to pull out of the matrix. Both processes consume energy, therefore increasing the fracture toughness. This application for fiber-reinforced composites requires poor bonding rather than good bonding between the two constituents. Glass, alumina, and silicon carbide can be reinforced with SiC fibers in this manner.

16–9 LAMINAR COMPOSITE MATERIALS

Laminar composites include very thin coatings, thicker protective surfaces, claddings, bimetallics, laminates, and a host of others. In addition, the fiber-reinforced composites produced from tapes or fabrics can also be considered as partly laminar. Many laminar composites are designed to improve corrosion resistance while retaining low cost, high strength, or light weight. Other important applications include superior wear or abrasion resistance, improved appearance, and unusual thermal expansion characteristics.

Rule of Mixtures.   Some properties of the laminar composite materials along the lamellae are estimated from the rule of mixtures. Density, electrical and thermal conductivity, and modulus of elasticity can be calculated with little error.

$$\text{Density} = \rho_c = \Sigma f_i \rho_i$$
$$\text{Electrical conductivity} = \sigma_c = \Sigma f_i \sigma_i$$
$$\text{Thermal conductivity} = K_c = \Sigma f_i k_i \tag{16-10}$$
$$\text{Modulus of elasticity} = E_c = \Sigma f_i E_i$$

The laminar composites are very anisotropic. The properties perpendicular to the lamellae are

$$\text{Electrical conductivity} = \frac{1}{\sigma_c} = \Sigma \frac{f_i}{\sigma_i}$$

$$\text{Thermal conductivity} = \frac{1}{K_c} = \Sigma \frac{f_i}{K_i} \tag{16-11}$$

$$\text{Modulus of elasticity} = \frac{1}{E_c} = \Sigma \frac{f_i}{E_i}$$

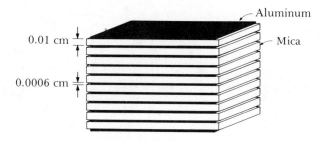

0.01 cm

0.0006 cm

FIGURE 16-33 A capacitor, composed of alternating layers of aluminum and mica, is one example of a laminar composite.

However, many of the really important properties, such as corrosion and wear resistance, depend primarily on only one of the components of the composite, so the rule of mixtures is inappropriate.

◇ | **Example 16-12**

Capacitors used to store electrical charge are essentially laminar composites built up from alternating layers of a conductor and an insulator (Figure 16-33). Suppose we construct a capacitor from 10 sheets of mica, each 0.01 cm thick, and 11 sheets of aluminum, each 0.0006 cm thick. The electrical conductivity of aluminum is 3.8×10^5 (ohm^{-1} · cm^{-1}) and the conductivity of mica is 10^{-13} (ohm^{-1} · cm^{-1}). Determine the electrical conductivity of the capacitor parallel and perpendicular to the sheets.

Answer:

Suppose the capacitor plates are 1 cm^2. Then the volume fractions are

$$V_{Al} = (11 \text{ sheets})(0.0006 \text{ cm})(1 \text{ cm}^2) = 0.0066 \text{ cm}^3$$
$$V_{mica} = (10 \text{ sheets})(0.01 \text{ cm})(1 \text{ cm}^2) = 0.1 \text{ cm}^3$$
$$f_{Al} = \frac{0.0066}{0.0066 + 0.1} = 0.062 \qquad f_{mica} = \frac{0.1}{0.0066 + 0.1} = 0.938$$

Parallel

$$\sigma = (0.062)(3.8 \times 10^5) + (0.938)(10^{-13}) = 0.24 \times 10^5 (\text{ohm}^{-1} \cdot \text{cm}^{-1})$$

Perpendicular

$$\frac{1}{\sigma} = \frac{0.062}{3.8 \times 10^5} + \frac{0.938}{10^{-13}} = 0.938 \times 10^{13}$$

$$\sigma = \frac{1}{0.938 \times 10^{13}} = 1.07 \times 10^{-13} (\text{ohm}^{-1} \cdot \text{cm}^{-1})$$

The composite, or capacitor, has high conductivity parallel to the plates, but acts as an insulator perpendicular to the plates.

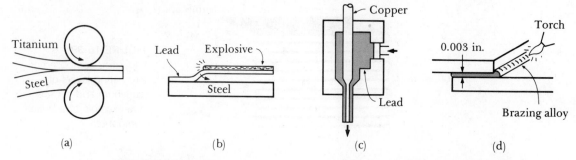

(a)   (b)   (c)   (d)

FIGURE 16-34 *Techniques for producing laminar composites. (a) Roll bonding, (b) explosive bonding, (c) coextrusion, and (d) brazing.*

Several methods are used to produce laminar composites, including a variety of deformation and joining techniques (Figure 16-34).

Rolling.   Most of the metallic laminar composites, such as claddings and bimetallics, are produced by hot or cold roll bonding. If the percent deformation is great enough, the pressure exerted by the rolls breaks up the oxides at the surface, brings the surfaces into atom-to-atom contact, and permits the two surfaces to be welded.

Explosive Bonding.   An explosive charge can provide the pressure required to join metals, as described in Chapter 7 and Figure 16-35. This process is particularly well suited for joining very large plate that will not fit into a rolling mill.

Coextrusion.   Very simple laminar composites, such as coaxial cable, are produced by coextruding two metals through a die in such a way that the soft

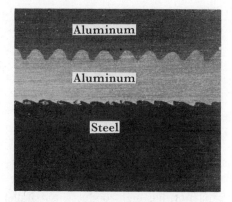

FIGURE 16-35 A laminar composite of two aluminum plates and a steel plate formed by explosive bonding.

material surrounds the harder material. Similarly, a thermoplastic polymer could be coextruded around a metal conductor wire.

Pressing.   For small components, high pressures at elevated temperatures provide welding. Hot pressing is frequently used to cure the adhesive in laminates.

Brazing.   Brazing can join composite plates. The metallic sheets are separated by a very small clearance, preferably about 0.003 in., and heated above the melting temperature of the brazing alloy. The molten brazing alloy is drawn into the thin joint by capillary action.

16–10   EXAMPLES AND APPLICATIONS OF LAMINAR COMPOSITES

The number of laminar composites is so varied and their applications and intentions are so numerous that we cannot make generalizations concerning their behavior. Instead we will examine the characteristics of a few commonly used examples.

Laminates.   Laminates are layers of materials joined by an organic adhesive. A familiar laminate is plywood, in which an odd number of wood veneer piles are stacked so that the grain is at right angles in each alternating ply. The piles are glued by an adhesive such as a phenolic or amine resin. Plywood permits wood products to be available in large sizes yet be inexpensive and resistant to splitting and warping.

Safety glass is a laminate in which a plastic adhesive, such as polyvinyl butyral, joins two pieces of glass; the adhesive prevents fragments of glass from flying about when the glass is broken. Laminates are used for insulation in motors, for gears, for printed circuit boards, and for decorative items such as Formica countertops and furniture.

A recently developed laminate, Arall (aramid aluminum laminate), has been developed as a possible skin material for aircraft. An aramid fiber, such as Kevlar, is woven into a fabric, impregnated with an adhesive, and laminated between layers of aluminum (Figure 16-36). The composite laminate has an unusual combination of strength, stiffness, corrosion resistance, and light weight. In addition, fatigue resistance is improved; cracks that initiate in the outer aluminum layers are arrested on reaching the aramid layer.

The adhesive laminates combine unusual characteristics including light weight, flame retardance, impact strength, corrosion resistance, easy forming and machining, good ablative ability, and good insulation characteristics.

Hard Surfacing.   Hard, wear-resistant surfaces can be deposited on softer, more ductile materials by fusion-welding techniques known as *hard surfacing*.

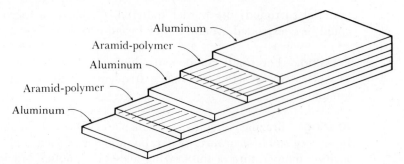

FIGURE 16-36   Schematic diagram of an aramid aluminum laminate, Arall, which has potential for aerospace applications.

Hard-surfacing alloys include hardenable grades of steel, irons and steels that produce hard carbides, cobalt-base alloys, and certain nonferrous alloys. Composite tungsten carbide rods can also be used to provide tungsten carbide at the wear surface. Similar welding procedures can improve corrosion resistance or heat resistance at surfaces.

Clad Metals.   Clad materials are metal-metal composites. A common example is United States silver coinage. A Cu-80% Ni alloy is bonded to both sides of a Cu-20% Ni alloy. The ratio of thicknesses is about $\frac{1}{6} : \frac{2}{3} : \frac{1}{6}$. The high-nickel alloy gives a silver color, while the predominantly copper core provides low cost.

Clad materials provide a combination of good corrosion resistance with high strength. *Alclad* is a clad composite in which commercially pure aluminum is bonded to higher strength aluminum alloys. The pure aluminum protects the higher strength alloy from corrosion. The thickness of the pure aluminum layer is about 1% to 15% of the total thickness. Alclad is used in aircraft construction, heat exchangers, building construction, and storage tanks, where combinations of corrosion resistance, strength, and light weight are desired.

Bimetallics.   Temperature indicators and controllers take advantage of the different coefficients of thermal expansion of the two metals in the laminar composite. If two pieces of metal are heated, the metal with the higher coefficient of thermal expansion becomes longer (Figure 16-37). If the two pieces of metal are rigidly bonded together, the difference in their coefficients causes the strip to bend and produce a curved surface. If one end of the strip is fixed, the free end moves. The amount of movement depends on the temperature; by measuring the curvature or deflection of the strip, we can determine the temperature. Likewise, if the free end of the strip activates a relay, the strip can turn on or off a furnace or air conditioner to regulate temperature.

Metals selected for bimetallics must have (a) very different coefficients of thermal expansion, (b) expansion characteristics that are reversible and repeatable, and (c) a high modulus of elasticity, so the bimetallic device can do work. Often the low-expansion strip is made from Invar, an iron-nickel alloy,

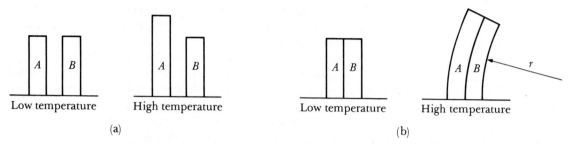

Low temperature   High temperature   Low temperature   High temperature

(a)   (b)

FIGURE 16-37   The effect of thermal expansion coefficient on the behavior of bimetallics. (a) Increasing the temperature increases the length of one metal more than the other. (b) If the two metals are joined, the difference in expansion causes a radius of curvature to be produced.

while the high-expansion strip may be brass, Monel, manganese-nickel-copper, nickel-chromium-iron, or pure nickel.

Bimetallics can act as circuit breakers as well as thermostats; if a current passing through the strip becomes too high, heating causes the bimetallic to deflect and break the circuit.

16–11   SANDWICH STRUCTURES

Sandwich materials have thin layers of a facing material joined to a lightweight filler material, such as a polymer foam. Neither the filler nor the facing material is strong or rigid, but the composite possesses both properties. A

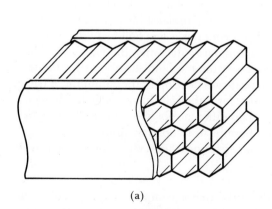

(a)   (b)

FIGURE 16-38   A honeycomb structure. Thin aluminum foil is glued at selected locations, then expanded into a cellular panel. Thicker aluminum facing sheets produce a strong, rigid structure. (a) Schematic drawing, (b) actual structure.

familiar example is corrugated cardboard. A corrugated core of paper is bonded on either side to flat, thick paper. Neither the corrugated core nor the facing paper is rigid, but the combination is.

Another important example is the honeycomb structure used in aircraft applications. A *honeycomb* is produced by gluing thin aluminum strip at selected locations. The honeycomb material is then expanded to produce a very low density cellular panel that by itself is unstable (Figure 16-38). When an aluminum facing sheet is adhesively bonded to either side of the honeycomb, however, a very stiff, rigid, strong, and exceptionally lightweight sandwich weighing as little as 1 pound per cubic foot is obtained.

The honeycomb cells can have a variety of shapes, including hexagonal, square, rectangular, and sinusoidal and can be made from a variety of materials other than aluminum, including fiberglass, paper, and aramid polymers. The honeycomb cells can be filled with foam or fiberglass to provide excellent sound and vibration absorption.

16–12   WOOD

Wood is one of the most familiar of materials. Although wood is not a "high-tech" material, we are literally surrounded by it in our homes and value it for its beauty. In addition, wood is a strong, lightweight material that still dominates much of the construction industry.

We can consider wood to be a complex fiber-reinforced composite composed of long, unidirectionally aligned tubular polymer cells in a polymer matrix. Furthermore, the polymer tubes in turn are composed of cellulose fibers also aligned nearly parallel to the axes of the tubes. This arrangement provides excellent tensile properties in the longitudinal direction of the wood.

Macrostructure of Wood.   A tree is composed of several layers (Figure 16-39). The outer layer, or *bark*, protects the tree; the *cambium*, just beneath the bark, contains new growing cells; the *sapwood* contains a few hollow living cells that store nutrients and serve as the conduit for water; and the *heartwood*, which contains only dead cells, provides most of the mechanical support for the tree.

The tree grows when new elongated cells develop in the cambium. Early in the growing season, the cells are large; later the cells have a smaller diameter, thicker walls, and a higher density. This difference between the early, or *spring*, wood and the late, or *summer*, wood permits us to observe annual growth rings. In addition, some cells grow in a radial direction; these cells, called *rays*, provide storage and transport of food.

Structure of the Cells.   The bulk of the tree is composed of elongated cells, often having an aspect ratio of 100 or more, which constitute about 95% of the solid material in wood. The fibrous cells are actually composed of several layers of long cellulose polymer chains. The cells begin to grow by forming a very thin, flexible primary wall containing randomly oriented cellulose fibers (Figure

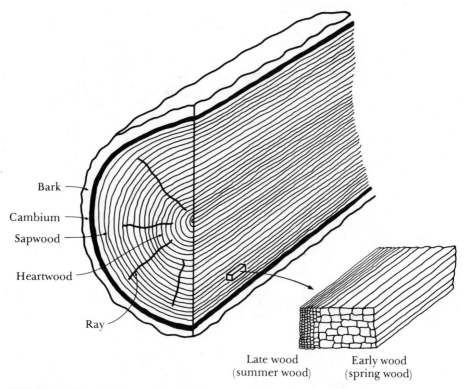

Bark

Cambium

Sapwood

Heartwood

Ray

Late wood
(summer wood)

Early wood
(spring wood)

FIGURE 16-39   *The macrostructure of wood, showing the different layers and the cell size within an annual growth ring.*

16-40). The wall then thickens as the cell grows, with three layers of differently oriented fibers making up this secondary wall. The middle layer of the secondary wall, which is the thickest, contains fibers most closely aligned with the axis of the cell.

The cell walls are composed of three primary polymer ingredients. The majority of the wall is made up of long, crystalline cellulose fibers [Figure 16-40(b)] having a degree of polymerization of as much as 30,000. Some additional fibers, called *hemicellulose*, which have a low degree of polymerization, are also present. The fibers are bonded by an amorphous polymer cement called *lignin*, which also binds the cells to one another.

Water is also present in the wood, sometimes weighing more than the solid material. The water is present in the hollow cells as well as in the cell walls. Other constituents present in the cells include various oils, called *extractives*, and minerals, including silica.

Hardwood versus Softwood.   Hardwoods are obtained from deciduous trees such as oak, elm, beech, birch, walnut, and maple. In the hardwoods, the elongated cells, or fibers, are relatively short, with a diameter of less than 0.1 mm and a length of less than 1 mm. Contained within the wood are longitudinal pores, or vessels, which carry water through the tree.

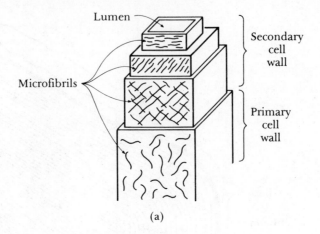

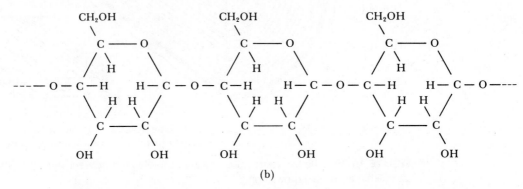

FIGURE 16-40 *The structure of a cell. (a) The structure of the cell, including the orientation of the cellulose fibers in each layer of the cell wall. (b) The structure of the cellulose fibers.*

Softwoods, obtained from evergreens such as pine, fir, spruce, and cedar, have similar structures. In softwoods, the cells, also called *tracheids*, tend to be somewhat longer than in the hardwoods. The hollow center of the cells, called the *lumen*, is responsible for transporting water. In general, the density of the softwoods tends to be lower than that of the hardwoods due to a greater percentage of void space.

Properties of Wood. The material making up the individual cells in virtually all woods has very similar properties. The cells, composed of the cellulose fibers and lignin, have a longitudinal tensile strength of about 100,000 psi and a density of about 1.45 g/cm^3. Several factors affect the final properties of the wood itself.

One important factor is the amount of water in the wood. The percent water is given by

$$\% \text{ Water} = \frac{\text{weight of water}}{\text{weight of dry wood}} \times 100 \qquad (16\text{-}12)$$

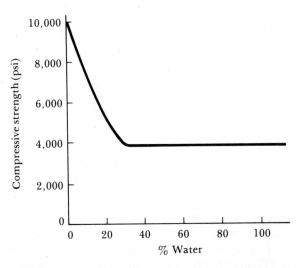

FIGURE 16-41 *Effect of percent water in a typical wood on the compressive strength parallel to the grain.*

Based on this definition, it is possible that a wood could be described as containing more than 100% water, with the sapwood containing considerably more water than the heartwood. As the wood dries, water is eliminated first from the lumen or vessels, where it is least tightly held. During this period, the strength of the wood remains nearly constant (Figure 16-41). On continued drying, giving less than about 30% water, water is lost from the cell walls; during this stage, the strength will increase.

As water is lost during drying, shrinkage of the wood also occurs, which can cause warping and cracking. Shrinkage is due primarily to loss of water from the cell walls, causing much greater dimensional changes perpendicular to the cells than parallel to the cells. Shrinkage can cause distortion and cracking of the wood if allowances are not made. In addition, when the wood is used, the water content in the wood can change, depending on the relative humidity in the environment. As the wood gains or loses water during use, shrinkage and swelling will continue to occur. If the wood construction does not allow movement due to moisture fluctuations, again warping and cracking can occur.

Loss of water also changes the overall density of the wood. The density of dried wood varies from 0.3 to 0.8 g/cm^3. The density is much lower than that of the cellulose wood substance, due to the empty lumen. As the moisture content in the wood decreases, the density will increase.

The type of wood is also important (Table 16-6). Because they contain less of the higher density late wood, softwoods typically are less dense than hardwoods, with densities ranging from 0.3 to 0.5 g/cm^3, compared with 0.3 to 0.8 g/cm^3 in hardwoods. The strength of the wood tends to increase as the density increases; hardwoods are normally stronger than softwoods.

The properties also depend on imperfections in the wood. Clear wood, free of imperfections such as knots, may have a longitudinal tensile strength of 10,000

TABLE 16-6   Properties of typical woods

Wood	Density (for 12% water) (g/cm^3)	Modulus of Elasticity (psi)	Compressive Strength (Parallel to Cells) (psi)
Cedar	0.32	1,100,000	4600
Pine	0.35	1,200,000	4800
Fir	0.48	2,000,000	7200
Maple	0.48	1,500,000	6000
Birch	0.62	2,000,000	8200
Oak	0.68	1,800,000	7400

to 20,000 psi. Less expensive construction lumber, which usually contains many imperfections, may have a tensile strength below 5000 psi.

Because of the anisotropic behavior of wood, its properties depend on how the log is cut (Figure 16-42). Most lumber is cut in a tangential-longitudinal or radial-longitudinal manner. Because of the orientation of the fibers, the strength in the longitudinal direction may be 25 to 50 times greater than the strength in the radial or tangential directions.

Due to its low density and good strength, clear wood has a specific strength and specific modulus that compared very well with those of other common construction materials (Table 16-7). Wood also has very good toughness, largely due to the slight misorientation of the cellulose fibers in the middle layer of the secondary wall. Under load, the fibers straighten, permitting some ductility and energy absorption.

Wood is normally used in compression or bending. Unfortunately, woods generally are very poor under these conditions. In compression, the fibers in the cells tend to buckle. When loads are applied perpendicular to the cells, strengths are even less.

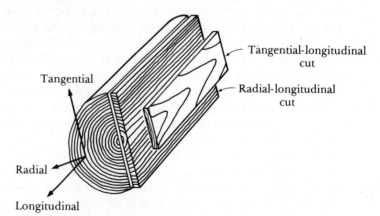

FIGURE 16-42   The different directions in a log; because of differences in cell orientation and the grain, wood has anisotropic behavior.

TABLE 16-7   Comparison of the specific strength and specific modulus of wood with those of other common construction materials

Material	Specific Strength ($\times 10^5$ psi)	Specific Modulus ($\times 10^7$ psi)
Clear wood	7	9.5
Aluminum	5	10.5
1020 steel	2	10.5
Copper	1.5	5.5
Concrete	0.6	3.5

After F. F. Wangaard, "Wood: Its Structure and Properties," *J. Educ. Models for Mat. Sci. and Engr.*, Vol. 3, No. 3, 1979.

◇   **Example 16-13**

A green wood has a density of $0.86 \, \text{g/cm}^3$ and contains 175% water. Calculate the density of the wood after it is completely dried.

Answer:

A 100-cm^3 sample of the wood would weigh 86 g. From Equation 16-12, we can calculate the weight of the dry wood to be

$$\% \text{ Water } = \frac{\text{weight of water}}{\text{weight of dry wood}} \times 100 = 175$$

$$= \frac{\text{green weight} - \text{dry weight}}{\text{dry weight}} \times 100 = 175$$

$$\text{Dry weight of wood } = \frac{(100)\,(\text{green weight})}{275}$$

$$= \frac{(100)\,(86)}{275} = 31.3 \, \text{g}$$

$$\text{Density of dry wood } = \frac{31.3 \, \text{g}}{100 \, \text{cm}^3} = 0.313 \, \text{g/cm}^3$$

Plywood.   The anisotropic behavior of wood can be reduced, and wood products can be made in larger sizes, by producing plywood. Thin layers of wood called *plies* are cut from logs. The plies are then stacked together with the grains between adjacent plies oriented at 90° angles. The plies are then adhesively bonded to one another.

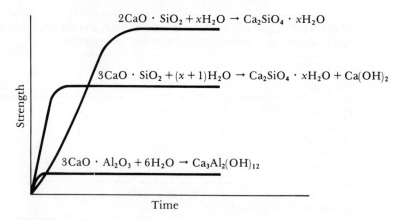

FIGURE 16-43 *The rate of hydration of the minerals in Portland cement.*
$3CaO \cdot Al_2O_3$ and $3CaO \cdot SiO_2$ set rapidly but produce low strength;
$2CaO \cdot SiO_2$ sets slowly but produces high strength.

16–13 CONCRETE

Concrete, another common construction material, is a particulate composite in which an aggregate, usually gravel and sand, is bonded in a matrix of Portland cement and water. A cementation reaction between water and the minerals in the cement provides the required strength.

The cement binder, which is very fine in size, is composed of various ratios of $3CaO \cdot Al_2O_3$, $2CaO \cdot SiO_2$, $3CaO \cdot SiO_2$, and other minerals. When water is added to the cement, a hydration reaction occurs in which the water is rigidly attached to the minerals to produce a solid gel. Water must continue to be made available to the concrete during curing; evolution of heat during the reaction sometimes makes this difficult.

The properties and behavior of the final concrete aggregate depend on a number of factors.

Composition of the Cement. The minerals in cement behave differently during hydration: $3CaO \cdot Al_2O_3$ and $3CaO \cdot SiO_2$ produce rapid setting but low strengths. The $2CaO \cdot SiO_2$ reacts more slowly but produces higher strengths (Figure 16-43). By controlling the relative amounts of the minerals in the cement, as well as other special accelerating or retarding agents, both the rate of setting and the final strength can be controlled. Normally several weeks are required before complete curing of the concrete is achieved.

Water-Cement Ratio. The water-cement ratio determines the workability and the final strength of the concrete. *Workability* refers to how easily the concrete slurry can fill all of the space in the form—air pockets trapped due to poor workability reduce the strength of the concrete structure. Workability can

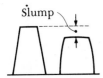

FIGURE 16-44
The slump test, in which deformation of a concrete shape under its own weight is measured, is used to describe the workability of the concrete mix.

be measured by the *slump test*. A wet concrete shape 12 in. tall is produced (Figure 16-44) and is permitted to stand under its own weight. After some period of time, it deforms. The reduction in height of the form is the slump. A larger slump, caused by a higher water-cement ratio, indicates greater workability.

Unfortunately, the strength of the concrete decreases when the water-cement ratio increases (Figure 16-45). High ratios also increase the amount of shrinkage of the concrete during curing. Consequently, some compromise must be made between strength, shrinkage, and workability when selecting the water-cement ratio. A ratio of 0.45 to 0.55 is typical.

Fortunately, workability can be improved by other methods. For example, small bubbles of entrained air and a variety of organic plasticizers may improve the workability with relatively little reduction in the strength of the concrete.

Aggregate.   The aggregate must be clean, strong, and durable, must have the appropriate size and distribution of sizes, and must produce a high packing factor. The size distribution is critical in minimizing the amount of open porosity in the finished concrete—interconnected porosity due to poor packing permits water to penetrate the concrete. When the water freezes and thaws, the resulting expansion can cause the concrete to disintegrate. Again, entrained air bubbles may help minimize these problems, provided the bubbles are well dispersed.

An angular rather than round aggregate provides better strength due to mechanical interlocking and a greater surface area for bonding.

Normally the aggregates are sand and gravel. However, in some cases special aggregates may be used. Lightweight concretes can be produced by using mineral slags produced during steel-making operations; these concretes provide better thermal insulation characteristics. Particularly heavy concretes can be produced using dense minerals or even metal shot; these heavy concretes might be used in building nuclear reactors to better absorb radiation.

Reinforced and Prestressed Concrete.   Concrete is capable of obtaining good compressive strengths but performs poorly in tension. When tall concrete structures (which might lead to buckling) or concrete beams (which might include tensile loads) are designed, the concrete may be reinforced with steel wire or bars. The steel reinforcement helps minimize cracking and crack openings when tensile loads are imposed.

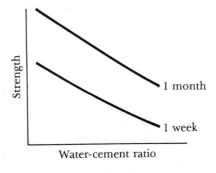

FIGURE 16-45 The water-cement ratio and the setting time are important factors in determining the strength of concrete.

(a)

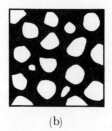
(b)

FIGURE 16-46 The ideal structure of asphalt (a) compared with the undesirable structure (b) in which round grains, a narrow distribution of grains, and excess binder all reduce the strength of the final material.

Even better reinforcement occurs when the steel wire or bars are prestressed. When the bars are placed into the form, a tensile force may be applied to cause elastic strain of the steel. Concrete is then poured around the reinforcing bars. After curing, the force on the steel is removed. As the steel attempts to recover its elastic strain, the surrounding concrete is placed in compression.

16–14 ASPHALT

Asphalt is a composite of aggregate and bitumen, which is a thermoplastic polymer. The properties of the asphalt are determined by the characteristics of the aggregate and binder and their relative amounts.

The aggregate, as in concrete, should be clean and angular and should have a distribution of grain sizes to provide a high packing factor and good mechanical interlocking between the aggregate grains (Figure 16-46).

The binder, composed of thermoplastic polymer chains, bonds the aggregate particles. The binder has a relatively narrow useful temperature range, being brittle at sub-zero temperatures and beginning to melt at relatively low temperatures. Additives, such as gasoline or kerosene, can be used to modify the binder, permitting it to liquefy more easily during mixing while causing the asphalt to cure more rapidly after application.

The ratio of binder to aggregate is important; just enough binder should be added so that the aggregate particles touch but voids are minimized. Excess binder permits viscous deformation of the asphalt under load.

SUMMARY

The strengthening mechanisms discussed in earlier chapters are difficult to apply to composite materials. Instead composites are designed to provide unusual combinations of properties that cannot be obtained by the typical techniques used to control microstructure and mechanical properties. This is especially true for the laminar and particulate composites, which are almost always designed to satisfy special service requirements other than strength. Fiber-reinforced composites are normally designed to produce unusual combinations of strength, stiffness, temperature resistance, and light weight. The rule of mixtures is suitable for determining the behavior of many simple fiber-reinforced materials.

GLOSSARY

Aspect ratio. The length of a fiber divided by its diameter.

Bimetallic. A laminar composite material produced by joining two strips of metal with different thermal expansion coefficients, making the material sensitive to temperature changes.

Cambium. The layer of growing cells in wood.

Carbonizing. Driving off the noncarbon atoms from a polymer fiber, leaving behind a carbon fiber of high strength. Also known as pyrolizing.

Cemented carbides. Particulate composites containing hard ceramic particles bonded with a soft metallic matrix. The composite combines high hardness and cutting ability yet still has good shock resistance.

Cladding. The good corrosion-resistant or high-hardness layer of a laminar composite formed onto a less expensive or higher strength backing.

Compocasting. Injection of a thixotropic mixture of an alloy and a filler material into a die at high pressures to form a composite.

Delamination. Separation of individual plies of a fiber-reinforced composite.

Honeycomb. A lightweight but stiff assembly of aluminum strip joined and expanded to form the core of a sandwich structure.

Lignin. Polymer cement in wood that bonds the cellulose fibers in the wood cells.

Lumen. The hollow center of a fibrous cell in wood.

Precursor. The polymer fiber that is carbonized to produce carbon fibers.

Prepregs. Layers of fibers in unpolymerized resins. After the prepregs are stacked to form a desired structure, polymerization joins the layers together.

Pultrusion. A method for producing composites containing mats or continuous fibers.

Rovings. Bundles of less than 10,000 filaments.

Rule of mixtures. The statement that the properties of a composite material are a function of the volume fraction of each material in the composite.

Sandwich A composite material constructed of a lightweight, low-density material surrounded by dense, solid layers. The sandwich combines overall light weight with excellent stiffness.

Sizing. Coating glass fibers with an organic material to improve bonding and moisture resistance in fiberglass.

Slump test. A test to measure the workability of a concrete mix.

Specific modulus. The modulus of elasticity divided by the density.

Specific strength. The strength of a material divided by the density.

Staples. Fibers chopped into short lengths.

Surfacing. A welding technique by which a hard or corrosion-resistant material is deposited on the surface of a second material, producing a laminar composite.

Thixotropic. The ability of a partly liquid–partly solid material to maintain its shape until a stress is applied, when it then flows like a liquid.

Tow. A bundle of more than 10,000 filaments.

Tracheids. Hollow, fibrous tubes that comprise the structure of softwoods.

Whiskers. Very fine fibers grown in a manner that produces single crystals with no mobile dislocations, thus giving nearly theoretical strengths.

Workability. The ease with w ' ch a concrete slurry can fill all of tł   ace in a form.

PRACTICE PROBLEMS

1. A dispersion-strengthened SAP aluminum alloy is produced from metal powder particles having a diameter of 0.01 mm. The Al_2O_3 film on the surface of each particle is spheroidized during the powder metallurgy manufacturing process. If the SAP contains 15 vol % Al_2O_3, calculate the original thickness of the oxide film on the aluminum powder particles.

2. The density of Al_2O_3 is about 3.85 g/cm^3. A SAP aluminum alloy is produced by powder metallurgy processing using powder particles having a diameter of 0.01 mm with an oxide coating of 0.0001 mm. A dispersion of spherical oxide particles 0.005 mm in diameter is produced. Calculate (a) the vol % Al_2O_3 present in the SAP, (b) the density of the SAP alloy, and (c) the number of oxide particles per 1000 g of alloy.

3. A tungsten alloy containing 2 wt % thorium is converted into a powder, compacted into a desirable shape, and oxidized during sintering. ThO_2 particles with a diameter of 1000 Å are produced. If all of the thorium is oxidized, calculate (a) the vol % ThO_2 produced in the alloy and (b) the number of oxide particles per cubic centimeter. (The density of ThO_2 is 9.86 g/cm^3.)

4. A nickel-yttrium alloy is internally oxidized to produce Y_2O_3 particles each 1500 Å in diameter. Careful measurements indicate there are 2×10^{12} oxide particles per cubic centimeter of the composite. If the density of Y_2O_3 is 5.01 g/cm^3, calculate (a) the vol % Y_2O_3 in the alloy and (b) the wt % Y originally in the alloy.

5. Explain why aluminum, which is a more ductile, tougher material than cobalt, is not used to produce cermets for high-speed machining tools.

6. A C-6 cemented carbide cutting tool contains 8 wt % Co, 10 wt % TaC, 10 wt % TiC, and 72 wt % WC. Calculate the density of the material. (See Example 16-2 for densities.)

7. A grinding wheel is 8 in. in diameter and 1 in. thick and weighs 5.5 lb. It is composed of SiC (density of 3.2 g/cm^3) bonded by silica glass (density of 2.5 g/cm^3), with the SiC particles approximating cubes of dimension 0.01 in. (a) Calculate the volume fraction of SiC particles in the wheel and (b) determine the number of SiC particles lost from the wheel after it is worn to a diameter of 7 in.

8. An abrasive grinding wheel contains 70 vol % Al_2O_3, a phenolic binder, and porosity. Determine the volume percent porosity if the grinding wheel has a density of 2.95 g/cm^3. The density of the phenolic is 1.28 g/cm^3 and that of Al_2O_3 is 3.96 g/cm^3.

9. An electrical contact material is produced by infiltrating copper into a porous tungsten compact. If the density of the finished composite is 15 g/cm^3, calculate (a) the volume fraction of copper in the composite, (b) the volume fraction of pores in the tungsten compact before infiltration, and (c) the original density of the tungsten compact before infiltration.

10. Suppose we use hollow glass beads as extenders in a thermosetting polyester. The glass beads have an outside diameter of 1 mm and a wall thickness of 0.01 mm. The glass has a density of $2.5 \, \text{g/cm}^3$ and the polyester has a density of $1.28 \, \text{g/cm}^3$. Calculate the number of beads required per cubic meter to produce a composite with a density of $0.85 \, \text{g/cm}^3$.

11. Calculate the number of grams of clay that must be added to $1000 \, \text{g}$ of polyethylene to produce a modulus of elasticity of $120,000 \, \text{psi}$. The density of clay is $2.4 \, \text{g/cm}^3$ and that of polyethylene is $0.95 \, \text{g/cm}^3$. (See Figure 16-7.)

12. Round sand grains coated with $1.5 \, \text{wt} \%$ phenolic resin are often used to make molds and cores for metal castings. (a) If the sand grains are $0.5 \, \text{mm}$ in diameter, determine the average thickness of the resin coating. The sand has a density of $2.2 \, \text{g/cm}^3$ and the resin has a density of $1.28 \, \text{g/cm}^3$. (b) The apparent density of the sand is $95 \, \text{lb/ft}^3$. Estimate the volume percent porosity in the sand.

13. We wish to compocast a Zn-27% Al alloy, introducing $30 \, \text{vol} \%$ glass beads during the processing. (a) From Figure 13-13, estimate the required casting temperature and (b) determine the density of the final product. The density of the glass is $2.2 \, \text{g/cm}^3$ and that of the zinc alloy is $5.03 \, \text{g/cm}^3$.

14. Using Figure 14-24, estimate the liquidus temperature of E glass.

15. A composite is produced by incorporating $30 \, \text{vol} \%$ HS carbon fibers in polyester. (a) Determine the volume percent polyethylene fibers required to produce the same modulus of elasticity and (b) compare the specific modulus of the polyethylene-reinforced composite with that of the carbon-reinforced composite. The density of the polyester is $1.28 \, \text{g/cm}^3$ and the modulus is $650,000 \, \text{psi}$.

16. There is some interest in producing magnesium reinforced with HM carbon fibers. Determine the density and modulus of elasticity both parallel and perpendicular to the fibers if $350 \, \text{g}$ of fibers are introduced into $300 \, \text{g}$ of magnesium. The density of the magnesium is $1.74 \, \text{g/cm}^3$ and its modulus is $6.5 \times 10^6 \, \text{psi}$.

17. An aluminum panel for an airplane measures $6 \, \text{ft} \times 8 \, \text{ft} \times 0.25 \, \text{in}$. (a) Determine the weight of the aluminum panel. (b) Determine the weight giving the same modulus of elasticity if the panel is made of polyethylene fiber–reinforced epoxy of the same thickness. The density of the epoxy is $1.3 \, \text{g/cm}^3$ and its modulus is $450,000 \, \text{psi}$.

18. Calculate (a) the vol $\%$ fibers and (b) the specific modulus of a unidirectionally aligned fiberglass boat hull having a density of $1.6 \, \text{g/cm}^3$. The matrix is polyester (density $= 1.28 \, \text{g/cm}^3$, modulus of elasticity $= 650,000 \, \text{psi}$) and S glass fibers are employed.

19. Nickel-base superalloys can be reinforced with tungsten fibers. (a) Determine the volume fraction of tungsten fibers needed to obtain a modulus of elasticity of $40 \times 10^6 \, \text{psi}$ perpendicular to the fibers. (b) Calculate the modulus of elasticity parallel to the fibers for this composite. The density of the superalloy is about $8.5 \, \text{g/cm}^3$ and the modulus of elasticity is $28 \times 10^6 \, \text{psi}$.

20. A hybrid composite is produced in which polyester is reinforced with HS carbon and Kevlar fibers. The volume percent of the carbon fibers is twice that of the Kevlar fibers. If the density of the composite is $1.4 \, \text{g/cm}^3$, estimate (a) the vol $\%$ of carbon, Kevlar, and polyester and (b) the wt $\%$ of carbon, Kevlar, and polyester. The density of the polyester is $1.28 \, \text{g/cm}^3$.

21. A 1 in. diameter cable is constructed of a $\frac{1}{8}$ in. diameter aluminum core coated with steel. Determine the thermal conductivity of the composite cable. The thermal conductivity of aluminum is $0.57 \, \text{cal/cm} \cdot \text{s} \cdot \text{K}$ and that of steel is $0.24 \, \text{cal/cm} \cdot \text{s} \cdot \text{K}$.

22. Two composites are produced; one composite contains $30 \, \text{vol} \%$ SiC particles, while a second contains an equal amount of SiC but in the form of fibers. Which will have the higher heat capacity? Will heat capacity be dependent on the orientation of the fibers?

23. Explain why bonding between carbon fibers and an aluminum matrix must be excellent, while bonding between silicon nitride fibers and a silicon carbide matrix should be poor.

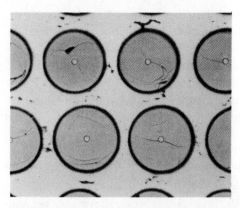

FIGURE 16-47 Borsic fiber–reinforced aluminum. The fibers are composed of a thick layer of boron deposited on a small-diameter tungsten filament (× 1000). Reprinted with permission from *Journal of Metals*. Vol. 24, No. 9, 1972, a publication of the Metallurgical Society, Warrendale, Pennsylvania 15086 USA.

24. Figure 16-47 shows the cross section of a Borsic fiber–reinforced aluminum composite material. Estimate (a) the vol % of tungsten, boron, and aluminum in the composite and (b) the modulus of elasticity parallel to the fibers. (c) If tungsten were not present in the boron fiber core, by what percent would the modulus of elasticity of the composite increase or decrease?

25. A composite of epoxy reinforced with 65 vol % carbon fibers is intended to have a minimum modulus of elasticity of 35×10^6 psi. (a) At what temperature should the carbon fibers be produced? (b) What tensile strength will these fibers possess?

26. Aramid fibers can be produced by a condensation reaction involving the two monomers shown below. Suppose 100 g of monomer *A* is reacted with monomer *B*. (a) What by-product is formed as a result of the polymerization reaction? (b) How many grams of monomer *B* are required? (c) How many grams of the by-product are formed? (d) What is the total weight of the aramid fibers produced?

27. A capacitor much like the one shown in Figure 16-33 is produced by stacking together 21 copper sheets 0.0008 cm thick and 20 BaTiO$_3$ sheets 0.05 cm thick. Determine the electrical conductivity both parallel and perpendicular to the sheets. The electrical conductivity of copper is 6×10^5 ohm$^{-1} \cdot$ cm^{-1} and that for BaTiO$_3$ is 1×10^{-12} ohm$^{-1} \cdot$ cm^{-1}.

28. Suppose a 0.25 in. thick iron sheet is coated with a 0.005 in. enamel of glass. (a) Calculate the percent decrease in thermal conductivity between the uncoated and coated iron. (b) Calculate the thickness of the glass enamel required to reduce the thermal conductivity of the steel by 50%. The thermal conductivity of iron is 0.19 cal/cm $\cdot$ s $\cdot$ K and that of glass is 0.003 cal/cm $\cdot$ s $\cdot$ K.

29. Suppose we sandwich a 0.1 cm thick layer of polycarbonate between two sheets of 0.5 cm thick glass. Calculate the modulus of elasticity of the composite perpendicular to the sheet. The modulus and density of polycarbonate are 400,000 psi and 1.2 g/cm^3, while those of the glass are 12×10^6 psi and 2.3 g/cm^3.

30. An aluminum honeycomb structure is produced. The honeycomb material is in the form of hexagonal cells, with each face of the cell 0.5 cm long. The aluminum walls of the cell are 0.008 cm thick. Separated by 1 in. of honeycomb are two aluminum sheets each 0.15 cm thick. (a) Estimate the density of the structure. (b) How many pounds would a 4 ft × 8 ft panel of the honeycomb weigh? Compare with a 4 ft × 8 ft × 1 in. thick solid aluminum plate.

31. For the honeycomb structure described in Problem 16–30, estimate the thermal conductivity of the honeycomb parallel to the walls of the cells. The thermal conductivity of aluminum is 0.57 cal/cm $\cdot$ s $\cdot$ K and that of air is about 0.00005 cal/cm $\cdot$ s $\cdot$ K.

32. A completely dried 2 in. × 4 in. × 8 in. sample of wood has a density of 0.3 g/cm^3. (a) How many liters of water are absorbed by the sample if it contains 125% water? (b) Calculate the density after the wood absorbs this amount of water.

33. The measured density of a 100-cm^3 sample of birch (see Table 16-6) is 0.98 g/cm^3. Calculate (a) the density of completely dry birch and (b) the percent water in the original sample.

34. A contractor calls for a slump of 2 in. for the concrete basement floor of a new house, while a slump of 5 in. is specified for the walls. Explain the difference in the two specifications.

35. The tar used in producing asphalt has a density of about 1.5 g/cm^3. Determine the typical weight of one yard (27 ft^3) of asphalt if the asphalt contains 65 vol % sand (density of 2.2 g/cm^3).

Part **IV**

PHYSICAL PROPERTIES OF ENGINEERING MATERIALS

The physical behavior of materials is described by a variety of electrical, magnetic, optical, and thermal properties. Most of these properties are determined by the atomic structure, atomic arrangement, and crystal structure of the material. In Chapter 17, we will find that the atomic structure, in particular the energy gap between the electrons in the valence and conduction bands, helps us to divide materials into conductors, semiconductors, and insulators. The atomic structure is responsible for the dielectric and ferromagnetic behavior discussed in Chapter 18 and explains many of the optical properties such as emission and transparency which are discussed in Chapter 19.

The physical properties can be altered to a significant degree by changing the short- and long-range order of the atoms as well as by introducing and controlling imperfections in the atomic structure and arrangement. We will find that strengthening mechanisms and metal processing techniques, for example, have a significant effect on electrical conductivity of metals. Improved magnets are obtained by introducing lattice defects or by controlling grain size. In this section, we will again demonstrate the importance of the structure-property-processing relationship.

17

ELECTRICAL CONDUCTIVITY

17–1 INTRODUCTION

In many applications, the electrical behavior of the material is more critical than the mechanical behavior. Metal wire used to transfer current over long distances must have a high electrical conductivity so that little power is lost by heating of the wire. Ceramic insulators must prevent arcing between conductors. Semiconductor devices used to convert solar energy to electrical power must be as efficient as possible to make solar cells a practical alternative energy source.

To select and use materials for electrical and electronic applications, we must understand how properties such as electrical conductivity are produced and controlled. We must also realize that electrical behavior is influenced by the structure of the material, the processing of the material, and the environment to which the material is exposed. To accomplish these goals, we must examine in greater detail the electronic structure of groups of atoms.

17–2 RELATIONSHIP BETWEEN OHM'S LAW AND ELECTRICAL CONDUCTIVITY

Most of us are familiar with the common form of Ohm's law

$$V = IR \tag{17-1}$$

where V is the voltage (volts, V), I is the current (amps, A), and R is the resistance (ohms, Ω) to the current flow. The resistance R is a characteristic of the size, shape, and properties of the materials that compose the circuit.

$$R = \rho \frac{l}{A} = \frac{l}{\sigma A} \tag{17-2}$$

where l is the length (cm) of the conductor, A is the cross-sectional area (cm²) of the conductor, ρ is the electrical *resistivity* (ohm · cm, $\Omega \cdot$ cm), and σ, which is the reciprocal of ρ, is the electrical *conductivity* (ohm$^{-1} \cdot$ cm^{-1}). We can use this equation to design resistors, since we can vary the length or the cross-sectional area of the device.

◇ | **Example 17-1**

Calculate the loss of power in a copper transmission line 1500 m long when a current of 50 A is flowing. The copper wire has a diameter of 0.1 cm and the electrical resistivity is $1.67 \times 10^{-6} \Omega \cdot$ cm.

$$R = \frac{\rho l}{A} = (1.67 \times 10^{-6}) \frac{(1500 \text{ m})(100 \text{ cm/m})}{(\pi/4)(0.1 \text{ cm})^2} = 31.9 \, \Omega$$

The power loss in the form of heating of the wire is

$$\text{Power} = VI = I^2 R = (50)^2 (31.9) = 7.98 \times 10^4 \text{ W}$$

A second form of Ohm's law is obtained if we combine Equations 17-1 and 17-2.

$$V = IR = \frac{Il}{\sigma A}$$

$$\frac{I}{A} = \sigma \frac{V}{l}$$

If we define I/A as the *current density* J (A/cm²) and V/l as the *electric field* ξ (V/cm), then

$$J = \sigma \xi \qquad\qquad (17\text{-}3)$$

We can also determine that the current density J is

$$J = nq\bar{v}$$

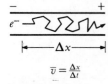

$$\bar{v} = \frac{\Delta x}{\Delta t}$$

FIGURE 17-1
Charge carriers, such as electrons, are deflected by atoms or lattice defects and take an irregular path through a conductor. The average rate at which the carriers move is the drift velocity $\bar{v}$.

where n is the number of charge carriers (carriers/cm³), q is the charge on each carrier (1.6×10^{-19} C), and $\bar{v}$ is the *average drift velocity* (cm/s) at which the charge carriers move (Figure 17-1). Thus

$$\sigma \xi = nq\bar{v} \quad \text{or} \quad \sigma = nq \frac{\bar{v}}{\xi}$$

The term $\bar{v}/\xi$ is called the *mobility* μ (cm²/V · s).

$$\mu = \frac{\bar{v}}{\xi}$$

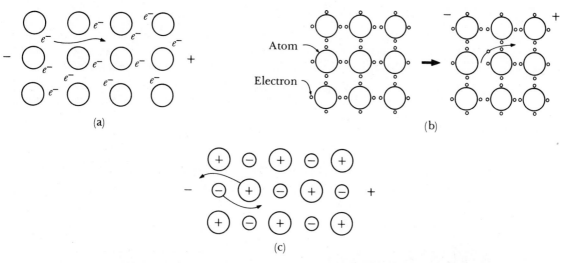

FIGURE 17-2 *Charge carriers in different materials. (a) Valence electrons in the metallic bond move easily, (b) covalent bonds must be broken in semiconductors and insulators for an electron to be able to move, and (c) entire ions must diffuse to carry charge in many ionically bonded materials.*

Finally,

$$\sigma = nq\mu \tag{17-4}$$

The charge q is a constant; from inspection of the equation, we find that we can control the electrical conductivity of a material by controlling the number of charge carriers in the material or by controlling the mobility, or ease of movement, of the charge carriers.

Electrons are the charge carriers in conductors (such as metals), semiconductors, and insulators whereas ions carry the charge in most ionic compounds (Figure 17-2). The mobility depends on atomic bonding, lattice imperfections, microstructure, and, in ionic compounds, diffusion rates.

Table 17-1 includes some useful units and relationships.

TABLE 17-1   *Some useful relationships and units*

Electron volt $= 1\,eV = 1.6 \times 10^{-19}$ joules $= 1.6 \times 10^{-12}$ ergs
1 amp $= 1$ coulomb/second
1 volt $= 1$ amp $\cdot$ ohm
$k =$ Boltzmann's constant $= 8.63 \times 10^{-5}\,eV/K = 1.38 \times 10^{-16}\,erg/K$
kT at room temperature $= 0.025\,eV$

◇ | **Example 17-2**

Calculate the mobility of an electron in copper, assuming that all of the valence electrons contribute to the current flow.

Answer:

The valence of copper is one; therefore the number of valence electrons equals the number of copper atoms in the material. The lattice parameter of copper is 3.6151×10^{-8} cm and, since copper is FCC, there are 4 atoms/unit cell.

$$n = \frac{(4 \text{ atoms/cell})(1 \text{ electron/atom})}{(3.6151 \times 10^{-8} \text{ cm})^3} = 8.467 \times 10^{22} \text{ electrons/cm}^3$$

$$q = 1.6 \times 10^{-19} \text{ C}$$

$$\mu = \frac{\sigma}{nq} = \frac{1}{\rho nq} = \frac{1}{(1.67 \times 10^{-6})(8.467 \times 10^{22})(1.6 \times 10^{-19})}$$

$$= 44.2 \text{ cm}^2/\Omega \cdot \text{C} = 44.2 \text{ cm}^2/\text{V} \cdot \text{s}$$

◇ | **Example 17-3**

Calculate the average drift velocity for the electrons in a 100-cm copper wire when 10 V are applied.

Answer:

The electric field is

$$\xi = \frac{V}{l} = \frac{10}{100} = 0.1 \text{ V/cm}$$

$$\bar{v} = \mu \xi = (44.2)(0.1) = 4.42 \text{ cm/s}$$

17-3 BAND THEORY

In Chapter 2 we found that the electrons in a single atom occupy discrete energy levels. The Pauli exclusion principle permits each energy level to contain only two electrons. For example, the $2s$ level of a single atom contains one energy level and two electrons. The $2p$ level contains three energy levels and a total of six electrons.

When N atoms come together to produce a solid, the Pauli principle still requires that only two electrons in the entire solid have the same energy. Each energy level broadens into a band (Figure 17-3). Consequently, the $2s$ band in a solid contains N discrete energy levels and $2N$ electrons, two in each energy level. Each of the $2p$ levels contains N energy levels and $2N$ electrons. Since the three $2p$ bands actually overlap, we could alternately describe a single, broad $2p$ band containing $3N$ energy levels and $6N$ electrons.

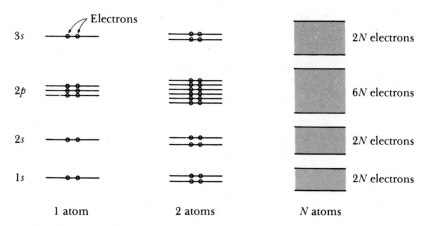

FIGURE 17-3 *The energy levels broaden into bands as the number of electrons grouped together increases.*

◇ | **Example 17-4**

How many energy levels are present in the 2s band of a pure aluminum crystal 1 cm × 1 cm × 1 cm in size?

Answer:

Aluminum is FCC, has 4 atoms/unit cell, and has a lattice parameter of 4.0496×10^{-8} cm.

$$N = \text{number of atoms} = \frac{4 \text{ atoms/cell}}{(4.0496 \times 10^{-8})^3} = 6 \times 10^{22} \text{ atoms/cm}^3$$

$2N$ = energy levels in the 2s band

$2N$ = 12×10^{22} energy levels in the 2s band/cm^3

Band Structure of Alkali Metals.   The alkali metals in column IA of the periodic table have only one electron in the outermost s level of their electronic structure. Figure 17-4 shows an idealized picture of the band structure in sodium, which has an electronic structure of $1s^2 2s^2 2p^6 3s^1$. The vertical line represents the equilibrium interatomic spacing of the atoms in solid sodium.

The shaded areas in Figure 17-4 indicate the portion of the band in which the energy levels are completely occupied by electrons. The 3s valence band of sodium is only half filled and only the lowest possible energy levels within the 3s band are occupied.

The *Fermi energy* E_f is the energy at which half of the possible energy levels in the band are actually occupied by electrons. At absolute zero, only the energy

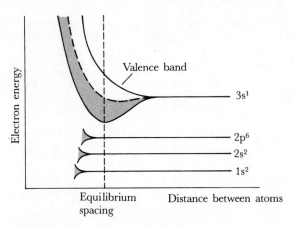

FIGURE 17-4 The simplified band structure for sodium. The energy levels broaden into bands. The 3s band, which is only half filled with electrons, is responsible for conduction in sodium.

levels in the bottom half of the 3s band of sodium are occupied (Figure 17-5), and the Fermi energy is the energy at the center of the 3s band.

When the temperature of the metal is increased, electrons near the Fermi energy gain enough energy to enter the conduction band. But an equal number of energy levels in the valence band are left unoccupied. Consequently, the Fermi energy is unchanged.

The Fermi distribution $f(E)$ gives the probability that a particular energy level E in the band is occupied by an electron. The function $f(E)$, which may vary from 0 to 1, is

$$f(E) = \frac{1}{1 + \exp\left(\dfrac{E - E_f}{kT}\right)} \tag{17-5}$$

where k is Boltzmann's constant (8.63×10^{-5} eV/K). At absolute zero, the probability $f(E)$ that an electron has an energy E less than E_f is one; the

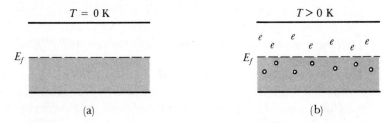

FIGURE 17-5 (a) At absolute zero, all of the electrons in the outer energy level have the lowest possible energy. (b) When the temperature is increased, some electrons are excited into unfilled levels. Note that the Fermi energy is unchanged.

probability $f(E)$ that an electron has an energy E greater than E_f is zero. At higher temperatures, some electrons enter energy levels in the conduction band.

The Fermi distribution shows that above absolute zero some electrons in a metal possess sufficient energy to exceed the Fermi level and enter the conduction band. At the same time, vacant energy levels within the valence band are created. When an electrical field is applied, the electrons in the conduction band are accelerated towards the positive terminal of the circuit, while electrons in the valence band decelerate towards the negative terminal by filling vacant sites in the valence band. Both of these events cause a net flow of current.

Sodium and other alkali metals have good electrical conductivity because of the half-filled outermost s band. Table 17-2 compares the electronic structures and conductivities of the alkali metals with those of other metals.

Band Structures of Other Metals. The band structure and electrical behavior vary for different groups of metals. The simplified band structures for four metals are shown in Figure 17-6 and conductivities are listed in Table 17-2.

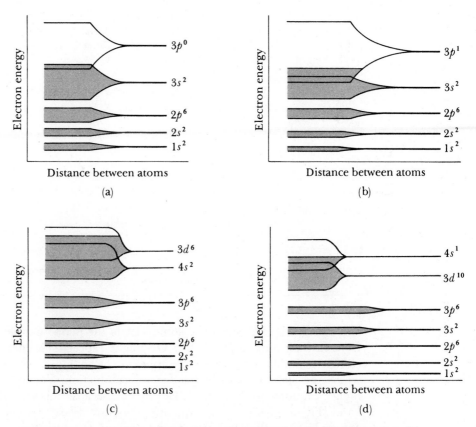

FIGURE 17-6 *Simplified band structures for selected metals. (a) Magnesium, (b) aluminum, (c) iron, and (d) copper.*

TABLE 17-2   Electronic structure and electrical conductivity of several groups of metals at 25°C

Metal	Electronic Structure	Electrical Conductivity $(ohm^{-1} \cdot cm^{-1})$
Alkali metals		
Li	$1s^2 2s^1$	1.07×10^5
Na	$1s^2 2s^2 2p^6 3s^1$	2.13×10^5
K	$1s^2 2s^2 2p^6 3s^2 3p^6 4s^1$	1.64×10^5
Rb	$\ldots \ldots 4s^2 4p^6 5s^1$	0.86×10^5
Cs	$\ldots \ldots 5s^2 5p^6 6s^1$	0.50×10^5
Alkali earths		
Be	$1s^2 2s^2$	2.50×10^5
Mg	$1s^2 2s^2 2p^6 3s^2$	2.25×10^5
Ca	$1s^2 2s^2 2p^6 3s^2 3p^6 4s^2$	3.16×10^5
Sr	$\ldots \ldots 4s^2 4p^6 5s^2$	0.43×10^5
Aluminum and Group IIIA		
B	$1s^2 2s^2 2p^1$	0.03×10^5
Al	$1s^2 2s^2 2p^6 3s^2 3p^1$	3.77×10^5
Ga	$\ldots 3s^2 3p^6 3d^{10} 4s^2 4p^1$	0.66×10^5
In	$\ldots 4s^2 4p^6 4d^{10} 5s^2 5p^1$	1.25×10^5
Tl	$\ldots 5s^2 5p^6 5d^{10} 6s^2 6p^1$	0.56×10^5
Transition metals		
Sc	$1s^2 2s^2 2p^6 3s^2 3p^6 3d^1 4s^2$	0.77×10^5
Ti	$\ldots \ldots \ldots 3d^2 4s^2$	0.24×10^5
V	$\ldots \ldots \ldots 3d^3 4s^2$	0.40×10^5
Cr	$\ldots \ldots \ldots 3d^5 4s^1$	0.77×10^5
Mn	$\ldots \ldots \ldots 3d^5 4s^2$	0.11×10^5
Fe	$\ldots \ldots \ldots 3d^6 4s^2$	1.00×10^5
Co	$\ldots \ldots \ldots 3d^7 4s^2$	1.90×10^5
Ni	$\ldots \ldots \ldots 3d^8 4s^2$	1.46×10^5
Copper and Group IB		
Cu	$1s^2 2s^2 2p^6 3s^2 3p^6 3d^{10} 4s^1$	5.98×10^5
Ag	$\ldots \ldots \ldots 4p^6 4d^{10} 5s^1$	6.80×10^5
Au	$\ldots \ldots \ldots 5p^6 5d^{10} 6s^1$	4.26×10^5

From *ASM Metals Handbook*, Vol. 2, 9th Ed., 1979.

The alkali earth metals, column IIA of the periodic table, have two electrons in their outermost s band. We might expect these metals to have poor electrical conductivity, since there appear to be no unoccupied energy levels into which the electrons can be excited for conduction. Yet these metals have a higher conductivity than the alkali metals because the p band overlaps the s band [Figure 17-6(a)]. Consequently, there are a large number of unoccupied energy levels in the combined $3s$ and $3p$ band to which a magnesium electron can be excited.

Aluminum, which follows sodium and magnesium in atomic number, has a partly filled $3p$ band, which serves as the conduction band [Figure 17-6(b)]. Electrons easily enter unoccupied levels in the $3p$ band.

The transition metals, including scandium through nickel, contain one or two electrons in their outermost s band but possess an unfilled d band. In iron [Figure 17-6(c)], the $4s$ band overlaps the $3d$ band, which contains only six electrons. Electrons may enter the conduction band in the upper half of the overlapping bands. However, complex interactions between the bands prevent the conductivity from being as high as in some of the better conductors.

Group IB metals—copper, silver and gold—contain one electron in their outermost s band and immediately follow the transition metal groups in atomic number. The inner d band electrons are tightly held by the atom core and do not interact with the electrons in the s band [Figure 17-6(d)]. Copper, silver, and gold have very high conductivities.

◇ | **Example 17-5**

From the electronic structure of tungsten, sketch the expected band structure and compare the conductivity, $1.77 \times 10^5 \, \text{ohm}^{-1} \cdot \text{cm}^{-1}$, with that of metals with a similar band structure.

Answer:

The electronic structure of tungsten, which has an atomic number of 74, is

$$1s^2 2s^2 2p^6 3s^2 3p^6 3d^{10} 4s^2 4p^6 4d^{10} 4f^{14} 5s^2 5p^6 5d^4 6s^2$$

The band structure, which is shown in Figure 17-7, has overlapping $5d$ and $6s$ levels. The structure resembles that of iron. Moreover, the conductivity of tungsten, $1.77 \times 10^5 \, \text{ohm}^{-1} \cdot \text{cm}^{-1}$, is similar to that of iron, $1.00 \times 10^5 \, \text{ohm}^{-1} \cdot \text{cm}^{-1}$.

17-4 CONTROLLING THE CONDUCTIVITY OF METALS

The conductivity of a pure, defect-free metal is determined by the electronic structure of the atoms. But we can significantly affect the conductivity by influencing the mobility μ of the carriers. The mobility is proportional to the drift velocity $\bar{v}$, which is low if the electrons collide with imperfections in the lattice. The *mean free path* is the average distance between collisions; a long mean free path permits high mobilities and high conductivities.

If the metal contains no lattice defects and performs at absolute zero degrees, the mean free path is infinite and the electrical resistivity is zero. However, no metals are perfect, nor do they operate at absolute zero.

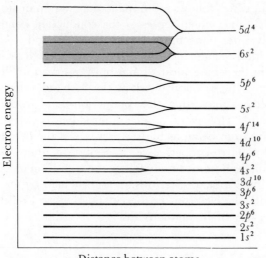

FIGURE 17-7 Simplified band structure of tungsten.

Temperature Effect.   When the temperature of a metal increases, thermal energy causes the atoms to vibrate (Figure 17-8). At any instant, the atom may not be in its equilibrium position and therefore interacts with and scatters electrons. The mean free path decreases, the mobility of electrons is reduced, and the resistivity increases. The change in resistivity with temperature can be estimated from the equation

$$\rho = \rho_r(1 + a\Delta T) \tag{17-6}$$

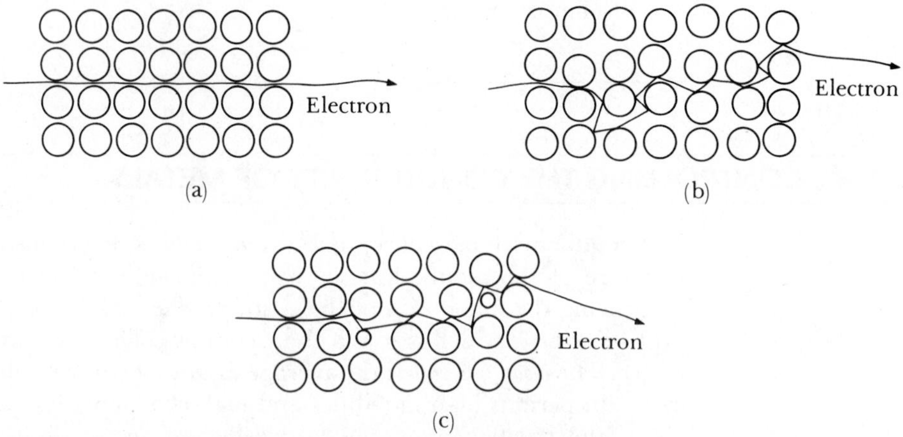

FIGURE 17-8 Movement of an electron through (a) a perfect crystal, (b) a crystal heated to a high temperature, and (c) a crystal containing lattice defects. Scattering of the electrons reduces the mobility and conductivity.

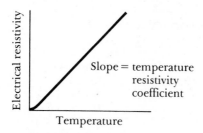

FIGURE 17-9 The effect of temperature on the electrical resistivity of a metal with a perfect lattice. The slope of the curve is the temperature resistivity coefficient.

where ρ_r is the resistivity at room temperature (25°C), ΔT is the temperature difference between the temperature of interest and room temperature, and a is the *temperature resistivity coefficient*. The relationship between resistivity and temperature is linear over a wide temperature range (Figure 17-9). Examples of the temperature resistivity coefficient are given in Table 17-3. The resistivity of the metal due only to thermal vibration of the atoms is ρ_T.

◇ | **Example 17-6**

Calculate the electrical conductivity of pure copper at (a) 400°C and (b) − 100°C.

Answer:

The resistivity of copper at room temperature is $1.67 \times 10^{-6} \Omega \cdot cm$ and the temperature resistivity coefficient is $0.0068 \, \Omega \cdot cm/°C$.

(a) At 400°C

$$\rho = \rho_r(1 + a\Delta T) = (1.67 \times 10^{-6})[1 + 0.0068(400 - 25)]$$
$$\rho = 5.929 \times 10^{-6} \Omega \cdot cm$$
$$\sigma = 1/\rho = 1.69 \times 10^5 \, ohm^{-1} \cdot cm^{-1}$$

(b) At − 100°C

$$\rho = (1.67 \times 10^{-6})[1 + 0.0068(-100 - 25)] = 0.251 \times 10^{-6} \Omega \cdot cm$$
$$\sigma = 39.9 \times 10^5 \, ohm^{-1} \cdot cm^{-1}$$

Effect of Lattice Defects.   Lattice imperfections scatter electrons and thus reduce the mobility and conductivity of the metal [Figure 17-8(c)]. Greater numbers of defects reduce the mean free path and have a pronounced effect on the conductivity. For example, the increase in the resistivity due to solid solution

TABLE 17-3 The temperature resistivity coefficient for selected metals

Metal	Room Temperature Resistivity $(\times 10^{-6}\,\Omega \cdot cm)$	Temperature Resistivity Coefficient $(\Omega \cdot cm/^\circ C)$
Be	4.0	0.0250
Mg	4.45	0.0165
Ca	3.91	0.0042
Al	2.65	0.0043
Cr	12.90	0.0030
Fe	9.71	0.0065
Co	6.24	0.0060
Ni	6.84	0.0069
Cu	1.67	0.0068
Ag	1.59	0.0041
Au	2.35	0.0040

From *ASM Metals Handbook*, Vol. 2, 9th Ed., 1979.

atoms is

$$\rho_d = b(1 - x)x \tag{17-7}$$

where ρ_d is the increase in resistivity due to the defects, x is the atomic fraction of the impurity or solid solution atoms present, and b is the *defect resistivity coefficient*. In a similar manner, vacancies, dislocations, and surface defects, such as grain boundaries, also reduce the conductivity of the metal. Each defect contributes to an increase in the resistivity of the metal. Thus, the overall resistivity is

$$\rho = \rho_T + \rho_d \tag{17-8}$$

where ρ_d equals the contributions from solid solution atoms, interstitial atoms, vacancies, grain boundaries, and other imperfections. The effect of the defects is independent of temperature (Figure 17-10).

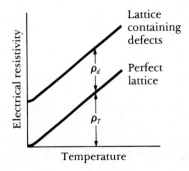

FIGURE 17-10 The electrical resistivity of a metal is composed of a constant defect contribution ρ_d and a variable temperature contribution ρ_T.

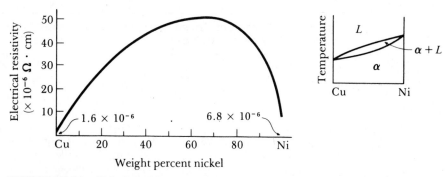

FIGURE 17-11   *Electrical resistivity in the copper-nickel system. Solid solution strengthening of either pure copper or nickel increases the electrical resistivity.*

Effect of Processing and Strengthening.   Strengthening mechanisms and metal processing techniques affect the electrical properties of a metal in different ways.

Solid solution strengthening is a poor way to obtain high strength in metals intended to have high conductivities. The mean free paths are very short due to the random distribution of the interstitial or substitutional atoms. Figure 17-11 shows the influence of alloying element in the isomorphous copper-nickel system. A maximum in the resistivity occurs near 60% Ni. The change in resistivity is very large—from about $1.6 \times 10^{-6}\,\Omega \cdot cm$ to more than $50 \times 10^{-6}\,\Omega \cdot cm$—as the composition changes from pure copper to Cu-60% Ni. Figure 17-12 shows

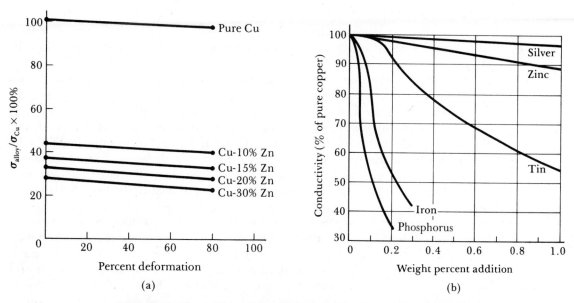

FIGURE 17-12   *(a) The effect of solid solution strengthening and cold working on the electrical conductivity of copper and (b) the effect of selected elements on the electrical conductivity of copper.*

TABLE 17-4   The effect of alloying, strengthening, and processing on the electrical conductivity of copper and its alloys

Alloy	$\dfrac{\sigma_{\text{alloy}}}{\sigma_{\text{Cu}}} \times 100$	Remarks
Pure annealed copper	101	Few lattice defects to scatter electrons; the mean free path is long.
Pure copper deformed 80%	98	Many dislocations, but because of the tangled nature of the dislocation networks, the mean free path is still long.
Dispersion-strengthened Cu-0.7% Al_2O_3	85	The dispersed phase is not as closely spaced as solid solution atoms, nor is it coherent, as in age hardening. Thus, the effect on conductivity is small.
Solution-treated Cu-2% Be	18	The alloy is single phase; however, the small amount of solid solution strengthening from the supersaturated beryllium greatly decreases conductivity.
Aged Cu-2% Be	23	During aging, the beryllium leaves the copper lattice to produce a coherent precipitate. The precipitate does not interfere with conductivity as much as the solid solution atoms.
Cu-35% Zn	28	This alloy is solid solution strengthened by zinc. The conductivity is low but not as low as when beryllium, which has an atomic radius much different from copper, is present.

the effect of zinc and other alloying elements on the conductivity of copper; as the amount of alloying element increases, the conductivity decreases substantially.

Age hardening and dispersion strengthening reduce the conductivity less than solid solution strengthening. Table 17-4 shows the effect of several strengthening mechanisms on the conductivity of copper and its alloys.

In the dispersion-strengthened two-phase alloys, such as eutectoid and eutectic alloys, the electrical conductivity or resistivity can be estimated from the rule of mixtures. Figure 17-13 shows the electrical resistivity in lead-tin alloys. The change in resistivity initially is very steep since the major effect is solid solution strengthening. However, the change is slower in the two-phase region, which suggests that the rule of mixtures might apply.

Strain hardening and grain size control have less effect on conductivity. Since dislocations and grain boundaries are further apart than solid solution atoms, there are large volumes of metal that have a long mean free path. Consequently, cold working is an effective way to increase the strength of a metallic conductor without seriously impairing the electrical properties of that material. In addition, the effects of cold working on conductivity can be eliminated by the low temperature recovery heat treatment, in which good conductivity is restored while the strength is retained. Both Figure 17-12 and

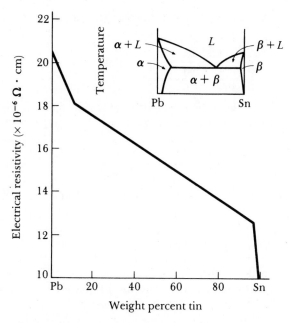

FIGURE 17-13 The electrical resistivity in the lead-tin eutectic system. The conductivity is linear in the two-phase region.

Table 17-4 illustrate the effect of cold working on conductivity; compared with solid solution strengthening, the effect of cold working is nearly negligible.

◇ | **Example 17-7**

Is Equation 17-7 valid for the copper-zinc system? If so, calculate the defect resistivity coefficient for zinc in copper.

Answer:

Let's look at the data in Figure 17-12 at 0% cold work. The conductivity of copper is $5.98 \times 10^5 \, \text{ohm}^{-1} \cdot \text{cm}^{-1}$.

% Zn	$\dfrac{\sigma_{\text{alloy}}}{\sigma_{\text{Cu}}} \times 100$	σ_{alloy}	ρ_{alloy}
0	101	6.00×10^5	0.167×10^{-5}
10	44	2.63×10^5	0.380×10^{-5}
15	37	2.21×10^5	0.452×10^{-5}
20	33	1.97×10^5	0.508×10^{-5}
30	28	1.67×10^5	0.599×10^{-5}

We can convert to atomic fraction x; a sample calculation is shown.

$$x_{Zn} = \frac{\dfrac{10\,\text{wt}\,\%}{65.38}}{\dfrac{10\,\text{wt}\,\%}{65.38} + \dfrac{90\,\text{wt}\,\%}{63.54}} = 0.0975$$

where 65.38 is the atomic weight of zinc and 63.54 is the atomic weight of copper. Now let's calculate the terms $x(1 - x)$ and $\Delta\rho = \rho_{\text{alloy}} - \rho_{Cu}$.

% Zn	x_{Zn}	$x(1 - x)$	$\Delta\rho = \rho_d$
0	0	0	0
10	0.0975	0.088	0.213×10^{-5}
15	0.146	0.125	0.285×10^{-5}
20	0.196	0.158	0.341×10^{-5}
30	0.294	0.208	0.432×10^{-5}

The results in this table are plotted in Figure 17-14. The straight line that results indicates that the relationship in Equation (17-7) is valid. The slope of the graph is the defect resistivity coefficient.

$$b = \frac{0.4 \times 10^{-5} - 0.2 \times 10^{-5}}{0.19 - 0.08} = 1.8 \times 10^{-5}\,\Omega \cdot \text{cm}$$

Anisotropic Electrical Behavior.   The electrical conductivity or resistivity in single crystals or in metals with a texture or preferred orientation may vary with direction, since the mean free path may vary. Typical examples of anisotropic behavior are given in Table 17-5.

TABLE 17-5   Anisotropic electrical resistivity of selected metals

Metal	Crystal Structure	Electrical Resistivity ($\times 10^{-6}\,\Omega \cdot \text{cm}$)			
		a_0	b_0	c_0	Polycrystalline
Mg	HCP	4.48		3.74	4.45
Ga	Orthorhombic	17.4	8.1	54.3	15.05

From *ASM Metals Handbook*, Vol. 2, 9th Ed., 1979.

17–5   THERMOCOUPLES

When we increase the temperature of a metal, more electrons are excited into higher energy levels. When one end of a wire is heated, more electrons will be excited there than at the other, cooler end. The large number of excited electrons

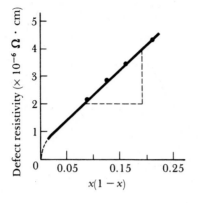

FIGURE 17-14 The resistivity due to lattice defects, or zinc substitution atoms, versus the parameter $x(1 - x)$. The slope of the graph is the defect resistivity coefficient for zinc in copper. (See Example 17-7.)

at the hot end then move toward the cold end, making the cold end more negatively charged, and a voltage difference between the ends of the conductor is produced [Figure 17-15(a)].

We cannot measure this voltage when only a single conductor is used. However, suppose we join two wires of different materials. The flow of electrons from the hot junction to a voltmeter is different in each material, giving two different potentials [Figure 17-15(b)]. This phenomenon, called the *Seebeck effect*,

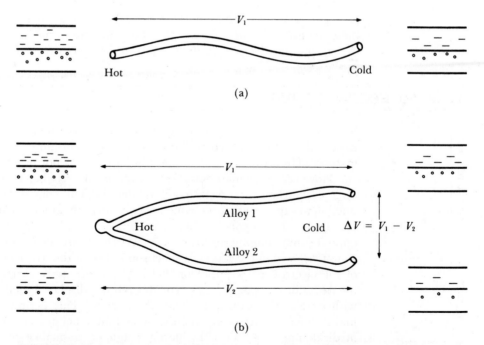

FIGURE 17-15 (a) An increase in the temperature will increase the number of electrons in the conduction band at the heated end of a conductor. (b) When two different conductors are joined, their different responses to the temperature create a voltage difference that is directly related to the temperature.

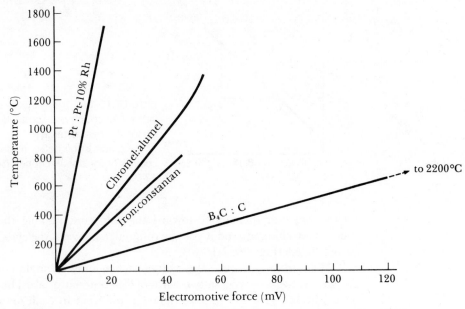

FIGURE 17-16   The relationship between voltage and temperature for several thermocouples.

produces a net voltage difference between the cold ends of the two wires. Because the potential difference increases as the temperature increases (Figure 17-16), we can use this device as a *thermocouple* to measure or control temperature.

17–6   SUPERCONDUCTIVITY

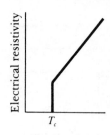

FIGURE 17-17
The electrical resistivity of a superconductor becomes zero below some critical temperature T_c.

Some crystals cooled to absolute zero behave as superconductors, since the electrical resistivity becomes zero and a current flows indefinitely in the material. Unfortunately, obtaining absolute zero is not practical.

However, some materials display superconductive behavior above absolute zero, even when the crystal contains defects. The change from normal conduction to superconduction, which occurs abruptly at a critical temperature T_c (Figure 17-17), is a complex phenomenon; electrons having the same energy but opposite spin combine to form pairs. When the frequency of atom vibrations (phonons) within the lattice is synchronized with the frequency or wavelength of the electron pairs, superconductivity can occur. A variety of metals and intermetallic compounds, in particular Nb_3Sn, display this effect (Table 17-6), with critical temperatures up to about $20\,K$. This permits materials to be made superconductive by cooling with liquid helium at $4\,K$. A number of medical diagnostic tools, including magnetic resonance imaging, use these superconductors.

The critical temperature depends on the magnetic field acting on the conductor [Figure 17-18(a)]. A magnetic field greater than H_c completely

TABLE 17-6   The critical temperature and magnetic field for superconductivity of selected metals and compounds

Metal	Critical Temperature (K)	Critical Magnetic Field H_o (oersted)
W	0.015	1.15
Ti	0.39	56–100
Al	1.18	105
Sn	3.72	305
Hg	4.15	411
Pb	7.23	803
Nb	9.25	1970
La_3Se_4	8.6	
$SnTa_3$	8.35	
Nb_3Sn	18.05	
GaV_3	16.8	
$AlNb_3$	18.0	
$(Al_{0.8}Ge_{0.2})Nb_3$	20.7	
$(La,Sr)_2CuO_4$	40.0	
$YBa_2Cu_3O_{7-x}$	93.0	

From *Handbook of Chemistry and Physics*, 56th Ed., CRC Press, 1975, and other sources.

suppresses superconduction. The temperature below which superconduction occurs in a magnetic field is given by

$$H_c = H_o\left[1 - \left(\frac{T}{T_c}\right)^2\right] \tag{17-9}$$

where H_o is defined in Figure 17-18. In addition, the current density flowing through the conductor affects superconductivity; if the current density J is too high, superconductivity may be lost, as shown in Figure 17-18(b).

Until about 1986, the extremely low critical temperatures required the use of liquid helium for cooling and consequently limited the applications for superconductivity. Fortunately, a new group of ceramic materials is being developed whose critical temperatures are above 90 K. These include the so-called 1-2-3 compounds of the form $YBa_2Cu_3O_{7-x}$, where ''x'' indicates that some oxygen ions are missing from the complicated perovskite crystal structure (Figure 17-19). Other ceramics, including $(La, Sr)_2CuO_4$ and $(La, Ba)_2CuO_4$, have a similar behavior. The ability to superconduct at temperatures above 77 K permits cooling to be done relatively inexpensively using liquid nitrogen. Unusually rapid development of these materials, including their manufacture into usable amounts and sizes, is occurring, although the brittleness of these ceramics makes it difficult to form the material into useful shapes such as wire. Potential applications include a number of electronic devices as well as more efficient power transmission lines, high-speed computers, magnetically levitated trains, improved batteries for nonpolluting electric automobiles, and fusion power plants.

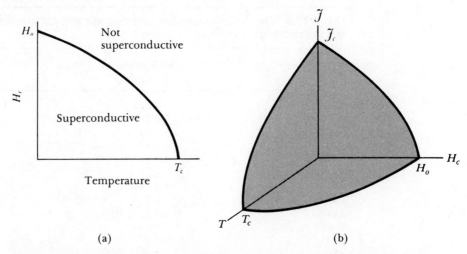

FIGURE 17-18 (a) The effect of a magnetic field on the temperature below which superconductivity occurs and (b) the superconducting envelope, showing the combined effects of temperature, magnetic field, and current density. Conditions within the envelope produce superconductivity.

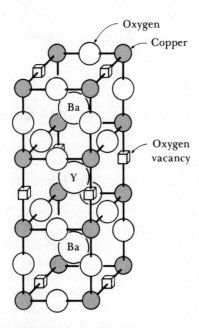

FIGURE 17-19 The crystal structure of the $YBa_2Cu_3O_{7-x}$ superconducting compound is composed of three perovskite cells in a layer form, giving an orthorhombic structure.

◇ **Example 17-8**

Niobium is to be used as a superconductor in a magnetic field of 500 oersted. What temperature must be obtained for niobium to be superconductive?

Answer:

From Table 17-6, $T_c = 9.25 \text{ K}$ and $H_o = 1970$ oersted.

$$H_c = H_o \left[1 - \left(\frac{T}{T_c} \right)^2 \right]$$

$$500 = 1970 \left[1 - \left(\frac{T}{9.25} \right)^2 \right] = 1970 \left(1 - \frac{T^2}{85.56} \right)$$

$$T = \sqrt{(0.746)(85.56)} = 7.99 \text{ K}$$

17-7 ENERGY GAPS—INSULATORS AND SEMICONDUCTORS

The elements in Group IVA of the periodic table contain two electrons in their outer p shell and have a valence of four. We expect the Group IVA elements to have a high conductivity due to the unfilled p band. However, this behavior is not observed.

These elements are covalently bonded; consequently, the electrons in the outer s and p bands are rigidly bound to the atoms. The restrictions caused by covalent bonding produce a complex change in the band structure, producing *hybridization*. The $2s$ and $2p$ levels of the carbon atoms in diamond can contain up to eight electrons, but there are only four valence electrons available. When N carbon atoms are brought together to form solid diamond, the $2s$ and $2p$ levels interact and produce two bands (Figure 17-20). Each of the hybrid bands can contain $4N$ electrons. Since there are only $4N$ electrons available, the lower, or valence, band is completely filled while the upper, or conduction, band is empty.

A large *energy gap* E_g separates the electrons from the conduction band in diamond. Few electrons possess sufficient energy to jump the forbidden zone to the conduction band. Consequently, diamond is an excellent insulator, with an electrical conductivity of less than $10^{-18} \text{ ohm}^{-1} \cdot \text{cm}^{-1}$.

Almost all of the covalently and ionically bonded materials have a band structure containing an energy gap between the valence and conduction bands and, like diamond, behave as *electrical insulators*. Table 17-7 shows the electrical conductivity of several polymer and ceramic materials at room temperature. Except for silicon carbide and boron carbide, the electrical conductivity of these materials is extremely low.

By increasing the temperature or the applied voltage, energy is supplied which makes it easier for electrons to gain enough energy to overcome the energy

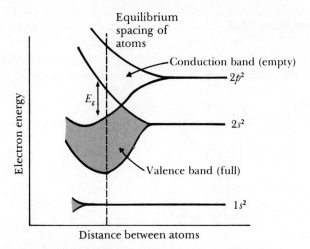

FIGURE 17-20 *The band structure of carbon in the diamond form. The 2s and 2p levels combine to form two hybrid bands separated by an energy gap E_g.*

gap. For example, the electrical conductivity of boron nitride increases from about 10^{-13} at room temperature to 10^{-4} at 800°C, while the conductivity of silicon nitride may increase from 10^{-15} to 10^{-7} in this same temperature range.

Although germanium, silicon, and tin have the same crystal and band structure as diamond, the energy gap is smaller. In fact, the energy gap in tin is so small that tin behaves as a metal. The energy gap is somewhat larger in silicon and germanium—these elements behave as *semiconductors*. Table 17-8 gives the energy gap and other properties for these four elements.

TABLE 17-7 Electrical conductivity of selected polymers and ceramics at room temperature

Material	Electrical Conductivity (ohm$^{-1} \cdot$ cm^{-1})
Polymers	
Polyethylene	10^{-15}
Cellulosic	10^{-12} to 10^{-16}
Polytetrafluoroethylene	10^{-18}
Polystyrene	10^{-17} to 10^{-19}
Polyester	10^{-15} to 10^{-16}
Epoxy	10^{-12} to 10^{-17}
Ceramics	
Alumina	10^{-14}
Silica glass	10^{-17}
Boron nitride	10^{-13}
Silicon nitride	10^{-11} to 10^{-17}
Silicon carbide	10^{-1} to 10^{-2}
Boron carbide	1 to 2

TABLE 17-8   Electronic structure and electrical conductivity of the Group IVA elements at 25°C

Metal	Electronic Structure	Electrical Conductivity $(\text{ohm}^{-1} \cdot \text{cm}^{-1})$	Energy Gap (eV)	Electron Mobility $(\text{cm}^2/\text{V} \cdot \text{s})$	Hole Mobility $(\text{cm}^2/\text{V} \cdot \text{s})$
C (diamond)	$1s^2 2s^2 2p^2$	$< 10^{-18}$	5.4	1800	1400
Si	$1s^2 2s^2 2p^6 3s^2 3p^2$	5×10^{-6}	1.107	1900	500
Ge	$\ldots\ldots 4s^2 4p^2$	0.02	0.67	3800	1820
Sn	$\ldots\ldots 5s^2 5p^2$	0.9×10^5	0.08	2500	2400

17–8   INTRINSIC SEMICONDUCTORS

Because the energy gap E_g is small in silicon and germanium, some electrons may possess enough thermal energy to be excited into the conduction band. The excited electrons leave behind unoccupied energy levels, or *holes*, in the valence band. When an electron moves to fill a hole, another hole is created from the original electron source; consequently the holes appear to act as positively charged electrons and can carry an electrical charge. When a voltage is applied to the material, the electrons in the conduction band accelerate towards the positive terminal, while holes in the valence band move towards the negative terminal (Figure 17-21). Current is therefore conducted by the movement of both the electrons and the holes. The material behaves as an *intrinsic semiconductor*.

The conductivity is determined by the number of electron-hole pairs.

$$\sigma = n_e q \mu_e + n_h q \mu_h \tag{17-10}$$

where n_e is the number of electrons in the conduction band, n_h is the number of holes in the valence band, and μ_e and μ_h are the mobilities of the electrons and holes (Table 17-8). Note that the mobility of the electrons is always greater than that of the holes. In intrinsic semiconductors

$$n = n_e = n_h$$

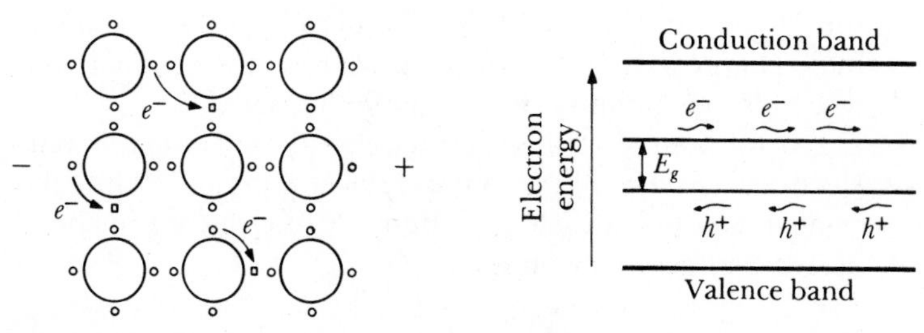

FIGURE 17-21   When a voltage is applied to a semiconductor, the electrons move through the conduction band while the electron holes move through the valence band in the opposite direction.

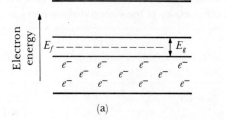
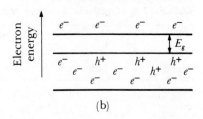

FIGURE 17-22 The distribution of electrons and holes in the valence and conduction bands (a) at absolute zero and (b) at an elevated temperature.

Therefore, the conductivity is

$$\sigma = nq(\mu_e + \mu_h) \tag{17-11}$$

In intrinsic semiconductors we control the number of charge carriers, and hence the electrical conductivity, by controlling the temperature. At absolute zero, all of the electrons are in the valence band, while all of the levels in the conduction band are unoccupied [Figure 17-22(a)]. As the temperature increases, there is a greater probability that an energy level in the conduction band is occupied (and an equal probability that a level in the valence band is unoccupied, or that a hole is present) [Figure 17-22(b)]. The number of electrons in the conduction band, which is equal to the number of holes in the valence band, is given by

$$n = n_e = n_h = n_0 \exp\left(\frac{-E_g}{2kT}\right) \tag{17-12}$$

where n_0 can be considered a constant, although it too actually depends on temperature. Higher temperatures permit more electrons to cross the forbidden zone and hence the conductivity increases.

$$\sigma = n_0 q(\mu_e + \mu_h) \exp\left(\frac{-E_g}{2kT}\right) \tag{17-13}$$

The behavior of the semiconductor is opposite that of metals (Figure 17-23). As the temperature increases, the conductivity of a semiconductor increases because more charge carriers are present, whereas the conductivity of a metal decreases due to lower mobility of the charge carriers.

If the source of the exciting energy or voltage is removed, the holes and electrons will recombine, but only over a period of time. For instance, when we remove an applied field, the number of electrons in the conduction band will decrease at a rate given by

$$n = n_0 \exp\left(\frac{-t}{\tau}\right) \tag{17-14}$$

where t is the time after the energy is removed, n_0 is a constant, and τ is a constant

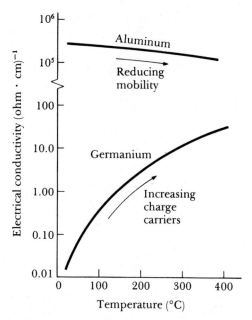

FIGURE 17-23 The electrical conductivity versus temperature for semiconductors compared with metals.

called the *recombination time*. This characteristic is important in the operation of a number of semiconductor devices.

◇ **Example 17-9**

For germanium at 25°C, estimate (a) the number of charge carriers, (b) the fraction of the total electrons in the valence band that are excited into the conduction band, and (c) the constant n_o.

Answer:

From Table 17-8

$$\sigma = 0.02 \, \text{ohm}^{-1} \cdot \text{cm}^{-1} \qquad E_g = 0.67 \, \text{eV}$$

$$\mu_e = 3800 \, \text{cm}^2/\text{V} \cdot \text{s} \qquad \mu_h = 1820 \, \text{cm}^2/\text{V} \cdot \text{s}$$

$$2kT = 2(0.025) = 0.05 \, \text{eV at } T = 25°C$$

(a) From Equation 17-11

$$n = \frac{\sigma}{q(\mu_e + \mu_h)} = \frac{0.02}{(1.6 \times 10^{-19})(3800 + 1820)} = 2.2 \times 10^{13}$$

There are 2.2×10^{13} electrons/cm³ and 2.2×10^{13} holes/cm³ helping to conduct charge in germanium at room temperature.

(b) The lattice parameter of diamond cubic germanium is 5.6575×10^{-8} cm. The total number of electrons in the valence band of germanium is

$$\text{Total electrons} = \frac{(8 \text{ atoms/cell}) (4 \text{ electrons/atom})}{(5.6575 \times 10^{-8} \text{cm})^3}$$

$$= 1.77 \times 10^{23}$$

$$\text{Fraction excited} = \frac{2.2 \times 10^{13}}{1.77 \times 10^{23}} = 1.24 \times 10^{-10}$$

(c) From Equation 17-12

$$n_o = \frac{n}{\exp\left(\dfrac{-E_g}{2kT}\right)} = \frac{2.2 \times 10^{13}}{\exp\left(\dfrac{-0.67}{0.05}\right)} = 1.45 \times 10^{19} \text{ carriers/cm}^3$$

17-9 EXTRINSIC SEMICONDUCTORS

We cannot accurately control the behavior of an intrinsic semiconductor since slight variations in temperature change the conductivity. However, by intentionally adding a small number of impurity atoms to the material, we can produce an *extrinsic semiconductor*. The conductivity of the extrinsic semiconductor depends primarily on the number of impurity, or *dopant*, atoms and in a certain temperature range may even be independent of temperature. Conductivity is therefore controllable and stable.

n-Type Semiconductors.   Suppose we add an impurity atom such as antimony, which has a valence of five, to silicon or germanium. Four of the electrons from the antimony atom participate in the covalent bonding process, while the extra electron enters an energy level in a donor state just below the conduction band (Figure 17-24). Since the extra electron is not tightly bound to the atoms, only a small increase in energy E_d is required for the electron to enter the conduction band.[1] The energy gap controlling conductivity is now E_d rather than E_g (Table 17-9). No corresponding holes are created when the donor electrons enter the conduction band.

Some intrinsic semiconduction still occurs, with a few electrons gaining enough energy to jump the large E_g gap. The total number of charge carriers is

$$n_{\text{total}} = n_e(\text{dopant}) + n_e(\text{intrinsic}) + n_h(\text{intrinsic})$$

[1] Many texts define E_d as the energy difference between the top of the valence band and the donor band. In this case, the energy increase required would be defined as $E_g - E_d$.

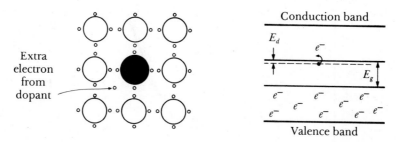

FIGURE 17-24 When a dopant atom with a valence greater than four is added to silicon, an extra electron is introduced and a donor energy state is created. Now electrons are more easily excited into the conduction band.

or

$$n_o = n_{od} \exp\left(\frac{-E_d}{kT}\right) + 2n_o \exp\left(\frac{-E_g}{2kT}\right) \tag{17-15}$$

where n_{od}, n_o are approximately constant. At low temperatures, few intrinsic electrons and holes are produced and the number of electrons is about

$$n_{\text{total}} \approx n_{od} \exp\left(\frac{-E_d}{kT}\right) \tag{17-16}$$

As the temperature increases, more of the donor electrons jump the E_d gap until, eventually, all of the donor electrons enter the conduction band. At this point, we have reached *donor exhaustion* (Figure 17-25). The conductivity is virtually constant, since no more donor electrons are available and the temperature is still too low to produce many intrinsic electrons and holes. The conductivity is

$$\sigma = n_d q \mu_e \tag{17-17}$$

where n_d is the maximum number of donor electrons, determined by the number of impurity atoms that are added.

TABLE 17-9 The donor and acceptor energy gaps in electron volts when silicon and germanium semiconductors are doped

Dopant	Silicon		Germanium	
	E_d	E_a	E_d	E_a
P	0.045		0.0120	
As	0.049		0.0127	
Sb	0.039		0.0096	
B		0.045		0.0104
Al		0.057		0.0102
Ga		0.065		0.0108
In		0.160		0.0112

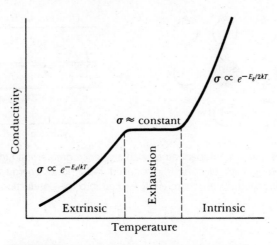

FIGURE 17-25 The effect of temperature on the conductivity of an extrinsic semiconductor. At low temperatures, the conductivity increases as more donor electrons enter the conduction band. At moderate temperatures, donor exhaustion has occurred. At high temperatures, intrinsic semiconduction becomes important.

We prefer to operate in the exhaustion, or plateau, temperature range. Generally, semiconductors with a large E_g have the broadest exhaustion plateau.

At high temperatures, the term $\exp\left(-E_g/2kT\right)$ becomes significant and the conductivity increases again according to

$$\sigma = qn_d\mu_e + q(\mu_e + \mu_h)n_o \exp\left(\frac{-E_g}{2kT}\right) \qquad (17\text{-}18)$$

If we plot σ versus $1/T$, we obtain an Arrhenius relationship from which E_g and E_d can be calculated.

◇ **Example 17-10**

Calculate the number of extrinsic charge carriers per cubic centimeter in an n-type semiconductor when one out of every 1,000,000 atoms in silicon is replaced by an antimony atom. Estimate the conductivity of the semiconductor in the exhaustion zone.

Answer:

The lattice parameter of diamond cubic silicon is 5.4307×10^{-8} cm.

$$n_d = \frac{\left(1\,\dfrac{\text{electron}}{\text{Sb atom}}\right)\left(10^{-6}\,\dfrac{\text{Sb atom}}{\text{Si atom}}\right)\left(8\,\dfrac{\text{Si atoms}}{\text{unit cell}}\right)}{(5.4307 \times 10^{-8}\,\text{cm})^3}$$

$$n_d = 5 \times 10^{16} \text{ electrons/cm}^3$$
$$\sigma = n_d q \mu_e = (5 \times 10^{16})(1.6 \times 10^{-19})(1900) = 15.2 \text{ ohm}^{-1} \cdot \text{cm}^{-1}$$

◇ **Example 17-11**

How many carriers are required to give a conductivity of $100 \text{ ohm}^{-1} \cdot \text{cm}^{-1}$ in the exhaustion region of silicon? How many antimony atoms would have to be added to silicon?

Answer:

$$\sigma = n_d q \mu_e = n_d (1900)(1.6 \times 10^{-19}) = 100$$

$$n_d = \frac{100}{(1900)(1.6 \times 10^{-19})} = 3.29 \times 10^{17} \text{ electrons/cm}^3$$

$$3.29 \times 10^{17} = \frac{\left(1 \dfrac{\text{electron}}{\text{Sb atom}}\right) \left(x \dfrac{\text{Sb atom}}{\text{Si atom}}\right) \left(8 \dfrac{\text{Si atoms}}{\text{unit cell}}\right)}{(5.4307 \times 10^{-8} \text{ cm})^3}$$

$$x = 6.59 \times 10^{-6} \text{ Sb atom/Si atom,}$$
$$\text{or } 6.59 \text{ Sb atoms/}10^6 \text{ Si atoms}$$

◇

---

p-**Type Semiconductors.**   When we add an impurity such as gallium, which has a valence of three, to a semiconductor, there are not enough electrons to complete the covalent bonding process. An electron hole is created in the valence band which can be filled by electrons from other locations in the band (Figure 17-26). The holes act as acceptors of electrons. These hole sites have a somewhat higher than normal energy and create an acceptor level of possible electron energies just above the valence band (Table 17-9). An electron must gain an

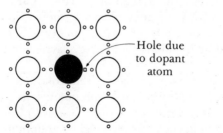

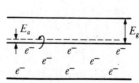

FIGURE 17-26 When a dopant atom with a valence of less than four is substituted into the silicon lattice, an electron hole is created in the structure and an acceptor energy level is created just above the valence band. Little energy is required to excite the electron holes into motion.

energy of only E_a in order to create a hole in the valence band. The hole then moves and carries the charge. Now we have a *p-type semiconductor*.

Again some intrinsic semiconduction may occur. The number of charge carriers is

$$n_t = n_h(\text{acceptor}) + n_e(\text{intrinsic}) + n_h(\text{intrinsic})$$

$$n_t = n_{oa} \exp\left(\frac{-E_a}{kT}\right) + 2n_o \exp\left(\frac{-E_g}{2kT}\right) \tag{17-19}$$

At low temperatures, the acceptor levels predominate.

$$n_t = n_{oa} \exp\left(\frac{-E_a}{kT}\right) \tag{17-20}$$

Eventually, the temperature is high enough to cause *acceptor saturation* and

$$\sigma = n_a q \mu_h \tag{17-21}$$

where n_a is the maximum number of acceptor levels, or holes, introduced by the dopant. At higher temperatures, intrinsic semiconduction becomes important and

$$\sigma = n_a q \mu_h + q(\mu_e + \mu_h)n_o \exp\left(\frac{-E_g}{2kT}\right) \tag{17-22}$$

Semiconducting Compounds.   Silicon and germanium are the only elements that have practical applications as semiconductors. However, a large number of ceramic and intermetallic compounds display the same effect. Examples are given in Table 17-10.

The *stoichiometric semiconductors*, usually intermetallic compounds, have crystal structures and band structures similar to DC silicon and germanium. Elements from Group III and Group V of the periodic table are classic examples. Gallium from Group III and arsenic from Group V combine to form a compound GaAs, with an average of four valence electrons per atom. The $4s^2 4p^1$ levels of gallium and the $4s^2 4p^3$ levels of arsenic produce two hybrid bands, each capable of containing $4N$ electrons. An energy gap of 1.35 eV separates the valence and conduction bands. The GaAs compound can be doped to produce either an *n*-type or a *p*-type semiconductor. The large energy gap E_g leads to a broad exhaustion plateau and high mobilities of charge carriers in the compound lead to high conductivities.

The energy gap of the stoichiometric semiconductors can be carefully tailored by alloying various Group III–V compounds, as illustrated in Figure 17-27, in which E_g is plotted versus the lattice parameter of the compound. For example, as gallium atoms in GaAs are replaced by aluminum atoms, the energy gap increases from 1.4 to 2.2 eV. If arsenic atoms in AlAs are replaced by antimony atoms, the energy gap decreases from 2.2 to 1.7 eV. Generally, substituting elements with a higher atomic number, such as Sb for As, will produce a smaller energy gap.

TABLE 17-10   Energy gaps and mobilities for semiconducting compounds

Compound	Energy Gap (eV)	Electron Mobility ($cm^2/V \cdot s$)	Hole Mobility ($cm^2/V \cdot s$)
ZnS	3.54	180	5
ZnTe	2.26	340	100
CdTe	1.44	1,200	50
GaP	2.24	300	100
GaAs	1.35	8,800	400
GaSb	0.67	4,000	1,400
InSb	0.165	78,000	750
InAs	0.36	33,000	460
ZnO	3.2	180	
CdS	2.42	400	
CdSe	1.74	650	
PbS	0.37	600	600
PbTe	0.25	1,600	600
$CdSnAs_2$	0.26	22,000	250

From *Handbook of Chemistry and Physics*, 56th Ed., CRC Press, 1975.

The *nonstoichiometric*, or *defect, semiconductors* are ionic compounds containing an excess of either anions (producing a *p*-type semiconductor) or cations (producing an *n*-type semiconductor). A number of oxides and sulfides have this behavior. For example, if an extra zinc atom is added to ZnO, the zinc atom enters the structure as an ion, Zn^{2+}, giving up two electrons which contribute to the number of charge carriers. These electrons can be activated by a small

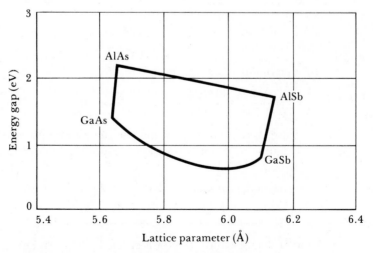

FIGURE 17-27   *Energy gaps and lattice parameters of semiconductors in the (Al, Ga) (As, Sb) system.*

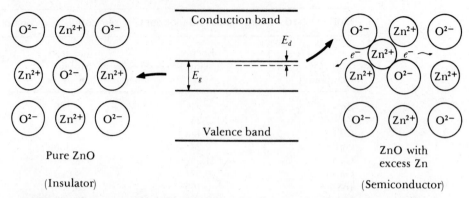

Pure ZnO

(Insulator)

ZnO with
excess Zn

(Semiconductor)

FIGURE 17-28 Extra interstitial zinc atoms can ionize and introduce extra electrons, creating an *n*-type defect semiconductor in ZnO.

increase in energy to carry current (Figure 17-28). The ZnO now behaves as an *n*-type semiconductor.

◇ | **Example 17-12**

ZnO has the zinc blende crystal structure [Figure 3-28]. (a) Calculate the lattice parameter and density of the ZnO unit cell. (b) If the ZnO is doped with one extra Zn ion per 100 Zn ions, estimate the number of charge carriers in one cubic centimeter of zinc blende.

Answer:

(a) The ionic radii are $r_{Zn} = 0.74\,Å$ and $r_O = 1.32\,Å$. From inspection of Figure 3-28, we find that the ions touch along the body diagonal, where

$$4r_{Zn} + 4r_O = \sqrt{3}\,a_0$$
$$4(0.74) + 4(1.32) = 8.24 = \sqrt{3}\,a_0$$
$$a_0 = 4.758\,Å = 4.758 \times 10^{-8}\,cm$$

(b) The number of zinc ions in one cubic centimeter of zinc blende is

$$\frac{4\ Zn\ ions/cell}{(4.758 \times 10^{-8}\,cm)^3} = 3.71 \times 10^{22}$$

Since there is one dopant ion per 100 Zn ions, there must be 3.71×10^{20} dopant ions. Each dopant ion contributes two electrons. Therefore, there are

$$Charge\ carriers = 2(3.71 \times 20^{20}) = 7.42 \times 10^{20}/cm^3$$

Another defect semiconductor is created when two Fe^{3+} ions are substituted for three Fe^{2+} ions in FeO, thereby creating a vacancy (Figure 17-29). The Fe^{3+} ions act as electron acceptors, and a *p*-type semiconductor is produced.

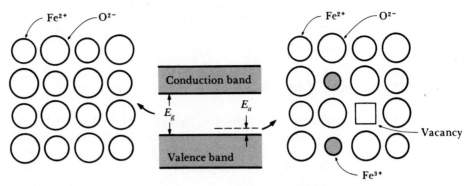

FIGURE 17-29   Two Fe^{3+} ions and a vacancy substitute for three Fe^{2+} ions, maintaining overall charge balance but creating an acceptor level, giving a p-type defect semiconductor.

Amorphous Semiconductors.   Semiconduction can also occur in a number of noncrystalline, or amorphous, materials. The traditional semiconducting materials, including silicon, germanium, and GaAs, can be produced as amorphous materials as well as crystalline materials. In addition, other materials such as sulfur, selenium, and tellurium behave as semiconductors in their amorphous states.

17–10   APPLICATIONS OF SEMICONDUCTORS TO ELECTRICAL DEVICES

Many electronic devices have been developed using the characteristics of semiconduction. A few of these are described here. Others, particularly those that interact with light, will be discussed in Chapter 19.

Thermistors.   The conductivity of semiconductors increases with temperature (Figure 17-30). By knowing the relationship between conductivity and temperature, we can use the semiconductor to measure temperature. Thermistors also

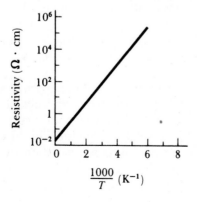

FIGURE 17-30   The electrical resistivity of an $Fe_3O_4 \cdot MgCr_2O_4$ thermistor versus temperature.

have other uses, including use as a fire alarm. When the thermistor heats, it passes a larger current through a circuit and activates the alarm. Thermistors have better sensitivities when the E_g is large.

◇ Example 17-13

A thermistor made from an $Fe_3O_4 \cdot MgCr_2O_4$ defect semiconductor is 0.1 cm in diameter and 1 cm long. Calculate (a) the current produced in the thermistor in a 16-V circuit at 27°C and (b) the temperature when a current of 12.5 mA flows in the 16-V thermistor circuit. (See Figure 17-30.)

Answer:

(a) First, let's determine the resistivity at $T = 27°C$ or 300 K. Since

$$\frac{1000}{T} = \frac{1000}{300} = 3.33$$

the resistivity is $3 \times 10^2 \, \Omega \cdot cm$. From Ohm's law, the current at 27°C is

$$V = IR = I\rho \frac{l}{A}$$

$$I = \frac{VA}{\rho l} = \frac{(16)(\pi/4)(0.1)^2}{(300)(1)} = 0.00042 \, A = 0.42 \, mA$$

(b) When the current is 12.5 mA (0.0125 A)

$$\rho = \frac{VA}{Il} = \frac{(16)(\pi/4)(0.1)^2}{(0.0125)(1)} = 10 \, \Omega \cdot cm$$

From Figure 17-30, $10 \, \Omega \cdot cm$ corresponds to $1000/T = 2.2$

$$T = \frac{1000}{2.2} = 455 \, K = 181°C$$

Pressure Transducers.   The band structure and the energy gap are a function of the spacing between the atoms in the material (Figure 17-31). When pressure is applied to the semiconductor, atoms are forced closer together, the energy gap decreases, and the conductivity increases. If we measure the conductivity, we can in turn calculate the pressure acting on the material.

Magnetometers.   Semiconductors measure the strength of a magnetic field by using the *Hall effect*. A charge carrier moving through a magnetic field is deflected to one side of the material through which the charge is moving (Figure 17-32). Electrons are deflected in one direction, holes in the other. This creates a voltage drop across the material which is related to the current and the magnetic field by

$$V_H = HJR_H \tag{17-23}$$

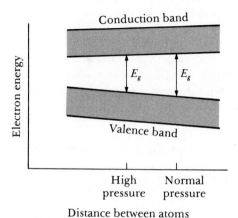

FIGURE 17-31   Pressure squeezes the
atoms closer together in a semiconductor,
reducing the energy gap and increasing
the electrical conductivity.

where V_H is the Hall voltage, J is the current density, H is the magnetic field, and R_H is the Hall coefficient, a constant given by

$$R_H \ = \ \frac{1}{n_h q} \ = \ - \ \frac{1}{n_e q} \tag{17-24}$$

If we apply a given current through a known cross section of the material and measure the voltage, we can calculate the strength of the magnetic field. The Hall effect also permits us to determine whether we have a p-type or an n-type semiconductor, since the voltage drop will have opposite signs.

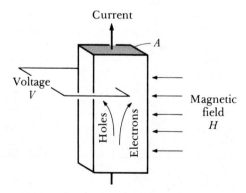

FIGURE 17-32   The Hall effect. Electrons
and holes are deflected by the magnetic
field. By measuring the voltage, the strength
of the magnetic field can be calculated.

◇ **Example 17-14**

A silicon rod 0.125 cm in diameter is doped with one antimony atom per million silicon atoms. A current of 10 A is passed through the rod when the rod is in a magnetic field. If 6 V is measured across the silicon semiconductor, calculate the strength of the magnetic field.

Answer:

From Example 17-10, $n_e = n_d = 5 \times 10^{16}$ electrons/cm^3.

$$R_H = -\frac{1}{n_e q} = \frac{-1}{(5 \times 10^{16})(-1.6 \times 10^{-19})} = 125 \text{ cm}^3/\text{C}$$

$$H = \frac{V}{\mathcal{J} R_H} = \frac{6 \text{ V}}{\left[\dfrac{10 \text{ A}}{(\pi/4)(0.125)^2}\right](125)} = 5.89 \times 10^{-5} \text{ V} \cdot \text{C}/\text{A} \cdot \text{cm}$$

◇

Rectifiers (*p-n* Junction Devices). Rectifiers are produced by joining an *n*-type semiconductor to a *p*-type semiconductor, forming a *p-n junction* (Figure 17-33). Electrons are concentrated in the *n*-type junction, while holes are concentrated in the *p*-type junction. The resulting electrical imbalance creates a voltage, or contact potential, across the junction.

If we place an external voltage on the *p-n* junction so that the negative terminal is at the *n*-type side, both the electrons and holes move towards the junction and eventually recombine. The movement of the electrons and holes causes a net current to be produced. This is called *forward bias* [Figure 17-33(b)]. By increasing the forward bias, the current passing through the junction increases [Figure 17-34(a)].

However, if the applied voltage is reversed, creating a *reverse bias*, both the holes and electrons move away from the junction [Figure 17-33(c)]. With no charge carriers in the depleted zone, the junction behaves as an insulator and virtually no current flows [Figure 17-34(a)].

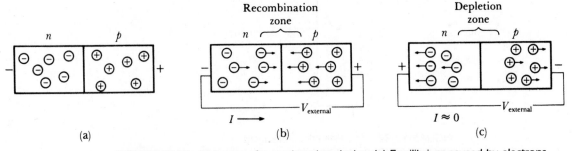

FIGURE 17-33 Behavior of a *p-n* junction device. (a) Equilibrium caused by electrons concentrating in the *n*-side and holes in the *p*-side, (b) forward bias causes a current to flow, and (c) reverse bias does not permit a current to flow.

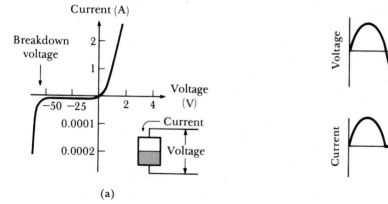

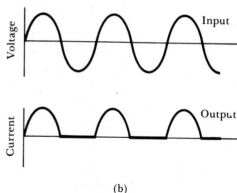

(a)

(b)

FIGURE 17-34   (a) The current-voltage characteristic for a *p-n* junction. Note the different scales in the first and third quadrants. (b) If an alternating signal is applied, rectification occurs and only half of the input signal passes the rectifier.

Because the *p-n* junction permits current to flow in only one direction, it passes only half of an alternating current, therefore converting the alternating current to direct current [Figure 17-34(b)]. These are called *rectifier diodes.* Typically a small leakage current is produced for reverse bias due to the movement of thermally activated electrons and holes.

When the reverse bias becomes too large, however, any carriers that do leak through the insulating barrier of the junction are highly accelerated, excite other charge carriers, and cause a high current in the reverse direction [Figure 17-34(a)]. We can use this phenomenon to design voltage-limiting devices. By proper doping and construction of the *p-n* junction, the breakdown or *avalanche voltage* can be preselected. When the voltage in the circuit exceeds the breakdown voltage, a high current flows through the junction and is diverted from the rest of the circuit. These devices are called *Zener diodes* and are used to protect circuitry from accidental high voltages.

Bipolar Junction Transistors.   A transistor can be used as a switch or an amplifier. One example is the *bipolar junction transistor* (BJT), which is often used in the central processing units of computers due to their rapid switching response. A bipolar junction transistor is a sandwich of either *n-p-n* or *p-n-p* semiconductor materials (Figure 17-35). There are three zones in the transistor —the *emitter*, the *base*, and the *collector*. As in the *p-n* junction, electrons are initially concentrated in the *n*-type material and holes are concentrated in the *p*-type material.

Figure 17-36 shows an *n-p-n* transistor and its electrical circuit, both schematically and as it might appear when implanted in a silicon chip. The electrical signal to be amplified is connected between the base and the emitter, with a small voltage between these two zones. The output from the transistor, or the amplified signal, is connected between the emitter and the collector and operates at a higher voltage. The circuit is connected so that a forward bias is

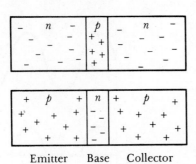

FIGURE 17-35 The *n-p-n* and *p-n-p* transistors.

Emitter   Base   Collector

produced between the emitter and the base (the positive voltage is at the *p*-type base), while a reverse bias is produced between the base and the collector (with the positive voltage at the *n*-type collector). The forward bias causes electrons to leave the emitter and enter the base.

Electrons and holes will attempt to recombine in the base; however, if the base is exceptionally thin and lightly doped, or if the recombination time τ is long, almost all of the electrons pass through the base and enter the collector. The reverse bias between the base and collector accelerates the electrons through the collector, the circuit is completed, and an output signal is produced. The current through the collector is given by

$$I_c = I_0 \exp\left(\frac{V_E}{B}\right) \tag{17-25}$$

where I_0 and B are constants and V_E is the voltage between the emitter and the base. If the input voltage V_E is increased, a very large current I_c is produced and amplification has occurred.

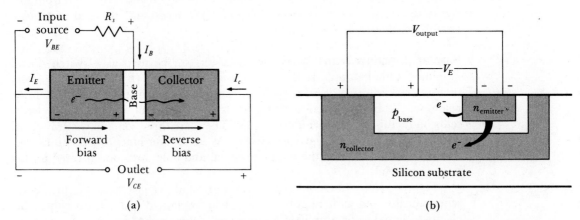

(a)   (b)

FIGURE 17-36 (a) A circuit for an *n-p-n* bipolar junction transistor. The input creates a forward and reverse bias that causes electrons to move from the emitter, through the base, and into the collector, creating an amplified output. (b) Sketch of the cross section of the transistor.

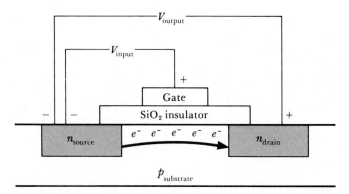

FIGURE 17-37 An *n-p-n* metal oxide semiconductor field effect transistor.

Field Effect Transistors.   A second type of transistor, which is more often used for storing data in computer memories, is the field effect transistor (FET), which behaves in a somewhat different manner from the bipolar junction transistors. Figure 17-37 shows an example of a metal oxide semiconductor (MOS) field effect transistor, in which two *n*-type regions are formed into a *p*-type substrate. One of the *n*-type regions is called the *source*; the second is called the *drain*. A third component of the transistor is a conductor, called a *gate*, that is separated from the semiconductor by a thin insulating layer of SiO_2. A potential is applied between the gate and the source, with the gate region being positive. The potential draws electrons to the vicinity of the gate, but the electrons cannot enter the gate because of the silica. The concentration of electrons beneath the gate makes this region more conductive so that a large potential between the source and drain permits electrons to flow from the source to the drain, producing an amplified signal. By changing the input voltage between the gate and the source, the number of electrons in the conductive path changes, thus also changing the output signal.

The field effect transistors are generally less expensive to produce than the bipolar junction transistors, and, because FETs occupy less space, they are preferred in microelectronic integrated circuits, where perhaps 100,000 transistors are to be present in a single silicon chip.

17–11   MANUFACTURE AND FABRICATION OF SEMICONDUCTOR DEVICES

In order to produce electronic components that require little power, operate very rapidly, yet are inexpensive, microelectronic integrated circuits formed on silicon chips may contain as many as 1,000,000 transistors or other devices, each having dimensions of as little as 10^{-4} cm. In order to produce these circuits, special technologies are required. The starting point for the most common devices is pure, single crystal silicon. The following steps might briefly

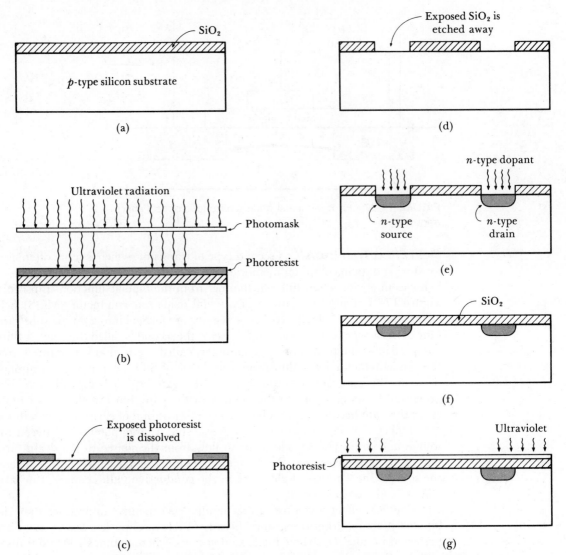

FIGURE 17-38 *Production of an FET semiconductor device. (a) A p-type silicon substrate is oxidized. (b) Ultraviolet radiation passing through a photomask exposes a portion of the photoresist layer. (c) The exposed photoresist is dissolved. (d) The exposed silica is removed by etching. (e) An n-type dopant is introduced to produce the source and drain. (f) The silicon is again oxidized. (g) Photolithography is repeated to introduce other components, including electrical leads, for the device.*

summarize the production of an FET transistor from silicon in a microelectronic integrated circuit (Figure 17-38).

   1. Pure silicon, perhaps containing a controlled amount of a *p*-type dopant material, is produced either by the Czochralski method for growing single

crystals from a liquid or by the zone refining technique; these methods were discussed in Chapter 9. The size of the silicon rods produced is limited to about 6 in. (150 mm) in diameter due to segregation problems, although at some future point processing in outer space may increase this limit.

2. The silicon rod is then sliced into wafers, each only about 0.01 in. (0.25 mm) thick, which will act as the *p*-type substrate for the transistor. The wafers are polished so that the variation in the height of the surface is only about 10 Å. The wafer is heated in the presence of oxygen to form a thin layer of SiO_2 over the surface.

3. A set of *photomasks* is prepared which contain the details of the integrated circuit. The photomasks perform the same function as a photographic negative in normal film processing. In addition, a polymer, called a *photoresist*, is used to coat the silica layer on the wafer. Ultraviolet radiation passing through the photomask exposes portions of the photoresist; the exposed areas are later dissolved in appropriate solvents, leaving photoresist covering those parts of the wafer that are not to be treated in the next step. This process is known as *photolithography*.

4. The silicon wafer, with the photoresist still present at locations not exposed by the ultraviolet radiation, is then immersed in acid; the acid removes the silica from the exposed areas, leaving behind a clean silicon surface. Finally, the remainder of the photoresist is removed chemically. At this point, the wafer consists of a surface that is part silica and part clean silicon.

5. Next, *n*-type dopants are introduced onto the exposed silicon areas, using a diffusion or ion implantation technique. In the *diffusion* process, the wafer is heated to over 1000°C in an atmosphere containing a dopant element. Because the silica acts as a diffusion barrier, the doping element diffuses only into the surface of the exposed silicon, changing the *p*-type substrate into an *n*-type region. By using appropriate processing temperatures and times, the necessary size and composition of doped regions can be produced.

In the *ion implantation* technique, dopant atoms are ionized at high temperatures, then fired at a high velocity into the wafer, becoming embedded into the silicon lattice at the surface. Again, the silica prevents these ions from reaching the masked-off regions of the wafer.

6. After doping, the wafer is again heated to produce another layer of silica over the doped surface. A portion of this silica layer will form the insulating layer at the gate of the FET. Photolithography can be repeated to produce the patterns that will permit the deposition of electrical contacts (usually aluminum) at the gate, source, and drain regions as well as insulating layers such as silica or silicon nitride. Techniques such as vacuum evaporation, sputtering, or chemical vapor deposition techniques are used to deposit these layers.

7. Finally, the chips are removed from the wafer, the external leads are bonded onto the chip, and the chip is encapsulated in an insulating ceramic or polymer material.

While many exotic materials and technologies are used to produce integrated circuits, the next generation of semiconducting devices must be even faster. One of the ways to produce faster response times is to make the individual devices even smaller; this may require using electron or X-ray beams rather than the ultraviolet light used in photolithography; producing better insulators for gate regions in FETs; producing conductors that will be stable; and perhaps even producing three-dimensional chips with alternating layers of silicon and silica rather than the current planar chips.

A second approach to achieving faster times is to use alternative materials. Although silicon is the normal basis for integrated circuits, electrons move more rapidly in semiconductors such as GaAs, permitting the compound devices to respond two to five times faster than silicon-based devices. New technologies are being developed to economically produce these materials, including *heteroepitaxy*. In this technique, thin films of alternating materials are grown one on top of the other, with the alternate layers perhaps varying from *p*-type to *n*-type behavior. This can be done by permitting the film to precipitate from a liquid solution during cooling (liquid-phase epitaxy), by condensing from a gas (chemical-vapor deposition), or by growing from a heated beam of atoms or molecules (molecular-beam epitaxy).

Finally, as will be discussed in Chapter 19, semiconducting devices may be designed to operate using light rather than electricity; semiconductor materials such as GaAs, which are photosensitive, and fiber-optic glasses may form the basis for future computers and communication systems.

17–12 CONDUCTIVITY IN OTHER MATERIALS

Electrical conductivity in most ceramics and polymers, as mentioned previously, is normally very low; however, in this section we will find that special conditions or materials will provide limited or even good conduction. In addition, electrical conductivity is often of importance in composite materials; we will find that conduction of electrical current depends on the type of composite as well as the materials present in the composite, providing extremes of excellent conduction or excellent insulation.

Conduction in Ionic Materials.   Conduction in ionic materials often occurs by movement of entire ions, since the energy gap is too large for electrons to enter the conduction band. Therefore, most ionic materials behave as insulators.

In ionic materials, the mobility of the charge carriers, or ions, is

$$\mu = \frac{ZqD}{kT} \tag{17-26}$$

where D is the diffusion coefficient, k is Boltzmann's constant, T is the absolute temperature, q is the charge, and Z is the valence of the ion. The mobility is many orders of magnitude lower than the mobility of electrons; hence, the conductivity is very small.

$$\sigma = nZq\mu \tag{17-27}$$

Impurities and vacancies increase conductivity; vacancies are necessary for diffusion in substitutional types of crystal structures, and impurities can also diffuse and help carry the current. High temperatures also increase conductivity because the rate of diffusion increases. The conductivity in molten ionic materials may be much higher than normal, due to the greater mobility of the ions in the liquid.

◇ | **Example 17-15**

Suppose that the electrical conductivity of MgO is determined primarily by the diffusion of the Mg^{2+} ions. Estimate the mobility of the Mg^{2+} ions and calculate the electrical conductivity of MgO at 1800°C.

Answer:

From Figure 5-7, the diffusion coefficient of Mg^{2+} ions in MgO at 1800°C is $10^{-10} cm^2/s$. For MgO, $Z = 2/ion$, $q = 1.6 \times 10^{-19} C$, $k = 1.38 \times 10^{-23} J/K$, and $T = 2073 K$.

$$\mu = \frac{ZqD}{kT} = \frac{(2)(1.6 \times 10^{-19})(10^{-10})}{(1.38 \times 10^{-23})(2073)} = 1.12 \times 10^{-9} C \cdot cm^2/J \cdot s$$

Since $C = A \cdot s$ and $J = A \cdot V \cdot s$

$$\mu = 1.12 \times 10^{-9} cm^2/V \cdot s$$

MgO has the NaCl structure, with four magnesium ions per unit cell. The lattice parameter is $3.96 \times 10^{-8} cm$, so the number of Mg^{2+} ions per cubic centimeter is

$$n = \frac{4\ Mg^{2+}\ ions/cell}{(3.96 \times 10^{-8}\ cm)^3} = 6.4 \times 10^{22}\ ions/cm^3$$

$$\sigma = nZq\mu = (6.4 \times 10^{22})(2)(1.6 \times 10^{-19})(1.12 \times 10^{-9})$$

$$= 22.94 \times 10^{-6} C \cdot cm^2/cm^3 \cdot V \cdot s$$

Since $C = A \cdot s$ and $V = A \cdot \Omega$

$$\sigma = 2.294 \times 10^{-5} \Omega^{-1} \cdot cm^{-1}$$

Conduction in Polymers. Because their electrons are involved in covalent bonding, polymers normally have a band structure that gives a large energy gap and a very low electrical conductivity. As a result, polymers are frequently used in applications that require electrical insulation, preventing short circuits, arcing, and safety hazards. Table 17-7 shows the very low conductivity of some typical polymers.

In some cases, however, the low conductivity may be a hindrance. For example, static electricity can accumulate on housings for electronic equipment and damage the internal solid-state devices. If lightning strikes the polymer matrix composite wing of an airplane, severe damage can occur. In addition to minimizing such problems, electrically conductive polymers could be decoratively finished by electroplating, or new polymer semiconducting devices could be developed for unique combinations of properties and applications.

Some techniques have been used to permit traditional polymers to become somewhat more conductive. Polymers have been coated with more conductive materials such as nickel and zinc. Conductive filler materials, either in particulate or fiber form, have been introduced to produce conductive polymer matrix composites; incorporation of nickel-plated carbon fibers provides a unique method of combining high stiffness with conductivity. Hybrid composites containing metal fibers along with normal carbon, glass, or aramid fibers have been considered for lightning-safe aircraft skins.

Other polymers can be made more conductive by doping or processing techniques. When acetal polymers are doped with agents such as arsenic pentafluoride, electrons are able to freely jump from one atom to another along the backbone of the chain, increasing the conductivity to near the range of metals. Because doping is easier for short chains, conductivity is higher for acetals with a low degree of polymerization.

Some polymers, such as polyphthalocyanine, can be cross-linked by special curing processes to raise the conductivity from 10^{-12} ohm^{-1} · cm^{-1} to between 10^{-4} and 10^2 ohm^{-1} · cm^{-1}, which permits the polymer to behave as a semiconductor (Figure 17-39). Because of the cross-linking, electrons can move more easily from one chain to another.

Conduction in Composites.   The electrical conductivity in composites varies from that of insulators to that of metals, depending on the individual properties of the matrix and reinforcement as well as the size and geometry of the reinforcement. A composite composed of aramid fibers in a polymer matrix will act as an insulator, since both constituents are insulators. A tungsten fiber–reinforced superalloy will behave as a conductor. Polymer or ceramic matrix composites reinforced with conductive fibers can provide good conductivity. The aspect ratio of fibers is important; for a constant volume fraction, the conductivity will be higher when the fibers have a large aspect ratio, a small diameter, and a low density. Fibers meeting these criteria will tend to overlap to a greater extent, permitting the higher conductivity.

Orientation of laminar composites is critical for electrical conductivity. A laminar composite of alternating layers of polyethylene and aluminum will be an insulator perpendicular to the sheets but a conductor parallel to the sheets.

SUMMARY

Electrical conductivity is particularly sensitive to atomic bonding, atomic structure, and processing of the material. Consequently, metals have a high conductivity, which is reduced when the temperature increases or lattice imper-

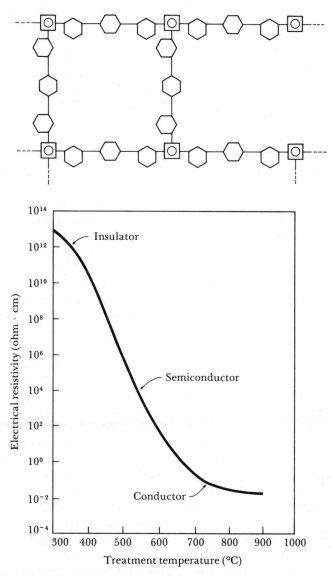

FIGURE 17-39 *Production of polyphthalocyanine gives a strong cross-linked structure that permits electrons to easily move from chain to chain. By appropriate heat treatment, the polymer may behave as an insulator, a semiconductor, or a conductor.*

fections are introduced by alloying or processing. Semiconductors and insulators have lower conductivities due to the nature of the ionic or covalent bonds. However, their conductivity can be increased by higher temperatures, by the introduction of the right type of lattice imperfections, or by doping. By taking advantage of the influence of temperature and structure on conductivity, a large variety of special electronic devices can be constructed and utilized.

GLOSSARY

Acceptor saturation. When all of the extrinsic acceptor levels in a p-type semiconductor are filled.

Avalanche voltage. The reverse bias voltage that causes a large current flow in a p-n junction.

Conduction band. The unfilled energy levels into which electrons can be excited to provide conductivity.

Current density. The current flowing through a given cross-sectional area.

Defect resistivity coefficient. Relates the effect of lattice imperfections to the conductivity.

Donor exhaustion. When all of the extrinsic donor levels in an n-type semiconductor are filled.

Doping. Addition of controlled amounts of impurities to increase the number of charge carriers in a semiconductor.

Drift velocity. The average rate at which electrons or other charge carriers move through the material.

Electric field. The voltage gradient, or volts per unit length.

Energy gap. The energy between the top of the valence band and the bottom of the conduction band that a charge carrier must obtain before it can transfer a charge.

Extrinsic semiconductor. A semiconductor prepared by adding impurities or dopants which determine the number of charge carriers.

Fermi energy. The energy midway between the valence and conduction bands.

Forward bias. Connecting a junction device so that holes and electrons flow towards the junction to produce a net current flow.

Hall effect. A charge carrier moving through a magnetic field is deflected. By measuring the voltage, the strength of the field can be measured.

Holes. Unfilled energy levels in the valence band. Because electrons move to fill these holes, the holes move and a current is carried.

Hybridization. The valence and conduction bands are separated by an energy gap. This leads to the semiconductive behavior of silicon and germanium.

Intrinsic semiconductor. A semiconductor in which the temperature determines the conductivity.

Mobility. The ease with which a charge carrier moves through a material.

Recombination time. A constant related to the time required for electrons and holes to recombine when an electric field is removed.

Rectifiers. p-n junction devices that permit current to flow in only one direction in a circuit.

Reverse bias. Connecting a junction device so that holes and electrons flow away from the junction, preventing a net current flow.

Seebeck effect. A temperature difference between two ends of a conductor will create a voltage.

Superconductivity. Flow of current through a material that has no resistance to that flow.

Temperature resistivity coefficient. Relates the effect of temperature to the conductivity.

Thermistor. A semiconductor device that is particularly sensitive to changes in temperature, permitting it to serve as an accurate measure of temperature.

Thermocouple. A pair of conductor wires. The voltage produced between the wires when the temperature is changed can be used to measure temperature.

Transistor. A semiconductor device that can be used to amplify electrical signals.

Valence band. The energy levels filled by electrons in their lowest energy states.

Zener diode. A *p-n* junction device which, with a very large reverse bias, causes a current to flow.

PRACTICE PROBLEMS

1. A current of 100 A is passed through a 0.125-in. diameter wire 1 mile long. Calculate the power loss if the wire is made of (a) copper, (b) germanium, and (c) silicon. (See Tables 17-2 and 17-8.)

2. A current density of 200,000 A/cm^2 is applied to a 500-m long aluminum wire. A resistance of 1 ohm is measured along the wire. Calculate (a) the diameter of the wire and (b) the voltage imposed on the wire.

3. The power loss in a 0.1-cm diameter gold wire is to be 100 W when a 4-A current is flowing in the circuit. Calculate the required length of the wire.

4. A 40-ohm resistor is produced from a 0.25-cm diameter iron wire. Calculate the required length of the wire.

5. An electric field of 10 V/cm is applied to a 50-cm length of magnesium wire. Calculate (a) the voltage and (b) the current density in the wire.

6. A current density of 10,000 A/cm^2 is applied to a silver wire. If all of the valence electrons serve as charge carriers, determine the average drift velocity of the electrons.

7. Suppose the mobility of an electron in nickel is 500 cm^2/V · s. Estimate the fraction of the valence electrons that are carrying an electrical charge.

8. The electrical conductivity of silicon is 5×10^{-6} ohm^{-1} · cm^{-1} and the mobility of the electrons is 1900 cm^2/V · s. What fraction of the valence electrons in silicon carry an electrical charge?

9. A voltage of 50 V is applied to a 10-cm length of 0.1-cm diameter copper wire. Deter-mine the average drift velocity of the electrons in miles per hour if 50% of the valence electrons are responsible for conducting the current.

10. A current of 200 A is passed through the arc in an arc-welding process. The diameter of the arc is about 5 mm and the arc extends 2.5 mm from the tip of the electrode to the metal being welded. A voltage of 40 V is applied across the arc. Calculate (a) the current density in the arc, (b) the electric field across the arc, and (c) the electrical conductivity of the gases in the arc during welding.

11. Calculate the number of energy levels present in the 3*p* level of 1 cm^3 of solid HCP titanium.

12. Calculate the electrical resistivity of pure aluminum at $-100°$C and $+200°$C.

13. Calculate the temperature of a pure nickel rod that has an electrical resistivity of 10×10^{-6} ohm · cm.

14. To what temperature must you raise pure magnesium to double its room temperature electrical resistivity?

15. Based on Figure 17-12(b), determine the defect resistivity coefficient for tin in copper.

16. To what temperature must you raise pure copper to obtain the same electrical resist-ivity that you would get by alloying with 25 wt % zinc? (See Example 17-7.)

17. Suppose the electrical conductivity of iron containing 3 at % impurity at 500°C is 0.02×10^6 ohm^{-1} · cm^{-1}. Determine the con-tribution to resistivity due to temperature and impurities by calculating (a) the expected resist-ivity of pure iron at 500°C, (b) the resistivity due to impurities, and (c) the defect resistivity coefficient.

18. A nickel alloy filament 0.1 mm in diameter and 1 m long operating on a 120-V circuit consumes 335 W at 500°C. Calculate (a) the current flowing in the filament during use and (b) the current that initially flows in the filament before it heats to the operating temperature, assuming that the alloy has the same temperature resistivity coefficient as pure nickel.

19. Based on the data in Figure 17-11, estimate the defect resistivity coefficient for nickel in copper.

20. The electrical resistivity of a Au-30 wt % Pt alloy is 22×10^{-6} ohm · cm at room temperature. Calculate the defect resistivity coefficient for platinum in gold.

21. (a) Determine the equations that relate millivolt reading to temperature for Pt : Pt − 10% Rh and iron : constantan thermocouples. (b) Suppose that at 600°C your accuracy in measuring voltage is ± 0.2 mV. What is the error in degrees Celsius for each of the thermocouples? Which would you prefer for best accuracy?

22. Suppose you would like to use a $B_4C : C$ thermocouple to monitor the temperature in a jet engine. What millivolt reading would you expect if the temperature were 2100°C?

23. Figure 17-40 shows the output from a chromel : alumel thermocouple immersed in an aluminum-silicon alloy during freezing. Assuming that the alloy is hypereutectic, estimate (a) the liquidus temperature, (b) the eutectic temperature, and (c) the composition of the alloy. (See Figure 10-19.)

24. Will niobium (Nb) be superconductive at 5 K when a magnetic field of 550 oersted is present? Explain.

25. Suppose an alumina (Al_2O_3) rod 1 mm in diameter and 1 cm in length is used as an insulator in a 240-V circuit at room temperature. Calculate (a) the current flowing in the circuit and (b) the number of electrons passing through the alumina rod per second. Repeat for the case in which the rod is made of aluminum instead.

26. If the electrical conductivity of dia-

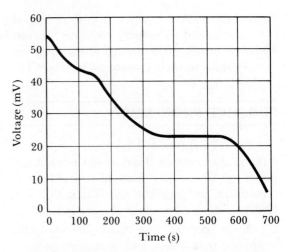

FIGURE 17-40 Cooling curve for a hypereutectic Al-Si alloy (Problem 17-23).

mond is 10^{-18} ohm · cm, determine (a) the number of charge carriers per cubic centimeter, (b) the fraction of the total electrons in the valence band that are excited into the conduction band, and (c) the constant n_0 at room temperature. Assume that the atomic radius for carbon in diamond is 0.77 Å.

27. For tin, determine (a) the number of charge carriers per cubic centimeter, (b) the fraction of the total electrons in the valence band that are excited into the conduction band, and (c) the constant n_0 at room temperature.

28. For pure silicon, estimate (a) the value of n_0 at room temperature and (b) the electrical conductivity at 500°C if the mobility remains constant.

29. Estimate the temperature at which the electrical conductivity of pure silicon will be twice its value at 25°C.

30. Suppose there are 2×10^{13} electrons/cm^3 serving as charge carriers in germanium when an electric field is applied. When the field is removed, 2×10^8 electrons/cm^3 remain after 10^{-6} s. Determine (a) the recombination time and (b) the time required for 99% of the electrons and holes to recombine.

31. Suppose silicon is doped with 0.0001 at % phosphorus. Estimate the electrical conductivity of the semiconductor in the exhaustion range.

32. Estimate the at % antimony required to produce an electrical conductivity of $1000 \, ohm^{-1} \cdot cm^{-1}$ in germanium in the exhaustion range.

33. Calculate the number of grams of gallium that must be added to 1 kg of silicon in order to produce an electrical conductivity of $30 \, ohm^{-1} \cdot cm^{-1}$ in the saturation range.

34. Suppose 0.472 kg of gallium are combined with 0.528 kg of arsenic to produce GaAs. (a) Will this produce a *p*-type or an *n*-type semiconductor? (b) Calculate the number of extrinsic charge carriers per cubic centimeter, using the lattice parameter from Figure 17-27. GaAs has the zinc blende crystal structure.

35. For which one of the semiconducting compounds in Figure 17-27 would you expect the broadest exhaustion or saturation plateau?

36. A ZnO crystal is produced in which one interstitial Zn atom is introduced for every 1000 Zn lattice sites. Estimate (a) the number of charge carriers per cubic centimeter and (b) the electrical conductivity at 25°C.

37. Each Fe^{3+} ion in FeO can serve as an acceptor site for an electron. If there is one vacancy per 500 unit cells of the FeO crystal (which has the sodium chloride structure), determine the number of possible charge carriers per

cubic centimeter. The lattice parameter of FeO is 4.29 Å.

38. Determine the energy gap E_g for the $Fe_3O_4 \cdot MgCr_2O_4$ thermistor in Figure 17-30.

39. Calculate the temperature when an $Fe_3O_4 \cdot MgCr_2O_4$ thermistor operating in a 12-V circuit produces a current of 15 mA. The thermistor is in the form of a wire 0.1 cm in diameter and 12 cm in length.

40. A germanium crystal 1 cm in diameter and 30 cm in length is doped with indium. A current of 10 A is passed through the rod with a voltage of 12 V. A magnetic field of 5 V · C/ A · cm is measured. Calculate the number of indium atoms per 10^6 Ge atoms in the semiconductor.

41. When a voltage of 10 mV is applied to the emitter of a transistor, a current of 3 mA is produced. If the voltage is increased to 15 mV, the current through the collector increases to 8 mA. By what percentage will the collector current be increased when the emitter voltage is doubled from 10 mV to 20 mV?

42. CaO has the sodium chloride crystal structure and a lattice parameter of 4.62 Å. Suppose the conductivity of CaO at 977°C is measured to be $2.408 \times 10^{-10} \, ohm^{-1} \cdot cm^{-1}$ and $1.445 \times 10^{-5} \, ohm^{-1} \cdot cm^{-1}$ at 1810°C. Calculate (a) the activation energy for diffusion of Ca^{2+} in CaO and (b) the mobility of Ca^{2+} at both temperatures.

18

DIELECTRIC AND MAGNETIC PROPERTIES

18-1 INTRODUCTION

The response of a material to an electric field can be used to advantage even when no charge is transferred. These effects are described by the dielectric properties of the material. Dielectric materials possess a large energy gap between the valence and conduction bands; thus the materials have a high electrical resistivity. Two important applications for dielectric materials include electrical insulators and capacitors. Insulators, used to prevent the transfer of charge in an electric circuit, include the plastic covering on electrical wires and the ceramic "bells" used in high-voltage power lines. Capacitors are used to store electric charge. Other characteristics of dielectrics include electrostriction, piezoelectricity, and ferroelectricity.

The effects of a magnetic field on a material are equally profound. Some magnetic materials possess a permanent magnetization for applications ranging from toys to computer storage; other magnetic materials are used in electric motors and transformers.

Although the responses of a material to an electric or magnetic field are based on very different phenomena, a similar approach is used to describe the two effects. For this reason, dielectric and magnetic behavior will be discussed together in this chapter and some of the similarities will become apparent in our discussion.

18-2 DIPOLES AND POLARIZATION

In both electrical and magnetic materials, the application of a field causes the formation and movement of dipoles. *Dipoles* are atoms or groups of atoms that have an unbalanced charge. In an imposed electric field, the dipoles become aligned in the material, causing *polarization*.

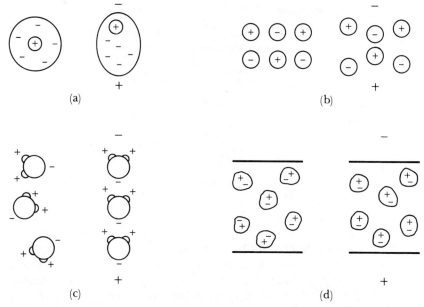

FIGURE 18-1 Polarization mechanisms in materials. (a) Electronic polarization, (b) ionic polarization, (c) molecular polarization, and (d) space charges.

Dipoles.   When an electric field is applied to a material, dipoles are induced within the atomic or molecular structure and become aligned with the direction of the field. In addition, any permanent dipoles already present in the material are aligned with the field. The material is polarized. The polarization P (C/m^2) is

$$P \;=\; \mathcal{Z}qd \tag{18-1}$$

where $\mathcal{Z}$ is the number of charge centers that are displaced per cubic meter, q is the electronic charge, and d is the displacement between the positive and negative ends of the dipole. Four mechanisms cause polarization—electronic polarization, ionic polarization, molecular polarization, and space charges (Figure 18-1).

Electronic Polarization.   When an electric field is applied to an atom, the electronic arrangement is distorted, with electrons concentrating on the side of the nucleus near the positive end of the field. The atom acts as a temporary, induced dipole. This effect, which occurs in all materials, is small and temporary.

◇ | **Example 18-1**

Suppose that the average displacement of the electrons relative to the nucleus in a copper atom is 1×10^{-8} Å when an electric field is imposed on a copper plate. Calculate the polarization.

Answer:

The atomic number of copper is 29, so there are 29 electrons in each copper atom. The lattice parameter of copper is 3.6151 Å. Thus

$$Z = \frac{(4\ \text{atoms/cell})(29\ \text{electrons/atom})}{(3.6151 \times 10^{-10}\,\text{m})^3} = 2.46 \times 10^{30}\,\text{electrons/m}^3$$

$$P = Zqd = \left(2.46 \times 10^{30}\,\frac{\text{electrons}}{\text{m}^3}\right)$$

$$\times \left(1.6 \times 10^{-19}\,\frac{\text{C}}{\text{electron}}\right)(10^{-8}\,\text{A})(10^{-10}\,\text{m/A})$$

$$= 3.94 \times 10^{-7}\,\text{C/m}^2$$

Ionic Polarization.   When an ionically bonded material is placed in an electric field, the bonds between the ions are elastically deformed. Consequently, the charge is minutely redistributed within the material. Depending on the direction of the field, cations and anions move either closer together or further apart. These temporarily induced dipoles provide polarization and may also change the overall dimensions of the material.

◇ | **Example 18-2**

The ionic polarization observed in a NaCl crystal is $4.3 \times 10^{-8}\,\text{C/m}^2$. Calculate the displacement between the Na^+ and Cl^- ions.

Answer:

In this example, there is one electric charge on each Na^+ ion. In the NaCl unit cell, which has a lattice parameter of 5.5 Å, there are four Na^+ ions.

$$Z = \frac{(4\ Na^+\ \text{ions/cell})(1\ \text{charge}/Na^+\ \text{ion})}{(5.5 \times 10^{-10}\,\text{m})^3}$$

$$= 2.4 \times 10^{28}\,\text{charges/m}^3$$

$$d = \frac{P}{Zq} = \frac{4.3 \times 10^{-8}}{(2.4 \times 10^{28})(1.6 \times 10^{-19})}$$

$$= 11.2 \times 10^{-18}\,\text{m} = 11.2 \times 10^{-8}\,\text{A}$$

Molecular Polarization.   Some materials contain natural dipoles. When a field is applied, the dipoles rotate to line up with the imposed field. Water molecules, shown schematically in Figure 18-1, represent a material that possesses molecular polarization. Many organic molecules, as illustrated in Example 18-3, also behave in a similar manner, as do a variety of organic oils and waxes.

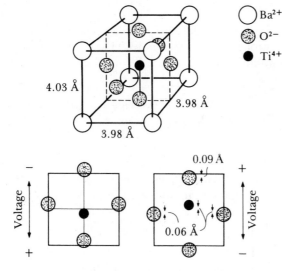

FIGURE 18-2 The crystal structure of barium titanate, $BaTiO_3$. Because of the displacement of O^{2-} and Ti^{4+} ions, the unit cell is a permanent dipole and produces excellent polarization.

In a number of materials, polarization occurs in this same manner but, when the electric field is removed, the dipoles remain in alignment, causing permanent polarization. In amorphous polymers, for example, segments of the chains possess sufficient mobility to cause polarization. Glass, an amorphous ceramic, behaves in much the same way. However, when the field is removed, the chain segments may not move back to the original position, leaving a permanent polarization.

Barium titanate, $BaTiO_3$, a crystalline ceramic, has an asymmetrical structure at room temperature (Figure 18-2). The titanium ion is displaced slightly from the center of the unit cell and the oxygen ions are displaced slightly in the opposite directions from their face-centered positions, causing the crystal to be tetragonal and permanently polarized. When an alternating current is applied to barium titanate, the titanium ion moves back and forth between its two allowable positions to assure that polarization is aligned with the field. In this particular material, polarization is highly anisotropic and the crystal must be properly aligned with respect to the applied field.

Space Charges.   A charge may develop at interfaces between phases within a material, normally as a result of the presence of impurities. The charge moves on the surface when the material is placed in an electric field. This type of polarization is not an important factor in most common dielectrics.

◇ | **Example 18-3**

Compare the tendencies of the following organic molecules to act as dipoles and produce molecular polarization—CH_4, CH_3Cl, CH_2Cl_2, $CHCl_3$, and CCl_4.

Answer:

The structure of each molecule is shown schematically below. The chlorine atoms are strong centers of negative charge and hydrogen atoms are weak centers of positive charge.

CH_4

H
|
H—C—H
|
H

Because the molecule is symmetrical, no molecular polarization is expected.

CH_3Cl

H
\
H—C—Cl
/
H

A large dipole effect is created, since the positive hydrogen atoms are displaced from the chlorine atom.

CH_2Cl_2

H   Cl
\ /
C
/ \
H   Cl

The dipole effect is smaller, since the large chlorine atoms become predominant in the structure.

$CHCl_3$

Cl
|
H—C—Cl
|
Cl

CCl_4

Cl
|
Cl—C—Cl
|
Cl

Again the molecule is symmetrical and no molecular polarization is expected.

The measured dipole moments for each of the molecules, given in $C \cdot m$, are: CH_4, 0; CH_3Cl, 6.2×10^{-30}; CH_2Cl_2, 5.0×10^{-30}; $CHCl_3$, 3.0×10^{-30}; CCl_4, 0.

◇

◇ | **Example 18-4**

Calculate the maximum polarization per cubic centimeter and the maximum charge that can be stored per square centimeter for barium titanate.

Answer:

The strength of the dipoles is given by the product of the charge and the distance between the charges. In $BaTiO_3$, the separations are the distances that the Ti^{4+} and O^{2-} ions are displaced from the normal lattice points (Figure 18-2). The charge on each ion is the product of q and the number of excess or missing electrons. Thus the dipole moments are

$$Ti^{4+}: (1.6 \times 10^{-19})(4 \text{ electrons/ion})(0.06 \times 10^{-8}\,cm)$$

$$= 0.384 \times 10^{-27}\,C \cdot cm/ion$$

$$O^{2-}_{(top)}: (1.6 \times 10^{-19})(2 \text{ electrons/ion})(0.09 \times 10^{-8}\text{cm})$$
$$= 0.288 \times 10^{-27}\text{C} \cdot \text{cm/ion}$$
$$O^{2-}_{(side)}: (1.6 \times 10^{-19})(2 \text{ electrons/ion})(0.06 \times 10^{-8}\text{cm})$$
$$= 0.192 \times 10^{-27}\text{C} \cdot \text{cm/ion}$$

Each oxygen ion is shared with another unit cell, so the total dipole moment in the unit cell is

$$\begin{aligned}
\text{Dipole moment} &= (1\ \text{Ti}^{4+}/\text{cell})(0.384 \times 10^{-27}) \\
&\quad + (1\ \text{O}^{2-} \text{ at top/cell})(0.288 \times 10^{-27}) \\
&\quad + (2\ \text{O}^{2-} \text{ at sides/cell})(0.192 \times 10^{-27}) \\
&= 1.056 \times 10^{-27}\text{C} \cdot \text{cm/cell}
\end{aligned}$$

The polarization per cubic centimeter is

$$\begin{aligned}
P &= \frac{1.056 \times 10^{-27}\text{C} \cdot \text{cm/cell}}{(3.98 \times 10^{-8}\text{cm})^2(4.03 \times 10^{-8}\text{cm})} \\
&= 1.65 \times 10^{-5}\text{C/cm}^2
\end{aligned}$$

The total charge on a $BaTiO_3$ crystal 1 cm $\times$ 1 cm is

$$\begin{aligned}
Q &= PA = (1.65 \times 10^{-5})(1)^2 \\
&= 1.65 \times 10^{-5}\text{C}
\end{aligned}$$

18-3 DIELECTRIC PROPERTIES AND THEIR CONTROL

Dielectric materials are used for a variety of electrical devices; one of their most important functions is to control the flow of an electrical charge, typically by acting as an electrical insulator. We will examine a number of applications for dielectrics in the following sections. First, we should define some of the important dielectric characteristics and how these characteristics are affected by service conditions.

Dielectric Constant. When a voltage is imposed on two conductive materials that are separated from one another by a vacuum (Figure 18-3) we would expect no current flow. Instead, the electrical charge produced by the voltage will remain stored in the circuit. The magnitude of the charge that can be stored between the conductors is called the *capacitance C* and is related to the imposed voltage by

$$Q = CV \tag{18-2}$$

where V is the voltage across the conductors and Q is the stored charge in coulombs. The units for capacitance are coulombs/volt, or farads (F).

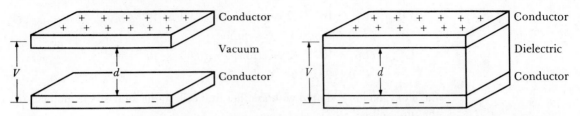

FIGURE 18-3 Capacitors are constructed by separating two or more conductor plates by a dielectric material, whose properties determine the effectiveness of the device.

The capacitance depends on the material between the conductors, the size and geometry of the conductors, and the separation between the conductors. For the geometry shown in Figure 18-3, the capacitance in a vacuum is given by

$$C = \varepsilon_0 \frac{A}{d} \tag{18-3}$$

where A is the area of each conductor and d is the distance between the plates. The constant ε_0 is the *permittivity* of a vacuum and is 8.85×10^{-12} F/m or 8.85×10^{-14} F/cm.

In a vacuum, some charge can be stored between the conductors (Figure 18-4). However, when a dielectric material replaces the vacuum between the conductors, polarization can occur in the dielectric and permit additional charge to be stored. The ability of the dipoles in the dielectric to polarize and store charge is reflected by the permittivity ε, which is a property of the dielectric material. Now the capacitance is given by

$$C = \varepsilon \frac{A}{d} \tag{18-4}$$

Normally we describe the ability of a material to polarize and store electrical charge by the *relative permittivity* or *dielectric constant* κ, which is simply the ratio of the permittivity of the material to the permittivity of a vacuum

$$\kappa = \frac{\varepsilon}{\varepsilon_0} \tag{18-5}$$

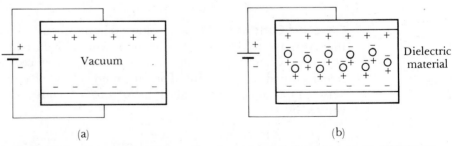

(a)   (b)

FIGURE 18-4 A charge can be stored at the conductor plates in a vacuum. However, when a dielectric is placed between the plates, the dielectric polarizes and additional charge is stored. (a) Shows a total of 12 units of charge and (b) shows a total of 22 units.

The dielectric constant κ is the normal way to describe the ability of a material to store a charge.

The dielectric constant, as expected, is related to the polarization that can be achieved in the material.

$$P = (\kappa - 1)\varepsilon_0 \xi \qquad (18\text{-}6)$$

where ξ is the strength of the electric field (V/m). For materials that polarize easily, both the dielectric constant and the capacitance will be large and, in turn, a large quantity of charge can be stored. In addition, Equation 18-6 suggests that polarization will increase, at least until all of the dipoles are aligned, as the voltage (expressed by the strength of the electric field) increases.

◇ | **Example 18-5**

Suppose sodium chloride has a polarization of $4.3 \times 10^{-8}\,\text{C/m}^2$ in an electric field of 1000 V/m. Calculate the dielectric constant for sodium chloride.

Answer:

From Equation (18-6)

$$P = (\kappa - 1)\varepsilon_0 \xi$$

$$\kappa - 1 = \frac{P}{\varepsilon_0 \xi} = \frac{4.3 \times 10^{-8}}{(8.85 \times 10^{-12})(1000)}$$

$$\kappa - 1 = 4.9$$

$$\kappa = 5.9$$

◇ | **Example 18-6**

Rank the expected dielectric constants for polyethylene, polystyrene, rubber, and epoxy based on their molecular structure and ability to be polarized. Compare your ranking with the measured values at 10^6 Hz in Table 18-1.

Answer:

Polyethylene contains a simple carbon backbone with only small hydrogen atoms attached. Consequently, the chains are easily rearranged, giving a low residual polarization and low dielectric constant of about 2.3.

Polystyrene is an asymmetrical chain with a benzene ring attached. The mobility of the chains is thus reduced and a slightly higher dielectric constant, 2.5, is observed.

Rubber is cross-linked to further reduce the mobility of the polymer chains, causing still higher dielectric constants, about 3.2.

TABLE 18-1   Properties of selected dielectric materials

Material	Dielectric constant (at 60 Hz)	(at 10^6 Hz)	Dielectric Strength (10^6 V/m)	tan δ (at 10^6 Hz)	Resistivity (ohm · m)
Polyethylene	2.3	2.3	20	0.0001	$> 10^{14}$
Teflon	2.1	2.1	20	0.00007	10^{16}
Polystyrene	2.5	2.5	20	0.0002	10^{16}
PVC	3.5	3.2	40	0.05	10^{10}
Nylon	4.0	3.6	20	0.04	10^{13}
Rubber	4.0	3.2	24		
Phenolic	7.0	4.9	12	0.05	10^{10}
Epoxy	4.0	3.6	18		
Paraffin wax		2.3	10		$10^{13}-10^{17}$
Fused silica	3.8	3.8	10	0.00004	10^9-10^{10}
Soda-lime glass	7.0	7.0	10	0.009	10^{13}
Al_2O_3	9.0	6.5	6	0.001	10^9-10^{12}
TiO_2		14–110	8	0.0002	$10^{11}-10^{16}$
Mica		7.0	40		10^{11}
$BaTiO_3$		3000	12		10^6-10^{13}
Water		78.3			10^{12}

Epoxy is a rigid, three-dimensional network polymer. Permanent dipoles are locked into place, giving high dielectric constants, about 3.6.

Dielectric Strength.   Unfortunately, we find that if the applied voltage is too high or the separation between the two conductors is too small, the dielectric device will break down and discharge and the electrical charge will be lost. Conduction of the charge through the dielectric will occur. The *dielectric strength* is the maximum electric field ξ that the dielectric material can maintain between the conductors. The dielectric strength therefore places an upper limit on both C and Q.

$$\text{Dielectric strength} = \xi_{\text{max.}} = \left(\frac{V}{d}\right)_{\text{max.}} \qquad (18\text{-}7)$$

In order to construct a small device capable of storing large charges in an intense field, we must select materials with both a high dielectric strength and a high dielectric constant. Dielectric strengths and dielectric constants for typical materials are shown in Table 18-1.

Electrical Conductivity.   In order for the dielectric to store energy, charge carriers such as electrons or ions must be prevented from moving through the material from one conductor to the other. As a consequence, dielectric materials always have a very high electrical resistivity, as shown in Table 18-1. Ceramic and polymer materials, which normally have electrical resistivities in excess of 10^{11} ohm · cm, are used as dielectric materials.

Effect of Material Structure. The polarization, and therefore the ability of the material to store charge, is closely related to the structure of the material. The material should possess permanent dipoles that move easily in an electric field and still produce high dielectric constants. In water, organic liquids, oils, and waxes, molecular polarization is accomplished easily, since the molecules making up the liquid or wax are very mobile and respond quickly to the application of the electric field. Unfortunately, these types of dielectrics are not easily incorporated into devices for electrical circuits.

Segments of the chains in amorphous polymers have sufficient mobility to polarize and, because they are solid, are easily made into electrical devices. Chains in more rigid structures, such as glassy or crystalline polymers, are less mobile and have lower dielectric constants and dielectric strengths than their amorphous counterparts. Amorphous polymers with asymmetrical chains have a higher dielectric constant, even though the chains may not easily align, because the strength of each molecular dipole is greater. Thus, polyvinyl chloride and polystyrene have dielectric constants greater than polyethylene.

Ceramic glasses, also amorphous structures, may permit some movement of segments of the glassy structure. Electronic and ionic polarization in crystalline ceramics also provide dielectric constants of the same order of magnitude as the polymer materials. However, certain ceramics, such as barium titanate ($BaTiO_3$) can provide exceptionally large dielectric constants because of the molecular polarization caused by the asymmetrical structure of the unit cell, as described earlier. Each unit cell contains a titanium ion that must move only a short distance in response to an electric field; consequently, polarization is very rapid and strong.

Imperfections in the structure are also critical. Often, breakdown of a dielectric is a result of a current flow through the material following cracks, grain boundary impurities, or moisture.

Frequency. Dielectric materials are often used in alternating-current circuits. The dipoles must therefore switch directions, often at a high frequency, in order for the electronic device to perform satisfactorily.

The energy losses are due primarily to two factors—current leakage and dipole friction. Losses due to current leakage are low if the electrical resistivity is high. Dipole friction occurs when reorientation of the dipoles is difficult, as in complex organic molecules. The greatest loss occurs at frequencies at which the dipoles can almost, but not quite, be reoriented (Figure 18-5). At lower frequencies, losses are low because the dipoles have time to move. At higher frequencies, losses are low because the dipoles do not move at all.

Electronic polarization occurs easily even at frequencies as high as 10^{16} Hz, since no rearrangement of atoms is necessary (Figure 18-6). Ionic polarization also occurs readily up to 10^{13} Hz; only a simple elastic distortion of the bonds

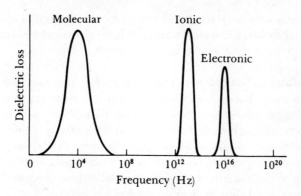

FIGURE 18-5 The influence of frequency on dielectric loss. Losses are greatest at frequencies at which one of the contributions to polarization is lost.

between the ions is required. However, materials that rely on molecular polarization are very sensitive to frequency, since entire atoms or groups of atoms must be rearranged. Consequently, the response of the dielectric material to an alternating field is reduced and complete polarization may not occur.

The structure also influences the frequency effect. Gases and liquids polarize at higher frequencies than solids. Amorphous polymers and ceramics polarize at higher frequencies than their crystalline counterparts. Polymers with bulky

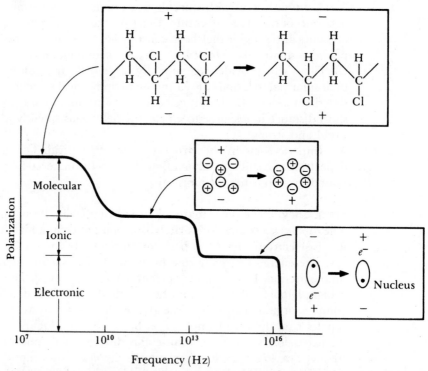

FIGURE 18-6 The response of the material to frequency depends on the polarization mechanism. Permanent molecular dipoles do not respond as rapidly as electronic and ionic dipoles.

asymmetrical groups attached to the chain polarize only at low frequencies. Therefore, polyethylene and polytetrafluoroethylene have the same dielectric constant at almost all frequencies, while the dielectric constant for polyvinyl chloride decreases as the frequency increases.

We can intentionally select a frequency so that materials with permanent dipoles have a high dielectric loss and materials that polarize only by electronic or ionic contributions have a low dielectric loss. Consequently, the permanent dipole materials heat, but the other materials remain cool. Microwave ovens are used to cure many polymer adhesives; the materials to be joined, including metals, have a low loss factor, while the adhesive has a high loss factor. The heat produced in the adhesive due to dielectric losses initiates the thermosetting reaction.

◇ | **Example 18-7**

Explain why microwave ovens can heat food without heating the container.

Answer:

The organic food substances contain highly complex, polar molecules with a high dielectric loss, while the cookware is composed of material with a lower loss.

Dissipation and Dielectric Loss Factors. When an alternating current is applied to a perfect dielectric, the current will lead the voltage by 90°. However, due to losses, the current leads the voltage by only $90° - \delta$, where δ is called the *dielectric loss angle*. When the current and voltage are out of phase by the dielectric loss angle, electrical energy or power is lost, often in the form of heat. The *dissipation factor* is given by

$$\text{Dissipation factor} = \tan \delta \qquad (18\text{-}8)$$

and the dielectric loss factor is

$$\text{Dielectric loss factor} = \kappa \tan \delta \qquad (18\text{-}9)$$

The total power lost P_L is related to the dissipation factor, the dielectric constant, the electric field, the frequency, and the volume of the dielectric material.

$$P_L = 5.556 \times 10^{-11} \kappa \tan \delta \xi^2 f\upsilon \qquad (18\text{-}10)$$

where the electric field is given in volts per meter, the frequency f in hertz, the volume υ in cubic meters, and the power loss in watts. We can minimize heating, even with a large dielectric constant, if a material with a small loss angle is selected. On the other hand, if we want to use dielectric heating, we can select a material with a large dissipation factor.

Temperature. When the temperature increases, permanent dipoles have a greater mobility, polarize more easily, and give a higher dielectric constant. However, the higher temperatures again permit the dielectric to break down and may cause the crystal structure to change to a less polar condition, which greatly reduces the polarization.

18–4 DIELECTRIC PROPERTIES AND CAPACITORS

A capacitor is an electrical device used to store charge received from a circuit. The capacitor may smooth out fluctuations in the signal, accumulate charge to prevent damage to the rest of the circuit, store charge for later distribution, or even change the frequency of the electric signal. Capacitors are designed so that the charge is stored in a polarized material between two conductors, as described in Figure 18-3. As we found earlier, the charge that can be stored depends on the capacitance, which in turn depends on the design of the capacitor and the dielectric material that is used. The material between the conductors must easily polarize, so the dielectric constant should be high yet have a high electrical resistivity to prevent the charge from passing from one plate to the next. In order to operate at high voltages and yet be made as small as possible, the dielectric strength should be high. The dielectric loss factor should be small to minimize heating.

The disc-shaped capacitor in Figure 18-7(a) is a common type of parallel plate capacitor, but, from a practical standpoint, only a limited charge can be stored. We could increase the stored charge by making the conductor plates larger, but the capacitor device would then become bulky. We could also increase the charge by reducing the distance between the conductor plates, but then the danger of a dielectric breakdown and discharge would be increased. One way to improve the performance of a capacitor is to increase the number of plates, as shown in Figure 18-7(b). For a capacitor containing n parallel conductor plates, the capacitance is

$$C = \varepsilon_0 \kappa (n - 1) \frac{A}{d} \tag{18-11}$$

The use of many large plates, a small separation between the plates, a high dielectric constant, and a high dielectric strength improve the ability of the capacitor to store a charge.

Capacitors can also be manufactured by producing rolled tubes [Figure 18-7(c)]. By rolling a dielectric film with a metal conductor into a tube, compact devices can be produced in an economical manner. Typical examples of capacitor construction and materials are included in Table 18-2. Liquid dielectrics can be used if the liquid is impregnated into special kraft paper, which itself is a dielectric. Polymers such as polyester (Mylar), polystyrene, polycarbonate, and cellulose (paper) are often used in the rolled capacitor tubes.

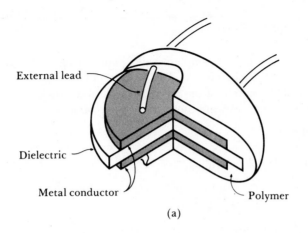

External lead

Dielectric

Metal conductor

Polymer

(a)

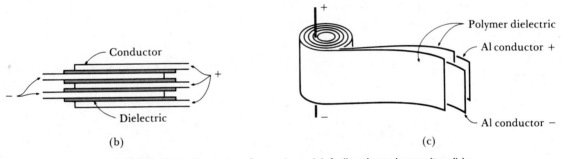

Conductor

Dielectric

(b)

Polymer dielectric

Al conductor +

Al conductor −

(c)

FIGURE 18-7   Examples of capacitors. (a) A disc-shaped capacitor, (b) a multiple-conductor parallel plate capacitor, and (c) a tube capacitor.

TABLE 18-2   Capacitor construction and materials

Types of Capacitors	Capacitor Shape and Characteristics
Mica (0.0001 in.) Lead foil (0.002 in.)	Plate capacitors with good temperature stability, good for radio frequencies, bulky
Mica—sprayed silver coating	Plate capacitors, same capacitance as other mica capacitors
Glass-metal foil	Plate capacitors, moisture resistant, same capacitance as mica capacitors
Kraft paper (with wax or oil)—foil or metallized with tin	Rolled tubes, usually with several layers of paper between each set of conductors
Plastic (such as polyester)—foil	Rolled tubes
Ceramic—sprayed silver	Plate or tube type

Electrolytic
   $Al-Al_2O_3$—liquid
   $Ta-TaO$—liquid	

◇ | **Example 18-8**

We want to make a simple parallel plate capacitor that can store 4×10^{-5} C at a potential of 10,000 V. The separation between the plates is to be 0.02 cm. Calculate the area of the plates required if the dielectric is (a) a vacuum, (b) polyethylene, (c) water, and (d) barium titanate.

Answer:

$$C = \frac{Q}{V} = \frac{4 \times 10^{-5}}{10,000} = 4 \times 10^{-9} \, \text{F}$$

$$A = \frac{Cd}{\kappa \varepsilon_0} = \frac{(4 \times 10^{-9})(0.02)}{(8.85 \times 10^{-14})\kappa} = \frac{900}{\kappa}$$

(a) For vacuum, $\kappa = 1$, $A = 900 \, \text{cm}^2$
(b) For polyethylene, $\kappa = 2.26$, $A = 398 \, \text{cm}^2$
(c) For water, $\kappa = 78.3$, $A = 11.5 \, \text{cm}^2$
(d) For barium titanate, $\kappa = 3000$, $A = 0.3 \, \text{cm}^2$

The benefit of a high dielectric constant on the size of capacitors is obvious.

◇ | **Example 18-9**

A mica capacitor $0.4 \, \text{in}^2$ and 0.0001 in. thick is to have a capacitance of $0.0252 \, \mu\text{F}$. (a) How many plates are needed? (b) What is the maximum allowable voltage?

Answer:

From Table 18-1, typical properties for mica include $\kappa = 7$ and a dielectric strength of 40×10^6 V/m.

(a) $C = \varepsilon_0 \kappa \dfrac{A}{d} (n - 1) = 0.0252 \times 10^{-6} \, \text{F}$

$$n - 1 = \frac{Cd}{\varepsilon_0 \kappa A} = \frac{(0.0252 \times 10^{-6})(0.0001 \, \text{in.})(0.0254 \, \text{m/in.})}{(8.85 \times 10^{-12})(7)(0.4 \, \text{in}^2)(0.0254 \, \text{m/in.})^2}$$

$$n - 1 = 4$$
$$n = 5$$

To be able to store the required charge, we need five conductor plates, with four layers of dielectric.

(b) $(40 \times 10^6 \, \text{V/m})(0.0254 \, \text{m/in.}) = 1.016 \times 10^6 \, \dfrac{V}{\text{in.}} = \dfrac{V}{d}$

$$V = (1.016 \times 10^6)(0.0001) = 101.6 \, \text{V}$$

18–5   DIELECTRIC PROPERTIES AND ELECTRICAL INSULATORS

Materials used to insulate an electric field from the surroundings must also be dielectric. Electrical insulators must possess a high electrical resistivity, a high dielectric strength, and a low loss factor. However, a high dielectric constant is not a necessary requirement for insulators and, in fact, may even be undesirable. Most polymer and ceramic materials, including glass, satisfy some or all of these requirements.

The high electrical resistivity, which results from the large energy gap between the valence and conduction bands, prevents current leakage.

A high dielectric strength prevents catastrophic breakdown of the insulator at high voltages. *Internal failure* of the insulator occurs if impurities provide donor or acceptor levels that permit electrons to be excited into the conduction band. *External failure* is caused by arcing along the surface of the insulator or through interconnected porosity within the insulator body. In particular, adsorbed moisture on the surface of ceramic insulators presents a problem. Glazes on ceramic insulators seal off porosity and reduce the effect of surface contaminants.

The small dielectric constant prevents polarization, so charge is not stored locally at the insulator. Low dielectric constants are desirable for insulators, but high constants are required for capacitors.

18–6   PIEZOELECTRICITY AND ELECTROSTRICTION

When an electric field is applied, polarization may change the dimensions of the material, an effect called *electrostriction*. This might occur as a result of atoms acting as egg-shaped particles rather than spheres, or the bonds between ions changing in length, or by distortion due to the orientation of the permanent dipoles in the material.

However, certain dielectric materials display a further property. When a dimensional change is imposed on the dielectric, polarization occurs and a voltage or field is created (Figure 18-8). Dielectric materials that display this reversible behavior are piezoelectric. Barium titanate, a solid solution of $PbZrO_3$–$PbTiO_3$, or PZT, and more complicated ceramics such as $(Pb, La)(Ti, Zr)O_3$, or PLTZ, are common materials that display this behavior. We can write the two reactions that occur in piezoelectrics as

$$\text{Field produced by stress } = \; \xi = g\sigma \tag{18-12}$$

$$\text{Strain produced by field } = \; \varepsilon = d\xi \tag{18-13}$$

where ξ is the electric field (V/m), σ is the applied stress (Pa), ε is the strain, and g and d are constants. Typical values for d are given in Table 18-3. The constant g is related to d through the modulus of elasticity E.

$$E \; = \; \frac{1}{gd} \tag{18-14}$$

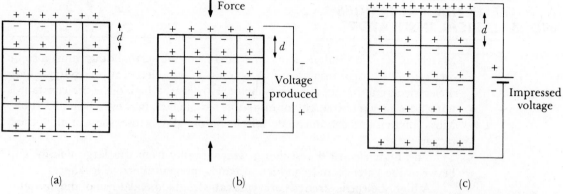

(a)   (b)   (c)

FIGURE 18-8   The piezoelectric effect. (a) Piezoelectric crystals have a charge difference due to permanent dipoles. (b) A compressive force reduces the distance between charge centers, changes the polarization, and introduces a voltage. (c) A voltage changes the distance between charge centers, causing a change in dimensions.

◇   **Example 18-10**

A 50-lb force is applied to a 0.1 in. × 0.1 in. wafer of barium titanate that is 0.01 in. thick. Calculate the strain produced by the force and the voltage that is created. The modulus of elasticity is 10×10^6 psi.

Answer:

First let's convert units.

$$\sigma = \frac{F}{A} = \frac{50\,\text{lb}}{(0.01)^2} = 5000\,\text{psi} \times 0.006895\,\text{Pa/psi} = 34.5\,\text{Pa}$$

$$E = (10 \times 10^6\,\text{psi})(0.006895\,\text{Pa/psi}) = 6.895 \times 10^4\,\text{Pa}$$

$$\varepsilon = \frac{\sigma}{E} = \frac{34.5}{6.895 \times 10^4} = 5 \times 10^{-4}\,\text{m/m}$$

From Table 18-3, $d = 100 \times 10^{-12}\,\text{m/V}$

$$\xi = \frac{\varepsilon}{d} = \frac{5 \times 10^{-4}}{100 \times 10^{-12}} = 5 \times 10^6\,\text{V/m}$$

$$V = (\xi)(\text{thickness}) = (5 \times 10^6\,\text{V/m})(0.01\,\text{in.})(0.0254\,\text{m/in.})$$

$$= 1270\,\text{V}$$

Materials which are permanently polarized display this effect. For example, we can measure a potential difference across a barium titanate crystal because the titanium ion is always shifted slightly from the body-centered position in the unit cell. The most common of these materials include quartz, barium titanate, lead titanate, lead zirconate, CdS, and ZnO.

TABLE 18-3 The piezoelectric
constant *d* for selected materials

Material	Piezoelectric Constant d $(C/Pa \cdot m^2 = m/V)$
Quartz	2.3×10^{-12}
$BaTiO_3$	100×10^{-12}
$PbZrTiO_6$	250×10^{-12}
$PbNb_2O_6$	80×10^{-12}

The piezoelectric effect is used in *transducers*, which convert acoustical waves (sound) into electric fields or electric fields into acoustical waves (Figure 18-9). Sound of a particular frequency produces a strain in a piezoelectric material. The dimensional changes polarize the crystal, creating an electric field. In turn, the electric field is transmitted to a second piezoelectric crystal; the electric field produces dimensional changes in the second crystal which produce an acoustical wave that is amplified. This description depicts the telephone. Similar electro-mechanical transducers are used for stereo record players and other audio devices. The piezoelectric effect is also employed in tuners for radios; by imposing the correct strain in the crystal, only the desired frequency is picked up and amplified.

18–7   FERROELECTRICITY

The presence of polarization in a material after the electric field is removed can be explained in terms of a residual alignment of permanent dipoles. Barium titanate is again an excellent example. Materials that retain a net polarization when the field is removed are called *ferroelectric*.

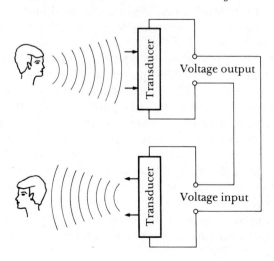

FIGURE 18-9 Schematic diagram showing the use of piezoelectric transducers for the telephone.

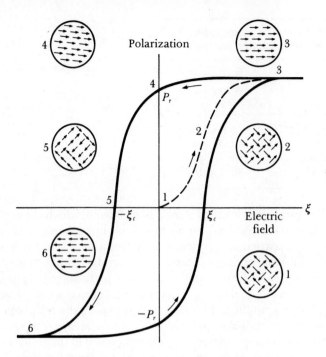

FIGURE 18-10 The ferroelectric hysteresis loop, showing the influence of the electric field on polarization and the alignment of the dipoles.

Many dielectric materials do not display this behavior, even though they have permanent dipoles. When the field is removed, the dipoles become randomly arranged, the polarization of each dipole is canceled by neighboring dipoles, and no net polarization results. However, in ferroelectric materials, the orientation of one dipole influences the surrounding dipoles to have an identical alignment. We can examine this behavior by describing the effect of an electric field on polarization (Figure 18-10).

Let's begin with a crystal whose dipoles are randomly oriented so there is no net polarization. When a field is applied, the dipoles begin to line up with the field (points 1 to 3 in Figure 18-10). Eventually the field aligns all of the dipoles and the maximum, or *saturation*, polarization P_s is obtained (point 3). When the field is subsequently removed, a *remanent* polarization P_r remains (point 4) due to the coupling between the dipoles. The material is permanently polarized. The ability to retain polarization permits the ferroelectric material to retain information, making the material useful in computer circuitry.

When a field is applied in the opposite direction, the dipoles must be reversed. A *coercive field* ξ_c must be applied to remove the polarization and randomize the dipoles (point 5). If the reverse field is increased further, saturation occurs with the opposite polarization (point 6). As the field continues to

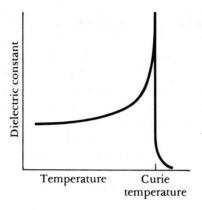

FIGURE 18-11   The effect of temperature on the dielectric constant of barium titanate. Above the Curie temperature, the molecular polarization is lost due to a change in crystal structure and barium titanate is no longer ferroelectric.

alternate, a *hysteresis loop* is described showing how the polarization of the ferroelectric varies with the field. The area contained within the hysteresis loop is related to the energy required to cause polarization to switch from one direction to the other.

The ferroelectric behavior depends on temperature. Above the critical *Curie temperature*, dielectric and consequently ferroelectric behavior are lost (Figure 18-11). In some materials, such as barium titanate, the Curie temperature corresponds to a change in crystal structure from the distorted tetragonal structure in Figure 18-2 to a normal cubic perovskite unit cell [Figure 14-2(a)]. Consequently, the permanent dipoles in each unit cell no longer exist. Curie temperatures of typical ferroelectrics are shown in Table 18-4.

Ferroelectric materials are always dielectric and piezoelectric and have a high dielectric constant which make them particularly suitable for use in capacitors and piezoelectric transducers.

TABLE 18-4   The Curie temperature for selected ferroelectrics

Material	Curie Temperature ($^\circ$C)
$SrTiO_3$	-200
$Cd_2Nb_2O_7$	-88
Rochelle salt	24
$BaTiO_3$	120
$PbZrO_3$	233
$PbTa_2O_6$	260
$KNbO_3$	435
$PbTiO_3$	490
$PbNb_2O_6$	570
$NaNbO_3$	640

18–8   MAGNETIC DIPOLES AND MAGNETIC MOMENTS

In dielectric materials, polarization occurs when induced or permanent electric dipoles are oriented by an interaction between the material and the electric field. In an analogous manner, *magnetization* occurs when induced or permanent magnetic dipoles are oriented by an interaction between the magnetic material and a magnetic field. Magnetization enhances the influence of the magnetic field, permitting larger magnetic energies to be stored than if the material were absent. This energy can be stored permanently or temporarily and can be used to do work.

Each electron in an atom has two magnetic moments. A *magnetic moment* is simply the strength of the magnetic field associated with the electron. This moment, called the *Bohr magneton*, is equal to

$$\text{Bohr magneton} = \frac{qh}{4\pi m_e} = 9.27 \times 10^{-24}\,\text{A} \cdot \text{m}^2 \qquad (18\text{-}15)$$

where q is the charge on the electron, h is Planck's constant, and m_e is the mass of the electron. The magnetic moments are due to the orbital motion of the electron around the nucleus and the spin of the electron about its own axis (Figure 18-12).

◇ Example 18-11

Magnetization is given in units of $\text{A} \cdot \text{m}^2/\text{m}^3$, A/m, or oersted. We will find later that each iron atom contains four electrons that act as magnetic dipoles. Calculate the maximum magnetization that we might expect in iron. The lattice parameter of BCC iron is 2.866 Å. (Note that this example is similar to Example 18-1, where we calculated polarization.)

Answer:

$$M = \left(\frac{2\ \text{atoms/cell}}{(2.866 \times 10^{-10}\,\text{m})^3}\right)\left(4\,\frac{\text{Bohr magnetons}}{\text{atom}}\right)$$

$$\times \left(9.27 \times 10^{-24}\,\frac{\text{A} \cdot \text{m}^2}{\text{magnetons}}\right)$$

$$= (0.085 \times 10^{30})(4)(9.27 \times 10^{-24}) = 3.15 \times 10^6\,\text{A/m}$$

or

$$M = (3.15 \times 10^6\,\text{A/m})(4\pi \times 10^{-3}\,\text{oersted/A/m}) = 3.96 \times 10^4\,\text{oersted}$$

When we discussed electronic structure and quantum numbers in Chapter 2, we pointed out that each discrete energy level could contain two electrons,

FIGURE 18-12 *Origin of magnetic dipoles. (a) The spin of the electron produces a magnetic field with a direction dependent on the quantum number m_s. (b) Electrons orbiting about the nucleus create a magnetic field about the atom.*

each having an opposite spin. The magnetic moments of each electron pair in an energy level are opposed and, consequently, whenever an energy level is completely full, there is no *net* magnetic moment.

Based on this reasoning, we expect any atom of an element with an odd atomic number to have a net magnetic moment from the unpaired electron. However, this is not the case. In most of these elements, the unpaired electron is a valence electron. Because the valence electrons from each atom interact, the magnetic moments, on average, cancel and no net magnetic moment is associated with the material.

However, certain elements, such as the transition metals, have an inner energy level that is not completely filled. The elements scandium through copper, whose electronic structures are shown in Table 18-5, are typical. Except for chromium and copper, the valence electrons in the $4s$ level are paired; the unpaired electrons in chromium and copper are canceled by interactions with other atoms. Copper also has a completely filled $3d$ shell and thus does not display a net moment.

The electrons in the $3d$ level of the remaining transition elements do not enter the shell in pairs. Instead, as in manganese, the first five electrons have the same spin. Only after half of the $3d$ level is filled do pairs with opposing spins form. Therefore, each atom in a transition metal has a permanent magnetic

TABLE 18-5 *The electron spins in the $3d$ energy level in transition metals, with arrows indicating the direction of spin*

Metal	3d					4s
Sc	↑					↑↓
Ti	↑	↑				↑↓
V	↑	↑	↑			↑↓
Cr	↑	↑	↑	↑	↑	↑
Mn	↑	↑	↑	↑	↑	↑↓
Fe	↑↓	↑	↑	↑	↑	↑↓
Co	↑↓	↑↓	↑	↑	↑	↑↓
Ni	↑↓	↑↓	↑↓	↑	↑	↑↓
Cu	↑↓	↑↓	↑↓	↑↓	↑↓	↑

moment, equal in strength to the number of unpaired electrons. Each atom behaves as a magnetic dipole.

The response of the atom to an applied magnetic field depends on how the magnetic dipoles represented by each atom react to the field. Most of the transition elements react in such a way that the sum of the individual atoms' magnetic moments is zero. However, the atoms in nickel, iron, and cobalt undergo an *exchange interaction*, whereby the orientation of the dipole in one atom influences the surrounding atoms to have the same dipole orientation, producing a desirable amplification of the effect of the magnetic field.

18–9 MAGNETIZATION, PERMEABILITY, AND THE MAGNETIC FIELD

Let's look at the relationship between the magnetic field and magnetization. Figure 18-13 depicts a coil having n turns; when an electric current is passed through the coil, a magnetic field H is produced, with the strength of the field given by

$$H = \frac{0.4\pi nI}{l} \tag{18-16}$$

where n is the number of turns, l is the length of the coil (m), and I is the current (A). The field is given in units of A/m, which can be converted to oersted by multiplying by $4\pi \times 10^{-3}$. When the magnetic field is applied in a vacuum,

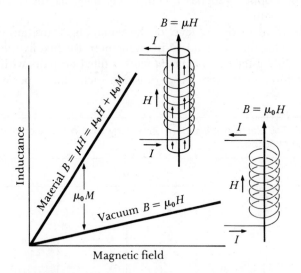

FIGURE 18-13 A current passing through a coil sets up a magnetic field H with a flux density B. The flux density is higher when a magnetic core is placed within the coil.

lines of magnetic flux are produced. A greater number of lines of flux increases the work that the magnetic field can accomplish. The flux density, or *inductance*, is related to the applied field by

$$B = \mu_0 H \qquad (18\text{-}17)$$

where B is the inductance (gauss), H is the magnetic field (oersted), and μ_0 is the *magnetic permeability* of a vacuum (gauss/oersted). The permeability in a vacuum is a constant, 1 gauss/oersted.

When we place a material within the magnetic field, the magnetic inductance is determined by the manner in which induced and permanent magnetic dipoles interact with the field. The magnetic inductance now is

$$B = \mu H \qquad (18\text{-}18)$$

where μ is the magnetic permeability of the material in the field. If the magnetic moments reinforce the applied field, $\mu > \mu_0$; but if the magnetic moments oppose the field, $\mu < \mu_0$.

We can describe the influence of the magnetic material by the *relative permeability* μ_r, where

$$\mu_r = \frac{\mu}{\mu_0} \qquad (18\text{-}19)$$

A large relative permeability means that the material has amplified the effect of the magnetic field. Thus, the relative permeability has the same importance that the dielectric constant, or relative permittivity, has in dielectrics.

The *magnetization M* represents the increase in the magnetic inductance due to the core material, so we can rewrite the equation for inductance as

$$B = \mu_0 H + \mu_0 M \approx \mu_0 M \qquad (18\text{-}20)$$

The *magnetic susceptibility* χ, which is the ratio between magnetization and the applied field, gives the amplification produced by the material.

$$\chi = \frac{M}{H} \qquad (18\text{-}21)$$

Because the term $\mu_0 M$ is often much greater than $\mu_0 H$, we can frequently equate $B = \mu_0 M$. We sometimes interchangeably refer to either inductance or magnetization. Normally, we are interested in producing a high inductance B or magnetization M. This is accomplished by selecting materials that have a high relative permeability or magnetic susceptibility.

◇ | **Example 18-12**

By combining Equations 18-18 and 18-20, derive the relationship between magnetic susceptibility and relative permeability.

Answer:

$$B = \mu H = \mu_0 H + \mu_0 M$$

$$\frac{\mu H}{\mu_0} = H + M$$

$$\mu_r H = H + M$$

$$\mu_r H - H = M$$

$$(\mu_r - 1)H = M$$

$$\mu_r - 1 = \frac{M}{H}$$

$$\mu_r = 1 + \chi$$

   Example 18-13

A magnetic field of 30 oersted is applied to a material with a relative permeability of 5000. Calculate the magnetization and inductance.

Answer:

$$\mu_r = 1 + \chi = 1 + \frac{M}{H} = 5000$$

$$\frac{M}{H} = 5000 - 1 = 4999$$

$$M = 4999H = 4999(30) = 149{,}970 \text{ gauss}$$

$$\mu = \mu_r \mu_0 = (5000)(1) = 5000 \text{ gauss/oersted}$$

$$B = \mu H = 5000(30) = 150{,}000 \text{ gauss}$$

18–10   INTERACTIONS BETWEEN MAGNETIC DIPOLES AND THE MAGNETIC FIELD

When a magnetic field is applied to a collection of atoms, several types of behavior may be observed (Figure 18-14).

Diamagnetic Behavior.   A magnetic field acting on any atom induces a magnetic dipole for the entire atom by influencing the magnetic moment due to the orbiting electrons. These dipoles oppose the magnetic field, causing the magnetization to be less than zero. This behavior, called *diamagnetism*, gives a

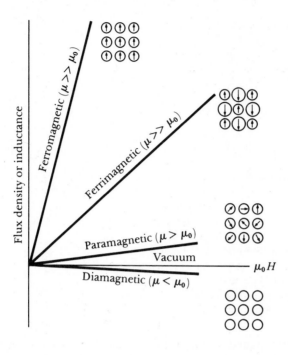

FIGURE 18-14 The effect of the core material on the flux density. No magnetic moment is produced in diamagnetic materials. Progressively stronger moments are present in paramagnetic, ferrimagnetic, and ferromagnetic materials for the same applied field.

relative permeability of about 0.99995. Diamagnetic behavior has no important applications for magnetic materials or devices and, in important magnetic materials, is overwhelmed by one of the following effects.

Paramagnetism.   When materials have unpaired electrons, a net magnetic moment due to electron spin is associated with each atom. When a magnetic field is applied, the dipoles line up with the field, causing a positive magnetization. However, because the dipoles do not interact, extremely large magnetic fields are required to align all of the dipoles. In addition, the effect is lost as soon as the magnetic field is removed. This effect, called *paramagnetism*, is also of little importance. The relative permeability is less than about 1.01.

Ferromagnetism.   Ferromagnetic behavior is due to the unfilled energy levels in the $3d$ level of iron, nickel, and cobalt. In ferromagnetic materials, the permanent unpaired dipoles easily line up with the imposed magnetic field due to the exchange interaction, or mutual reinforcement of the dipoles. Large magnetizations are obtained even for small magnetic fields, giving relative permeabilities as high as 10^6.

Antiferromagnetism.   In materials such as manganese, chromium, MnO, and NiO, the magnetic moments produced in neighboring dipoles line up in opposition to one another in the magnetic field, even though the strength of each dipole is very high. This effect is illustrated for MnO in Figure 18-15. These materials are *antiferromagnetic* and have zero magnetization. The difference

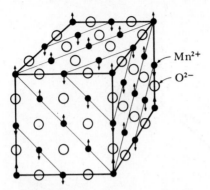

FIGURE 18-15 The crystal structure of MnO consists of alternating layers of {1 1 1} type planes of oxygen and manganese ions. The magnetic moments of the manganese ions in every other (1 1 1) plane are oppositely aligned. Consequently, MnO is antiferromagnetic.

between ferromagnetism and antiferromagnetism is in the interactions between neighboring dipoles—whether they reinforce or oppose one another.

Ferrimagnetism.   In ceramic materials, different ions have different magnetic moments. In a magnetic field, the dipoles of ion A may line up with the field while dipoles of ion B oppose the field. But because the strengths of the dipoles are not equal, a net magnetization results. The *ferrimagnetic* materials can provide good amplification of the imposed field. We will look at a group of ceramics called ferrites, which display this behavior, in a later section.

18–11   DOMAIN STRUCTURE AND THE HYSTERESIS LOOP

Ferromagnetic materials have their powerful influence on magnetization because of the positive interaction between the dipoles of neighboring atoms. Within the grain structure of a ferromagnetic material, a substructure composed of magnetic domains is produced, even in the absence of an external field. *Domains* are regions in the material in which all of the dipoles are aligned. In a material that has never been exposed to a magnetic field, the individual domains have a random orientation. The net magnetization in the material as a whole is zero.

Boundaries, called *Bloch walls*, separate the individual domains, much like grain boundaries. The Bloch walls are narrow zones in which the direction of the magnetic moment gradually and continuously changes from that of one domain to that of the next (Figure 18-16). The domains are typically very small, about 0.005 cm or less, while the Bloch walls are about 1000 Å thick.

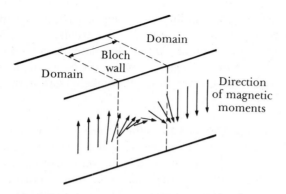

FIGURE 18-16 The magnetic moments in adjoining atoms change direction continuously across the boundary between domains.

Movement of Domains in a Magnetic Field. When a magnetic field is imposed on the material, domains that are nearly lined up with the field grow at the expense of unaligned domains. In order for the domains to grow, the Bloch walls must move; the field provides the force required for this movement. Initially the domains may grow with difficulty and relatively large increases in the field may be required to produce even a little magnetization; this is indicated in Figure 18-17 by a shallow slope, which is the initial permeability of the material. As the field increases in strength, favorably oriented domains grow

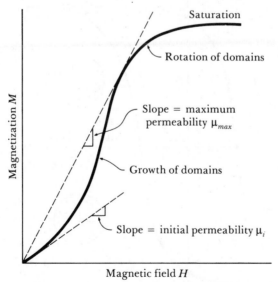

FIGURE 18-17 When a magnetic field is first applied to a magnetic material, magnetization initially increases slowly, then more rapidly as the domains begin to grow. Later, magnetization slows, as domains must rotate eventually to reach saturation.

more easily, with the permeability increasing as well. A maximum permeability can be calculated, as shown in the figure. Eventually, the unfavorably oriented domains disappear and domain rotation completes the alignment of the domains with the field. The *saturation magnetization*, produced when all of the domains are properly oriented, is the greatest amount of magnetization that the material can obtain.

Effect of Removing the Field. When the field is removed, the resistance offered by the domain walls prevents regrowth of the domains into random orientations. As a result, many of the domains remain oriented near the direction of the original field and a residual magnetization, known as the *remanence*, is present in the material. The material acts as a permanent magnet. Figure 18-18 shows this effect in the magnetization-field curve.

Effect of an Alternating Field. If we now apply a field in the reverse direction, the domains grow with an alignment in the opposite direction. A *coercive*

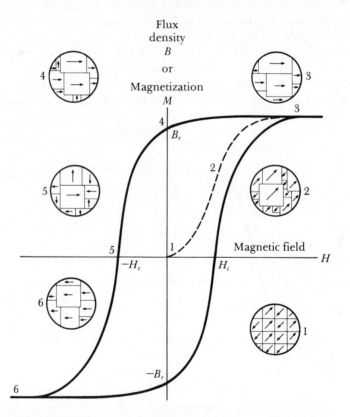

FIGURE 18-18 The ferromagnetic hysteresis loop, showing the effect of the magnetic field on inductance or magnetization. The dipole alignment leads to saturation magnetization (point 3), a remanence (point 4), and a coercive field (point 5).

field H_c is required to force the domains to be randomly oriented and cancel one another's effect. Further increases in the strength of the field eventually align the domains to saturation in the opposite direction.

As the field continually alternates, the magnetization versus field relationship traces out a *hysteresis loop*. The hysteresis loop describes the strength and direction of the magnetization in an alternating magnetic field.

Magnetostriction.   A less obvious effect that occurs when the field is applied is that the material changes dimensions, either expanding or contracting in the direction of the field. This phenomenon, similar to electrostriction in dielectrics, is related to attraction or repulsion between the dipoles. Unfortunately, this behavior does not lend itself as readily to producing transducers, which is a common application for electrostriction.

18–12 APPLICATION OF THE MAGNETIZATION-FIELD CURVE

The behavior of a material in a magnetic field is related to the size and shape of the hysteresis loop (Figure 18-19). Let's look at three applications for magnetic materials.

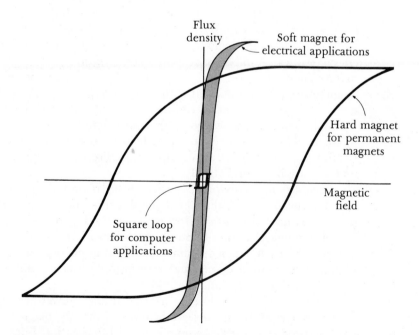

FIGURE 18-19 *Comparison of the hysteresis loops for three applications of ferromagnetic materials—electrical applications, computer applications, and permanent magnets.*

TABLE 18-6 Properties of selected soft or electrical magnetic materials

Material	Maximum Relative Permeability	Saturation Magnetization (gauss)	Coercive Field (oersted)
99.91% pure iron	5,000	21,500	1
99.95% pure iron	180,000	21,500	0.05
Fe-3% Si (oriented)	30,000	20,000	0.15
Fe-4% Si (not oriented)	7,000	19,700	0.5
4S Permalloy (54.7% Fe, 45% Ni, 0.3% Mn)	25,000	16,000	0.3
Supermalloy (79% Ni, 15.7% Fe, 5% Mo, 0.3% Mn)	800,000	8,000	2
A6 Ferroxcube (Mn, Zn) Fe_2O_4		4,000	
B2 Ferroxcube (Ni, Zn) Fe_2O_4		3,000	

From *Handbook of Chemistry and Physics*, 56th Ed., CRC Press, 1975, and other sources.

Magnetic Materials for Electrical Applications. Ferromagnetic materials are used to enhance the magnetic field produced when an electric current is passed through the material. The magnetic field is then expected to do work. Applications include cores for electromagnets, electric motors, transformers, generators, and other electrical equipment. These devices utilize an alternating field, so that the core material is continually cycled through the hysteresis loop.

Electrical magnetic materials, often called *soft* magnets, should have several characteristics. A high saturation magnetization is desired, permitting the material to do the most work, while a high permeability permits this saturation magnetization to be obtained with small imposed fields. A small coercive force, indicating that the domains can be reoriented with small fields, is also desired. A small remanence is desired so that no magnetism remains when the field is removed. These characteristics also lead to a small hysteresis loop, therefore minimizing energy losses during operation. Properties for several important soft magnets are listed in Table 18-6.

In addition, the frequency at which the material operates is important. If the frequency is so high that the domains cannot be realigned in each cycle, the device may heat, just as in dielectric materials, due to *dipole friction*. In addition, higher frequencies naturally produce more heating because the material cycles through the hysteresis loop more often, losing energy during each cycle. For high-frequency applications, materials must be selected that permit the dipoles to be aligned at exceptionally rapid rates.

Energy can also be lost by heating if *eddy currents* are produced. During operation, electrical currents can be induced into the magnetic material. These currents produce power losses and joule, or I^2R, heating. Eddy current losses are particularly severe when the material operates at high frequencies. If the electrical resistivity is high, eddy current losses can be held to a minimum. Soft magnets produced from ceramic materials have a high resistivity and therefore are less likely to heat than metallic magnets.

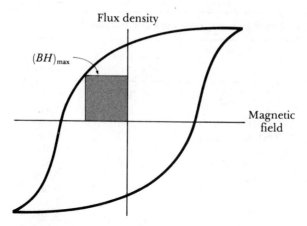

FIGURE 18-20 *The largest rectangle drawn in the second or fourth quadrant of the B-H curve gives the maximum BH product. $(BH)_{max}$ is related to the power, or energy required to demagnetize the permanent magnet.*

Magnetic Materials for Computer Memories.   Magnetic materials are used to store bits of information in computers. Memory is stored by magnetizing the material in a certain direction. For example, if the north pole is up, the bit of information stored is 1. If the north pole is down, then a 0 is stored.

For this application, materials with a square hysteresis loop, a low remanence, a low saturation magnetization, and a low coercive field are preferable. Ferrites containing manganese, magnesium, or cobalt may satisfy these requirements. The square loop assures that a bit of information placed in the material by a field remains stored; a steep and abrupt change in magnetization is required to remove the information from storage in the ferromagnet. Furthermore, the magnetization is produced by small external fields, so the coercive field, saturation magnetization, and remanence should be low.

Magnetic Materials for Permanent Magnets.   Finally, magnetic materials are used to make strong permanent magnets (Table 18-7). Permanent magnets require high remanence, high permeability, high coercive fields, and high power.

The *power* of the magnet is related to the size of the hysteresis loop, or the maximum product of B and H. The area of the largest rectangle that can be drawn in the second or fourth quadrants of the B-H curve is related to the energy required to demagnetize the magnet (Figure 18-20). For the product to be large, both the remanence and the coercive field should be large.

Development of strong permanent magnets, often said to be magnetically *hard*, is aimed at improving both the magnetic permeability and the stability of the domains. We will see in the following sections how this is achieved.

TABLE 18-7   Selected properties of hard or permanent magnetic materials

Material	Remanence (gauss)	Coercive field (oersted)	$(BH)_{max.}$ (gauss · oersted)
Steel (0.9% C, 1.0% Mn)	10,000	50	200,000
Alnico I (21% Ni, 12% Al, 5% Co, bal Fe)	7,100	440	1,400,000
Alnico V (24% Co, 14% Ni, 8% Al, 3% Cu, bal Fe)	13,100	640	6,000,000
Alnico XII (35% Co, 18% Ni, 8% Ti, 6% Al, bal Fe)	5,800	950	1,600,000
Cunico (50% Cu, 29% Co, 21% Ni)	3,400	660	800,000
Cunife (60% Cu, 20% Fe, 20% Ni)	5,400	550	1,500,000
Silmanal (86.6% Ag, 8.8% Mn, 4.4% Al)	550	6,000	75,000
Co_5Sm	9,500	9,500	25,000,000
$BaO \cdot 6Fe_2O_3$	4,000	2,400	2,500,000
$SrO \cdot 6Fe_2O_3$	3,400	3,300	—
Neodymium-iron-boron	—	—	45,000,000

From *Handbook of Chemistry and Physics*, 56th Ed., CRC Press, 1975, and other sources.

◇   **Example 18-14**

Determine the power, or *BH* product, for the magnetic material whose properties are shown in Figure 18-21.

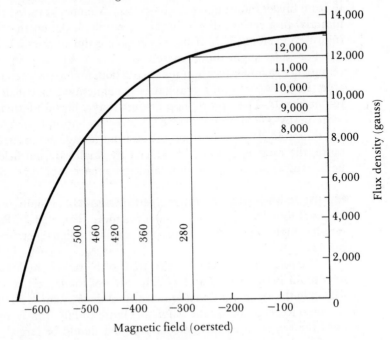

FIGURE 18-21   The fourth quadrant of the *B–H* curve for a permanent magnetic material. (See Example 18-14.)

Answer:

Several rectangles have been drawn in the fourth quadrant of the *B-H* curve. The *BH* product in each is

$$BH_1 = (12,000)(280) = 3.4 \times 10^6 \text{ gauss} \cdot \text{oersted}$$
$$BH_2 = (11,000)(360) = 4.0 \times 10^6 \text{ gauss} \cdot \text{oersted}$$
$$BH_3 = (10,000)(420) = 4.2 \times 10^6 \text{ gauss} \cdot \text{oersted}$$
$$BH_4 = (9,000)(460) = 4.1 \times 10^6 \text{ gauss} \cdot \text{oersted}$$
$$BH_5 = (8,000)(500) = 4.0 \times 10^6 \text{ gauss} \cdot \text{oersted}$$

Thus, the power is about 4.2×10^6 gauss · oersted.

18–13 THE CURIE TEMPERATURE

When the temperature of a ferromagnetic material is increased, the added thermal energy increases the mobility of the domains, making it easier for them to become aligned but also preventing them from remaining aligned when the field is removed. Consequently, saturation magnetization, remanence, and the coercive field are all reduced at high temperatures (Figure 18-22). If the temperature exceeds the *Curie temperature*, ferromagnetic behavior is no longer observed. The Curie temperature (Table 18-8) depends on the type of magnetic material and can be changed by alloying elements.

The dipoles can still be aligned in a magnetic field above the Curie temperature, but the dipoles become randomly aligned when the field is removed. Above the Curie temperature, the material displays paramagnetic behavior.

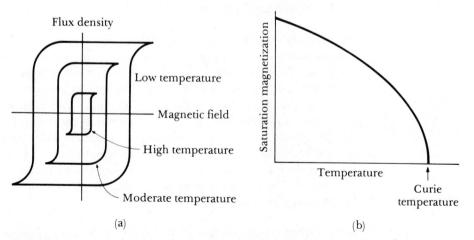

FIGURE 18-22 *The effect of temperature on (a) the hysteresis loop and (b) the saturation remanence. Ferromagnetic behavior disappears above the Curie temperature.*

TABLE 18-8 Curie temperatures for selected magnetic materials

Material	Curie Temperature (°C)
Gd	16
Ni	358
Fe	770
Co	1131

From *Handbook of Chemistry and Physics*, 56th Ed., CRC Press, 1975.

18–14 MAGNETIC MATERIALS

Let's look at typical metallic alloys and ceramic materials used in magnetic applications and discuss how their properties and behavior can be enhanced.

Magnetic Metals. Pure iron, nickel, and cobalt are not usually used for electrical applications; they have high electrical conductivities and relatively large hysteresis loops, leading to excessive power losses. On the other hand, they are relatively poor permanent magnets; the domains are easily reoriented and both the remanence and the *BH* product are small compared with those of more complex alloys. Some improvement in the magnetic properties is gained by introducing defects into the structure. Dislocations, grain boundaries, boundaries between multiple phases, and point defects may help to pin the domain boundaries, therefore keeping the domains aligned when the original magnetizing field is removed. However, complex alloys are more suitable for producing powerful permanent magnets.

Iron-Nickel Alloys. Some iron-nickel alloys, such as Permalloy, have high permeabilities, making them useful as soft magnets. One example of an application for these magnets is the "head" that stores or reads information on a computer disk (Figure 18-23). As the disk rotates beneath the head, a current produces a magnetic field in the head. The magnetic field in the head in turn magnetizes a portion of the disk. The direction of the field produced in the head determines the orientation of the magnetic particles embedded in the disk and consequently stores information. The information can be retrieved by again spinning the disk beneath the head. The magnetized region in the disk induces a current in the head; the direction of the current depends on the direction of the magnetic field in the disk.

Silicon Iron. Introduction of 3% to 5% Si into iron produces an alloy that, after proper processing, is useful in electrical applications such as motors and generators.

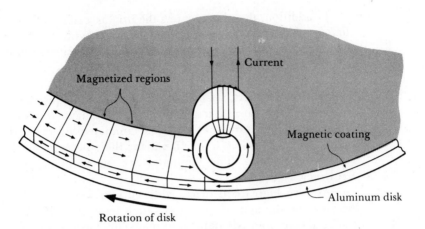

Rotation of disk

FIGURE 18-23   Information can be stored or retrieved from a magnetic disk by use of an electromagnetic head. A current in the head magnetizes domains in the disk during storage; the domains in the disk induce a current in the head during retrieval.

We take advantage of the anisotropic magnetic behavior of silicon iron to obtain the best performance. Unusually small hysteresis loops and coercive fields are obtained when the crystal structure of the silicon iron is lined up with the field in the most easily magnetized direction (Figure 18-24). As a result of rolling and subsequent annealing, a sheet texture is formed in which the $\langle 100 \rangle$ directions in each grain are aligned. Because the silicon iron is most easily magnetized in $\langle 100 \rangle$ directions, the field required to give saturation magnetization is very small and both a small hysteresis loop and a small remanence are observed.

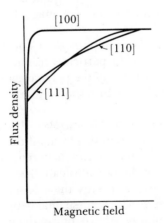

FIGURE 18-24   The initial magnetization curve for silicon iron is highly anisotropic; magnetization is easiest when the $\langle 100 \rangle$ directions are aligned with the field.

In addition, the silicon enters the iron as a solid solution alloying element, therefore reducing the electrical conductivity. While the low conductivity helps to reduce eddy current losses, the silicon also reduces the saturation magnetization and the Curie temperature.

Composite Magnets.   Composite material techniques have been used to further reduce eddy current losses. Thin sheets of silicon iron are laminated around a layer of an insulating material. The laminated layers are then built up to the desired overall thickness. The laminant thereby increases the resistivity of the composite magnets. The laminates are successful at low and intermediate frequencies.

At very high frequencies, eddy current losses are even more significant because the domains do not have time to realign. In this case, a composite material containing domain-sized magnetic particles in a polymer matrix may be used. The particles, or domains, rotate easily in the soft polymer, while eddy current losses are minimized because of the high resistivity of the polymer.

Metallic Glasses.   Amorphous metallic glasses, often complex iron-boron alloys, are produced by employing extraordinarily high cooling rates during solidification (rapid solidification processing). The metallic glasses can be produced in the form of thin tapes, which can be stacked together to produce larger materials. These materials behave as soft magnets with a high magnetic permeability; the absence of grain boundaries may permit easy movement of the domains, while a high electrical resistivity minimizes eddy current losses.

Magnetic Tape.   Magnetic materials for information storage must have a square loop and a low coercive field, permitting very rapid transmission of information. Magnetic tape for audio or video applications is produced by evaporating, sputtering, or plating particles of a magnetic material such as Fe_2O_3 onto a polymer tape.

Both floppy disks and hard disks for computer data storage are produced in a similar manner. In a hard disk, magnetic particles are embedded in a polymer film on a flat aluminum substrate. Because of the polymer matrix and the small particles, the domains can rotate quickly in response to a magnetic field.

Complex Metallic Alloys for Permanent Magnets.   Improved permanent magnets are produced by making the grain size so small that only one domain is present in each grain. Now the boundaries between domains are grain boundaries rather than Bloch walls; the domains can change their orientation only by rotating, which requires greater energy than domain growth.

Two techniques are used to produce these magnetic materials—phase transformations and powder metallurgy. Alnico, one of the most common of the complex metallic alloys, has a single-phase BCC structure at high temperatures. But when Alnico slowly cools below 800°C, a second BCC phase rich in iron and cobalt precipitates. This second phase is so fine that each precipitate particle is a single domain, producing a very high remanence, coercive field, and power.

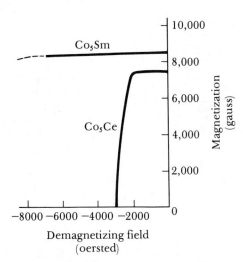

FIGURE 18-25 Demagnetizing curves for Co_5Sm and Co_5Ce, representing a portion of the hysteresis loop.

Often the alloys are permitted to cool and transform when in a magnetic field to align the domains as they form.

A second technique—powder metallurgy—is used for a group of rare earth metal alloys, including samarium-cobalt. A composition giving Co_5Sm, an intermetallic compound, has excellent magnetic properties (Figure 18-25) due to unpaired magnetic spins in the $4f$ electrons of samarium. The brittle intermetallic is crushed and ground to produce a fine powder in which each particle is a domain. The powder is then compacted when in an imposed magnetic field to align the powder domains. Careful sintering to avoid growth of the particles produces a solid powder metallurgy magnet. Another rare earth magnet based on neodymium, iron, and boron has a BH product of 45×10^6 gauss · oersted.

Ferrimagnetic Ceramic Materials.   Common magnetic ceramics are the ferrites, which have a spinel crystal structure (Figure 18-26). Each metallic ion in the crystal structure behaves as a dipole. Although the dipole moments of each type of ion may oppose one another, the strengths of the dipoles are different, a net magnetization develops, and ferrimagnetic behavior is observed.

We can understand the behavior of these ceramic magnets by looking at magnetite, Fe_3O_4. Magnetite contains two different iron ions, Fe^{2+} and Fe^{3+}, so we could rewrite the formula for magnetite as $Fe^{2+} Fe_2^{3+} O_4^{2-}$. The magnetite, or spinel, crystal structure is based on an FCC arrangement of oxygen ions, with iron ions occupying selected interstitial sites. Although the spinel unit cell actually contains eight of the FCC arrangements, we need only examine one of the FCC subcells.

1. The oxygen ions are in the FCC positions of the subcell; thus there are a total of four oxygen ions.

2. Octahedral sites, which are surrounded by six oxygen ions, are present at each edge and the center of the subcell. The sites at the 12 edges are each shared by four subcells; thus the number of octahedral sites belonging uniquely

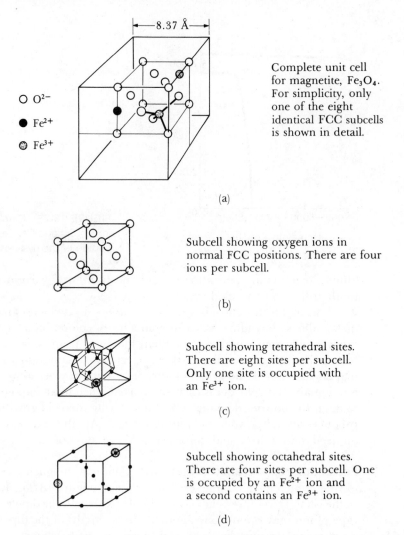

FIGURE 18-26 (a) The structure of magnetite, Fe_3O_4. Details of the complex arrangement of the Fe^{2+}, Fe^{3+}, and O^{2-} ions are shown in (b), (c), and (d).

to each subcell is 12 edges/4 subcells plus one center site, giving a total of four sites. One Fe^{2+} and one Fe^{3+} ion occupy octahedral sites.

3. Tetrahedral sites, surrounded by four oxygen ions, have indices in the subcell such as $\frac{1}{4}, \frac{1}{4}, \frac{1}{4}$ and represent the centers of oxygen tetrahedra. There are eight tetrahedral sites in the subcell. One Fe^{3+} ion occupies one of the tetrahedral sites.

4. When Fe^{2+} ions form, the two $4s$ electrons of iron are removed but all of the $3d$ electrons remain. Because there are four unpaired electrons in the $3d$ level of iron, the magnetic strength of the Fe^{2+} dipole is four Bohr magnetons. However, when Fe^{3+} forms, both $4s$ electrons and one of the $3d$ electrons are removed. The Fe^{3+} ion has five unpaired electrons in the $3d$ level and thus has a strength of five Bohr magnetons.

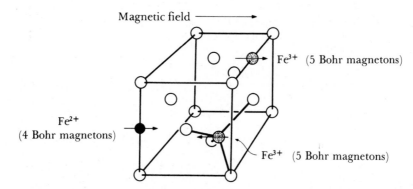

FIGURE 18-27 *The subcell of magnetite. The ions in the octahedral sites line up with the magnetic field, but ions in tetrahedral sites oppose the field. A net magnetic moment is produced by this ionic arrangement.*

5. The ions in the tetrahedral sites of the magnetite line up so that their magnetic moments oppose the applied magnetic field, but the ions in the octahedral sites reinforce the field (Figure 18-27). Consequently, the Fe^{3+} ion in the tetrahedral site neutralizes the Fe^{3+} ion in the octahedral site—the Fe^{3+} ions have antiferromagnetic behavior. However, the Fe^{2+} ion in the octahedral site is not opposed by any other ion and therefore reinforces the magnetic field. In a reversing magnetic field, magnetite displays a hysteresis loop, just as in ferromagnetic materials.

◇ | **Example 18-15**

Calculate the total magnetic moment per cubic centimeter in magnetite.

Answer:

In the subcell (Figure 18-27), the total magnetic moment is four Bohr magnetons, obtained from the Fe^{2+} ion, since the magnetic moments from the two Fe^{3+} ions are canceled by each other.

In the unit cell overall, there are eight subcells, so the total magnetic moment is 32 Bohr magnetons per cell.

The size of the unit cell, with a lattice parameter of 8.37×10^{-8} cm, is

$$V_{cell} = (8.37 \times 10^{-8})^3 = 5.86 \times 10^{-22} \, cm^3$$

The magnetic moment per cubic centimeter is

$$\frac{Total}{moment} = \frac{32 \text{ Bohr magnetons/cell}}{5.86 \times 10^{-22} \, cm^3/cell} = 5.46 \times 10^{22} \text{ magnetons/cm}^3$$

$$= (5.46 \times 10^{22})(9.27 \times 10^{-24} \, A \cdot m^2/\text{magneton})$$

$$= 0.51 \, A \cdot m^2/cm^3 = 5.1 \times 10^5 \, A/m$$

This represents the magnetization M at saturation.

TABLE 18-9 *Magnetic moments for ions in the spinel structure*

Ion	Bohr Magnetons
Fe^{3+}	5
Mn^{2+}	5
Fe^{2+}	4
Co^{2+}	3
Ni^{2+}	2
Cu^{2+}	1
Zn^{2+}	0

When ions are substituted for Fe^{2+} ions in the spinel structure, the magnetic behavior may be changed. Ions that may not produce ferromagnetism in a pure metal may contribute to ferrimagnetism in the spinels, as shown by the magnetic moments in Table 18-9. Soft electrical magnets are obtained when the Fe^{2+} ion is replaced by various mixtures of manganese, zinc, nickel, and copper. The nickel and manganese ions have magnetic moments that partly cancel the effect of the two iron ions, but a net ferrimagnetic behavior, with a small hysteresis loop, is obtained. The high electrical resistivity of these ceramic compounds helps minimize eddy currents and permits the materials to operate at high frequencies. Ferrites used in computer applications may contain additions of manganese, magnesium, or cobalt to produce a square loop hysteresis behavior.

Another group of soft ceramic magnets is based on garnets, which include yttria iron garnet, $Y_3Fe_5O_{12}$ (YIG). These complex oxides, which may be modified by substituting aluminum or chromium for iron or by replacing yttrium with lanthanum or praesydium, behave much like the ferrites. Another garnet, based on gadolinium and gallium, can be produced in the form of a thin film; tiny magnetic domains can be produced in the garnet film. These domains, or magnetic bubbles, can then serve as storage units for computers; once magnetized, the domains do not lose their memory in case of a sudden power loss.

Hard ceramic magnets selected as permanent magnets include another complex oxide family. Typical examples of these materials include $SrO \cdot 6Fe_2O_3$, $BaO \cdot 6Fe_2O_3$, and $PbO \cdot 6Fe_2O_3$.

SUMMARY

Dielectric and magnetic properties are related to the ability of a material to store an electric charge or to retain magnetization. In either case, the behavior of the material is enhanced when it contains permanent dipoles that can be controlled. The permanent dipoles can be influenced by temperature, microstructure, and processing to produce a wide range of beneficial properties.

GLOSSARY

Antiferromagnetism. Opposition of adjacent magnetic dipoles, causing zero net magnetization.

Bloch walls. The boundaries between magnetic domains.

Bohr magneton. The strength of a magnetic moment.

Capacitor. An electrical device, constructed from alternating layers of a dielectric and a conductor, which is capable of storing a charge.

Coercive field. The strength of the electric or magnetic field required to eliminate polarization or magnetization from a material.

Curie temperature. The temperature above which ferroelectric or ferromagnetic behavior is lost.

Diamagnetism. The effect caused by the magnetic moment due to the orbiting electrons, which produces a slight opposition to the imposed magnetic field.

Dielectric constant. The ratio of the permittivity of a material to the permittivity of a vacuum, thus describing the relative ability of a material to polarize and store a charge.

Dielectric loss. The fraction of energy lost each time an electric field in a material is reversed.

Dielectric strength. The maximum electric field that can be maintained between two conductor plates.

Dipole friction. Loss in energy when a cyclic electric field is applied to a dielectric material due to the inability of the dipoles to realign rapidly.

Dipoles. Atoms or groups of atoms having an unbalanced charge or moment.

Domains. Small regions within a material in which all of the dipoles are aligned.

Eddy currents. Currents induced into a material by the imposition of an electric field.

Electronic polarization. Polarization of an atom when the electrons are displaced to one side of the atom.

Electrostriction. The dimensional change that occurs in a material when an electric field is acting on it.

Ferrimagnetism. Magnetic behavior obtained when two types of dipoles, having different strengths, oppose one another but a net magnetization remains.

Ferroelectricity. Alignment of domains so that a net polarization remains after the electric field is removed.

Ferromagnetism. Alignment of domains so that a net magnetization remains after the magnetic field is removed.

Hard magnet. Ferromagnetic material that has a large hysteresis loop and remanence.

Hysteresis loop. The loop traced out by the polarization or magnetization as the electric or magnetic field is cycled.

Inductance. The flux density produced by a magnetic field.

Ionic polarization. Polarization of an ionic material by the relative displacement of the anions and cations.

Magnetic moment. The strength of the magnetic field associated with an electron.

Magnetic permeability. The ratio between the magnetic field and the inductance or magnetization.

Magnetic susceptibility. The ratio between magnetization and the applied field.

Magnetization. The sum of all of the magnetic moments per unit volume.

Molecular polarization. Polarization caused by the asymmetrical nature of certain molecules or crystal structures.

Paramagnetism. The net magnetic moment caused by alignment of the electron spins when a magnetic field is applied.

Permittivity. The ability of the material to

polarize and store a charge within the material.

Piezoelectricity. The ability in some materials for a change in electric field to change the dimensions of the material, while a change in dimensions produces an electric field.

Polarization. Alignment of dipoles so that a charge can be permanently stored.

Power. The strength of a permanent magnet as expressed by the maximum product of the inductance and magnetic field.

Remanence. The polarization or magnetization that remains in a material after

it has been removed from the field due to permanent alignment of the dipoles.

Saturation. When all of the dipoles have been aligned by the field, producing the maximum polarization or magnetization.

Soft magnet. Ferromagnetic material that has a small hysteresis loop and little energy loss in an alternating field.

Space charge. An electrical charge, which can move in the presence of an electric field, that develops on surfaces or at interfaces within a material, thus contributing to polarization.

PRACTICE PROBLEMS

1. The electronic polarization in a tungsten crystal is determined to be $4 \times 10^{-7} \, C/m^2$. Determine the average displacement of the electrons from the nucleus.

2. The electronic polarization in a platinum crystal is determined to be $7 \times 10^{-8} \, C/m^2$. Determine the average displacement of the electrons from the nucleus.

3. Calculate the polarization when an electric field causes an average displacement of the electrons from the nucleus in a gold crystal of $5 \times 10^{-8} \, \text{Å}$.

4. Calculate the polarization when an electric field causes an average displacement of the electrons from the nucleus in a silicon crystal of $3 \times 10^{-16} \, cm$.

5. Calculate the ionic polarization expected for KCl if an electric field causes a displacement of $6 \times 10^{-8} \, \text{Å}$. KCl has the cesium chloride structure.

6. Calculate the ionic polarization expected for CaO if an electric field causes a displacement of $4 \times 10^{-16} \, cm$. CaO has the sodium chloride structure.

7. Calculate the displacement between the ions in ZnS if the ionic polarization is $6 \times 10^{-8} \, C/m^2$. ZnS has the zinc blende crystal structure.

8. A 1-mm thick alumina (Al_2O_3) dielectric is used in a 60-Hz circuit. Determine the voltage required to produce polarization of $6 \times 10^{-8} \, C/m^2$. Will this voltage cause the dielectric to break down?

9. A diamond cube $2 \, mm \times 2 \, mm \times 2 \, mm$ is introduced as a dielectric into a circuit. The total charge Q on one face of the crystal is $3 \times 10^{-5} \, C$. Calculate the voltage acting on the crystal. The dielectric constant of diamond is 5.5.

10. Calcium fluoride has the fluorite crystal structure with a lattice parameter of $5.43 \, \text{Å}$ and a dielectric constant of 7.36. Calculate the thickness of the crystal if 50 V causes a displacement of $2 \times 10^{-8} \, \text{Å}$ between the calcium and fluoride ions.

11. KCl has the cesium chloride structure. When 12 V are applied to a 0.1-cm thick crystal, a displacement of $1.186 \times 10^{-6} \, \text{Å}$ is observed between the ions. Calculate the dielectric constant for KCl.

12. Suppose you would like to select a dielectric for use in a 60-Hz circuit. Calculate the ratio of Teflon thickness to Al_2O_3 thickness required to obtain the same polarization and charge.

13. Calculate the ratio of polyethylene thickness to phenolic thickness if each dielectric is to store the maximum charge in a 12,000-V circuit without breakdown.

14. A $BaTiO_3$ dielectric 4 mm thick is used in a 24,000-V circuit. The dielectric breaks down and a current flows through the barium titanate. (a) Is this breakdown expected? (b) If not, explain why breakdown may have occurred.

15. Suppose polyvinyl chloride is used as a dielectric in a 220-V circuit operating at 10^6 Hz. The polymer is in the form of a sheet 2 cm × 2 cm × 0.01 cm. Calculate the power loss due to the dielectric.

16. A nylon dielectric in the form of a 1-cm diameter cylinder 0.001 mm thick is used in a circuit operating at 10^6 Hz. Calculate the maximum voltage in the circuit if the power loss is to be less than 0.15 W.

17. Calculate the capacitance of a parallel plate capacitor containing 20 layers of Teflon, where each Teflon sheet has the dimensions 2 cm × 2 cm × 0.02 cm.

18. A parallel plate capacitor having a capacitance of $0.034 \mu F$ is to be constructed of fused silica sheets each 1 cm × 1 cm × 0.001 cm. Determine the number of fused silica sheets and the number of conductor sheets that are required.

19. Determine the surface area of a 0.025-μF parallel plate capacitor containing three layers of dielectric each 0.0015 cm thick if the dielectric is (a) polyvinyl chloride and (b) barium titanate. Assume that the capacitor is operating in a 10^6-Hz circuit.

20. What force must be applied to a 1 mm × 1 mm × 0.01 mm barium titanate crystal if a voltage of 12 V is to be produced? How much strain is required?

21. A strain of 5×10^{-4} in./in. is produced in a barium titanate crystal having an area of $0.25 \, cm^2$, resulting in a voltage of 75 V. Calculate (a) the thickness of the crystal and (b) the applied force.

22. Determine the ratio of the forces required to produce a voltage of 125 V in quartz versus barium titanate. The modulus of elasticity of quartz is 10.4×10^6 psi.

23. Determine the voltage required to eliminate polarization in a 0.015-cm thick dielectric made from material A in Figure 18-28.

24. If a voltage of 25 V is required to eliminate polarization in material B in Figure 18-28, determine the thickness of the dielectric.

25. Material B in Figure 18-28, which originally is not polarized, is placed in an electric field that causes polarization of $4 \times 10^{-8} \, C/m^2$. Determine (a) the field required to do this and (b) the dielectric constant of the material at this point.

26. An electric field of 4000 V/m is applied to material A in Figure 18-28. Assuming that the material originally was not polarized, determine (a) the polarization and (b) the dielectric constant at that point.

27. Plot a graph showing how the dielectric constant of originally unpolarized material B in Figure 18-28 changes as the electric field increases.

28. Calculate and compare the maximum magnetization we would expect in iron, nickel, cobalt, and gadolinium. There are seven electrons in the $4f$ level of gadolinium.

29. A Ni-25% Co alloy, which has a lattice parameter of 3.52 Å, is prepared. Assuming no interactions between the cobalt and nickel atoms, calculate (a) the at% Co in the alloy and (b) the maximum magnetization of the alloy.

30. A nickel-copper alloy is prepared which has a lattice parameter of 3.54 Å. The maximum magnetization of the alloy is measured as 1.46×10^6 A/m. Assuming no interactions between the Ni and Cu atoms, estimate the wt% Cu in the alloy.

31. Supermalloy, a soft magnetic material, is surrounded by a 20-m long, 30-turn coil of a conductor through which a current of 5 A is passed. Calculate (a) the magnetic field H in

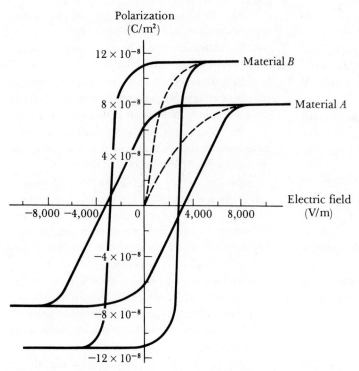

FIGURE 18-28   The hysteresis loop for two ferroelectric materials.

oersted, (b) the magnetization M, and (c) the inductance B.

32.   Using a core of 4S Permalloy, you would like to produce a 19-m long, 300-turn coil of a conductor, giving an inductance of 75,000 gauss. What current must you use in the coil?

33.   Calculate the strength of the magnetic field required to produce saturation magnetization in Supermalloy.

34.   Using Figure 18-29, (a) calculate and plot the BH product as a function of magnetic field and (b) determine the power of the magnetic material.

35.   Using Figure 18-30, (a) calculate and plot the BH product as a function of magnetic field and (b) determine the power of the magnetic material.

36.   Using Figure 18-29, calculate (a) the initial permeability and (b) the maximum permeability of the material.

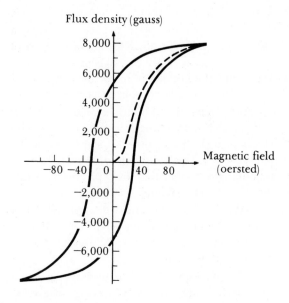

FIGURE 18-29   The hysteresis loop for a ferromagnetic material.

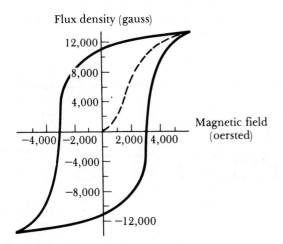

FIGURE 18-30 The hysteresis loop for a ferromagnetic material.

37. Using Figure 18-30, calculate (a) the initial permeability and (b) the maximum permeability of the material.

38. Determine the power of the Co_5Ce material shown in Figure 18-25.

39. A magnetic field obtained from a 100-turn coil 12 m in length produces a magnetization of 3000 gauss in the material shown in Figure 18-29. Determine (a) the required magnetic field, (b) the relative permeability of the material in the magnetic field, and (c) the current required to produce the magnetization.

40. A magnetic field of 2000 oersted is obtained in a rod made from the material shown in Figure 18-30. Determine (a) the magnetization produced and (b) the relative permeability at that field.

41. Suppose we replace 5% of the Fe^{2+} ions in magnetite with Cu^{2+} ions. Determine the total magnetic moment per cubic centimeter.

42. The total magnetic moment per cubic centimeter in a spinel structure in which Mn^{2+} ions have replaced a portion of the Fe^{2+} ions is 5.5×10^5 A/m. (a) Calculate the fraction of the Fe^{2+} ions that have been replaced and (b) determine the wt% Mn in the spinel.

19

OPTICAL AND
THERMAL PROPERTIES

19–1 INTRODUCTION

Optical and thermal properties are related to the interaction of a material with radiation in the form of waves or particles of energy. The frequency, wavelength, and energy of the radiation are determined by the source. For example, gamma rays are produced by changes in the structure of the nucleus of the atom; X rays, ultraviolet radiation, and the visible spectrum are produced by changes in the electronic structure of the atom. Infrared radiation, microwaves, and radio waves are low-energy, long-wavelength radiation caused by vibration of the atoms or crystal structure. When radiation interacts with a material, a variety of effects are produced, including absorption, colors, fluorescence, conduction of heat, and electronic behavior. By examining these phenomena, not only can we better understand the behavior of materials, we can also use this behavior to produce aircraft that cannot be detected by radar; lasers; fiber-optic devices; light-emitting diodes; solar cells; analytical instruments for determining crystal structure or material composition; and many more critical devices.

Optical Properties

19–2 EMISSION OF CONTINUOUS AND CHARACTERISTIC RADIATION

Energy, or radiation in the form of waves or particles called *photons*, can be emitted from a material. The important characteristics of the photons—their energy E, wavelength λ, and frequency v—are related by the equation

Incoming electron

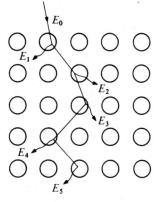

$$E_1 + E_2 + E_3 + E_4 + E_5 = E_0$$

FIGURE 19-1   When an accelerated electron strikes and interacts with a material, its energy may be reduced in a series of steps. In the process, several photons of different energies E_1 to E_5 are emitted, each with a unique wavelength.

$$E \ = \ h\nu \ = \ \frac{hc}{\lambda} \tag{19-1}$$

where c is the speed of light (3×10^{10} cm/s) and h is Planck's constant (6.62×10^{-27} erg $\cdot$ s or 6.62×10^{-34} J $\cdot$ s). This equation permits us to consider the photon either as a particle of energy E or as a wave with a characteristic wavelength and frequency.

Continuous Spectrum.   A stimulus such as a high-energy electron is decelerated when it strikes a material. As the electron decelerates, energy is given up and emitted as photons. Each time the electron strikes an atom, more of its energy is given up. Each interaction, however, may be more or less severe, so the electron gives up a different fraction of its energy each time, producing photons of different wavelengths (Figure 19-1). A *continuous spectrum*, or *white radiation*, is produced (Figure 19-2).

If the electron were to lose all of its energy in one impact, the minimum wavelength of the emitted photons would be equivalent to the original energy of the stimulus. Thus the continuous spectrum has a *short wavelength limit* λ_{SWL}. When the energy of the stimulus increases, the short wavelength limit decreases and the number and energy of the emitted photons increase, giving a more intense continuous spectrum.

◇ | **Example 19-1**

A voltage of 10,000 V is applied to a heated tungsten filament, which then emits electrons. Calculate the short wavelength limit that the electron can stimulate if it strikes a target material.

Answer:

First, let's calculate the energy of the emitted electrons, noting that there are 1.6×10^{-12} ergs/volt.

$$E = (10{,}000\,\text{V})(1.6 \times 10^{-12}\,\text{erg/V}) = 1.6 \times 10^{-8}\,\text{erg}$$

$$hc = (6.62 \times 10^{-27}\,\text{erg} \cdot \text{s})(3 \times 10^{10}\,\text{cm/s}) = 19.86 \times 10^{-17}\,\text{erg} \cdot \text{cm}$$

$$\lambda_{SWL} = \frac{hc}{E} = \frac{19.86 \times 10^{-17}}{1.6 \times 10^{-8}} = 1.24 \times 10^{-8}\,\text{cm} = 1.24\,\text{Å}$$

Characteristic Spectrum.   If the incoming stimulus has sufficient energy, an electron from an inner energy level is excited into an outer energy level. To restore equilibrium, the empty inner level is filled by electrons from a higher level.

There are discrete energy differences between any two energy levels. When an electron drops from one level to a second level, a photon having that particular energy and wavelength is emitted. Photons with this energy and wavelength comprise the *characteristic spectrum* and are X rays. The characteristic spectrum appears as a series of peaks superimposed on the continuous spectrum (Figure 19-2).

This effect is illustrated in Figure 19-3. We typically refer to the energy levels by the K, L, M, . . . designation, as described in Chapter 2. If an electron is excited from the K shell, electrons may fill that vacancy from any outer shell. Normally, electrons in the next closest shell fill the vacancies. Thus, photons with energies $\Delta E = E_{\text{K}} - E_{\text{L}}$ (K_α X rays) or $\Delta E = E_{\text{K}} - E_{\text{M}}$ (K_β X rays) are emitted. If an electron from the L shell fills the K shell, then an electron from the M shell may fill the L shell, giving a photon with energy $\Delta E = E_{\text{L}} - E_{\text{M}}$ (L_α X rays) which has a longer wavelength or lower energy. Note that we need a more energetic stimulus to produce K_α X rays than that required for L_α X rays.

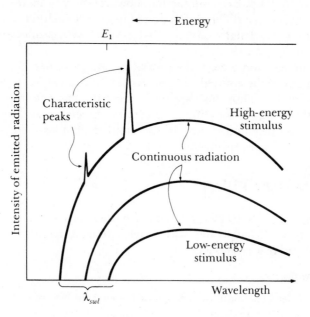

FIGURE 19-2 The continuous and characteristic spectra of radiation emitted from a material. Low-energy stimuli produce a continuous spectrum of low-energy, long-wavelength photons. A more intense, higher energy spectrum is emitted when the stimulus is more powerful until eventually characteristic radiation is observed.

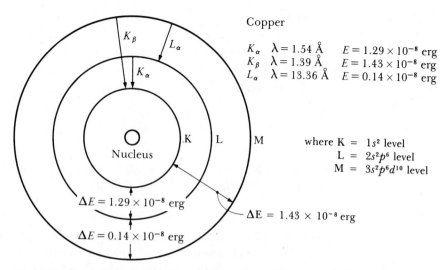

Copper

K_α $\lambda = 1.54$ Å $E = 1.29 \times 10^{-8}$ erg
K_β $\lambda = 1.39$ Å $E = 1.43 \times 10^{-8}$ erg
L_α $\lambda = 13.36$ Å $E = 0.14 \times 10^{-8}$ erg

where K = $1s^2$ level
L = $2s^2p^6$ level
M = $3s^2p^6d^{10}$ level

FIGURE 19-3   Characteristic X rays are produced when electrons change from one energy level to a lower energy level, as illustrated here for copper. The energy and wavelength of the X rays are fixed by the energy differences between the energy levels.

TABLE 19-1   Characteristic emission lines and absorption edges for selected elements

Metal	K_α (Å)	K_β (Å)	L_α (Å)	Absorption Edge (Å)
Al	8.337	7.981		7.951
Si	7.125	6.768		6.745
S	5.372	5.032		5.018
Cr	2.291	2.084		2.070
Mn	2.104	1.910		1.896
Fe	1.937	1.757		1.743
Co	1.790	1.621		1.608
Ni	1.660	1.500		1.488
Cu	1.542	1.392	13.357	1.38
Mo	0.711	0.632	5.724	0.620
W	0.211	0.184	1.476	0.178

From B. Cullity, *Elements of X-ray Diffraction*, 2nd Ed., Addison-Wesley, 1978.

Examples of a portion of the characteristic spectra for several elements are included in Table 19-1. The absorption edge in the table will be explained in a later section.

◇   **Example 19-2**

Suppose an electron accelerated at 5000 V strikes a copper target. Will K_α, K_β, or L_α X rays be emitted from the copper target?

Answer:

The electron must possess enough energy to excite an electron to a higher level, or its wavelength must be less than that corresponding to the energy difference between the shells.

$$E = (5000)(1.6 \times 10^{-12}) = 0.8 \times 10^{-8}\,\text{erg}$$

$$\lambda = \frac{hc}{E} = \frac{(6.62 \times 10^{-27})(3 \times 10^{10})}{0.8 \times 10^{-8}}$$

$$= 2.48 \times 10^{-8}\,\text{cm} = 2.48\,\text{Å}$$

For copper, K_α is 1.542 Å, K_β is 1.392 Å, and L_α is 13.357 Å. Therefore, the L_α peak may be produced but K_α and K_β will not.

◇

We have so far looked primarily at the emission of rather short wavelength photons, ordinarily classified as X rays, that are emitted when electrons near the nucleus of the atom change energy levels. Depending on the source of the photons, we could have different wavelengths of radiation. As we will see, electronic reactions in the outer energy level of certain materials may produce visible light. The entire spectrum of electromagnetic radiation is shown in Figure 19-4. The continuous spectrum produced when a stimulus strikes a material may

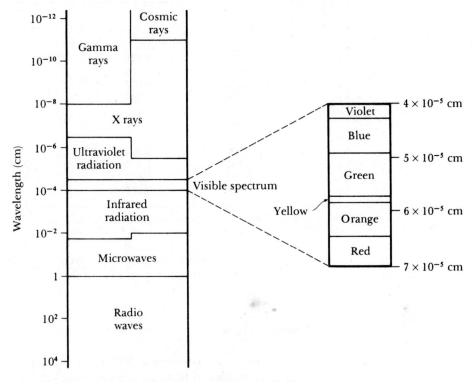

FIGURE 19-4   The electromagnetic spectrum of radiation.

contain all radiation having a longer wavelength (smaller energy) than that of the stimulus. Thus, a stimulus producing X rays could also produce ultraviolet radiation. However, the longer wavelength radiation is often absorbed by the material.

19-3 EXAMPLES OF EMISSION PHENOMENA

Let's look at some particular examples of emission phenomena which are of some familiarity and importance.

X rays. As we have already noted, X rays are produced when the inner shell electrons are stimulated, normally by an incident electron. The most useful of the X rays are the most energetic, shortest wavelength photons produced by filling the K and L levels. These X rays can be used to determine the composition of the material.

If we bombarded an unknown material with very high energy electrons, the material emits X rays in both the characteristic and the continuous spectra. We can then measure the wavelength or energy of the characteristic radiation. If we match the emitted characteristic wavelengths with those expected for various materials, the identity of the material can be determined. We can also measure the intensity of the characteristic peaks. By comparing measured intensities with standard intensities, we can estimate the amount of each emitting atom and determine the composition of the material. We can perform this test on large samples of the material using X-ray fluorescent analysis or on a microscopic scale using the electron microprobe or the scanning electron microscope (SEM), permitting us to identify individual phases or even inclusions in the microstructure.

$\Diamond$ | **Example 19-3**

An unknown metal is bombarded with high-energy X rays, producing the emission spectrum shown in Figure 19-5. (a) Determine the identity of the material from the spectrum and Table 19-1. (b) What is the energy of the exciting X rays?

Answer:

(a) From Figure 19-5, we find three characteristic peaks. The wavelengths are about 0.62 Å, 0.71 Å, and 5.75 Å. From Table 19-1, we find that molybdenum has $K_\beta = 0.632$ Å, $K_\alpha = 0.711$ Å, and $L_\alpha = 5.724$ Å, providing a good match. The material is most likely molybdenum.

(b) The energy of the stimulus corresponds to the energy giving the short wavelength limit of the spectrum. The $\lambda_{SWL} = 0.4$ Å.

$$E = \frac{hc}{\lambda_{SWL}} = \frac{(6.62 \times 10^{-27})(3 \times 10^{10})}{0.4 \times 10^{-8}}$$

$$= 4.965 \times 10^{-8} \, \text{erg}$$

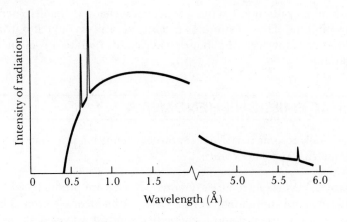

FIGURE 19-5 The emission spectrum for Example 19-3.

The X rays are produced by a voltage of

$$E = \frac{4.965 \times 10^{-8}}{1.6 \times 10^{-12}} = 31,000\,V$$

◇ | **Example 19-4**

The photograph in Figure 19-6 was obtained using a scanning electron microscope (SEM) at a magnification of 1000. The beam of electrons in the SEM was directed at the three different phases, producing the characteristic peaks. From the energy spectra, determine the probable identity of each phase.

Answer:

Phase A

The energy of the single peak is about $1.5\,keV = 1500\,eV$

$$\lambda = \frac{hc}{E} = \frac{(6.62 \times 10^{-27}\,erg \cdot s)(3 \times 10^{10}\,cm/s)}{(1500\,eV)(1.6 \times 10^{-12}\,erg/eV)(10^{-8}\,cm/Å)} = 8.275\,Å$$

Phase B

The energies of the two peaks are $1.5\,keV = 1500\,eV$ and $1.7\,keV = 1700\,eV$. The wavelength of the second, large peak is

$$\lambda = \frac{(6.62 \times 10^{-27})(3 \times 10^{10})}{(1700)(1.6 \times 10^{-12})(10^{-8})} = 7.30\,Å$$

Phase C

We observe five peaks with the following energies and wavelengths.

　　$1.5\,keV$:　$8.275\,Å$

　　$1.7\,keV$:　$7.30\,Å$

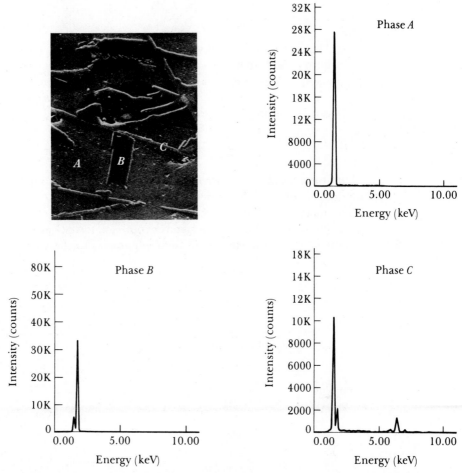

FIGURE 19-6   Scanning electron micrograph of a multiple-phase material. The energy distribution of the emitted radiation from the three phases marked *A*, *B*, and *C* are shown. The identity of each phase is determined in Example 19-4.

$$5.8\,\text{keV:} \quad \frac{(6.62 \times 10^{-27})(3 \times 10^{10})}{(5800)(1.6 \times 10^{-12})(10^{-8})} = 2.14\,\text{Å}$$

$$6.4\,\text{keV:} \quad \frac{(6.62 \times 10^{-27})(3 \times 10^{10})}{(6400)(1.6 \times 10^{-12})(10^{-8})} = 1.94\,\text{Å}$$

$$7.1\,\text{keV:} \quad \frac{(6.62 \times 10^{-27})(3 \times 10^{10})}{(7100)(1.6 \times 10^{-12})(10^{-8})} = 1.75\,\text{Å}$$

From Table 19-1

$$\lambda = 8.275 \simeq 8.337 = K_\alpha\,\text{Al}$$
$$\lambda = 7.30 \ \ \simeq 7.125 = K_\alpha\,\text{Si}$$
$$\lambda = 2.14 \ \ \simeq 2.104 = K_\alpha\,\text{Mn}$$

$$\lambda = 1.94 \simeq 1.937 = K_\alpha \text{ Fe}$$
$$\lambda = 1.75 \simeq 1.757 = K_\beta \text{ Fe}$$

Thus, phase A appears to be an aluminum matrix, phase B appears to be a silicon needle, perhaps containing some aluminum, and phase C appears to be an Al-Si-Mn-Fe compound. Actually, this is an aluminum-silicon alloy. The stable phases are aluminum and silicon, with inclusions forming when manganese and iron are present as impurities.

Luminescence.   *Luminescence* is the conversion of radiation or other forms of energy to visible light caused by reactions in the outer energy levels of an atom. The incident radiation excites electrons from the valence band into the conduction band. When the electrons drop back to the valence band, photons are emitted. If the wavelength of these photons is in the visible light range, luminescence occurs (Figure 19-7).

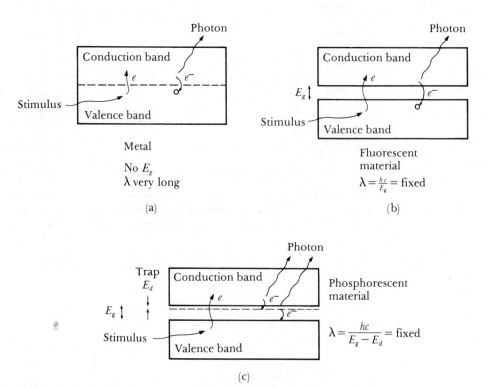

FIGURE 19-7   Luminescence occurs when photons have a wavelength in the visible spectrum. (a) In metals, there is no energy gap so luminescence does not occur. (b) Fluorescence occurs when there is an energy gap. (c) Phosphorescence occurs when the photons are emitted over a period of time due to donor traps in the energy gap.

In metals, the energy of the emitted photon is very small, since the valence and conduction bands overlap, and the wavelength is longer than the visible light spectrum. Therefore, luminescence does not occur. However, in certain ceramic materials, the energy gap between the valence and conduction bands is such that an electron dropping through this gap produces a photon in the visible range. One wavelength, corresponding to the energy gap E_g, predominates.

Two different effects may be observed in luminescent materials—fluorescence and phosphorescence. In *fluorescence*, luminescence ceases when the stimulus is removed. All excited electrons drop back to the valence band and the photons are emitted within about 10^{-8} s [Figure 19-7(b)].

However, *phosphorescent* materials have impurities which introduce a donor level within the energy gap [Figure 19-7(c)]. The stimulated electrons first drop into the donor level and are trapped. The electrons must then escape the trap before returning to the valence band. There is a delay of more than 10^{-8} s before the photons are emitted. When the source is removed, electrons in the traps gradually escape and emit light over some additional period of time. The intensity of the luminescence is given by

$$\ln\left(\frac{I}{I_0}\right) = -\frac{t}{\tau} \tag{19-2}$$

where τ is the *relaxation time*, a constant for the material. After time t following removal of the source, the intensity of the luminescence is reduced from I_0 to I. Phosphorescent materials are very important in the operation of television screens. In this case, the relaxation time must not be too long or the images begin to overlap. In color televisions, three types of phosphorescent materials are used —the energy gaps are engineered so that red, green, and blue colors are produced.

◇ | **Example 19-5**

The wavelength of the luminescent radiation is determined by impurities in the material, which in turn determine the energy gap. (a) Determine the wavelength of a photon required to excite an electron in ZnS. (b) An impurity in ZnS gives an energy trap 1.38 eV below the conduction band. Calculate the wavelength and determine the type of radiation produced during luminescence.

Answer:

(a) From Table 17-10, the energy gap for pure ZnS is 3.54 eV. A photon must have a maximum wavelength of

$$\lambda = \frac{hc}{E} = \frac{(6.62 \times 10^{-27})(3 \times 10^{10})}{(3.54)(1.6 \times 10^{-12})(10^{-8})} = 3506\,\text{Å}$$

to excite an electron into the conduction band. This wavelength corresponds to ultraviolet radiation.

(b) When the excited electron returns to the valence band, it first enters the trap. This causes a photon to be emitted with the wavelength

$$\lambda = \frac{(6.62 \times 10^{-27})(3 \times 10^{10})}{(1.38)(1.6 \times 10^{-12})(10^{-8})} = 8995\,\text{Å}$$

This wavelength, which is in the infrared spectrum, is not visible.

However, the electron then drops from the trap to the valence band. The energy of the photon, $3.54 - 1.38 = 2.16\,\text{eV}$, corresponds to a wavelength of

$$\lambda = \frac{(6.62 \times 10^{-27})(3 \times 10^{10})}{(2.16)(1.6 \times 10^{-12})(10^{-8})} = 5747\,\text{Å}$$

This photon is in the visible spectrum and gives yellow light.

◇ **Example 19-6**

In a television screen the intensity of the phosphorescent material decreases by a factor of 10 within $10^{-3}\,\text{s}$. What is the relaxation time?

Answer:

If $I_0 = 10$, then $I = 1$.

$$\ln\left(\frac{I}{I_0}\right) = \ln\frac{1}{10} = \frac{-t}{\tau} = \frac{-10^{-3}}{\tau}$$

$$-2.3 = \frac{-10^{-3}}{\tau}$$

$$\tau = 4.3 \times 10^{-4}\,\text{s}$$

Light-Emitting Diodes.   Luminescence can be used to advantage in creating light-emitting diodes (LEDs), which are often used to display data. LEDs are used to provide the display for watches, clocks, calculators, and other electronic devices. These devices are *p-n* junction devices engineered so that the E_g is in the visible spectrum (often red). A voltage applied to the diode in the forward-bias direction causes holes and electrons to recombine at the junction. Photons are produced as a result of the recombination (Figure 19-8).

Lasers.   Lasers are another example of a special application of luminescence. In certain materials, electrons excited by a stimulus (such as the flash tube shown in Figure 19-9) produce photons which in turn excite additional photons of identical wavelength. Consequently, a large amplification of the photons emitted in the material occurs. By proper choice of stimulant and material, the wavelength of the photons can be in the visible range. The output of the laser

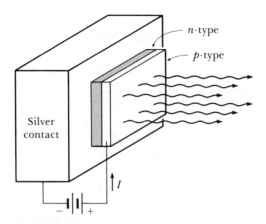

FIGURE 19-8 Diagram of a light-emitting diode (LED). A forward-bias voltage across the *p-n* junction produces photons.

is a beam of photons that are parallel, of the same wavelength, and coherent. In a coherent beam, the wavelike nature of the photons is in phase so that destructive interference does not occur. Lasers are useful in heat treating and melting of metals, welding, surgery, mapping, transmission and processing of information, and a variety of other applications, including reading of the compact disks used to produce noise-free stereo recordings.

A variety of materials are used to produce lasers. Ruby, which is Al_2O_3 doped with a small amout of Cr_2O_3, and yttrium aluminum garnet (YAG) doped with neodymium are two common solid-state lasers, while other lasers are based on CO_2 gas. The solid-state lasers typically are less penetrating than the gas laser.

Semiconductor lasers such as GaAs, which have an energy gap corresponding to a wavelength in the visible range, are also used. Figure 19-10 illustrates how a semiconductor laser might operate. When the semiconductor is excited by a voltage applied to the device, electrons jump from the valence band to the conduction band, leaving behind holes in the valence band. When an electron collapses back to the valence band and recombines with a hole, a photon having an energy and wavelength equivalent to the energy gap is produced. This photon stimulates another electron to drop from the conduction band to the valence band, creating a second photon having an identical

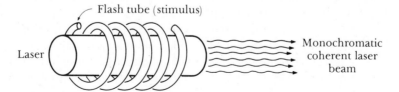

FIGURE 19-9 The laser converts a stimulus into a beam of coherent photons.

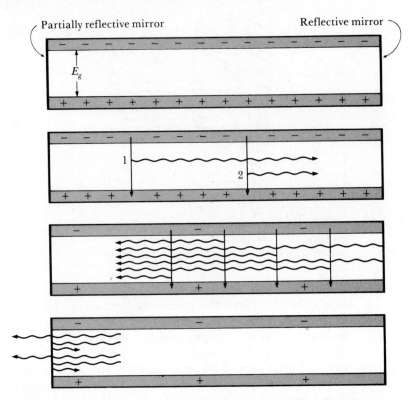

FIGURE 19-10 Creation of a laser beam from a semiconductor. (a) Electrons are excited into the conduction band by an applied voltage. (b) Electron 1 recombines with a hole to produce a photon. The photon stimulates the emission of photon 2 by a second recombination. (c) Photons reflected from the mirrored end stimulate even more photons. (d) A fraction of the photons are emitted as a laser beam, while the rest are reflected to stimulate more recombinations.

wavelength and a frequency which is in phase with the first photon. A mirror at one end of the laser crystal completely reflects the photons, trapping the photons within the semiconductor. The reflected photons stimulate even more recombinations until an intense wave of photons is produced. The photons then reach the other end of the crystal, which is only partly mirrored. A fraction of the photons emerge from the crystal as a monochromatic, coherent laser beam, while the rest of the photons remain in the crystal to stimulate further recombinations. The applied voltage assures that a steady source of excited electrons is available to produce additional photons. A continuous laser beam is therefore produced. Figure 19-11 schematically depicts one design for a semiconductor laser.

Although doped silicon or germanium could be used as lasers, they tend to produce an excessive amount of heat.

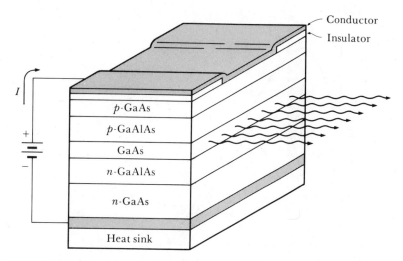

FIGURE 19-11   *Schematic cross section of a GaAs laser. Because the surrounding p- and n-type GaAlAs layers have a higher energy gap and a lower index of refraction than GaAs, the photons are trapped in the active GaAs layer.*

Thermal Emission.   When a material is heated, electrons are thermally excited to higher energy levels, particularly in the outer energy levels where the electrons are less tightly bound to the nucleus. The electrons immediately drop back to their normal levels and release photons with low energy and long wavelengths.

As the temperature increases, the thermal agitation increases and the maximum energy of the emitted photons increases. A continuous spectrum of radiation is emitted, with a minimum wavelength and an intensity distribution dependent on the temperature. The photons may include wavelengths in the visible spectrum; consequently the color of the material changes with temperature. At low temperatures, the wavelength of the radiation is too long to be visible. As the temperature increases, emitted photons have shorter wavelengths. At 700°C we begin to see a reddish tint; at 1500°C the orange and red wavelengths are emitted (Figure 19-12). Higher temperatures produce all wavelengths in the visible range, and the emitted spectrum is white light. Eventually, the blue end of the spectrum is predominant. By measuring the intensity of a narrow band of the emitted wavelengths with a pyrometer, we can estimate the temperature of the material.

19–4   INTERACTION OF PHOTONS WITH A MATERIAL

Photons, either characteristic or continuous, cause a number of optical phenomena when they interact with the electronic or crystal structure of a material. Figure 19-13 illustrates several things that can happen to a beam of photons when it strikes a material, including reflection, refraction, absorption,

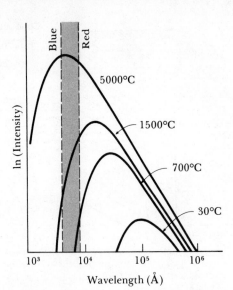

FIGURE 19-12 The intensity versus wavelength of photons emitted thermally from a material. As the temperature increases, more photons are emitted from the visible spectrum.

and transmission. Several factors are important in determining the behavior of the photons in the material, with the energy required to cause an electron to jump from the valence band to the conduction band being of particular importance. As a first approximation, we can say that if the energy of the photons is greater than the energy gap between the valence band and the conduction band, reflection or absorption will occur as the photon's energy is absorbed by exciting the electron. However, if the energy of the photons is less than the energy gap between the valence and conduction bands, the photons are transmitted through the material. The following sections treat these interactions in greater detail.

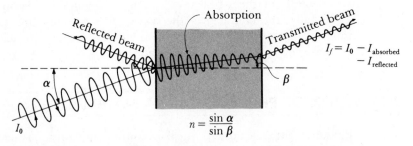

FIGURE 19-13 Interaction of photons with a material. In addition to reflection, absorption, and transmission, the beam changes direction, or is refracted. The change in direction is given by the index of refraction n.

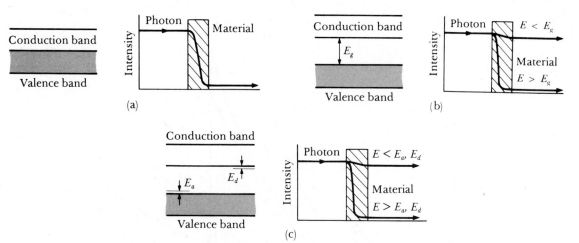

FIGURE 19-14 Relationships between absorption and the energy gap. (a) Metals, (b) insulators and intrinsic semiconductors, and (c) extrinsic semiconductors.

Absorption and Transmission. If the incoming photons interact with the valence electrons, the photons give up their energy and may be absorbed. This will occur if the energy gap in the material is small. On the other hand, if the energy gap is larger than the energy of the incident photons, the photons may be transmitted.

In metals, the valence and conduction bands overlap; therefore, radiation of almost any wavelength will be absorbed and the metal is generally considered to be opaque [Figure 19-14(a)]. However, if the metal is very thin and the wavelength of the radiation is very short, not all of the photons are absorbed. The amount of absorption is given by

$$\ln\left(\frac{I}{I_0}\right) = -\mu x \tag{19-3}$$

where x is the path through which the photons move (usually the thickness of the material) and μ is the *linear absorption coefficient* of the material for photons. The absorption coefficient is related to the density of the material, the wavelength of the radiation, and the energy required to stimulate an electron (either from the valence band to the conduction band or from one energy level to another energy level). The absorption coefficient of X rays in three metals is shown as a function of the wavelength of the X rays in Figure 19-15. Note that the absorption coefficient changes abruptly at a particular wavelength corresponding to the energy required to remove an electron from the K shell of the atom; this *absorption edge* is important in certain X-ray analytical techniques. Also note that as the wavelength increases beyond the absorption edge, the absorption coefficient increases dramatically; by the time we reach the wavelength of visible light, absorption is virtually complete unless the metal is exceptionally thin.

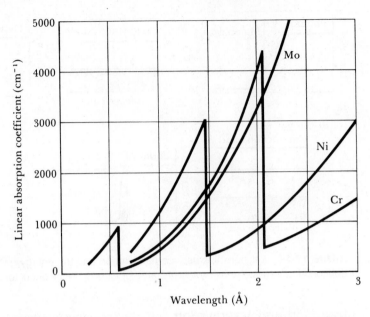

FIGURE 19-15 The linear absorption coefficient versus wavelength for several metals. Note the sudden decrease in absorption coefficient for wavelengths greater than the absorption edge.

However, in most insulators there is a large energy gap between the valence and conduction bands. If the energy of the photons is less than the energy gap, no electrons will gain enough energy to escape the valence band and therefore absorption will not occur [Figure 19-14(b)]. Unless the photons interact with imperfections in the material, the material is transparent. This is the case for glass, many high-purity crystalline ceramics, and amorphous polymers such as acrylics, polycarbonates, polysulfones, and many others.

In semiconductors, the energy gap is smaller than that of insulators, particularly in extrinsic semiconductors which have been doped to produce donor or acceptor energy levels. In intrinsic semiconductors, absorption will occur when the photons have energies exceeding the energy gap E_g, while transmission may occur for less energetic photons [Figure 19-14(b)]. In extrinsic semiconductors, absorption will occur when photons have energies greater than E_a or E_d [Figure 19-14(c)]. Semiconductors are therefore opaque to short-wavelength radiation but transparent to long-wavelength photons; for example, silicon and germanium appear opaque to visible light but are transparent to the longer wavelength infrared radiation.

While insulators and many semiconductors are expected to be transparent to long-wavelength photons based on the energy gap criterion, impurities may reduce transmission of the photons. Some impurities may create acceptor or donor levels in the crystal structure; other impurities, such as porosity or grain boundaries, may cause the photons to be scattered, making the material translucent or even opaque. Consequently, a crystalline polymer is expected to be more absorptive than an amorphous polymer.

◇ **Example 19-7**

Suppose a source of zinc K_α X rays is shielded from the surroundings by aluminum foil. If 95% of the energy of the X-ray beam is to penetrate the foil, determine the maximum thickness of the aluminum. The linear absorption coefficient in this case is $108\,\text{cm}^{-1}$.

Answer:

The final intensity of the beam must be at least $0.95I_0$. Thus

$$\ln\left(\frac{0.95I_0}{I_0}\right) = -(108)(x)$$

$$\ln(0.95) = -0.051 = -108x$$

$$x = \frac{-0.051}{-108} = 0.00047\,\text{cm}$$

◇ **Example 19-8**

Determine the shortest wavelength you would expect to be transmitted in the following materials: intrinsic silicon, phosphorus-doped silicon, diamond, tin.

Answer:

From the energy gaps in Tables 17-8 and 17-9

Intrinsic silicon : $\qquad E_g = 1.107\,\text{eV}$

$$\lambda = \frac{(6.62 \times 10^{-27})(3 \times 10^{10})}{(1.107)(1.6 \times 10^{-12})(10^{-8})} = 11,200\,\text{Å}$$

Phosphorus-doped silicon : $\quad E_d = 0.045\,\text{eV}$

$$\lambda = \frac{(6.62 \times 10^{-27})(3 \times 10^{10})}{(0.045)(1.6 \times 10^{-12})(10^{-8})} = 276,000\,\text{Å}$$

Diamond: $\qquad E_g = 5.4\,\text{eV}$

$$\lambda = \frac{(6.62 \times 10^{-27})(3 \times 10^{10})}{(5.4)(1.6 \times 10^{-12})(10^{-8})} = 2,300\,\text{Å}$$

Tin: $\qquad E_g = 0.08\,\text{eV}$

$$\lambda = \frac{(6.62 \times 10^{-27})(3 \times 10^{10})}{(0.08)(1.6 \times 10^{-12})(10^{-8})} = 155,000\,\text{Å}$$

Of these four materials, only diamond transmits visible light.

TABLE 19-2 Index of refraction of selected materials for photons of wavelength 5890 Å

Material	Index of Refraction
Ice	1.309
NaCl	1.544
Quartz	1.544
Diamond	2.417
TiO_2	2.7
Water	1.333
Plastics	1.5
Glasses	1.5
Leaded glass	2.5
Air	1.0

From *Handbook of Chemistry and Physics*, 56th Ed., CRC Press, 1975.

Refraction. Even when photons are transmitted by the material, the photon loses some of its energy and therefore has a slightly longer wavelength. The photon then behaves as though the speed of light in the material has been reduced and the beam of photons changes directions (Figure 19-13). Suppose photons traveling in a vacuum impinge on a material. If α and β, respectively, are the angles that the incident and refracted beams make with the normal to the surface of the material, then

$$n = \frac{c}{v} = \frac{\lambda_{vacuum}}{\lambda} = \frac{\sin \alpha}{\sin \beta} \qquad (19\text{-}4)$$

The ratio n is the index of refraction, c is the speed of light in a vacuum, and v is the speed of light in the material. Typical values of the index of refraction for several materials are listed in Table 19-2.

If the photons are traveling in material 1 and then pass into material 2, the velocities of the incident and refracted beams will depend on the ratio between their indices of refraction, again causing the beam to change directions

$$\frac{v_1}{v_2} = \frac{n_2}{n_1} = \frac{\sin \alpha}{\sin \beta} \qquad (19\text{-}5)$$

We can use this latter expression to determine whether the beam will be transmitted as a refracted beam or reflected. A beam traveling through material 1 will be reflected rather than transmitted if the angle β becomes 90°.

◇ | **Example 19-9**

Suppose a beam of light is passing through a glass fiber, making an angle of 60° to the axis of the fiber. The fiber is in air. The index of refraction of the glass

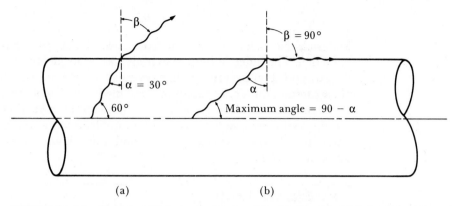

FIGURE 19-16   Diagram of light beam in glass fiber for Example 19-9. (a) Calculation of angle β and (b) calculation of maximum angle α.

is 1.5 and the index of refraction for air is 1.0. (a) Will the beam of light remain in the glass fiber? (b) What is the maximum angle between the beam of light and the axis of the fiber that will prevent leakage of the beam from the fiber?

Answer:

(a) Using Figure 19-16, we find that $\alpha = 90 - 60 = 30°$. If we let the glass be material 1, then from Equation 19-5

$$\frac{n_2}{n_1} = \frac{\sin \alpha}{\sin \beta}$$

$$\frac{1}{1.5} = \frac{\sin 30}{\sin \beta}$$

$$\sin \beta = 1.5 \sin 30 = 1.5(0.50) = 0.75$$

$$\beta = 48.6°$$

The angle β is less than 90° and the beam of light will be lost from the fiber.

(b) To reflect the beam, the angle β must be 90°. Thus

$$\frac{1}{1.5} = \frac{\sin \alpha}{\sin 90} = \sin \alpha$$

$$\sin \alpha = 0.6666 \quad \text{or} \quad \alpha = 41.8°$$

If the angle between the beam and the axis of the fiber is $90 - 41.8 = 48.2°$ or less, the beam will be reflected.

More interaction and refraction of the photons occurs when the electrons in the material are more easily polarized. Consequently, we expect to find a relationship between the index of refraction and the dielectric constant of the material.

$$n = \sqrt{\kappa} \tag{19-6}$$

Materials that polarize easily, such as dielectric materials, have a high index of refraction. The index of refraction is also larger for denser materials, leading to the exceptional appearance of fine crystal glassware or diamonds. However, n is not a constant for a particular material; the frequency of the photons will affect the index of refraction. In addition, the index of refraction is anisotropic in certain crystalline materials, leading to problems with birefringence, or double refraction. However, ceramic glasses and amorphous polymers typically have isotropic behavior.

◇ | **Example 19-10**

What type of polarization—electronic, ionic, or molecular—is normally involved when visible light is refracted?

Answer:

The wavelength of visible light is about 6000 Å or 6×10^{-5} cm. The frequency of the light is therefore

$$v = \frac{c}{\lambda} = \frac{3 \times 10^{10}\,\text{cm/s}}{6 \times 10^{-5}\,\text{cm}} = 5 \times 10^{14}\,\text{Hz}$$

From Figure 18-6, we find that the frequency is too high for either ionic or molecular polarization but will cause electronic polarization.

◇ | **Example 19-11**

Based on the indices of refraction in Table 19-2, what type of glass would you recommend for use in making fine crystal?

Answer:

The leaded glasses have the highest index of refraction and therefore would be expected to give the best appearance.

◇ | **Example 19-12**

Suppose a beam of photons in a vacuum strikes a sheet of polyethylene at an angle of 10° to the normal to the surface of the polymer. Calculate the index of refraction of polyethylene and find the angle between the incident beam and the beam as it passes through the polymer.

Answer:

The index of refraction is related to the dielectric constant. From Table 18-1, $\kappa = 2.3$.

$$n = \sqrt{\kappa} = \sqrt{2.3} = 1.52$$

The angle β is

$$n = \frac{\sin \alpha}{\sin \beta}$$

$$\sin \beta = \frac{\sin \alpha}{n} = \frac{\sin 10°}{1.52} = \frac{0.174}{1.52} = 0.114$$

$$\beta = 6.55°$$

Reflection.   When a beam of photons strikes a material such as a metal, electrons are excited into a higher energy level in the conduction band. If we were using visible light, we would expect that the photons would be totally absorbed, no light would be reflected, and the metal would appear to be black. However, as soon as the electrons are excited into the higher energy level, they immediately drop back to their stable levels, emitting photons of the same wavelength as the incident photons. Therefore, the material is reflective.

Even in materials that are not opaque, some reflection of an incident beam of photons may occur. The *reflectivity R*, which gives the percentage of the beam that is reflected, is related to the index of refraction. In a vacuum

$$R = \left(\frac{n-1}{n+1}\right)^2 \times 100 \tag{19-7}$$

Materials with a high index of refraction have a higher reflectivity than materials with a low index. In addition, a smoother surface will provide less scattering and a higher reflectivity.

◇ | **Example 19-13**

What fraction of the photons striking polyethylene are reflected?

Answer:

The index of refraction for polyethylene, as calculated in Example 19-12, is 1.52. Thus, the reflectivity is

$$R = \left(\frac{n-1}{n+1}\right)^2 \times 100 = \left(\frac{1.52-1}{1.52+1}\right)^2 \times 100 = \left(\frac{0.52}{2.52}\right)^2 \times 100$$

$$= 4.3\%$$

Example 19-14

A beam of photons passes through a 2-cm thick glass, with 60% of the original beam intensity being transmitted. If the dielectric constant of the glass is 3.2, calculate the linear absorption coefficient of the photons in the glass.

Answer:

A portion of the beam is reflected. The index of refraction is

$$n = \sqrt{\kappa} = \sqrt{3.2} = 1.79$$

$$R = \left(\frac{n-1}{n+1}\right)^2 \times 100 = \left(\frac{1.79-1}{1.79+1}\right)^2 \times 100 = 8.0\%$$

Thus, the intensity of the beam entering the glass is $100 - 8 = 92\%$. We will let $I_0 = 92\%$. The final beam has an intensity of 60%. The rest of the beam is absorbed.

$$\ln\left(\frac{I}{I_0}\right) = -\mu x$$

$$\ln\left(\frac{60}{92}\right) = -2\mu \qquad \ln(0.652) = -2\mu$$

$$\mu = \frac{0.428}{2} = 0.214\,\text{cm}^{-1}$$

Selective Absorption, Transmission, or Reflection.   Unusual optical behavior is observed when photons are selectively absorbed, transmitted, or reflected. We have already found that semiconductors transmit long-wavelength photons but absorb short-wavelength radiation. There are a variety of other cases in which similar selectivity produces unusual optical properties.

The colors of certain metals, such as copper and gold, are related to their ability to more easily reflect, or reemit, only a certain range of wavelengths and absorb the remaining photons. Copper absorbs the shorter wavelength photons on the blue or violet end of the visible spectrum but reflects photons with longer wavelengths at the red end of the spectrum. Since we see only the reflected light, copper and gold appear reddish. Other metals, such as aluminum and silver, reflect photons of all wavelengths in the visible spectrum; therefore, they appear white.

In certain materials, replacement of normal ions by transition or rare earth elements may produce a *crystal field* which creates new energy levels within the structure. This occurs when Cr^{3+} ions replace Al^{3+} ions in Al_2O_3. The new energy levels absorb visible light in the violet and green-yellow portions of the spectrum (Figure 19-17). Red wavelengths in particular are transmitted, with a tint of blue also passing through the crystal. This produces the reddish color

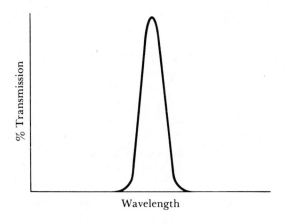

FIGURE 19-17 Selective absorption and transmission of certain wavelengths due to crystal fields or *F*-centers can produce colors in materials.

in ruby. In addition, the chromium ion replacement creates an energy level that permits luminescence to occur when the electrons are excited by a stimulus; lasers produced from chromium-doped ruby produce a characteristic red beam because of this.

Glasses can also be doped with ions that produce selective absorption and transmission. The effects of several types of ions in glasses are shown in Table 19-3.

Electron or hole traps, called *F*-centers, can also be present in crystals. When fluorite (CaF_2) is produced so that there is excess calcium, a fluoride ion vacancy is produced. To maintain electrical neutrality, an electron is trapped in the vacancy, producing energy levels that absorb all visible photons with the exception of purple. In a similar manner, replacement of Si^{4+} ions with Al^{3+} ions in quartz (SiO_2), followed by irradiation with gamma or X rays, produces a hole in the energy band structure.

Polymers, particularly those containing an aromatic ring in the backbone, can have complex covalent bonds that produce an energy level structure which causes selective absorption. For this reason, chlorophyll in plants appears green and hemoglobin in blood appears red.

TABLE 19-3 Effect of ions on colors produced in glasses

Ion	Color
Cr^{2+}	Blue
Cr^{3+}	Green
Cu^{2+}	Blue-green
Mn^{2+}	Orange
Fe^{2+}	Blue-green
U^{6+}	Yellow

◇ | **Example 19-15**

Polychromic glass, used for sunglasses, contains silver atoms. The glass darkens in sunlight but becomes transparent in darkness. Explain this phenomenon.

Answer:

In bright light, the silver ions in the glass gain an electron by excitation by the photons and are reduced from Ag^+ to metallic silver atoms. Thus, absorption of photons occurs. When the incoming light diminishes in intensity, the silver reverses to silver ions and no absorption occurs.

X-ray Filters.   A material may selectively absorb X rays that have sufficient energy to excite electrons from inner orbitals (Figure 19-18). Very little absorption of X rays having slightly longer wavelengths is observed. The wavelength that just causes excitation of the electrons is the absorption edge, as discussed in Section 19–4.

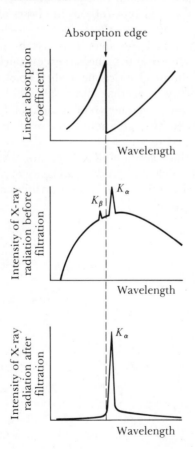

FIGURE 19-18 Elements have a selective lack of absorption of certain wavelengths. If a filter is selected with an absorption edge between the K_α and K_β peaks of an X-ray spectrum, all X rays except K_α are absorbed.

In X-ray diffraction, we can take advantage of this selective absorption to filter out almost all of the continuous spectrum, permitting only a single desired characteristic wavelength to pass. This characteristic wavelength is then used in X-ray diffraction experiments. Note that the K_β peak is eliminated along with most of the continuous spectrum if the filter has an absorption edge lying between K_α and K_β. Absorption edges for several elements are included in Table 19-1.

◇ | **Example 19-16**

What filter material should you use to isolate the K_α peak of nickel?

Answer:

From Table 19-1, $K_\alpha = 1.660 \, \text{Å}$ and $K_\beta = 1.500 \, \text{Å}$ for nickel. We need a filter with an absorption edge between these characteristic peaks. Cobalt, with an absorption edge of $1.608 \, \text{Å}$, will work.

Photoconduction.   Photoconduction occurs in semiconducting materials if the semiconductor is part of an electrical circuit. In this case, the stimulated electrons produce a current rather than emission (Figure 19-19). If the energy of an incoming photon is sufficient, an electron is excited into the conduction band or a hole is created in the valence band and the electron or hole then carries a charge through the circuit. The maximum wavelength of the incoming photon that is required to produce photoconduction is related to the energy gap in the semiconductive material.

$$\lambda_{\text{max.}} = \frac{hc}{E_g} \tag{19-8}$$

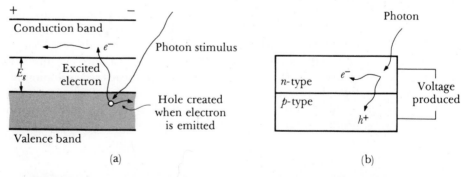

FIGURE 19-19   (a) Photoconduction in semiconductors involves absorption of a stimulus by exciting electrons from the valence band to the conduction band. Rather than dropping back to the valence band to cause emission, the excited electrons carry a charge through an electrical circuit. (b) A solar cell takes advantage of this effect.

We can use this principle for "electric eyes" that open or close doors or switches when a beam of light focused on a semiconductive material is interrupted. Note that photoconduction is the inverse of luminescence and LEDs. In the present case, photons produce a voltage and current, whereas in an LED a voltage produces photons and light.

◇ | **Example 19-17**

What is the maximum wavelength of light energy that should be used in a photoconductive electric eye circuit if intrinsic silicon is used as the semiconductor at room temperature?

Answer:

Photons must have an energy greater than the energy gap in silicon, 1.107 eV.

$$E = 1.107 = \frac{hc}{\lambda} = \frac{(6.62 \times 10^{-27} \text{ erg} \cdot \text{s})(3 \times 10^{10} \text{ cm/s})}{\lambda}$$

$$\lambda = \frac{(6.62 \times 10^{-27} \text{ erg} \cdot \text{s})(3 \times 10^{10} \text{ cm/s})}{(1.107 \text{ eV})(1.6 \times 10^{-12} \text{ erg/eV})}$$

$$= 1.12 \times 10^{-4} \text{ cm}$$

Solar cells are *p-n* junctions designed so that photons excite electrons into the conduction band. The electrons move to the *n*-side of the junction, while holes move through the valence band to the *p*-side of the junction. This produces a contact voltage due to the charge imbalance. If the junction device is connected to an electric circuit, the junction acts as a battery to power the circuit.

Diffraction. In X-ray diffraction, the wavelike nature of X rays is used to determine information about the crystal structure and atom location within the material. The wavelike X-ray radiation interacts with the electronic dipoles and is scattered in all directions. The radiation scattered from one atom interacts destructively with radiation from other atoms except in certain directions (Figure 19-20). In these directions, the scattered radiation is reinforced rather than destroyed.

The radiation, which is of the identical wavelength as the original X-ray beam, is reinforced in the angles given by *Bragg's law*.

$$\sin \theta = \frac{\lambda}{2d_{hkl}} \tag{19-9}$$

where the angle θ is half of the angle between the diffracted beam and the original beam direction, λ is the wavelength, and d_{hkl} is the interplanar spacing that causes constructive reinforcement of the beam.

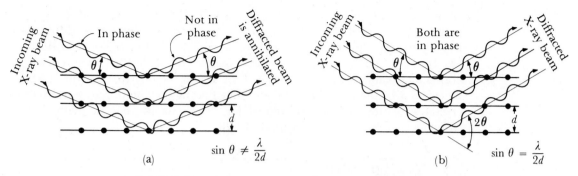

FIGURE 19-20   (a) Destructive and (b) reinforcing interactions between X rays and the crystal structure of a material. Reinforcement occurs at angles that satisfy Bragg's law.

The *Debye–Scherrer X-ray diffraction* technique is one method used to determine interplanar spacings. X rays having a known wavelength (usually filtered K_α radiation) strike a powdered specimen. If the wavelength is fixed, atoms on any particular plane cause the X rays to be diffracted at a particular angle. The diffracted X-ray beam intersects and exposes a film surrounding the specimen (Figure 19-21). Only cones of radiation from planes whose interplanar spacings satisfy Bragg's law intersect the film. By analyzing the film and calculating the 2θ angle, we can determine the interplanar spacing.

We can determine the type of crystal structure and the lattice parameter of a material using the Debye–Scherrer technique. To identify the crystal structure of a cubic material, we would note the pattern of the diffracted lines, typically by creating a table of $\sin^2 \theta$ values. By combining Equation 19-9 with the interplanar spacing equation (Equation 3-6), we find that

$$\sin^2 \theta = \frac{\lambda^2}{4a_0^2} (h^2 + k^2 + l^2) \tag{19-10}$$

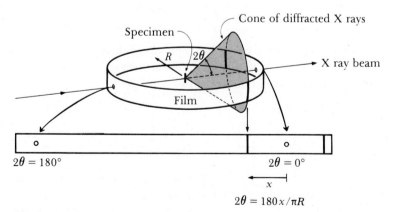

FIGURE 19-21   The Debye–Scherrer X-ray diffraction technique uses a film that intercepts a cone of X rays diffracted by the powdered specimen.

In simple cubic metals, all possible planes will diffract, giving an $h^2 + k^2 + l^2$ pattern of 1, 2, 3, 4, 5, 6, 8, ... In body-centered cubic metals diffraction occurs only from planes having an even $h^2 + k^2 + l^2$ sum of 2, 4, 6, 8, 10, 12, 14, 16, ... For face-centered cubic metals, more destructive interference occurs and planes having $h^2 + k^2 + l^2$ of 3, 4, 8, 11, 12, 16, ... will diffract. By calculating the values of $\sin^2 \theta$ and then finding the appropriate pattern, the crystal structure can be determined for metals having one of these simple structures, as illustrated in Example 19-20.

◇ | **Example 19-18**

Determine the angle 2θ between the transmitted and diffracted K_α X-ray beam of nickel when the (111) planes in copper $(a_0 = 3.6151\,\text{Å})$ are responsible for diffraction.

Answer:

From the interplanar spacing equation (Equation 3-6)

$$d_{111} = \frac{a_0}{\sqrt{h^2 + k^2 + l^2}} = \frac{3.6151}{\sqrt{1^2 + 1^2 + 1^2}} = 2.087\,\text{Å}$$

From Table 19-1, the wavelength of K_α X rays of nickel is 1.660 Å.

$$\sin \theta = \frac{\lambda}{2d_{111}} = \frac{1.660}{2(2.087)} = 0.398$$

$$\theta = 23.45°$$

$$2\theta = 46.9°$$

◇ | **Example 19-19**

Suppose chromium radiation of wavelength 2.291 Å produces a diffracted line on a film at $2\theta = 58.2°$. What is the interplanar spacing of the crystallographic plane that produces the diffracted line?

Answer:

$$d_{hkl} = \frac{\lambda}{2 \sin \theta} = \frac{2.291}{2 \sin 29.1} = 2.355\,\text{Å}$$

If we were able to determine that the line represents a (111) plane, we might identify the specimen as gold.

◇   **Example 19-20**

The results of an X-ray diffraction experiment using molybdenum X rays with $\lambda = 0.7107$ Å show that diffracted peaks occur at the following 2θ angles.

Peak	2θ	Peak	2θ
1	20.20	5	46.19
2	28.72	6	50.90
3	35.36	7	55.28
4	41.07	8	59.42

Determine the crystal structure, the indices of the plane producing each peak, and the lattice parameter of the material.

Answer:

We can first determine the $\sin^2 \theta$ value for each peak, then divide through by the lowest denominator.

Peak	2θ	$\sin^2 \theta$	$\sin^2 \theta/0.0308$	$h^2 + k^2 + l^2$	(hkl)
1	20.20	0.0308	1	2	(110)
2	28.72	0.0615	2	4	(200)
3	35.36	0.0922	3	6	(211)
4	41.07	0.1230	4	8	(220)
5	46.19	0.1539	5	10	(310)
6	50.90	0.1847	6	12	(222)
7	55.28	0.2152	7	14	(321)
8	59.42	0.2456	8	16	(400)

When we do this, we find a pattern of $\sin^2 \theta/0.0308$ values of 1, 2, 3, 4, 5, 6, 7, and 8. If the material were simple cubic, the 7 would not be present because no planes have an $h^2 + k^2 + l^2$ value of 7. Therefore, the pattern must really be 2, 4, 6, 8, 10, 12, 14, 16, . . . and the material must be body-centered cubic. The (hkl) values listed give these required $h^2 + k^2 + l^2$ values.

We could then use 2θ values for any of the peaks to calculate the interplanar spacing and thus the lattice parameter. Picking peak 8,

$$2\theta = 59.42 \quad \text{or} \quad \theta = 29.71$$

$$d_{400} = \frac{\lambda}{2 \sin \theta} = \frac{0.7107}{2 \sin (29.71)} = 0.71699 \text{ Å}$$

$$a_0 = d_{400} \sqrt{h^2 + k^2 + l^2} = (0.71699)(4) = 2.866 \text{ Å}$$

This is the lattice parameter for body-centered cubic iron.

19–5   PHOTONIC SYSTEMS AND MATERIALS

Photonic systems use light to transmit information. Telephone communication systems, for instance, use fiber optics as a means for transmitting larger numbers of messages than can be done using conventional techniques. Supercomputers may rely on photonic systems rather than the present electronic systems based largely on electronic transport through silicon semiconducting devices.

A photonic system must generate a light signal from some other source, such as an electrical signal, transmit the light to a receiver, process the data received, and convert the data to a usable form (Figure 19-22). In order to do this, photonic materials are required. Most of the principles and materials presently utilized in photonic systems have already been introduced in the previous sections. Let us now review these materials in the context of an actual system, pointing out some of the special requirements that are needed.

Generating the Signal.   In order to best transmit and process information, the light should be coherent and monochromatic. Thus a laser is an ideal method for generating the photons. The Group III-V semiconductors such as GaAs, GaAlAs, and InGaAsP have energy gaps that provide emitted photons in the visible spectrum. Lasers built from these materials can be energized by a voltage, therefore generating a laser beam. Light-emitting diodes are a second device that can be used to produce the photons.

By varying the voltages applied to these devices, the intensity of the photon beam can be varied as well. The intensity of the beam can then be used to convey information.

Transmitting the Beam.   Waveguides composed of glass fibers are used to transmit the light from the source to the receiver. In order for the optical fibers efficiently to transmit light long distances, the glass must have exceptional transparency and must not leak any of the light.

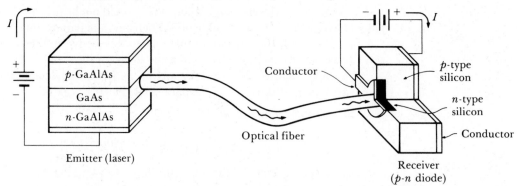

FIGURE 19-22 A photonic system for transmitting information involves a laser or LED to generate photons from an electrical signal, optical fibers to transmit the beam of photons efficiently, and an LED receiver to convert the photons back into an electrical signal.

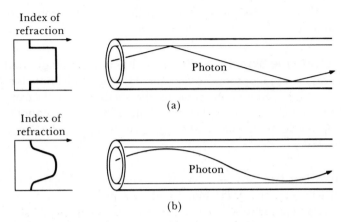

FIGURE 19-23 *Two methods for controlling leakage from an optical fiber. (a) A composite glass fiber, in which the index of refraction is slightly different in each glass, and (b) a glass fiber doped at the surface to lower the index of refraction.*

The index of refraction is important when considering losses due to leakage. In a previous section, we found that light will be reflected if the index of refraction of the surroundings is less than that of the fiber, provided that the angle between the beam and the length of the fiber is sufficiently small. Simple optical fibers are often a composite glass material, with the central core having a higher index of refraction than the glass sheath. The difference in index of refraction helps keep the beam within the central core, and the beam travels the path shown in Figure 19-23(a).

Even if the angle between the beam and the fiber is small, the distance that the beam must travel through a composite fiber is considerably longer than the fiber itself, due to the sharp reflections that are involved at the distinct boundary between the two types of glass. More complex fibers contain a core glass that is doped at the surface with B_2O_3 or GeO_2; these dopants gradually lower the index of refraction near the surface of the inner fiber. The gradual change in index allows the beam of light to change direction gradually, rather than sharply, reducing the total path it must follow. The more complex fibers help to transmit a less distorted signal, particularly in long fibers.

Light is also scattered internally by the glass. Some scattering due to the disordered structure of the glass will occur for ultraviolet photons; excitation of electrons in the atoms of the glass can occur and produce new photons of a different wavelength for ultraviolet beams, while infrared beams can cause vibrational excitations of the bonds between the atoms; any kind of impurity or porosity in the fiber can cause scattering or refraction of the beam. To minimize internal losses, photons of an appropriate wavelength and exceptionally pure silica glass are used. Even better materials would include certain crystalline materials, such as ZrF_2, As_2Se_3, or KI. Because these materials are crystalline, scattering due to a disordered structure would not occur. Because the atoms in these materials are more massive, less excitation of electrons would occur.

Because the bonding is weaker than in SiO_2, vibrational excitations would be minimized.

Receiving the Signal. Either the light-emitting diodes that might be used to generate a signal or more conventional silicon semiconductor diodes can receive the signal. When a photon reaches the *p-n* diode, an electron is excited into the conduction band, leaving behind a hole. If a voltage is applied to the diode, the electron-hole pair will create a current that can be amplified and further processed.

Processing the Signal. Normally the received signal is immediately converted into an electronic signal and then processed using conventional silicon-based semiconductor devices. However, certain materials, such as $LiNbO_3$, have a nonlinear optical response. When a beam of photons is received by such a material, the material can act as a transistor and amplify the signal or act as a switch (or a logic gate in a computer) and control the path of the beam. Development of photonic transistors could lead to an optically based computer.

Thermal Properties

19–6   HEAT CAPACITY

Thermal properties, including heat capacity, thermal conductivity, and thermal expansion, are influenced by atomic vibration and, in the case of thermal conductivity, transfer of energy by electrons. The vibration may be expressed as an energy or the wavelike nature of the energy can be utilized.

At absolute zero, the atoms have a minimum energy. But when heat is supplied to the material, the atoms gain thermal energy and vibrate at a particular amplitude and frequency. This vibration produces an elastic wave called a *phonon*. The energy of the phonon can be expressed in terms of wavelength or frequency, as in Equation 19-1.

$$E = hc/\lambda = h\nu \tag{19-11}$$

The material gains or loses heat by gaining or losing phonons. The energy, or number of phonons, required to change the temperature of the material one degree is of interest. We can express this energy as the heat capacity or the specific heat.

The *heat capacity* is the energy required to raise the temperature of one mole of a material one degree. The heat capacity can be expressed either at constant

TABLE 19-4 The specific heat of selected materials at 27°C

Material	Specific Heat (cal/g · K)
Al	0.215
Cu	0.092
B	0.245
Fe	0.106
Pb	0.038
Mg	0.243
Ni	0.106
Si	0.168
Ti	0.125
W	0.032
Zn	0.093
Water	1.0
He	1.24
N	0.249
Silica	0.265
Al_2O_3	0.20
SiC	0.25
Si_3N_4	0.17
Polymers	0.20–0.35
Diamond	0.124

pressure, C_p, or at a constant volume, C_v. At high temperatures, the heat capacity for a given volume of material approaches

$$C_p = 3R = 6\,\text{cal/mol} \cdot \text{K} \tag{19-12}$$

where R is the gas constant (1.987 cal/mol). However, as shown in Figure 19-24, the heat capacity is not a constant. The heat capacity of metals approaches 6 cal/mol · K near room temperature, but in ceramics, this value is not reached until near 1000°C.

The *specific heat* is the energy required to raise the temperature of a particular weight or mass of a material one degree. The relationship between specific heat and heat capacity is

$$\text{Specific heat} = c = \frac{\text{heat capacity}}{\text{atomic weight}} \tag{19-13}$$

In most engineering calculations, the specific heat is more conveniently used. The specific heat of typical materials is given in Table 19-4. Neither the heat capacity nor the specific heat depends significantly on the structure of the material; thus, changes in dislocation density, grain size, or vacancies have little effect.

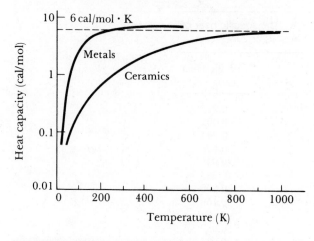

FIGURE 19-24 The heat capacity as a function of temperature for metals and ceramics.

◇ | **Example 19-21**

Calculate the temperature after 5000 cal are introduced to 250 g of tungsten originally at 25°C.

Answer:

The specific heat of tungsten is 0.032 cal/g · K. Thus

$$\Delta T = \frac{5000\,\text{cal}}{(250\,\text{g})\,(0.032\,\text{cal/g}\cdot\text{K})} = 625\,\text{K} = 625°\text{C}$$

$$T_{\text{final}} = 25 + 625 = 650°\text{C}$$

◇ | **Example 19-22**

Suppose the temperature of 50 g of niobium increases 75°C when heated for a period of time. Estimate the specific heat and determine the heat in calories required.

Answer:

The atomic weight of niobium is 92.91 g/g · mole. We can use Equation 19-13 to estimate the heat required to raise the temperature of one gram one °C.

$$c \approx \frac{6}{92.91} = 0.0646\,\text{cal/g}\cdot°\text{C}$$

Thus the total heat required is

$$\text{heat} = (0.0646\,\text{cal/g}\cdot°\text{C})\,(50\,\text{g})\,(75°\text{C})$$

$$= 242\,\text{cal}$$

◇ | **Example 19-23**

Calculate the expected specific heat of aluminum, copper, iron, and silicon based on the classical heat capacity of 6 cal/mol · K and compare with the measured values in Table 19-4.

Answer:

The weight in grams of one mole of each metal is the atomic weight.

Metal	Atomic Weight (g/g · mole)	$\dfrac{6\,\text{cal/mol}\cdot\text{K}}{\text{Atomic Weight}}$ (cal/g · K)	Measured Specific Heat (cal/g · K)
Al	26.981	0.222	0.215
Cu	63.54	0.094	0.092
Fe	55.847	0.107	0.106
Si	28.08	0.214	0.168

The results, although not precise, do closely correspond to the measured values for specific heat.

19–7   THERMAL EXPANSION

An atom that gains thermal energy and begins to vibrate behaves as though it has a larger atomic radius. The average distance between the atoms and the overall dimensions of the material increase. The change in the dimensions of the material Δl per unit length is given by the *linear coefficient of thermal expansion α*.

$$\alpha \;=\; \frac{1}{l}\frac{\Delta l}{\Delta T} \tag{19-14}$$

where ΔT is the increase in temperature and l is the initial length.

The linear coefficient of thermal expansion is related to the strength of the atomic bonds (Figure 2-15). In order for the atoms to move from their equilibrium separation, energy must be introduced into the material. If a very deep energy trough caused by strong atomic bonding is characteristic of the material, the atoms separate to a lesser degree and the material has a low linear coefficient of thermal expansion. This relationship also indicates that materials having a high melting temperature, also due to strong atomic attractions, have low linear coefficients of thermal expansion (Figure 19-25). Materials like lead and polyethylene expand a great deal compared with diamond, tungsten, and ceramics (Table 19-5).

TABLE 19-5 The linear coefficient of thermal expansion at room temperature for selected materials

Material	Linear Coefficient of Thermal Expansion ($\times 10^{-6}$ cm/cm $\cdot$ °C)
Al	25
Cu	16.6
Fe	12
Pb	29
Mg	25
Ni	13
Si	3
Ti	8.5
W	4.5
1020 steel	12
Gray iron	12
Stainless steel	17.3
3003 aluminum alloy	23.2
Yellow brass	18.9
Invar (Fe-36% Ni)	1.54
Polyethylene	100
Polystyrene	70
Polyethylene—30% glass fiber	48
Epoxy	55
6,6-Nylon	80
6,6-Nylon—33% glass fiber	20
Fused quartz	0.55
Al_2O_3	6.7
Si_3N_4	3.3
SiC	4.3
Partially stabilized ZrO_2	10.6

From *Handbook of Chemistry and Physics*, 56th Ed., CRC Press, 1975, and other sources.

Several precautions must be taken when calculating dimensional changes in materials.

1. The expansion characteristics of some materials, particularly single crystals or materials having a preferred orientation, may be anisotropic.

2. Allotropic materials have abrupt changes in their dimensions when the phase transformation occurs (Figure 19-26). These abrupt changes contribute to cracking of refractories on heating or cooling and quench cracks in steels.

3. The linear coefficient of expansion is not constant at all temperatures. Normally, α either is listed in handbooks as a complicated temperature-dependent function or is given as a constant for only a particular temperature range.

4. Interaction of the material with electric or magnetic fields produced by magnetic domains may prevent normal expansion until temperatures above the

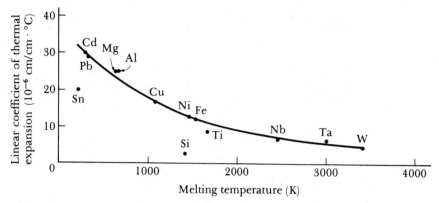

FIGURE 19-25 The relationship between the linear coefficient of thermal expansion and the melting temperature in metals at 25°C. Higher melting point metals tend to expand to a lesser degree.

Curie temperature are reached. This is the case for Invar, an Fe-36% Ni alloy, which undergoes practically no dimensional changes at temperatures below the Curie temperature, about 200°C. This makes Invar attractive as a material for bimetallics (Figure 19-26).

◇ | **Example 19-24**

Explain why, in Figure 19-25, the linear coefficients of thermal expansion for silicon and tin do not fall on the curve. How would you expect germanium to fit into this figure?

Answer:

Both silicon and tin are covalently bonded. The strong covalent bonds are more difficult to stretch than the metallic bonds (a deeper trough in the energy-separation curve), so these elements have a lower coefficient. Since germanium also is covalently bonded, its thermal expansion should be less than that predicted by Figure 19-25.

◇ | **Example 19-25**

An aluminum casting solidifies at 660°C. At that point, the casting is 25 cm long. What is the length after the casting cools to room temperature?

Answer:

The linear coefficient of thermal expansion for aluminum is 25×10^{-6} cm/cm · °C. The temperature change from solidification to room temperature is $660 - 27 = 633$°C.

$$\Delta l = \alpha l \Delta T = (25 \times 10^{-6})(25)(633) = 0.4 \, \text{cm}$$
$$l_f = l - \Delta l = 25 - 0.4 = 24.6 \, \text{cm}$$

The contraction in castings is normally compensated for by making the original pattern oversized by an amount calculated from the linear coefficient of thermal expansion. In this case, the original pattern should be 25.4 cm long to permit a final aluminum casting exactly 25 cm long.

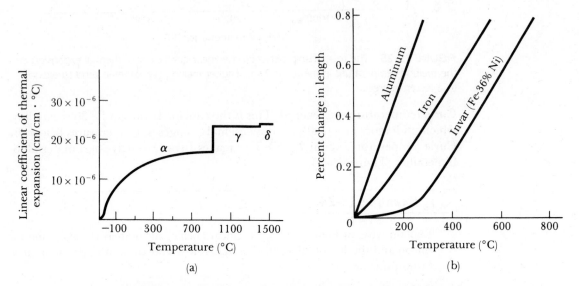

(a)   (b)

FIGURE 19-26 (a) The linear coefficient of thermal expansion of iron changes abruptly at temperatures where an allotropic transformation occurs. (b) The expansion of Invar is very low due to the magnetic properties of the material at low temperatures.

19-8   THERMAL CONDUCTIVITY

The *thermal conductivity* K is a measure of the rate at which heat is transferred through a material. The conductivity relates the heat Q transferred across a given plane of area A per second when a temperature gradient $\Delta T/\Delta x$ exists (Figure 19-27).

$$\frac{Q}{A} = K \frac{\Delta T}{\Delta x} \qquad (19\text{-}15)$$

Note that the thermal conductivity K plays the same role in heat transfer that the diffusion coefficient D does in mass transfer.

If valence electrons are easily excited into the conduction band, thermal energy can be transferred by the electrons. The amount of energy transferred

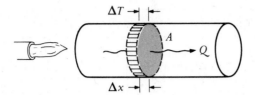

FIGURE 19-27 When one end of a bar is heated, a heat flux Q/A flows towards the cold end at a rate determined by the temperature gradient produced in the bar.

depends on the number of excited electrons and their mobility; thus we expect a relationship between thermal and electrical conductivity.

$$\frac{K}{\sigma T} \;=\; L \;=\; 5.5 \times 10^{-9}\,\text{cal} \cdot \Omega/\text{s} \cdot \text{K}^2 \tag{19-16}$$

where L is the Lorentz constant.

Thermally induced vibrations of the atoms cause the emission of phonons which also transfer energy through the material. We expect higher temperatures to increase the rate of heat transfer due to higher energy phonons.

◇ | **Example 19-26**

A window glass 1 cm thick and 4 ft × 4 ft square separates a room at 25°C from the outside, at 40°C. Calculate the amount of heat entering the room through the window each day.

Answer:

From Table 19-6, the thermal conductivity of a soda-lime glass, typical of windows, is 0.0023 cal/cm · s · K.

$$\frac{\Delta T}{\Delta x} \;=\; \frac{40 - 25}{1} \;=\; 15\,\text{K/cm}$$

$$\frac{Q}{A} \;=\; K\frac{\Delta T}{\Delta x} \;=\; (0.0023)\,(15) \;=\; 0.0345\,\text{cal/cm}^2 \cdot \text{s}$$

$$A \;=\; [(4\,\text{ft})\,(12\,\text{in./ft})\,(2.54\,\text{cm/in.})]^2 \;=\; 1.486 \times 10^4\,\text{cm}^2$$

$$t \;=\; (1\;\text{day})\,(24\,\text{h/day})\,(3600\,\text{s/h}) \;=\; 8.64 \times 10^4\,\text{s}$$

$$\frac{\text{Heat}}{\text{Day}} \;=\; \left(\frac{Q}{A}\right)(t)\,(A) \;=\; (0.0345)\,(8.64 \times 10^4)\,(1.486 \times 10^4)$$

$$=\; 4.36 \times 10^7\,\text{cal/day}$$

Thermal Conductivity in Metals. The electronic contributions are the dominant factor in the conduction of thermal energy in metals and alloys, since the valence band is not completely filled. But the thermal conductivity also depends on lattice defects, microstructure, and processing of the metal. Thus cold-worked

TABLE 19-6 Thermal conductivity of selected materials at 27°C

Material	Thermal Conductivity (cal/cm · s · K)	Material	Thermal Conductivity (cal/cm · s · K)
Al	0.57	Yellow brass	0.53
Cu	0.96	Cu-30% Ni	0.12
Fe	0.19	Ar	0.000043
Mg	0.24	Carbon (graphite)	0.80
Pb	0.084	Carbon (diamond)	5.54
Si	0.36	Soda-lime glass	0.0023
Ti	0.052	Vitreous silica	0.0032
W	0.41	Vycor glass	0.0030
Zn	0.28	Fireclay	0.00064
Zr	0.054	Al_2O_3	0.038
1020 steel	0.24	ZrO_2	0.012
Ferrite	0.18	Si_3N_4	0.035
Cementite	0.12	Silicon carbide	0.21
304 stainless steel	0.072	Polyimide	0.0005
Gray iron	0.19	6,6-Nylon	0.0006
3003 aluminum alloy	0.67	Polyethylene	0.0008

metals, solid solution-strengthened metals, and two-phase alloys might display lower conductivities compared with their defect-free counterparts.

We expect higher temperatures to reduce the mobility and the thermal conductivity of metals. However, higher temperatures also increase the energy of the electrons and permit heat to be transferred by lattice vibration. In metals, the thermal conductivity often initially decreases with temperature, becomes nearly constant, and then increases slightly, as in iron (Figure 19-28). However, conductivity decreases continuously when aluminum is heated but increases continuously when platinum is heated.

Thermal Conductivity in Ceramics. Lattice vibrations, or phonons, are responsible for transfer of heat in ceramic or insulating materials. In these materials, the energy gap is too large for many electrons to be excited into the conduction band except at very high temperatures. Many ceramics, including glass, have higher thermal conductivities at higher temperatures due to higher energy phonons and some electronic contributions. However, other ceramics, such as SiC and Al_2O_3, tend to conduct heat more slowly at high temperatures.

Other factors influence the thermal conductivity of ceramics. Materials with a close-packed structure, low density, and high modulus of elasticity produce high-energy phonons that encourage high thermal conductivities. Crystalline solids have higher conductivities than their amorphous or glassy counterparts because less scattering of phonons occurs. Porosity reduces thermal conductivity. The best insulating brick, for example, contains a large fraction of porosity. We can consider this material as a composite material, where porosity is one of the materials.

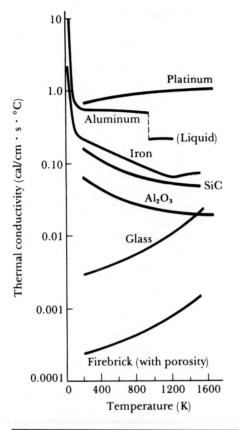

FIGURE 19-28   The effect of temperature on the thermal conductivity of selected materials.

◇   **Example 19-27**

Suppose we are able to introduce 30 vol % porosity into a soda-lime glass. If we assume that the thermal conductivity of the porosity is zero, how much heat do we save in Example 19-26 per day?

Answer:

Let's use the rule of mixtures to estimate the thermal conductivity of our glass-porosity composite.

$$K_c = K_g f_g + K_p f_p = (0.0023)(0.7) + (0)(0.3)$$
$$= 0.0016 \, \text{cal/cm} \cdot \text{s} \cdot \text{K}$$

$$\text{Total heat} = \left(\frac{Q}{A}\right)(t)(A) = (K)\left(\frac{\Delta T}{\Delta x}\right)(t)(A)$$

$$= (0.0016)(15)(8.64 \times 10^4)(1.486 \times 10^4) = 3.08 \times 10^7 \, \text{cal/day}$$

$$\text{Heat savings} = 4.36 \times 10^7 - 3.08 \times 10^7 = 1.28 \times 10^7 \, \text{cal/day}$$

Thermal Conductivity in Semiconductors. Heat is conducted in semiconductors by both phonons and electrons. At low temperatures, phonons are the principal carriers of energy, but at higher temperatures, electrons are excited through the small energy gap into the conduction band and the thermal conductivity increases significantly.

19–9   THERMAL SHOCK

Cracking due to volume changes may occur when there is a sudden change in temperature, particularly in brittle materials like glass and ceramics. *Thermal shock* is a combination of (a) expansion or contraction constraints, (b) temperature gradients due to the thermal conductivity, and (c) phase transformations.

When a piece is cooled quickly, a temperature gradient is produced. This can lead to different amounts of contraction in different areas. Residual tensile stresses are produced, which can cause a Griffith flaw to produce brittle fracture.

We often measure the resistance to thermal shock by determining the maximum temperature difference that can be tolerated during a quench without affecting the mechanical properties of the material. Fused silica has a thermal shock resistance of about 3000°C. Figure 19-29 shows the effect of quenching temperature difference on the modulus of rupture in sialon $(Si_3Al_3O_3N_5)$; no cracks and no change in the properties of the ceramic are evident until the quenching temperature approaches 950°C. Other ceramics have poorer resistance; partially stabilized zirconia (PSZ) and Si_3N_4 have a thermal shock resistance of 500°C; SiC has a shock resistance of 350°C, and Al_2O_3 and ordinary glass have a resistance of about 200°C.

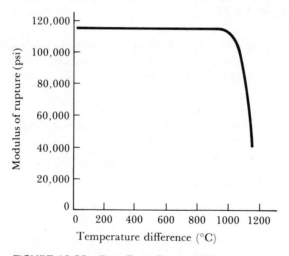

FIGURE 19-29 The effect of quenching temperature difference on the modulus of rupture of sialon. The thermal shock resistance of the ceramic is about 950°C.

If a material has a high thermal conductivity, a low coefficient of thermal expansion, and no polymorphic transformation, we expect fewer problems with thermal shock. This is the case in fused silica but not in crystalline silica. This is also a reason why zirconia is stabilized with CaO or MgO to produce a cubic crystal structure at all temperatures.

We do not expect thermal shock to be a problem in most metals; metals normally have sufficient ductility to permit deformation rather than fracture.

SUMMARY

The optical and thermal properties of a material depend on interactions between radiation or energy and the atomic structure and arrangement. As a consequence, we can influence many of these properties by altering the atomic arrangement through changes in the composition and processing of the material. This in turn permits us to create myriad devices, such as light-emitting diodes, lasers, and photonic systems, that perform many functions currently satisfied by electronic devices.

GLOSSARY

Absorption edge. The wavelength at which the absorption characteristics of a material abruptly change.

Bragg's law. The relationship describing the angle at which a beam of X rays of a particular wavelength diffracts from crystallographic planes of a given interplanar spacing.

Characteristic spectrum. The spectrum of radiation emitted from a material which occurs at fixed wavelengths corresponding to particular energy level differences within the atomic structure of the material.

Continuous spectrum. Radiation emitted from a material having all wavelengths longer than a critical short wavelength limit.

Debye–Scherrer technique. A technique using the interaction between X-ray radiation and planes of atoms in a crystal to obtain information concerning the identity and characteristics of a material.

Fluorescence. Emission of radiation from a material only when the material is actually being stimulated.

Heat capacity. The energy required to raise the temperature of one mole of a material one degree.

Index of refraction. Relates the change in velocity and direction of radiation as it passes through a transparent material.

Laser. A beam of monochromatic coherent radiation produced by the controlled emission of photons.

Light-emitting diodes (LEDs). Electronic p-n junction devices that convert an electrical signal into visible light.

Linear absorption coefficient. Describes the ability of a material to absorb radiation.

Linear coefficient of thermal expansion. Describes the amount by which each unit length of a material changes when the temperature of the material changes by one degree.

Lorentz constant. Relates thermal and electrical conductivity.

Luminescence. Conversion of radiation to visible light.

Phonon. An elastic wave that transfers energy through a material.

Phosphorescence. Emission of radiation from a material after the material is stimulated.

Photoconduction Production of a current due to stimulation of electrons into the conduction band by light radiation.

Photons. Energy or radiation produced from atomic, electronic, or nuclear sources that can be treated as particles or waves.

Reflectivity. The percent of incident radiation that is reflected.

Relaxation time. The time required for $1/e$ of the electrons to drop from the conduction band to the valence band in luminescence.

Short wavelength limit. The shortest wavelength or highest energy radiation emitted from a material under particular conditions.

Specific heat. The energy required to raise the temperature of one gram of a material one degree.

Thermal conductivity. Measures the rate at which heat is transferred through a material.

Thermal emission. Emission of photons from a material due to excitation of the material by heat.

Thermal shock. Cracking of a brittle material when subjected to stresses caused by a sudden temperature change.

X rays. Electromagnetic radiation produced by changes in the electronic structure of atoms.

PRACTICE PROBLEMS

1. A tungsten filament emits a continuous spectrum having a minimum wavelength of 1.02 Å. Calculate the voltage on the filament.

2. Determine the highest frequency radiation produced when a tungsten filament is heated using a 10,500-V supply.

3. Calculate the minimum voltage required to produce the K_α peak in cobalt.

4. A voltage of 2017 V is required to produce the K_α characteristic peak in potassium. Calculate the wavelength of the K_α peak.

5. Determine the difference in energy between electrons in (a) the K and L shells, (b) the K and M shells, and (c) the L and M shells of tungsten.

6. Figure 19-30 shows the results of an X-ray fluorescence analysis. Determine (a) the accelerating voltage and (b) the identity of the elements in the sample.

7. Figure 19-31 shows the results of an X-ray fluorescence analysis. Determine (a) the accelerating voltage and (b) the identity of the elements in the sample.

8. Figure 19-32 shows the energies of photons produced from an energy-dispersive analysis of radiation emitted from a specimen in a scanning electron microscope. Determine the identity of the elements in the sample.

9. Figure 19-33 shows the energies of photons produced from an energy-dispersive analysis of radiation emitted from a specimen in a scanning electron microscope. Determine the identity of the elements in the sample.

10. If the relaxation time is 3×10^{-4} s, determine the time required for the intensity of a phosphorescent material to decrease to 1% of the original intensity.

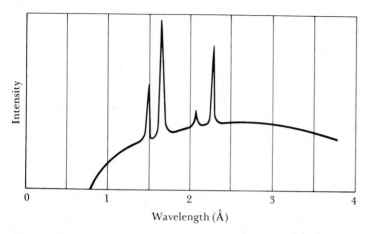

FIGURE 19-30 Results from an X-ray fluorescence analysis of an unknown metal sample (Problem 19-6).

11. The intensity of a phosphorescent material is reduced to 90% of the original intensity after 8.43×10^{-7} s. Calculate the relaxation time for the material.

12. The intensity of a phosphorescent material is reduced to 95% of its original intensity after 2.56×10^{-6} s. Determine the time required for the intensity to decrease to 0.1% of the original intensity.

13. When europium is added to ZnS, donor bands are introduced which provide for emission of photons in the red spectrum. What is

the energy difference between (a) the donor level and the conduction band and (b) the donor level and the valence band? (See Table 17-10.)

14. Will an incident beam of photons having a wavelength of 8000 Å cause luminescence in (a) CdS, (b) CdSe, and (c) PbTe? What is the maximum wavelength of incident photons required for each material? (See Table 17-10.)

15. Using the data in Table 17-10, determine (a) the wavelength of photons produced from pure GaAs, (b) whether the photons are in

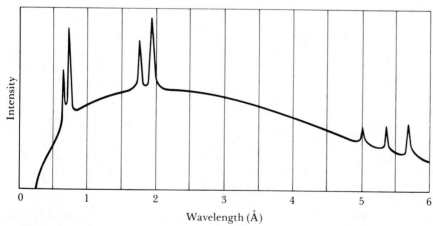

FIGURE 19-31 Results from an X-ray fluorescence analysis of an unknown metal sample (Problem 19-7).

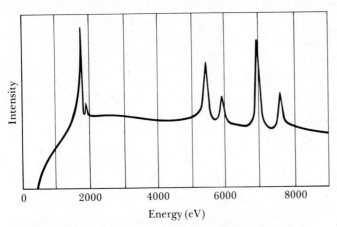

FIGURE 19-32 X-ray emission spectrum for Problem 19-8.

the infrared, visible, or ultraviolet portion of the spectrum, and (c) the location of the electron trap with respect to the valence band if GaAs is to produce photons in the yellow portion of the spectrum.

16. By appropriately doping yttrium aluminum garnet with neodymium, electrons are excited within the 4*f* energy band of the Nd atoms. Determine the approximate energy transition if the Nd:YAG serves as a laser, producing a wavelength of 5320 Å. What color would the laser beam possess?

17. Which, if any, of the semiconducting compounds listed in Table 17-10 could produce a green laser beam with a wavelength of 5130 Å?

18. Determine the wavelength of photons emitted from (a) pure silicon and (b) silicon doped with indium. Are the photons in the infrared, visible, or ultraviolet portion of the spectrum? (See Tables 17-8 and 17-9.)

19. Based on the data in Figure 17-27, determine how the wavelength of the emitted photons changes as aluminum replaces gallium, producing AlAs rather than GaAs.

20. A beam of X rays having a wavelength of 2.291 Å is passed through a platinum foil having a linear absorption coefficient of $6540 \, \text{cm}^{-1}$. If only 2% of the original intensity of the beam is transmitted, determine the thickness of the foil.

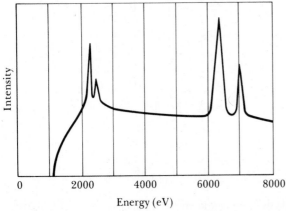

FIGURE 19-33 X-ray emission spectrum for Problem 19-9.

21. Only 32% of an X-ray beam passes through a 0.001-cm foil of titanium. Determine (a) the linear absorption coefficient and (b) the thickness required to filter out all but 1% of the X-ray beam.

22. Using Figure 19-15, determine the exponent in the relationship between the absorption coefficient and incident wavelength for nickel. (The exponent should be a whole number.) Based on this relationship, estimate the absorption coefficient of a nickel foil for X rays having a wavelength of 13 Å.

23. From Figure 17-27, determine whether any of the solid solutions between GaAs and GaSb will transmit visible light.

24. Which, if any, of the following semiconducting compounds are capable of transmitting visible light—InSb, InAs, GaAs, and GaP? (See Table 17-10.)

25. Describe how the index of refraction will change when the wavelength or frequency of the incident radiation changes for (a) polyvinyl chloride and (b) polyethylene.

26. Suppose a laser beam strikes a 1-cm thick sheet of leaded glass at an angle of 15° to the normal of the glass. Determine the angle between the beam and the normal as the laser passes through the glass. By what distance will the transmitted beam be displaced from its original path by the glass?

27. Suppose a beam of light from an LED passes through air and impinges at an angle of 30° to the normal of a composite sheet of TiO_2 and polyethylene. Determine the angle between the beam and the normal of the composite as the beam of light passes (a) through the TiO_2 and (b) through the polyethylene.

28. A beam of light passing through a glass sheet ($n = 1.5$) at an angle of 26° to the normal of the sheet then enters a second sheet of a transparent dielectric. If the beam of light in the dielectric makes an angle of 18° to the normal, determine (a) the index of refraction and (b) the dielectric constant of the dielectric.

29. Suppose we construct a composite glass fiber, with the inner portion of the fiber having an index of refraction of 1.5 and the outer sheath of the fiber having an index of refraction of 1.48. Calculate the maximum angle that a beam of light can deviate from the axis of the fiber without escaping from the inner portion of the fiber.

30. Suppose that 25% of the intensity of a beam of photons entering a material at a 90° angle to the surface is transmitted through a 1-cm thick material with a dielectric constant of 1.44. Determine the fraction of the beam that is (a) reflected and (b) absorbed. (c) Calculate the linear absorption coefficient of the photons in the material.

31. The linear absorption coefficient of a 0.25-cm thick material is $0.8\,cm^{-1}$. The intensity of a transmitted beam is 10% of the original beam. Determine (a) the reflectivity, (b) the index of refraction, and (c) the dielectric constant of the material.

32. What filter material would you use to isolate the K_α peak of the following X rays—iron, manganese, nickel?

33. Figure 19-34 shows the intensity of the radiation obtained from a copper X-ray generating tube as a function of wavelength. The accompanying table shows the linear absorption coefficient for a manganese filter for several wavelengths. If the Mn filter is 0.002 cm thick, calculate the intensity of the transmitted X-ray beam for each wavelength.

Wavelength (Å)	Linear Absorption Coefficient (cm^{-1})
0.711	249
1.436	1739
1.542	2110
1.659	2586
1.79	3202
1.937	473
2.103	591
2.291	739

34. The results of an X-ray diffraction analysis using copper radiation ($\lambda = 1.5418\,Å$)

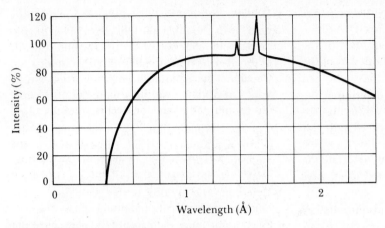

FIGURE 19-34 Intensity of the initial spectrum from a copper X-ray source before filtering (Problem 19-33).

are given as 2θ values below. Determine (a) the crystal structure of the metal, (b) the Miller indices of each of the peaks, and (c) the lattice parameter of the material.

Peak	2θ
1	26.65
2	38.05
3	47.06
4	54.81
5	62.04
6	68.71
7	81.33

35. Figure 19-35 shows the results from an X-ray diffraction experiment as the intensity of the diffracted peak versus the 2θ angle. Determine (a) the crystal structure of the metal, (b) the indices of the planes that produce each of the peaks, (c) the lattice parameter of the metal, and (d) the probable identity of the metal. Copper radiation ($\lambda = 1.5418\,\text{Å}$) is used.

36. An X-ray film with a 180-mm diameter is used in an X-ray camera. A diffracted line 40 mm from the exit port of the beam is produced. Determine the interplanar spacing of

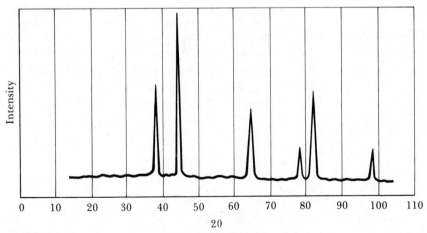

FIGURE 19-35 Results of an X-ray diffraction study of an unknown metal (Problem 19-35).

the plane that produced the diffracted line if copper radiation ($\lambda = 1.5418\,\text{Å}$) is used.

37. In a Debye–Scherrer X-ray film obtained from a powdered chromium specimen with copper radiation ($\lambda = 1.5418\,\text{Å}$), the first three diffraction lines are located at $2\theta = 44.4°$, $64.6°$, and $81.8°$. Determine the Miller indices corresponding to each line if the lattice parameter of BCC chromium is $2.8845\,\text{Å}$.

38. Calculate the temperature of a 1-kg sample of nickel originally at 25°C after 4000 calories are introduced.

39. Calculate the heat in calories required to raise the temperature of 50 g of silica by 50°C.

40. Suppose a 5 cm × 5 cm block of magnesium 2 cm thick is used as a heat sink; determine the thickness of a 5 cm × 5 cm block of copper required to remove the same amount of heat with only a 5°C temperature increase.

41. Liquid copper is poured into a mold 1 m in length and allowed to solidify. Calculate the length of the copper casting after cooling to room temperature.

42. A gray iron mold is produced by machining a cavity 15 in. long at room temperature. The mold is then heated to 600°F and liquid aluminum is poured into it. Assuming that the temperature of the mold does not change, calculate the final length of the aluminum casting after it cools to room temperature (75°F).

43. A 50-cm aluminum conductor wire is coated with a protective layer of epoxy. If the coated wire is heated from 25°C to 100°C, determine the final length of both the aluminum and the epoxy. What is likely to happen to the epoxy coating as a result of the expansion?

44. A 10 cm × 10 cm plate of magnesium is coated with a thin layer of fused quartz; the composite is then heated from 25°C to 350°C. Calculate the expected dimensions of both the magnesium and the quartz after heating. What is likely to happen to the quartz layer?

45. A casting is to be made from a 1020 steel. A sand core having a diameter of 10 cm is placed into the mold, the steel is poured around the sand core, and solidification occurs. After freezing, the sand core is knocked out of the casting, leaving behind a hole in the casting. Calculate the diameter of the hole. The melting temperature for the steel is about 1500°C.

46. Suppose an aluminum plate 10 cm × 10 cm × 1 cm thick separates a heat source at 300°C from a bath containing 1000 ml of water at 25°C. Calculate (a) the heat Q transferred to the water each second and (b) the time required to warm the water to 26°C.

47. We would like to heat 100 ml of water from 10°C to 11°C in 60 s. We will do this by just touching the end of a 1-cm long cylinder of copper in the water while the other end of the copper cylinder is in contact with a heat source operating at 800°C. Calculate the required diameter of the cylinder if half of the heat is lost from the cylinder during transfer.

48. A window glass (soda-lime) 5 mm thick and 1 m square separates a room at 25°C from the outside, at 0°C. (a) Calculate the amount of heat lost from the room through the window each day. (b) Suppose we produce a window composed of two layers of glass each 2 mm thick separated by a 1-mm thick layer of 6,6 nylon. Recalculate the amount of heat lost from the room each day.

49. From the data in Table 19-6, use the thermal conductivity to calculate the room temperature electrical conductivity of aluminum and compare your answer with the value given in Table 17-2.

V

PROTECTION AGAINST DETERIORATION AND FAILURE OF MATERIALS

In this section, the failure of materials by corrosion, wear, and fracture is discussed. We will again find that deterioration or failure is related to the structure, properties, and processing of the materials. In Chapter 20, corrosion and wear will be examined; electrochemical corrosion will be found to be particularly important. In addition to determining the mechanism for corrosion, we will look at techniques for controlling and preventing damage to the material by these processes.

In Chapter 21, we will review the mechanical failure of materials and pick up some hints on how to identify the cause for fracture. We will also examine a number of techniques by which we can nondestructively test a material to determine if it is subject to fracture.

20

CORROSION AND WEAR

20-1 INTRODUCTION

The composition and physical integrity of a solid material are altered in a corrosive environment. In chemical corrosion, the material is dissolved by a corrosive liquid. In electrochemical corrosion, metal atoms are removed from the solid material due to an electric circuit that is produced. Metals and certain ceramics react with a gaseous environment, usually at elevated temperatures, and the material may be destroyed by formation of oxides or other compounds. Polymers undergo cross-linking or degradation when exposed to oxygen at elevated temperatures. Materials may also be altered when exposed to radiation or even bacteria. Finally, a variety of wear and wear-corrosion mechanisms alter the shape of materials. Billions of dollars are required to repair the damage done by corrosion each year.

20-2 CHEMICAL CORROSION

In *chemical corrosion*, or direct solution, the material dissolves in a corrosive liquid medium. The material continues to dissolve in the liquid until either the material is consumed or the liquid is saturated. A simple example is salt dissolving in water.

Liquid Metal Attack. Liquid metals first attack a solid at high-energy locations, such as grain boundaries. If these regions continue to be attacked preferentially, cracks eventually grow (Figure 20-1). Often this form of corrosion is complicated by wetting, formation of compounds or fluxes that accelerate the attack, or electrochemical corrosion.

Selective Leaching. One particular element in an alloy may be selectively dissolved, or leached, from the solid. *Dezincification* occurs in brass containing

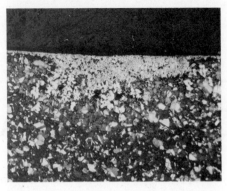

FIGURE 20-2   Photomicrograph of a copper deposit in brass, showing the effect of dezincification (× 50).

FIGURE 20-1
Molten lead is held in thick steel pots during refining. In this case, the molten lead has attacked a weld in a steel plate and cracks have developed. Eventually, the cracks propagate through the steel and molten lead leaks from the pot.

more than 15% Zn. Both copper and zinc are dissolved by aqueous solutions at high temperatures; the zinc ions remain in solution while the copper ions are replated onto the brass (Figure 20-2). Eventually, the brass becomes porous and weak.

Graphitic corrosion of gray cast iron occurs when iron is selectively dissolved in water or soil, leaving behind interconnected graphite flakes and a corrosion product. Localized graphitic corrosion often causes leakage or failure of buried gray iron gas lines, leading to explosions.

Fluxing of Ceramics.   Ceramic refractories used to contain molten metal during melting or refining may be dissolved by the slags that are produced on the metal surface. Frequent replacement of the refractories is thus required.

Solvents for Polymers.   Polymers dissolve in liquid solvents having similar structures. Polyethylene, which has a straight chainlike structure, dissolves readily in organic solvents whose molecules resemble the ethylene molecule. Polystyrene dissolves more easily in organic solvents such as benzene that have a similar molecular structure. Some polymers, such as polytetrafluoroethylene, are exceptionally resistant to chemical attack by almost all solvents.

Often chemical attack of polymers does not involve solution but, instead, the solvent may be absorbed into the polymer structure. This may cause swelling, cracking, or plasticization of the polymer, reducing the mechanical properties of the material. In many polymers, including nylon, water is a severe offender.

General Principles of Chemical Corrosion.   Several common features are observed in chemical corrosion.

1. Small ions or molecules dissolve faster than more complicated structures. A wax dissolves more rapidly in a liquid organic solvent than does polyethylene. Ionic salts dissolve faster in a flux than do complex silicate ions.

2. Solution occurs more rapidly when the solid and the dissolving liquid have similar structures.

3. Solution is accelerated at higher temperatures due to more rapid dissociation rates and higher solubilities.

Prevention of chemical corrosion is relatively difficult. Low temperatures and protective coatings may be helpful. However, the best way to prevent chemical corrosion is to avoid contact between the solid material and the liquid solvent or to select combinations of materials that are not soluble in one another. Thus, a refractory material selected to contain molten metal should not react with the slag that is produced on the molten metal.

◇ | **Example 20-1**

Suppose a slag that is produced during melting and refining of steel at 1600°C contains 50% SiO_2 and 50% FeO. The refractory containing the steel could be made from Al_2O_3 or MgO. From the ternary phase diagrams (Figure 20-3), which refractory would appear to work best for these conditions?

Answer:

In each plot, we can draw a line that runs from the 50% SiO_2–50% FeO point to pure Al_2O_3 or MgO. When we do this, we find that, at 1600°C, the solubility of Al_2O_3 or MgO in the slag is

Al_2O_3: Up to 40% Al_2O_3 dissolves in the SiO_2-FeO liquid

MgO: Up to 24% MgO dissolves in the SiO_2-FeO liquid

As the refractory is used numerous times, each time with fresh slag, the Al_2O_3 refractory will erode much more rapidly than the MgO refractory.

20–3 THE ELECTROCHEMICAL CELL

An *electrochemical cell* is formed when two pieces of metal in contact with one another are placed in a conducting liquid medium, or *electrolyte*. The complete electric circuit that is produced permits either *electroplating* or *electrochemical corrosion*.

Components of an Electrochemical Cell.   There are four components in an electrochemical cell (Figure 20-4).

1. Anode: The *anode* gives up electrons to the circuit and corrodes.
2. Cathode: The *cathode* receives electrons from the circuit by means of a chemical, or cathode, reaction. Ions that combine with the electrons produce a by-product at the cathode.
3. Physical contact: The anode and cathode must be electrically connected, usually by physical contact, to permit the electrons to flow from the anode to the cathode.
4. Electrolyte: A liquid electrolyte must be in contact with both the anode and the cathode. The electrolyte is conductive, thus completing the circuit. The

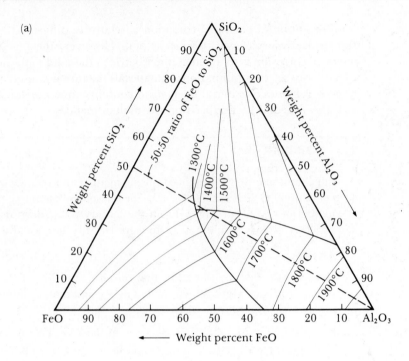

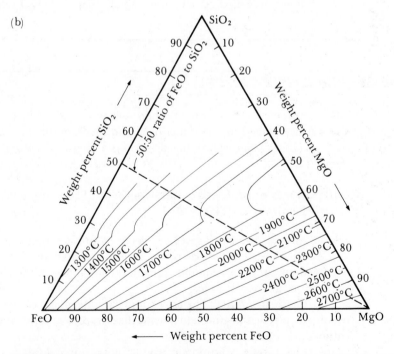

FIGURE 20-3   The liquidus plots for (a) SiO_2-FeO-Al_2O_3 and (b) SiO_2-FeO-MgO

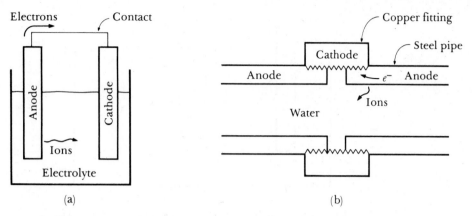

FIGURE 20-4   *The components in an electrochemical cell. (a) A possible electroplating set-up. (b) A corrosion cell between a steel water pipe and a copper fitting.*

electrolyte provides the means by which metallic ions leave the anode surface and assures that ions move to the cathode to accept the electrons.

This description of an electrochemical cell defines either electrochemical corrosion or electroplating. The two processes differ depending on the purpose of the cell, the source of the electric potential by which a current is caused to flow in the circuit, and the type of reaction that occurs at the cathode.

Anode Reaction.   The anode, which is a metal, undergoes an *oxidation reaction* by which metal atoms are ionized. The metal ions enter the electrolytic solution while the electrons leave the anode through the electrical connection.

$$M \rightarrow M^{n+} + ne^- \tag{20-1}$$

Because metal atoms leave the anode, the anode corrodes.

Cathode Reaction in Electroplating.   In electroplating, a *reduction reaction*, which is the reverse of the anode reaction, occurs at the cathode.

$$M^{n+} + ne^- \rightarrow M \tag{20.2}$$

The metal ions, either intentionally added to the electrolyte or formed by the anode reaction, combine with electrons at the cathode. The metal then plates out and covers the cathode surface.

Cathode Reactions in Corrosion.   Except in unusual conditions, plating of a metal does not occur during electrochemical corrosion. Instead, the reduction reaction forms a gas, solid, or liquid by-product at the cathode (Figure 20-5).

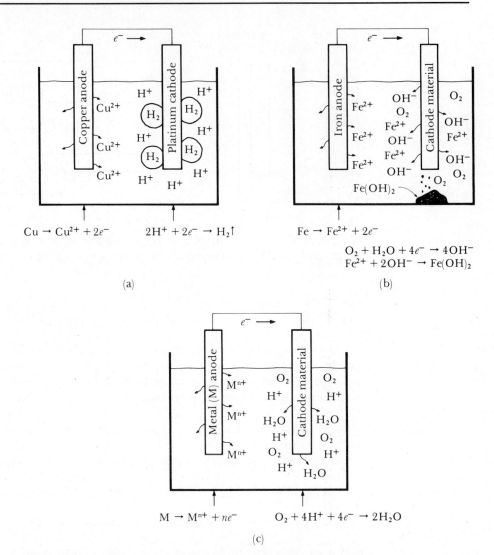

FIGURE 20-5 The anode and cathode reactions in typical electrolytic corrosion cells. (a) The hydrogen electrode, (b) the oxygen electrode, and (c) the water electrode.

1. The hydrogen electrode: In oxygen-free liquids, such as hydrochloric acid (HCl) or stagnant water, the most common cathode reaction is the evolution of hydrogen.

$$2H^+ + 2e^- \rightarrow H_2\uparrow \qquad (20\text{-}3)$$

This is the *hydrogen electrode.*

2. The oxygen electrode: In aereated water, oxygen is available to the cathode and hydroxyl, or OH^-, ions are formed.

$$O_2 + 2H_2O + 4e^- \rightarrow 4OH^- \qquad (20\text{-}4)$$

The *oxygen electrode* enriches the electrolyte in OH^- ions. These ions react with positively charged metallic ions, such as Fe^{2+}, finally producing a solid product, such as $Fe(OH)_2$, or rust.

3. The water electrode: In oxidizing acids, the cathode reaction produces water as a by-product.

$$O_2 + 4H^+ + 4e^- \rightarrow 2H_2O \qquad (20\text{-}5)$$

If a continuous supply of both oxygen and hydrogen is available, the *water electrode* produces neither a buildup of solid rust nor a high concentration or dilution of ions at the cathode.

20–4   THE ELECTRODE POTENTIAL IN ELECTROCHEMICAL CELLS

In electroplating, an imposed voltage is required to cause a current to flow in the cell. But in corrosion a potential naturally develops when a material is placed in a solution. Let's see how the potential required to drive the corrosion reaction develops.

Electrode Potential.   When a perfect ideal metal is placed in an electrolyte, an *electrode potential* is developed which is related to the tendency of the material to give up its electrons. However, the driving force for the oxidation reaction is offset by an equal but opposite driving force for the reduction reaction. No net corrosion occurs. Consequently, we cannot measure the electrode potential for a single electrode material.

Electromotive Force Series.   To determine the tendency of a metal to give up its electrons, we measure the *potential difference* between the metal and a standard electrode using a half-cell (Figure 20-6). The metal electrode to be tested is placed in a 1-M solution of its ions. A reference standard electrode is also placed in a 1-M solution of its ions. The two electrolytes are in electrical contact but are not permitted to mix with one another. Each electrode establishes its own electrode potential. By measuring the voltage between the two electrodes when the circuit is open, we obtain the potential difference. The potential of the hydrogen electrode, which is taken as our standard reference electrode, is arbitrarily set equal to zero volts. If the metal has a greater tendency to give up electrons than hydrogen, then the potential of the metal is negative—the metal is anodic with respect to the hydrogen electrode.

The *electromotive force*, or emf, *series* shown in Table 20-1 compares the electrode potential E_0 for each metal with the hydrogen electrode under standard conditions of 25°C and 1-M solution of ions in the electrolyte. Note that the measurement of the potential is made when the electric circuit is open. The voltage difference begins to change as soon as the circuit is closed.

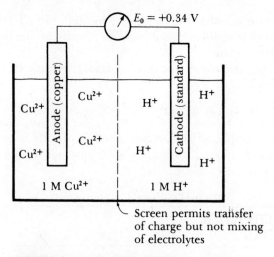

$E_0 = +0.34$ V

Screen permits transfer
of charge but not mixing
of electrolytes

FIGURE 20-6 The half-cell used to measure the electrode potential of copper under standard conditions. The electrode potential of copper is the potential difference between it and the standard hydrogen electrode in an open circuit. Since E_0 is greater than zero, copper is cathodic compared with the hydrogen electrode.

TABLE 20-1 The electromotive force (emf) series for selected elements

	Metal	Electrode Potential (V)
Anodic	$Li \rightarrow Li^+ + e^-$	-3.05
	$Mg \rightarrow Mg^{2+} + 2e^-$	-2.37
	$Al \rightarrow Al^{3+} + 3e^-$	-1.66
	$Ti \rightarrow Ti^{2+} + 2e^-$	-1.63
	$Mn \rightarrow Mn^{2+} + 2e^-$	-1.63
	$Zn \rightarrow Zn^{2+} + 2e^-$	-0.76
	$Cr \rightarrow Cr^{3+} + 3e^-$	-0.74
	$Fe \rightarrow Fe^{2+} + 2e^-$	-0.44
	$Ni \rightarrow Ni^{2+} + 2e^-$	-0.25
	$Sn \rightarrow Sn^{2+} + 2e^-$	-0.14
	$Pb \rightarrow Pb^{2+} + 2e^-$	-0.13
	$H_2 \rightarrow 2H^+ + 2e^-$	0.00
	$Cu \rightarrow Cu^{2+} + 2e^-$	$+0.34$
	$4(OH)^- \rightarrow O_2 + 2H_2O + 4e^-$	$+0.40$
	$Ag \rightarrow Ag^+ + e^-$	$+0.80$
	$Pt \rightarrow Pt^{4+} + 4e^-$	$+1.20$
	$2H_2O \rightarrow O_2 + 4H^+ + 4e^-$	$+1.23$
Cathodic	$Au \rightarrow Au^{3+} + 3e^-$	$+1.5$

Effect of Concentration on the Electrode Potential.   The electrode potential depends on the concentration of the electrolyte. The *Nernst equation* permits us to estimate the electrode potential in nonstandard solutions.

$$E \;=\; E_0 + \frac{0.0592}{n} \log\,(C_{ion}) \tag{20-6}$$

where E is the electrode potential in a solution containing a concentration C_{ion} of the metal in molar units, n is the valence of the metallic ion, and E_0 is the standard electrode potential in a 1-M solution. Note that when $C_{ion} = 1$, $E = E_0$.

◇ | Example 20-2

Suppose 1 g of copper as Cu^{2+} is dissolved in 1000 g of water to produce an electrolyte. Calculate the electrode potential of the copper half-cell in this electrolyte.

Answer:

From chemistry, we know that a standard 1-M solution of Cu^{2+} is obtained when we add 1 mol of Cu^{2+} (an amount equal to the atomic mass of copper) to 1000 g of water. The atomic mass of copper is 63.54 g/g · mole. The concentration of the solution when only 1 g of copper is added must be

$$C_{ion} \;=\; \frac{1}{63.54} = 0.0157\,M$$

From the Nernst equation, with $n = 2$ and $E_0 = +0.34\,V$,

$$E \;=\; E_0 + \frac{0.0592}{n} \log\,(C_{ion})$$

$$E \;=\; 0.34 + \frac{0.0592}{2} \log\,(0.0157)$$

$$\;=\; 0.34 + (0.0296)(-1.8)$$

$$\;=\; 0.29\,V$$

Rate of Corrosion or Plating.   The amount of metal plated on the cathode in electroplating or removed from the metal by corrosion can be determined from *Faraday's equation.*

$$w \;=\; \frac{ItM}{nF} \tag{20-7}$$

where w is the weight plated or corroded (g/s), I is the current (A), M is the atomic mass of the metal, n is the valence of the metal ion, t is the time (s), and

F is Faraday's constant, 96,500 C. Often the current is expressed in terms of *current density*, $i = I/A$, so Equation 20-7 becomes

$$w = \frac{iAtM}{nF} \tag{20-8}$$

where the area A (cm^2) is the surface area of the anode or cathode.

◇ | **Example 20-3**

Copper is electroplated onto one side of a 1 cm × 1 cm cathode using a current of 10 A. Calculate (a) the weight of copper plated per hour and (b) the time required to make a copper plate 0.1 cm thick.

Answer:

(a) From Faraday's equation, using $n = 2$ and $M = 63.54\,\text{g/g}\cdot\text{mole}$

$$w = \frac{ItM}{nF} = \frac{(10)(3600\,\text{s/h})(63.54)}{(2)(96{,}500)}$$

$$= 11.85\,\text{g/h}$$

(b) The time required to produce a 0.1-cm thick layer on a 1 cm^2 cathode is

$$\rho_{Cu} = 8.96\,\text{g/cm}^3 \qquad A = 1\,\text{cm}^2$$

$$\text{Volume/h} = \frac{11.85}{8.96} = 1.323\,\text{cm}^3/\text{h}$$

$$\text{Volume required} = \text{thickness} \times \text{area} = 0.1\,\text{cm} \times 1\,\text{cm}^2 = 0.1\,\text{cm}^3$$

$$\text{Time} = \frac{0.1}{1.323} = 0.076\,\text{h} = 4.6\,\text{min}$$

◇ | **Example 20-4**

An iron container 10 cm × 10 cm at its base is filled to a height of 20 cm with a corrosive liquid. A current is produced as a result of an electrolytic cell, and after 4 weeks, the container has decreased in weight by 70 g. Calculate (a) the current and (b) the current density involved in the corrosion of the iron.

Answer:

(a) The total exposure time is

$$t = (4\,\text{wk})(7\,\text{d/wk})(24\,\text{h/d})(3600\,\text{s/h}) = 2.42 \times 10^6\,\text{s}$$

From Faraday's equation, using $n = 2$ and $M = 55.847\,\text{g/g}\cdot\text{mole}$

$$I = \frac{wnF}{tM} = \frac{(70)(2)(96{,}500)}{(2.42 \times 10^6)(55.847)}$$

$$= 0.1\,\text{A}$$

(b) The total surface area of iron in contact with the corrosive liquid and the current density are

$$A = (4 \text{ sides})(10 \times 20) + (1 \text{ bottom})(10 \times 10) = 900 \text{ cm}^2$$

$$i = \frac{I}{A} = \frac{0.1}{900} = 1.11 \times 10^{-4} \text{A/cm}^2$$

  Example 20-5

Suppose that in a corrosion cell composed of copper and zinc, the current density at the copper cathode is 0.05 A/cm^2. The area of both the copper and zinc electrodes is 100 cm^2. Calculate (a) the corrosion current, (b) the current density at the zinc anode, and (c) the zinc loss per hour and the copper plated per hour.

Answer:

(a) The corrosion current is

$$I = i_{Cu} A_{Cu} = (0.05 \text{ A/cm}^2)(100 \text{ cm}^2) = 5 \text{ A}$$

(b) The current in the cell is the same everywhere. Thus

$$i_{Zn} = \frac{I}{A_{Zn}} = \frac{5}{100} = 0.05 \text{ A/cm}^2$$

(c) From Faraday's equation,

$$W_{(zinc \, loss)} = \frac{ItM}{nF} = \frac{(5)(3600 \text{ s/h})(65.38)}{(2)(96,500)}$$
$$= 6.1 \text{ g/h}$$
$$W_{(copper \, plated)} = \frac{(5)(3600 \text{ s/h})(63.54)}{(2)(96,500)}$$
$$= 5.9 \text{ g/h}$$

20–5 THE CORROSION CURRENT AND POLARIZATION

In electroplating, we wish to use high currents to increase the rate at which metal ions are deposited at the cathode. We can, within limits, independently control the current and thus the rate of electroplating by controlling the potential in the electrochemical cell.

On the other hand, to protect metals from corrosion, we wish to make the current as small as possible. Unfortunately, the corrosion current is very difficult to measure, control, or predict. Part of this difficulty can be attributed to various changes that occur during operation of the corrosion cell. A change in the potential of an anode or cathode, which in turn affects the current in the cell, is called *polarization*. There are three important sources of polarization.

Activation Polarization.   *Activation polarization* is related to the energy required to cause the anode or cathode reactions to occur. If, by proper selection of the anode and cathode materials, we can increase the degree of polarization, these reactions will occur with greater difficulty and the rate of corrosion will be reduced. Under ideal conditions, there are mathematical methods by which we can calculate the current density i produced during corrosion based on the degree of activation polarization for a given electrochemical cell. The current density can then be used to calculate the current and rate of corrosion.

Unfortunately, we often do not know the exact areas of the anode and cathode. Furthermore, small differences in composition and structure in the anode and cathode materials may dramatically change the activation polarization. Finally, segregation effects in the electrodes may cause the activation polarization to vary from one location to another. All of these factors make it more difficult to predict the corrosion current.

Concentration Polarization.   After corrosion begins, the concentration of ions at the anode or cathode surface may change. For example, a higher concentration of metal ions may be produced at the anode if the ions are unable to diffuse rapidly into the electrolyte. Hydrogen ions may be depleted at the cathode in a hydrogen electrode, or a high OH^- concentration may develop at the cathode in an oxygen electrode. When this occurs, either the anode or cathode reaction is stifled because fewer electrons are released at the anode or accepted at the cathode.

In any of these examples, the current density, and thus the rate of corrosion, decreases because of *concentration polarization*. Normally, polarization is less pronounced when the electrolyte is highly concentrated, the temperature is increased, or the electrolyte is vigorously agitated. Each of these factors increases the current density and encourages electrochemical corrosion.

Resistance Polarization.   *Resistance polarization* is a consequence of the electrical resistivity of the electrolyte. If a greater resistance to the flow of the current is offered, the rate of corrosion is reduced. Again the degree of resistance polarization may change as the composition of the electrolyte changes during the corrosion process.

20–6   TYPES OF ELECTROCHEMICAL CORROSION

In this section we will look at some of the more common forms that electrochemical corrosion takes.

TABLE 20-2   The galvanic series in seawater

Anodic	Magnesium
	Magnesium alloys
	Zinc
	Galvanized steel
	5052 aluminum
	3003 aluminum
	1100 aluminum
	6053 aluminum
	Alclad
	Cadmium
	2017 aluminum
	2024 aluminum
	Low-carbon steel
	Cast iron
	410 stainless steel (active)
	50% Pb-50% Sn solder
	316 stainless steel (active)
	Lead
	Tin
	Cu-40% Zn brass
	Manganese bronze
	Nickel-base alloys (active)
	Cu-35% Zn brass
	Aluminum bronze
	Copper
	Cu-30% Ni alloy
	Nickel-base alloys (passive)
	Stainless steels (passive)
	Silver
	Titanium
	Graphite
	Gold
Cathodic	Platinum

After *ASM Metals Handbook*, Vol. 10, 8th Ed., 1975.

Uniform Attack.   When a metal is placed in an electrolyte, some regions are anodic to other regions. However, the location of these regions moves and even reverses from time to time. Since the anode and cathode regions continually shift, the metal corrodes uniformly even without contact with a second material.

Galvanic Attack.   Galvanic attack occurs when certain areas always act as anodes, while other areas always act as cathodes. These electrochemical cells are called *galvanic cells* and can be separated into three types—composition cells, stress cells, and concentration cells.

Composition Cells.   Composition cells, or *dissimilar metal* corrosion, develop when two metals or alloys, such as copper and iron, form an electrolytic cell.

Because of the effect of alloying elements and electrolyte concentration on polarization, the emf series may not tell us which regions corrode and which are protected. Instead, we use a *galvanic series*, in which the different alloys are ranked according to their anodic or cathodic tendencies in a particular environment (Table 20-2). We may find a different galvanic series for seawater, freshwater, and industrial atmospheres.

◇ | **Example 20-6**

A brass fitting used in a marine application is joined by soldering with lead-tin solder. Will the brass or the solder corrode?

Answer:

From the galvanic series, we find that all of the copper-base alloys are more cathodic than a 50% Pb-50% Sn solder. Thus, the solder is the anode and corrodes. Corrosion of the solder can contaminate water in plumbing systems with lead.

◇

Composition cells develop in two-phase alloys, where one phase is more anodic than the other. Since ferrite is anodic to cementite in steel, small microcells cause steel to galvanically corrode (Figure 20-7). Almost always, a two-phase alloy has less resistance to corrosion than a single-phase alloy of a similar composition.

Intergranular corrosion occurs when precipitation of a second phase or segregation at grain boundaries produces a galvanic cell. In zinc alloys, for example,

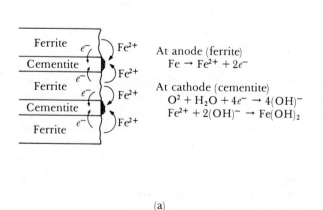

At anode (ferrite)
$$Fe \rightarrow Fe^{2+} + 2e^-$$

At cathode (cementite)
$$O^2 + H_2O + 4e^- \rightarrow 4(OH)^-$$
$$Fe^{2+} + 2(OH)^- \rightarrow Fe(OH)_2$$

(a)

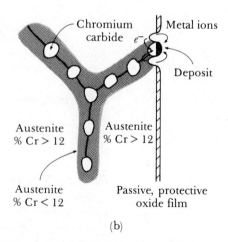

(b)

FIGURE 20-7   *Example of microgalvanic cells in two-phase alloys. (a) In steel, ferrite is anodic to cementite. (b) In austenitic stainless steel, precipitation of chromium carbide makes the austenite in the grain boundaries anodic.*

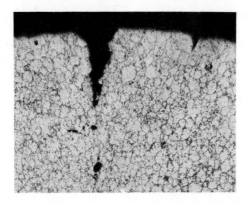

FIGURE 20-8 Photomicrograph of intergranular corrosion in a zinc die casting. Segregation of impurities to the grain boundaries produces microgalvanic corrosion cells (× 50).

impurities such as cadmium, tin, and lead segregate at the grain boundaries during solidification. The grain boundaries are anodic compared with the remainder of the grains, and corrosion of the grain boundary metal occurs (Figure 20-8). In austenitic stainless steels, chromium carbides can precipitate at grain boundaries [Figure 20-7(b)]. The formation of the carbides removes chromium from the austenite adjacent to the boundaries. The low-chromium austenite at the grain boundaries is anodic to the remainder of the grain and corrosion occurs at the grain boundaries.

Stress Cells.   Stress cells develop when a metal contains regions with different local stresses. The most highly stressed, or high energy, regions act as anodes to the less stressed cathodic areas (Figure 20-9). Regions with a finer grain size, or a higher density of grain boundaries, are anodic to coarse-grained regions of the

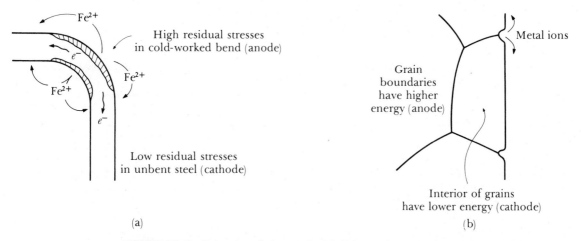

(a)   (b)

FIGURE 20-9 Examples of stress cells. (a) Cold work required to bend a steel bar introduces high residual stresses at the bend, which then is anodic and corrodes. (b) Because grain boundaries have a high energy, they are anodic and corrode.

same material. Highly cold worked areas are anodic to less cold worked areas. High applied stresses accelerate corrosion.

Stress corrosion occurs by galvanic action, but other mechanisms, such as adsorption of impurities at the tip of an existing crack, may also occur. Failure occurs as a result of corrosion and the applied stress. Higher applied stresses reduce the time required for failure.

Fatigue failures are also initiated or accelerated when corrosion occurs. *Corrosion fatigue* can reduce fatigue properties by initiating cracks, perhaps by producing pits or crevices, and by increasing the rate at which the cracks propagate.

◇ | **Example 20-7**

A cold-drawn steel wire is formed into a nail by additional deformation, producing the point at one end and the head at the other. Where will the most severe corrosion of the nail occur?

Answer:

Since the head and point have been cold worked an additional amount compared with the shank of the nail, the head and point serve as anodes and corrode most rapidly.

Concentration Cells.   Concentration cells develop due to differences in the electrolyte (Figure 20-10). A difference in metal ion concentration causes a difference in electrode potential, according to the Nernst equation. The metal in contact with the most concentrated solution is the cathode; the metal in contact with the dilute solution is the anode.

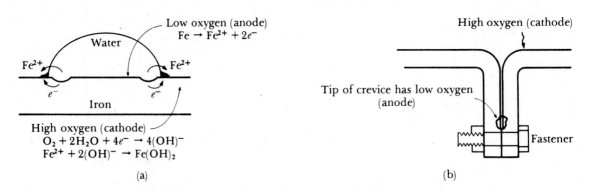

FIGURE 20-10 *Concentration cells. (a) Corrosion occurs beneath a water droplet on a steel plate due to low oxygen concentration in the water. (b) Corrosion occurs at the tip of a crevice due to limited access of oxygen.*

The *oxygen concentration* cell (often referred to as oxygen starvation) occurs when the cathode reaction is the oxygen electrode, $2H_2O + O_2 + 4e^- \rightarrow 4OH^-$. Electrons flow from the low-oxygen region, which serves as the anode, to the high-oxygen region, which serves as the cathode.

Deposits, such as rust or water droplets, shield the underlying metal from oxygen. Consequently, the metal *under* the deposit is the anode and corrodes. This causes one form of *pitting corrosion*. *Waterline corrosion* is similar—metal above the waterline is exposed to oxygen, while metal beneath the waterline is deprived of oxygen. Hence, the metal under the water corrodes. Normally, the metal far below the surface corrodes more slowly than metal just below the waterline due to differences in the distance that electrons must travel. Because cracks and crevices have a lower oxygen concentration than the surrounding base metal, the tip of a crack or crevice is the anode, causing *crevice corrosion*.

Pipe buried in soil may corrode because of differences in the composition of the soil. Velocity differences may cause concentration differences—stagnant water contains low oxygen concentrations, while fast-moving, aerated water contains higher oxygen concentrations; metal near the stagnant water is anodic and corrodes.

◇ | **Example 20-8**

Two pieces of steel are joined mechanically by crimping the edges. Why would this be a bad idea if the steel is then exposed to water? If the water contains salt, would corrosion be affected?

Answer:

By crimping the steel edges, we produce a crevice. The region in the crevice is exposed to less air and moisture, so it behaves as the anode in a concentration cell. The steel in the crevice corrodes.

Salt in the water increases the conductivity of the water, permitting electrical charge to be transferred at a more rapid rate. This causes a higher current density and thus faster corrosion due to less resistance polarization.

20–7 PROTECTION AGAINST ELECTROCHEMICAL CORROSION

The problem of corrosion of metals is serious but not hopeless. A number of techniques combat corrosion, including design, coatings, inhibitors, cathodic protection, passivation, and materials selection.

Design. By properly designing metal structures, corrosion can be slowed or even avoided. Some of the factors that should be considered are as follows.

1. Prevent the formation of galvanic cells. For example, steel pipe is frequently connected to brass plumbing fixtures, producing a galvanic cell that causes the steel to corrode. By using intermediate plastic fittings to electrically insulate the steel and brass, this problem can be minimized.

2. Make the anode area much larger than the cathode area. For example, copper rivets can be used to fasten steel sheet. Because of the small area of the copper rivets, a limited cathode reaction occurs. The copper accepts few electrons and the steel anode reaction proceeds slowly. If, on the other hand, steel rivets are used for joining copper sheet, the small steel anode area gives up many electrons, which are accepted by the large copper cathode area. Corrosion of the steel rivets is then very rapid.

◇ Example 20-9

Consider a copper-zinc corrosion couple. If the current density at the copper cathode is $0.05\,A/cm^2$, calculate the weight loss of zinc per hour if (a) the copper cathode area is $100\,cm^2$ and the zinc anode area is $1\,cm^2$ and (b) the copper cathode area is $1\,cm^2$ and the zinc anode area is $100\,cm^2$.

Answer:

(a) For the small zinc anode area,

$$I = i_{Cu} A_{Cu} = (0.05\,A/cm^2)(100\,cm^2) = 5\,A$$

$$w_{Zn} = \frac{ItM}{nF} = \frac{(5)(3600)(65.38)}{(2)(96,500)} = 6.1\,g/h$$

(b) For the large zinc anode area,

$$I = i_{Cu} A_{Cu} = (0.05\,A/cm^2)(1\,cm^2) = 0.05\,A$$

$$w_{Zn} = \frac{ItM}{nF} = \frac{(0.05)(3600)(65.38)}{(2)(96,500)} = 0.061\,g/h$$

The rate of corrosion of the zinc is reduced significantly when the zinc anode is much larger than the cathode.

---

3. Design components so that fluid systems are closed rather than open and so that stagnant pools of liquid do not collect. Partly filled tanks undergo waterline corrosion. Open systems continuously dissolve gas, providing ions that participate in the cathode reaction and encourage concentration cells.

4. Avoid crevices between assembled or joined materials (Figure 20-11). Welding may be a better joining technique than brazing, soldering, or mechanical fasteners. Galvanic cells develop in brazing or soldering, since the filler metals have a different composition from the metal being joined. Mechanical fasteners produce crevices that lead to concentration cells. However, if the filler metal is closely matched to the base metal, welding may prevent these cells from developing.

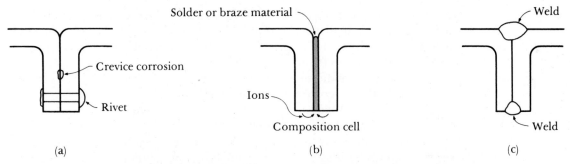

Solder or braze material

Weld

Crevice corrosion

Rivet

Ions

Composition cell

Weld

(a)

(b)

(c)

FIGURE 20-11 *Alternative methods for joining two pieces of steel. (a) Fasteners may produce a concentration cell, (b) brazing or soldering may produce a composition cell, and (c) welding with a filler metal that matches the base metal may avoid galvanic cells.*

5. In some cases, the rate of corrosion cannot be reduced to a level that will not interfere with the expected lifetime of the component. In such cases, the assembly should be designed in such a manner that the corroded part can easily and economically be replaced.

Coatings. Coatings are used to isolate the anode and cathode regions. Temporary coatings, such as grease or oil, provide some protection but are easily disrupted. Organic coatings, such as paint, or ceramic coatings, such as enamel or glass, provide better protection. However, if the coating is disrupted, a small anodic site is exposed that undergoes rapid, localized corrosion.

Metallic coatings include tin-plated and galvanized (zinc-plated) steel (Figure 20-12). A continuous coating of either metal isolates the steel from the

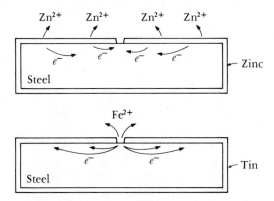

Zn^{2+} Zn^{2+} Zn^{2+} Zn^{2+}

e^- e^- e^- e^-

Steel

Zinc

Fe^{2+}

e^- e^-

Steel

Tin

FIGURE 20-12 *Zinc-plated steel and tin-plated steel are protected differently. Zinc protects steel even when the coating is scratched, since zinc is anodic to steel. Tin does not protect steel when the coating is disrupted, since steel is anodic with respect to tin.*

electrolyte. However, when the coating is scratched, exposing the underlying steel, the two coatings behave differently. The zinc continues to be effective, since zinc is anodic to steel. Since the area of the exposed steel cathode is small, the zinc coating corrodes at a very slow rate and the steel remains protected. However, steel is anodic to tin, so a tiny steel anode is created when tinplate is scratched. Rapid corrosion of the steel subsequently occurs. Composite materials, such as Alclad, are used to improve corrosion resistance of high-strength two-phase aluminum alloys.

Chemical conversion coatings are produced by a chemical reaction with the surface. Liquids such as zinc acid orthophosphate solutions form an adherent phosphate layer on the metal surface. The phosphate layer is, however, rather porous and is more often used to improve paint adherence. Stable, adherent, nonporous, nonconducting oxide layers form on the surface of aluminum, chromium, and stainless steel. These oxides exclude the electrolyte and prevent the formation of galvanic cells.

Inhibitors.   Some chemicals, when added to the electrolyte solution, migrate preferentially to the anode or cathode surface and produce concentration or resistance polarization (Figure 20-13). Chromate salts perform this function in automobile radiators. However, if insufficient inhibitor is added, small local anode areas are unprotected and corrosion is accelerated.

Cathodic Protection.   We can protect against corrosion by supplying the metal with electrons and forcing the metal to be a cathode (Figure 20-14). Cathodic protection can be produced by using a sacrificial anode or an impressed voltage.

A *sacrificial anode* is attached to the material to be protected, forming an electrochemical circuit. The sacrificial anode corrodes, supplies electrons to the metal, and thereby prevents an anode reaction at the metal. The sacrificial anode, typically zinc or magnesium, is consumed and must eventually be

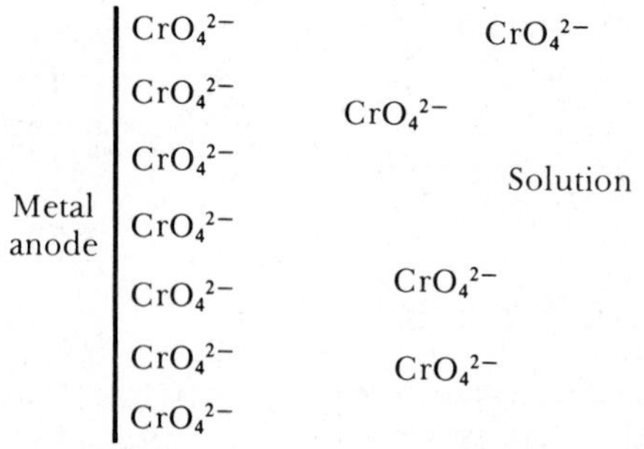

$$CrO_4^{2-}$$

CrO_4^{2-}

CrO_4^{2-}

CrO_4^{2-}

CrO_4^{2-}

Metal anode $\quad CrO_4^{2-}$

Solution

CrO_4^{2-}

CrO_4^{2-}

CrO_4^{2-}

CrO_4^{2-}

CrO_4^{2-}

FIGURE 20-13 Inhibitors may concentrate at the anode, causing severe concentration polarization, and significantly reduce the rate of corrosion of the anode.

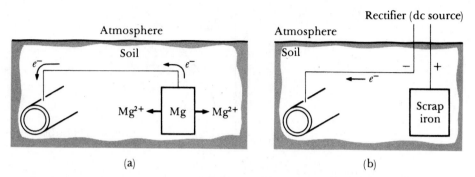

FIGURE 20-14 Cathodic protection of a buried steel pipeline. (a) A sacrificial magnesium anode assures that the galvanic cell makes the pipeline the cathode. (b) An impressed voltage between a scrap iron auxiliary anode and the pipeline assures that the pipeline is the cathode.

replaced. Applications include preventing corrosion of buried pipelines, ships, off-shore drilling platforms, and water heaters.

An *impressed voltage* is obtained from a direct current source connected between an auxiliary anode and the metal to be protected. Essentially, we have a battery connected so that electrons flow to the metal, causing the metal to be the cathode. The auxiliary anode, such as scrap iron, corrodes.

Passivation or Anodic Protection.   Metals near the anodic end of the galvanic series are *active* and serve as anodes in most electrolytic cells. However, if these metals are made *passive*, or more cathodic, they corrode at slower rates than normal. *Passivation* is accomplished by producing strong anodic polarization, preventing the normal anode reaction. Thus the term *anodic protection*.

We can produce passivation by exposing the metal to highly concentrated oxidizing solutions. If iron is dipped in very concentrated nitric acid, the iron rapidly and uniformly corrodes to form a thin, protective iron hydroxide coating. The coating protects the iron from subsequent corrosion in nitric acid.

We can also cause passivation by increasing the potential on the anode above a critical level. A passive film forms on the metal surface, causing strong anode polarization, and the current decreases to a very low level. Passivation of aluminum is called *anodizing* and a thick oxide coating is produced. This oxide layer can be dyed to produce attractive colors. The coating remains passive even when the potential is removed.

Materials Selection and Treatment.   Corrosion can be prevented or minimized by selecting appropriate materials and heat treatments. In castings, for example, segregation causes tiny, localized galvanic cells that accelerate corrosion. We can improve corrosion resistance with a homogenization heat treatment. When metals are formed into finished shapes by bending, differences in the amount of cold work and residual stresses cause local stress cells. These may be minimized by a stress relief anneal or a full recrystallization anneal.

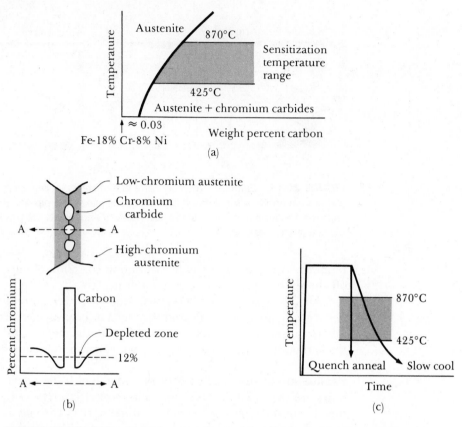

FIGURE 20-15 (a) Intergranular corrosion in austenitic stainless steel. (b) Slow cooling permits chromium carbides to precipitate at grain boundaries. (c) A quench anneal to dissolve the carbides may prevent intergranular corrosion.

The heat treatment is particularly important in austenitic stainless steels (Figure 20-15). When the steel cools slowly from 870°C to 425°C, chromium carbides precipitate at the grain boundaries. The austenite at the grain boundaries may contain less than 12% chromium, which is the minimum required to produce a passive oxide layer. The steel is sensitized. Because the grain boundary regions are small and highly anodic, rapid corrosion of the austenite at the grain boundaries occurs. There are several techniques by which we can minimize the problem.

1. If the steel contains less than 0.03% C, the chromium carbides do not form.

2. If the percent chromium is very high, the austenite may not be depleted to below 12% Cr, even if the chromium carbides form.

3. Addition of titanium or niobium ties up the carbon as TiC or NbC, preventing the formation of chromium carbide. The steel is said to be *stabilized*.

4. The sensitization temperature range—425°C to 870°C—should be avoided during manufacture and service.

5. In a *quench anneal* heat treatment, the stainless steel is heated above 800°C, causing the chromium carbides to dissolve. The structure, now containing 100% austenite, is rapidly quenched to prevent formation of carbides.

◇ | **Example 20-10**

The peak temperatures obtained near a weld on a 304 stainless steel are shown in Figure 20-16. Where will corrosion most likely occur as the cooling rate after welding increases?

Answer:

Figure 20-16 shows the region of the weld that will heat above the sensitization range and the region that will be held in the sensitization range. The region that is held in the sensitization range may eventually contain precipitated carbides. Carbides will precipitate in the other region only if the weld cools slowly. Thus, for slow cooling, the entire heat-affected area may be sensitized and will corrode. For faster cooling, only the region that heated into the sensitization range will corrode.

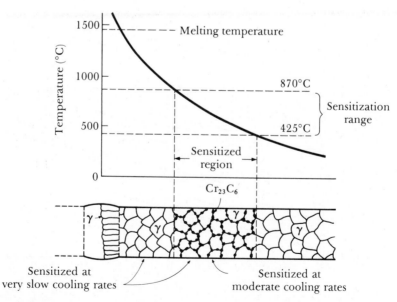

FIGURE 20-16   *The peak temperature surrounding a stainless steel weld and the sensitized structure produced when the weld slowly cools.*

20–8   OXIDATION AND OTHER GAS REACTIONS

Materials of all types may react with oxygen and other gases. These reactions can, like corrosion, alter the composition, properties, or integrity of the material.

Oxidation of Metals.   Metals may react with oxygen to produce an oxide at the surface. We are interested in three aspects of this reaction—the ease with which the metal oxidizes, the nature of the oxide film that forms, and the rate at which oxidation occurs.

The ease with which oxidation occurs is given by the *free energy of formation* for the oxide (Figure 20-17). There is a large driving force for the oxidation of magnesium and aluminum, but there is little tendency for the oxidation of nickel or copper.

◇   **Example 20-11**

Explain why we should not add alloying elements such as chromium to pig iron before the pig iron is converted to steel in a basic oxygen furnace at 1700°C.

Answer:

In a basic oxygen furnace, we lower the carbon content of the metal from about 4% to much less than 1% by blowing pure oxygen through the molten metal. If chromium were already present before the steel making began, Figure 20-17

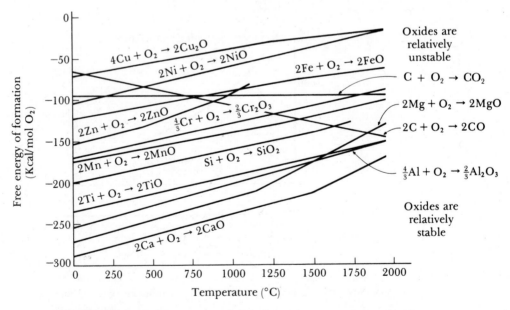

FIGURE 20-17   The free energy of formation of selected oxides as a function of temperature. A large negative free energy indicates a more stable oxide.

tells us that chromium would oxidize before the carbon, since chromium oxide has a lower free energy of formation, or is more stable, than carbon monoxide (CO). All of the expensive chromium that we add is lost before the carbon is removed from the pig iron.

The type of oxide film determines the rate at which oxidation occurs and whether the oxide causes the metal to be passive. Three types of behavior are observed, depending on the relative volumes of the oxide and the metal (Figure 20-18). We can determine this ratio from the Pilling-Bedworth equation for the following oxidation reaction.

$$nM + mO_2 \rightarrow M_nO_{2m} \tag{20-9}$$

The *Pilling-Bedworth ratio* is

$$\text{P-B ratio} = \frac{\text{oxide volume per metal atom}}{\text{metal volume per atom}}$$

$$= \frac{(M_{oxide})(\rho_{metal})}{n(M_{metal})(\rho_{oxide})} \tag{20-10}$$

where M is the atomic or molecular mass, ρ is the density, V is the volume, and n is the number of metal atoms in the oxide, as defined in Equation 20-9. Examples of the Pilling-Bedworth ratio for several metal-oxide combinations are shown in Table 20-3.

If the Pilling-Bedworth ratio is less than one, the oxide occupies a smaller volume than the metal from which it formed. Tensile stresses develop in the

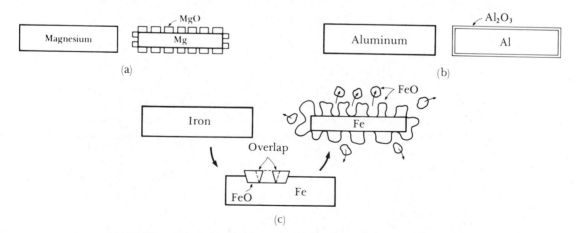

FIGURE 20-18   Three types of oxides may form, depending on the volume ratio between the metal and the oxide. (a) Magnesium produces a porous oxide film, (b) aluminum forms a protective, adherent, nonporous oxide film, and (c) iron forms an oxide film that spalls off the surface and provides poor protection.

TABLE 20-3 The Pilling-Bedworth ratio for selected metal–metal oxide systems

Metal and Oxide	Density of Oxide (g/cm^3)	Pilling-Bedworth Ratio
Mg-MgO	3.6	0.8
Al-Al$_2$O$_3$	4.0	1.3
Ti-TiO$_2$	5.1	1.5
Zr-ZrO$_2$	5.6	1.5
Fe-Fe$_2$O$_3$	5.3	2.1
Cr-Cr$_2$O$_3$	5.1	2.1
Cu-Cu$_2$O	6.2	1.6
Ni-NiO	6.9	1.6
Si-SiO$_2$	2.7	1.9
U-UO$_2$	11.1	1.9
W-WO$_3$	7.3	3.3

From J. West, *Basic Corrosion and Oxidation*, 1980, Ellis Horwood.

oxide film, causing the film to crack and become porous. Oxidation can then continue rapidly, which is typical of metals such as magnesium.

If the ratio is equal to one, the volumes of the oxide and metal are equal and an adherent, nonporous, protective film forms, typical of aluminum and titanium. In most cases, the film tends to be protective until the Pilling-Bedworth ratio exceeds about two.

If the ratio is greater than two, the oxide volume is greater than that of the metal and, initially, the oxide forms a protective layer. However, as the thickness of the film increases, high compressive stresses develop in the oxide. The oxide may flake from the surface, exposing fresh metal, which continues to oxidize. This nonadherent oxide layer is typical of iron.

◇ | **Example 20-12**

The density of aluminum is 2.7 g/cm^3 and that of Al$_2$O$_3$ is about 4 g/cm^3. Describe the characteristics of the aluminum oxide film. Compare with the oxide film that forms on tungsten; the density of tungsten is 19.254 g/cm^3 and that of WO$_3$ is 7.3 g/cm^3.

Answer:

For $2Al + \frac{3}{2}O_2 \rightarrow Al_2O_3$, the molecular weight of Al$_2$O$_3$ is 101.96 and that of aluminum is 26.981.

$$\text{P-B} = \frac{M_{Al_2O_3}\rho_{Al}}{nM_{Al}\rho_{Al_2O_3}} = \frac{(101.96)(2.7)}{(2)(26.981)(4)} = 1.28$$

For tungsten, $W + \frac{3}{2}O_2 \rightarrow WO_3$, the molecular weight of WO$_3$ is 231.85 and that of tungsten is 183.85.

$$P\text{-}B = \frac{M_{WO_3}\rho_W}{nM_W\rho_{WO_3}} = \frac{(231.85)(19.254)}{(1)(183.85)(7.3)} = 3.33$$

Since P-B $\simeq$ 1 for aluminum, the Al_2O_3 film is nonporous and adherent, providing protection to the underlying aluminum. However, P-B > 1 for tungsten, so the WO_3 should be nonadherent and nonprotective.

The rate at which oxidation occurs depends on the access of oxygen to the metal atoms. A *linear* rate of oxidation occurs when the oxide is porous (as in magnesium) and oxygen has continued access to the metal surface.

$$y = kt \qquad (20\text{-}11)$$

where y is the thickness of the oxide, t is the time, and k is a constant that depends on the metal and temperature.

A *parabolic* relationship is observed when diffusion of ions or electrons through a nonporous oxide layer is the controlling factor. This relationship is observed in iron, copper, and nickel.

$$y = \sqrt{kt} \qquad (20\text{-}12)$$

Finally, a *logarithmic* relationship is observed for the growth of thin oxide films that are particularly protective, as for aluminum and possibly chromium.

$$y = k \ln (ct + 1) \qquad (20\text{-}13)$$

where k and c are constants for a particular temperature, environment, and composition.

◇ Example 20-13

At 1000°C, pure nickel follows a parabolic oxidation curve given by the constant $k = 3.9 \times 10^{-12}\,cm^2/s$ in an oxygen atmosphere. If this relationship is not affected by the thickness of the oxide film, calculate the time required for a 0.1-cm nickel sheet to completely oxidize.

Answer:

Assuming the sheet oxidizes from both sides

$$y = \sqrt{kt} = \sqrt{(3.9 \times 10^{-12})(t)} = \frac{0.1\,cm}{2\ sides} = 0.05\,cm$$

$$t = \frac{(0.05)^2}{3.9 \times 10^{-12}} = 6.4 \times 10^8\,s = 20.3\ years$$

The temperature also affects the rate of oxidation. In many metals, the rate of oxidation is controlled by the rate of diffusion of oxygen or metal ions through the oxide. If oxygen diffusion is more rapid, the oxidation reaction will occur between the oxide and the metal; if metal ion diffusion is more rapid, oxidation occurs at the oxide-atmosphere interface. Consequently, we would expect that oxidation rates would follow an Arrhenius relationship, increasing exponentially as the temperature increases.

Another factor that influences the stability and protectiveness of the oxide film is the difference in coefficient of thermal expansion between the oxide and the underlying metal. Typically, the oxide film has a lower expansion coefficient than the metal. When a metal is oxidized at a high temperature and is then allowed to cool, the metal will contract by a greater amount than the oxide; this in turn will impose compressive stresses on the oxide that may cause it to fail, particularly if the Pilling-Bedworth ratio is already high.

Selective Oxidation.   The curves in Figure 20-17 show that different metals have different tendencies to oxidize. The metal having the largest negative free energy of formation oxidizes first when more than one metal is present. This behavior is used to refine metals. In steel making, oxygen blown into liquid pig iron reacts first with carbon, rather than reacting with iron. The carbon content of pig iron is lowered until the desired steel is produced. We utilize the same principle in designing oxidation-resistant alloys. When chromium is added to steel, the chromium oxidizes first, produces a protective chromium oxide film, and protects the underlying metal.

Another example of selective oxidation is the *decarburization* of steel during hot working or heat treatment. Oxygen reacts with the carbon at the surface of the steel, leaving behind a low-carbon surface layer that reduces the mechanical properties of the steel at the surface.

Oxidation of Ceramics.   Most oxide ceramics are not significantly affected by oxygen, even at high temperatures. However, the high-temperature usefulness of boron nitride and graphite is limited by their oxidation.

Oxidation of Polymers.   Exposure of polymers to oxygen, usually at elevated temperatures, alters the polymerized structure. In rubber, for example, oxygen provides additional cross-linking, causing the rubber to become harder. Oxygen may also cause depolymerization, or chain scission, reducing the degree of polymerization and permitting small molecules to escape as a gas, or cause charring or even burning of the polymer at high temperatures. Polymers based on silicon rather than carbon backbones are more resistant to oxidation and can be used at higher temperatures.

20–9   RADIATION DAMAGE

A variety of problems occur in materials exposed to radiation, causing changes in the structure and properties of the material. A tiny sample of these problems include the following.

Metals.   High-energy radiation, such as neutrons, may knock an atom out of its normal lattice site, creating interstitials and vacancies. These point defects reduce electrical conductivity and cause ductile materials to become harder and more brittle. Annealing may help eliminate radiation damage.

Ceramics.   Radiation also creates point defects in ceramic materials. Normally, little effect on mechanical properties is observed, since ceramics are brittle, but physical properties, such as thermal conductivity and optical properties, may be impaired.

Polymers.   Even low-energy radiation can alter the structure in polymers. Chains can be broken, reducing the degree of polymerization or causing branching to occur, thus reducing the strength of the polymer. Exposure to ultraviolet radiation from sunlight is a serious problem in many polymers, including polystyrene, and often requires that stabilizers be added to the polymer to minimize long-term damage.

20–10   WEAR AND EROSION

Wear and erosion remove material from a component by mechanical attack of solids or liquids. Corrosion and mechanical failure also contribute to this type of attack.

Adhesive Wear.   Adhesive wear, also known as scoring, galling, or seizing, occurs when two solid surfaces slide over one another under pressure. Surface projections, or *asperities*, are plastically deformed and eventually welded together by the high local pressures (Figure 20-19). As sliding continues, these bonds are broken, producing cavities on one surface, projections on the second surface, and frequently tiny, abrasive particles, all of which contribute to further wear of the surfaces.

Wear rates are often difficult to predict or quantify. One expression that has been used to correlate sliding wear to some of the contributing factors is

$$\text{Volume loss} \ = \ kFvt \tag{20-14}$$

where wear rate is expressed by the volume of lost material in cubic inches, F is the load in pounds, v is the velocity in feet per minute, t is the time in hours, and k is a constant for a given material and conditions.

Many factors may be considered in trying to improve the wear resistance of materials. Designing components so that loads are small, surfaces are smooth, and continual lubrication is possible will help prevent adhesions that cause the loss of material.

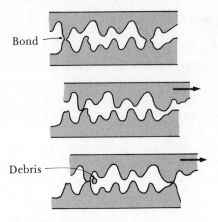

Bond

Debris

FIGURE 20-19 The asperities on two rough surfaces may initially be bonded. A sufficient force breaks the bonds and the surfaces slide. As they slide, asperities may be fractured, wearing away the surfaces and producing debris.

The properties and microstructure of the material are also important. Normally, if both surfaces have high hardnesses that are approximately the same, the wear rate is low. The presence, size, shape, and distribution of inclusions, second-phase particles, and grain boundaries in the material will affect the hardness. High strength, to help resist the applied loads, and good toughness and ductility, which prevent tearing of material from the surface, may be beneficial. Ceramic materials, with their exceptional hardness, are expected to provide good adhesive wear resistance.

Selection of the material is often critical. In some cases, even low-strength materials may provide acceptable wear resistance. For example, pearlitic gray cast iron, which contains graphite flakes that provide excellent self-lubrication at sliding surfaces, is frequently used for the ways in machine tools such as lathes.

Polymers are also used in applications in which adhesive wear might be expected. When polymers rub against metal surfaces, particularly at low temperatures and loads, adhesive wear may be slow due to the low coefficient of friction between the polymer and the metal. Wear resistance can be improved if the coefficient of friction is reduced still further by the addition of 15% to 20% polytetrafluoroethylene (PTFE or Teflon) to the polymer or by the introduction of reinforcing fibers such as glass, carbon, or aramid (Table 20-4).

TABLE 20-4 The wear rate factor for 6,6 nylon sliding against steel for various additives

Polymer	k = Wear Factor ($\times 10^{-10} in^3 \cdot min/ft \cdot lb \cdot h$)
6,6 Nylon	200
6,6 Nylon with 30% glass fibers	75
6,6 Nylon with 20% aramid fibers	62
6,6 Nylon with 18% PTFE	6

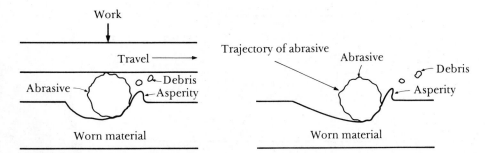

FIGURE 20-20 Abrasive wear, caused by either trapped or free-flying abrasives, produces troughs in the material, piling up asperities that may fracture into debris.

Abrasive Wear. Abrasive wear occurs when material is removed from the surface by contact with hard particles, which either may be present at the surface of a second material or may be present as loose particles between two surfaces (Figure 20-20). Abrasive wear is unlike adhesive wear in that no bonding occurs. This type of wear is common in machinery such as plows, scraper blades, crushers, and grinders used to handle abrasive materials and may also occur when hard particles are unintentionally introduced into moving parts of machinery. Abrasive wear is also used for grinding operations to remove material intentionally.

Materials with a high hardness, good toughness, and high hot strength are most resistant to abrasive wear. Typical materials used for abrasive wear applications include quenched and tempered steels; carburized or surface-hardened steels; austenitic manganese steels (containing 3% to 5% Cr, 2% to 4% Ni, 10% to 20% Mn, and up to 1% C), which strengthen by work hardening during use; cobalt alloys such as Stellite; composite materials, including tungsten carbide cermets; white cast irons; and hard surfaces produced by welding. Most ceramic materials also resist wear effectively due to their high hardness; however, their brittleness may sometimes limit their usefulness in abrasive wear conditions.

Liquid Erosion. The integrity of a material may be destroyed by erosion caused by high pressures associated with a moving liquid. The liquid causes strain hardening of the metal surface, leading to localized deformation, cracking, and loss of material. Two types of liquid erosion deserve mention.

Cavitation occurs when a liquid containing a dissolved gas enters a low-pressure region. Gas bubbles, which precipitate and grow in the liquid, collapse when the pressure subsequently increases (Figure 20-21). The high-pressure, local shock wave that is produced may exert a pressure of thousands of atmospheres against the surrounding material. Cavitation is frequently encountered in propellors, dams and spillways, and hydraulic pumps.

Liquid impingement occurs when liquid droplets carried in a rapidly moving gas strike a metal surface (Figure 20-22). High localized pressures develop because of the initial impact and the rapid lateral movement of the droplets from

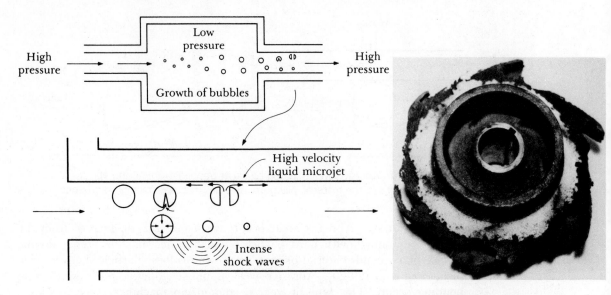

FIGURE 20-21 Cavitation occurs when gas bubbles grow from the liquid in a low-pressure region, then collapse on reentering a high-pressure region. Collapse due to the implosion of the gas bubbles creates high-intensity shock waves or high-velocity microjets of liquid, which erode the material surface. The photograph illustrates the effect of cavitation on an impeller used in a municipal water system.

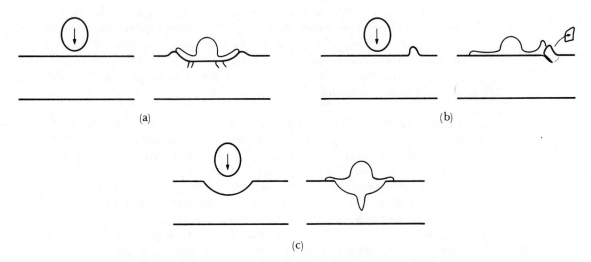

FIGURE 20-22 Effects of impingement corrosion. (a) A water droplet may create a crater and even cracks in a ductile material. (b) Asperities on the surface arrest the spreading droplet, fracture, and pull out of the surface. (c) Impingement on existing pits accelerates the growth of the pit.

the impact point along the metal surface. Water droplets carried by steam may erode turbine blades in steam generators and nuclear power plants.

Liquid erosion can be minimized by proper materials selection and design, including the following.

1. Minimizing velocity. Cavitation erosion increases at an exponential rate with velocity, and impingement erosion increases with v^5 to v^6.

2. Control of the liquid. Assuring that the liquid is deaereated so that bubbles cannot form or assuring that excess moisture is removed from steam helps prevent liquid erosion.

3. Selecting hard, tough materials, such as Stellite, tool steels, titanium, or nickel-base alloys, which absorb the impact of the gas or liquid droplets.

4. Organic coatings. Coating the material with an elastomer permits the organic polymer to absorb the shock of cavitation or impingement and protect the underlying material from erosion.

SUMMARY

Deterioration of a material by corrosion is primarily a chemical process which is determined by the mutual solubility of materials or the tendency of a material to give up its electrons in an electrochemical cell. To a certain extent, we can control this deterioration by excluding the corrosive environment, including the use of coatings. However, the structure of the material also plays an important role in many corrosive situations; producing more homogeneous structures free of stresses and high-energy locations by controlling the composition, processing, and heat treatment will improve corrosion resistance.

Deterioration by wear is reduced by improved control over the structure and properties of the material. In particular, alloy selection and materials processing which provide high hardness lead to lower rates of wear.

GLOSSARY

Abrasive wear. Removal of material from surfaces by the cutting action of hard particles.

Adhesive wear. Removal of material from surfaces of moving equipment by momentary local bonding, then bond fracture, at the surfaces.

Anode. The location at which corrosion occurs as electrons and ions are given up.

Anodizing. An anodic protection technique in which a thick oxide layer is deliberately produced on a metal surface.

Asperities. Small projections on the surface of moving parts that participate in adhesive wear.

Cathode. The location at which electrons are accepted and a by-product is produced during corrosion.

Cavitation. Erosion of a material surface by the pressures produced when a gas bubble collapses within a moving liquid.

Chemical corrosion. Removal of atoms from a material by virtue of the solubility or chemical reaction between the material and the surrounding liquid.

Composition cells. Electrochemical corrosion cells produced between two materials having a different composition. Also known as galvanic cells.

Concentration cells. Electrochemical corrosion cells produced between two locations on a material at which the composition of the electrolyte is different.

Corrosion fatigue. Accelerated failure of a material by combined corrosion and a cyclic load.

Crevice corrosion. A special concentration cell in which corrosion occurs in crevices because of the low concentration of oxygen.

Decarburization. Selective oxidation of carbon from the surface of a steel.

Dezincification. A special chemical corrosion process by which both zinc and copper atoms are removed from brass but the copper is replated back onto the metal.

Electrochemical cell. A cell in which electrons and ions can flow by separate paths between two materials, producing a current which in turn leads to corrosion or plating.

Electrochemical corrosion. Corrosion produced by the development of a current in an electrochemical cell which removes ions from the material.

Electrode potential. Related to the tendency of a material to corrode. The potential is the voltage produced between the material and a standard electrode.

Electrolyte. The conductive medium through which ions move to carry current in an electrochemical cell.

Electroplating. The precipitation of ions on the cathode in an electrochemical cell.

Emf series. The arrangement of elements according to their electrode potential, or their tendency to corrode.

Galvanic series. The arrangement of alloys according to their tendency to corrode in a particular environment.

Graphitic corrosion. A special chemical corrosion process by which iron is leached from cast iron, leaving behind a weak, spongy mass of graphite.

Hydrogen electrode. The cathode reaction by which electrons and hydrogen ions from the corrosion circuit combine to produce hydrogen gas.

Impingement. Erosion of a material surface due to collisions with a rapidly moving liquid.

Impressed voltage. A cathodic protection technique by which a direct current is introduced into the material to be protected, thus preventing the anode reaction.

Inhibitors. Additions to the electrolyte that preferentially migrate to the anode or cathode, cause polarization, and reduce the rate of corrosion.

Intergranular corrosion. Corrosion at grain boundaries because grain boundary segregation or precipitation produce local galvanic cells.

Oxidation. Reaction of a metal with oxygen to produce a metallic oxide. This normally occurs most rapidly at high temperatures.

Oxidation reaction. The anode reaction by which electrons are given up to the electrochemical cell.

Oxygen electrode. The cathode reaction by which electrons and ions combine to produce OH^- ion groups in the electrolyte.

Oxygen starvation. The concentration cell in which low-oxygen regions of the electrolyte cause the underlying material to behave as the anode and corrode.

Passivation. Producing strong anodic polarization by causing a protective coating to form on the anode surface and interrupt the electric circuit.

Pilling-Bedworth ratio. Describes the type of oxide film that forms on a metal surface during oxidation.

Polarization. Changing the voltage between the anode and cathode to reduce the rate of corrosion. Activation polarization is related to the energy required to cause the anode or cathode reaction; concentration polarization is related to changes in the composition of the electrolyte; and resistance polarization is related to the electrical resistivity of the electrolyte.

Quench anneal. The heat treatment used to dissolve carbides and prevent intergranular corrosion in stainless steels.

Reduction reaction. The cathode reaction by which electrons are accepted from the electrochemical cell.

Sacrificial anode. Cathodic protection by which a more anodic material is connected electrically to the material to be protected. The anode corrodes to protect the desired material.

Sensitization. Precipitation of chromium carbides at the grain boundaries in stainless steels, making the steel sensitive to intergranular corrosion.

Stabilization. Addition of titanium or niobium to a stainless steel to prevent intergranular corrosion.

Stress cells. Electrochemical corrosion cells produced by differences in imposed or residual stresses at different locations in the material.

Stress corrosion. Deterioration of a material in which an applied stress accelerates the rate of corrosion.

Water electrode. The cathode reaction by which electrons and ions combine to produce water in the electrolyte.

PRACTICE PROBLEMS

1. A brass plumbing fitting produced from a Cu-30% Zn alloy operates in the hot water system of a building. After some period of use, cracking and leaking occur, although on visual examination no metal appears to have been corroded. Offer an explanation for why the fitting failed.

2. A gray cast iron pipe is used in the water distribution system for a city. The pipe fails and leaks, even though it is well below the freezing line of the soil and no corrosion noticeable to the naked eye has occurred. Offer an explanation for why the fitting failed.

3. The slag covering liquid steel in a furnace is composed of 4 parts by weight SiO_2 to 1 part FeO. Because the melting point of the slag is less than the temperature of the liquid steel (1600°C), the slag is also a liquid and begins to attack the furnace refractory. Determine how much (a) MgO or (b) Al_2O_3 would have to be added to make the liquidus temperature of the slag equal to or greater than the steel temperature so that slag-refractory problems are reduced.

4. Which of the following coatings would be expected to protect iron even if the coating is scratched to expose the iron—polyethylene, glass enamel, zinc, nickel, tin, lead?

5. Suppose 5 g of Ni^{2+} are dissolved in 1000 ml of water to produce an electrolyte. Calculate the electrode potential of the nickel half-cell.

6. A half-cell produced by dissolving lead in water produces an electrode potential of -0.15 V. Calculate the amount of lead that must be added to 1000 ml of water to produce this potential.

7. An electrode potential in a silver half-cell is found to be 0.78 V. Determine the concentration of Ag^+ ions in the electrolyte.

8. A current density of 0.03 A/cm^2 is applied to a 125-cm^2 cathode. What period of time is required to plate out a 0.05-cm coating of gold onto a cathode?

9. Determine the plating current density required to deposit 50 g of tin per 1000 cm^2 onto a cathode in 1.5 h.

10. Suppose a copper water pipe having an inside diameter of 2 cm is accidentally connected to the power system of a building, causing a current of 0.2 A to flow through the pipe. Determine the rate of corrosion of the pipe. If the wall thickness of the pipe is 0.1 cm, estimate the time required before the pipe begins to leak.

11. An aluminum probe 1 in. in diameter is submerged to a depth of 10 in. in an electrolyte. After one year, the diameter of the probe is reduced to $\frac{7}{8}$ in. Calculate the current density at the aluminum surface that caused the corrosion.

12. In a corrosion cell composed of copper and iron, a current density of $0.008\,A/cm^2$ is applied to the cathode. The area of the copper cathode is $100\,cm^2$ and the area of the iron anode is $200\,cm^2$. Calculate the amount of iron that is lost from the anode each hour.

13. Would corrosion of iron be higher in lake water or seawater? Explain in terms of polarization.

14. An Alclad composite material is produced in which one layer of 2024 aluminum alloy is sandwiched between two layers of 1100 aluminum alloy. (a) Would you expect the corrosion resistance of the composite to be good or poor? Explain. (b) Suppose that one of the 1100 alloy sheets were deeply scratched. Would the corrosion resistance change? Explain.

15. A 1080 steel rod 0.25 in. in diameter is to be formed into a coiled spring. For best corrosion resistance, should the rod be formed hot or cold? Explain.

16. A steel nut is very securely tightened onto a bolt used in a marine environment. After several months, the nut contains numerous cracks. Explain why cracking might have occurred.

17. An austenitic stainless steel is found to corrode in the heat-affected zone a short distance from the fusion zone. (a) Why does corrosion occur at this location? (b) What does this tell you about the carbon content of the steel? (c) What does this tell you about the type of welding process that must have been done? (d) What could you do to prevent the occurrence of this corrosion?

18. An aircraft wing composed of carbon fiber–reinforced epoxy is connected to a titanium forging on the fuselage. Will corrosion occur in the carbon (graphite) fibers, the epoxy, or the titanium? Explain.

19. The inside surface of a cast iron pipe is coated with tar, which provides a protective coating. Acetone in a chemical laboratory is routinely drained through the pipe. Explain why, after some period of time, the pipe begins to corrode.

20. A frequent coating for a low-carbon steel is cadmium. Does cadmium provide corrosion protection to the steel even if the coating is scratched?

21. Almost pure tin is soemtimes used to solder copper for electrical uses. Will the tin solder or the copper corrode in a corrosive environment?

22. The axle of an automobile is normally sealed in a housing and, consequently, the axle should survive the lifetime of the automobile. However, if the housing leaks, the axle may fail. Explain why a leaky housing may greatly reduce the life of the axle.

23. A sheet of annealed 1040 steel is exposed to the weather. Although the steel is not in contact with any other material that could act as a cathode, corrosion gradually occurs. Explain why the steel will corrode.

24. A steel column partly submerged in lake water is found to corrode severely just beneath the water surface. How would the corrosion pattern change if the steel column were submerged in a small pond instead of a large lake?

25. An annealed copper sheet, a cold-worked copper sheet, and a recrystallized copper sheet are submerged in an electrolyte. Which would be expected to be most resistant to corrosion? Which would be least resistant to corrosion? Explain.

26. A steel pipe used to transport crude oil is buried in various types of soil. Different rates of corrosion of the pipe are noted, depending on the nature of the soil. Explain why the soil might affect the corrosion rate of the pipe.

27. In which case would you expect less severe corrosion problems—joining two steel sheets with aluminum rivets or joining two aluminum sheets with steel rivets? Explain.

28. A steel plate can be protected by a coating of paint. Will the type of protection provided by the paint resemble the type of protection provided by tin or that provided by zinc? Explain.

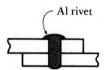

FIGURE 20-23
Two steel sheets
joined by an
aluminum rivet
(Problem 20-29).

29. Two sheets of 1020 steel are joined together with an aluminum rivet (Figure 20-23). During the joining process, the aluminum rivet is severely deformed. Discuss the possible corrosion cells that might be created as a result of this process.

30. A corrosion cell is set up between iron and zinc. The current density at the iron sheet is $0.02 \, A/cm^2$. Calculate the weight loss of zinc per week (a) if the zinc has a surface area of $10 \, cm^2$ and the iron has a surface area of $100 \, cm^2$ and (b) if the zinc has a surface area of $100 \, cm^2$ and the iron has a surface area of $10 \, cm^2$.

31. A cadmium coating is used to protect a $1 \, m \times 1 \, m$ steel sheet. A scratch 0.1 cm wide and 10 cm long is made in the cadmium coating. If 2 g of cadmium are lost uniformly from the coating per week, determine the corrosion current produced in the sheet.

32. Anodized aluminum sheet is joined by steel rivets. Would you expect the steel or the anodized aluminum to act as the anode? Explain.

33. Figure 20-24 shows a cross section through a plastic-encapsulated integrated circuit, including a small microgap between the lead frame and the polymer. If chloride ions from the manufacturing environment are able to penetrate the package, determine what types of corrosion cells might develop and what portions of the integrated circuit might corrode.

34. Liquid pig iron containing 4% C, 0.5% Ni, 1.2% Si, and 0.8% Mn is held at 1625°C. If oxygen is blown into the liquid pig iron, describe the order in which the different elements in the liquid bath will be oxidized.

35. Determine the Pilling-Bedworth ratio for the following metals and predict the behavior of the oxide that forms on the surface.

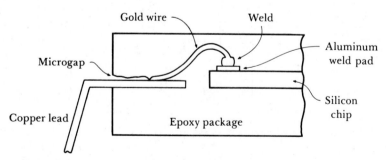

FIGURE 20-24 Cross section through an integrated circuit showing the external lead connection to the chip (Problem 20-33).

Metal	Metal Density (g/cm^3)	Oxide	Oxide Density (g/cm^3)
Ag	10.49	Ag_2O	7.143
Ba	3.5	BaO	5.72
Ca	1.55	CaO	3.30
Ce	6.689	Ce_2O_3	6.86
Co	8.832	CoO	6.45
Mo	10.22	MoO_2	6.47
Na	0.967	Na_2O	2.27
Nb	8.57	Nb_2O_5	4.47
Sn	5.765	SnO	6.446
Zr	6.505	ZrO_2	5.6

36. When palladium is oxidized at $1480°C$, 0.1 mg of Pd metal is lost per square centimeter per hour. At $604°C$, only 0.0001 mg is lost per square centimeter per hour. (a) Assuming that oxidation is controlled by the diffusion of Pd ions through the oxide, calculate the activation energy for the diffusion of Pd in PdO. (b) Suppose a 10 cm × 10 cm foil of Pd is placed into an oxidizing atmosphere at $1000°C$ for 10 days. Calculate the amount of Pd that is oxidized. Remember that oxidation occurs from both sides of the foil.

37. Suppose a $100 cm^2$ surface area of copper is exposed to an oxidizing atmosphere at an elevated temperature. After 1 h, 1×10^{-7} cm of copper has oxidized; after 1000 h, 3.164×10^{-6} cm has oxidized; and after 100,000 h, 3.164×10^{-5} cm has oxidized. (a) Determine from the data whether oxidation follows a linear, parabolic, or logarithmic pattern. (b) Calculate the total weight loss of copper after 1 year.

38. A newly developed stainless steel was tested at $500°C$ in an oxygen atmosphere. After 10,000 s, a weight *gain* of $0.62 g/m^2$ was measured; after 40,000 s, the weight gain was $1.24 g/m^2$; and after 90,000 s, the gain was $1.86 g/m^2$. (a) Determine from the data whether oxidation follows a linear, parabolic, or logarithmic pattern. (b) Explain why there is a weight gain.

39. Fiber-reinforced polymers have better resistance to sliding wear than unreinforced polymers. Would you expect the orientation of the fibers to have an effect on wear resistance? (See Figure 20-25.) Which would be most effective—normal, longitudinal, or transverse fibers? Explain. Would the aspect ratio of the fibers be important? Explain.

40. A nylon cylinder 10 in. in diameter slides continuously against a steel housing under a load of 25 lb. The cylinder rotates at 250 rpm. (a) Calculate the length of time before 0.01 in. of 6,6 nylon is worn from the surface of the cylinder. (b) Calculate the time if 18% PTFE is added to the nylon.

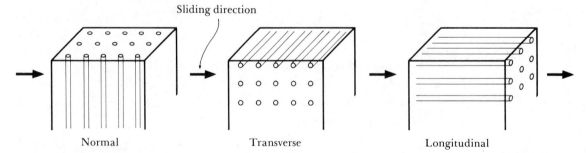

FIGURE 20-25 Three orientations of fibers in a polymer intended for wear resistance (Problem 20-39).

21

FAILURE—ORIGIN, DETECTION, AND PREVENTION

21–1   INTRODUCTION

In spite of our understanding of the behavior of materials, failures frequently occur. The sources of these failures include improper design, materials selection, and materials processing and abuse. The engineer must anticipate potential failures and consequently exercise good design, materials and processing selection, quality control, and testing to prevent failures. When failures do occur, the engineer must determine the cause so that failures can be prevented in the future.

Although the topic of failure analysis is far too complex to cover in one chapter, we will discuss a few general principles. First, we will identify the fracture mechanism by which failure occurs. Then we will discuss some considerations that may help us prevent failures, including nondestructive testing techniques.

21–2   DETERMINING THE FRACTURE MECHANISM IN METAL FAILURES

Failure analysis requires a combination of technical understanding, careful observation, detective work, and common sense. Knowledge of the history of the failed component, including the applied stress, the environment, the temperature, the intended structure and properties, and unusual changes in any of these factors, helps make identification of the cause of failure much easier.

An understanding of fracture mechanisms may also lead to the cause of failure. In this section we will concentrate on identifying the mechanism by which a metal fails when subjected to a stress. We will consider five common fracture mechanisms—ductile, brittle, fatigue, creep, and stress corrosion failures.

Ductile Fracture.   Ductile fracture normally occurs in a *transgranular* manner (through the grains) in metals that have good ductility and toughness. Often a considerable amount of deformation, including necking, is observed in the failed component. The deformation occurs before the final fracture. Ductile fractures are usually due to simple overloads, or applying too high a stress to the material.

Ductile fracture in a simple tensile test begins by the nucleation, growth, and coalescence of microvoids at the center of the test bar (Figure 21-1). *Microvoids* form when a high stress causes separation of the metal at grain boundaries or interfaces between the metal and inclusions. As the local stress continues to increase, the microvoids grow, connect, and produce larger cavities. Eventually, the metal-to-metal contact area is too small to support the load and final fracture occurs.

Deformation by slip also contributes to the ductile fracture of a metal. We

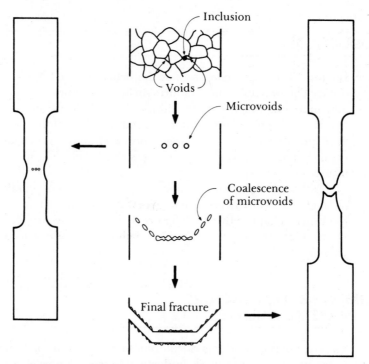

FIGURE 21-1 When a ductile material is pulled in a tensile test, necking begins and voids form, starting near the center of the bar, by nucleation at grain boundaries or inclusions. As deformation continues, a 45° shear lip may form, producing a final cup and cone fracture.

FIGURE 21-2 The cup and cone fracture observed when a ductile material, in this case an annealed 1018 steel, fractures in a tensile test. The original diameter of the test bar was 0.505 in.

know that slip occurs when the resolved shear stress reaches the critical resolved shear stress and that the resolved shear stresses are highest at a 45° angle to the applied tensile stress (Schmid's law).

These two aspects of ductile fracture give the failed surface characterisic features that, like clues, help us to determine whether a metal has failed by ductile fracture. In thick metal sections, we expect to find evidence of necking, with a significant portion of the fracture surface having a flat face where microvoids first nucleated and coalesced, and a small *shear lip*, where the fracture surface is at a 45° angle to the applied stress. The shear lip, indicating that slip occurred, gives the fracture a cup and cone appearance (Figure 21-2). Simple macroscopic observation of this fracture may be sufficient to identify the ductile fracture mode.

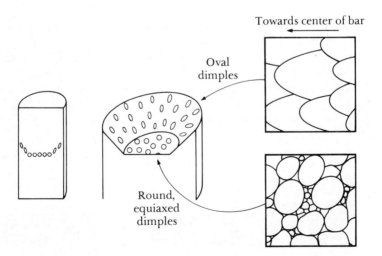

FIGURE 21-3 Dimples form during ductile fracture. Equiaxed dimples form in the center where microvoids grow. Elongated dimples, pointing towards the origin of failure, form on the shear lip.

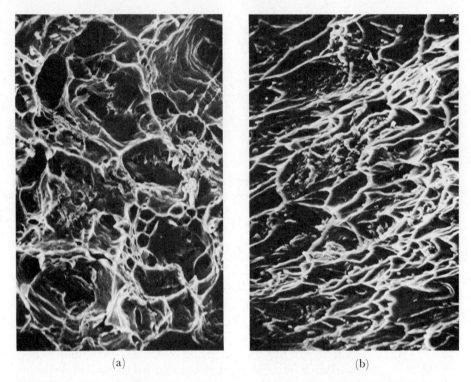

(a)   (b)

FIGURE 21-4 Scanning electron micrographs of an annealed 1018 steel exhibiting ductile fracture in a tensile test. (a) Equiaxed dimples at the flat center of the cup and cone, and (b) elongated dimples at the shear lip (× 1250).

Examination of the fracture surface at a high magnification, perhaps using a scanning electron microscope, reveals a dimpled surface (Figure 21-3). The dimples are traces of the microvoids produced during fracture. Normally, these microvoids are round, or equiaxed, when a normal tensile stress produces the failure [Figure 21-4(a)]. However, on the shear lip, the dimples are oval shaped, or elongated, with the ovals pointing towards the center of the tensile bar where fracture began [Figure 21-4(b)].

In thin plate, less necking is observed and the entire fracture surface may be a shear face (Figure 21-5). Microscopic examination of the fracture surface

FIGURE 21-5 Ductile fracture of a thin annealed 1080 steel plate in a tensile test. Necking is observed, but the entire fracture is a shear lip rather than a cup and cone. The sample is 0.125 in. thick.

shows elongated dimples rather than equiaxed dimples, indicating a greater proportion of 45° slip than in the thicker metals.

◇ **Example 21-1**

A chain used to hoist heavy loads fails. Examination of the failed link indicates considerable deformation and necking prior to failure. List some of the possible reasons for failure.

Answer:

This description suggests that the chain failed in a ductile manner by a simple tensile overload. Two factors could be responsible for this failure.

1. The load exceeded the hoisting capacity of the chain. Thus, the stress due to the load exceeded the yield strength of the chain, permitting failure. Comparison of the load to the manufacturer's specifications will indicate that the chain was not intended for such a heavy load. This is the fault of the user!

2. The chain was of the wrong composition or was improperly heat treated. Consequently, the yield strength was lower than intended by the manufacturer and could not support the load. This is the fault of the manufacturer!

Brittle Fracture.   Brittle fracture occurs in high-strength metals or metals with poor ductility and toughness. Furthermore, even metals that are normally ductile may fail in a brittle manner at low temperatures, in thick sections, at high strain rates (such as impact), or when flaws play an important role. Brittle fractures are frequently observed when impact rather than overload causes failure.

In brittle fracture, little or no plastic deformation is required. Initiation of the crack normally occurs at small flaws, which cause a concentration of stress. The crack may move at a rate approaching the velocity of sound in the metal. Normally, the crack propagates most easily along specific crystallographic planes, often the {100} planes, by *cleavage*. In some cases, however, the crack may

FIGURE 21-6 Brittle fracture of a quenched 1080 steel tensile bar. Because the microstructure is entirely martensite, a flat, brittle fracture surface is obtained. The sample is 0.125 in. thick.

FIGURE 21-7 Scanning electron micrograph of a brittle fracture surface on a quenched 1080 steel (× 5000).

take an *intergranular* (along the grain boundaries) path, particularly when segregation or inclusions weaken the grain boundaries.

Brittle fracture can be identified by observing the features on the failed surface. Normally, the fracture surface is flat and perpendicular to the applied stress in a tensile test (Figure 21-6). If failure occurred by cleavage, each fractured grain is flat, differently oriented, and gives a crystalline or "rock candy" appearance to the fracture surface (Figure 21-7). Often the layman claims that the metal failed because it crystallized. Of course, we know that the

FIGURE 21-8 The chevron pattern in a 0.5-in. diameter quenched 4340 steel. The steel failed in a brittle manner by an impact blow.

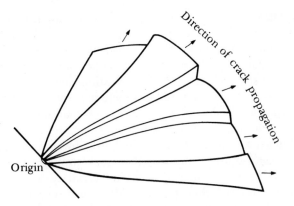

FIGURE 21-9 The chevron pattern forms as the crack propagates from the origin at different levels. The pattern points back to the origin.

metal was crystalline to begin with and the surface appearance is due to the cleavage faces.

Another common fracture feature is the *chevron pattern* (Figure 21-8), produced by separate crack fronts propagating at different levels in the material. A radiating pattern of surface markings, or ridges, fans away from the origin of the crack (Figure 21-9). The chevron pattern is visible with the naked eye or a magnifying glass and helps us identify both the brittle nature of the failure process as well as the origin of the failure.

◇ **Example 21-2**

An engineer investigating the cause of an automobile accident finds that the right rear wheel has broken off at the axle. The axle is bent. The fracture surface reveals a chevron pattern pointing towards the surface of the axle. Suggest a possible cause for the fracture.

Answer:

The evidence suggests that the axle did not break prior to the accident. The deformed axle means that the wheel was still attached when the load was applied. The chevron pattern indicates that the wheel was subjected to an intense impact blow, which was transmitted to the axle, causing failure. The preliminary evidence suggests that the driver lost control and crashed and the force of the crash caused the axle to break. Further examination of the fracture surface, microstructure, composition, and properties may verify that the axle was properly manufactured.

Fatigue Fracture.   A metal fails by fatigue when an alternating stress greater than the endurance limit is applied. Fracture occurs by a three-step process involving (a) nucleation of a crack, (b) slow, cyclic propagation of the crack, and (c) catastrophic failure of the metal. Cracks nucleate at locations of highest stress and lowest local strength. Normally, nucleation sites are at or near the surface, where the stress is at a maximum, and include surface defects such as scratches or pits, sharp corners due to poor design or manufacture, inclusions, grain boundaries, or dislocation concentrations.

Once nucleated, the crack grows towards lower stress regions. Because of the stress concentration at the tip, the crack propagates a little bit further during each cycle until the load-carrying capacity of the remaining metal is approached. The crack then grows spontaneously, usually in a brittle manner.

Fatigue failures are often easy to identify. The fracture surface, particularly near the origin, is typically smooth. The surface becomes rougher as the original crack increases in size and finally may be fibrous during final crack propagation.

Microscopic and macroscopic examination reveal a fracture surface including a beach mark pattern and striations (Figure 21-10). *Beach marks* are normally formed when the load is changed during service or when the loading is intermittent, perhaps permitting time for oxidation inside the crack. *Striations*, which are on a much finer scale, may show the position of the crack tip after each cycle. Observation of beach marks always suggests a fatigue failure; unfortunately, the absence of beach marks does not rule out fatigue failure.

◇ **Example 21-3**

A crankshaft in a diesel engine fails. Examination of the crankshaft reveals no plastic deformation. The fracture surface is smooth. In addition, several other cracks appear at other locations in the crankshaft. What type of failure mechanism would you expect?

Answer:

Since the crankshaft is a rotating part, the surface experiences cyclical loading. We should immediately suspect fatigue. The absence of plastic deformation supports our suspicion. Furthermore, the presence of other cracks is consistent with fatigue—the other cracks didn't have time to grow to the size that produced catastrophic failure. Examination of the fracture surface will probably reveal beach marks or fatigue striations.

Creep and Stress Rupture.   At elevated temperatures, a metal undergoes thermally induced plastic deformation even though the applied stress is below the nominal yield strength. The typical creep curve shows three regions— primary creep, as dislocations and other lattice imperfections are rearranged to cause rapid plastic deformation; secondary or steady-state creep, when

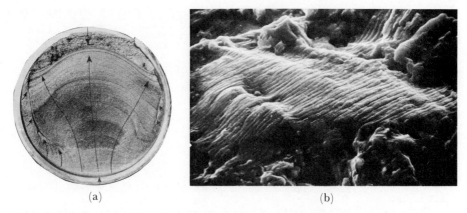

(a)   (b)

FIGURE 21-10   Fatigue fracture surface. (a) At low magnifications, the beach mark pattern indicates fatigue as the fracture mechanism and points to the origin of the failure. (b) At very high magnifications, closely spaced striations formed during fatigue are observed (× 1000). (a) From C. A. Cottell, "Fatigue Failures with Special Reference to Fracture Characteristics," *Failure Analysis: The British Engine Technical Reports*, American Society for Metals, 1981, p. 318.

dislocation climb and cross-slip cause a steady, continuous plastic deformation; and tertiary creep, when necking, void nucleation and coalescence, or grain boundary sliding cause rapid deformation and failure (Figure 21-11). *Creep failures* are defined as excessive deformation or distortion of the metal part, even if fracture has not occurred. *Stress-rupture failures* are defined as the actual fracture of the metal part.

Normally, ductile stress-rupture fractures include necking of the metal during tertiary creep and the presence of many cracks that did not have an

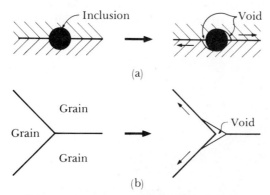

FIGURE 21-11   Grain boundary sliding during creep causes (a) the creation of voids at an inclusion trapped at the grain boundary and (b) the creation of a void at a triple point where three grains are in contact.

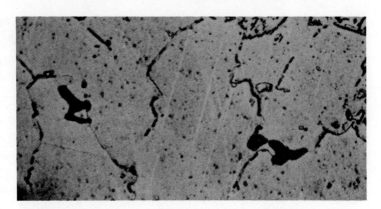

FIGURE 21-12 Creep cavities formed at grain boundaries in an austenitic stainless steel (× 500). From *Metals Handbook*, Vol. 7, 8th Ed., American Society for Metals, 1972.

opportunity to produce final fracture. Furthermore, grains near the fracture surface tend to be elongated. Ductile stress-rupture failures are generally transgranular and occur at high creep rates, short rupture times, and relatively low exposure temperatures.

Brittle stress-rupture failures usually are intergranular, show little necking, and occur more often at slow creep rates and high temperatures. Equiaxed grains are observed near the fracture surface. Brittle failure typically occurs by formation of voids at the intersection of three grain boundaries and precipitation of additional voids along grain boundaries by diffusion processes (Figure 21-12).

Stress-Corrosion Fractures.   Stress-corrosion fractures occur at stresses well below the yield strength of the metal due to attack by a corrosive medium. Deep, fine corrosion cracks are produced even though the metal as a whole shows little uniform attack. The stresses either can be externally applied or can be stored

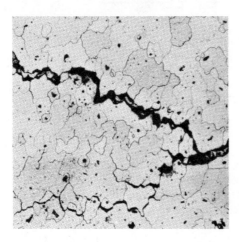

FIGURE 21-13 Photomicrograph of a metal near a stress-corrosion fracture, showing the many intergranular cracks formed as a result of the corrosion process (× 200). From *Metals Handbook*, Vol. 7, 8th Ed., American Society for Metals, 1972.

residual stresses. Stress-corrosion failures are often identified by microstructural examination of the nearby metal. Ordinarily, extensive branching of the cracks along grain boundaries is observed (Figure 21-13). The location at which cracks initiated may be identified by the presence of a corrosion product.

◇ | **Example 21-4**

A titanium pipe used to transport a corrosive material at 400°C is found to fail after several months. How would you determine the cause for the failure?

Answer:

Since a period of time at a high temperature was required before failure occurred, we might first suspect a creep or stress-corrosion mechanism for failure. Microscopic examination of the material near the fracture surface would be advisable. If many tiny, branched cracks leading away from the surface are noted, stress-corrosion is a strong possibility. However, if the grains near the fracture surface are elongated, with many voids between the grains, creep is a more likely culprit.

21–3 FRACTURE IN NONMETALLIC MATERIALS

In ceramic materials, the ionic or covalent bonds permit little or no slip. Consequently, failure is a result of brittle rather than ductile fracture. Most

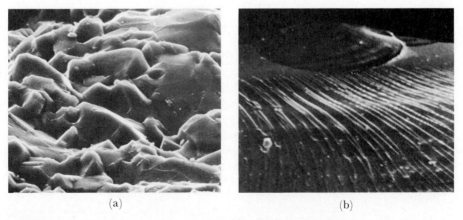

(a)   (b)

FIGURE 21-14 Scanning electron micrographs of fracture surfaces in ceramics. (a) The fracture surface of Al_2O_3, showing the cleavage faces ($\times$ 1250). (b) The fracture surface of glass, showing the mirror zone and tear lines characteristic of conchoidal fracture ($\times$ 300).

crystalline ceramics fail by cleavage along widely spaced, close-packed planes. The fracture surface typically is smooth and frequently no characteristic surface features point to the origin of the fracture [Figure 21-14(a)].

Glasses also fracture in a brittle manner. Frequently, a *conchoidal* fracture surface is observed. This surface contains a very smooth mirror zone near the origin of the fracture, with tear lines comprising the remainder of the surface [Figure 21-14(b)]. The tear lines point back to the mirror zone and the origin of the crack, much like the chevron pattern in metals.

Polymers can fail by either a ductile or a brittle mechanism. Below the glass transition temperature, thermoplastic polymers fail in a brittle manner, much like a ceramic glass. Likewise, the hard thermosetting polymers fail by a brittle mechanism. Thermoplastics, however, fail in a ductile manner above the glass transition temperature, giving evidence of extensive deformation and even necking prior to failure. The ductile behavior is a result of sliding of the polymer chains, which is not possible in glassy or thermosetting polymers.

Fracture in fiber-reinforced composite materials is more complex. Typically, these composites contain strong, brittle fibers surrounded by a soft, ductile matrix, as in boron-reinforced aluminum. When a tensile stress is applied along the fibers, the soft aluminum deforms in a ductile manner, with void formation and coalescence eventually producing a dimpled fracture surface. As the aluminum deforms, the load is no longer transmitted effectively to the fibers; the fibers break in a brittle manner until there are too few fibers left intact to support the final load.

Fracture becomes easier if bonding between the fibers is poor. Voids can form between the fibers and the matrix, causing *pull-out*. Voids can also form between layers of the matrix if composite tapes or sheets are not properly bonded, causing *delamination* (Figure 21-15).

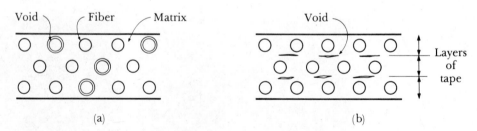

FIGURE 21-15   Fiber-reinforced composites can fail by several mechanisms. (a) Due to weak bonding between the matrix and fibers, fibers can pull out of the matrix, creating voids. (b) If the individual layers of the matrix are poorly bonded, the matrix may debond, creating voids.

◇ | **Example 21-5**

Describe the difference in fracture mechanism between a boron-reinforced aluminum composite and a glass fiber–reinforced epoxy composite.

Answer:

In the boron-aluminum composite, the aluminum matrix is soft and ductile; thus we expect the matrix to fail in a ductile manner, while the boron fibers fail in a brittle manner.

Both glass fibers and epoxy are brittle; thus the composite as a whole should display little evidence of ductile fracture.

21-4   SOURCE AND PREVENTION OF FAILURES IN METALS

We can prevent metal failures by several approaches—design of components, selection of appropriate materials and processing techniques, and consideration of the service conditions.

Design.   Components must be designed to (a) permit the material to withstand the maximum stress that is expected to be applied during service, (b) avoid stress raisers that cause the metal to fail at lower than expected loads, and (c) assure that deterioration of the material during service does not cause failure at lower than expected loads.

Creep, fatigue, and stress-corrosion failures occur at stresses well below the yield strength. The design of the component must be based on the appropriate creep, fatigue, and stress-corrosion data, not yield strength. Designers must not introduce galvanic cells when components are fabricated, particularly from different materials.

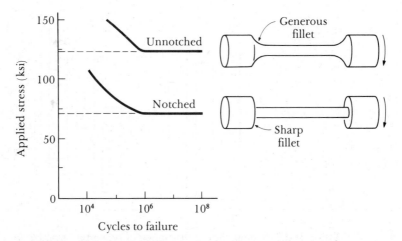

FIGURE 21-16   *Notches, introduced by poor design, significantly reduce fatigue failure, as shown here for heat-treated 2340 steel.*

Stress raisers produced by designed-in notches such as sharp fillets or keyways must be avoided. These sharp corners concentrate stresses so that fatigue or corrosion cracks can more easily nucleate (Figure 21-16).

Materials Selection.   A tremendous variety of materials are available to the engineer for any application, many of which are capable of withstanding high applied stresses (Figure 21-17). Selection of a material is based both on the ability of the material to serve and on the cost of the material and its processing.

The engineer must consider the condition of the material. For example, age-hardened, cold-worked, or quenched and tempered alloys lose their strength at high temperatures. Figure 21-18 shows the temperature ranges over which a number of groups of alloys can operate as a function of the applied stress.

Materials Processing.   All finished components are at one time passed through some type of processing—casting, forming, machining, joining, or heat treatment—producing the appropriate shape, size, and properties. However, a variety of flaws can be introduced. The engineer must design to compensate for the flaws or must detect their presence and either reject the material or correct

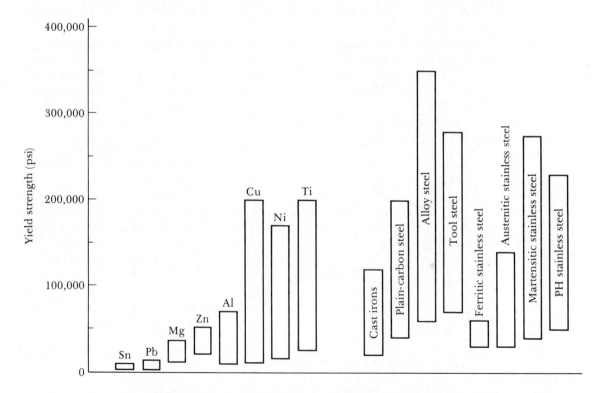

FIGURE 21-17 Comparison of the range of properties available for many important metals and alloys. A wide range of properties is possible for each alloy system, depending on composition and treatment.

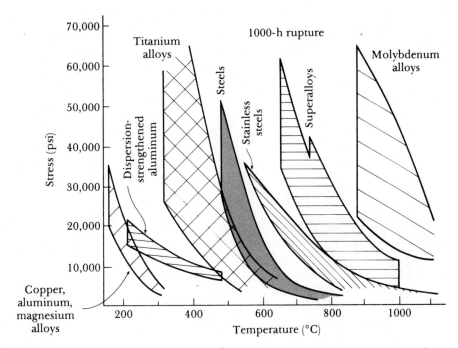

FIGURE 21-18 *Stress-rupture plots showing the range of suitable temperatures for several alloys.*

the flaw. Detection of flaws will be discussed in the next section. Figure 21-19 illustrates some of the typical flaws that might be introduced.

Service Conditions.   The performance of a material is influenced by the service conditions, including the type of loading, the corrosive or atmospheric environment, and the temperature to which an assembly is exposed.

Another source of failure is abuse of the material in service. This includes overloading a material—for example, using a chain to lift a tank whose weight exceeds the capacity of the chain. An ordinary carpenter's hammer should not be used as a crowbar by striking the hammer with another metallic instrument. Flakes of metal may spall off of the face of the hammer, which is heat treated to a high hardness, causing injury.

Improper maintenance, such as inadequate lubrication of moving parts, can lead to adhesive wear, overheating, and oxidation. If overheated, the microstructure changes and decreases the strength or ductility of the metal.

◇ | **Example 21-6**

An alloy steel, which is welded using an electrode that produces a high-hydrogen atmosphere, is found to fail in the heat-affected zone near the weld. What factors may have contributed to the failure?

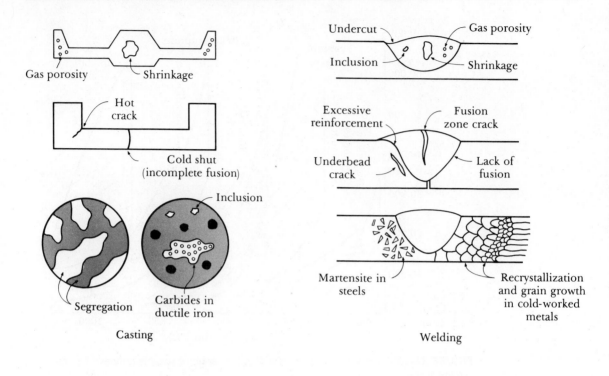

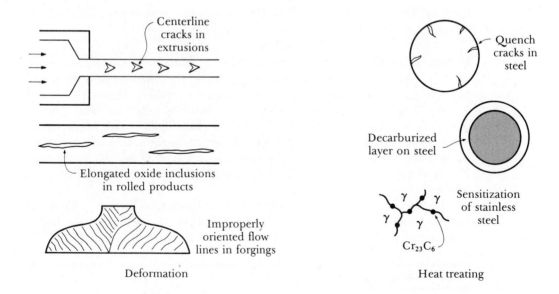

FIGURE 21-19 Typical defects introduced into a metal during processing.

Answer:

If we assume that the joint is designed so that the maximum stresses can be withstood, failure probably occurred as a result of flaws introduced during welding. Since the steel failed in the heat-affected zone rather than the fusion zone, possible problems include the following.

(a) Severe grain growth may have weakened the metal near the fusion zone.

(b) Since an alloy steel has good hardenability, martensite may have formed during cooling. A large grain size also increases the likelihood of martensite formation.

(c) Hydrogen, dissolved in the fusion zone, may have diffused to the heat-affected zone and caused hydrogen embrittlement, or underbead cracking.

(d) Because of temperature differences in the steel during welding, residual stresses may build up in the weld. Tensile residual stresses reduce the local load-bearing ability of the weld.

(e) Elongated inclusions in the steel could act as stress raisers and initiate cracks.

Additional information obtained concerning the initial state of the steel, the welding parameters, and the microstructure would permit us to pin down the cause for failure.

 Example 21-7

A steel rope, composed of many tiny strands of steel, passes over a 2-in. diameter pulley. After several months, the steel rope fails, dropping its heavy load. Suggest a possible reason for failure.

Answer:

The long period of time required before failure occurred suggests that fatigue might be the culprit. Each time the steel rope passes over the pulley, the outer strands of the steel experience a high stress. This stress may exceed the endurance limit. After passing over the pulley enough times, the strands begin to fail by fatigue. This increases the stress on the remaining strands, accelerating their failure, until the rope is overloaded and breaks. A larger diameter pulley reduces the stress on the fibers to below the endurance limit, so the rope will not fail.

◇ | **Example 21-8**

After a helicopter crash, the teeth on a gear in the transmission are found to have worn away. The gear, which is a carburized alloy steel, is intended to have a surface hardness of $R_c 60$. However, the hardness measured on a portion of one tooth that is still intact reveals a hardness of $R_c 30$. Suggest possible causes for the failure.

Answer:

We know that the rate of wear increases when the hardness decreases. Thus, we confidently blame the failure on the soft gear. However, we still need to determine why the gear was soft. One explanation would be that the gear was not carburized or heat treated. Microscopic examination would reveal this. Suppose that examination shows the presence of a proper case depth and an overtempered martensitic structure. This would suggest that the gear, originally properly manufactured, overheated during use, softened, and then began to wear. In this case, failure may have been due to loss of oil from the gear box, permitting the gear to overheat.

21–5 DETECTION OF POTENTIALLY DEFECTIVE MATERIALS

Care in design, materials selection, materials processing, and service conditions help prevent failure of metals. However, how do we determine whether our engineering process has been successful? Often a simple visual examination will reveal defects. However, to assure that our final part performs successfully, we may conduct either destructive or nondestructive tests.

Destructive Testing. Special tests are used to determine the properties of the finished product. For example, chain manufacturers routinely pull their product to failure to determine if the chain can withstand the rated load. By so doing, the effectiveness of the manufacturing process can be monitored. We use statistical methods to show that if a certain sample is good, all other parts will also be good.

Hardness Test. In some situations, a hardness test can be used to assure proper heat treatment. The hardness test does produce an indentation on the surface of the inspected part; depending on whether this indentation is harmful, hardness testing may be considered either destructive or nondestructive. We may be able to determine whether an aluminum alloy has been aged to the required properties, or a steel has been adequately tempered, or a gray cast iron has been fully annealed. However, the hardness test does not tell us the microstructure, or if a crack is present at the surface, or if a large shrinkage cavity lies within a casting.

Proof Test.   In many instances a *proof test* can be designed. We load a part to its rated capacity and see if the part remains intact. If the part is never abused or used to support loads higher than the proof test, we can usually be confident that the part will perform properly. The proof test, and the remaining tests in this section, are nondestructive.

Radiography.   In radiographic testing, the transmission and absorption characteristics of a material are utilized to produce a visual image of any flaws in a material. There are several components in a radiographic technique.

1. A source of a penetrating radiation is required. X rays emitted from a tungsten target are most common, but occasionally gamma rays or neutrons are required.

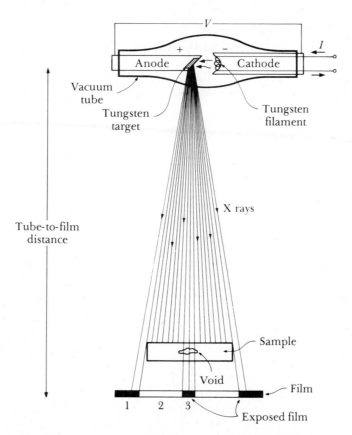

FIGURE 21-20   The elements in an X-ray radiographic nondestructive testing set-up.

2. A detection system is needed. Normally, special film is used to detect the amount of radiation that is transmitted through the material. Other detectors include fluorescent screens, Geiger counters, Xerographs, or television screens.

3. The flaw, or discontinuity, in the material must have a different absorption characteristic than the material.

In *X-ray radiography* (Figure 21-20), an X-ray tube provides the source of the radiation. Electrons are emitted from a tungsten filament cathode and accelerated at a high voltage onto an anode material, also tungsten. The beam excites electrons in the inner shells of the tungsten target and a continuous spectrum of X rays is emitted when the electrons fall back to their equilibrium states. The emitted X rays are directed to the material to be inspected. A small fraction of the X rays are transmitted through the material to expose the film.

The intensity I_0 of the incoming X-ray radiation depends on three factors.

1. High tube voltages produce higher energy X rays.

2. Longer tube-to-film distances permit the X rays to spread out. The intensity decreases with the square of the tube-to-film distance.

3. Increasing the tube current increases the intensity of the emitted X-ray beam by increasing the number of X rays.

TABLE 21-1   Absorption coefficients of selected elements for tungsten X rays and for neutrons

Element	Density (g/cm^3)	μ_m (X rays) $\lambda = 0.098\,\text{Å}$ (cm^2/g)	μ_m (neutrons) $\lambda = 1.08\,\text{Å}$ (cm^2/g)
H	—	0.280	0.11
Be	1.85	0.131	0.0003
B	2.3	0.138	24
C	2.2	0.142	0.00015
N	0.00116	0.143	0.048
O	0.00133	0.144	0.00002
Mg	1.74	0.152	0.001
Al	2.7	0.156	0.003
Si	2.33	0.159	0.001
Ti	4.54	0.217	0.001
Fe	7.87	0.265	0.015
Ni	8.9	0.310	0.028
Cu	8.96	0.325	0.021
Zn	7.133	0.350	0.0055
Mo	10.2	0.79	0.009
Sn	7.3	1.17	0.002
W	19.3	2.88	0.036
Pb	11.34	3.5	0.0003

After W. McGonnagle, *Nondestructive Testing*, McGraw-Hill, 1961.

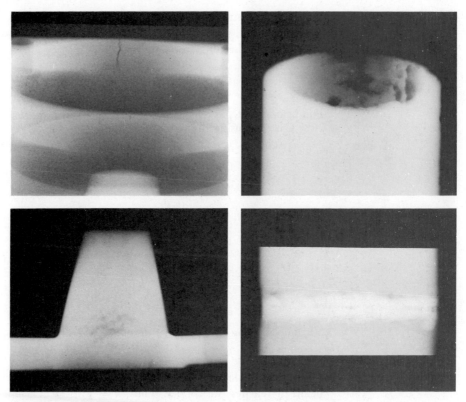

FIGURE 21-21 Radiographs of (a) magnesium casting containing crack and hydrogen porosity; (b) graphitic corrosion of a cast iron pipe; (c) solidification shrinkage in an aluminum casting; (d) incomplete penetration of a steel weld.

The intensity I of the transmitted X-ray beam depends on the absorption coefficient and the thickness of the material.

$$I = I_0 \exp(-\mu x) = I_0 \exp(-\mu_m \rho x) \qquad (21\text{-}1)$$

where I_0 is the intensity of the incoming beam, μ is the linear absorption coefficient (cm^{-1}), μ_m is the mass absorption coefficient (cm^2/g), ρ is the density (g/cm^3), and x is the thickness of the material (cm). The mass absorption coefficients for selected materials are included in Table 21-1. If a large shrinkage cavity is present inside a casting, the absorption of the X rays by the cavity is lower than in the solid metal and consequently the intensity of the transmitted beam is greater. The film is exposed to more radiation and, after developing, is much darker (Figure 21-21).

◇ **Example 21-9**

In Figure 21-20, let's assume that the material being inspected is a copper plate 1 in. thick and that a discontinuity $\frac{1}{4}$ in. thick is present at point 3. The discontinuity contains air. Estimate the intensity I at locations 1, 2, and 3 if the copper plate is placed on the film 30 in. from the source. The average wavelength of the tungsten X rays is 0.098 Å.

Answer:

Point 1

The beam must pass through air only. If we assume that air is 80% N_2 and 20% O_2, then using the rule of mixtures

$$\mu_m, \text{air} = f_{O_2}\mu_{m,O_2} + f_{N_2}\mu_{m,N_2}$$
$$= (0.2)(0.144) + (0.8)(0.143)$$
$$= 0.143 \text{ cm}^2/\text{g}$$
$$\rho_{\text{air}} = f_{O_2}\rho_{O_2} + f_{N_2}\rho_{N_2}$$
$$= (0.2)(1.33 \times 10^{-3}) + (0.8)(1.16 \times 10^{-3})$$
$$= 1.19 \times 10^{-3} \text{g/cm}^3$$
$$\frac{I}{I_0} = \exp(-\mu_m\rho x)$$
$$= \exp(-0.143)(1.19 \times 10^{-3})(30 \text{ in.})(2.54 \text{ cm/in.})$$
$$= \exp(-0.013) = 0.987$$

Virtually no absorption occurs in air.

Point 2

Since virtually no absorption occurs in air, let's neglect absorption of the beam before it reaches the copper. Thus, $x = 1$ in. $= 2.54$ cm. From Table 21-1, $\mu_{m,\text{Cu}} = 0.325 \text{ cm}^2/\text{g}$.

$$\frac{I}{I_0} = \exp(-\mu_m\rho x) = \exp(-0.325)(8.96)(2.54) = \exp(-7.40)$$
$$= 6.1 \times 10^{-4}$$

Practically none of the beam is transmitted.

Point 3

Since the $\frac{1}{4}$-in. cavity absorbs no X rays compared with the copper, we can ignore it and consider the thickness to be 0.75 in.

$$\frac{I}{I_0} = \exp(-\mu_m\rho x) = \exp(-0.325)(8.96)(0.75)(2.54)$$
$$= \exp(-5.55) = 3.89 \times 10^{-3}$$

The X-ray beam intensity is about one order of magnitude greater when it passes through the discontinuity. Thus, the film will be darker beneath the discontinuity.

Obtaining a radiograph that sensitively detects the presence of a discontinuity in the material requires optimization of the film, accelerating voltage,

---

TABLE 21-2   Factors producing good sensitivity
and definition in X-ray radiography

Fine-grained film emulsion
Low X-ray energy or long wavelength
Low accelerating voltage
Long tube-to-film distance
Thin metal samples
Large difference in relative absorption coefficients
   of material and discontinuity
Orienting part so discontinuities are near the film

---

exposure, and various geometric factors (Table 21-2). The *sensitivity* is defined as

$$\text{Percent sensitivity} \;=\; \frac{\Delta x}{x} \times 100 \tag{21-2}$$

where x is the total thickness of the material and Δx is the smallest change in the thickness of the material that is visible on the film. The sensitivity tells us the minimum size of flaw that can be detected. If the sensitivity is very small, we can detect the presence of very small discontinuities.

◇ | **Example 21-10**

The sensitivity of a particular X-ray radiographic set-up is 3%. What is the thickness of the smallest discontinuity that we can detect in a casting 1 in. thick? In a casting 3 in. thick?

Answer:

$$\text{Percent sensitivity} \;=\; 3 \;=\; \frac{\Delta x}{x} \times 100$$

If $x = 1$ in., then $\Delta x = (3\%)(1)/100 = 0.03$ in.
If $x = 3$ in., then $\Delta x = (3\%)(3)/100 = 0.09$ in.

The geometry and physical properties of the discontinuity also affect how easily it can be detected (Figure 21-22).

1. Shrinkage voids or gas porosity in castings can be detected if their size falls within the sensitivity of the radiographic technique. The voids or pores are normally round, so the direction of the X rays compared with the orientation of the casting is not critical. The voids or pores are filled with either a gas or a vacuum which have very different absorption coefficients than the metal.

2. Nonmetallic inclusions in rolled products may be flattened during deformation processing. If the X rays are perpendicular to the inclusion, the thin inclusions may not cause a noticeable change in intensity. Either the

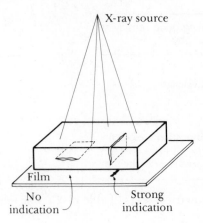

FIGURE 21-22 The importance of orientation on the detection of discontinuities in a material by radiography.

geometry between the X-ray source and the flaw must be changed or some other nondestructive testing technique may be required.

3. Detection of cracks caused by casting, welding, incipient fatigue, or stress-corrosion is dependent on the crack geometry. If the crack is perpendicular to the incident radiation, it may not be detected, but the crack may be easily observed if it is parallel to the radiation.

◇   **Example 21-11**

A crack 3 cm long and 0.01 cm wide is present in a 5 cm × 5 cm × 5 cm block of aluminum. Determine the relative intensity of the transmitted tungsten X-ray beam if the X rays are parallel to the crack. Repeat for the case in which the X rays are perpendicular to the crack.

Answer:

Let's assume that the crack does not absorb X rays. From Table 21-1, $\mu_{m,\mathrm{Al}} = 0.156 \, \mathrm{cm^2/g}$ and $\rho_{\mathrm{Al}} = 2.7 \, \mathrm{g/cm^3}$.

No crack: $x = 5 \, \mathrm{cm}$

$$\frac{I}{I_0} = \exp\left(-\mu_m \rho x\right) = \exp\left(-0.156\right)(2.7)(5) = \exp\left(-2.1\right) = 0.12$$

Perpendicular to crack: $x = 5 - 0.01 = 4.99 \, \mathrm{cm}$

$$\frac{I}{I_0} = \exp\left(-\mu_m \rho x\right) = \exp\left(-0.156\right)(2.7)(4.99) = \exp\left(-2.1\right) = 0.12$$

Parallel to crack: $x = 5 - 3 = 2 \, \mathrm{cm}$

$$\frac{I}{I_0} = \exp\left(-\mu_m \rho x\right) = \exp\left(-0.156\right)(2.7)(2) = \exp\left(-0.84\right) = 0.43$$

There is a large change in the intensity of the transmitted beam if the beam is parallel to the crack, but no change if the beam is perpendicular to the crack.

Improvements in radiographic techniques have made possible better resolutions and even three-dimensional representations of flaws. Semiconductor materials such as *p-n* diodes receive the transmitted X-ray beam and convert the beam to a current pulse; the current pulses are then electronically and computer enhanced to provide improved sensitivity and sharpness of the radiograph.

(a)

(b)

(c)

FIGURE 21-23 X-ray radiographs of an integrated circuit at three magnifications obtained by electronic reception of the transmitted beam. The radiograph shows the 14 copper lead pads, gold wire and weld balls, and aluminum weld pad. (a) × 5, (b) × 20, (c) × 50. From *Nondestructive Testing Handbook*, Vol. 3: *Radiography and Radiation Testing*, 2nd Ed., American Society for Nondestructive Testing, 1985.

Inspection of very small parts, including integrated circuits (Figure 21-23), can be accomplished. By using computerized axial tomography (the CAT scan), either the X-ray source and detector or the sample is rotated during radiography. Several exposures at different locations can be made. Again, the transmitted beam is received by electronic means and analyzed by computer to provide information in three dimensions. Two-dimensional radiographs can then be generated by the computer for any plane within the sample. This technique is also used frequently for medical purposes.

Gamma ray radiography employs an intense radiation of a single wavelength produced by nuclear disintegration of radioactive materials. The intensity of the radiation depends on the type and size of the radioactive source. Cobalt 60 produces gamma rays having an average energy of about 1.33×10^6 eV (or a wavelength of about 0.009 Å); the gamma rays are much more energetic than the X rays obtained from tungsten (about 0.125×10^6 eV or 0.098 Å). Consequently, Co^{60} is used for thick, absorptive materials. The mass absorption coefficient of most elements for Co gamma rays is about 0.055 cm²/g. Cesium 137, which produces gamma rays of 0.66×10^6 eV, and indium 192, which produces gamma rays of 0.31 to 0.60×10^6 eV, are used when less energetic radiation is needed.

The intensity of the gamma ray source decreases with time.

$$I = I_0 \exp(-\lambda t) \tag{21-3}$$

where λ is the decay constant for the material and t is the time. The *half-life* is the time required for I to decrease to $0.5I_0$. The half-life of cobalt 60 is about 5.27 years and that for iridium 192 is 74 days. Usually, the source is no longer suitable after about two half-lives have elapsed.

◇ | **Example 21-12**

Determine the decay constant for cobalt, which has a half-life of 5.27 years.

Answer:

In Equation 21-3, $I = 0.5I_0$ when $t = 5.27$ years.

$$0.5I_0 = I_0 \exp(-5.27\lambda)$$
$$0.5 = \exp(-5.27\lambda)$$
$$\ln(0.5) = -5.27\lambda$$
$$\lambda = \frac{-0.693}{-5.27} = 0.131 \text{ years}^{-1} = 4.15 \times 10^{-9} \text{s}^{-1}$$

Neutron radiography is occasionally used because the neutrons are absorbed by nuclear rather than electronic interactions. Materials that have similar absorption coefficients for X rays or gamma rays may be easily distinguished by neutron radiography (Table 21-1). However, the source for the neutrons is a nuclear reactor, which is not widely available.

Ultrasonic Testing.   A material may both transmit and reflect elastic waves. An ultrasonic transducer composed of quartz, barium titanate, or lithium sulfate uses the piezoelectric effect to introduce a series of elastic pulses into the material at a high frequency, usually greater than 100,000 Hz. The pulses create a compressive strain wave that propagates through the material. The elastic wave, or phonon, is transmitted through the material at a rate that depends on the modulus of elasticity and the density of the material. For a thin rod

$$v = \sqrt{\frac{Eg}{\rho}} \tag{21-4}$$

where E is the modulus of elasticity, g is the acceleration due to gravity (384 in./s^2), and ρ is the density of the material. More complicated expressions are required for pulses propagating in bulkier material; the velocity is higher in the bulk materials than in thin rods because of the greater transverse restraint experienced by the compressive wave. Examples of the bulk ultrasonic velocity v for several materials are given in Table 21-3.

Three common techniques are used to ultrasonically inspect a material.

1. The pulse-echo or reflection method: An ultrasonic pulse is generated and transmitted through the material (Figure 21-24). When the elastic wave

TABLE 21-3   Ultrasonic velocities for selected materials

Material	Velocity (in./s × 10^5)	Modulus of Elasticity (psi × 10^6)	Density (g/cm^3)
Al	2.46	10.4	2.7
Cu	1.82	16	8.96
Pb	0.77	2.4	11.34
Mg	2.27	6.6	1.74
Ni	2.37	30	8.9
60% Ni-40% Cu	2.1	26	8.9
Ag	1.43	10.9	10.49
Stainless steel	2.26	28.5	7.91
Sn	1.33	8	7.3
W	2.04	58.9	19.25
Air	0.13		1.3 × 10^{-3}
Glass	2.22	10.4	2.32
Lucite	1.05	0.5	1.18
Polyethylene	0.77	11.6	0.9
Quartz	2.26	10.0	2.65
Water	0.59		1.00

From W. McGonnagle, *Nondestructive Testing*, McGraw-Hill, 1961.

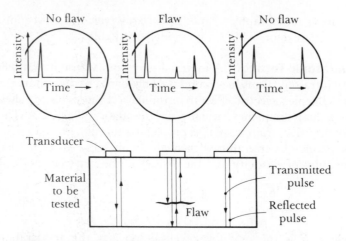

FIGURE 21-24 The pulse-echo ultrasonic test. The time required for a pulse to travel through the metal, reflect off a discontinuity on the opposite side, and return to the transducer is measured with an oscilloscope.

strikes an interface, a portion is reflected and returns to the transducer. Both the initial and the reflected pulse can be displayed on an oscilloscope.

From the oscilloscope display, we measure the time required for the pulse to travel from the transducer to the reflecting interface back to the transducer. If we know the velocity at which the pulse travels in the material, we can determine how far the elastic wave traveled and can calculate the distance below the surface at which the reflecting interface is located. If there are no flaws in the material, the beam reflects from the opposite side of the material and our measured distance corresponds to twice the wall thickness.

If a discontinuity is present and properly oriented beneath the transducer, at least a portion of the pulse reflects from the discontinuity and registers at the transducer in a shorter period of time. Now our calculations show that a discontinuity lies within the material and even tell us the depth of the discontinuity below the surface. By moving the transducer we can estimate the size of the discontinuity. Automatic scanning techniques can move the transducer over the surface and display the results of the entire scan, showing the exact location of flaws. If we also combine holography with ultrasonics, we can obtain a three-dimensional picture of the discontinuities.

◇ | **Example 21-13**

In a pulse-echo ultrasonic inspection of an aluminum rod, the oscilloscope shows three peaks. The first, at time zero, is the initial transmitted pulse; the second, at time 1.63×10^{-5} s, is the reflection from an internal discontinuity; and the third peak, at 2.44×10^{-5} s, is the reflection from the opposite surface of the material. Calculate the thickness of the material and the depth below the surface of the flaw.

Answer:

From Table 21-3, we expect the ultrasonic velocity in aluminum to be 2.46×10^5 in./s. Thus, the total distance that the pulse traveled in each case is

$$\text{Discontinuity:} \quad \text{Distance} = (2.46 \times 10^5)(1.63 \times 10^{-5})$$
$$= 4 \text{ in.}$$
$$\text{Back surface:} \quad \text{Distance} = (2.46 \times 10^5)(2.44 \times 10^{-5})$$
$$= 6 \text{ in.}$$

Since the total distance includes the return path, the actual depth of the discontinuity is $4/2 = 2$ in. and the actual thickness of the rod is $6/2 = 3$ in.

2. Through-transmission method: In this method an ultrasonic pulse is generated at one transducer and detected at the opposite surface by a second transducer (Figure 21-25). The initial and transmitted pulses are displayed on an oscilloscope. The loss of energy from the initial to the transmitted pulse depends on whether or not a discontinuity is present in the material.

3. Resonance method: In both the reflection and transmission methods, the phonon, or elastic wave, is treated as an energy particle. In the resonance method, we utilize the wavelike nature of the phonon. A continuum of pulses is generated and travels as an elastic wave through the material (Figure 21-26). By selecting a wavelength or frequency so that the thickness of the material is a whole number of half-wavelengths, a stationary elastic wave is produced and reinforced in the material. A discontinuity in the material prevents resonance

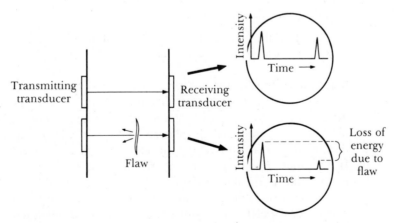

FIGURE 21-25 *The through-transmission ultrasonic test. The presence of a discontinuity reflects a portion of the transmitted beam, thus reducing the intensity of the pulse at the receiving transducer.*

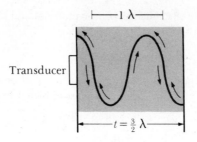

FIGURE 21-26 In the resonance method of ultrasonic testing, a stationary elastic wave is produced by changing the wavelength or frequency of the ultrasonic pulse until a whole number of half-wavelengths matches the thickness of the material.

from occurring. However, this technique is used more frequently to determine accurately the thickness of the material.

  Example 21-14

The thickness of copper is to be determined by the resonance method. A frequency of 1,213,000 Hz is required to produce 12 half-wavelengths at resonance. Determine the thickness of the copper.

Answer:

From Table 21-3, the ultrasonic velocity in copper is 1.82×10^5 in./s.

$$\lambda = \frac{v}{\nu} = \frac{1.82 \times 10^5}{1,213,000} = 0.15 \text{ in.}$$

$$\frac{\lambda}{2} = 0.075 \text{ in.}$$

$$\frac{\lambda}{2}(12) = (0.075)(12) = 0.9 \text{ in.} = \text{thickness of the copper}$$

In ultrasonic testing, the discontinuity should be perpendicular to the beam. A crack parallel to the beam is not detected.

Magnetic Particle Inspection. Magnetic particle testing detects discontinuities near the surface of ferromagnetic materials. A magnetic field is induced in the material to be tested (Figure 21-27), producing lines of flux. The magnetic field can be introduced by placing the material in a coil, passing a current through the metal, or touching the material with a yoke or probe (Figure 21-28). If a discontinuity is present in the material, the reduction in the magnetic permeability of the material due to the discontinuity alters the flux density of the magnetic field. Leakage of the lines of flux into the surrounding atmosphere creates local north and south poles which attract magnetic powder particles. The particles may be added dry or in a fluid such as water or light oil for better movement; they may be dyed or coated with a fluorescent material to aid in their detection.

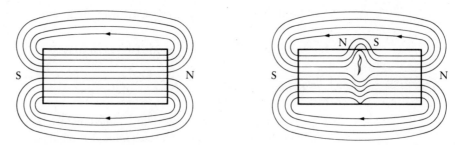

FIGURE 21-27 A flaw in a ferromagnetic material causes a disruption of the normal lines of magnetic flux. If the flaw is at or near the surface, lines of flux leak from the surface. Magnetic particles are attracted to the flux leakage and indicate the location of the flaw.

Several requirements must be satisfied in order to detect discontinuities by magnetic particle inspection.

1. The discontinuity must be perpendicular to the lines of flux. Thus, different methods of introducing the magnetic field detect differently oriented discontinuities.

2. The discontinuity must be near the surface or the flux lines will merely crowd together rather than leak from the material. Magnetic particle testing is well suited for locating quench cracks, fatigue cracks, or cracks induced by grinding, all of which occur at the surface.

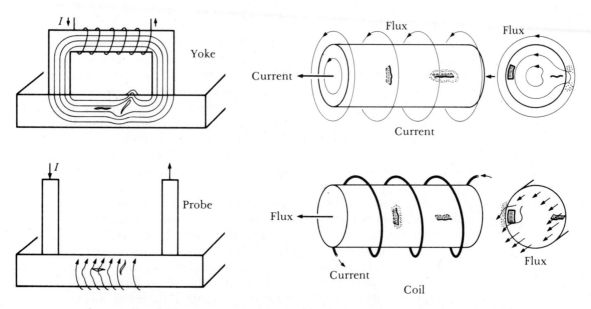

FIGURE 21-28 Magnetic particle testing techniques and the relationship between the flux pattern and the orientation of the flaw.

3. The discontinuity must have a lower magnetic permeability than the metal.

4. Only ferromagnetic materials can be tested.

◇ | **Example 21-15**

Two cylindrical bars of a steel are joined by friction welding. Describe a magnetic particle test that will determine if the bars are properly joined.

Answer:

The potential weld defect is perpendicular to the cylindrical bars. Thus, we want to produce longitudinal magnetization to detect any flaw. Either a yoke or a coil technique produces the proper orientation of the magnetic flux lines with the weld.

Eddy Current Testing.   In eddy current testing, we rely on the interaction between the material and an electromagnetic field. An alternating current flowing in a conductive coil produces an electromagnetic field. If a conductive material is placed near or within the coil (Figure 21-29), the field of the coil will induce eddy currents and additional electromagnetic fields in the sample, which in turn interact with the original field of the coil. By measuring the effect of the sample on the coil, we can deduce information concerning the structure and properties of the sample.

In the eddy current test, we can optimize the size and geometry of the coil, the proximity of the coil to the sample, and the frequency and magnitude of the current in the coil for a particular application. Once the optimum conditions are determined, we can detect changes in electrical conductivity or magnetic permeability of the sample. Changes in conductivity or permeability are often due to differences in composition, microstructure, and properties. If properly calibrated, the eddy current test can determine whether a material has been properly heat treated. Because discontinuities in the sample will alter the electromagnetic fields, potentially harmful defects can be detected. Even changes in the size of the sample or the thickness of the plating on a sample might be detected by the test.

Eddy current testing, like magnetic particle inspection, is best suited for detecting flaws near the surface of the sample. Particularly at high frequencies, the eddy currents do not penetrate deeply below the surface.

The eddy current test is particularly rapid compared with most other nondestructive testing techniques. Therefore, large numbers of parts can be tested quickly and economically. Often the eddy current test is set up as a "go or no go" test that is standardized on good parts. If the interaction between the coil and the part is the same when other parts are tested, the parts may be assumed to be good.

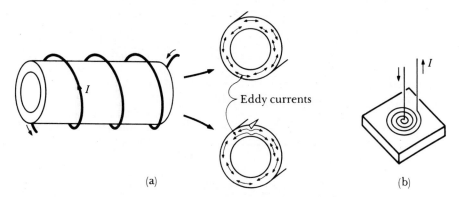

FIGURE 21-29 (a) The through-coil method and (b) the probe method for eddy current inspection.

Electromagnetic Testing. This process is often used for sorting materials but also may be used for analysis of microstructure and heat treatment. Electromagnetic testing can analyze the hysteresis loop characteristics of ferromagnetic materials to sort steels based on composition, heat treatment, hardness, residual stress, or case depth or can be used to determine amounts of ferrite present in austenitic stainless steels (Figure 21-30).

Liquid Penetrant Inspection. Discontinuities such as cracks that penetrate to the surface can be detected by the *dye penetrant* technique. A liquid dye is drawn by capillary action into a thin crack that might otherwise be invisible. A four-step process is involved (Figure 21-31). The surface is first thoroughly cleaned. A liquid dye is sprayed onto the surface and permitted to stand for a period of time, during which the dye is drawn into any surface discontinuities. Excess dye is then cleaned from the metal surface. Finally, a developing solution is sprayed onto the surface. The developer reacts with any dye that remains, drawing the dye from the cracks. The dye then can be observed, either because

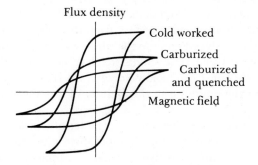

FIGURE 21-30 Hysteresis loop analysis is an electromagnetic testing method used to evaluate the condition of steels.

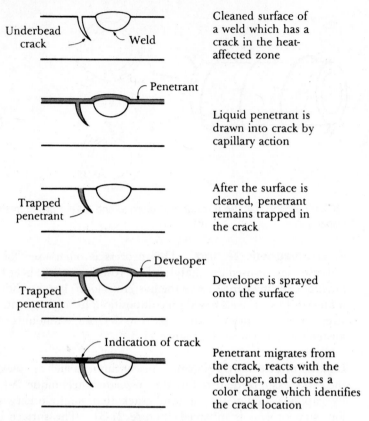

FIGURE 21-31 The elements of the dye penetrant test, used to detect cracks that penetrate to the surface.

it changes the color of the developer or because it fluoresces under an ultraviolet lamp.

A similar process uses a krypton gas penetrant rather than a liquid dye, therefore allowing penetration and detection of smaller flaws. The sample is first exposed to a vacuum to draw adsorbed air from the flaw: the mildly radioactive krypton gas is then introduced and the gas enters the flaws. A radiation-sensitive emulsion containing silver halide similar to that used for radiographic film is then sprayed onto the surface. The low-energy beta rays emitted by the krypton expose and darken the emulsion in the vicinity of any defects.

Thermography. Often imperfections in a material will alter the rate of heat flow in the vicinity, leading to high temperature gradients or hot spots. In *thermography*, a temperature-sensitive coating is applied to the surface of a material, then the material is uniformly heated and cooled. The temperature will be higher in the vicinity of an imperfection than at other locations; therefore, the color of the coating at this location will be different and easily detected.

A variety of coatings can be used. Heat-sensitive paints, heat-sensitive papers, organic compounds or phosphors that produce visible light when excited by infrared radiation, and crystalline organic materials known as liquid crystals are commonly used. Paints are available for use at temperatures as high as 1600°C.

One important use of thermography is in detecting lack of bonding between or delamination of the individual monolayers or tapes that make up many of the fiber-reinforced composite material structures, particularly in the aerospace industry.

Acoustic-Emission Inspection. Accompanying many microscopic phenomena, such as crack growth or phase transformations, is the release of strain energy in the form of high-frequency elastic stress waves, much like those produced during an earthquake. In the *acoustic-emission test*, a stress below the nominal yield strength of the material is applied. Due to stress concentrations at the tip of a preexisting crack, the crack may enlarge, releasing the strain energy surrounding the crack tip. The elastic stress wave associated with the movement of the crack can be detected by a piezoelectric sensor, amplified, and analyzed. Cracks as small as 10^{-6} in. long can be detected by this technique. By using several sensors simultaneously, the location of the crack can also be determined.

Acoustic emission testing can be used for all materials. The technique is used to detect microcracks in aluminum aircraft components even before the microcracks are large enough to imperil the safety of the aircraft. Cracks in polymers and ceramics can be detected. The test will detect breakage of fibers in fiber-reinforced composite materials as well as debonding between the fibers and the matrix.

21–6 NONDESTRUCTIVE TESTING AND FRACTURE MECHANICS

In Chapter 6, we discussed fracture mechanics and pointed out that if we know the size of the flaws present in a material, we can predict whether the flaws will cause fracture for a given applied stress. The nondestructive tests described in the previous section can be used to find these flaws, and in many cases either the actual size or the maximum size of the flaws can be determined. For example, the percent sensitivity in the X-ray radiography technique can be used to estimate the smallest detectable flaw. With experience and experimentation, maximum flaw sizes could also be determined for other tests, including ultrasonic inspection, magnetic particle inspection, and others. Particularly small flaws might be detected by acoustic emission.

By combining nondestructive testing and fracture mechanics, we have the ability not only to locate flaws but to determine whether they pose any threat to the serviceability of the part. If the flaws are exceptionally small, they may not propagate under the expected loads and, consequently, the part may continue in service. By periodically monitoring the flaw size and charting its growth, we can determine when a part must be scrapped or repaired. This approach is often used for aerospace components.

SUMMARY

We often are able to determine the manner in which a material fails by examining the characteristic features on the fracture surface or the microstructure adjacent to the fracture. This analysis helps us to improve our materials selection or engineering. By combining nondestructive testing and fracture toughness techniques, we may be able to prevent failures. The nondestructive testing techniques, which rely on the physical properties of the material, are consequently very sensitive to the material's structure, including incipient cracks or other discontinuities.

GLOSSARY

Acoustic emission testing. Detection of elastic stress waves produced when a small applied stress causes enlargement of a flaw in a material.

Beach marks. Marks on the surface of a fatigue fracture which represent the position of the crack front at various times during failure.

Brittle fracture. Fracture of a material with little or no deformation.

Chevron pattern. Markings caused by the merging of crack fronts in brittle fracture. The markings form arrows which point back towards the origin of the brittle fracture.

Cleavage. Brittle fracture along particular crystallographic planes in the grains of the material.

Conchoidal fracture. A characteristic fracture surface in glass, with a mirror zone near the origin of the fracture and tear lines pointing towards the origin.

Creep failure. Excessive deformation or distortion of a material at high temperatures, without fracture occurring.

Ductile fracture. Fracture of a material with significant deformation required.

Eddy current testing. A nondestructive testing technique to detect flaws or evaluate structure and properties by determining the reaction between the material and an electric field.

Electromagnetic testing. Evaluating the structure or treatment of a ferromagnetic material by determining its response to a magnetic field.

Fatigue fracture. Fracture of a material due to a cyclical application of a load.

Intergranular fracture. Fracture of a material along the grain boundaries.

Liquid penetrant inspection. A nondestructive testing technique in which a liquid, drawn into a surface imperfection by capillary action, is exposed by a dye or ultraviolet light.

Magnetic particle inspection. A nondestructive testing technique that relies on the interruption of lines of magnetic flux by imperfections near the surface.

Mass absorption coefficient. Related to the absorption of X rays or other radiation by a material.

Microvoids. Tiny voids at the fracture surface formed by separation of the material at grain boundaries or other interfaces during ductile failure.

Proof test. Loading a material to its designed capacity to determine if it is capable of proper service.

Radiography. A nondestructive testing

technique that relies on a difference between the absorption of radiation by the material and flaws in the material.

Sensitivity. A measure of the minimum size of a flaw that can be detected in a material by a particular radiographic set-up.

Shear lip. The surface formed by ductile fracture that is at a 45° angle to the direction of the applied stress.

Stress-corrosion fracture. Fracture caused by a combination of corrosion and a stress below the yield strength.

Stress-rupture fracture. Fracture of a material due to prolonged exposure at a high temperature.

Striations. Microscopic traces of the location of a fatigue crack.

Thermography. Detection of discontinuities in a material by the change in color of a coating caused by temperature differences induced in a material.

Transgranular fracture. Fracture of a material through the grains rather than along the grain boundaries.

PRACTICE PROBLEMS

1. Figure 21-32 is a photograph of a 1.25-in. diameter axle from an automobile; after an accident, the axle was found to be broken. From the photograph, determine the mechanism by which the axle fractured and the location at which fracture began (express by the angle measured clockwise from the top of the photograph). What is the probable cause for the accident?

2. Figure 21-33 shows the fracture surface of a coil spring made from a 0.25-in. diameter steel rod. From the photograph, determine the mechanism by which the axle fractured and the location at which fracture began (express by the angle measured clockwise from the top of the photograph).

3. Figure 21-34 shows the end of a steel wire used to reinforce an automobile tire. From

FIGURE 21-32 Photograph of the fracture surface of an automobile axle (Problem 21-1).

FIGURE 21-33 Photograph of the fracture surface of a coil spring (Problem 21-2).

FIGURE 21-34 Photograph of the fractured ends of several steel wires used to reinforce an automobile tire (Problem 21-3).

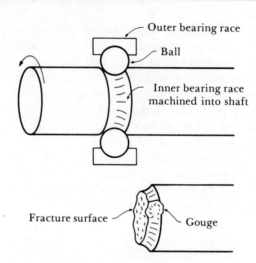

FIGURE 21-36 Sketch of a shaft for an automotive water pump, showing the relationship between the balls and inner and outer races and the fracture surface after failure of the shaft (Problem 21-5).

the appearance of the wire, determine the mechanism by which the wire fractured.

4. Figure 21-35 shows a diagram of a pin and hanger strap assembly used on a bridge. The bridge connection is designed in such a manner that as the ambient temperature changes, the resulting expansion or contraction of the bridge can occur by rotation of the pin. The pin failed,

as indicated, and a cross section of the pin is also shown. Discuss the probable mechanism for fracture and how this fracture came about.

5. A steel shaft in an automotive water pump rotates on steel ball bearings. The shaft fractured, causing the fan blade that was attached to the shaft to injure a mechanic adjusting the

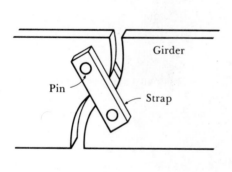

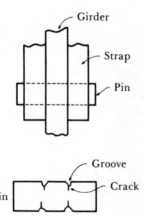

FIGURE 21-35 Sketch of a strap-pin-girder connection for a highway bridge (Problem 21-4). (a) Overall picture of the assembly, (b) section through the connection, and (c) cross section of pin showing grooves and cracks just prior to fracture.

engine of the automobile. On inspection, the shaft was found to have fractured at a groove in which the balls rotated (Figure 21-36). Severe deformation of the metal had occurred in this groove. Discuss some of the possible causes for this fracture.

6. Determine (a) the mass absorption coefficient and (b) the linear absorption coefficient for a Cu-30% Ni alloy. Be sure to use atomic fraction.

7. A gas bubble 2 mm in diameter is located in a 2-cm thick aluminum casting. Compare the intensity of an X-ray beam transmitted through the section containing the bubble with that of a beam transmitted through the section containing no bubble.

8. A 1-mm diameter tungsten inclusion is located in a 1-cm thick aluminum weld. Compare the intensity of an X-ray beam transmitted through the section containing the inclusion with that of a beam transmitted through the section containing no inclusion.

9. The sensitivity when radiographing a 0.75-in. thick iron plate is 2.5%. Calculate (a) the smallest air-filled flaw that can be detected and (b) the ratio formed by the intensity of the beam transmitted through the smallest flaw divided by the intensity of the beam passing through the solid plate.

10. Based on Table 21-1, determine the relationship between mass absorption coefficient and atomic number Z for X rays with $\lambda = 0.098\,Å$. Is a similar relationship obtained for the absorption of neutrons?

11. What percentage of the incident X-ray beam is transmitted through (a) a 3-cm thick magnesium plate, (b) a 3-cm thick magnesium plate containing a 2-cm long, 0.01-cm thick crack perpendicular to the beam, and (c) a 3-cm thick magnesium plate containing a 2-cm long, 0.01-cm thick crack parallel to the beam?

12. Aluminum is reinforced with 50 vol % boron fibers each 0.005 cm in diameter. If the fiber-reinforced composite is 0.5 cm thick, deter-

mine the ratio of the intensities of an X-ray beam transmitted through a cracked fiber and through the remainder of the composite. Would you expect that a broken fiber could be easily detected by X-ray radiography?

13. A laminar composite material composed of a 3-cm thick copper plate bonded to a 0.5-cm thick lead sheet is then rolled to a final thickness of 0.25 cm. Calculate (a) the intensity of a transmitted X-ray beam before rolling and (b) the intensity after rolling. Describe how a radiographic technique could be used to control the thickness of a rolled material.

14. Often the intensity of a radiation source and the absorption capability of a material are related by the *half value layer* (HVL). This is the thickness of the material that will reduce the intensity of a radiation beam by half. Calculate the HVL for aluminum and iron using (a) tungsten X rays, (b) cobalt 60 gamma rays, and (c) neutrons.

15. Radium has a half-life of 1620 years. Calculate (a) the decay constant for radium and (b) the percentage of the radium source intensity remaining after 25 years.

16. The intensity of a thulium 170 source is 76% of its original value after 50 days. Calculate (a) the decay constant and (b) the half-life of thulium.

17. Calculate the thin-rod ultrasonic velocity in aluminum, copper, and lead and compare with the velocity in bulk materials given in Table 21-3.

18. An ultrasonic pulse introduced into a 5-in. thick lead block returns to the transducer in $1.0 \times 10^{-4}\,s$. Is there a flaw in the lead block? If so, at what depth beneath the surface is the flaw located?

19. An ultrasonic pulse introduced into a 4-in. thick polymer material returns to the transducer in $9.4 \times 10^{-5}\,s$. Calculate the ultrasonic velocity in the material.

20. An ultrasonic pulse is introduced into a thin rod of a material 10 in. long. A pulse returns in $3.3 \times 10^{-4}\,s$. If the density of the material is

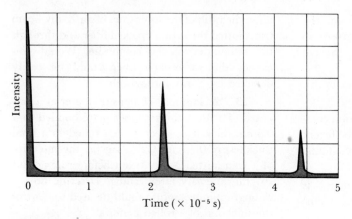

FIGURE 21-37 Sketch of the oscilloscope trace for a pulse-echo ultrasonic test (Problem 21-21).

measured to be $1.33 \, \mathrm{g/cm^3}$, estimate the modulus of elasticity of the material.

21. Figure 21-37 shows the oscilloscope trace of an ultrasonic inspection of the 5-in. wall of a stainless steel pressure vessel. (a) Calculate the ultrasonic velocity in the stainless steel. (b) Is a flaw present? If so, at what depth beneath the surface of the steel is the flaw located?

22. Two sheets of nickel are sandwiched about a core of copper, giving a laminar composite that is 1.5 in. thick. An ultrasonic pulse is introduced into the composite plate; return pulses are observed at $0.844 \times 10^{-5} \mathrm{s}$, $0.981 \times 10^{-5} \mathrm{s}$, and $1.297 \times 10^{-5} \mathrm{s}$. Estimate the thicknesses of the copper core and each of the nickel sheets.

23. A laminar composite is produced from 2-in. thick stainless steel and a 0.2-in. thick polymer. In a through-transmission ultrasonic test, a pulse is received after $1.135 \times 10^{-5} \mathrm{s}$. Calculate the ultrasonic velocity in the polymer.

24. We would like to monitor the thickness of 0.002 cm of silver foil using a resonance ultrasonic test. What frequency must we select if we want to produce 9 half-wavelengths in the foil?

25. A resonance ultrasonic test is used to determine the thickness of a glass enamel on cast iron. A frequency of $22.2 \times 10^6 \mathrm{Hz}$ is required to produce 3 half-wavelengths in the enamel. What is the enamel thickness?

26. Figure 21-38 shows the results of ultrasonic inspection of ductile cast iron (see Chapter 12) for different heat treatments and for different degrees of nodularity, or roundness, of the graphite. Expain why the heat treatment and nodularity might affect the ultrasonic velocity.

27. Describe how the shape of the graphite in a cast iron treated with different amounts of magnesium might affect the attenuation in a through-transmission ultrasonic test.

28. Describe how thermography might detect lack of bonding between the cover sheets and the cells in a honeycomb structure.

29. Which of the following nondestructive tests would not be suitable for inspecting alumina $(\mathrm{Al_2O_3})$ for surface cracks oriented perpendicular to the surface—X-ray radiography, ultrasonic inspection, eddy current inspection, magnetic particle inspection, dye penetrant inspection? Explain.

30. Describe a test that might be used to count the number of wires in a wire rope that will break when a given load is applied.

31. The center of an aluminum casting 2 in. thick contains shrinkage in the form of many small, disconnected pores. Would the following

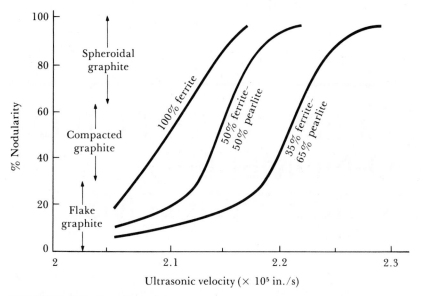

FIGURE 21-38 *The effect of nodularity and heat treatment of cast irons on the ultrasonic velocity in cast irons (Problem 21-26).*

inspection techniques locate the porosity? Explain for each case.

> Through-transmission ultrasonic inspection
> Pulse-echo ultrasonic inspection
> Eddy current inspection
> Magnetic particle inspection
> X-ray radiography
> Rockwell hardness testing

32. Which of the following inspection techniques might locate quench cracks produced when a high-carbon steel plate is rapidly quenched? Explain for each case.

> Pulse-echo ultrasonic inspection
> Eddy current inspection
> Magnetic particle inspection with the probe method
> X-ray radiography
> Acoustic emission

33. Which of the following inspection techniques might be effective as a means to assure that a 0.002-cm thick nickel plating is present on a zinc casting? Explain for each case.

> Pulse-echo ultrasonic inspection
> X-ray radiography

> Acoustic emission
> Eddy current inspection
> Electromagnetic testing

34. Which of the following inspection techniques might be effective in detecting whether an austenitic stainless steel contains between 5% and 10% ferrite? Explain for each case.

> Through-transmission ultrasonic inspection
> Acoustic emission
> Electromagnetic testing
> Magnetic particle testing
> Eddy current inspection

35. Safety glass might be produced by joining two sheets of glass with a thin layer of thermoplastic polymer. Which of the following inspection techniques might be effective in determining if bonding between the glass sheets is complete? Explain for each case.

> Electromagnetic testing
> Eddy current inspection
> Thermography
> Magnetic particle inspection
> X-ray radiography

ANSWERS TO SELECTED ODD-NUMBERED PRACTICE PROBLEMS

CHAPTER 1

1. Absorbs radiation, lightweight   **3.** Ceramics
5. Ceramic; composite   **7.** Ceramic
9. Powder metallurgy   **11.** Joining
13. Composites   **15.** It is an electrical insulator.
17. Semiconductor; metal; ceramic; polymer
19. Brass; thermoplastic polymer

CHAPTER 2

1. $28.1088 \, g/g \cdot mole$ **3.** 54.55% Br^{79} **5.** 4
7. 5.58×10^{23} atoms; 5.58×10^{23} electrons
9. 2.53×10^{21} U/g vs. 5.57×10^{22} B/g;
4.82×10^{22} U/cm^3 vs. 1.28×10^{23} B/cm^3
11. $10.2 \, cm^3$ **13.** 10.2×10^{23} atoms
15. Decreases   **17.** Al_2O_3 **19.** 0.039
21. Metallic

CHAPTER 3

1. 2   **3.** Simple tetragonal   **5.** BCC
7. (a) $4.9486 \, \text{Å}$; (b) $1.749 \, \text{Å}$
9. 2.3×10^{22} cells/cm^3 **11.** BCC   **13.** 6
15. 0.75   **17.** -1.71%; expands   **19.** 0.5
21. $[2\,\bar{2}\,\bar{1}]$; $[\bar{1}\,0\,1]$; $[1\,\bar{1}\,1]$; $[\bar{3}\,12\,\bar{8}]$
23. $(\bar{1}\,2\,1)$; $(0\,0\,3)$; $(0\,\bar{3}\,\bar{2})$
25. $[2\,1\,0]$; $[\bar{4}\,\bar{8}\,3]$; $[0\,\bar{1}\,2]$; $[1\,2\,\bar{1}]$
27. $(1\,\bar{1}\,0\,1)$; $(\bar{1}\,0\,1\,1)$; $(1\,\bar{2}\,1\,0)$

31. $[1\,0\,1]$ $[\bar{1}\,0\,\bar{1}]$ $[0\,1\,1]$ $[0\,\bar{1}\,\bar{1}]$ $[1\,\bar{1}\,0]$ $[\bar{1}\,1\,0]$
33. 12; 8; 4   **35.** 48; 16; 8
37. $(\bar{4}\,1\,2)$; $(4\,\bar{1}\,2)$; $(4\,1\,\bar{2})$; $(4\,1\,2)$
39. $112.6°$ **41.** $(1\,0\,1)$ **43.** $(0\,2\,1)$
45. Yes   **47.** Yes
49. (a) 1.443×10^7 points/cm;
(b) 2.887×10^7 points/cm;
(c) 1.443×10^7 points/cm
51. (a) 0.577; (b) 1.0; (c) 0.408; BCC
53. (a) 0.453; (b) 0.34; (c) 0.907; FCC
55. (a) 3.61×10^{14} points/cm^2;
(b) 3.61×10^{14} points/cm^3;
(c) 14.43×10^{14} points/cm^3
57. 3.54×10^7 points/cm; 2.774×10^7 points/cm
59. 3.33×10^7 points/cm; 2×10^7 points/cm;
3.43×10^7 points/cm; 1.25×10^7 points/cm
61. (a) $0.890 \, \text{Å}$; (b) $1.824 \, \text{Å}$; (c) $1.665 \, \text{Å}$; (d) $1.090 \, \text{Å}$
65. 3   **67.** (a) 4; (b) 6; (c) 6; (d) 4; (e) 8
69. Sodium chloride
71. (a) 4 Te, 8 O; (b) 6; (c) $7.92 \, \text{Å}$; (d) 0.472
73. Cesium chloride; (a) $4.1917 \, \text{Å}$; (b) 0.693;
(c) $4.8 \, g/cm^3$
75. 0.5; 0.34
77. 8 Si, 16 O; (a) $8.037 \, \text{Å}$; (b) 0.302;
(c) $1.538 \, g/cm^3$

CHAPTER 4

1. Mg, Cd; Rh, Be

3. $[\bar{1}\,\bar{1}\,1]$, $[1\,1\,\bar{1}]$, $[\bar{1}\,1\,1]$, $[1\,\bar{1}\,\bar{1}]$
5. $(\bar{1}\,1\,0)$, $(1\,\bar{1}\,0)$, $(1\,0\,\bar{1})$, $(\bar{1}\,0\,1)$, $(0\,\bar{1}\,1)$, $(0\,1\,\bar{1})$
7. No   **9.** $[0\,1\,0]$, $[0\,\bar{1}\,0]$ **11.** 2.7411 Å
13. (a) 1267 g; (b) 0.00127 g
15. 3.86×10^9 cm/cm³   **17.** ∞; 552 psi; 552 psi
19. ∞; 20,836 psi; 20,836 psi   **21.** 5507 psi; no
23. (a) 0.001675; (b) 1×10^{20}/cm³
25. (a) 5.43×10^{18}/cm³; (b) 1.9157 g/cm³
27. 0.2   **29.** 3.924 g/cm³   **31.** (a) 0.005; (b) 50
33. 2.191 g/cm³   **35.** 9   **37.** 9.23; fine   **39.** 3.8
41. 0.038°   **43.** Create vacancy

CHAPTER 5

1. (a) 10,132 cal/mole; (b) 7.325 ipm
3. 176,132 cal/mole; 1.28×10^{12} cm²/s
5. 70,000 cal/mole
7. (a) -0.0999 at%/cm;
   (b) -4.408×10^{11} atoms/cm³ · cm
9. (a) -9.99994×10^{24} atoms/cm³ · cm;
   (b) 10.737×10^{20} atoms/cm² · s
11. -4.5095×10^9 atoms/cm³ · cm   **13.** 85°C
15. 1.005% C; 0.53%; 0.245% C   **17.** 6.67 h
19. 0.327% C   **21.** 0.018% N   **23.** 28.3 h
25. 6.56 h   **27.** Ni in Cu   **31.** Na; 46,490 cal/mole

CHAPTER 6

1. (a) Will deform; (b) will not neck
3. 16.3×10^6 psi   **5.** Yes   **7.** 0.2495 in.
9. 7.5%; 11.89%
11. (a) 27,000 psi; (b) 31,000 psi; (c) 6.5×10^6 psi;
   (d) 12.25%; (e) 14.8%; (f) 30,000 psi;
   (g) 35,179 psi
13. (a) 320 MPa; (b) 470.59 MPa; (c) 82.68 GPa;
   (d) 6.2%; (e) 5.91%; (f) 470.59 MPa;
   (g) 500.15 MPa
15. 79,754 psi = 549.9 MPa   **17.** FCC
19. 81.86 lb   **21.** 1.365 in.   **23.** 64.752 lb
25. 2.31 d   **29.** (a) 416.7 d; (b) 83.3 d; (c) 12.5 d
31. 2.82 in.   **33.** 980°C   **35.** 601.4°C
37. 25.92 in.   **39.** 11,045 lb   **41.** 2.05 mm
43. 49.5 MPa · m$^{1/2}$ **45.** 1,075,000 psi
47. (a) 0.07074 in.; (b) 0.03537 in.   **49.** No

CHAPTER 7

1. (a) 0.25   **3.** 0.0109 mm; 32,700%
5. (a) 97.2%; (b) 97.2%

7. (a) 92.2%; (b) 110,000 psi TS, 70,000 psi YS,
   0% elongation
9. 0.318 in.
11. (a) 20% to 30%; (b) 0.3125 in. to 0.3571 in.
13. No
15. Need 25% to 40% final cold work;
   0.2143 in. to 0.25 in.
17. 0.05977 in.
19. (a) 0.1195 in.; (b) 89.73 lb; (c) will survive
21. (a) ~ 500°C, ~ 900°C, ~ 1300°C; (b) ~ 700°C;
   (c) ~ 1100°C; (d) ~ 2660°C
23. Smaller grain size; higher grain growth
   temperature
27. (a) 0.066 in.; (b) 28,000 psi;
   (c) 800 lb/in. for sheet, 1848 lb/in. for joint
29. More difficult to produce bonding
31. (a) Cold; (b) hot; (c) cold work and anneal;
   (d) cold; (e) hot

CHAPTER 8

3. 13.9 Å   **5.** 715°C   **7.** 547   **9.** 527,160
11. (a) 0.0149; (b) 0.149; (c) 0.514   **13.** 114°C
15. (a) 16.9 min/cm²; (b) 16.9 min
17. 40 min/in²; 1.6; Styrofoam decreases time
19. (a) 0.0107 cm (b) 10 s   **21.** Cylinder
23. Sphere   **25.** 46.47 in³; 3.404 in.; 5.106 in.
27. Will not   **29.** 0.000033 cm, 0.0123 cm
31. Surface; 360 s, 0.04 cm; mid-radius: 720 s,
   0.06 cm; center: 1200 s, 0.10 cm
33. (a) 2100°C; (b) 1750°C; (c) 350°C; (d) 9°C/s;
   (e) 350 s; (f) 300 s; (g) 100°C
35. 4.29%   **37.** 6.93%; 2.38×10^6 pores
39. (a) -5.97%; (b) expands   **41.** 0.003 s

CHAPTER 9

3. Au-Ag   **5.** Ge; no
7. 1310°C; 1260°C; 50°C
9. 1250°: α, 50% Ni in α, 100% α;
   1300°: $\alpha + L$, 58% Ni in α, 46% Ni in L,
      33% α–67% L;
   1350°: L, 50% Ni in L, 100% L
11. 2200°: S, 50% MgO in S, 100% S;
   2400°: $S + L$, 62% MgO in S,
      37% MgO in L, 52% S–48% L;
   2600°: L, 50% MgO in L, 100% L
13. Sodium chloride; yes
15. (a) 2520°C; (b) 2150°C;

(c) 40% FeO in L, 18% FeO in S;
(d) 54.5% L–45.5% S

17. 55% FeO   19. 2300°C

21. (a) 3020°C; (b) 2820°C;
(c) 83% W in α, 68% W in L;
(d) 13.3% α-86.7% L

23. 54.6% W

25. (a) 17.8 at% W; (b) 21 wt% W or 11.8 at% W
in L, 40 wt% W or 25.2 at% W in α;
(c) 47.4% α-52.6% L

27. 66.4% W; $L + \alpha$

29. (a) 34% FeO; (b) 82% FeO

31. (a) 84% W
(b) 2900°: 59% W in L, 76% W in α,
35.3% L-64.7% α;
2800°: 70% W in α, 100% α;
2700°: 70% W in α, 100% α
(c) 2900°: 59% W in L, 81% W in α,
50% L-50% α;
2800°: 48% W in L, 77% W in α,
24% L-76% α;
2700°: 37% W in L, 72% W in α,
5.7% L-94.3% α
(d) 50% W in last L, 70% W in last α
(e) 32% W in last L, 54% W in last α
(f) 2820°C vs 2660°C

33. (a) 2970°C; (b) 2760°C; (c) 210°C; (d) 3200°C;
(e) 230°C; (f) 8 min; (g) 12 min; (h) 65% W

CHAPTER 10

1. (a) μ, nonstoichiometric; (b) α, β, γ, δ;
(c) peritectic at 1150°:
$$L \ (30\% \ B) + \delta \ (5\% \ B) \to \gamma \ (15\% \ B);$$
monotectic at 950°:
$$L_1 \ (40\% \ B) \to L_2 \ (60\% \ B) + \gamma \ (25\% \ B);$$
eutectic at 750°:
$$L \ (67\% \ B) \to \gamma \ (25\% \ B) + \beta \ (90\% \ B);$$
eutectoid at 450°:
$$\gamma \ (20\% \ B) \to \alpha \ (10\% \ B) + \beta \ (92\% \ B);$$
peritectoid at 300°:
$$\alpha \ (5\% \ B) + \beta \ (95\% \ B) \to \mu \ (50\% \ B)$$

3. (a) χ, γ_1, γ_2, γ, β, all nonstoichiometric; (b) α;
(c) eutectic at 1020°: $L \to \alpha + \beta$;
peritectic at 1020°: $L + \beta \to \chi$;
eutectoid at 960°: $\chi \to \beta + \gamma_1$;
eutectoid at 780°: $\gamma_1 \to \beta + \gamma_2$;
eutectoid at 550°: $\beta \to \alpha + \gamma_2$;
peritectoid at 380°: $\alpha + \gamma_2 \to \gamma$

5. (a) $MgSiO_3$, Mg_2SiO_4, fixed; (b) no, no;

(c) monotectic at 1690°:
$$L_1 \to SiO_2 + L_2;$$
eutectic at 1850°:
$$L \to Mg_2SiO_4 + MgO;$$
peritectic at 1550°:
$$L + Mg_2SiO_4 \to MgSiO_3$$
eutectic at 1530°:
$$L \to SiO_2 + MgSiO_3$$

7. (a) 5% Sn;
(b) 15% Sn in L, 8% Sn in α, 57% L, 43% α;
(c) 300°, 270°, 150°;
(d) 12% Sn in α, 100% α;
(e) 2% Sn in α, 100% Sn in β, 90% α, 10% β;
(f) 10.7% eutectic

9. 31% α, 69% β

11. At 1540°: 12.5% SiO_2, 87.5% $MgSiO_3$;
At 1850°: 92.9% Mg_2SiO_4, 7.1% MgO;
both are brittle

13. (a) Hypereutectic; (b) 98% Sn;
(c) 61.9% Sn in L, 97.5% Sn in β,
63% L, 37% β;
(d) 19% Sn in α, 97.5% Sn in β, 28.7% α,
71.3% β;
(e) 97.5% Sn in primary β, 61.9% Sn in
eutectic, 37% primary β, 63% eutectic

15. (a) Hypereutectic; (b) 100% Si; (c) 99.83% Si
in β, 12.6% Si in L, 71.5% β, 28.5% L;
(d) 99.83% Si in β, 1.65% Si in α, 74.7% β,
25.3% α; (e) 99.83% Si in primary β, 12.6% Si
in eutectic, 71.5% primary β, 28.5% eutectic

17. 74.5% Sn, hypereutectic

19. 28.3% Si   21. 73.5% B

23. (a) 1400°C; (b) 200°C; (c) 1200°C; (d) 570°C;
(e) 630°C; (f) 220 s; (g) 240 s; ~60% Si

27. 0.75%   29. 43.1% Li; more

31. (a) 21.75% Pb; (b) 60.4% L_1; 0% L_2

33. 3; 4   37. (a) 270°C; (b) γ; (c) α and γ

CHAPTER 11

1. 4.72% θ; 6.48% θ

5. 5% β in conventional; 35% β in RSP

7. 81.53% α, 18.47% Fe_3C, 91.86% pearlite,
8.14% primary Fe_3C

9. 0.215% C   11. 1.20% C, hypereutectoid

13. 0.52% C   15. 830°C, 2.134% C

17. 7.846 g/cm^3 19. 1.637% C

21. 68% ε, 32% β; brittle; alloys with primary β are
brittle

23. 86%   25. (a) 3.5 × 10^{-5} cm; (b) 300 MPa

27. (a) bainite; (b) 375°C
29. Austenitize at 800°C, quench to 300°C for > 2000 s, cool
31. Fine pearlite   **33.** Bainite
35. Tempered martensite
37. 0.92%C, 91.7% martensite
39. (a) 790°C; (b) 1.77% C   **41.** 0.55% C
43. 500°C to 590°C

CHAPTER 12

1. 95.82% α, 4.18% Fe_3C, 62.82% primary α, 37.18% pearlite
3. 1010   **5.** 10100   **7.** 865°C
9. 22,222 cm^2/cm^3 for pearlitic; 240 cm^2/cm^3 for spheroidite
11. (a) Fe_3C + martensite;
(b) Fe_3C, pearlite, + martensite;
(c) bainite; (d) Fe_3C + bainite;
(e) tempered martensite; (f) Fe_3C + pearlite
13. 400°C for > 200 s   **15.** 300°C
17. (a) 8 s; (b) 10 s; (c) 2 s   **19.** Yes, if at 460°C
21. (a) 0.60% C; (b) 750°C; 870°C to 890°C
25. 320°C   **27.** Pearlite
29. (a) Hypereutectoid; (b) Fe_3C + α; (c) no
31. (a) 3 s at 300°C; (b) 30,000 s at 540°C
33. 1095 and 4340   **35.** 6.5 to 9°C/s
37. > 0.4; agitated oil
39. 2.1 in.   **43.** 2 in.
45. R_c41
47. 22.4% martensite; 77.6% α; 0.60% C; R_c65
49. 4.1%; hypoeutectic; austenite
51. 1040 steel: ferrite + coarse pearlite; gray iron: all ferrite
53. (a) 4.3%; (b) 20; (c) 20,000 psi; (d) ~12,000 psi
55. Reduce FSG time; faster cooling rate
57. All lost; no   **59.** (a) Better; (b) better
61. SSG   **63.** 4.485% C

CHAPTER 13

1. Al-Li: 1.39 × 10^8 in.; Al: 1.02 × 10^8 in.
3. Al_6Mn **5.** 64% α, 36% β
7. Stronger   **9.** 55,500 psi   **11.** 23%   **13.** No
15. (a) 98.125 lb; (b) 412 lb
17. 400°C: 107; 650°C: 267; 800°C: 667   **19.** H12
21. 54.5%   **23.** High   **27.** $\beta \rightarrow \alpha + \eta$; 270°C
29. Solid solution strengthening   **31.** 850°C
33. (a) ~500°C; (b) $\alpha + \beta_{ss}$;

(c) 27.8% α, 72.2% β;
(d) α precipitate in β matrix
35. FCC alloys

CHAPTER 14

1. 3   **3.** (a) 2 Ti, 4 O; (b) 0.66; (c) yes
5. (a) 6   **7.** Metasilicate   **9.** 1.074 g
11. 13.82% Mg, 32.89% Fe, 16.25% Si, 37.04% O
13. 2.32   **15.** 1833 g PbO; 330.5 g MgO
17. 3578 lb   **19.** 55,505 cal/mole
21. (a) 17.1% Na_2O; (b) 1100°C; (c) 19% SiO_2
23. (a) 55% MgO; (b) 83.3% forsterite; (c) 1850°C
25. 1805°C   **27.** 1680°C
29. (a) 147.3 kg; 28.2% Al_2O_3, 41.0% SiO_2, 30.8% CaO; (b) yes
31. (a) 25.4% Na_2O, 64.9% SiO_2, 9.7% CaO; (b) 1020°C
33. (a) 33%; (b) 52.56%; (c) 0.372   **35.** 39.96 lb
37. 50% feldspar, 33% clay, 17% silica; translucent porcelain

CHAPTER 15

1. (b) Addition   **3.** 24,495 Å   **5.** 4762
7. (a) 172.8 g; (b) 163 cm^3
9. (a) 0.068 g; (b) 1.204 × 10^{21} chains
11. (a) 16.01 g · mole; (b) 64.04 kcal; (c) 267°C
13. (a) 15.267 kcal; (b) 76.335°C
15. (a) 567; (b) 20,413 g/g · mole
17. Slower
19. 17,500; 14,950   **25.** 0.9966 g/cm^3
27. Rolling   **29.** 1.936 kg   **31.** 0.02125
37. 17,088 cal/mole
39. (a) 2500 g; (b) 1500 g; (c) 6000 g (d) 5000 g
41. 134.48 g   **43.** 42.552 g   **45.** 51.98 cm^3

CHAPTER 16

1. 0.00028 mm
3. (a) 4.34% ThO_2; (b) 8.3 × 10^{13} particles/cm^3
7. (a) 0.76; (b) 8.95 × 10^6 particles
9. (a) 0.412; (b) 0.412; (c) 11.321 g/cm^3
11. 1083 g   **13.** (a) 475°C; (b) 4.181 g/cm^3
15. (a) 72.2%; (b) 3.27 × 10^8 in. vs 2.56 × 10^8 in.
17. (a) 167.6 lb; (b) 69.12 lb
19. (a) 0.575; (b) 45.54 × 10^6 psi
21. 0.245 cal/cm · s · K
25. (a) 2400°C; (b) 260,000 psi
27. 9.9 × 10^3 ohm^{-1} · cm^{-1} parallel;

1.017×10^{-12} ohm^{-1} · cm^{-1} perpendicular

29. 3.29×10^6 psi   **31.** 0.01188 cal/cm · s · K

33. (a) 0.5536 g/cm^3; (b) 77% H$_2$O   **35.** 3292 lb

CHAPTER 17

1. (a) 3.399×10^4 W; (b) 1.016×10^{12} W;
(c) 4.065×10^{15} W

3. 209 m   **5.** (a) 500 V; (b) 2.25×10^6 A/cm^2

7. 0.099 **9.** 9.876 mph   **11.** 1.7×10^{23}/cm^3

13. $91.96°$C   **15.** 2.58×10^{-4}

17. (a) 3.969×10^{-5} ohm · cm; (b) 11.31×10^{-6};
(c) 3.89×10^{-4}

19. 1.7×10^{-4}

21. (b) $40°$C for Pt: Pt-Rh; $7°$C for iron: constantan

23. (a) $1000°$C; (b) $570°$C; (c) 44% Si

25. Al$_2$O$_3$: (a) 1.89×10^{-14} A; (b) 1.18×10^5/s; Al:
(a) 7.11×10^5 A; (b) 4.44×10^{24}/s

27. (a) 1.148×10^{20}; (b) 9.81×10^{-4};
(c) 5.44×10^{20}

29. $35°$C   **31.** 15.17 ohm^{-1} · cm^{-1}

33. 0.0001862 **35.** AlAs   **37.** 5.1×10^{19}/cm^3

39. $727°$C   **41.** 610%

CHAPTER 18

1. 0.535×10^{-8} Å   **3.** 3.726×10^{-6} C/m^2

5. 2×10^{-8} C/m^2 **7.** 9.91×10^{-8} Å

9. 3.08×10^8 V   **11.** 4.75 **13.** 0.6

15. 1.72 W   **17.** 0.000743 μF

19. (a) 44.1 cm^2; (b) 0.047 cm^2

21. (a) 0.015 mm (b) 861.9 N $= 194$ lb

23. 0.48V   **25.** (a) 800 V/m; (b) 6.65

29. (a) 24.9%; (b) 1.912×10^6 A/m

31. (a) 0.118 oersted; (b) $94{,}399.9$ oersted;
(c) $94{,}400$ G

33. 0.796 A/m   **35.** (b) 17×10^6 oersted · G

37. (a) 1.6; (b) 4

39. (a) 2387 A/m; (b) 100; (c) 228 A

41. 4.87×10^5 A/m

CHAPTER 19

1. $12{,}169$ V   **3.** 6934 V

5. (a) 9.41×10^{-8} erg; (b) 10.79×10^{-8} erg;
(c) 1.35×10^{-8} erg

7. (a) $49{,}650$ V; (b) Mo, Fe, S   **9.** Fe, S

11. 8×10^{-6} s   **13.** (a) 1.83 eV; (b) 1.71 eV

15. 9194 Å; (b) infrared; (c) 2.14 eV below
conduction band

17. CdS   **19.** From 9194 Å to 5640 Å

21. (a) 1139 cm^{-1}; (b) 0.004 cm   **23.** None

25. (a) Increases with frequency;
(b) remains constant

27. (a) $10.67°$; (b) $19.47°$ **29.** $9.37°$

31. (a) 8.13%; (b) 1.79; (c) 3.2

35. (a) FCC; (c) 4.072 Å; (d) gold

37. ($1\,1\,0$), ($2\,0\,0$), ($2\,1\,1$)   **39.** 662.5 cal

41. 0.98241 m

43. Al: 50.09375 cm; epoxy; 50.20625 cm

45. 9.823 cm   **47.** 0.5795 cm

49. 3.478×10^5 ohm^{-1} · cm^{-1}

CHAPTER 20

1. Dezincification   **3.** (a) 35%; (b) 27%

5. -0.282 V   **7.** 0.4594 mole

9. 0.015 A/cm^2 **11.** 1.37×10^{-4} A/cm^2 **15.** Hot

17. (a) Intergranular corrosion; (b) $> 0.03\%$ C;
(c) slow cooling process

19. Tar dissolved by acetone   **21.** Tin

23. Ferrite-cementite galvanic couple

25. Annealed; cold worked   **27.** Steel rivets

31. 0.00568 A   **37.** (a) Parabolic; (b) 0.00836 g

CHAPTER 21

1. Brittle   **3.** Ductile   **7.** 1.088

9. (a) 0.01875 in.; (b) 1.0399

11. (a) 45.23%; (b) 45.35%; (c) 76.76%

13. (a) 3.87×10^{-13}; (b) 0.1298

15. (a) 4.279×10^{-4} y^{-1}; (b) 98.9%

17. Al: 1.986×10^5 in./s; Cu: 1.379×10^5 in./s; Pb:
0.475×10^5 in./s; thin rod velocities are lower

19. 0.85×10^5 in./s

21. (a) 2.273×10^5 in./s; (b) 2.5 in.

23. 0.8×10^5 in./s   **25.** 0.015 in.

27. Attenuation increases as Mg decreases

29. Ultrasonic, eddy current, magnetic particle

31. Both ultrasonic tests and radiography would.

33. Eddy current and electromagnetic

35. Thermography

APPENDIX A Selected physical properties of metals

Metal		Atomic Number	Crystal Structure	Lattice Parameter (Å)	Atomic Mass (g/g · mole)	Density (g/cm³)	Melting Temperature (°C)
Aluminum	Al	13	FCC	4.04958	26.981	2.699	660.4
Antimony	Sb	51	hex	$a = 4.307$	121.75	6.697	630.7
				$c = 11.273$			
Arsenic	As	33	hex	$a = 3.760$	74.9216	5.778	816
				$c = 10.548$			
Barium	Ba	56	BCC	5.025	137.3	3.5	729
Beryllium	Be	4	hex	$a = 2.2858$	9.01	1.848	1290
				$c = 3.5842$			
Bismuth	Bi	83	hex	$a = 4.546$	208.98	9.808	271.4
				$c = 11.86$			
Boron	B	5	rhomb	$a = 10.12$	10.81	2.3	2300
				$\alpha = 65.5°$			
Cadmium	Cd	48	HCP	$a = 2.9793$	112.4	8.642	321.1
				$c = 5.6181$			
Calcium	Ca	20	FCC	5.588	40.08	1.55	839
Cerium	Ce	58	HCP	$a = 3.681$	140.12	6.6893	798
				$c = 11.857$			
Cesium	Cs	55	BCC	6.13	132.91	1.892	28.6
Chromium	Cr	24	BCC	2.8844	51.996	7.19	1875
Cobalt	Co	27	HCP	$a = 2.5071$	58.93	8.832	1495
				$c = 4.0686$			
Copper	Cu	29	FCC	3.6151	63.54	8.93	1084.9
Gadolinium	Gd	64	HCP	$a = 3.6336$	157.25	7.901	1313
				$c = 5.7810$			
Gallium	Ga	31	ortho	$a = 4.5258$	69.72	5.904	29.8
				$b = 4.5186$			
				$c = 7.6570$			
Germanium	Ge	32	FCC	5.6575	72.59	5.324	937.4
Gold	Au	79	FCC	4.0786	196.97	19.302	1064.4
Hafnium	Hf	72	HCP	$a = 3.1883$	178.49	13.31	2227
				$c = 5.0422$			
Indium	In	49	tetra	$a = 3.2517$	114.82	7.286	156.6
				$c = 4.9459$			
Iridium	Ir	77	FCC	3.84	192.9	22.65	2447
Iron	Fe	26	BCC	2.866	55.847	7.87	1538
			FCC	3.589	(> 912°C)		
			BCC		(> 1394°C)		
Lanthanum	La	57	HCP	$a = 3.774$	138.91	6.146	918
				$c = 12.17$			
Lead	Pb	82	FCC	4.9489	207.19	11.36	327.4
Lithium	Li	3	BCC	3.5089	6.94	0.534	180.7
Magnesium	Mg	12	HCP	$a = 3.2087$	24.312	1.738	650
				$c = 5.209$			
Manganese	Mn	25	cubic	8.931	54.938	7.47	1244
Mercury	Hg	80	rhomb		200.59	13.546	− 38.9
Molybdenum	Mo	42	BCC	3.1468	95.94	10.22	2610
Nickel	Ni	28	FCC	3.5167	58.71	8.902	1453
Niobium	Nb	41	BCC	3.294	92.91	8.57	2468
Osmium	Os	76	HCP	$a = 2.7341$	190.2	22.57	2700
				$c = 4.3197$			

APPENDIX A   Selected physical properties of metals (Continued)

Metal		Atomic Number	Crystal Structure	Lattice Parameter (Å)	Atomic Mass (g/g · mole)	Density (g/cm³)	Melting Temperature (°C)
Palladium	Pd	46	FCC	3.8902	106.4	12.02	1552
Platinum	Pt	78	FCC	3.9231	195.09	21.45	1769
Potassium	K	19	BCC	5.344	39.09	0.855	63.2
Rhenium	Re	75	HCP	a = 2.760	186.21	21.04	3180
				c = 4.458			
Rhodium	Rh	45	FCC	3.796	102.99	12.41	1963
Rubidium	Rb	37	BCC	5.7	85.467	1.532	38.9
Ruthenium	Ru	44	HCP	a = 2.6987	101.07	12.37	2310
				c = 4.2728			
Selenium	Se	34	hex	a = 4.3640	78.96	4.809	217
				c = 4.9594			
Silicon	Si	14	FCC	5.4307	28.08	2.33	1410
Silver	Ag	47	FCC	4.0862	107.868	10.49	961.9
Sodium	Na	11	BCC	4.2906	22.99	0.967	97.8
Strontium	Sr	38	FCC	6.0849	87.62	2.6	768
			BCC	4.84	(> 557°C)		
Tantalum	Ta	73	BCC	3.3026	180.95	16.6	2996
Technetium	Tc	43	HCP	a = 2.735	98.9062	11.5	2200
				c = 4.388			
Tellurium	Te	52	hex	a = 4.4565	127.6	6.24	449.5
				c = 5.9268			
Thorium	Th	90	FCC	5.086	232	11.72	1755
Tin	Sn	50	FCC	6.4912	118.69	5.765	231.9
Titanium	Ti	22	HCP	a = 2.9503	47.9	4.507	1668
				c = 4.6831			
			BCC	3.32	(> 882°C)		
Tungsten	W	74	BCC	3.1652	183.85	19.254	3410
Uranium	U	92	ortho	a = 2.854	238.03	19.05	1133
				b = 5.869			
				c = 4.955			
Vanadium	V	23	BCC	3.0278	50.941	6.1	1900
Yttrium	Y	39	HCP	a = 3.648	88.91	4.469	1522
				c = 5.732			
Zinc	Zn	30	HCP	a = 2.6648	65.38	7.133	420
				c = 4.9470			
Zirconium	Zr	40	HCP	a = 3.2312	91.22	6.505	1852
				c = 5.1477			
			BCC	3.6090	(> 862°C)		

APPENDIX B The atomic and ionic radii of selected elements

Element	Atomic Radius (Å)	Valence	Ionic Radius (Å)
Aluminum	1.432	+ 3	0.51
Antimony		+ 5	0.62
Arsenic		+ 5	2.22
Barium	2.176	+ 2	1.34
Beryllium	1.143	+ 2	0.35
Bismuth		+ 5	0.74
Boron	0.46	+ 3	0.23
Bromine	1.19	− 1	1.96
Cadmium	1.49	+ 2	0.97
Calcium	1.976	+ 2	0.99
Carbon	0.77	+ 4	0.16
Cerium	1.84	+ 3	1.034
Cesium	2.65	+ 1	1.67
Chlorine	0.905	− 1	1.81
Chromium	1.249	+ 3	0.63
Cobalt	1.253	+ 2	0.72
Copper	1.278	+ 1	0.96
Fluorine	0.6	− 1	1.33
Gallium	1.218	+ 3	0.62
Germanium	1.225	+ 4	0.53
Gold	1.442	+ 1	1.37
Hafnium		+ 4	0.78
Hydrogen	0.46	+ 1	1.54
Indium	1.570	+ 3	0.81
Iodine	1.35	− 1	2.20
Iron	1.241 (BCC)	+ 2	0.74
	1.269 (FCC)	+ 3	0.64
Lanthanum	1.887	+ 3	1.016
Lead	1.75	+ 4	0.84
Lithium	1.519	+ 1	0.68
Magnesium	1.604	+ 2	0.66
Manganese	1.12	+ 2	0.80
		+ 3	0.66
Mercury	1.55	+ 2	1.10
Molybdenum	1.363	+ 4	0.70
Nickel	1.243	+ 2	0.69
Niobium	1.426	+ 4	0.74
Nitrogen	0.71	+ 5	0.15
Oxygen	0.60	− 2	1.32
Palladium	1.375	+ 4	0.65
Phosphorus	1.10	+ 5	0.35
Platinum	1.387	+ 2	0.80
Potassium	2.314	+ 1	1.33
Rubidium	2.468	+ 1	0.70
Selenium		− 2	1.91
Silicon	1.176	+ 4	0.42
Silver	1.445	+ 1	1.26
Sodium	1.858	+ 1	0.97

APPENDIX B   The atomic and ionic radii of selected elements (Continued)

Element	Atomic Radius (Å)	Valence	Ionic Radius (Å)
Strontium	2.151	+ 2	1.12
Sulfur	1.06	− 2	1.84
Tantalum	1.43	+ 5	0.68
Tellurium		− 2	2.11
Thorium	1.798	+ 4	1.02
Tin	1.405	+ 4	0.71
Titanium	1.475	+ 4	0.68
Tungsten	1.371	+ 4	0.70
Uranium	1.38	+ 4	0.97
Vanadium	1.311	+ 3	0.74
Yttrium	1.824	+ 3	0.89
Zinc	1.332	+ 2	0.74
Zirconium	1.616	+ 4	0.79

INDEX